Atomic Masses of the Naturally Occurring Isotopic Mixtures of the Elements

Element	Symbol	Atomic Number	Atomic Mass[a] (u)	Element	Symbol	Atomic Number	Atomic Mass[a] (u)
hydrogen	H	1	1.00797	iodine	I	53	126.9044
helium	He	2	4.0026	xenon	Xe	54	131.30
lithium	Li	3	6.939	cesium	Cs	55	132.905
beryllium	Be	4	9.0122	barium	Ba	56	137.34
boron	B	5	10.811	lanthanum	La	57	138.91
carbon	C	6	12.01115	cerium	Ce	58	140.12
nitrogen	N	7	14.0067	praseodymium	Pr	59	140.907
oxygen	O	8	15.9994	neodymium	Nd	60	144.24
fluorine	F	9	18.9984	promethium	Pm	61	(145)[b]
neon	Ne	10	20.183	samarium	Sm	62	150.35
sodium	Na	11	22.9898	europium	Eu	63	151.96
magnesium	Mg	12	24.312	gadolinium	Gd	64	157.25
aluminum	Al	13	26.9815	terbium	Tb	65	158.924
silicon	Si	14	28.086	dysprosium	Dy	66	162.50
phosphorus	P	15	30.9738	holmium	Ho	67	164.930
sulfur	S	16	32.064	erbium	Er	68	167.26
chlorine	Cl	17	35.453	thulium	Tm	69	168.934
argon	Ar	18	39.948	ytterbium	Yb	70	173.04
potassium	K	19	39.102	lutetium	Lu	71	174.97
calcium	Ca	20	40.08	hafnium	Hf	72	178.49
scandium	Sc	21	44.956	tantalum	Ta	73	180.948
titanium	Ti	22	47.90	tunsten	W	74	183.85
vanadium	V	23	50.942	rhenium	Re	75	186.2
chromium	Cr	24	51.996	osmium	Os	76	190.2
manganese	Mn	25	54.9380	iridium	Ir	77	192.2
iron	Fe	26	55.847	platinum	Pt	78	195.09
cobalt	Co	27	58.9332	gold	Au	79	196.967
nickel	Ni	28	58.71	mercury	Hg	80	200.59
copper	Cu	29	63.54	thallium	Tl	81	204.37
zinc	Zn	30	65.37	lead	Pb	82	207.19
gallium	Ga	31	69.72	bismuth	Bi	83	208.980
germanium	Ge	32	72.59	polonium	Po	84	(210)[b]
arsenic	As	33	74.9216	astatine	At	85	(210)[b]
selenium	Se	34	78.96	radon	Rn	86	(222)[b]
bromine	Br	35	79.909	francium	Fr	87	(223)[b]
krypton	Kr	36	83.80	radium	Ra	88	(226)[b]
rubidium	Rb	37	85.47	actinium	Ac	89	(227)[b]
strontium	Sr	38	87.62	thorium	Th	90	232.038
yttrium	Y	39	88.905	protactinium	Pa	91	(231)[b]
zirconium	Zr	40	91.22	uranium	U	92	238.03
niobium	Nb	41	92.906	neptunium	Np	93	(237)[b]
molybdenum	Mo	42	95.94	plutonium	Pu	94	(242)[b]
technetium	Tc	43	(99)[b]	americium	Am	95	(243)[b]
ruthenium	Ru	44	101.07	curium	Cm	96	(248)[b]
rhodium	Rh	45	102.905	berkelium	Bk	97	(247)[b]
palladium	Pd	46	106.4	californium	Cf	98	(249)[b]
silver	Ag	47	107.870	einsteinium	Es	99	(254)[b]
cadmium	Cd	48	112.40	fermium	Fm	100	(253)[b]
indium	In	49	114.82	mendelevium	Md	101	(256)[b]
tin	Sn	50	118.69	nobelium	No	102	(253)[b]
antimony	Sb	51	121.75	lawrencium	Lw	103	(259)[b]
tellurium	Te	52	127.60	(unnamed)	?	104	(260)[b]

[a] Based on $^{12}C = 12$ (exactly). Some of the masses are poorly known, primarily because of lack of knowledge of the relative isotopic abundances.

[b] Radioactive element; the number in parentheses is the mass number of the isotope with the longest known half-life.

TO THE STUDENT: A Study Guide for the textbook is available through your college bookstore under the title Study Guide to accompany GENERAL PHYSICS WITH BIOSCIENCE ESSAYS, Second Edition, by Jerry B. Marion and William F. Hornyak. The Study Guide, prepared by Edward E. Beasley of Gallaudet College in Washington, D.C., can help you with course material by acting as a tutorial, review and study aid. If the Study Guide is not in stock, ask the bookstore manager to order a copy for you.

General Physics
with Bioscience Essays

John Wiley & Sons

New York
Chichester
Brisbane
Toronto
Singapore

Jerry B. Marion
William F. Hornyak

Department of Physics
University of Maryland

SECOND EDITION

General Physics with Bioscience Essays

Cover photo: Arthur D'Arazien/The Image Bank

Production supervised by Linda Indig
Manuscript edited by Bruce Safford
Cover designed by Ann Marie Renzi
Photo researched by Elyse Rieder
Illustrations by John Balbalis

Copyright © 1985, by John Wiley & Sons, Inc.

All rights reserved. Published simultaneously in Canada.

Reproduction or translation of any part of
this work beyond that permitted by Sections
107 and 108 of the 1976 United States Copyright
Act without the permission of the copyright
owner is unlawful. Requests for permission
or further information should be addressed to
the Permissions Department, John Wiley & Sons.

Library of Congress Cataloging in Publication Data:
Marion, Jerry B.
    General physics with bioscience essays.

    Includes index.
    1. Physics.   2. Life sciences.   I. Hornyak, William F.
(William Frank), 1922–   . II. Title.
QC21.2.M363 1985   642'.5   84-15278
ISBN 0-471-89878-3

Printed in the United States of America

10 9 8 7 6 5 4 3 2 1

In Memorium

It is with great sorrow that we note the untimely death of Jerry B. Marion. His loss to the teaching community in physics and the cessation of his prodigious output of excellent textbooks will be sadly felt for many years to come. Regrettably, he was unable to see this printed version of the second edition we began together some years earlier.

William F. Hornyak

Preface

The wide acceptance of the first edition of General Physics with Bioscience Essays has also provided a valuable source of suggestions for improvements by its many users. In addition, several detailed reviews have been solicited resulting in further comments. In response to these suggestions, this new edition has been prepared.

The present edition continues as a basic text for a general introductory course in physics for students whose main interests and careers lie in other areas. The mathematical ability necessary to master the material presented here is not great—high school algebra is used extensively and simple trigonometry is used where necessary—the methods of calculus are not used at all. The Appendix summarizes all of the mathematical techniques required to understand the discussions and to solve the problems.

For most of the college and university students who take an introductory physics course at the level of this book, it will be the only formal course in physics that they will take. Consequently, we have again presented a comprehensive and balanced overview of the subject. Some sections have been partially rewritten, some new sections added, and the motion of rigid bodies expanded into a new chapter. There is a strong emphasis on classical physics, with discussions of all the important topics, but there is also a generous amount of material on modern physics. The present edition has brought this material up-to-date, adding the most recent findings in physics and astronomy. Brief discussions are now included of such topics as: black holes, pulsars, neutron stars, the big bang theory, nucleogenesis, quarks and gluons, color forces, and the recent progress in a unified field theory. (If insufficient time is available to cover all the topics here, Chapter 17, Relativity, Chapter 20, The Structure of Matter, and the section in Chapter 21 on elementary particles can be omitted.)

To understand and appreciate the various concepts and applications of physics, considerable drill in problem solving is required. Each chapter contains a collection of worked examples covering all of the important points. Moreover, at the end of each chapter there is a list of questions to test the student's comprehension of the concepts and a generous number of problems to test his or her problem-solving abilities. In the present edition the problems have also been identified by the relevant section. The more difficult problems are indicated by an asterisk, and the answers to the odd-numbered problems are at the back of the book. Altogether, with many new additions, there are now approximately 1200 questions and problems in the 21 chapters of this edition.

For the student who wishes assistance in a planned program of study, a new Study Guide is available to accompany this text.

Many of the students who take an introductory physics course are looking toward careers in medicine, dentistry, nursing, medical technology, microbiology, chemistry, or a variety of other professions in or related to the life sciences. These students, in particular, sometimes wonder how the subject of physics plays a role in the behavior of living things. The attempt to provide a partial answer to this curiosity has led to the development of the unique aspect of both the first edition and of its continuation in the present edition, that is, the inclusion of a number of essays on bioscience topics that emphasize the importance of physical principles in the operation of living systems. These essays—generally, two or three pages in length and identified by a colored border on the pages—will be found in every chapter (except the chapter on relativity theory). Each essay is related directly to the material in the chapter of which it is a part. But the essays are supplementary to and separate from the material in the chapters themselves. That is, the essays are optional and can be completely skipped without a loss in the flow of physics ideas. However, by omitting the essays, some of the most interesting physics in the book will be lost!

One of the ideas in preparing this collection of bioscience essays was to provide, for every main physics topic, some *quantitative* life-science application. Therefore, every essay contains some numerical discussion of the topic and some numerical problems so the student can see how calculations are carried out in a different area of science.

These essays do not represent a complete survey of the many ways that physics touches upon living systems. Nor do they represent even the most profound of the topics in the biological sciences that involve physics concepts. Instead, the selection of material has been made to illustrate the impact of physics in a variety of areas, with applications that have an interest or appeal to most students and which do not require mathematics beyond the level of the main text. A final consideration was to limit the amount of material to that which could be incorporated into a typical one year physics course without overbalancing the content of the course. (The essays here amount to about 12 percent of the text.) Consequently, some of the topics that might have been included were omitted in favor of others in different areas or because of space or mathematics limitations.

We would like to thank the several persons who read the new revised text with diligence and care and who offered so many suggestions that have resulted in a significant improvement in the text: Professors Fred Becchetti (University of Michigan), Charles E. Brient (Ohio University), Irvin G. Clator (University of North Carolina), Susanta K. Ghorai (Alabama State University), Eastman N. Hatch (Utah State University), Thomas R. Manney (Kansas State University), James L. Monroe (Pennsylvania State University), and Richard M. Prior (University of Arkansas). Finally, we thank Cheryl Connor for her usual excellent work in preparing the manuscript.

Jerry B. Marion
William F. Hornyak
College Park, Maryland

Contents

Chapter 1, The Physical View of the World Around Us,   1

1.1 Physics as an Experimental Science,   1
1.2 The Fundamental Units of Measure,   2
  • Scaling and the Sizes of Things,   7
1.3 Properties of Matter,   10

Chapter 2, Motion,   21

2.1   Displacement and Path Length,   21
2.2   Average Speed and Average Velocity,   22
2.3   Distance-Time Graphs,   23
2.4   Instantaneous Velocity,   26
2.5   Acceleration,   28
  • Running Speeds,   30
2.6   The Motion of Falling Bodies,   32
2.7   Vectors,   36
2.8   Motion in Two Dimensions,   39
  • The Running Long Jump,   42
2.9   Uniform Circular Motion,   44
2.10 Accelerated Circular Motion,   47

Chapter 3, Force and Linear Momentum,   55

3.1 The Idea of Force,   55
3.2 Newton's Laws,   56
3.3 The Dimensions of Force,   59
3.4 Mass and Weight,   59
  • Bone-Breaking Forces in Jumping,   61
3.5 Applications of Newton's Laws,   62
3.6 Some Implications of Newtonian Dynamics,   67
3.7 Momentum,   69

Chapter 4, Torque and Angular Momentum,   79

4.1 Torque and Angular Momentum,   79
4.2 Center of Mass,   83
  • Distribution of Mass in the Human Body,   85
4.3 Rotation of Rigid Bodies,   86

Chapter 5, Forces in Equilibrium,   93

5.1 Static Equilibrium of Translation,   93
  • Traction Systems,   97
5.2 Friction,   99
5.3 Static Equilibrium of Rotation,   101
  • Forces in Muscles and Bones,   103

Chapter 6, Gravitation, 113

6.1 The Gravitational Force, 113
- Artificial Gravity, 117

6.2 Planetary Motion, 119
6.3 Satellite Orbits, 123

Chapter 7, Energy, 127

7.1 Work, 127
7.2 Kinetic and Potential Energy, 130
- The Energetics of Running, 131

7.3 Conservation of Energy, 135
- The Energetics of Jumping, 137

7.4 Elastic and Inelastic Collisions, 141
7.5 Rotational Kinetic Energy, 144
7.6 Gravitational Potential Energy, 145

Chapter 8, Fluids, 153

8.1 Pressure, 153
8.2 Pressure with Static Fluids, 155
8.3 Buoyancy, 158
8.4 Surface Tension, 161
8.5 Fluids in Motion, 164
8.6 The Viscous Flow of Fluids, 167
- The Flow of Blood in the Circulatory System, 169

Chapter 9, Temperature and Heat, 179

9.1 Temperature, 179
9.2 Thermal Expansion, 181
9.3 Heat, 185
- Energy and Metabolic Rates of Animals, 187

9.4 Heat Transfer, 191
- Metabolic Rates of Humans, 195

Chapter 10, Gas Dynamics 203

10.1 The Gas Laws, 203
10.2 Kinetic Theory, 207
- Diffusion, 211
- Osmosis, 214

10.3 The First Law of Thermodynamics, 216
10.4 The Second Law of Thermodynamics, 217
10.5 Heat Engines, 220
10.6 Changes of Phase, 221

Chapter 11, Elasticity and Vibrations, 231

11.1 Basic Elastic Properties of Materials, 232
- Elastic Properties of Biological Materials, 235

11.2 Simple Harmonic Motion, 238

Chapter 12, Electric Fields and Currents, 249

12.1 The Electrostatic Force, 249
12.2 Electrostatic Potential Energy, 254
12.3 Potential Difference and the Electron Volt, 255
12.4 The Electric Field, 257
12.5 Capacitance, 265
- Membrane Potentials and Nerve Impulses, 271

12.6 Electric Current, 276
12.7 Electrical Resistance, 278
12.8 Electric Power, 281

Chapter 13, Electromagnetism, 289

13.1 Magnetism, 289
13.2 Electromagnetism, 291
13.3 Effects of Magnetic Fields on Moving Charges, 293
13.4 Orbits of Charged Particles in Magnetic Fields, 296
- Electromagnetic Blood Flowmeters, 299

13.5 Magnetic Fields Produced by Electric Currents, 302
13.6 Magnetic Materials, 304
13.7 Electromagnetic Induction, 307

Chapter 14, Applied Electromagnetism, 319

14.1 Electric Circuits Containing Batteries and Resistors, 319
14.2 Capacitance and Inductance in DC Currents, 325
- Effects of Electric Current in the Human Body, 328

14.3 Electric Devices, 332

14.4 Alternating Current, 335
14.5 Electronics, 342

Chapter 15, Waves, Sound, and Radiation, 351

15.1 Traveling Waves on Strings and Springs, 351
15.2 Standing Waves, 356
15.3 Sound, 358
- The Ear and Hearing, 369

15.4 Diffraction and Interference, 372
15.5 Electromagnetic Radiation 381

Chapter 16, Light and Optics, 391

16.1 The Speed of Light, 391
16.2 Reflection and Refraction, 393
16.3 The Formation of Images by Lenses and Mirrors, 397
16.4 Color and Spectra, 407
- The Eye and Vision, 410

16.5 Polarized Light, 413
16.6 Optical Instruments, 416
- Visual Acuity, 421

Chapter 17, Relativity, 429

17.1 Light Signals—The Basis of Relativity, 430
17.2 Time and Length in Special Relativity, 432
17.3 Variation of Mass with Velocity, 437
17.4 Mass and Energy, 439
17.5 General Relativity, 441

Chapter 18, The Quantum, 449

18.1 Radiation and Quanta, 449
18.2 The Photoelectric Effect, 451
18.3 Waves or Particles? Two Crucial Experiments, 455
- The Electron Microscope, 456

18.4 The Basis of Quantum Theory, 464
18.5 The Quantization of Momentum and Energy, 466
18.6 The Uncertainty Principle, 468

Chapter 19, Atoms and Atomic Radiations, 475

19.1 Atomic Models, 475
19.2 The Hydrogen Atom, 478
19.3 Angular Momentum and Spin, 483
19.4 Quantum Theory of the Hydrogen Atom, 487
19.5 The Exclusion Principle and Atomic Shell Structure, 488
19.6 Atomic Radiations, 492
- Medical Uses of Lasers, 498

Chapter 20, The Structure of Matter, 505

20.1 Bonds between Atoms, 505
- Molecular Spectroscopy, 509

20.2 Crystals, 514
20.3 Theory of Solids, 517
20.4 Low Temperature Phenomena, 523

Chapter 21, Nuclei and Particles, 529

21.1 Nuclear Masses, 529
21.2 Radioactivity, 533
- Radiation Exposures of Humans, 541

21.3 Nuclear Reactions, 545
21.4 Nuclear Fission, 546
21.5 Nuclear Reactors, 550
- Biological Effects of Radiation Exposure in Humans, 554

21.6 Nuclear Fusion, 558
21.7 Particles and Antiparticles, 559
21.8 Hadrons: The Strongly Interacting Particles, 560
21.9 Leptons and the Weak Interaction, 563

Appendix, 571

Answers to Odd-Numbered Problems, 577

Answers to Selected Essay Exercises, 583

Chapter Opening Photo Credits 585

Index, 587

General Physics
with Bioscience Essays

The Physical View of the World Around Us

The science of physics is a growing, changing body of knowledge about the way in which Nature behaves. The physicist seeks to describe this behavior with the simplest possible models. In this chapter we examine the tools that the physicist uses in making these descriptions—mathematics and standards of measurement. Without an organized system for reporting the results of experiments (measurement standards) and without a precise language for formulating theories (mathematics), an accurate description of natural phenomena would not be possible. Therefore, we begin with a discussion of the methods we use for describing things.

Throughout this book we are concerned with the behavior of *matter*. Accordingly, in this chapter we also take a first look at the properties and structure of matter, the forms in which it occurs, and its ultimate composition. As we proceed with our discussions, we will build on these ideas and examine the workings of Nature in the large scale of the macroscopic world as well as in the small scale of the microscopic domain.

1.1 Physics as an Experimental Science

The Final Test Is in the Laboratory

The scientist seeks to learn the "truth" about Nature. In physics we can never learn "absolute truth" because physics is basically an experimental science; experiments are never perfect and, therefore, our knowledge of Nature must always be imperfect. We can only state at a certain moment in time the extent and the precision of our knowledge of Nature, with the full realization that both the extent and the precision will increase in the future. Our understanding of the physical world has as its foundation experimental measurements and observations; on these are based our theories that organize our facts and deepen our understanding of Nature.

Physics is not an armchair activity. The ancient Greek philosophers debated the nature of the physical world, but they would not test their conclusions, they would not experiment. Real progress was made only centuries later, when it was finally realized that the key to scientific knowledge lay in observation and experiment, combined with logic and reason. Of course, the formulation of ideas in physics involves a certain amount of just plain *thinking*, but when the final analysis is made, the crucial questions can only be answered by experiments.

The Philosophy of Discovery

The mere accumulation of facts does not constitute good science. Certainly, facts are a necessary ingredient in any science, but facts alone are of limited value. In order to utilize our facts fully, we must understand the relationships among them; we must systematize our information and discover how one event produces or influences another event. In doing this, we follow the *scientific method:* the coupling of observation, reason, and experiment.

The scientific method is not a formal procedure or a detailed map for the exploration of the unknown. In science we must always be alert to a new idea and prepared to take advantage of an unexpected opportunity. Progress in science occurs only as the result of the symbiotic relationship that exists between observational information and the formulation of ideas that correlate the facts and allow us to appreciate the interrelationships among the facts. The scientific method is actually not a "method" at all; instead, it is an attitude or philosophy concerning the way in which we approach the real physical world and attempt to gain an understanding of the way Nature works.

Johannes Kepler (1571–1630), the greatest of the early astronomers, followed the scientific method when he analyzed an incredible number of observations of the positions of planets in the sky. From these facts he was able to deduce the correct description of planetary motion: the planets move in elliptical orbits around the Sun.

Kepler's procedure—amassing facts and trying various hypotheses until he found one that accounted for all the information—is not the only way to utilize the scientific method. When Erwin Schrödinger was working on the problems associated with the new experiments in atomic physics in the 1920s, he set out to find a description of atomic events that could be formulated in a mathematically beautiful way. Schrödinger deviated from the "normal" procedure of the scientific method. Instead of closely following the experimental facts and attempting to relate them, he sought only to find an aesthetically pleasing mathematical description of the general trend of the results. This pursuit of mathematical beauty led Schrödinger to develop modern quantum theory. In the realm of atoms, where quantum theory applies, Nature does indeed operate in a beautiful way. At essentially the same time, Werner Heisenberg followed the more conventional approach and formulated an alternative version of quantum theory, which is equivalent to the theory constructed by Schrödinger in every respect that can be experimentally tested.

The Language of Physics

One of the significant steps forward in our understanding of the behavior of Nature was the realization that it is the Earth that moves around the Sun, and not the Sun that moves around the Earth. The simple statement that "the Earth moves around the Sun" represents a new dimension in physical thinking. As important as this idea is, nevertheless, it is incomplete. We cannot say that we really understand a physical phenomenon until we have reduced the description to a statement involving *numbers*. Physics is a precise science and its natural language is mathematics. Only when Johannes Kepler gave a mathematical description of planetary motion and Isaac Newton derived the same results on the basis of his theory of universal gravitation could it be said that a proper analysis of the motion of the Earth and the planets had finally been made.

Although the most sophisticated mathematics can be used in developing physical theories, in this book we restrict the use of mathematics to algebra, geometry, and trigonometry. The reader will find a mathematical review in the appendices at the end of this book.

1.2 The Fundamental Units of Measure

Units and Standards

In our subsequent discussions we will encounter a variety of physical quantities—for example, length, time, mass, force, momentum, energy, and so forth. These quantities not only have magnitudes but they have *dimensions* and *units* as well. It makes no sense to state that a certain length is 12—we must also specify the units in which the magnitude has this value. Whether the length is 12 centimeters or 12 miles makes a considerable difference!

It is also necessary to have *standards* for the units of physical measure. If we state that the size of a building lot is 30 paces by 60 paces, we have only a crude idea of the area. But if we state that the size is 20 meters by 40 meters, we know the area precisely because the *meter* is a well-defined and standard unit of length.

In this book we will use the metric system of physical measure. The particular metric units and abbreviations used will be those that have been recommended by the commission on "Le Système International d'Unités." We will refer to these as SI units.

The Standard of Length

Although an enormous number of units for the specification of length have been invented, only those of the British system and metric systems survive today. The unit of length in the British system is the *yard;* the derived units are the *foot* ($\frac{1}{3}$ yard), the *inch* ($\frac{1}{36}$ yard), and the *statute mile* (1760 yards). In the metric system the unit of length is the *meter,* which was originally conceived as 10^{-7} of the distance from the equator to the North Pole along a meridian passing through Paris. In order to provide a more practical standard, in 1889 the meter was officially defined as the distance between two parallel scribe marks on a specially constructed bar of platinum-iridium.

The meter-bar definition of the meter suffers from two disadvantages: not only is the precision inadequate for many scientific purposes, but comparisons of lengths with a bar that is kept in a standards laboratory are quite inconvenient. These difficulties have been overcome with the definition, by international agreement in 1961, of a *natural* unit of length based on an atomic radiation. Because all atoms of a given species are identical, their radiations are likewise identical. Therefore, an atomic definition of length is reproducible everywhere. We now accept as the standard of length the wavelength of a particular orange radiation emitted by krypton atoms. The standard was arrived at by carefully measuring the length of the standard meter bar in terms of the wavelength of krypton light. It was then decided that exactly 1,650,763.73 wavelengths would constitute 1 meter. This definition is then consistent with the previous definition in terms of the distance between the scribe marks on the meter bar, but is has the advantage of being approximately 100 times as precise. Now the standard can be reproduced in many laboratories throughout the world, not in standards laboratories alone.

In the early part of this book we will occasionally refer to the British system of units for length (with which you are probably more familiar). When we have completed the introductory material, we will forgo completely the use of British units.

Table 1.1 gives the conversion factors connecting some of the units of length in the metric and British systems. Notice that the inch, the foot, and the yard are now defined *exactly* in terms of centimeters. Thus, the krypton wavelength is the standard of length for both systems.

Table 1.2 shows in a schematic way the enormous range of distances we encounter in the Universe. Notice that there is a factor of 1000 between successive marks on the vertical scale. Between the smallest and the largest things about which we have any comprehension, the span is more than 40 factors of 10!

Table 1.3 gives the values of some of the distances that we will find useful in our discussions.

Table 1.1 Conversion Factors for Length

1 cm = 0.3937 in.	1 in. = 2.54 cm
1 m = 3.281 ft	1 ft = 30.48 cm } exactly
1 km = 0.6214 mi	1 yd = 91.44 cm
= 3281 ft	1 mi = 5280 ft
	= 1.609 km

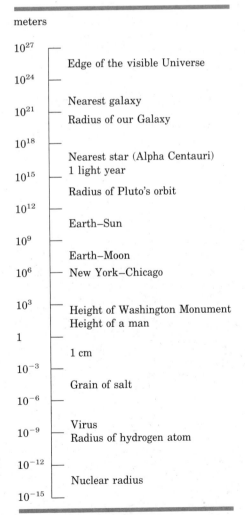

Table 1.2 The Range of Distances in the Universe

meters	
10^{27}	
	Edge of the visible Universe
10^{24}	
	Nearest galaxy
10^{21}	Radius of our Galaxy
10^{18}	
	Nearest star (Alpha Centauri)
10^{15}	1 light year
	Radius of Pluto's orbit
10^{12}	
	Earth–Sun
10^{9}	
	Earth–Moon
10^{6}	New York–Chicago
10^{3}	
	Height of Washington Monument
	Height of a man
1	
	1 cm
10^{-3}	
	Grain of salt
10^{-6}	
	Virus
10^{-9}	Radius of hydrogen atom
10^{-12}	
	Nuclear radius
10^{-15}	

4 The Physical View of the World Around Us

Table 1.3 Some Useful Distances

To Alpha Centauri (nearest star)	4.04×10^{16} m
1 light year (L.Y.)	9.460×10^{15} m
1 astronomical unit (A.U.) (Earth–Sun distance)	1.496×10^{11} m
Radius of Sun	6.960×10^{8} m
Earth–Moon distance	3.844×10^{8} m
Radius of Earth	6.378×10^{6} m
Wavelength of yellow sodium light	5.89×10^{-7} m
1 Ångstrom (Å)	10^{-10} m
Radius of hydrogen atom	5.292×10^{-11} m
Radius of proton	1.2×10^{-15} m

Table 1.4 Relative Precision of Various Types of Clocks

	Precision	
Type of Clock	1 s in	1 part in
Hour glass	1.5 min	10^{2}
Pendulum clock	3 h	10^{4}
Tuning fork	1 day	10^{5}
Quartz crystal oscillator	3 y	10^{8}
Ammonia resonator	30 y	10^{9}
Cesium resonator	3×10^{4} y	10^{12}
Hydrogen maser	3×10^{6} y	10^{14}

The Standard of Time

The unit of time in both the British and metric systems is the *second,* which, until recently, was defined as 1/86,400 of the mean solar day. The determination of time by observation of the rotation of the Earth is inadequate for high precision work because of minor but quite perceptible changes in the speed of the Earth's rotation.

In order to improve the precision of time measurements, in 1967 we adopted a *natural* unit for time just as we had done previously for a length standard. Our present-day *atomic clocks* (Fig. 1.1) depend on the characteristic vibrations of cesium atoms. The *second* is defined as the time required for 9,192,631,770 complete vibrations to occur in cesium. With this definition of the second, it is possible to compare time intervals to 1 part in 10^{12}, which corresponds to 1 second in 30,000 years. Current research with other atomic vibrations indicates that we will soon have a clock that will be precise to 1 part in 10^{14} or to 1 second in 3 million years!

Time standards for practical working purposes are provided by radio station WWV, located in Fort Collins, Colorado, and operated by the National Bureau of Standards. WWV operates on frequencies of 2.5, 5, 10, 15, 20, and 25×10^{6} cycles per second, which are controlled to 1 part in 10^{10} by comparison with a cesium clock. A beat is given every second and 10 times per hour the time is given by voice. Several other countries also maintain radio stations that broadcast time signals.

Table 1.5 shows the range of time intervals that we encounter in the Universe. Notice that the span from the shortest to the longest time interval is greater than 40 orders of magnitude, about the same as the range of distances shown in Table 1.2.

The Standard of Mass

The international standard of mass is a cylinder of platinum-iridium,[1] which is defined as 1 kilogram = 10^{3} grams.

It would, of course, be highly desirable to have an atomic standard for mass just as we have for length and time. We do have such a standard that is used in the comparison of the masses of atoms and molecules, but, unfortunately, we have no precision method at present of utilizing this standard above

National Bureau of Standards

Figure 1.1 A cesium clock constructed at the National Bureau of Standards (Boulder Laboratories). This clock can measure time intervals to a precision equivalent to 1 second in 30,000 years.

---

[1] Originally, the kilogram was defined as the mass of 1000 cm³ of water.

Table 1.5 The Range of Time Intervals in the Universe

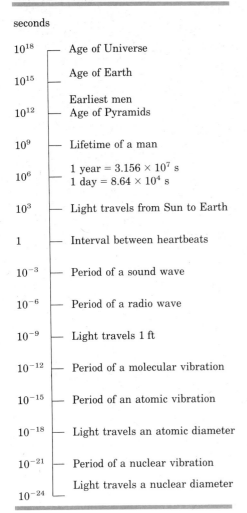

Table 1.6 The Range of Masses in the Universe

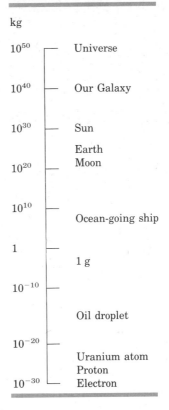

Table 1.7 Some Important Masses

Object	Mass (kg)
Sun	1.991×10^{30}
Earth	5.997×10^{24}
Moon	7.35×10^{22}
Proton	1.672×10^{-27}
Electron	9.108×10^{-31}

the level of individual atoms and molecules. When the technology has developed to the point that we can determine precisely the mass of the standard kilogram in terms of the atomic mass standard, we will certainly adopt an atomic unit of mass as our standard.

Table 1.6 shows some of the marker points on the gigantic range of masses that we find in the Universe. Values of some of the more important masses are given in Table 1.7.

In this book we will use only the metric system of units for mass. However, common and legal usage still refers to the British system of units for mass. Although certain ambiguities exist for this unit, it is sufficient for our purposes to point out that an object stated to have a weight of a pound (lb) has, by legal definition, a mass of exactly 0.45359237 kilograms (kg). Thus a student of weighing 150 lb would have a mass of 68 kg.

The Use of Units

All physical quantities have *dimensions* and *units*. When we make numerical statements or write numerical equations concerning physical quantities, we must include the units of the quantities. If we make the statement, "The distance traveled is equal to the speed multiplied by the time," we could express this more briefly as

$$\text{distance} = \text{speed} \times \text{time}$$

or, by giving arbitrary symbols to the quantities, as

$$d = s \times t$$

This equation does not explicitly contain the units of the various quantities; the equation is valid for any system of units as long as they are used consistently. For example,

$$120 \text{ mi} = 60 \frac{\text{mi}}{\text{h}} \times 2 \text{ h}$$

Not only must the *numbers* balance in such an equation, but so must the *units*. On the right-hand side of the equation, the time unit "hour" occurs in both numerator and denominator and therefore cancels, leaving "miles" as the unit on both sides of the equation.

We can alter the unit of any quantity by using *conversion factors*. In order to convert 60 mi/h to the corresponding number of m/s, we use the conversions

1 h = 3600 s, 1 mi = 1.609 km, and
1 km = 1000 m

Thus, we can form the ratios

$$\frac{1 \text{ h}}{3600 \text{ s}} = 1, \quad \frac{1{,}609 \text{ km}}{1 \text{ mi}} = 1, \text{ and}$$

$$\frac{1000 \text{ m}}{1 \text{ km}} = 1$$

Because we can multiply any quantity by a factor of *unity* without altering the value (we only change the *scale*), we can write

$$60 \frac{\text{mi}}{\text{h}} = 60 \frac{\text{mi}}{\text{h}} \times \frac{1 \text{ h}}{3600 \text{ s}} \times \frac{1.609 \text{ km}}{1 \text{ mi}} \times \frac{1000 \text{ m}}{1 \text{ km}}$$

$$= \frac{60 \times 1.609 \times 1000 \text{ m}}{3600 \text{ s}}$$

$$= 26.8 \text{ m/s}$$

Note how "miles," "kilometers," and "hours" have canceled in this expression.

Any unit that appears in the denominator can always be written with a negative exponent; for example,

$$30 \text{ m/s} = 30 \text{ m s}^{-1}$$

An exponent $^{-1}$ always means *per*. It is sometimes convenient to use this notation.

Always give units when writing the numerical values of physical quantities. Be alert for the need in some instances to specify the algebraic sign (+ or −) for the calculated quantity. *Always* check equations to ensure that the units on both sides are the same (or are equivalent in the sense that they are related by a conversion factor such as demonstrated above); if the units are *not* the same, something is wrong!

Your pocket calculator is probably capable of giving the results of calculations to a large number of significant figures. For our purposes it will never be necessary to go beyond carrying calculations to a final result quoted to three significant figures, as in the above calculation. Assume that when a problem quotes a length to be 2 m for example, it is intended to be an exact value.

The Three Basic Units

The units of all mechanical quantities can be expressed in terms of the basic units of length, mass, and time. When we introduce such quantities as *force* and *energy*, for convenience we shall give special names to the units (newtons and joules), but these units are defined as certain combinations of length, mass, and time. These three units—meter, kilogram, and second—are all that we require; every mechanical quantity can be expressed in terms of these units.

Later, when treating heat and electricity, other basic units will have to be introduced.

Approximations and Estimates

Although the science of physics attempts to describe natural phenomena in terms that are as precise as possible, there are many occasions on which an approximation or even a crude estimate is quite adequate. For example, if we wish to describe the motion of the Earth around the Sun, it is not necessary to take account of the Earth's geologic features or its internal structure. Indeed, we obtain highly accurate results by making the approximation that the Earth is a *particle*, that its size is unimportant for the effects under consideration. If we shift our attention to the study of volcanoes or earthquakes, then, clearly, the internal structure of the Earth is of crucial importance.

Every situation should be examined to determine what approximations can be made to simplify the problem. We should always ask, "What features of this event or phenomenon can we neglect so that the calculations will be easier?" The features we find to be negligible in one situation may not be so in another. In one problem we may include the effects of gravity but neglect friction. In another problem friction may be important and the gravitational effect may be negligible.

Numerical approximations are also useful in judging whether a calculation is proceeding properly. Before carrying out a lengthy computation, it is often profitable to obtain an estimate of the answer by making such approximations as $3.7 + 5.4 \cong 10$, $\pi \cong 3$, $\sqrt{2} \cong 1.5$, and so forth. By quickly calculating the numerical coefficients and combining the powers of 10, you will be able to see whether your answer is going to be reasonable. If your result turns out to be that the size of an atom is 2×10^{-3} m, there is no point in improving the accuracy of the calculation until you correct the gross error!

You will also find it useful to develop your powers of estimation. Frequently, just a little thought will allow you to obtain an estimate (even a crude estimate) for something you may never have considered before. Try these:

How many cars are involved in a 2-mile traffic jam on a three-lane highway?

How many golf balls could be placed in your bedroom?

How many miles of interstate highway are there in the United States?

How many barrels of oil are required to supply the American public with gasoline for their automobiles for a year? (You need to know how many gallons of gasoline can be produced from a barrel of oil. The yield is about 20 gallons per barrel. Could you have estimated that figure?)

(Answers will be found at the end of this chapter.)

Scaling and the Sizes of Things

We see in the world around us a variety of living things that have vastly different sizes. The smallest cells have a size of about 10^{-6} m, whereas the largest living things, the giant sequoia trees, grow to heights in excess of 100 m. This range in size of living matter amounts to a factor of 10^8 or 100 million.

The characteristics and the function of an organism are related to its size. A rabbit scaled up to the size of an elephant would not be a viable creature. Nor could humans exist if they were the size of a mouse. To see how the life-style of an organism is determined by its size, we must examine the idea of *scaling*.

The biological properties of an organism depend to a remarkable extent on its geometrical properties of length, surface area, and volume. For regular geometrical objects, it is easy to relate these properties. To do this in a general way, we use the idea of a *characteristic length L* for an object. For a cube, the characteristic length is the length of a side; for a sphere, the characteristic length is the radius. Then, we know that

$$\text{surface area} \propto L^2; \quad \text{volume} \propto L^3$$

Moreover, if the object has a uniform density, the mass of the object is also proportional to the cube of the characteristic length.

To *scale* an object means to change its characteristic length by some factor. The area and the volume then scale according to the relations above. If the characteristic length is doubled, then the surface area increases by a factor of 4 and the volume increases by a factor of 8. Many biological properties depend upon the ratio of the surface area of an organism to its volume. This ratio is determined by the organism's characteristic length:

$$\frac{\text{surface area}}{\text{volume}} \propto \frac{L^2}{L^3} = \frac{1}{L}$$

The characteristic length of a regular geometrical object, such as a cube or a sphere, is easy to define. But what is the characteristic length associated with an ant, or a dog, or a human? Because we are interested only in approximate, not precise comparisons of the properties and functions of different organisms, we can use for the characteristic length any obvious or convenient length. Thus, the value of L for a human will be height (about 2 m); for a dog, $L \cong 1$ m; and for an ant, $L \cong 0.5$ cm.

Now, let us look at some of the biological properties of organisms that depend on the characteristic length.

Strength

How can we compare the strengths of organisms of different sizes? A full-grown human (mass $\cong$ 80 kg) can lift an object with a mass about equal to his or her own mass. A grasshopper (mass $\cong$ 1 g), on the other hand, can lift an object with a mass about 15 times its own mass. Does this mean that a grasshopper is "stronger" than a human? To answer this question, we need to refer to the features of the source of strength, that is, muscles. All muscle is composed of bundles of muscle fibers. These fibers are all quite similar and are packed with about the same density in the muscles of different organisms. The strength of a muscle is, to a sufficient approximation, directly proportional to the number of fibers in the muscle, that is, to the cross-sectional area of the muscle. The muscle area of an organism, again to a sufficient approximation, is proportional to a characteristic cross-sectional area, that is, to the square of its characteristic length. Thus,

$$\text{strength} \propto L^2$$

In order to compare the strengths of two organisms with different sizes, we should use the strength per unit mass. We can call this quantity the *specific strength*:

$$\text{specific strength} = \frac{\text{strength}}{\text{mass}} \propto \frac{L^2}{L^3} = \frac{1}{L}$$

where we have used the fact that the mass is proportional to the cube of the characteristic length. Now, to compare the specific strengths of the grasshopper ($L \cong 2$ cm) and a human ($L \cong 2$ m = 200 cm), we write

$$\frac{\text{specific strength of grasshopper}}{\text{specific strength of human}} = \frac{\frac{1}{L(\text{grasshopper})}}{\frac{1}{L(\text{human})}}$$

$$= \frac{L(\text{human})}{L(\text{grasshopper})} = \frac{200 \text{ cm}}{2 \text{ cm}}$$

$$= 100$$

That is, simply because of its smaller size, the grasshopper should have a specific strength 100 times that of a human. But, as pointed out above, a human can lift a mass about equal to his or her own mass, whereas a grasshopper can lift a mass about 15 times its own mass. Thus, the actual strength of a grasshopper falls short of that expected on the basis of its size. A human is able to use his or her muscular capacity more efficiently than can a grasshopper.

Food Requirements

The rate at which heat is generated in an animal's body is proportional to the amount of food intake which, in turn, is proportional to the mass (or the volume) of the animal. The rate at which heat is lost by an animal to its surroundings is proportional to the surface area of the animal's body. (Moreover, for warm-blooded animals whose body temperature remains constant, the rate of heat loss is greater in cold climates than in hot climates). The heat loss of an animal per unit body mass is therefore inversely proportional to the animal's characteristic length. A very small animal must replace the loss of body heat by an almost continual intake of food. Such an animal must daily eat an amount of food that is a large fraction of its body mass. A mouse each day eats food with a mass equal to about one-quarter of its own mass. The tiny shrew will die of starvation if it goes without food for more than about 3 hours. A giant elephant, on the other hand, has the opposite problem. Its rate of heat generation taxes the ability of its skin area to dispose of the excess. For this reason, elephants take every opportunity to cool themselves at waterholes. Insects, whose ratio of surface area to volume is relatively much larger than that of any warm-blooded animal, cannot take in food at a rate sufficient to maintain a constant body temperature. For insects this does not constitute a problem, however, because they are cold-blooded creatures with a body temperature always equal to the ambient temperature. By this mechanism the heat losses, and consequently the food intake requirements, are greatly reduced. (We will examine this problem with a more sophisticated approach in the essay on page 187).

Terminal Velocity

If you fall to the ground from a third-floor window, you will probably suffer a serious injury. However, if a flea or other small insect experiences such a fall it will survive uninjured. Even a small animal, such as a mouse, will probably be able to walk away after a fall from this height. The reason that a small animal or insect can negotiate a substantial fall without serious effects is due to its relatively large surface-area-to-volume ratio. When an object or an organism falls through the air, the frictional resistance of the air retards the motion and prevents the speed of fall from increasing beyond a certain value.

The downward force on a falling object is due to gravity and is proportional to the mass (or the volume) of the object. The effect of air resistance is to produce a retarding force that depends on the cross-sectional area of the object and on its velocity. Therefore, as an object falls, its velocity increases until the retarding force becomes equal to the downward force. When this point is reached, the object continues its fall at constant velocity. This maximum velocity of free fall is called the *terminal velocity*. The greater the area-to-volume ratio of an object or organism, the smaller will be the terminal velocity. The terminal velocity of a falling human is about 65 m/s (about 145 mi/h) if he or she is "spread eagled" the way sky divers fall; if he or she assumes a ball-like shape, the terminal velocity will increase to about 105 m/s (about 235 mi/h). The maximum rate of fall for a small insect is a few meters per second; no significant injury will result from an impact with such a small velocity. (See Section 8.6 for more details concerning terminal velocity.)

References
R. McNeill Alexander, *Size and Shape,* Edward Arnold, London, 1971.

J.B.S. Haldane, *Possible Worlds,* Harper and Brothers, New York, 1928. (Reprint of essay, "On Being the Right Size" in *The World of Mathematics,* J. R. Newman, ed., Simon and Schuster, New York, 1956, p. 952.)

F. R. Hallett, P. A. Speight, and R. H. Stinson, *Introductory Biophysics,* Halsted (Wiley), New York, 1977, Section 7.9.

■ Exercises

1. Some science fiction stories have been written about giant mutant insects that terrorize the populace. Could an ant that had grown to the size of a human actually move about? Explain.

2. Could a 20-story building be built in the same way as a 2-story frame dwelling? Explain.

3. If a dog were scaled up to the size of an elephant, the diameter of the elephant's legs would still be greater than that of the dog's legs. Why is this so?

4. Why are small animals such as mice and shrews not found in the arctic regions?

5. Cells absorb nutrients through their surfaces. Why do cells grow only to a certain size before they divide?

1.3 Properties of Matter

As we proceed with our discussions of the behavior of physical systems, we will encounter *matter* in a variety of forms. Before we begin these discussions, we take this opportunity to describe briefly some of the important features of matter.

Conservation of Mass

Mass is the first *property of matter* (as distinct from the basically geometrical concepts of length and time) that we have introduced. We have come to appreciate the permanence and the immutability of mass. A block of metal has a certain mass. This mass does not change with time. Nor does it change when the dimensions are changed; we can alter the shape by hammering or forging, but the mass does not change. We can even dissolve it in acid, but the combined mass of the acid and the metal (together with any gases that might be evolved in the process) remains unchanged.

These facts concerning mass constitute the first of a series of important fundamental statements called the *conservation laws* of physics. Instead of emphasizing the *differences* between various physical processes, in order to understand the fundamentals of Nature it is more profitable to seek those properties that are the *same* in any process. The conversion of water into ice, and the conversion of water into its constituent gases, oxygen and hydrogen, are two different physical processes. But a common feature of these processes is that the *mass* of the material remains constant. A long series of experiments, all of which exhibit this feature, has led us to conclude that *mass is conserved* in all physical processes. This is the first of several *conservation laws* that we will discuss.

Having argued that the conservation of mass is an important physical law, it must now be pointed out that mass is not really a conserved quantity! We consider mass to be conserved because in all ordinary processes we can detect, even with the most sensitive instruments, no change in the mass of an object. However, Einstein's relativity theory has shown us (and countless experiments have verified) that mass and energy are intimately connected and one can be exchanged for the other in the proper circumstances. Therefore, the conservation law properly refers, not to mass (or energy) alone, but to *mass-energy.* The conditions under which exchanges of mass and energy are detectable do not occur in ordinary processes. Consequently, we usually treat mass and energy as separately conserved quantities. (We will discuss mass-energy in Chapter 17.)

Mass Density

One of the important features of a particular type of matter that distinguishes it from other types of matter is its *density.* We say that a piece of lead is "heavy" and that a piece of wood is "light." By such a statement we mean that the mass of a certain volume of lead is much greater than the mass of the same volume of wood. That is, we are comparing the *mass per unit volume* of the two substances. This quantity, (mass)/(volume), is the *mass density*:

$$\text{density} = \frac{\text{mass}}{\text{volume}}$$

$$\rho = \frac{M}{V} \qquad (1.1)$$

In the metric system, densities are expressed either in kg/m³ or in g/cm³ (see Table 1.8). It is probably easier to think of densities in g/cm³ because in these units the density of water is 1. Some densities we will find useful are given in Table 1.9.

● *Example 1.1*

What is the average density of the Earth?
The radius of the Earth is 6.38×10^6 m. Therefore, the volume is

$$V = \tfrac{4}{3}\pi R^3 = \tfrac{4}{3}\pi \times (6.38 \times 10^6 \text{ m})^3$$
$$= 1.09 \times 10^{21} \text{ m}^3$$

Table 1.8 Some Representative Mass Densities

Type of Matter	Density g/cm³	kg/m³
Nuclear matter	10^{14}	10^{17}
Center of Sun	10^2	10^5
Water	1	10^3
Air	10^{-3}	1
Laboratory high vacuum	10^{-18}	10^{-15}
Interstellar space	10^{-24}	10^{-21}
Intergalactic space	10^{-30}	10^{-27}

Table 1.9 Densities of Different Substances

Substance	Density g/cm³	kg/m³
Gold	19.3	1.93×10^4
Mercury	13.6	1.36×10^4
Lead	11.3	1.13×10^4
Copper	8.93	8.93×10^3
Iron	7.86	7.86×10^3
Aluminum	2.70	2.70×10^3
Water	1.00	1.00×10^3
Ice	0.92	9.2×10^2
Alcohol (methyl)	0.81	8.1×10^2
Air (standard conditions)	1.29×10^{-3}	1.29
Helium (standard conditions)	1.78×10^{-4}	0.178
Hydrogen (standard conditions)	0.90×10^{-4}	0.090

The mass of the Earth is 5.98×10^{24} kg, so the average density is

$$\rho = \frac{M}{V} = \frac{5.98 \times 10^{24} \text{ kg}}{1.09 \times 10^{21} \text{ m}^3}$$
$$= 5.49 \times 10^3 \text{ kg/m}^3 \text{ (or } 5.49 \text{ g/cm}^3\text{)}$$

This is the density averaged over the entire Earth; actually, the core has a density of about 12 g/cm³ and the mantle (the region near the surface) has a density of about 3 g/cm³.

● *Example 1.2*

A piece of machinery can be constructed either from steel ($\rho = 7.88$ g/cm³) or from a magnesium-aluminum alloy ($\rho = 2.45$ g/cm³). When steel is used, the mass of the piece is 46 kg. What would be the mass if the alloy were used?

The volume of the piece can be determined from the mass when steel is used:

$$V = \frac{M}{\rho} = \frac{46 \text{ kg}}{7.88 \times 10^3 \text{ kg/m}^3} = 5.84 \times 10^{-3} \text{ m}^3$$

(Notice that the density must be expressed in kilograms per cubic meter if the mass is expressed in kilograms.) Now, the mass of the piece when the alloy is used is obtained from

$$M = \rho V = (2.45 \times 10^3 \text{ kg/m}^3) \times (5.84 \times 10^{-3} \text{ m}^3)$$
$$= 14.3 \text{ kg}$$

Thus, a considerable reduction in mass would result if the magnesium-aluminum alloy were used instead of steel. ■

The States of Matter

The matter that we see around us occurs in three different forms: solid, liquid, and gas. One of the Earth's most abundant substances—water—is familiar to us in all three forms. At ordinary room temperature, water is a liquid, but at low temperatures it freezes to become solid ice and at high temperatures it vaporizes to become gaseous steam. Many other substances behave in the same way if the temperature is changed by a sufficient amount. For example, we usually consider iron to be a solid, but at very high temperatures iron will melt and at still higher temperatures it will vaporize. Also, oxygen is a gas under ordinary conditions, but as the temperature is decreased to very low values, oxygen will first liquefy and then solidify.

What determines whether a particular substance is a solid or a liquid or a gas? In any sample of matter, forces of attraction exist between the atoms that make up the substance. In solids these forces are strong and the atoms are held firmly in place. In gases the interatomic forces are extremely weak and each gas atom moves freely among the other atoms. In liquids the forces are sufficiently strong to maintain the matter in a condensed state but they are not strong enough to prevent the atoms from sliding easily through the surrounding atoms.

When heat is added to a substance, the temperature increases and the constituent atoms move more rapidly. For a solid, this increased motion is in the form of vibrations around the fixed average positions of the atoms. If a sufficient amount of heat is added, the atoms will vibrate so violently that the interatomic bonds will be broken. The atoms will then be able to move past one another; that is, the substance begins to flow. The temperature at which a substance changes from the solid to the liquid state is called the *melting point* (which is the same as the *freezing point* for the liquid-to-solid transition).

In many types of solids, the atoms are arranged in regular repeating patterns and form *crystals* (see Fig. 1.2). In a bulk sample of a crystalline material, quartz crystals. Sodium chloride (ordinary table salt) occurs as crystals, and the tiny cubes are easy to see with a low-power magnifier. Many other everyday substances are also crystalline, although the crystals are often so small that we do not recognize them and therefore do not think of the material as crystalline. An iron rod, for example, does not appear to be a crystal; yet, if you clean the surface and examine the iron with a microscope, you will see

Figure 1.3 Natural quartz crystals.

that it consists of an enormous number of tiny crystalline pieces that are bonded tightly together.

When the temperature of a crystalline material is increased, melting occurs at a well-defined temperature because all of the interatomic bonds are the same and are therefore broken by the same degree of atomic agitation. On the other hand, substances such as glass, tar, and plastics do not exist in crystalline form. The interatomic bonds in these materials are not all the same and they do not all break at the same temperature. Consequently, noncrystalline materials do not exhibit sharp melting points. If you heat a piece of plastic, you will find that it softens gradually instead of changing suddenly from solid to liquid.

Any change of state that involves an increase in the mobility of the atoms requires the addition of heat to break the interatomic bonds. Thus, liquid-to-gas, solid-to-liquid, and solid-to-gas changes all require that heat be added to the substance. The reverse changes take place when heat is removed from the substance. The various changes of state are summarized in Fig. 1.4.

Most everyday substances exist in one (or more) of the three ordinary states of matter. However, there is one additional state of matter that is extremely important even though it is in evidence in only a few of our everyday materials. This fourth state of matter is *plasma*. If we begin with a solid substance and continually add heat, the substance first changes into a liquid and eventually into a gas. Further heating causes the atoms to move even more rapidly and the collisions between the moving

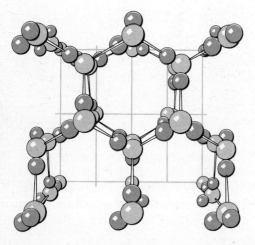

Figure 1.2 The crystal structure of ice. The gray spheres represent oxygen atoms and the brown spheres represent hydrogen atoms. The sticks represent the electrical bonds that exist between the oxygen atom of one molecule and a hydrogen atom of a neighboring molecule. (Adapted from A. M. Buswell and W. H. Rodebush, Scientific American, April, 1956.)

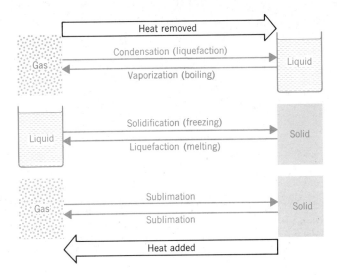

Figure 1.4 Heat transfers involved in the changes of state of matter. (Notice that the gas-to-solid and solid-to-gas transitions are both called *sublimation*.)

atoms become more frequent and more violent. When the atomic speeds have increased sufficiently, the collisions can cause electrons to be knocked off the atoms. At extremely high temperatures, only a few, or perhaps no electrons remain attached to the atomic nuclei. The substance is then a plasma—highly charged atoms (*ions*) moving in a sea of electrons. On Earth, plasmas are produced only in a limited number of special situations (such as in fluorescent lighting tubes and in lightning strokes). But in stars, plasma is the normal, not the exceptional form of matter. Indeed, plasma is the most common state of matter in the Universe!

Atoms

By the end of the eighteenth century, chemists had formulated a reasonably precise idea of what constitutes a pure, chemically identifiable substance. Moreover, the chemists had been able to identify a number of substances that we call chemical *elements*. An element is a substance that cannot be decomposed into simpler constituents by any chemical means. Carbon, oxygen, sulfur, and copper are elements. The various elements can combine with one another to produce chemical *compounds*. The early chemists recognize that there are relatively few elements (we now know just over a hundred), whereas the number of possible chemical compounds is enormous.

Modern atomic theory began with the enunciation by Joseph Louis Proust (1754–1826) of the *law of definite proportions*. This law states that when two elements combine to form a distinct chemical compound, they do not combine indiscriminately but with a definite ratio of masses that is characteristic of the combining elements and the compound that is formed. Thus, when water is formed from the elements hydrogen and oxygen, 1 g of hydrogen always combines with 8 g of oxygen to form 9 g of water. The same is true for every other chemical compound—there is always a definite proportion of each element present in the compound regardless of the origin of the particular sample.

It was John Dalton (1766–1844), an English chemist, who first clearly recognized that Proust's law of definite proportions implies the existence of some fundamental unit for each chemical element. How can it be that 1 g of hydrogen always combines with exactly 8 g of oxygen to produce water? The answer must be that $\frac{1}{2}$ g of hydrogen combines with 4 g of oxygen, that $\frac{1}{4}$ g of hydrogen combines with 2 g of oxygen, that $\frac{1}{8}$ g of hydrogen combines with 1 g of oxygen, and so on. That is, there is some fundamental unit of hydrogen that combines with some fundamental unit of oxygen in such a way that the ratio of hydrogen mass to oxygen mass is 1:8. Dalton therefore concluded that Proust's law was sufficient evidence to suppose that every element is composed of individual fundamental units of matter—these units are *atoms*.

Methods have been developed for measuring with high precision the mass of one species of atom relative to another. Thus, we know that the mass of a uranium atom is 238 times that of a hydrogen atom and that the mass of a carbon atom is 12 times that of a hydrogen atom. In order to specify atomic masses we have adopted a special unit called the *unified atomic mass unit* (u). We define 1 u to be exactly $\frac{1}{12}$ of the mass of the carbon atom (the nucleus of which, as we will see later in this section, consists of 12 particles, six protons and six neutrons). In terms of u the masses of all atoms are approximately (but not exactly) integer numbers. The mass of a gold atom is 196.967 u and that of an aluminum atom is 26.9815 u. The atomic mass of the lightest element, hydrogen, is 1.007825 u. It is often sufficient for our purposes to give a particular atomic mass as an integer number of u instead of listing the precise value (but see the discussion in Section 21.1).

When two or more atoms combine to form a *molecule*, we specify the mass of the molecule in u by adding together the atomic masses of the constituent atoms. Table 1.10 lists the masses of some common atoms and molecules in u. Notice that both atomic and molecular masses are given for hydrogen and oxygen. We can measure the masses of individual atoms of these elements, but in Nature hydrogen

Table 1.10 Approximate Masses of Some Common Atoms and Molecules

Substance	Symbol	Mass u
Atoms		Atomic mass[a]
Hydrogen	H	1
Helium	He	4
Carbon	C	12
Oxygen	O	16
Aluminum	Al	27
Iron	Fe	56
Uranium	U	238
Molecules		Molecular mass[a]
Hydrogen	H_2	2
Methane	CH_4	16
Water	H_2O	18
Nitrogen	N_2	28
Oxygen	O_2	32
Methyl alcohol	CH_3OH	32
Butane	C_4H_{10}	58

[a] The actual masses differ slightly from these integer values. Some individual isotopic atomic masses are given in Table 1.11.

and oxygen occur as two-atom (*diatomic*) molecules, H_2 and O_2.

A variety of experimental techniques have been used to determine the connection between the u and the kilogram:

$$1 \text{ u} = 1.6605 \times 10^{-27} \text{ kg} \qquad (1.2)$$

Thus, the mass of a hydrogen atom is

$$m_H = (1.007825 \text{ u}) \times (1.6605 \times 10^{-27} \text{ kg/u})$$
$$= 1.6735 \times 10^{-27} \text{ kg}$$

The *relative* masses of atoms can be measured with high precision. We can, for example, state the mass in u of an iron atom or a gold atom to a precision of better than 1 part in 10^7. However, the connection between the u and the kilogram (Eq. 1.2) is known to a precision of only about 1 part in 10^4. This precision is not sufficient to permit the use of an atomic standard for mass in the way that we use atomic standards for length and time. When our methods of measurement have been improved, we will be able to use atomic standards for all of the fundamental physical quantities.

Avogadro's Number and the Mole

One of the important ideas in molecular theory is the concept of the *mole* as a means for measuring the quantity of a substance. We can reason in the following way. The mass of an iron atom is 56 u and the mass of a methane molecule is 16 u (see Table 1.10). Therefore, we can write

$$\frac{\text{mass of iron atom}}{\text{mass of methane molecule}} = \frac{56}{16}$$

Now, if we take a sample of N atoms of iron and N molecules of methane, the mass ratio will be the same:

$$\frac{\text{mass of } N \text{ atoms of iron}}{\text{mass of } N \text{ molecules of methane}} = \frac{56 N}{16 N} = \frac{56}{16}$$

Suppose that we make N sufficiently large that the mass of the iron sample becomes 56 g. This means that the mass of the methane sample becomes 16 g. Because N is the same for both samples, we can write

(no. of atoms in a 56-g sample of iron)
= (no. of molecules in a 16-g sample of methane)

When a sample has a mass in grams equal to the mass in u of the smallest unit of the substance, we say that the sample is *one mole* of the substance. (The abbreviation of "mole" is *mol*.) Thus, 1 mol of iron has a mass of 56 g and 1 mol of methane has a mass of 16 g. Now, we can write

(number of atoms in 1 mol of iron)
= (number of molecules in 1 mol of methane)

It is clear that the relationship we have obtained regarding the equality of the number of units of matter (atoms or molecules) in 1-mol samples does not apply exclusively to iron and methane. In fact, 1 mol of any substance contains the same number of basic matter units as does 1 mol of any other substance. This important idea was first realized in 1811 by the Italian Count Amedeo Avogadro (1776–1856), and in his honor the number of matter units in one mole is called *Avogadro's number*, N_0:

$$N_0 = 6.022 \times 10^{23} \text{ molecules/mol} \qquad (1.3)$$

Notice that we use the term "molecule" in presenting Avogadro's number. This term covers all cases if we understand that it means the smallest unit of the substance that occurs in Nature. For methane, the smallest unit is the methane *molecule*. But for iron, the smallest unit is the iron *atom*; thus, the iron "atom" is the same as the iron "molecule." Notice also that there is no significance to Avogadro's number if it is stated in terms of *atoms* instead of *molecules*. For example, 1 mol of methane does *not* contain 6×10^{23} *atoms*.

(Even though *kilograms* are used instead of *grams* in the International System of Units, the SI Commission has decided to retain the old definition of the mole in terms of *grams*. Chemists sometimes use the term *gram molecular weight* for the mole.)

●*Example 1.3*

Use the fact that the molecular mass of water is 18 u (Table 1.10) to estimate the size of a water molecule.

The mass of 1 mol of water is 18 g and water has a density of 10^3 kg/m³. Therefore, a mole of water will occupy a volume V, where

$$V = \frac{M}{\rho} = \frac{18 \times 10^{-3} \text{ kg/mol}}{10^3 \text{ kg/m}^3} = 1.8 \times 10^{-5} \text{ m}^3/\text{mol}$$

This volume contains N_0 molecules, so the number of molecules in 1 m³ is

$$N = \frac{N_0}{V} = \frac{6.02 \times 10^{23} \text{ molecules/mol}}{1.8 \times 10^{-5} \text{ m}^3/\text{mol}}$$
$$= 3.34 \times 10^{28} \text{ molecules/m}^3$$

The volume occupied by a single molecule is the inverse of N:

$$\frac{1}{N} = \frac{1}{3.34 \times 10^{28} \text{ molecules/m}^3}$$
$$= 3 \times 10^{-29} \text{ m}^3/\text{molecule}$$

If we consider each molecule to occupy a cubical box, the length d of a side of such a box will be the cube root of $1/N$:

$$d = \left(\frac{1}{N}\right)^{1/3} = (30 \times 10^{-30} \text{ m}^3)^{1/3}$$
$$\cong 3 \times 10^{-10} \text{ m}$$

Thus, the size of a water molecule is a few times 10^{-10} m (i.e., a few angstroms). A water molecule bears approximately the same relationship to 1 cm as the solar system does to our entire galaxy, the Milky Way! ■

Atoms, Electrons, and Ions

During the 1890s it became clear that atoms, which are the smallest units of matter that can be identified as definite chemical elements, must consist of particles of even smaller size. In 1897 Sir Joseph Thomson discovered the electron and showed that this particle is a fundamental piece of matter common to all atoms (see Chapter 18). Within a few years of this important discovery, it was realized that the atoms of every chemical element contain a particular number of electrons (when the atoms are in the normal, electrically neutral condition). For example, a hydrogen atom contains 1 electron, a carbon atom contains 6 electrons, and an iron atom contains 26 electrons. Every chemical element is characterized by the number of electrons contained in its atoms; this number is called the *atomic number* of the element and is denoted by the letter Z. Thus, hydrogen has $Z = 1$, carbon has $Z = 6$, and iron has $Z = 26$. The table inside the front cover gives the atomic number (and the atomic mass) of every known element.

An electron carries a negative electric charge. Therefore, if an electrically neutral atom contains a certain number of electrons, it must also contain an equal amount of positive charge. This positive charge resides in the atomic nucleus in the form of *protons*. A proton carries a positive electric charge that is exactly equal in magnitude to the electron charge. Thus, the nucleus of an atom with an atomic number Z contains Z protons.

It is relatively easy to remove from an atom one of the electrons that exist in the outer part of the atom. This process is called *ionization* and produces a "free" (i.e., an unattached) electron which can then participate in some physical or chemical process. The removal of an electron from an atom leaves the atom with an insufficient number of negative charges to balance the positive charge of the nucleus. Such an atom is called a positively charged *ion*. A carbon atom from which one electron has been removed would be designated a C^+ ion; if two electrons were removed, we would have a C^{++} or C^{2+} ion. Some types of atoms tend to acquire electrons. For example, when a chlorine atom picks up an extra electron it becomes a negatively charged ion, Cl^-. In later chapters we will examine some of the important processes in which ions participate.

Protons and Neutrons

In 1911 experiments performed under the direction of Ernest Rutherford revealed for the first time the existence of the atomic *nucleus* (see Chapter 19). Rutherford succeeded in showing that a nucleus is very much smaller than the atom of which it is a part (Fig. 1.5), but, nevertheless, the nucleus contains most of the mass of the atom. The basic unit of positively charged matter in all nuclei is the *proton*, which has a mass 1837 times that of an electron.

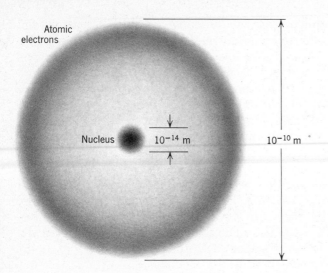

Figure 1.5 Schematic comparison of atomic and nuclear sizes. Atoms are about 10,000 times larger than nuclei.

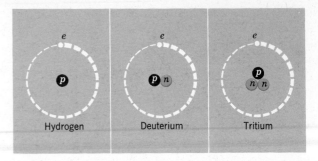

Figure 1.6 The three isotopes of hydrogen. Natural hydrogen consists predominantly of the $A = 1$ isotope. The $A = 3$ isotope (tritium) is unstable and undergoes radioactive decay. These simple schematic representations of atoms and nuclei are not realistic (see Section 18.4).

The simplest atom, hydrogen, has a nucleus that consists of a single proton ($Z = 1$). The element with $Z = 2$, helium, has two protons in its nucleus. However, measurements show that the mass of a helium atom is *four* times that of the hydrogen atom. The reason for this mass difference was unknown until, in 1932, James Chadwick discovered the *neutron*. The neutron is a particle that has a mass almost equal to that of a proton but it carries no electric charge. The nucleus of a helium atom contains two neutrons in addition to its two protons so that the total mass of a helium atom is four times that of a hydrogen atom. In fact, the nuclei of *all* atoms, with the single exception of hydrogen, contain neutrons.

The nucleus of an atom with atomic number Z contains Z protons. If the nucleus also contains N neutrons, the total number of particles in the nucleus is $A = Z + N$. This number A we call the *mass number* of the nucleus.

The proton and the neutron each have a mass of approximately 1 u:

$$\left. \begin{array}{l} \text{Proton mass} = m_p = 1.007276 \text{ u} \\ \text{Neutron mass} = m_n = 1.008665 \text{ u} \end{array} \right\} \quad (1.4)$$

Therefore, the mass of nucleus with mass number A is approximately equal to A u.

For comparison, the mass of an electron is

$$\text{Electron mass} = m_e = \tfrac{1}{1837} m_p = 0.000548 \text{ u} \quad (1.5)$$

Isotopes

Nuclei of the same chemical element do not all have the same mass. Although most hydrogen atoms have nuclei that consist of a single proton, a small fraction of natural hydrogen atoms (about 0.015 percent) have one proton and one neutron in their nuclei. This "heavy hydrogen" is called *deuterium* (see Fig. 1.6) and is often denoted by the symbol D. Another form of hydrogen atoms have nuclei with *two* neutrons: hydrogen with $A = 3$ is called *tritium* (T). The series of nuclei with a given value of Z but different values of A are called *isotopes* of the element.

Isotopes are classified as either *stable* or *radioactive*. A stable isotope will exist permanently in Nature but a radioactive isotope will eventually undergo a spontaneous transmutation that will form an isotope of a different element (see Chapter 2). Radioactive isotopes are known for all of the elements.

Some elements have a single stable isotope, but most have two or more. Different isotopes of a given element are distinguished by using the mass number as a superscript. Thus, the stable isotopes of helium are ^{3}He and ^{4}He (Fig. 1.7) and the stable isotopes of oxygen are ^{16}O, ^{17}O, and ^{18}O. A list of some of the isotopes of the lightest elements is given in Table 1.11.

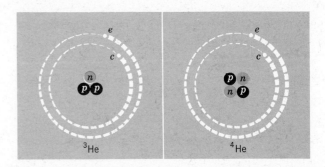

Figure 1.7 The stable isotopes of helium, ^{3}He and ^{4}He. All other isotopes of helium are unstable. The abundance of ^{3}He in natural helium is only 1.5 parts per million.

Table 1.11 Properties of Some Light Elements

Element	Z	A	Symbol	Remarks[a]
Hydrogen	1	1	^{1}H	Stable (99.985%)
	1	2	^{2}H or ^{2}D (deuterium)	Stable (0.015%)
	1	3	^{3}H or ^{3}T (tritium)	Radioactive
Helium	2	3	^{3}He	Stable (0.00015%)
	2	4	^{4}He	Stable (99.99985%)
	2	6	^{6}He	Radioactive
Lithium	3	6	^{6}Li	Stable (7.52%)
	3	7	^{7}Li	Stable (92.48%)
	3	8	^{8}Li	Radioactive
Beryllium	4	7	^{7}Be	Radioactive
	4	8	^{8}Be	Radioactive
	4	9	^{9}Be	Stable (100%)
	4	10	^{10}Be	Radioactive
Boron	5	10	^{10}B	Stable (18.7%)
	5	11	^{11}B	Stable (81.3%)
	5	12	^{12}B	Radioactive
Carbon	6	11	^{11}C	Radioactive
	6	12	^{12}C	Stable (98.89%)
	6	13	^{13}C	Stable (1.11%)
	6	14	^{14}C	Radioactive

[a] The numbers in parentheses are the relative natural abundances of the isotopes.

Summary of Important Ideas

The fundamental physical units in mechanics are those of *length, time,* and *mass.*

The basic unit of *length* is the *meter.* The *standard* of length is the wavelength of light from krypton gas.

The basic unit of *time* is the *second.* The *standard* of time is the vibration period of cesium atoms.

The basic unit of *mass* is the *kilogram.* The *standard* of mass cannot yet be stated in terms of atomic quantities with sufficient precision to be generally useful; therefore, the operational standard is a certain block of metal maintained in the international standards depository.

All physical quantities have *units,* and in equations relating various physical quantities, the *numbers* as well as the *units* on the two sides of the equation must be the same.

Mass is conserved; that is, mass can neither be created nor destroyed—only rearranged. (We shall later see that mass and energy are intimately connected and that the conservation law properly refers to *mass-energy.*)

Mass Density is mass per unit volume and is measured in kg/m^3 or g/cm^3.

Atomic and molecular masses are measured in terms of the *atomic mass unit* (u). 1 u is $\frac{1}{12}$ of the mass of the ^{12}C atom.

A sample of a substance that has a mass in grams equal to the molecular mass of the substance in u is called *one mole* of the substance and contains Avogadro's number N_0 of molecules. $N_0 = 6.022 \times 10^{23}$ molecules/mol.

Atoms consist of tiny, massive cores (*nuclei*) surrounded by *electrons.* A typical atomic size is about 10^{-10} m whereas nuclei have sizes about 10^4 times smaller.

All nuclei (except the hydrogen nucleus) contain *neutrons* as well as *protons.* The number of protons in the nucleus determines the atomic *species* and the number of neutrons determines the *isotope.*

◆Questions

1.1 Write down as many different units of length (modern or ancient) as you can remember or can find in any encyclopedia or dictionary. (You should have no difficulty in finding 20 or so.)

1.2 Look up the definition of *time* in a dictionary. Ignoring those definitions that do not deal with the physical concept of time, comment on the definitions that pertain to time as we use the word in physics. Do these definitions give you a clear understanding of time? Try to devise a better definition of physical time.

1.3 You suspect that your stopwatch runs 0.01 s slow every minute. Explain how you would use the WWV time signals to check this suspicion.

1.4 Suppose you are faced with the task of selecting a material from which the standard kilogram is to be made. What properties would be desirable for the material and what properties would be undesirable? What procedures would you establish for the use and preservation of the standard?

1.5 The density of a long, thin piece of wire is to be determined. What measurements must be made? Which measurement must be made with the greatest care and why?

1.6 How many electrons are there in the ion Cu^{4+}? (The table inside the front cover may be useful.)

1.7 How many neutrons are contained in a nucleus of ^{238}U?

1.8 The atomic number of the gas *neon* is 10. There are three isotopes of neon that occur in nature: ^{20}Ne, ^{21}Ne, and ^{22}Ne. How many protons and how many neutrons are there in each of these isotopes?

★Problems

Note: Those problems that are above average difficulty or require somewhat more in the way of computation are indicated by an asterisk*.

Section 1.2

1.1 How many feet are there in 1 km?

1.2 Convert 2.5 mi to meters.

1.3 One of the running distances in international competition is 1500 m, the so-called "metric mile." What fraction of a mile is a "metric mile"?

1.4 Machinists sometimes work in units of *mils* (10^{-3} in.) or *microinches* (10^{-6} in.). Express both of these units in centimeters.

1.5 A certain athlete runs the 100-yd dash in 9.4 s. What would be his expected time for the 100-m dash?

1.6 The distances to the outfield fences in the Cincinnati baseball park, Riverfront Stadium, are listed both in feet and in meters. The markings on the left, center, and rightfield fences that indicate the distances in feet are, respectively, 330, 404, and 375. What other numbers appear on these fences?

1.7 A jet airliner can make the trip from San Francisco to Washington, D.C. (3800 km) in 4 h. At this same speed, how long would be required for a round trip from the Earth to the Moon? (The Earth-Moon distance is given in Table 1.3.)

1.8* What is the conversion factor that relates cubic centimeters to cubic inches? Cubic feet to cubic meters?

1.9* A certain watch is claimed by the manufacturer to have an accuracy of 99.995 percent. By how many minutes might such a watch be in error after running a month?

1.10 A certain electronic computer can perform 350,000 arithmetical operations per second. How many microseconds are required (on the average) for each operation?

1.11 How many seconds are there in one year?

1.12 Express 1 microcentury in the more conventional unit of seconds.

1.13* Express the age of the Earth (4.5×10^9 y) in seconds.

1.14 A power blackout puts out of operation the elevators in a 50-story building. Estimate the number of steps that must be negotiated by an athletically inclined person who decides to climb to the top floor of the building.

1.15 The largest cells in living matter are just a bit smaller than the smallest thing that can be discerned with a low-power magnifying glass (about 0.1 mm). How many of these cells could be contained in 1 cm^3?

Section 1.3

1.16 A certain block is measured and found to have the following dimensions: length, 16.2 cm; width, 4.12 cm; height, 0.89 cm. The mass of the block is found to be 206.35 g. What is the mass density of the block?

1.17* An aluminum beverage can is made of sheet metal 0.5 mm thick. If the cylindrical can has a height of 11 cm and a diameter of 6 cm, calculate its mass.

1.18* Use the fact that 1 qt = 946.3 cm^3, and comment on the accuracy of the old saying, "A pint's a pound the world around."

1.19 A rectangular aluminum plate has dimensions 1.5 m × 2.5 m and has a mass of 253 kg. How thick is the plate?

1.20 What is the mass of air in a classroom that has dimensions 6 m × 20 m × 3 m?

1.21 Express the weight density of water in lb/ft^3.

1.22* Assume that the Sun consists entirely of hydrogen. (This is approximately correct.) How many hydrogen atoms are there in the Sun? (Use the information in Table 1.7.)

1.23* How many grams of hydrogen are there in 1 kg of methyl alcohol? (Refer to Table 1.10.)

1.24 The mass of 1 mol of hydrogen sulfide (H_2S) is 34 g. What is the approximate mass of a sulfur atom in u?

1.25 The density of octane (C_8H_{18}, a component of gasoline) is 0.703 g/cm^3. How many molecules of octane are there in 1 cm^3?

1.26* In order to obtain an estimate of atomic and molecular sizes, Lord Rayleigh (1842–1919), an English physicist, performed the following experiment. He took an oil droplet (mass = 8×10^{-4} g, density = 0.9 g/cm^3) and allowed it to spread out over a water surface. The area of the film was found to be 0.55 m^2. The oil would cover no greater area; any attempt to spread the film further resulted in tearing the film. The conclusion is that the layer of oil is one molecule thick. Oil consists of chainlike molecules having the property that only one end has an affinity for water. Therefore, when oil is spread on water, the oil molecules "stand on their heads." The film thickness therefore corresponds to the *length* of the oil molecule. Calculate this length from Lord Rayleigh's data. Each molecule of the type of oil used contains 16 atoms. Estimate the size of an individual atom.

1.27* Air consists primarily of nitrogen gas (N_2, molecular mass = 28 u). What is the average distance between the molecules in air under normal conditions of temperature and pressure? (Follow the method in Example 1.3. You will need the density of air; see Table 1.9.)

1.28* Air consists of a mixture of nitrogen (N_2) 75.5%, oxygen (O_2) 23.2%, and argon (Ar) 1.3% (percentages by weight). Determine the equivalent molecular mass of an "air molecule" assumed to be diatomic?

1.29* The atomic masses of naturally occurring elements are sometimes quite different from an integer number of atomic mass units because the element is actually a mixture of isotopes. Copper, for example, consists of two isotopes, ^{63}Cu (69.1%) and ^{65}Cu (30.9%). What do you expect the atomic mass of natural copper to be (approximately)? Compare your result with that given in the table inside the front cover.

1.30* Using the atomic mass for naturally occuring lithium given in the table inside the front cover and the isotopic masses for ^{6}Li: 6.015123 and ^{7}Li = 7.016004 determine the percentage abundance of these two isotopes in nature.

Answers to Estimates (p. 7)

- If the cars average 20 ft in length and are bumper to bumper, the number is approximately,

$N = (3 \times 2 \text{ mi}) \times (5000 \text{ ft/mi})/(20 \text{ ft/car})$
$= 1500$ cars.

- If the room measures 3m × 3m × 2m and taking the diameter of a golf ball to be 3 cm, the number is approximately

$$\frac{(18 \text{ m}^3 \times (100^3 \text{ cm}^3/\text{m}^3)}{27 \text{ cm}^3} = 7 \times 10^5$$

- Assume that the conterminous states are approximately rectangular in shape, 3000 mi × 1000 mi; assume that each interstate highway stretches completely across the country, north-south and east-west; guess that the average separation between adjacent highways is 150 mi. Then, the approximate total mileage is $(20 \times 1000) + (8 \times 3000) = 44,000$ mi. (The actual mileage is approximately 42,500 mi.)

- Guess that there are 100×10^6 automobiles in the country; assume that, on average, each is driven 15,000 mi/y and gets 20 mi/gal. Then, the number of barrels of oil needed is approximately $(10^8) \times (15,000 \text{ mi})/[(20 \text{ mi/gal}) \times (20 \text{ gal/bbl})] = 3.8 \times 10^9$ bbl. (This is reasonably close to the best estimates of the actual figure.)

2
Motion

If there is one feature that is characteristic of all forms of matter everywhere in the Universe, it is *motion*. The gas molecules in the air, even the atoms in solid matter, are continually in motion; the electric current that operates our appliances and motors is due to the motion of electrons along wires; the planets are in motion around the Sun; and even the gigantic galaxies both revolve and move through space. Historically, the study of simple motions constituted the first extensive and systematic application of the scientific method to a problem of the real physical world. Because the development of other physical ideas depends so heavily on *motion*, we begin our detailed study with a discussion of this important topic. The discussion is divided into two parts. In this chapter we concentrate on the techniques for *describing* motion—this is the subject of *kinematics*. In Chapter 3 we examine the *causes* of motion—this is the subject of *dynamics*.

2.1 Displacement and Path Length

Suppose you walk from one place to another following a certain path. We distinguish between the total distance you have walked along the path which we call the *path length,* and the straight-line distance between the starting point and the final or end point

Fratelli Alinari

Figure 2.1 Galileo Galilei (1564–1642). Through a series of ingenious experiments and well-constructed logical arguments, Galileo correctly formulated the laws governing the motion of falling bodies. Later, Newton incorporated these ideas into his all-encompassing theory of dynamics.

of your trip, called the *displacement*. Thus if you walked along a circular path on level ground to a final position which is very close to your starting point the path length would be nearly the circumfer-

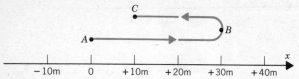

Figure 2.2 An object moves from point A to point B and retraces its motion back to point C.

ence of the circle while the displacement would be close to zero.

At first we consider only the simple case of *straight-line* motion. However, we allow the possibility of moving in *either direction* along a given straight line. Later, we shall be concerned with the direction of more general motion.

In order to specify positions along a straight line we introduce an origin, a distance scale, and define positive and negative directions. In Fig. 2.2 we show the motion of an object that started at point A (the origin, $x = 0$), traveled to point B (coordinate location $x = +30$ m), and returned to point C (coordinate location $x = +10$ m). The path length of the motion (i.e. total distance traveled from point A to point C) is $\ell = 30$ m $+ 20$ m $= 50$ m. The displacement due to the motion (i.e. the end point C relative to the starting point A) is $d = +10$ m. Had the object instead started at point C, gone to point B, and then reversed its motion arriving at point A, the path length would still be $\ell = 50$ m, however, its displacement would now be $d = -10$ m. From this example we see that the path length is always a positive definite quantity while displacement relating the final position to the initial position may be a positive or a negative quantity. It is possible to write a simple algebraic expression for the displacement d.

$$d = x(\text{endpoint}) - x(\text{beginning point}) \quad (2.1)$$

2.2 Average Speed and Average Velocity

The Rate of Movement

The concept of speed is familiar to everyone; it is the time rate at which something moves in space. The speedometer of an automobile registers the speed at which the vehicle is moving (although this instrument is frequently unreliable as any recipient of a speeding ticket will testify.) If a certain automobile trip of 30 mi along a straight stretch of highway is to be completed in 1 hr, the speed required is, clearly, 30 mi/h.

Note that no mention is made of whether a constant speed of 30 mi/h is maintained or whether the trip is made by stop-and-go driving. That is, the statement specifies only the *average* speed, or to be more specific, the time average speed. If the first 15 mi were covered in 15 minutes (at a constant speed of 60 mi/h) and, if because of heavy traffic, it required 45 minutes to negotiate the second 15 mi (so that the speed was 20 mi/h for this part of the trip), the average speed for the entire 30-mi journey would still be

$$\text{average speed} = \frac{\text{total distance traveled}}{\text{total elapsed time}} \quad (2.2)$$

or 30 mi/h. We denote speed by the symbol u and *average speed* by $\bar{u}$. Also, ℓ stands for the total distance traveled and t for the elapsed time. Therefore, in symbols, Eq. 2.2 becomes

$$\bar{u} = \frac{\ell}{t} \quad (2.3)$$

In general, for any interval of distance $\Delta \ell$ that is traversed along some path in the time interval Δt the average speed is given by [1]

$$\bar{u} = \frac{\Delta \ell}{\Delta t} \quad (2.4)$$

Note that u and $\bar{u}$ are always positive definite quantities.

● *Example 2.1*

In the two-part, 30-mi trip referred to above, we calculate $\bar{u}$ in the following way. We have for the two segments,

1. $\ell_1 = 15$ mi at $u_1 = 60$ mi/h
2. $\ell_2 = 15$ mi at $u_2 = 20$ mi/h

The average speed for the entire trip is

$$\bar{u} = \frac{\text{total distance}}{\text{total time}} = \frac{\ell_1 + \ell_2}{t_1 + t_2}$$

The individual times are

$$t_1 = \frac{\ell_1}{u_1} = \frac{15 \text{ mi}}{60 \text{ mi/h}} = \frac{1}{4} \text{ h}$$

$$t_2 = \frac{\ell_2}{u_2} = \frac{15 \text{ mi}}{20 \text{ mi/h}} = \frac{3}{4} \text{ h}$$

---

[1] The symbol $\Delta \ell$ does *not* imply the product of Δ and ℓ, but means "an interval of ℓ" or "an increment of ℓ." In general, a Greek delta, Δ, in front of a quantity means an *increment* of that quantity (see Fig. 2.3).

Thus,

$$\bar{u} = \frac{\ell_1 + \ell_2}{t_1 + t_2} = \frac{15 \text{ mi} + 15 \text{ mi}}{\frac{1}{4}\text{h} + \frac{3}{4}\text{h}} = \frac{30 \text{ mi}}{1 \text{ h}} = 30 \text{ mi/h}$$

Notice that the average speed is the time average and is *not* given by the average of u_1 and u_2:

$$\bar{u} \neq \tfrac{1}{2}(u_1 + u_2) = \tfrac{1}{2}(60 \text{ mi/h} + 20 \text{ mi/h}) = 40 \text{ mi/h}$$

● *Example 2.2*

Suppose that the first half of the distance between two points is covered at an average speed $\bar{u}_1 = 10$ km/h and that during the second half the speed is $\bar{u}_2 = 40$ km/h. What is the average speed for the entire trip?

Let 2ℓ be the total distance traveled and let t_1 and t_2 denote the times necessary for the two parts of the trip. Then,

$$\bar{u} = \frac{2\ell}{t_1 + t_2}$$

$$t_1 = \frac{\ell}{\bar{u}_1}; \quad t_2 = \frac{\ell}{\bar{u}_2}$$

$$t_1 + t_2 = \frac{\ell}{\bar{u}_1} + \frac{\ell}{\bar{u}_2} = \frac{\ell(\bar{u}_1 + \bar{u}_2)}{\bar{u}_1 \bar{u}_2}$$

Therefore,

$$\bar{u} = \frac{2\ell}{\dfrac{\ell(\bar{u}_1 + \bar{u}_2)}{\bar{u}_1 \bar{u}_2}} = \frac{2\bar{u}_1 \bar{u}_2}{\bar{u}_1 + \bar{u}_2}$$

$$= \frac{2 \times (10 \text{ km/h}) \times (40 \text{ km/h})}{10 \text{ km/h} + 40 \text{ km/h}} = \frac{800}{50} \text{ km/h}$$

$$= 16 \text{ km/h}$$

Again, notice that $\bar{u}$ is *not* obtained by averaging the individual speeds, $\bar{u}_1$ and $\bar{u}_2$. ■

A physically more meaningful description of the rate at which something moves along a straight line would also tell us about the *direction* of the motion as well as the magnitude of the rate. Such a quantity is the *average velocity*. We denote the average velocity by the symbol $\bar{v}$ and define it in terms of the displacement d.

$$\bar{v} = \frac{x(\text{end}) - x(\text{beginning})}{t(\text{elapsed time})} = \frac{d}{t} \quad (2.5)$$

In general, for any displacement Δx resulting during an elapsed time Δt, the average velocity is given by

$$\bar{v} = \frac{\Delta x}{\Delta t} \quad (2.6)$$

Note that since the displacement may be either a positive or negative quantity, the average velocity also carries a corresponding algebraic sign.

2.3 Distance-Time Graphs

The Geometrical Representation of Velocity

Suppose we know that an object moving in a straight line was at a point labeled by x_0 at a time t_0 and at a point x_1 at a later time t_1. We can indicate these facts in a graph of position *versus* time, as in Fig. 2.3. The net displacement is $\Delta x = x_1 - x_0$ and the time interval during which the motion occurred is $\Delta t = t_1 - t_0$. Therefore, according to Eq. 2.4, the average velocity for the motion is the increment of distance moved divided by the corresponding increment of time:

$$\bar{v} = \frac{\Delta x}{\Delta t} = \frac{x_1 - x_0}{t_1 - t_0} \quad (2.7)$$

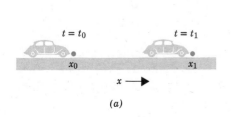

(a)

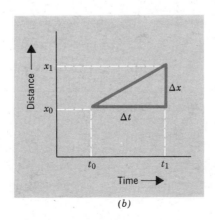

(b)

Figure 2.3 The motion in (a) is represented in the distance-time graph (b).

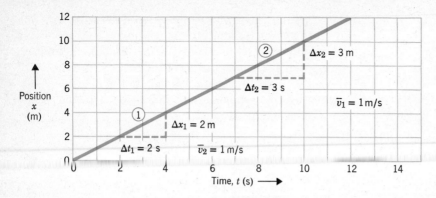

Figure 2.4 Graph of the position versus time for an object moving at a constant velocity of 1 m/s.

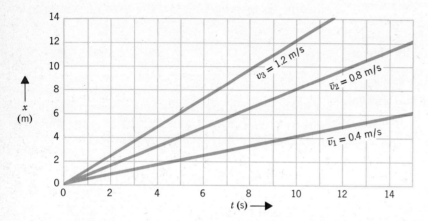

Figure 2.5 The greater the slope of the distance-time graph, the greater is the average velocity.

If the velocity with which an object moves is *constant*, then, clearly, the average velocity is always the same regardless of the particular time interval that is chosen for the computation. An object that moves with a constant velocity of 1 m/s will have traveled 1 m after 1 s, 5 m after 5 s, and 20 m after 20 s. The distance traveled is therefore a *linear function* of the time; that is, the graph of distance traveled *versus* time is a *straight line*. Figure 2.4 is a distance-time graph for an object moving with a constant velocity of +1 m/s. It is a simple matter to extract the average velocity for any given time interval from such a graph. Consider the case labeled ① in Fig. 2.4. The time interval for this case is $\Delta t_1 = 4\,s - 2\,s = 2\,s$, and the displacement is $\Delta x_1 = 4\,m - 2\,m = 2\,m$. The average velocity is therefore $\bar{v}_1 = \Delta x_1/\Delta t_1 = 2\,m/2\,s = 1\,m/s$. Geometrically, the average velocity is just the length of the vertical dotted line (indicated by Δx_1 in Fig. 2.4) divided by the length of the horizontal dotted line (Δt_1). (Do not be confused by the fact that the lengths of these lines, as shown in Fig. 2.4, are not equal; time and distance are different physical quantities and we have chosen different numerical scales for the two axes of the graph.) Similarly, the average velocity for the case labeled ② is $\bar{v}_2 = \Delta x_2/\Delta t_2 = 3\,m/3\,s = 1\,m/s$. No matter what time interval is chosen, the average

Table 2.1 Some Typical Speeds

Growth of hair (human head)	5×10^{-9} m/s = 15 cm/y
Rapidly moving glacier	3×10^{-6} m/s = 0.25 m/day
Tip of sweep-second hand on wrist watch	10^{-3} m/s = 1 mm/s
Sprinter	10 m/s
Batted baseball	50 m/s
Racing car	70 m/s = 250 km/h
Sound in air	330 m/s
X-15 rocket-plane	2×10^3 m/s = 2 km/s
Earth in orbit	3.0×10^4 m/s = 30 km/s
Electron in hydrogen atom	2.2×10^6 m/s
Light in vacuum	3×10^8 m/s

velocity will always be 1 m/s because the speed is *constant*.

Velocity and the Slope of the Distance-Time Graph

Figure 2.5 shows three different distance-time graphs, all of which are straight lines; each corresponds to a different average velocity. As the steepness of the line increases, the average velocity becomes greater. This is a general result: *the slope of a distance-time graph determines the average velocity*.

In Fig. 2.6 we show a distance-time graph that is composed of several straight-line segments. Here

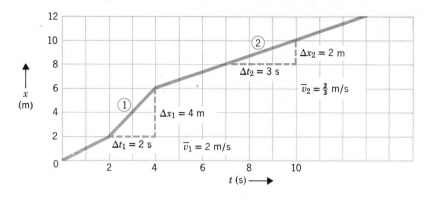

Figure 2.6 Distance-time graph for the motion of an object whose velocity is not constant.

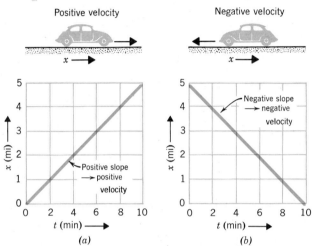

Figure 2.7 If a distance increases to the right, motion to the right takes place with positive velocity and motion to the left takes place with negative velocity. An *upward* slope is *positive* and denotes a *positive* velocity, a *downward* slope is *negative* and denotes a *negative* velocity.

the average velocity will depend on the time interval chosen. But the procedure for the computation of the average speed is exactly the same as before—the displacement is divided by the corresponding time interval. For the first interval we obtain $\bar{v}_1 = 2$ m/s, whereas for the second interval we have $\bar{v}_2 = \frac{2}{3}$ m/s. (What is $\bar{v}$ for the interval between $t = 2$ s and $t = 10$ s?)

In discussing straight-line motion we must distinguish between the two possible directions of motion. Whenever an object moves in the direction of positive x (for example, to the right in Fig. 2.3), the x values increase with time. This means that when we calculate $\bar{v} = \Delta x/\Delta t$, both Δx and Δt are positive quantities, and so $\bar{v}$ is *positive*. Then, the distance-time graph shows a line or curve that increases with time; that is, the slope of the distance-time graph is *positive* (Fig. 2.7a). However, if the object moves in the direction of negative x (Fig. 2.7b), the x values decrease with time and Δx is negative, thereby making $\bar{v}$ negative. The slope of the distance-time graph is then also negative (Fig. 2.7b).

● Example 2.3

Consider a certain round trip that is made by automobile. The first figure below shows a 20-km round trip between points A and B. The first 2 km were covered in 0.1 h; the driver found himself at kilometer marker 6 after 0.2 h; and he arrived at B after 0.3 h. The driver remained at B for 0.1 h before returning. He accomplished the return trip in 0.2 h. The distance-time graph for the complete journey is shown in the second diagram. What were the average velocities for the various sections of the trip?

$$\bar{v}_1 = \frac{2 - 0}{0.1 - 0} = \frac{2}{0.1} = 20 \text{ km/h}$$

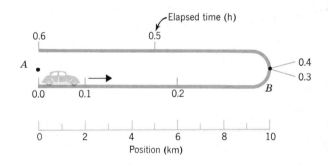

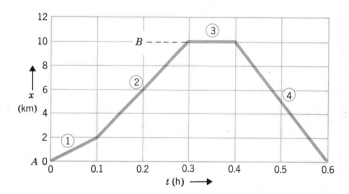

$$\bar{v}_2 = \frac{10-2}{0.3-0.1} = \frac{8}{0.2} = 40 \text{ km/h}$$

$$\bar{v}_3 = \frac{10-10}{0.4-0.3} = \frac{0}{0.1} = 0 \text{ km/h}$$

$$\bar{v}_4 = \frac{0-10}{0.6-0.4} = \frac{-10}{0.2} = -50 \text{ km/h}$$

Notice the following points:

1. The slope of the distance-time graph is steeper in interval ② than in interval ①, and the average velocity is correspondingly greater.

2. The graph is flat (that is, it has *zero* slope) in interval ③ when the automobile was stopped. The formula for computing the average velocity automatically gives zero.

3. In interval ④ the end of the interval (at 0.6 h) is at $x = 0$ km whereas the beginning (at 0.4 h) is at $x = 10$ km. Therefore, $\Delta x = 0 - 10 = -10$ km, a *negative* quantity which gives a *negative* average velocity. The *downward* slope of the graph in interval ④ indicates negative velocity.

4. The average velocity over the entire round trip is zero because the total displacement is zero. ∎

2.4 Instantaneous Velocity

The Limiting Average Velocity

If the motion of an object does not take place with constant velocity then, in general, the average velocity depends on the particular time interval chosen for the calculation. For the trip shown schematically in Example 2.3, if we choose the first 0.3 h, we obtain $\bar{v} = 10$ km/0.3 h $= 33\frac{1}{3}$ km/h, but if we choose the 0.3-h interval from 0.1 h to 0.4 h, we find $\bar{v} = 8$ km/0.3 h $= 26\frac{2}{3}$ km/h. It seems clear that it would be advantageous to have a method of specifying velocity that gives a unique answer without the necessity of always stating the time interval involved. For this purpose we need the concept of *instantaneous velocity*.

Figure 2.8 shows a distance-time graph that is *curved*. Start with point A, which corresponds to the position x_0 at the time t_0. If we take for the final position the position x_2, which occurs at time t_2

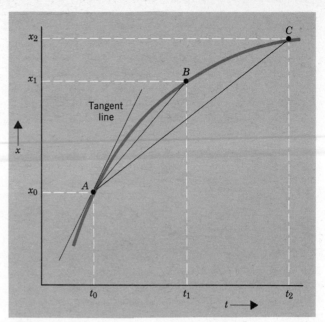

Figure 2.8 The average velocity between the initial point A and the final points C and B is greater for shorter time intervals. The instantaneous velocity at A results when the time interval is infinitesimally small, and is equal to the slope of the tangent line at that point. For this case, note that $v > \bar{v}$ for $\bar{v}$ calculated with any finite time interval starting at t_0.

(point C), we have for the average velocity in this interval,

$$\bar{v}_{02} = \frac{x_2 - x_0}{t_2 - t_0}$$

If we next reduce the time interval to $t_1 - t_0$, we find an average velocity

$$\bar{v}_{01} = \frac{x_1 - x_0}{t_1 - t_0} > \bar{v}_{02}$$

so that $\bar{v}_{01}$ is *greater* than $\bar{v}_{02}$; that is, the slope of the line AB is greater than the slope of the line AC. If we continue to reduce the time interval (always starting at x_0), the average velocity will increase further.[2] We could, in fact, continue this process indefinitely; we could take smaller time intervals and obtain greater and greater average velocities. But if the average velocity for a very small time interval is calculated and this interval is decreased still further, very little change will be produced in the aver-

---

[2] This *increase* of $\bar{v}$ as Δt is decreased is due to the fact that the distance-time graph has the particular curvature shown. What will happen to $\bar{v}$ as Δt is decreased if the curvature of the distance-time graph is *upward*?

2.4 Instantaneous Velocity

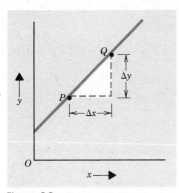

Figure 2.9 Graph of a line $y = mx + b$. The slope m is defined as the ratio $\Delta y/\Delta x$. Note that the slope does not depend on the selection of the points P and Q.

age velocity. We can therefore imagine a time interval so small that any further velocity reduction will not alter the average velocity. This limiting average velocity we call the *instantaneous* velocity v. Mathematically, we express this result as

$$v = \lim_{t_1 \to t_0} \frac{x_1 - x_0}{t_1 - t_0} = \lim_{\Delta t \to 0} \frac{\Delta x}{\Delta t} \qquad (2.8)$$

This equation states: "The instantaneous velocity v is given by the ratio of Δx to Δt in the limit that Δt approaches zero." Alternatively, "The instantaneous velocity is equal to the average velocity in the limit that the time interval becomes infinitesimally small."[3]

Geometrically, the instantaneous velocity is equal to the slope of the line that is tangent to the distance-time graph at the point in question. Recall from your study of algebra that the slope of any line $y = mx + b$ (including the tangent line) is defined as the ratio

$$\text{slope} = m = \frac{\Delta y}{\Delta x}$$

(see Figure 2.9).

Now that we have defined *average velocity* and *instantaneous velocity* the qualifier "instantaneous" is really unnecessary. We will use the term *velocity* to mean the instantaneous value. When we need the velocity averaged over a time interval we will use the term *average velocity*.

---

[3]Students familiar with calculus will recognize the instantaneous velocity to be the time derivative of displacement.

● *Example 2.4*

Suppose the position of an object moving along the x-axis is given by

$$x = \tfrac{1}{3}t^2$$

where x is in meters when t is in seconds. See graph. What is the velocity of the object at time $t_0 = 3s$?

At time t_0 the position is

$$x = \tfrac{1}{3}t_0^2$$

And at time $t_1 = t_0 + \Delta t$, the position is

$$x_1 = \tfrac{1}{3}t_1^2 = \tfrac{1}{3}(t_0 + \Delta t)^2$$
$$= \tfrac{1}{3}t_0^2 + \tfrac{2}{3}t_0\Delta t + \tfrac{1}{3}\Delta t^2$$

Therefore,

$$\frac{\Delta x}{\Delta t} = \frac{x_1 - x_0}{t_1 - t_0}$$
$$= \frac{(\tfrac{1}{3}t_0^2 + \tfrac{2}{3}t_0\Delta t + \tfrac{1}{3}\Delta t^2) - \tfrac{1}{3}t_0^2}{(t_0 + \Delta t) - t_0}$$
$$= \tfrac{2}{3}t_0 + \tfrac{1}{3}\Delta t$$

Thus we have

$$v = \lim_{\Delta t \to 0}(\tfrac{2}{3}t_0 + \tfrac{1}{3}\Delta t) = \tfrac{2}{3}t_0$$

Finally when $t_0 = 3s$,

$$v = \tfrac{2}{3} \times 3 = 2 \text{ m/s}$$

The tangent line to the distance-time graph at time $t_0 = 3s$ is also shown. Note that its slope is also 2m/s as it should be. ∎

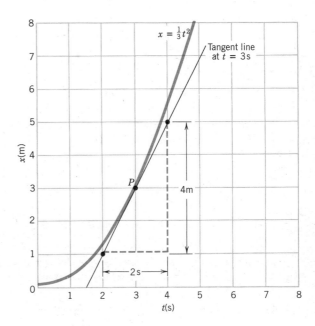

An object moving along a straight line, that sometimes reverses its sense of motion before continuing onward may have an average speed $\bar{u}$ and an average velocity $\bar{v}$ during certain time intervals that differ in magnitude. However, the instantaneous speed u and the instantaneous velocity v will always be equal in magnitude, although the velocity might be either a positive or negative quantity. That is: "Speed is simply the magnitude of velocity." We shall see later that for even general motion along curved paths in space we may consider speed to be the magnitude of the velocity.

2.5 Acceleration

Changing Velocity

When you "step on the gas" in an automobile, you do so in order to increase the speed, that is, you *accelerate*. When you apply the brakes, you do so in order to decrease the speed, or you *decelerate*.[4] In either case, the essential feature of the motion is that there is a *change* of speed and hence a change in velocity.

Figure 2.10a shows a case in which the velocity does *not* change; the motion is unaccelerated. Figure 2.10b shows a case in which the velocity increases linearly (that is, as a straight line) with the time, starting from $v = v_0$ at $t = t_0$. This is therefore a case of *accelerated* motion. Labeling the acceleration with the symbol a, we can express the average acceleration in a way analogous to that for the average velocity:

$$\bar{a} = \frac{\Delta v}{\Delta t} = \frac{v_1 - v_0}{t_1 - t_0} \qquad (2.9)$$

The unit of acceleration must be the unit of velocity divided by the unit of time, in other words, (m/s)/s, usually written as m/s², and read as "meters per second per second."

Just as velocity is the rate of change of distance with time, $\Delta x/\Delta t$, acceleration is the rate of change of velocity with time, $\Delta v/\Delta t$. Compare Eq. 2.7 for $\bar{v}$ with Eq. 2.9 for $\bar{a}$.

Constant Acceleration

In Fig. 2.10b the velocity time graph is a straight line, so that $\bar{a} = a =$ constant. Although there are

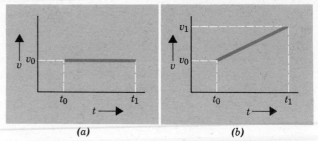

Figure 2.10 (a) A case of unaccelerated motion. (b) A case of constant acceleration. Notice that the graph is a straight line.

many physically interesting cases in which the acceleration changes with time, *we shall consider here only cases in which the acceleration is constant (or uniform)*.

If we displace the velocity time graph (Fig. 2.10b) to the left until the initial time occurs at $t_0 = 0$, the velocity time graph has the form shown in Fig. 2.11. Then, the acceleration is given by[5]

$$a = \frac{\Delta v}{\Delta t} = \frac{v - v_0}{t} \qquad (2.10)$$

from which

$$v - v_0 = at$$

or,

$$v = v_0 + at \qquad (2.11)$$

That is, the velocity at the time t is equal to the initial velocity v_0 plus the additional velocity ac-

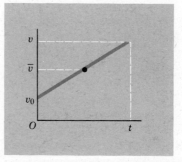

Figure 2.11 A velocity-time graph for motion with an initial velocity v_0. The average velocity during the time interval from 0 to t is $\bar{v}$.

---

[4]In physics we sometimes use the term *deceleration*, but usually we say that there has been a *negative acceleration*, with respect to the direction of the velocity.

[5]The speed or position that one usually wishes to determine is the *final* value; we will suppress the subscripts on these values and use a subscript zero for the initial values. We will also usually set $t = 0$ at the initial point.

quired by virtue of the constant acceleration that acts during the time t.

Next, we seek an expression for the distance traveled in the event that there is a constant acceleration. If we consider the initial position on the distance scale to be x_0, the additional distance traveled is given, as always, by the product of the *average velocity* and the time; thus

$$x = x_0 + \bar{v}t \qquad (2.12)$$

Referring to Fig. 2.11, it is easy to see that the average velocity for the case of constant acceleration, is just the average of v_0 and v. Hence,

$$\bar{v} = \tfrac{1}{2}(v_0 + v)$$

Substitution of this expression for $\bar{v}$ into Eq. 2.12 gives

$$x = x_0 + \tfrac{1}{2}(v_0 + v)t = x_0 + \tfrac{1}{2}v_0 t + \tfrac{1}{2}vt$$

Using Eq. 2.11 for v, we have

$$x = x_0 + \tfrac{1}{2}v_0 t + \tfrac{1}{2}(v_0 + at)t$$

or, finally,

$$x = x_0 + v_0 t + \tfrac{1}{2}at^2 \qquad (2.13)$$

That is, the distance traveled is equal to $v_0 t$ (the distance that would be traveled in the *absence* of acceleration, as we found previously) plus a term that depends on the acceleration and is proportional to the *square* of the elapsed time (Fig. 2.12).

For the case of *constant acceleration* and for the initial conditions, $x = x_0$ and $v = v_0$ at $t = 0$, we can summarize the results as follows:

Acceleration: $a = \text{const.}$
Velocity: $v = v_0 + at$ $\qquad (2.14)$
Distance: $x = x_0 + v_0 + \tfrac{1}{2}at^2$

● *Example 2.5*

A drag racer accelerates at a constant rate, starting from rest, and reaches a velocity of 240 mi/h in a distance of $\tfrac{1}{4}$ mi. What is the acceleration?

We have not derived a formula that allows us to calculate directly the acceleration from a knowledge of the final velocity and the distance traveled. We must therefore solve the problem in two steps, starting with a computation of the time required to go $\tfrac{1}{4}$ mi. We use

$$x = \bar{v}t$$

where $x_0 = 0$, $x = \tfrac{1}{4}$ mi and where the average velocity is

$$\bar{v} = \tfrac{1}{2}(v_0 + v) = \tfrac{1}{2}(0 + 240 \text{ mi/h})$$
$$= 120 \text{ mi/h}$$

The time required is, therefore,

$$t = \frac{x}{\bar{v}} = \frac{\tfrac{1}{4} \text{ mi}}{120 \text{ mi/h}}$$
$$= \frac{1}{480} \text{ h} \times \frac{3600 \text{ s}}{1 \text{ h}}$$
$$= 7.5 \text{ s}$$

Because the initial speed is zero, the acceleration is given by

$$a = \frac{v}{t}$$

where v is the final velocity, 240 mi/h. The result is then

$$a = \frac{240 \text{ mi/h}}{7.5 \text{ s}}$$
$$= 32 \text{ (mi/h)/s}$$

That is, the velocity increases by 32 mi/h each second.

In this case we have given the result in *mixed* units, (mi/h)/s, instead of in mi/h^2 or mi/s^2 or ft/s^2, because these mixed units seem to be the easiest to appreciate. There is no sense in establishing arbitrary rules regarding the use of units; we use those units that are most convenient or that convey the

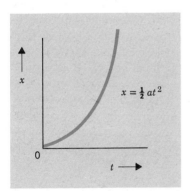

Figure 2.12 For the case of uniformly accelerated motion, the distance traveled is proportional to t^2 and, hence, the distance-time graph is a *parabola*. (Here, $v_0 = 0$, $x_0 = 0$.)

Example 2.6

An automobile traveling at a velocity of 40 km/h accelerates uniformly to a velocity of 90 km/h in 10 s. How far does the automobile travel during the time of acceleration?

$$v_0 = 40 \, \frac{km}{h} = 40 \, \frac{km}{h} \times 1000 \, \frac{m}{km} \times \frac{1 \, h}{3600 \, s}$$
$$= 11.1 \, m/s$$

$$a = \frac{\Delta v}{\Delta t} = \frac{90 \, km/h - 40 \, km/h}{10 \, s} = 5 \, (km/h)/s$$

$$= \frac{5 \, km}{h\text{-}s} \times \frac{1000 \, m}{km} \times \frac{h}{3600 \, s} = 1.39 \, m/s^2$$

$$x = v_0 t + \tfrac{1}{2} a t^2$$
$$= (11.11 \, m/s) \times (10 \, s) + \tfrac{1}{2} \times (1.39 \, m/s^2) \times (10 \, s)^2$$
$$= 111.1 \, m + 69.5 \, m$$
$$= 180.6 \, m$$

Suppose next that the automobile, traveling at 90 km/h, slows to 30 km/h in a period of 20 s. What was the acceleration?

$$a = \frac{v_2 - v_1}{\Delta t} = \frac{30 \, km/h - 90 \, k/h}{20 \, s}$$
$$= -3(km/h)/s.$$

The automobile was *slowing down* during this period so the acceleration was *negative* in this case. ∎

Running Speeds

How fast can a man run? It seems reasonably clear that the average speed for a run of a certain distance must depend on that distance. A runner can maintain top speed only for a limited time. Consequently, the average speed for long-distance runs must be less than that for sprints. We might expect, therefore, that the average speeds for world's records at different distances would show a steady decrease with increasing distance. This is not quite the case, however, as Fig. 1 shows. This diagram was prepared by drawing a smooth curve through the average speeds computed from all of the record times for men for races from 50 yards (45.7 m) to 1 mile (1609 m). Notice that the average speed actually increases with distance for distances less than about 200 m and thereafter decreases steadily.

We can understand the shape of the curve in Fig. 1 by examining the *instantaneous* speed of a runner during a race. A typical case for a race of 200 m is illustrated in Fig. 2. The runner begins from rest and accelerates at his maximum rate until he reaches his top speed. For a male sprinter, this acceleration time is about 2 s or slightly less, and his top speed is about 10.5 m/s. The average speed for the entire distance, however, is less than 10.5 m/s because the average speed during the

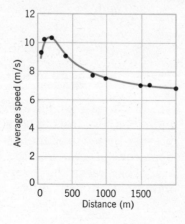

Figure 1 Average running speeds for records at different distances. A smooth curve has been drawn through the points corresponding to record times for races from 50 yd to 2000 m.

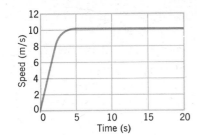

Figure 2 A typical speed-versus-time graph for a 200-m sprinter.

acceleration period was only about one-half the maximum speed. For this reason, the average speed for a 200-m race will be greater than that for a 100-m race because a greater fraction of the total time was run at the top speed in the longer race. If a runner could maintain his top speed indefinitely, the average speed would continue to increase with distance. But we know that a runner cannot do this, and Fig. 1 shows that the average speed begins to decrease for races at distances greater than about 200 m or 220 yd. (It is not possible to determine the distance at which the average speed begins to decrease because there are no records for distances between 220 yd and 400 m.)

The reason that a runner cannot maintain top speed indefinitely is that his oxygen supply is inadequate. The body contains stored oxygen in the muscles and gains further oxygen through respiration. While running at top speed, the oxygen requirement is not met by intake through the lungs, even though the respiration rate is increased. Therefore, a sprinter can maintain top speed only until his or her supply of stored oxygen is exhausted. This point of oxygen exhaustion probably occurs at a distance near 300 m. Consequently, for races longer than this distance, a runner must pace him or herself to run at a speed less than top speed. The longer the race, the lower this speed must be to ensure adequate oxygen for the entire race. This has been realized. All races of 220 yd or shorter are *sprints,* and longer races are called *middle-distance* or *long-distance* races. Only sprints are run at top speed throughout. (It might be interesting to have races at the exhaustion distance of 300 m.)

When racing against other individuals, a runner usually strives simply to *win,* and this may dictate a running strategy that is different from that which would produce the best time. If a runner is seeking to break a record, the optimum strategy would be to run at that steady speed which would result in complete oxygen exhaustion just prior to crossing the finish line.

References

J. B. Keller, "A Theory of Competitive Running," *Physics Today,* September 1973, p. 43.

H. W. Ryder, H. J. Carr, and P. Herget, "Future Performance in Footracing," *Scientific American,* June 1976, p. 109.

R. A. R. Tricker and B. J. K. Tricker, *The Science of Movement,* Mills and Boon, London, 1966, Chapter 21.

■ Exercises

1 Suppose that a sprinter can accelerate uniformly from rest to a top speed of 10.2 m/s in a time of 1.8 s. If the sprinter maintains this top speed for the remainder of a 200 m race, what will be his finishing time?

2 Suppose that another sprinter can reach top speed by accelerating uniformly for 1.7 s. What top speed must this sprinter maintain in order to equal the world's record of 19.5 s for 200 m?

3. When running long distance races (e.g., 1 mi), most runners maintain a relatively slow pace during the middle part of the race and then "kick" to the finish line. Is this the proper strategy to set a record? Could it be the proper strategy to optimize the likelihood of winning the race?

2.6 The Motion of Falling Bodies

Galileo's Experiments

When Galileo attacked the problem of motion of falling bodies, he sought to find a simple relationship connecting quantities that he could measure. By dropping objects of different weights from high places (though probably not from the Tower of Pisa as legend would have it), Galileo quickly concluded that the weight of an object was not a factor in its falling motion. (But see the comments on air resistance at the end of this section.)

Galileo began his quantitative experiments by rolling balls down inclined planes (see Fig. 2.13). In this way he was able to "dilute" the effect (gravity) that produced the motion of a freely falling body whose motion was too rapid for him to make accurate measurements. Because he lacked a clock to measure the short time intervals involved, he invented a *water clock* for the purpose. He used a large tank from which water was allowed to escape through a small pipe at the bottom. At the start of the motion to be studied, he began collecting the escaping water in a vessel and he removed the vessel from the stream at the end of the motion. By weighing the water collected he could compare the various short time intervals that were involved in his experiments to a precision of about 0.1 s.

Galileo's hypothesis was that a falling object (or one rolling down an inclined plane) would acquire equal increments of speed in equal intervals of time, that is, the motion would be one of *constant acceleration*. But he could not test his hypothesis directly because to do so would necessitate measuring the speed in several very short intervals during the motion. (A moment's reflection will reveal that this is a rather difficult experiment.)

By using a clever geometrical argument, Galileo reasoned that a body undergoing constant acceleration starting from rest would move, during any interval of time, a distance proportional to the square of that time (compare Eq. 2.13). This conclusion can be tested by simple experiments because it involves measuring *distances* and times instead of speeds and times. Using his water clock, Galileo showed that $x \propto t^2$ for balls rolling down his inclined plane. Furthermore, he showed that this relation held for *all* the angles of inclination of the plane for which he could make measurements. By extrapolating his results to an angle of 90°, at which point the plane is no longer involved in the motion, Galileo concluded that a freely falling body obeys the same relation, namely, $x \propto t^2$, and therefore that the body undergoes constant acceleration when falling freely.

The Acceleration Due to Gravity

Present-day techniques permit the verification of Galileo's hypothesis regarding falling bodies to be made with high precision. An experiment using a stroboscopic flash is shown in Fig. 2.14. From measurements made with such techniques we find that the acceleration experienced by any body falling freely (and without air resistance) near the surface of the Earth is approximately 9.80 m/s² or 32.2 ft/s². We give this important number the symbol g:

$$g = 9.80 \text{ m/s}^2 = 32.2 \text{ ft/s}^2 \tag{2.15}$$

Although we will always use the approximate values for g that are given above, it is of interest to note that because the Earth is not a homogeneous sphere (and also because it rotates), g varies from

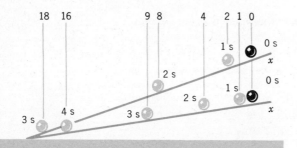

Figure 2.13 Galileo measured the times required for a ball to roll various distances down an inclined plane. He found that $x \propto t^2$ for all angles of inclination and thereby verified that the balls were undergoing constant acceleration.

2.6 The Motion of Falling Bodies

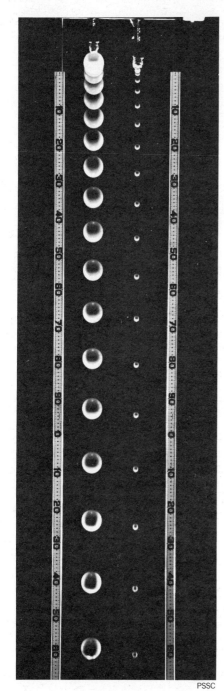

Figure 2.14 Two balls of unequal mass fall at the same rate. This is a stroboscopic photograph taken by opening the camera lens and flashing a light source at regular intervals. From the photograph, verify that the balls fall 4 times as far in 16 time units as they fall in 8 time units. The scales are marked in centimeters. Use the fact that $g = 9.80$ m/s² and calculate the time interval between successive flashes of the stroboscopic light. (Ans.: approximately 1/40 s.)

place to place on the surface of the Earth. At sea level, the variation as a function of latitude λ is shown in Fig. 2.15. The variation of g between $\lambda = 0°$ and $\lambda = 90°$ is approximately 0.5 percent.

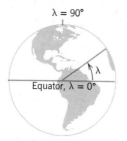

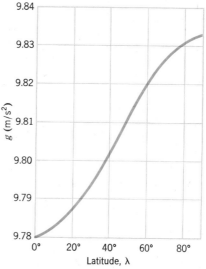

Figure 2.15 Variation of g over the surface of the Earth at sea level. The maximum variation is about 0.5 percent. The diagram above shows the definition of the latitude λ.

• *Example 2.7*

A ball is released from rest at a certain height. What is its velocity after falling 100 m?

Because the initial speed is zero, we use Eq. 2.13 with $y_0 = 0$, $v_0 = 0$, and $a = g$; the distance y is measured vertically *downward*. Following custom, we let vertical displacements be represented by the symbol y. Thus,

$$y = \tfrac{1}{2}gt^2$$

Solving for the time of fall, we have

$$t = \sqrt{\frac{2y}{g}}$$

The velocity after falling for a time t is

$$v = gt = g \times \sqrt{\frac{2y}{g}}$$

$$= \sqrt{2gy}$$

Substituting for g and x, we find the velocity to be

$$v = \sqrt{2 \times (9.80 \text{ m/s}^2) \times (100 \text{ m})}$$

$$= 44.3 \text{ m/s} \quad \blacksquare$$

Notice the important result we have obtained in this example. When an object falls, starting from rest, through a height h, the final velocity is

$$v = \sqrt{2gh} \qquad (2.16)$$

● *Example 2.8*

A ball is thrown upward with an initial speed of 15 m/s from the top of a building. Calculate the velocity and the position as a function of the time.

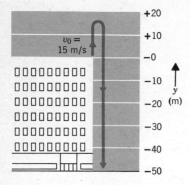

In this example we have *two* directions of motion to consider: first, the upward motion to the maximum height and then the downward motion toward the ground. Therefore, we must be careful to use the proper *signs* in our equations. We arbitrarily choose the positive direction to be *up*. Thus, the initial speed will be $v_0 = +15$ m/s. But the acceleration is *downward*, so $a = -g = -9.80$ m/s². We choose the origin for distance ($y = 0$) at the point from which the ball is thrown. The equations for velocity and distance therefore become

$$v = v_0 + at = (15 \text{ m/s}) - (9.80 \text{ m/s}^2) \times t$$
$$y = v_0 t + \tfrac{1}{2}at^2 = (15 \text{ m/s}) \times t - \tfrac{1}{2}(9.80 \text{ m/s}^2) \times t^2$$

From these equations we find, at one-second intervals, the following values:

t(s)	v(m/s)	y(m)
0	+15	0
1	+5.2	+10.1
2	−4.6	+10.4
3	−14.4	+0.9
4	−24.2	−18.4
5	−34.0	−47.5

Examining this table reveals that the ball reaches its maximum height at some time between one and two seconds after being thrown up into the air. It also appears that it returns to its initial starting point sometime just after three seconds have elapsed. Can we determine these times more accurately?

The ball reaches its maximum height when its velocity has become zero. (Why is this so?) Thus, this occurs at time $t = T$ found by solving.

$$0 = v_0 + aT \quad \text{or} \quad T = \frac{v_0}{g} = \frac{15 \text{ m/s}}{9.80 \text{ m/s}^2} = 1.53 \text{ s}$$

The maximum height reached is therefore

$$y_m = (15 \text{ m/s}) \times 1.53 \text{ s} - \tfrac{1}{2}(9.80 \text{ m/s}^2)(1.53 \text{ s}^2)$$
$$= +11.5 \text{ m}$$

The ball returns to its starting point ($y = 0$) when

$$0 = v_0 T' \times \tfrac{1}{2}a(T')^2 \quad \text{or} \quad T'(v_0 - \tfrac{1}{2}gT') = 0$$

Thus at the time $T' = \dfrac{2v_0}{g} = 2T = 3.06$ s

the ball is again at $y = 0$. (Since y is a quadratic function of time there is another solution to $y = 0$, namely the time $T' = 0$. What is the significance of this solution?)

In summary, after 1.53 s, the velocity of the ball has become zero; that is, the maximum height has been reached (11.5 m) and the subsequent motion is downward. All velocities for $t > 1.53$ s are therefore *negative*. At t 3.06 s the ball has returned to its starting point ($y = 0$) and for all subsequent times, y is *negative*.

The following diagrams show the velocity and the distance as functions of the time.

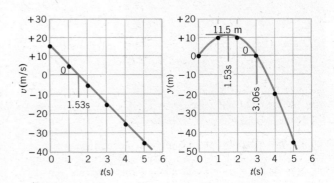

● *Example 2.9*

In Example 2.5 the drag racer achieved an acceleration of 32 (mi/h)/s. Compare this value with g.

$$a = 32 \frac{\text{mi}}{\text{h-s}}$$

$$= 32 \frac{\text{mi}}{\text{h}} \times \frac{1}{\text{s}} \times \frac{5280 \text{ ft}}{1 \text{ mi}} \times \frac{1 \text{ h}}{3600 \text{ s}}$$

$$= 32 \times \frac{5280 \text{ ft}}{3600 \text{ s}^2}$$

$$= 1.46 \times 32 \text{ ft/s}^2$$

$$= 1.46 \, g$$

This acceleration is about the maximum that can be achieved by a vehicle that travels on wheels and depends on the friction between the wheels and the road for its thrust. Attempts to surpass this maximum value by using a more powerful engine will result merely in spinning tires. (Rocket-powered cars and sleds can, of course, achieve much greater accelerations.) ∎

The Effects of Friction

We have thus far been discussing motion from an *idealized* viewpoint; we have considered none of the ever-present effects that will produce deviations from the results we have derived. For example, in any real physical case, there are frictional effects that will retard the motion of a ball rolling down a plane; air resistance will retard falling objects. In addition, the value of g decreases with increasing height above the surface of the Earth (see Section 7.6). Thus, our results are really valid only for the case of an object falling in a vacuum through a small distance near the surface of the Earth (so that g is essentially constant for the motion).

The expression for the velocity of a falling object, Eq. 2.11, indicates that the velocity will continue to increase as long as the object is falling. However, in reality this is not the case for motion through air; the resistance to the motion by the air increases with the velocity so that for any falling object there is a limiting velocity for which there is no longer any net acceleration and the velocity ceases to increase. This effect is illustrated in Fig. 2.16.

The limiting velocity of a falling object is called its *terminal* velocity and depends on the size and shape of the object as well as on its mass. An object that has a large surface area with which to contact the air will experience a relatively large retardation effect. A fluffy object, such as a loosely wadded paper ball, has a very low terminal speed. A spherical object that has a smooth surface and a large mass-to-surface-area ratio, such as a cannon ball, will suffer only a small retardation effect and therefore will have a high terminal velocity. If a sky diver falls freely for several thousand feet before opening his or her parachute, the diver can reach a terminal velocity as high as 235 mi/h (380 km/h or 10.5 m/s) (see Fig. 2.17).

The fact that a terminal velocity exists is indeed a fortunate circumstance. Otherwise a raindrop falling from a height of 10,000 ft would acquire a velocity of about 600 mi/h by the time it reached ground level. A heavy rain storm would then be able to cause enormous damage.

In Chapter 8 we will learn more about the frictional effects experienced by objects moving through fluids.

Figure 2.17 Because of air friction, a sky diver will attain a maximum terminal velocity of free fall of 235 mi/h (10.5 m/s). When "spread-eagled," the terminal velocity is reduced to about 145 mi/h (6.5 m/s).

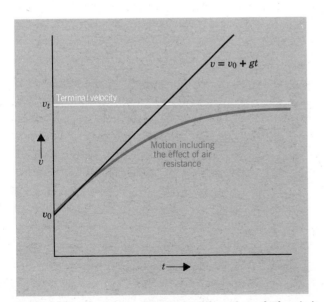

Figure 2.16 The motion of an object falling through the air is retarded by air resistance and its velocity does not increase linearly with the time. Instead, the velocity gradually approaches a limiting value, the *terminal* velocity v_t. (In this diagram, the velocity of fall is plotted as a positive quantity.)

2.7 Vectors

Direction and Magnitude

If you state that point B is 4 m from point A, you have not completely specified the location of B with respect to A. It is necessary to state that B is 4 m *east* of A in order to provide the instructions necessary to move from A to B. The *displacement* $A \rightarrow B$ requires both a *magnitude* and a *direction*. If you wish to describe the motion of an automobile, a statement that the speed is 60 km/h is not a complete description. Again, a *direction* is also required. Physical quantities that require both magnitude and direction for their specification are called *vectors*. *Displacement* is a vector quantity. The quantity that gives the speed of an object as well as its direction of motion is the *velocity vector* of the object.

Quantities that are completely specified by magnitude alone are called *scalars*. Mass, time, and temperature, for example, are scalar quantities. As we use the terms in physics, *speed* and *velocity* are not identical: speed is a scalar, velocity is a vector. We will use the term "speed" when we are interested only in the rate at which an object moves and are not concerned with the direction of the motion. When we wish to convey the impression that direction as well as magnitude is important, the term "velocity" will be used.

In order to distinguish vectors from scalars, we will use boldface type for vectors. Thus, the velocity vector will be denoted by **v**. If we are interested only in the magnitude of the vector **v**, we will write this as v. (Of course, the magnitude of **v** is just the *speed*.)

Some Simple Properties of Vectors

We can write equations connecting vectors just as we can for scalars. Thus, the equation

A = n**B**

means that the magnitude of **A** in n times the magnitude of **B** and, furthermore, that the direction of **A** is the same as that of **B**. In some cases, n will be a scalar quantity with dimensions; then, the product n**B** will have units different from those of **B** and can represent a different physical quantity. In the next chapter we shall see that the *force* **F** acting on a body (a vector quantity) is equal to the product of the mass m of the body (a scalar) and the acceleration **a** of the body (also a vector); that is, **F** = m**a**, which is Newton's famous equation.

In diagrams we represent vectors by arrows. The length of an arrow will be proportional to the

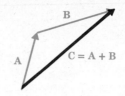

Figure 2.18 The addition of vectors **A** and **B** produces the vector **C**.

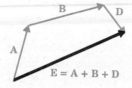

Figure 2.19 The addition of three vectors follows the same rule as for the addition of two vectors.

magnitude of the vector and the direction of an arrow will be the direction of the vector. We will need only a few simple rules regarding the manipulation of vectors.

In Fig. 2.18 we indicate the rule for vector addition. Graphically, in order to add **A** to **B**; we place the origin of the vector **B** at the head of the vector **A**. Then, the line connecting the origin of **A** with the head of **B** is the sum vector: **C** = **A** + **B**. More than two vectors can be added by simply continuing this procedure, as indicated in Fig. 2.19.

The negative of a certain vector **A** is another vector, $-$**A**, which has the same magnitude as **A** but the opposite direction.

The subtraction of one vector from another is the same as adding the negative vector. Figure 2.20 shows the subtraction of **B** and **A** according to two equivalent diagrams. Notice that the vector **B** is placed in different positions in the two diagrams, but that is has the same magnitude and same direction in each case. Of course, if one vector is added to another that has the same magnitude but opposite direction, the result is zero: **A** + ($-$**A**) = 0.

A useful concept that simply specifies the direction of some vector **A** is the *unit vector* written as $\hat{\mathbf{u}}_A$ and defined as

$$\hat{\mathbf{u}}_A = \mathbf{A}/A \qquad (2.17)$$

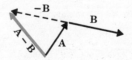

Figure 2.20 The subtraction of **B** from **A** according to two equivalent diagrams. Notice that (**A** $-$ **B**) + **B** = **A**.

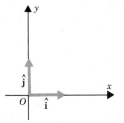

Figure 2.21 The unit vectors $\hat{\mathbf{i}}$ and $\hat{\mathbf{j}}$ defining the directions of the x- and y-axes.

Note that the unit vector $\hat{\mathbf{u}}_A$ is dimensionless even though **A** might have dimensions. The unit vector in the direction of the x-axis in a rectangular x-y coordinate frame is written $\hat{\mathbf{u}}_x = \hat{\mathbf{i}}$ while that in the y-direction is written $\hat{\mathbf{u}}_y = \hat{\mathbf{j}}$, see Fig. 2.21.

It is often required to determine the *component of a vector* **A** in a preferred direction, say in the direction of the x-axis (sometimes the component is referred to as the *projection of* **A** onto this axis.) To do this, construct a unit vector $\hat{\mathbf{u}}$ at the origin of the vector **A** and in the preferred direction. Then, the component of **A** in the direction of $\hat{\mathbf{u}}$ is defined to be

$$A_u = A \cos \theta_u \tag{2.18}$$

where θ_u is the angle between $\hat{\mathbf{u}}$ and **A**. We also define the *vector component* of **A** in the direction of $\hat{\mathbf{u}}$ to be

$$\mathbf{A}_u = (A \cos \theta_u)\hat{\mathbf{u}} \tag{2.19}$$

Note carefully that A_u has associated with it an algebraic sign: A_u is *not* simply the magnitude of $\mathbf{A}_u$; refer to Fig. 2.22.

Just as we can add two vectors and find a single vector that represents the sum, we can also *decompose* a given vector into two perpendicular components, vectors the sum of which yields the original vector. Figure 2.23b shows two mutually perpendicular vectors, $\mathbf{A}_x$ and $\mathbf{A}_y$, whose sum is **A**; that is, $\mathbf{A}_x + \mathbf{A}_y = \mathbf{A}$. When $\mathbf{A}_x$ and $\mathbf{A}_y$ lie along the x- and y-axes, respectively, we say that these vectors are the *rectangular component vectors* of **A**.

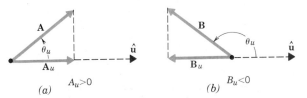

Figure 2.22 The vector components in the direction of $\hat{\mathbf{u}}$, illustrated for two different vectors.

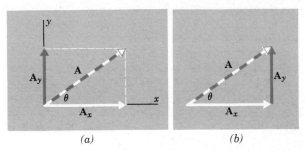

Figure 2.23 (a) The vector **A** is decomposed into the component vectors, $\mathbf{A}_x$ and $\mathbf{A}_y$. (b) The sum of the vectors $\mathbf{A}_x$ and $\mathbf{A}_y$ equals **A**.

Figure 2.23a shows the way in which the two vector components $\mathbf{A}_x$ and $\mathbf{A}_y$ are obtained. Using Eqs. 2.18 and 2.19, and the definitions of the unit vectors $\hat{\mathbf{i}}$ and $\hat{\mathbf{j}}$, we have

$$A_x = A \cos \theta$$
$$A_y = A \cos \left(\frac{\pi}{2} - \theta\right) = A \sin \theta \tag{2.20}$$

and

$$\mathbf{A}_x = A \cos \theta \, \hat{\mathbf{i}}$$
$$\mathbf{A}_y = A \sin \theta \, \hat{\mathbf{j}} \tag{2.21}$$

It is also evident that we can reverse the process and given A_x and A_y we can find A and the angle θ, and thus the vector **A**. From Fig. 2.23, we can also see that the magnitude of **A** is given by the Pythagorean Theorem (Appendix A.5):

$$A = \sqrt{A_x^2 + A_y^2} \tag{2.22}$$

and the angle θ is

$$\theta = \tan^{-1}(A_y/A_x) \tag{2.23}$$

In determining the angle θ (measured counterclockwise from the $+x$ direction) careful attention must be given to the algebraic signs of both A_x and A_y. Figure 2.24 should be studied carefully; for example, note that A_y/A_x is a positive quantity in both the first and third quadrants, however the angle θ is less than 90° in the first quadrant and between 180° and 270° in the third quadrant. The inverse trigonometric functions $\sin^{-1}$, $\cos^{-1}$, and $\tan^{-1}$ are all double valued in the range $0 < \theta < 360°$. Pocket calculators, however, only give one result for these operations.

(A review of the essentials of trigonometry is in Appendix A.5.)

38 Motion

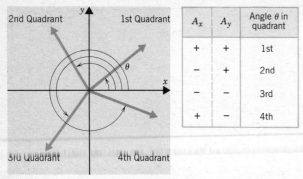
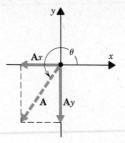

Figure 2.24 The angle θ with the x-axis may lie in any of the four quadrants of the x-y plane.

● *Example 2.10*

An airplane, whose ground speed in still air is 200 mi/h is flying with its nose pointed due north. If there is a cross wind of 50 mi/h in an easterly direction, what is the ground speed of the airplane and what is the direction of flight relative to the ground?

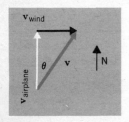

$v = \sqrt{v^2_{\text{airplane}} + v^2_{\text{wind}}}$
$v = \sqrt{(200)^2 + (50)^2}$
$= \sqrt{42,500}$
$= 206 \text{ mi/h}$

The direction of flight is specified by the angle θ:

$\theta = \tan^{-1}(v_{\text{wind}}/v_{\text{airplane}})$
$= \tan^{-1}(50/200)$
$= \tan^{-1} 0.25$
$= 14°$

The airplane actually moves in the direction 14° east of north.

● *Example 2.11*

Given that $A_x = -3$m and $A_y = 4$m, find the magnitude of A and the angle it makes with the x-axis.

First we show (an approximate sketch will do) the vector components $\mathbf{A}_x$ and $\mathbf{A}_y$ in the x-y plane.

Thus $\theta = \tan^{-1} \dfrac{-4}{-3} = \tan^{-1} 1.333$
$= 53.1 \text{ or } 233.1°$

We clearly want $\theta = 233.1°$ since $\mathbf{A}$ is in the third quadrant. The magnitude of $\mathbf{A}$ is

$A = \sqrt{(-3)^2 + (-4)^2} \text{ m}$
$= \sqrt{9 + 16} \text{ m} = 5 \text{ m}$ ∎

Acceleration is a Vector

In addition to displacement and velocity, we also require for the complete description of motion, the vector property of *acceleration*. The *magnitude* of the acceleration of an object is the rate of change of the velocity, and the *direction* of the acceleration is the direction of the *change* in the velocity. If an object is moving in a straight line in the $+x$ direction at the time $t_0 = 0$ with a velocity of 10 m/s and at a later time $t_1 = 1$ s has a velocity of 20 m/s in the same direction, then the x-component of velocity has increased and the acceleration vector is in the $+x$ direction: $a = +10$ m/s². However, if the velocity at t_0 is 20 m/s and is 10 m/s at t_1, then the velocity has decreased in the $+x$ direction and therefore the acceleration vector is in the $-x$ direction: $a = -10$ m/s². Figure 2.25 illustrates this point.

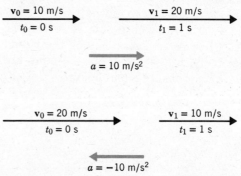

Figure 2.25 The direction of the acceleration vector is in the direction of the *change* in velocity. Uniform acceleration sumed.

● *Example 2.12*

At $t_0 = 0$ an automobile is moving eastward with a velocity of 30 mi/h. At $t_1 = 1$ min the automobile is moving northward at the same speed. What average acceleration has the automobile experienced?

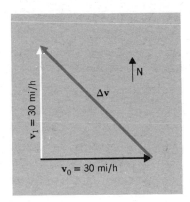

The figure shows the velocity vectors, $\mathbf{v}_0$ and $\mathbf{v}_1$, and the vector $\Delta\mathbf{v}$ that represents the *change* in velocity. We have

initial velocity + change in velocity = final velocity

That is,

$\mathbf{v}_0 + \Delta\mathbf{v} = \mathbf{v}_1$

or,

$\Delta\mathbf{v} = \mathbf{v}_1 - \mathbf{v}_0$

The magnitude of $\Delta\mathbf{v}$ is

$\Delta v = \sqrt{(30 \text{ mi/h})^2 + (30 \text{ mi/h})^2}$
$= \sqrt{1800 \text{ (mi/h)}^2}$
$= 42.4 \text{ mi/h}$

The magnitude of the average acceleration is

$\bar{a} = \dfrac{\Delta v}{\Delta t}$

$= \dfrac{42.4 \text{ mi/h}}{60 \text{ s}}$

$= 0.71 \text{ (mi/h)/s}$

The direction of $\Delta\mathbf{v}$, and hence the direction of $\mathbf{a}$ is, from the figure, in the direction *northwest*.

Note in particular that an acceleration is present if the velocity $\mathbf{v}$ changes either its magnitude as in the example shown in Fig. 2.25, *or* its direction as in the present example. ∎

2.8 Motion in Two Dimensions

Separation into Vector Components

If we drop an object from a certain height, we know that the object will undergo accelerated motion straight downward. What will happen if, just as we drop the object, we also give it some initial velocity (v_{0x}) in the *horizontal* direction? Clearly, the motion will no longer be straight downward but will be at some angle to the vertical. Now, we know that velocity is a vector quantity, so we can decompose the velocity vector in this case into a vertical (y) component and a horizontal (x) component. What equations describe the variation with time of these components? If we choose the upward direction as the direction of positive y, the acceleration due to gravity g acts only in the $-y$ direction. Therefore, the y- component of the velocity is

$v_y = a_y t = -gt$

Because there is no horizontal component of the acceleration, the x-motion is simply

$v_x = v_{0x}$

These two equations are summarized by the important statement that *the resultant velocity vector consists of two components which act independently*. Only the vertical component of the motion undergoes acceleration, whereas the horizontal component proceeds at the constant initial velocity v_{0x}.

Figure 2.26 shows the way in which the horizontal and vertical velocity components combine to

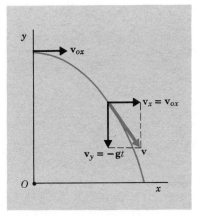

Figure 2.26 If an object is dropped and simultaneously given an initial horizontal velocity v_{ox}, this horizontal velocity component remains constant while the vertical component increases linearly with the time. Thus, the motion follows a curved (actually, *parabolic*) path.

40 Motion

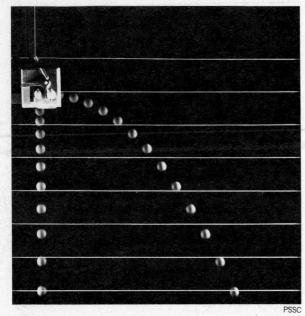

Figure 2.27 The two balls were released simultaneously; the one on the left was merely dropped while the other was given an initial horizontal velocity. The vertical components of the motion of both balls are exactly the same. The stroboscopic photograph was taken with a flash interval of 1/30 s. The distance between the horizontal bars is 0.15 m. From the photograph verify that $g = 9.80$ m/s². What is the horizontal component of the velocity of the projected ball?

give the instantaneous velocity vector **v**. Figure 2.27 is a stroboscopic photograph of two balls that are dropped simultaneously, one with a horizontal velocity component. The picture reveals that the vertical motions are indeed identical. The path followed by the projected ball is a *parabola*.

As the motion proceeds the velocity vector at any instant is always tangent to the path.

If we also allow the vertical motion to have an initial velocity v_{0y}, then the equations that describe the motion are:

Acceleration: $a_x = 0 \qquad a_y = -g$

Velocity: $v_x = v_{0x} \qquad v_y = v_{0y} + a_y t$
$\qquad\qquad\qquad\qquad = v_{0y} - gt$ (2.24)

Displacement: $x = x_0 + v_{0x}t \qquad y = y_0 +$
$\qquad\qquad\qquad\qquad v_{0y}t + \tfrac{1}{2}a_y t^2$

● **Example 2.13**

An object is released from a height of 20 m and is given an initial horizontal velocity of 10 m/s. When the object strikes the ground, how far has it traveled in the horizontal direction?

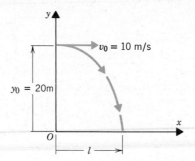

In the coordinate frame selected we have
$y_0 = +20$ m and $x_0 = 0$.

Because $v_{0y} = 0$, we have
$$y = y_0 - \tfrac{1}{2}gt^2$$

The final value of y is zero, so
$$0 = y_0 - \tfrac{1}{2}gt^2 \text{ or } 20 = \tfrac{1}{2}(9.80)t^2$$

Thus,
$$t = \sqrt{\frac{20 \text{ m}}{4.90 \text{ m/s}^2}} = 2.02 \text{ s}$$

Therefore,
$$\ell = v_{0x}t$$
$$= (10 \text{ m/s}) \times (2.02 \text{ s})$$
$$= 20.2 \text{ m} \quad \blacksquare$$

Ballistic Trajectory

The path followed by an unpowered and unguided projectile is called a ballistic trajectory. Figure 2.28 shows the path followed by such a projectile fired with an initial velocity $\mathbf{v}_0$ at an angle θ with respect to the horizontal. The components of $\mathbf{v}_0$ are

$$v_{0x} = v_0 \cos \theta \qquad (2.25\text{a})$$
$$v_{0y} = v_0 \sin \theta \qquad (2.25\text{b})$$

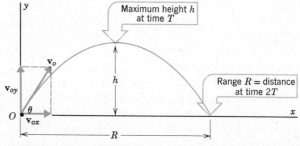

Figure 2.28 The path of a ballistic projectile fired with an initial velocity $\mathbf{v}_0$ at an angle θ with respect to the horizontal. The origin of the x-y coordinate frame is located at the launch point.

In the selected coordinate frame we have $x_0 = 0$ and $y_0 = 0$, and hence

$$x = v_{0x}t \tag{2.26a}$$
$$y = v_{0y}t + \tfrac{1}{2}a_y t^2 = v_{0y}t - \tfrac{1}{2}qt^2 \tag{2.26b}$$

The velocity components at any instant are given by

$$v_x = v_{0x} \tag{2.27a}$$
$$= v_0 \cos\theta$$
$$v_y = v_{0y} - gt \tag{2.27b}$$
$$= v_0 \sin\theta - gt$$

Let us now use these equations to find the maximum height to which the projectile will rise and the overall range of the projectile along the horizontal surface.

In order to find the maximum height, we calculate the value of y when the upward component of the velocity has decreased to zero, that is, $v_y = 0$ when $y = h$; we let $t = T$ be the time when this occurs. Substituting $v_y = 0$ and $t = T$ into Eq. 2.27b, we find

$$T = \frac{v_{0y}}{g}$$

Next, we use Eq. 2.26b and substitute $y = h$ when $t = T$; we then obtain

$$h = v_{0y}T - \tfrac{1}{2}gT^2$$
$$= v_{0y}\left(\frac{v_{0y}}{g}\right) - \tfrac{1}{2}g\left(\frac{v_{0y}}{g}\right)^2$$
$$= \frac{v_{0y}^2}{g} - \frac{1}{2}\frac{v_{0y}^2}{g}$$

or,

$$h = \frac{1}{2g}v_{0y}^2 = \frac{1}{2g}v_0^2\sin^2\theta \tag{2.28}$$

which gives the maximum height in terms of the initial velocity.

In order to find the horizontal range R we use the fact that the total flight requires a time that is *twice* the time necessary to reach maximum height. (Can you see why this is so? Hint: see Example 2.8.) That is, we set $x = R$ when $t = 2T$. Then, using Eq. 2.26a, we find

$$R = v_{0x} \times 2T$$

substituting for T,

$$R = \frac{2}{g}v_{0x}v_{0y}$$

Making use of Eqs. 2.25 for v_{0x} and v_{0y}, we obtain

$$R = \frac{2v_0^2}{g}\sin\theta\cos\theta \tag{2.29}$$

When Galileo was studying projectile motion[6] he showed by calculating a number of cases that maximum range occurs when the elevation angle is $\theta = 45°$ (see Problem 2.53). We can easily obtain this result by using the trigonometric relationship that states $\sin 2\theta = 2\sin\theta\cos\theta$. Then, Eq. 2.29 can be written as

$$R = \frac{v_0^2}{g}\sin 2\theta \tag{2.29}$$

The maximum value of the sine function is 1 and this occurs when $2\theta = 90°$ or $\theta = 45°$; then

$$R_{\max} = \frac{v_0^2}{g} \quad \text{at } \theta = 45° \tag{2.30}$$

● *Example 2.14*

An artillery piece is pointed upward at an angle of 35° with respect to the horizontal and fires a projectile with a muzzle velocity of 600 m/s. If air resistance is unimportant, to what height will the projectile rise and what will be its range?

The height is given by Eq. 2.28.

$$h = \frac{1}{2}\frac{v_{0y}^2}{g} = \frac{v_0^2}{2g}\sin^2\theta$$

$$= \frac{(600 \text{ m/s})^2}{2 \times 9.80 \text{ m/s}^2} \times \sin^2 35°$$

$$= 6040 \text{ m} = 6.04 \text{ km}$$

or about 3.8 mi.

The range is given by Eq. 2.29a:

$$R = \frac{v_0^2}{g}\sin 2\theta$$

---

[6]Although Galileo is usually credited with the discovery that projectiles move along parabolic paths, Leonardo da Vinci discussed parabolic motion in his notes about 1500.

$$= \frac{(600 \text{ m/s})^2}{9.80 \text{ m/s}^2} \times \sin 70°$$

$$= 34{,}500 \text{ m} = 34.5 \text{ km}$$

or about 21 mi. A rifle bullet fired with the same initial conditions would not travel nearly this far. Because a rifle bullet has a much larger surface-to-mass ratio than does an artillery shell, air resistance effects are much more severe and drastically reduce the range. ∎

The Running Long Jump

How far can a man jump using a running start? We can use the same type of analysis that we used in the study of projectile motion. At the point of departure, the center of the jumper's body has velocity components v_{0x} and v_{0y}. For v_{0x} we will use the maximum sprinting speed of 10.5 m/s (see the essay on Running Speeds earlier in this chapter). In order to find v_{0y}, we use a result that we will obtain later (see the essay on page 125), namely, that the height to which an individual can raise himself by jumping vertically is about 0.6 m. That is, we assume that the jumper maintains his horizontal velocity through the jump and that at the departure point he propels himself upward with the same velocity that his leg muscles could produce in the standing high jump. We will describe the motion of the jumper in terms of the motion of the center of his body. (The definition of *center of mass* is given later, in Section 3.9.) Thus, we can write the initial values of the jumper's velocity components as

$$v_{0x} = 10.5 \text{ m/s}$$
$$v_{0y} = \sqrt{2gh} = \sqrt{2 \times (9.8 \text{ m/s}^2) \times (0.6 \text{ m})}$$
$$= 4.85 \text{ m/s}$$

Therefore, the initial speed v_0 and direction of motion θ are

$$v_0 = \sqrt{v_{0x}^2 + v_{0y}^2} = \sqrt{(10.5 \text{ m/s})^2 + (4.85 \text{ m/s})^2}$$
$$= 11.6 \text{ m/s}$$

$$\theta = \tan^{-1} \frac{v_{0y}}{v_{0x}} = \tan^{-1} \frac{4.85}{10.5} = \tan^{-1} 0.46$$
$$= 25°$$

(The value we have obtained here for the angle at which the jumper projects himself, namely, 25°, is in close agreement with the angle seen in photographs of competitive long jumpers.) Using these values for v_0 and θ, we can compute the range using Eq. 2.29a:

$$R = \frac{v_0^2}{g} \sin 2\theta = \frac{(11.6 \text{ m/s})^2}{9.8 \text{ m/s}} \sin 50°$$

$$= 10.5 \text{ m}$$

To obtain this figure, we used the formula for the range of a projectile over a horizontal surface. In the case of a long jumper, however, he extends his range by finishing the jump with his body center *lower* than at the starting point. From an erect start to a sitting finish, the center of his body will be lowered by about 0.6 m. We can take this into account in the following way (see Fig. 1). At the distance R, the jumper's body center will be at the same height as at the departure point (1.0 m) and his horizontal velocity will be the same, that is, v_{0x}. Furthermore, the vertical

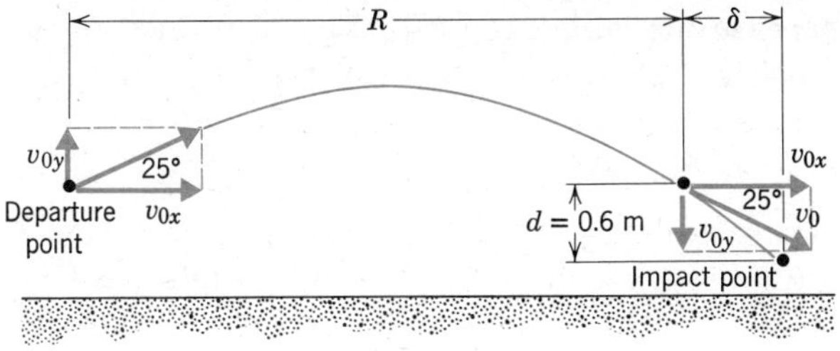

Figure 1 The running long jump.

velocity will be the same in magnitude as the departure velocity, that is, v_{0y}. With an initial downward velocity $v_{0y} = 4.85$ m/s, the time required to move through the distance $d = 0.6$ m is obtained from

$$d = v_{0y}t + \tfrac{1}{2}gt^2$$

Using the quadratic formula (see Appendix A.3), and expressing all quantities in metric units, we have

$$t = -\frac{v_{0y}}{g} \pm \frac{1}{g}\sqrt{v_{0y}^2 + 2gd}$$

$$= -\frac{4.85}{9.8} \pm \frac{1}{9.8}\sqrt{(4.85)^2 + (2)(9.8)(0.6)}$$

$$= -0.50 \pm 0.61$$

Clearly, the positive sign is required in this case, so

$t = 0.11$ s

Then, the horizontal distance moved during this time is

$\delta = v_{0x}t = (10.5 \text{ m/s}) \times (0.11 \text{ s})$
$\quad = 1.2$ m

Therefore, the total distance of the jump is

jump distance $= R + \delta = 10.5$ m $+ 1.2$ m $= 11.7$ m

Notice that by landing in the sitting position, the jumper extends his range by more than 10 percent. In addition, the jumper can add a further increment to his distance by arm and leg movements that change the position of the effective center of his body, but we have neglected this aspect of the motion.

The world record for the running long jump is about 9 m. Thus, the degree to which the maximum theoretical value has been attained in this event is about 77 percent.

References
G. Dyson, *The Mechanics of Athletics,* University of London Press, London, 1968, Chapter 8.

P. Davidovits, *Physics in Biology and Medicine,* Prentice-Hall, Englewood Cliffs, N.J., 1975, Chapter 3.

■ *Exercises*
1   What is the duration of the long jump illustrated in Fig. 1?

2. Describe how a jumper can increase his long-jump range by arm and leg motions that change the position of the effective center of his body. Make stickman sketches of the way a long jumper leaves the ground and the way that he lands.

3. If a long jumper could, by training, increase either his vertical or his horizontal takeoff speed, which would be more advantageous? Calculate the jump distance for a 10 percent increase in v_{0x} and the jump distance for the same fractional increase in v_{0y}.

2.9 Uniform Circular Motion

Angular Velocity

Another very important case of two-dimensional motion is that of motion in a circular path. For simplicity, we assume that the motion is *uniform*, that is, that the *speed* of the object is constant. If it takes 1 s for the object to make a complete revolution, we say that the object moves at an *angular* rate of 1 rev/s. But one complete revolution corresponds to 2π radians (360°), so we can alternatively state that the object moves with an *angular velocity* of 2π rad/s. It is customary to denote angular velocity (measured in radians per second) by the symbol ω. (Radian measure is reviewed in Appendix A4.)

Suppose that an object moves uniformly in a circle, as indicated in Fig. 2.29. The position of the object at any time is specified by the angle θ. The rate at which θ changes with time is the angular velocity:

$$\omega = \lim_{\Delta t \to 0} \frac{\Delta \theta}{\Delta t} \qquad (2.31)$$

(Notice that the *radian* is not a quantity that carries physical dimensions of length, time, or mass. Consequently, the *physical* unit of angular velocity is s^{-1}. We sometimes carry the abbreviation *rad* in calculations so that we are reminded that angles are being measured in radians, but this "unit" is always dropped in the final answer. As we will see in Chapter 11, the unit s^{-1} is the unit of *frequency*. Therefore, it is equally appropriate to refer to ω as angular *frequency* or as angular *velocity*. This is often done.)

The Period

For uniform circular motion θ increases linearly with time and thus ω given by Eq. 2.31 is a constant. The *period* of this circular motion is the time required for one complete revolution or cycle of the motion. If the period is denoted by the symbol τ, then $\omega \tau = 2\pi$ or:

$$\tau = \frac{2\pi}{\omega} \qquad (2.32)$$

Clearly, the period and the angular velocity are inversely related because the greater the angular velocity, the shorter the time required to make a revolution.

If an object moves with uniform speed in a circular path with radius r, the distance traveled in one period is the circumference of the circle, $2\pi r$. The time required for this motion is τ. Therefore, the speed is

$$v = \frac{\text{distance}}{\text{time}}$$
$$= \frac{2\pi r}{2\pi/\omega}$$

Thus,

$$v = r\omega \qquad (2.33)$$

For this case of circular motion, the velocity vector is continually changing in direction (see Fig. 2.30). However, if the motion is uniform, the *magni-*

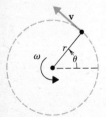

Figure 2.29 The position of an object moving in a circular path is given by the angle θ. The rate of change of θ is the angular velocity of the object. The velocity vector **v** is always tangent to the circular path.

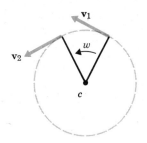

Figure 2.30 The velocity vector of an object moving in a circular path changes direction continually. If the motion is *uniform,* the *speed* of the object remains constant while **v** changes direction.

Figure 3.32 The two triangles are similar. In the triangle at the right, v stands for v_0 and v_1 which are equal.

tude of the velocity vector (the *speed*) is everywhere the same.

● *Example 2.15*

An automobile moves with a constant speed of 50 mi/h around a track of 1 mi diameter. What is the angular velocity and the period of the motion?

$$\omega = \frac{v}{r} = \frac{50 \text{ mi/h}}{0.5 \text{ mi}} = 100 \text{ rad/h (or } 100 \text{ h}^{-1})$$

$$\tau = \frac{2\pi}{\omega} = \frac{2\pi}{100 \text{ rad/h}} = 0.0628 \text{ h} = 3.77 \text{ min}$$

Notice that the "unit" *rad* is dropped in writing the final result. ■

Centripetal Acceleration

If an object moves without acceleration, there is no change of velocity and the velocity vector is therefore constant in magnitude and direction. Conversely, if there is any change of velocity, then there must have been acceleration. This change need not be in magnitude; a change in the *direction* of the velocity vector, even if the magnitude remains constant, requires acceleration. Thus, an object moving uniformly in a circular path is continually accelerated.

We can derive an expression for the acceleration in circular motion by referring to Fig. 2.31. At the time t_0 the velocity vector of the moving object is $\mathbf{v}_0$. At the time t_1 the motion has progressed by an angle $\Delta\theta$ and the velocity vector is $\mathbf{v}_1$. If the motion is uniform, $v_0 = v_1$, even though the directions of $\mathbf{v}_0$ and $\mathbf{v}_1$ are different. In moving from point A to point B, the object moves a distance Δs along the circumference of the circle. The length of the chord connecting the two points is Δx. In order to change the velocity vector from $\mathbf{v}_0$ to $\mathbf{v}_1$, the acceleration has produced a "change-in-velocity vector" which we label $\Delta\mathbf{v}$.

Because the velocity vectors, $\mathbf{v}_0$ and $\mathbf{v}_1$, are each perpendicular to radius lines at points A and B, respectively, it follows from the geometry that the triangle formed by the two radius lines and Δx is similar to the triangle formed by $\mathbf{v}_0$, $\mathbf{v}_1$, and $\Delta\mathbf{v}$ (see Fig. 2.31). That is, both triangles are isosceles triangles with the same angle $\Delta\theta$ between the equal sides (Fig. 2.32). Therefore, the ratios of the lengths of the sides of the two triangles are equal:

$$\frac{\Delta v}{v} = \frac{\Delta x}{r}$$

where v stands for either v_0 or v_1 which are equal. Thus, the magnitude of the velocity change is

$$\Delta v = \frac{v \times \Delta x}{r}$$

Now, the acceleration is given by the standard expression

$$a_c = \lim_{\Delta t \to 0} \frac{\Delta v}{\Delta t}$$

Substituting for Δv,

$$a_c = \lim_{\Delta t \to 0} \frac{v \times \Delta x / r}{\Delta t} = \lim_{\Delta t \to 0} \left(\frac{v}{r} \times \frac{\Delta x}{\Delta t} \right)$$

Because the quantities v and r are constants, they are not affected by the limiting process and we can rewrite the expression for a_c in the form

$$a_c = \frac{v}{r} \times \lim_{\Delta t \to 0} \frac{\Delta x}{\Delta t}$$

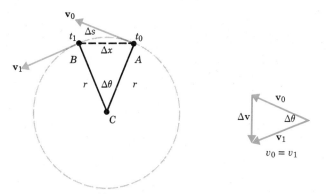

Figure 2.31 Velocity diagram of accelerated motion in a circle.

As Δt becomes infinitesimally small, the chord Δx becomes equal to the arc length Δs, and the limit of $\Delta s/\Delta t$ is the velocity v. Therefore,

$$a_c = \frac{v^2}{r} \qquad (2.34)$$

Referring to Fig. 2.31 we see that as Δt becomes small, $\Delta \theta$ also becomes small, and $\mathbf{v}_1$ almost coincides with $\mathbf{v}_0$. Then, $\Delta \mathbf{v}$ is a vector essentially perpendicular to both $\mathbf{v}_0$ and $\mathbf{v}_1$; that is, $\Delta \mathbf{v}$ points toward the center of the circle. Because the direction of the acceleration vector is the same as the direction of the change in velocity, the vector $\mathbf{a}_c$ is also directed toward the center of the circle. The acceleration is "center seeking" and is termed *centripetal* acceleration.

Using our previous result that $v = r\omega$ (Eq. 2.33), we can also express a_c as

$$a_c = \frac{v^2}{r} = \frac{(r\omega)^2}{r}$$

or,

$$a_c = r\omega^2 \qquad (2.35)$$

•*Example 2.16*

We know that if we drop an object while giving it a horizontal velocity component, the object will fall toward the surface of the Earth with the horizontal velocity remaining constant. The object will impact the Earth some distance away as in Example 2.13. With what velocity must an object be projected so that the curvature of its path is just equal to the curvature of the Earth?

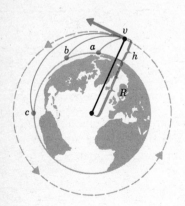

Suppose we start with the object at a distance h above the surface of the Earth and impart to it a "horizontal" velocity $\mathbf{v}$, as shown in the diagram. If the magnitude of $\mathbf{v}$ is small, the object will fall towards the Earth and impact at some point such as a. As the magnitude of $\mathbf{v}$ is increased the impact point moves further and further away, to points b, and c, and so forth. Eventually, when some critical speed is reached, the object will literally fall *around* the Earth, never hitting its surface. The height of the object above the surface of the Earth would therefore never decrease and the object would become a satellite of the Earth traveling in a circular orbit. The radius of the Earth is R so that the radius of the desired circular path of the object is $R + h$. The centripetal acceleration required to maintain the circular motion is

$$a_c = \frac{v^2}{r} = \frac{v^2}{R + h}$$

This centripetal acceleration is furnished by gravity, so we can substitute g for a_c; thus,

$$g = \frac{v^2}{R + h}$$

As we will see later when we study gravitation in more detail, the value of g depends on the distance from the center of the Earth. If h is small compared to R (even if $h = 100$ mi, it would still be only about 1/40 of the Earth's radius), the value of g will be essentially that appropriate for the surface of the Earth, that is, $g = 9.80$ m/s^2. Therefore, neglecting h compared to R, solving for v, we find

$$v = \sqrt{gR}$$
$$= \sqrt{(9.80 \text{ m/s}^2) \times (6.38 \times 10^6 \text{ m})}$$
$$= 7.91 \times 10^3 \text{ m/s} = 7.91 \text{ km/s}$$

or approximately 17,700 mi/hr.

The period of the motion is

$$\tau = \frac{2\pi}{\omega} = \frac{2\pi r}{v}$$
$$= \frac{2\pi \times (6.38 \times 10^6 \text{ m})}{7.90 \times 10^3 \text{ m/s}}$$
$$= 5070 \text{ s} = 84.6 \text{ min}$$

Therefore, a satellite moving in a circular orbit at a height under say 100 mi will require approximately an hour and a half to circle the Earth. Many of the artificial satellites that have been launched during the past few years have orbit characteristics similar to those in this example. ∎

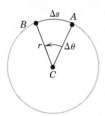

Figure 2.33 An object moves from point A to point B along a circular path in the time Δt.

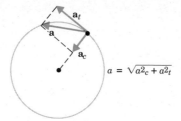

Figure 2.34 The linear acceleration **a** and its radial and tangential components.

2.10 Accelerated Circular Motion

An important generalization of the circular motion we have been considering up to this point is the case of nonuniform or accelerated circular motion. For this type of motion both the speed v and the angular velocity ω vary with time. However, the relationship between the instantaneous value of these two quantities remains the same as that given by Eq. 2.33, namely that $v = r\omega$. To see why this is so, consider the application of the basic definition of the angular velocity (Eq. 2.31) to the situation shown in Fig. 2.33. The object moves in its circular path from point A to point B in a time Δt. This motion represents an angular displacement $\Delta\theta$ and an arc length $\Delta s = r\Delta\theta$. Multiplying both the left and right-hand sides of Eq. 2.31 by the radius r we have

$$r\omega = \lim_{\Delta t \to 0} \frac{r\Delta\theta}{\Delta t} = \lim_{\Delta t \to 0} \frac{\Delta s}{\Delta t} = v$$

This desired result now refers to instantaneous values of v and ω.

While the instantaneous centripetal or radial acceleration of the object still remains the same as that given by Eqs. 2.34 and 2.35, namely

$$a_c = \frac{v^2}{r} = r\omega^2$$

An additional acceleration is present when v is not a constant. This new acceleration is referred to as the *tangential* acceleration a_t:

$$a_t = \lim_{\Delta t \to 0} \frac{\Delta v}{\Delta t} \qquad (2.36)$$

Since a_c and a_t are always mutually perpendicular, they may be considered to be the two components of a single vector, the linear acceleration **a**. Figure 2.34 shows the instantaneous linear acceleration, the magnitude of which is

$$a = \sqrt{a_c^2 + a_t^2} \qquad (2.37)$$

It is often convenient to relate the tangential acceleration a_t to its angular counterpart called the *angular* acceleration, denoted by the symbol α and defined to be

$$\alpha = \lim_{\Delta t \to 0} \frac{\Delta \omega}{\Delta t} \qquad (2.38)$$

Since for circular motion, r is a constant, it follows that a small change Δv in v corresponds to a small change $\Delta \omega$ in ω given by $\Delta v = r\Delta\omega$. Thus, Eq. 2.36 may be written

$$a_t = \lim_{\Delta t \to 0} \frac{\Delta v}{\Delta t} = r \lim_{\Delta t \to 0} \frac{\Delta \omega}{\Delta t}$$

or

$$a_t = r\alpha \qquad (2.39)$$

That is, the tangential acceleration is equal to the angular acceleration times the path radius.

A common situation met in practice is the case of a *constant* angular acceleration α. Obtaining the relationship between the angular position θ, the angular velocity ω, the angular acceleration α, and t follows the same reasoning that related x, v, and a for linear motion when a was a constant. If at time $t = 0$ the angular position of an object traveling in a circular path is θ_0 and the angular velocity is ω_0, then with constant angular acceleration α, we have a complete analogy with Eq. 2.14 for linear motion:

$$\alpha = \text{const}$$
$$\omega = \omega_0 + \alpha t \qquad (2.40)$$
$$\theta = \theta_0 + \omega_0 t + \tfrac{1}{2}\alpha t^2$$

● *Example 2.17*

An object initially at rest is subjected to a constant angular acceleration $\alpha = 5$ rad/s^2 while following a circular path of radius $r = 0.4$ m. Find the angle through which it has rotated in the time $t = 2/5$ s and determine the linear acceleration at this point.

Since we are informed that $\omega_0 = 0$, using Eq. 2.40 gives

$$\theta - \theta_0 = \tfrac{1}{2}\alpha t^2 = \tfrac{1}{2} \times (5 \text{ rad/s}^2)(2/5 \text{ s})^2$$
$$= 2/5 \text{ rad or } 22.9°$$

To determine **a** we must calculate both a_c and a_t at the time $t = 2/5$ s. The value of ω after an elapsed time of 2/5 s is

$$\omega = \alpha t = (5 \text{ rad/s}^2)(2/5 \text{ s}) = 2 \text{ rad/s}.$$

Thus the centripetal acceleration is

$$a_c = r\omega^2 = (0.4 \text{ m})(2 \text{ s}^{-1})^2 = 1.6 \text{ m/s}^2,$$

and the tangential acceleration is

$$a_t = r\alpha = (0.4 \text{ m})(5 \text{ s}^{-2}) = 2.0 \text{ m/s}^2.$$

Finally, the linear acceleration is obtained using Eq. 2.37:

$$a = \sqrt{(1.6)^2 + (2.0)^2} \text{ m/s}^2 = 2.56 \text{ m/s}^2$$

Summary of Important Ideas

Speed is the rate at which distance is traveled without regard to direction; speed is a scalar quantity.

Velocity is a vector quantity that specifies both the magnitude and the direction of the motion; the magnitude of the velocity vector is the speed.

On a distance-time graph, the *slope* of the curve at any point is the velocity at that instant.

Acceleration is the rate of change of velocity. Acceleration is a *vector*.

For objects falling near the surface of the Earth, the horizontal and vertical motions are *independent*. The vertical motion undergoes an acceleration $g = 9.80$ m/s$^2 = 32.2$ ft/s^2.

Ballistic motion near the surface of the Earth is *parabolic* (in the absence of air friction).

An object moving in a circle has a *centripetal acceleration* that is directed toward the center of the circle.

When the circular motion is accelerated there is, in addition to centripetal acceleration, a *tangential* acceleration.

◆Questions

2.1 Is it possible for a moving body that experiences a constant acceleration **a** to have **v** and **a** *always* in *opposite* directions?

2.2 The acceleration applied to a certain body is constant in magnitude and direction. (a) Describe a situation in which the velocity vector always has the same direction. (b) Describe a situation in which the body never moves in a straight line.

2.3 Why is it not possible to take account of air resistance effects in describing the motion of falling bodies by simply decreasing the value of g?

2.4 In the game of roulette, it is customary to set the ball into motion in the direction opposite to that of the revolving wheel. What is the difference between the motion of the ball relative to the wheel and the motion of the ball relative to the table? Can the ball ever have *zero* instantaneous velocity relative to the table even if the wheel is moving?

2.5 At what point or points in its path is the *speed* of a projectile a minimum? At what point or points is the speed a maximum? (Neglect air resistance.)

2.6 Describe three different kinds of situations that involve acceleration in the driving of an automobile.

2.7 A particle moves uniformly in a circular path. Describe the way in which the acceleration vector changes with time.

★Problems

Section 2.1

2.1 A bead is confined to slide along a taut string. It moves 80 cm to the right from its starting point, then reverses its motion and moves to the left 30 cm before resuming its motion to the right by an additional 50 cm. How far is the bead from its starting point? What is its displacement? What is the path length traversed?

2.2 A hiker, while following a straight-line path, drops his watch at some point. After proceeding along some distance he realizes the loss and returns to find the watch and then resumes his hike. He finds at the end of the hike that he is 10 km from the starting point yet his portable odometer shows he walked a total of 15 km. How far did he proceed after dropping the watch before turning back to recover it?

2.3 An object follows a circular path having a radius of 10 m. If the end point of the trip is diametrically opposite the starting point, what is the magnitude of the displacement? What is the path length?

Sections 2.2 and 2.3

2.4 A driver on an Interstate highway notices the mileage markers at the following times:

Mile 120	11:30 A.M.
140	11:50 A.M.
150	12:40 P.M.
200	1:40 P.M.
208	1:46 P.M.
208	1:50 P.M.
218	2:05 P.M.

(a) Plot a graph of distance *versus* time for the trip.
(b) Indicate on the graph the average speeds for the various straight-line portions of the graph.
(c) What apparently happened near noon and at 1:46 P.M.?

2.5 During a certain automobile trip the speeds at different time intervals were as follows:

1:00 P.M.–2:00 P.M.	$v = 30$ mi/h
2:00 P.M.–2:30 P.M.	$v = 40$ mi/h
2:30 P.M.–3:00 P.M.	stopped
3:00 P.M.–4:30 P.M.	$v = 20$ mi/h
4:30 P.M.–5:00 P.M.	$v = 60$ mi/h

(a) What was the total length of the trip?
(b) What was the average speed?
(c) How long did it take to go the first 55 mi?
(d) What was the average speed for the first two hours? For the last two hours?

2.6 An ocean liner makes a 3600 mi voyage in 8 days, 8 hours. What was the average speed during the trip?

2.7* What is the speed of the Earth in its orbit around the Sun? (The radius of the orbit is 1.50×10^{11} m.) Express the result in km/s.

2.8 An automobile moves at constant speed a distance of 20 m in 2 s; during the next 3 s it moves at constant speed only 10 m. What was the average speed (a) during the first 2 s, (b) during the next 3 s, (c) during the 5-s interval? (d) Compute the average speed for the first 3 s.

2.9 A car makes a 200-km trip at an average speed of 40 km/h. A second car starting 1 h later arrives at their mutual destination at the same time after following the same route. What was the average speed of the second car?

Section 2.4

2.10* Suppose the position of an object moving along the x-axis is given by

$$x = 5t + 3$$

where x is in meters when t is in seconds. What is the velocity of the object at any time t? (Hint: Refer to Example 2.4.)

2.11 An object is traveling along the x-axis with a constant velocity. If it is known to be at position $x = +3$m when $t = 2$ s, and to be at $x = +7$ m when $t = 3$ s, what is the velocity? What is the speed? Write the position of the object in the form

$$x = v_0 t + x_0$$

where x and x_0 are to be in meters and t in seconds.

2.12 Repeat Problem 2.11 with the change that at time $t = 3$ s the position is $x = -7$ m.

2.13* The distance-time graph for an object moving in a straight line is the semicircular curve shown in the diagram. Determine the velocity of the object when $t = 5$, 10, and 15 seconds. [Hint: In determining the tangent to the distance-time graph make use of the geometric properties of a circle.]

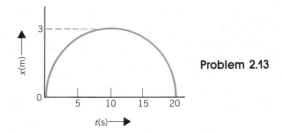

Problem 2.13

Section 2.5

2.14 An object moving along a straight path with an initial velocity of 12 m/s is uniformly accelerated at a rate of 3 m/s². What is the velocity of the object after 8 s of acceleration?

2.15 An automobile is traveling along a straight road at a velocity of 50 mi/h. At $t = 0$ the driver applies the brakes and decelerates uniformly to a velocity of 30 mi/h in 5 s. What was the value of the acceleration? Express the result in (mi/h)/s and in ft/s².

2.16 A sky diver is falling with a terminal velocity of 120 mi/h. It requires 2 s for her parachute to deploy and to decrease her downward velocity to 20 mi/h? What (constant) acceleration does the sky diver experience? Express the result in units of g. Is sky diving safe for a person who "blacks out" at $5\,g$?

2.17 An automobile starts from rest and after 3 s is moving with a velocity of 20 m/s along a straight course. If the acceleration to this velocity was constant, how far did the automobile move in the first 2 s of motion? How far did it move during the third second?

2.18 What must be the value of the constant acceleration of a light aircraft to reach liftoff velocity of 80 mi/h in 1000 ft of runway?

2.19 The driver of a car traveling at 90 mi/h observes a hazard on the road and applies the brakes, giving a constant deceleration of 7.5 ft/s². If the driver's reaction time is 0.15 s, how long does it take to stop after sighting the hazard? What distance does the automobile travel before coming to rest?

2.20* An automobile is cruising at a speed of 100 km/h on a straight stretch of highway with a posted speed limit of 80 km/h. As the automobile passes a parked highway patrol car, the officer accelerates his car at a uniform rate, reaching 60 km/h in 10 s; he continues to accelerate at the same rate until he catches the speeding car. How long did the chase last? How far from the parked position of the patrol car was the speeder overtaken? What was the speed of the patrol car as it overtook the speeder? [Hint: Sketch the distance-time graph for both vehicles and note the significance of the patrol car having overtaken the speeding car.]

2.21* A truck traveling at a constant speed of 80 km/h along a straight stretch of road passes a more slowly moving car. The instant the truck passes the car, the car begins to accelerate at a constant rate of 1.2 m/s² and passes the truck 0.5 km farther down the road. What was the speed of the car when it was passed by the truck?

Section 2.6

2.22 An object is released from rest and falls downward. Make a table of the distance moved (in meters during each second of fall for the first 8 s. Next, make a graph of the table entries. How would you characterize the curve that passes through the points?

2.23 A ball is thrown upward from the top of a tower with an initial speed of 45 m/s. What is the downward speed of the ball 3 s after it has passed the top of the tower on its way down?

2.24 A ball is thrown upward with an initial velocity of 30 m/s. How high will it rise? What will be the velocity of the ball when it returns to its original position?

2.25* An object is projected straight downward with an initial speed v_0. Show that after falling a distance h, the speed of the object is $v = \sqrt{2gh + v_0^2}$.

2.26* Plot a graph (v vs. t) of the velocity of an object thrown upward with an initial speed of 49 m/s. Determine *graphically* the approximate speed after 3 s and after 5 s. At what time will the speed be downward at 49 m/s?

2.27 On the Moon the acceleration due to gravity is only $\frac{1}{6}$ as large as on the Earth. An object is given an initial upward speed of 30 m/s at the surface of the Moon. How long will it take for the object to reach maximum height? How high above the surface of the Moon will the object rise?

2.28 A rocket is launched from the surface of the Earth with an upward acceleration of $4\,g$. After 10 s, what is the speed of the rocket and to what height has it risen?

2.29 Two automobiles, each traveling with a speed of 60 mi/h, crash head-on. The impact speed (that is, the relative speed at collision) is the same as if the cars had been dropped from what height? If it requires $\frac{1}{50}$ s for the cars to come to rest after the initial impact, what deceleration did each experience?

2.30 The use of *Mach numbers* is one way to specify speed. Mach 1 corresponds to the speed of sound in air and at sea level is approximately equal to 330 m/s; Mach 2 corresponds to twice the speed of sound, and so forth. A certain high-acceleration rocket can reach Mach 1.2 by the time it has traveled 300 m. What is the acceleration? Express the result as a multiple of g.

2.31* A ball is thrown directly downward with a speed of 7 m/s from a height of 20 m. When will the ball strike the ground? (To find t, you must solve a quadratic equation. This yields two possible answers. Argue on physical grounds which answer is appropriate.)

Section 2.7

2.32* The vector **A** points north and has a magnitude of 2 units. The vector **B** points west and has a magnitude of 3 units. Use a graphical construction and find the sum **A** + **B**. What is the *magnitude* of **A** + **B**? What is the *direction* of **A** + **B**? First, use a ruler and a protractor to obtain the result; then, perform a calculation to check the original result.

2.33 The vector **C** has a magnitude of 2 units and is directed along the $+x$ axis. The vector **D** has a magnitude of 2 units and is directed along the $-y$ axis. What is the magnitude and direction of the vector **C** − **D**?

2.34 A vector **A** has a magnitude of 40 units and is directed at a counterclockwise angle of 25° with respect to the x axis. What are the magnitudes of the component vectors, $\mathbf{A}_x$ and $\mathbf{A}_y$?

2.35 A certain vector **A** has vector components $\mathbf{A}_x = 3\hat{\mathbf{i}}$ and $\mathbf{A}_y = 4\hat{\mathbf{j}}$ so that we may write

$$\mathbf{A} = \mathbf{A}_x + \mathbf{A}_y = 3\hat{\mathbf{i}} + 4\hat{\mathbf{j}}.$$

Write an expression for the unit vector in the direction of the vector **A**.

2.36 An object has a constant velocity $\mathbf{v} = (10\hat{\mathbf{i}} + 5\hat{\mathbf{j}})$ m/s. If it is at the origin of the x-y plane at a certain time where is it after 5 s have elapsed? Where was it 3 s before arriving at the origin?

2.37* A point in the x-y plane is 50 cm from the origin. If its x coordinate is +30 cm find its y coordinate. Why are there two possible answers?

2.38* A vector **A** is in the third quadrant of the x-y plane. Its magnitude is 40 units and the magnitude of its y component is twice that of its x component. Write an expression for **A** in terms of the unit vectors $\hat{\mathbf{i}}$ and $\hat{\mathbf{j}}$.

2.39 Two forces $F_x = 15$ units and $F_y = 40$ units, act on an object. What is the resultant force (magnitude and direction)?

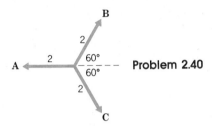
Problem 2.40

2.40 Each vector in the diagram has a length of 2 units.
(a) What is the sum **B** + **C**?
(b) What is the sum **B** + **C** + $\frac{1}{2}$**A**?
(c) What is the sum **A** + **B** + **C**?

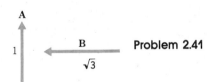
Problem 2.41

2.41 The vector **A** has unit magnitude and the magnitude of **B** is $\sqrt{3}$. What is the direction and the magnitude of **A** + **B**?

2.42 In still water a certain boat can travel at a speed of 10 km/h. The helmsman wishes to proceed *straight across* a river in which the current flows uniformly at a velocity of 5 km/h. In order to do this, the helmsman finds that he must steer the boat *upstream* at a certain angle. What is the angle?

2.43 At a certain instant an object has the following velocity components: $v_x = 300$ m/s, $v_y = 400$ m/s. What is the *speed* of the object at that instant? What is the *velocity* of the object in terms of the unit vectors $\hat{\mathbf{i}}$ and $\hat{\mathbf{j}}$?

2.44 An airplane is flying due north in still air at a velocity of 240 km/h. Suddenly the aircraft encounters a cross wind from the west whose velocity is 70 km/h. If the pilot does not alter his controls, what will be his new direction and new ground speed? Has the aircraft been accelerated?

2.45 Lay out the following vectors and *graphically* find the sum, first by adding **B** to **A** and then adding **C** to the sum, and next by adding **C** to **B** and then adding **A**. What is the difference in the two results?

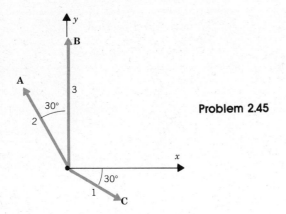

Problem 2.45

2.46 Find the sum of the three vectors in Problem 2.45 by adding *components*.

2.47 Consider the two vectors: **A** has a magnitude fo 6 units and is directed at an angle of 30° above the +x axis: **B** has a magnitude of 10 units and is directed at an angle of 45° above the +x axis. Find the resultant vector, **C** = **A** + **B**, by adding components.

2.48 In a children's game the instructions are to start from the school flag pole and go 30 paces north, 25 paces at 30° south of west, and 10 paces due east. How far is the student from the flag pole? What is the student's compass bearing?

2.49 Given the three vectors $\mathbf{A} = 3\hat{\mathbf{i}} + 4\hat{\mathbf{j}}$, $\mathbf{B} = 2\hat{\mathbf{i}} - 7\hat{\mathbf{j}}$, and $\mathbf{C} = -\hat{\mathbf{i}}$. What is the resultant vector of sum **A** + **B** + **C**? What is its magnitude and what counterclockwise angle does it make with the x-axis? Repeat the calculations for the sum **A** − **B** + **C**.

Section 2.8

2.50 An object is projected horizontally and falls through a vertical distance of 256 ft while moving horizontally a distance of 100 ft. What is its horizontal velocity?

2.51 A well-hit baseball will travel with a speed of about 50 m/s. Calculate the maximum range of such a baseball, neglecting air resistance. The effect of air resistance is to decrease the range to about 70 percent of the maximum value. Calculate the "practical" range in this case and compare with the distance of Mickey Mantle's record home run in 1953 (565 ft or 172 m).

2.52* During World War I, the Germans constructed an enormous railway gun (called Big Bertha by the Allies) that was used to bombard Paris from a distance of 120 km. What was the *minimum* muzzle velocity necessary for the gun to have had such a range? (Recall that the effect of air resistance is always to reduce the range, so the actual velocity must have been somewhat greater than that computed with the simple formula). What was the minimum time between firing and impact when the gun was adjusted for maximum range? What height did the shells reach?

2.53* By computing elaborate tables, Galileo showed that the maximum range of a projectile fired over level ground occurs for a muzzle elevation of 45°. He also showed that the range will be the same for muzzle angles of 45° + ϕ and 45° − ϕ, for any value of ϕ (between 0° and 45°). Show that this conclusion is correct by calculating the ranges for several values of ϕ. Sketch the two possible paths for some value of ϕ.

2.54 A projectile is launched at an angle of 45° above the horizontal. Show that the range of the projectile is equal to four times the maximum height reached.

2.55 A stone is thrown horizontally outward with an initial velocity of 20 m/s from the top of a cliff; 4 s later the stone is seen to strike the water below the cliff. What is the height of the cliff above the water?

2.56* A monkey hunter (probably a poacher) locates an animal hanging from a limb in a tree. The hunter points the barrel of his gun directly at the monkey and fires. The monkey sees the flash from the gun barrel and, sensing that he is being shot at, instantly releases his grip on the limb. Show that the bullet from the hunter's gun will strike the

Problem 2.56

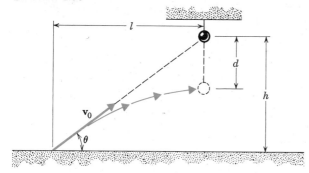

monkey, regardless of the speed of the bullet (assuming, of course, that the bullet reaches the monkey's position before he lands on the ground). Use the simplified diagram shown for the geometry of the situation.

2.57* A golfer attempts to drive the ball directly over a 15 m high tree that is 100 m from the tee. The ball leaves the tee with a velocity of 40 m/s with an elevation angle of 35° to the horizontal. Does the ball clear the tree? If so, by what amount? Is the ball rising or falling when it reaches the tree?

2.58* Water leaves the nozzle of a ground-level garden sprinkler with a speed of 10 m/s. At what angle above the horizontal is the nozzle pointed if the water falls on a flower bed 10 m away? Assume the water stream behaves in the manner of a projectile. (You might also wish to use the relationship $\sin 2\theta = 2 \sin\theta \cos\theta$).

2.59* If a ball player can throw a ball straight up in the air to a maximum height of 20 m, what is the furthest he can throw the ball on a flat playing field?

2.60* A target shooter fires a bullet with a muzzle velocity of 450 m/s at the bull's-eye of a target 100 m away and at the same elevation as the rifle. At what angle of elevation must the rifle barrel be set? With this setting, how high above the target is the rifle aimed?

Section 2.9

2.61 A belt-driven pulley 50 cm in diameter is rotating at a rate of 150 rev/s. Through what angle (in radian measure) will it have turned in 5 s? If the belt runs over the pulley without slipping, what length of belt traveled over the pulley?

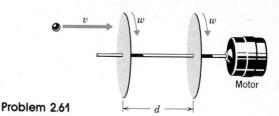

Problem 2.61

2.62* The muzzle velocity of a bullet can be measured by firing the bullet through two rotating discs that are separated by a certain distance d and then measuring the angular displacement $\Delta\theta$ between the two holes. If the discs are rotating together with the same angular velocity ω, show that the muzzle velocity is given by $v = \omega d/\Delta\theta$. If the discs are 1 m apart and are driven by a 1000-rpm motor and if the angular displacement between the holes is found to be 29°, what is the velocity of the bullet?

2.63 What is the angular velocity of the Earth's motion around the Sun? Around its own axis?

2.64 What is the angular velocity and the period of a four-speed phonograph turntable when operating at each of its possible speeds ($16\frac{2}{3}$, $33\frac{1}{3}$, 45, and 78 revolutions per minute or rpm)? What is the centripetal acceleration on a particle at the outer edge of a moving $33\frac{1}{3}$ rpm 12-in. diam. LP record?

2.65 An object is moving in a circular path of radius 3 m and is experiencing a centripetal acceleration of $3g$. What is the speed of the object in m/s? What is the period of the motion?

2.66 By what percentage should the speed of an object traveling in a circular path be reduced in order to decrease the centripetal acceleration by 5 percent? If the speed is kept the same, by what percentage should the radius be changed to produce the same change in the acceleration? Should the radius be increased or decreased?

Section 2.10

2.67 A wheel increases its rotational speed from 100 rev/s to 250 rev/s at a uniform rate over a 30 s time interval. What is the angular acceleration α in rad/s^2? What was the average angular velocity $\bar{\omega}$ during this acceleration period?

2.68 The motor of a record turntable rotating at $33\frac{1}{3}$ rev/min is turned off and the turntable coasts to rest in 25 s. What is the average angular acceleration $\bar{\omega}$ in rad/s²?

2.69 What angular velocity is reached after 10 s by a wheel starting from rest if it is undergoing a uniform angular acceleration of 10 rev/s²?

2.70 What angular acceleration is required to increase the angular velocity from 50 rad/s to 75 rad/s in 10 complete revolutions?

2.71 A wheel rotating counterclockwise at a rate of 100 rev/s is subject to a constant clockwise angular acceleration of 4 rad/s². How long will it take for the wheel to be rotating clockwise at a rate of 100 rev/s? Through what net angle has the wheel turned in this time? Discuss any similarity between this problem and that referring to a ball being thrown straight up into the air.

2.72 A disc with a radius of 50 cm starts from rest and is given a constant counterclockwise angular acceleration $\alpha = \pi$ rad/s². A particle glued to a point on the circumference has an initial counterclockwise angular location $\phi_0 = \pi/2$ rad, measured from a fixed reference line. After an elapsed time of 3 s find the following quantities for the particle: (a) the angular position, (b) the angular velocity, and (c) the tangential velocity.

2.73* A wheel starting from rest is subjected to a constant angular acceleration of 4 rad/s². At a certain elapsed time it is found that the tangential and centripetal accelerations of any point on the wheel are equal. Explain how this situation is possible and determine the time involved.

2.74 An object moving in a circular path with a radius of 75 cm is undergoing a uniform angular deceleration of 10 rev/s². When the angular velocity has been reduced to 2π rad/s find: (a) the tangential acceleration, (b) the centripetal acceleration, and (c) the magnitude and direction with respect to a radial line, of the total acceleration.

3

Force and Linear Momentum

We now turn to the subject of *dynamics* and examine the ways in which motion is influenced by the presence of *forces*. We know from experience that to set an object into motion requires a push or a pull; and we know that we can change the direction of motion of an object or stop it completely by the application of a push or a pull. In all of these situations, the crucial feature is that a *force* (represented by a push or a pull) is required to produce a *change* in the state of motion of an object. About 300 years ago, these qualitative facts were incorporated into a precise theory of dynamics formulated by Isaac Newton. In this chapter and the next we will study Newton's laws of motion and their application in a variety of situations.

During this century it has been found that the Newtonian laws need modification when exceedingly small distances or extremely high velocities are encountered. The discrepancies between prediction and observation under these circumstances paved the way for the development of the theories of quantum mechanics and of relativity. These limitations of Newtonian theory do not in any way imply that the theory is obsolete or unimportant. Indeed, under almost all ordinary circumstances the Newtonian laws are a correct description of the dynamics of physical systems.

3.1 The Idea of Force

Force and Inertia

We usually think of *force* in terms of the muscular action of a push or a pull. We appreciate that a bowling ball can be set into motion with a "small push" and that a "large push" is required to move an automobile. The automobile has a much greater mass than the bowling ball, and if we exert the same push on each object for the same time, we know that the state of motion of the automobile will be changed much less than will that of the bowling ball. Clearly, there must be some relationship connecting *force, mass,* and the *change in the state of motion*. As we will see, Newton's laws provide us with this relationship.

We know one other property of force from our experience. If we push in a certain direction on an object at rest, the object will begin to move in the same direction as the applied force. That is, force has *direction* as well as *magnitude:* force is a *vector* quantity.

We use the term *inertia* for the property of matter that causes it to resist a *change* in its state of motion, That is, the inertia of an automobile must be overcome in order to set it into motion. The measure of an object's inertia is its *mass*.

Frictional Forces

In addition to any force that we may apply to an object, there will in general be opposing *frictional forces* in operation as well. Even a well-polished ball rolling over a smooth surface will eventually come to rest because of friction. If the ball is not well-polished and if the surface is not smooth, the motion will cease much more quickly. Friction comes about when the irregularities in the surface of an object tend to bond to similar irregularities on the surface of another object against which it is sliding or over which it is rolling. Friction can never be eliminated completely, although it can be reduced to exceedingly small values in certain circumstances.

The frictional force on an object always acts in a direction *opposite* to the direction of motion. That is, a frictional force is a *retarding* force and always acts to slow down a moving object.

3.2 Newton's Laws

In 1687, Isaac Newton (Fig. 3.1) produced his most important work, the famous *Principia,* in which he presented a new theory of motion. Newton did not "invent" the subject of dynamics; on the contrary, he made the maximum use of previous work, especially the detailed experiments and analyses of Galileo. Newton's great contribution was to synthesize, from his own work and all that went before him, a complete description of the dynamics of bodies in motion. In this section we summarize Newtons' results by stating his famous three laws in modern terms, and we briefly discuss the meaning of each law.

Figure 3.1 Sir Isaac Newton (1642–1727), the great English mathematician and physicist who, as a young man, formulated a complete theory of dynamics, discovered the law of universal gravitation, and invented the calculus.

Newton's First Law (or the Law of Inertia)

If the net force on an object is zero (i.e., if the vector sum of all forces acting on an object is zero), then the acceleration of the object is zero and the object moves with constant velocity. That is, if no force is applied to an object at rest, the object remains at rest; if the object is in motion, it maintains a constant velocity. In mathematical terms,

$$\mathbf{F}_{\text{net}} = 0 \text{ implies } \mathbf{a} = 0 \quad \text{or} \quad \mathbf{v} = \text{const.} \quad (3.1)$$

A hockey puck, sliding over smooth ice, will move a great distance at nearly constant velocity even though there is no force pushing it along. Of course, the puck will eventually come to rest due to the friction that exists between the puck and the ice. The smoother we make the ice, the smaller will be the frictional drag and the farther the puck will slide. If the friction could be eliminated completely, we are led to believe that the puck would slide indefinitely far without any change in speed or direction.

Newton's first law, by itself, gives us only a crude notion regarding force. In fact, we have here only a definition of *zero* force. However, because zero force means zero acceleration, there is the strong implication that *force* is somehow intimately connected with *acceleration*. Indeed, we can define force as follows: *A force is any influence that is capable of producing a change in the state of motion of an object.* The second law states the connection between force and acceleration.

Newton's Second Law

The accelerated motion of a body can only be produced by the application of a force to that body. The direction of the acceleration is the same as the direction of the force and the magnitude of the acceleration is proportional to the magnitude of the force. The constant of proportionality is the *inertia* or *mass* of the body. In equation form, this law is

$$\mathbf{F} = m\mathbf{a} \quad (3.2)$$

Because **a** is a well-defined quantity, this law expresses the relationship between force and mass but it really defines neither. It remains for the third law to resolve the situation, as we will see presently.

Notice that Eq. 3.2 is a *vector* equation. That is, the vectors **F** and **a** are related by the scalar quantity *m*. If the force **F** is in a certain direction, then

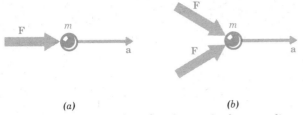

Figure 3.2 (a) Force and acceleration are in the same direction. (b) The acceleration is in the direction of the total (or net) force applied to the object.

the acceleration **a** is necessarily in that same direction.

It is essential to realize that whenever we wish to use Eq. 3.2 to describe the motion of an object, we must be certain that **F** in this equation is the *total (or net) force acting on the object in question*. If there are several individual forces acting on an object, **F** is the *vector sum* of all of these individual forces (Fig. 3.2). Whether the object is exerting forces on other bodies is irrelevant. We need only know the forces that are exerted *on* a body in order to calculate its motion.

Some examples will serve to illustrate these points. (We assume that friction is unimportant in all of these examples.) Figure 3.3 illustrates three basic points: (a) a force is necessary to impart acceleration; (b) the forces applied to an object can cancel so that there is no acceleration; and (c) the net force is always the *vector* sum of the individual forces and may not be in the direction of any one force. In Section 3.5 we examine some additional cases in detail.

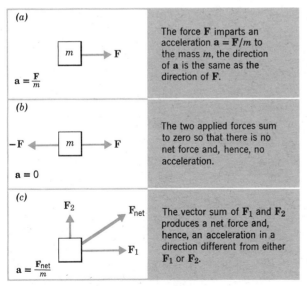

Figure 3.3 Three different force situations.

When considering objects in the Earth's environment a force that is *always* acting on the object is the gravitational force exerted by Earth on the object. This force cannot be shielded and is independent of the state of motion of the object. It is but one example of the universal law of gravitation, a subject we treat in detail in Chapter 6. For present purposes, we can determine the value of this force, designated by the symbol $\mathbf{F}_G$, by noting the observations of Galileo on freely falling bodies (Section 2.6).

If we allow an object to fall freely near the surface of the Earth, we know the acceleration that results is a constant given by $g = 9.80$ m/s². It is common experience that this acceleration worldwide is "downward" directed, that is it is directed toward the center of the Earth. Let us therefore introduce the acceleration due to the gravitational force on an object in free fall: **g** a vector directed towards the center of the Earth and approximately equal in magnitude to 9.80 m/s² at sea level. For an object in free fall the only force acting on the object (neglecting air resistance) is $\mathbf{F}_G$ the gravitational force. Using Newton's second law with $\mathbf{a} = \mathbf{g}$ and $\mathbf{F} = \mathbf{F}_G$

$$\mathbf{F}_G = m\mathbf{g} \qquad (3.3)$$

where m is the mass of the object. We therefore conclude that the gravitational force is just $m\mathbf{g}$. It is understood, however, that **g** is the acceleration due to gravity *appropriate for the place at which its measurement is made*.

Although we have determined the Earth's gravitational force $\mathbf{F}_G = m\mathbf{g}$ acting on an object with mass m by considering it to be in free fall, this is the value of $\mathbf{F}_G$ *no matter what the state of motion of the object*, even if it is at rest.

Newton's Third Law

Newton's third law of dynamics tells us that *single forces cannot occur; forces always act in pairs*. In any situation, we usually designate one force as the *force* and the other as the *reaction force*, but this distinction is entirely unnecessary. The third law states:

If object 1 exerts a force on object 2, then object 2 exerts an equal force, oppositely directed, on object 1. That is, a force is always paired with an equal reaction force. In equation form,

$$\mathbf{F}_{12} = -\mathbf{F}_{21} \qquad (3.4)$$

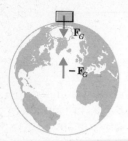

Figure 3.4 The Earth exerts a gravitational force F_G on a block at its surface, and the block exerts a gravitational force $-F_G$ on the Earth.

where the first subscript denotes the object receiving the force and the second subscript denotes the object exerting the force.

It is important to understand what the third law means and what it does *not* mean. Consider, first, the simple case of a block at rest on the surface of the Earth (Fig. 3.4). What forces are acting here? The block experiences a force $\mathbf{F}_G$ due to the gravitational pull of the Earth. The reaction force is $-\mathbf{F}_G$ and this is the gravitational force exerted on the Earth by the block. The force pair, $\mathbf{F}_G$ and $-\mathbf{F}_G$, consists of one force that acts on the block and one force that acts on the Earth. But there must be other forces also acting in this case. Otherwise, the block, subject to the single force $\mathbf{F}_G$, would accelerate toward the center of the Earth. This would be the situation if the block were released from some height above the Earth. In the case illustrated in Fig. 3.4, however, the block rests on the surface. To see the other forces that are acting, look at Fig. 3.5. The block exerts a force $\mathbf{w}$ on the surface of the Earth. (We call $\mathbf{w}$ the weight of the block. A complete discussion of *weight* is given in Section 3.4.) The reaction to $\mathbf{w}$ is $\mathbf{N} = -\mathbf{w}$, the force that the Earth's surface exerts on the block. (We call $\mathbf{N}$ the *normal* force; *normal* means acting *perpendicular* to the surface.) Thus, there are two forces acting on the block, $\mathbf{F}_G$ downward and $\mathbf{N}$ upward. Since the block experiences no acceleration (it is at rest) the net force on the block is zero. Therefore, $\mathbf{F}_G$ and $\mathbf{N}$ are equal in magnitude and opposite in direction. Similarly, there are two equal and oppositely directed forces acting on the Earth, $-\mathbf{F}_G$ and $\mathbf{w}$, and the Earth experiences no acceleration.

Notice that in this example the four forces that are acting all the same magnitude. These forces constitute two action-reaction pairs: $\mathbf{F}_G$ and $-\mathbf{F}_G$, $\mathbf{w}$ and $\mathbf{N}$. Although $\mathbf{F}_G = \mathbf{w}$, it is not correct to say that the normal force $\mathbf{N}$ is the reaction to the gravitational force $\mathbf{F}_G$; $\mathbf{N}$ is the reaction to $\mathbf{w}$. The normal force $\mathbf{N}$ cannot be the reaction to $\mathbf{F}_G$ because both $\mathbf{N}$ and $\mathbf{F}_G$ act *on* the block.

It is important always to distinguish carefully between those forces exerted *on* an object and those forces exerted *by* the object. An object moves in accordance with the forces acting *on* the object. An action-reaction pair always consists of one force exerted *on* the object and one force exerted *by* the object. The latter force has nothing to do with the motion of the object.

To illustrate this point, consider two blocks that are resting on a frictionless surface, as in Fig. 3.6. A force $\mathbf{F}$ is applied to m_1. If m_1 and m_2 are in contact, is this force transmitted to m_2? The third law tells us that whatever force m_1 exerts on m_2, then m_2 must exert this same force on m_1. But the third law does *not* specify the magnitude of this force; only the *second* law can be used to find this quantity. The total mass of the two blocks is $M = m_1 + m_2$ and the only externally applied force is $\mathbf{F}$; therefore, the acceleration of the two blocks together is

$$\mathbf{a} = \frac{\mathbf{F}}{M}$$

The acceleration of m_2 is, of course, also $\mathbf{a}$, so the force applied to m_2 that gives rise to this acceleration must be

$$\mathbf{F}_2 = m_2 \mathbf{a} = m_2 \frac{\mathbf{F}}{M}$$

which is smaller than the external force $\mathbf{F}$ by the fraction m_2/M. Similarly, the net force acting on m_1 is

$$\mathbf{F}_1 = m_1 \mathbf{a} = m_1 \frac{\mathbf{F}}{M}$$

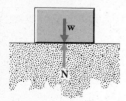

Figure 3.5 The block exerts a force $\mathbf{w}$ (its *weight*) on the surface, and the surface exerts an equal but oppositely directed force $\mathbf{N}$ (the *normal* force) on the block.

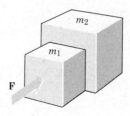

Figure 3.6 A force $\mathbf{F}$ is applied to m_1. Is this force transmitted to m_2?

which is just the sum of the externally applied force **F** and the reaction force applied on m_1 by m_2, namely, $-\mathbf{F}_2$:

$$\mathbf{F}_1 = \mathbf{F} - \mathbf{F}_2 = \mathbf{F} - \frac{\mathbf{F}m_2}{M} = \mathbf{F}\left(1 - \frac{m_2}{M}\right)$$

$$= \mathbf{F}\left(1 - \frac{m_2}{m_1 + m_2}\right) = \frac{m_1}{m}\mathbf{F}$$

(From the expression for $\mathbf{F}_1$ and $\mathbf{F}_2$, notice that $\mathbf{F} = \mathbf{F}_1 + \mathbf{F}_2$.)

The third law tells us that forces always occur in pairs. But the two forces in every action-reaction pair always act on *different* objects. To determine the motion of an object, we need to know only the forces acting *on* that object; it does not matter that the object exerts forces on other objects.

3.3 The Dimensions of Force

The Newton

Newton's second law specifies the dimensions of force;

$$F = ma$$

Therefore, the unit of force is

(unit of *mass*) × (unit of *acceleration*)
$$= (\text{kg}) \times (\text{m/s}^2)$$
$$= \text{kg-m/s}^2$$

If a net force **F**, acting on a mass of 1 kg, causes an acceleration of 1 m/s², the magnitude of the force is 1 kg-m/s². We give the special name *newton* (N) to the unit of force in the metric system:

$$1 \text{ N} = 1 \text{ kg-m/s}^2 \tag{3.5}$$

● *Example 3.1*

An object of mass 10 kg is at rest in deep space, far from any gravitating masses. A net force of 20 N is applied for 10 s. What is the final velocity? How far will the object have moved in the 10-s interval?

$$a = \frac{F}{m} = \frac{20 \text{ N}}{10 \text{ kg}} = 2 \text{ m/s}^2$$

The acceleration is constant, so

$$v = at = (2 \text{ m/s}^2) \times (10 \text{ s}) = 20 \text{ m/s}$$
$$x = \tfrac{1}{2}at^2 = \tfrac{1}{2}(2 \text{ m/s}^2) \times (10 \text{ s})^2 = 100 \text{ m}$$

This same object is brought to the Earth's surface and sits at rest. What is the magnitude of the gravitational force $\mathbf{F}_G$ exerted on it by the Earth?

$$F_G = mg = (10 \text{ kg}) \times (9.80 \text{ m/s}^2) = 98.0 \text{ N} \quad \blacksquare$$

3.4 Mass and Weight

Weight Is a Force

In discussing the forces involved in the case of a block at rest on the surface of the Earth in Section 3.2, we called the force the block exerted on the surface of the Earth, its *weight* **w**. When the analysis of the situation was completed, we found that **w** was equal to the gravitational force acting on the block $\mathbf{F}_G$ and hence equal to $m\mathbf{g}$. Evidently moving the block from one point on the Earth's surface to another (refer to Fig. 2.15) or to the surface of the Moon (where the gravitational force is only about $\tfrac{1}{6}$ that at the Earth's surface) would result in ascribing different weights to the same object. The concept of weight is thus of very limited utility.

It is important to distinguish between the concepts of mass and weight. Mass is an intrinsic property of matter. The simplest idea of mass is that it is a measure of the amount of matter in a body; that is, the mass of a body depends on the number and type of atoms in the body. These atoms do not change if we move the body from one location to another. Thus, the mass of a body remains the same whether it is on the Earth, or on the Moon, or in deep space.

In Newtonian dynamics we define the mass of a body as a measure of its inertia—that is, a measure of the body's ability to resist a change in its state of motion when acted on by a force (see Section 3.6). A net force of 1 N applied to a 1-kg object will produce an acceleration of 1 m/s² regardless of where the object is located. The property of a body that we call its *mass* is the same (by either definition) wherever the body is located. But this is not true of its *weight*. A change in the gravitational force acting on a body means a change in its *weight* but not in its mass.

In particular, in this text we define the weight of a body to be the force that the body exerts on another object which is supporting the body in the presence of a gravitational force. For an object that is not accelerating, the weight of the object and the gravitational force acting on it are equal: $\mathbf{w} = \mathbf{F}_G$. Because of this fact, we often say that the gravitational force *is* the weight. This is the definition you will find in many books, but is not entirely satisfactory. Equating **w** and $\mathbf{F}_G$ causes no difficulty in cases of zero acceleration, but in other situations it is more revealing to consider weight to be the force exerted by the object on other supporting objects.

How do we determine the weight of an object? Practical methods for weighing an object employ spring scales. When an object is placed on the platform of a scale the spring in the device is compressed and this deformation is communicated to a calibrated dial (or digital) read out. In using a scale from which the object is hung the spring is stretched. In either case the reading obtained is a measure of the force the object exerts on a spring, which we call its weight.

In Fig. 3.7a we have a block with mass $m = 2$ kg suspended from a spring scale attached to the ceiling of a box that is at rest. There are two forces acting on the block, the gravitational force $\mathbf{F}_G$ and the spring tension $\mathbf{T}_0$ which is the reaction to the weight of the block $\mathbf{w}$ (that is $\mathbf{T}_0 = -\mathbf{w}$.) Since the block is not accelerating, the net force on it must vanish, that is

$$\mathbf{F}_G + \mathbf{T}_0 = 0$$

Taking components, the upward direction is considered positive;

$$-mg + T_0 = 0 \quad \text{or} \quad T_0 = mg = 19.6 \text{ N}$$

Since $\mathbf{w}$ and $\mathbf{T}_0$ are action-reaction pairs, the magnitude of w (the weight) is also 19.6 N.

Now, suppose that the rope attached to the box is pulled upward so that the box and its contents undergo an acceleration $a = 2.2$ m/s² (Fig. 3.7b). The spring tension $\mathbf{T}$ acting on the block is now greater than $\mathbf{T}_0$ because an upward force greater in magnitude than F_G is necessary to produce the acceleration of the block, that is

$$\mathbf{F}_G + \mathbf{T} = \mathbf{ma}$$

Again, taking components with the upward direction considered positive,

$$-mg + T = ma \quad \text{or} \quad T = ma + mg = m(a + g)$$

Therefore, the spring tension is

$$T = (2 \text{ kg})(2.20 \text{ m/s}^2 + 9.80 \text{ m/s}^2)$$
$$= 24.0 \text{ N}$$

Because the block exerts a force on the string equal to T, we say that the block weighs 24.0 N. When not accelerating (Fig. 3.7a), the weight of the block was 19.6 N. The upward acceleration has caused the weight to increase by 4.4 N.

The same effect takes place when you accelerate upward in an elevator. Suppose that you stand on a bathroom-type scale in an elevator. When the elevator is at rest, the scale indicates your normal weight, mg. If the elevator now accelerates upward at a rate a, you feel an increased downward pressure, an increased weight of your body. In order for you to accelerate at the same rate as the elevator, the scale must exert on you an upward force equal to $mg + ma$. At the same time, you are exerting a force of the same magnitude on the scale. That is, your weight is $m(g + a)$, and the scale indicates the corresponding value. Notice carefully that what the scale reads is exactly your *weight;* the scale reading includes the accelerating force applied to you by the scale and the reaction to this force is part of your weight.

What happens if the elevator accelerates *downward* at the rate a? By the same reasoning, it is easy to see that the force you exert on the scale is now $m(g - a)$, and the scale indicates a weight that is *less* than your normal weight mg. If the elevator cable breaks and the elevator falls freely, the downward acceleration becomes g and your weight is $m(g - g) = 0$; that is, you are "weightless." In this situation there is a gravitational force $F_G = mg$ acting on you, but, nevertheless, you have zero weight. The condition of weightlessness simply means that in free fall a body does not exert a force on any other object in contact with it. (otherwise, the object would not be falling freely!) Astronauts in space vehicles orbiting the Earth are in a condition of free fall and

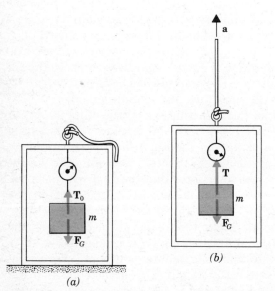

Figure 3.7 The upward acceleration of the box causes the tension in the spring of the scale (and therefore the weight of the block) to increase.

they are also weightless. Whenever an object is undergoing an acceleration in which **a** has a component in the direction of **g**, we must be careful to distinguish between the terms "gravitational force" and "weight." These two quantities are equal only in the case of zero acceleration along the direction of **g**.

Were it not for shopkeepers and diet conscious individuals we could totally dispense with referring to the weight of objects. Efforts to extend the concept of weight to situations involving complicated motion is not only unnecessary but confusing. We should note that chemists and physicists use an analytic balance to determine the mass of an object. This is a method of direct mass comparison against calibrated "weights" and is independent of gravitational or accelerational effects. All we ever need to specify for objects in the environment of large gravitating masses such as the Earth or Moon is the gravitational for $\mathbf{F}_G$ and the mass of the object m. We shall take this point of view in all that follows.

Bone-Breaking Forces in Jumping

A person who jumps or falls from a height and lands on his or her feet on hard ground places a great stress on the long bones of the leg. The most vulnerable bone is the tibia and the stress on this bone will be greatest at the point where the cross-sectional area is least, just above the ankle. The tibia will fracture if a compressive force (such as that exerted in landing from a jump) of more than about 50 000 N is applied (see Exercise 1 in the essay on page 238). If the jumper lands squarely on both feet, the maximum force that the leg bones can tolerate is twice this value, or 10^5 N. (This force corresponds to about 130 times the weight of a 75-kg human.)

The average force exerted on the leg bones is equal to the mass of the individual multiplied by the average acceleration in coming to rest from the velocity of fall immediately before landing:

$$\overline{F} = m\overline{a}$$

Now, we know that the velocity v acquired in a fall through a height H, starting from rest, is given by (see Eq. 2.16)

$$v^2 = 2gH$$

Similarly, the average acceleration $\overline{a}$ necessary to bring an object to rest from a velocity v in a distance h is given by

$$v^2 = 2\overline{a}h$$

Equating these two expressions for v^2 and solving for $\overline{a}$, we find

$$\overline{a} = g\frac{H}{h}$$

Therefore, the average force exerted during the deceleration is

$$\overline{F} = m\overline{a} = mg\frac{H}{h}$$

The quantity H/h is the ratio of the height of fall to the distance through which the deceleration to rest occurs.

If a person lands on both feet stiffly, without bending his or her knees, the distance h will be about 1 cm. Then, if $\overline{F}$ is to be no more than 130 mg (the fracture value), the maximum height of fall H is

$$H = \frac{\overline{F}h}{mg} = \frac{(130 \text{ mg}) \times (0.01 \text{ m})}{mg} = 1.3 \text{ m}$$

We therefore see that a fall through the relatively small height of 1.3 m (or 4.3 ft) can result in the fracture of the tibia if the landing is made stiffly.

By bending the knees during the landing, the deceleration distance can be made larger. If the individual collapses from an erect position to a crouching position during the landing, h will be about 0.6 m. This value for h is 60 times greater than that used in the computation above and would indicate that a jump from a height $H = 60 \times 1.3$ m $= 78$ m could be negotiated by bending during the landing. In this case, however, the deceleration force is exerted almost entirely by the tendons and ligaments instead of the leg bones, and these muscles are capable of withstanding only about $\frac{1}{20}$ the force that will fracture the bones. Thus, the maximum height is reduced to about 4 m or 13 ft. To land safely from such a height requires that the collapsing motion be made uniformly so that the instantaneous forces exerted on the muscles do not exceed the failure level. Therefore, jumping from a height of 4 m is still a risky business.

If an individual lands in a yielding substance such as water or soft snow, then much higher impact velocities can be tolerated. There are several documented cases of military personnel falling (without parachutes) from aircraft in flight and surviving because they landed in soft snow.

Reference

G. B. Benedek and F. M. H. Villars, *Physics with Illustrative Examples from Medicine and Biology*, Addison-Wesley, Reading, Mass., 1973, Vol. 1, Chapter 4.

■*Exercises*

1. In the example of the fall from 1.3 m, what is the velocity immediately before landing? How long does the deceleration process require?

2. Write Newton's second law as $\overline{F} \Delta t = m \Delta v$ and express Δv in terms of the height of fall. From what height can an individual fall safely (landing stiffly on both feet) if he or she lands on a spring-supported platform that causes deceleration time to be 0.01 s?

3.5 Applications of Newton's Laws

In Fig. 3.3 we showed three simple examples of the vector character of forces. We now look at some additional cases that summarize most of the important aspects of the types of problems we will consider.

Refer to the five cases in Fig. 3.8. In each example we identify all of the forces acting on the block and then sum these individual forces to obtain the net force. In these examples we assume frictional effects to be negligible.

(a) Here there are two forces acting on the object—the downward gravitational force $\mathbf{F}_G = m\mathbf{g}$ and the upward force due to the tension $\mathbf{T}$ in the string. In this case the block is stationary, the acceleration is zero, and the net force is zero: $\mathbf{F}_{\text{net}} = \mathbf{F} + \mathbf{F}_G = 0$.

(b) In this example the block is also stationary. The individual forces acting on the block are the gravitational force $\mathbf{F}_G = m\mathbf{g}$ and the upward force provided by the surface in reaction to the weight of the block. This force is perpendicular (or *normal*) to the surface and we call this the *normal force* $\mathbf{N}$. Again, $\mathbf{F}_{\text{net}} = \mathbf{N} + \mathbf{F}_G = 0$.

(c) Next, we attach a rope to the block and pull with a horizontal force $\mathbf{F}$. This force produces a tension $\mathbf{T}$ in the rope and this tension acts on the block. In the force diagram we see the sum, $\mathbf{F}_{\text{net}} = \mathbf{N} +$

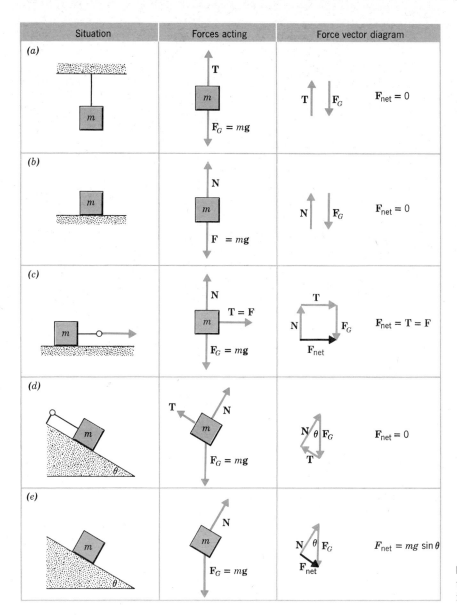

Figure 3.8 Force diagrams for five situations. The effect of friction is ignored in all cases.

$F_G + T = T = F$. The block therefore undergoes an acceleration $a = F/m$.

(d) The block is held on a slope by means of a restraining rope. The tension T in the rope, the normal force N, and the gravitational force F_G, sum to zero.

(e) If we remove the restraining rope in case (d) and neglect friction, there is a net force acting down the slope, which produces an acceleration of the block. This net force is equal to the sum $F_G + N$ and is just the negative of the tension T in case (d). Then, $F_{net} = mg \sin \theta$.

Note the common features in all these examples:

1 A diagram or sketch is made of the situation to be analyzed.

2 The specific object under investigation is isolated for special attention (in all the above cases it is the block) and shown in a *separate* diagram. Such diagrams are referred to as *free body* diagrams.

3 *All* the forces acting *on* the isolated object are shown. In making sure that you have not overlooked some relevant forces note that: all other objects in physical contact with the isolated object may exert forces on it (for example, connecting ropes or strings, supporting surfaces etc.); and if the object is in the Earth's environment then the gravitational force $F_G = mg$ also acts.

4 Newton's second law (a *vector* relationship) is then employed to obtain either the acceleration of the isolated object or if it is known to be at

rest, the second law is used to determine the unknown forces acting.

Let us now look at some other cases in detail.

● *Example 3.2*

Consider two blocks with masses m_1 and m_2 connected by a rope that passes over a pulley, as shown in the diagram. We assume that friction is unimportant. Notice that the tension is the same in both parts of the rope and the pulley serves merely to change the direction of T. What is the acceleration of the pair of blocks?

We solve this problem by writing $F = ma$ for each block. The net force acting on m_1 is the vector sum of $m_1\mathbf{g}$ and $\mathbf{T}$; similarly for m_2. Let us consider the *downward* direction to be positive. Then, we can write for the vertical components of the net forces,

On m_1: $m_1 g - T = m_1 a_1$
On m_2: $m_2 g - T = m_2 a_2 = -m_2 a_1$

In the last equation we have used the fact that the rope connecting m_1 and m_2 remains taut so that the accelerations of the two blocks are equal in magnitude but opposite in direction; that is, $a_2 = -a_1$.

Solving the two equations for T and equating the results, we find

$$m_1 g - m_1 a_1 = m_2 g + m_2 a_1$$

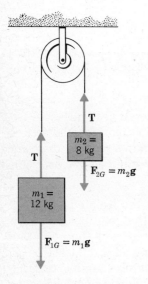

from which

$$a_1 = \frac{m_1 - m_2}{m_1 + m_2} g$$

Substituting the numerical values,

$$a_1 = \frac{12 \text{ kg} - 8 \text{ kg}}{12 \text{ kg} + 8 \text{ kg}} \times (9.80 \text{ m/s}^2)$$
$$= 1.96 \text{ m/s}^2$$

If we solve the first equation above for the tension T, we find

$$T = m_1(g - a_1) = (12 \text{ kg}) \times (9.80 \text{ m/s}^2 - 1.96 \text{ m/s}^2)$$
$$= 94.1 \text{ N}$$

The tension in the rope is therefore *greater* than $m_2 g = (8 \text{ kg}) \times (9.80 \text{ m/s}^2) = 78.4 \text{ N}$, so that m_2 undergoes an upward acceleration. The tension is *less* than $m_1 g = (12 \text{ kg}) \times (9.80 \text{ m/s}^2) = 117.6 \text{ N}$, so that m_1 accelerates downward. (The fact that a_1 was found to be *positive* agrees with this conclusion.)

Examine the case that m_1 approaches zero while m_2 remains 8 kg. What happens if m_1 approachs zero while m_2 remains 12 kg? Does the expression for a_1 correctly account for these cases as well?

There is, of course, no reason why the downward direction has to be taken as positive for both blocks in this example. We can recognize from the outset that the two blocks are required to move in opposite directions. Therefore, taking the downward direction as positive for m_1 we can set the upward direction as positive for m_2 and simply refer to the acceleration of either as a. Then we have for the new forces,

On m_1: $m_1 g - T = m_1 a$
On m_2: $T - m_2 g = m_2 a$

Adding these two equations gives the same results as before for $a = a_1$

● *Example 3.3*

Two masses, m_1 and m_2, are connected by a string (considered massless). The mass m_2 is placed on a flat frictionless surface and m_1 is suspended over the side of the surface on a frictionless pulley. What is the acceleration of the system?

The figure illustrates the situation and also shows the forces acting on each of the masses. Notice that the tension $\mathbf{T}$ acting on each mass has the same magnitude; the pulley serves only to change the direction of the tension vector.

This is a two-dimensional problem, and we choose the positive directions of the x- and y-axes as indicated in the diagram. There is only an x-component of $\mathbf{T}_2$ acting on m_2 and only a y-component of $\mathbf{T}_1$ acting on m_1; these two components are equal in

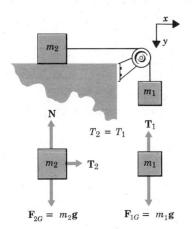

magnitude: $T_{2x} = T_{1y} = T$. Similarly, because the string is taut, the blocks move together. The x-component of the acceleration of m_2 has the same magnitude as the y-component of the acceleration of m_1:

$$a_{2x} = a_{1y} = a$$

Now, for each block we write Newton's second law:

For m_1:
$$\mathbf{F}_{net,1} = m_1 \mathbf{a}_1$$
$$m_1 \mathbf{g} + \mathbf{T}_1 = m_1 \mathbf{a}_1$$

Because we have only y-components, this vector equation reduces to

$$m_1 g - T_{1y} = m_1 a_{1y}$$

Notice that the negative sign is necessary because $\mathbf{T}_1$ in this case is *upward* (i.e., in the negative y-direction).

For m_2:
$$\mathbf{F}_{net,2} = m_2 \mathbf{a}_2$$
$$m_2 \mathbf{g} + \mathbf{N} + \mathbf{T}_2 = m_2 \mathbf{a}_2$$

The pair of vectors, $m_2 \mathbf{g}$ and $\mathbf{N}$, sum to zero. (Explain why this is so.) The component equation becomes

$$T_{2x} = m_2 a_{2x}$$

If we rewrite the two component equations using $T_{2x} = T_{1y} = T$ and $a_{2x} = a_{1y} = a$ we have

$$m_1 g - T = m_1 a$$
$$T = m_2 a$$

Now, we *add* these equations and obtain

$$m_1 g = (m_1 + m_2) a$$

Finally, we solve for the acceleration a:

$$a = \frac{m_1}{m_1 + m_2}$$

We could attack the problem in another way. First, we recognize that m_1 and m_2 are connected by a taut string and therefore can be considered to constitute a *system*. The mass of the system is $M = m_1 + m_2$ and Newton's second law is therefore written as

$$F_{net} = Ma$$

The net force on the system is just the gravitational force on m_1, in other words, $m_1 g$. That is, considering m_1 and m_2 as a *system*, we do not need to discuss the *internal* forces that hold it together (in this case, the tension in the string). The string remains taut and so there is always a fixed relationship between m_1 and m_2, which is determined by the length of the string; the pulley merely serves to change the direction of motion. Therefore, applying the second law, we have

$$Ma = m_1 g$$

or,

$$a = \frac{m_1}{M} g = \frac{m_1}{m_1 + m_2} g$$

which is the same result as before.

The solution to a problem is always simpler if all of the bodies involved act together as a system. Of course, before electing to use this approach in a problem, one must be careful to ensure that there is indeed a *system* that acts together as a whole.

● *Example 3.4*

As a final example without the presence of friction, consider the block on the inclined frictionless plane shown in the diagram. The block has a mass $m = 10$ kg and is subject to a constant force $F = 30$ N acting up the plane. What is the motion of the block?

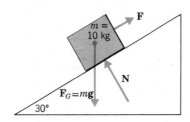

Three forces act on the block: (1) the externally applied force $\mathbf{F}$ acting up the plane, (2) the normal force $\mathbf{N}$ acting perpendicular to the plane, and (3) the gravitational force $m\mathbf{g}$ acting downward. The net force acting on the block is the vector sum of these three forces.

In order to simplify the vector summation, let us

decompose the force $m\mathbf{g}$ into components that act *parallel* to the plane and *perpendicular* to the plane. This is done in diagram (a) following. Now, we can represent all of the forces acting on the block as in diagram (b). Notice that the normal force $\mathbf{N}$ must exactly balance the component of $m\mathbf{g}$ that acts perpendicular to the plane because there can be no net force and no acceleration in this direction. The magnitude of the component of $m\mathbf{g}$ that acts down the plane (49 N) is greater than the magnitude of the restraining force $\mathbf{F}$ (30 N). Therefore, the net force acting on the block is 19 N directed down the plane, and the block undergoes an acceleration in this direction:

$$a = \frac{F_{\text{net}}}{m} = \frac{19 \text{ N}}{10 \text{ kg}} = 1.90 \text{ m/s}^2 \quad \blacksquare$$

Kinetic Friction

Up to this point we concentrated on examples in which we imagined friction to be a negligible effect. When a net force is applied to an object in a frictionless situation, the object accelerates. Much more common, however, are cases in which a force is applied to an object by some sort of push or pull and yet the object only moves with constant velocity if it moves at all. For example, when you push a large box across the floor at constant velocity, you can readily sense in your muscles the fact that you are exerting a force. The reason that the box does not accelerate even though you exert a force on it is that there is a retarding force also acting on the box, and this retarding force has the same magnitude as the applied force. The net force acting on the box is therefore zero, and the box moves with constant velocity. The retarding force that acts against your push is, of course, due to *friction*. The frictional force *always acts to retard* whatever motion is present.

We can identify two different types of frictional forces—those that are effective between surfaces that are at rest with respect to one another and those that are effective between surfaces that are in

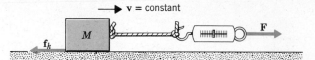

Figure 3.9 If the force $\mathbf{F}$ pulls the block at constant velocity, the scale reading is exactly equal to the frictional force. (Notice that $\mathbf{f}_k$ acts along the surface of contact and is not directed through the center of mass of the block.)

relative motion. We call the first of these the forces of *static* friction and we use the term *kinetic* friction for forces of the second type. We consider first the effects of kinetic (or *sliding*) friction.

How can we measure the magnitude of the kinetic friction that exists between a box and the floor over which it slides? Actually, such a measurement is easy to make. Figure 3.9 shows a block being pulled at constant velocity across a horizontal surface by a force $\mathbf{F}$. The spring scale indicates the magnitude of $\mathbf{F}$ directly in newtons. Because the block moves with constant velocity, the net force on the block must be zero. Therefore, the magnitude of the retarding frictional force $\mathbf{f}_k$ is exactly equal to the magnitude of the applied force $\mathbf{F}$. Thus, the scale reading is equal to the frictional force (whenever $\mathbf{v} = $ constant).

We can measure the frictional force between a block and a particular surface for a variety of conditions. We can alter the mass of the block or the surface area over which contact is made, and we can change the angle of the surface along which the block is pulled. These experiments tell us that the frictional force between the block and the surface depends on only one factor—the normal force exerted by the surface on the block. Remember that the normal force is the force exerted by a surface on the object and always acts perpendicular to the surface. That is, we can write

$$f_k = \mu_k N \tag{3.6}$$

The quantity μ_k is called the *coefficient of kinetic friction*.

The value of μ_k depends strongly on the details of the pair of surfaces that slide past one another. Generally, the addition of a lubricant between the surfaces tends to decrease the kinetic friction. (There is some dependence of μ_k on the velocity of sliding, but we shall ignore such effects.) Some values of μ_k are listed in Table 3.1.

When friction is present, as a block slides along a surface, it is convenient to replace the force the surface exerts on the block by its two vector components; the normal force $\mathbf{N}$ which acts perpendicular

Table 3.1 Typical Coefficients of Kinetic Friction

Materials	μ_k
Steel on steel, dry	0.4
Steel on steel, lubricated	0.08
Steel on ice	0.06
Copper on cast iron	0.3
Oak on oak (grains parallel)	0.5
Rubber on concrete, dry	0.7
Rubber on concrete, damp	0.5
Teflon on teflon	0.04

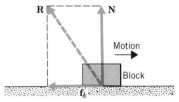

Figure 3.10 The force exerted by the surface of contact on a sliding block may be represented *either* by the vector **R** *or* its vector components **N** and $\mathbf{f}_k$.

to the surface, and $\mathbf{f}_k$ which acts along the surface of contact. In Fig. 3.10 the two vector components **N** and $\mathbf{f}_k$ are shown adding to the resultant force **R** the total force exerted by the surface on the block.

● *Example 3.5*

A block with a mass $M = 10$ kg is pulled along a horizontal surface by a force **F** whose magnitude is 50 N and which makes an angle of 25° with the horizontal. The coefficient of kinetic friction between the block and the plane is $\tfrac{2}{5}$. Determine the acceleration of the block.

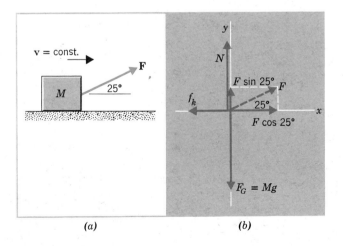

The force diagram for the situation in (a) is shown in (b). Because there is no motion in the y direction we have

$$N + F \sin 25° - Mg = 0$$

In the x direction we have

$$F \cos 25° - f_k = Ma$$

In addition the relationship between f_k and N is given by Eq. 3.6. From the first of the above equations the normal force N is

$$N = Mg - F \sin 25°$$
$$= (10 \text{ kg}) \times (9.80 \text{ m/s}^2) - (50 \text{ N}) \times \sin 25°$$
$$= 76.9 \text{ N}$$

The corresponding frictional force f_k is therefore

$$f_k = \mu_k N = (\tfrac{2}{5}) \times (76.9 \text{ N}) = 30.8 \text{ N}$$

Finally, the acceleration of the block is

$$a = \frac{1}{M}(F \cos 25° - f_k)$$
$$= \frac{1}{(10 \text{ kg})} \times (76.9 \cos 25° - 30.8) \text{ N}$$
$$= 3.89 \text{ m/s}^2 \quad \blacksquare$$

3.6 Some Implications of Newtonian Dynamics

The Third Law and the Definition of Mass

Newton's first law relates to the motion of a free body, that is, a body subject to no net force. The second law specifies the relationship that connects *force* and *mass* with the kinematical concept of acceleration. The second law does not provide us with a rigorous physical definition of either force or mass. It remains for the third law to give a proper definition of mass; then, with mass and acceleration defined, the second law completes the sequence by providing a proper definition of force.

Suppose that we have two objects that are isolated, that is, they interact only between themselves and with nothing else. An example of such a pair of isolated objects would be a star that has a single planet. Even the Sun and Earth are, to a good approximation, an "isolated" pair of objects because for many purposes we can neglect the presence of the Moon and other planets. Or we could consider two spacemen in deep space; if Spaceman 1 pushes on Spaceman 2 with a force $\mathbf{F}_{21}$, then 2 pushes on 1 with an equal and opposite force. That is, the third law states

$$\mathbf{F}_{12} = -\mathbf{F}_{21}$$

where $\mathbf{F}_{12}$ means the force *on* 1 *due* to 2 and $\mathbf{F}_{21}$ means the force *on* 2 *due* to 1. Now, using the second law, we can write

$$m_1 \mathbf{a}_1 = -m_2 \mathbf{a}_2$$

Considering only the magnitudes of the accelerations,

$$m_2 = \left(-\frac{a_1}{a_2}\right) m_1$$

Therefore, if we select m_1 as our standard mass, we can determine m_2 by measuring the acceleration ratio, a_1/a_2. (The negative sign indicates only that the two accelerations have opposite directions.) Thus, we can, in principle, compare masses by making only geometrical measurements on a pair of interacting objects. Notice that the objects can interact via *any* type of force; the gravitational force is not necessary. The concept of mass can therefore be given a proper definition without reference to the gravitational force or to any process that involves "weighing."

Inertial Reference Frames

Newton's laws involve the concept of *acceleration*. In order to measure acceleration we must specify some reference marks with respect to which the measurement is made. These reference marks constitute a *frame of reference* in the same way that a set of coordinate axes serves as a reference system for the plotting of graphical data. Not all reference frames are equally useful. If we set out to study the dynamics of planetary motion, we would not choose a coordinate system that is fixed with respect to the Earth; in such a system the planets wander in a complicated manner across the sky and undergo apparent motions that are not indicative of the forces actually acting on them. That is, in an Earth-fixed coordinate system, Newton's laws cannot be used directly to describe the motion of planets. However, if we choose a reference frame that is fixed with respect to the distant stars, the motion of the planets is found to conform to Newton's laws. *Any* reference frame in which Newton's laws are a correct description of the dynamics of moving bodies is called an *inertial reference frame*. (We use the term *inertial* reference frame because in such a frame a free body obeys the law of *inertia*, Newton's first law.) It is important to note that an accelerated reference frame is *not* an inertial frame.

To Newton, the distant stars, which appeared to have fixed positions in space, satisfied the need for a basic inertial reference frame. We now know that these stars are not "fixed" with respect to the Earth (or even with respect to the Sun) but undergo continual and complicated motions. The specification of an inertial reference frame is therefore not a simple problem. However, for all but the most sophisticated analyses, we can consider the distant stars to be fixed and to constitute an acceptable inertial reference frame.

The distant stars do not specify the only possible inertial reference frame. We can find many other reference frames in which Newton's laws are also valid. The Earth undergoes a complicated motion against the background of the distant stars. But if we confine our attention to small-scale phenomena that take place over relatively short periods of time, the motion of the Earth will not influence the phenomena to any appreciable extent. Therefore, in many practical situations, Newton's laws will be valid in a coordinate system fixed with respect to the Earth. Indeed, for most everyday applications of Newton's laws we find an Earth-fixed coordinate system to be quite adequate.

Suppose that we set up a laboratory in a large box that is at rest on the Earth. We equip ourselves with suitable meter sticks, clocks and spring balances so that we can test Newton's laws. By making various measurements we can verify, for example, that $\mathbf{F} = m\mathbf{a}$ is a valid equation. The laboratory coordinate system therefore is an inertial reference frame. Next, someone removes our box laboratory and places it on a train that is moving with constant velocity. (Let us suppose that our box contains no windows so that we cannot measure the velocity of the train.) If we repeat the measurements that were made when the laboratory was at rest, what will we find? We will find exactly the same results! That is, Newton's laws are also valid in the moving reference frame. The reason is easy to see. A measurement of a certain acceleration requires the measurement of a time interval and the difference between two velocities:

$$\mathbf{a} = \frac{\Delta \mathbf{v}}{\Delta t} = \frac{\mathbf{v}_1 - \mathbf{v}_0}{t_1 - t_0}$$

If we add a constant velocity $\mathbf{v}$ to the coordinate system (by transferring it to the moving train), we have

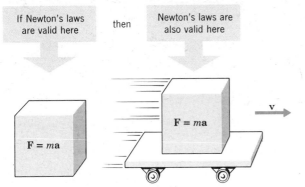

Figure 3.11 The addition of a constant velocity does not invalidate Newton's laws.

for our new velocities, $\mathbf{v}_1' = \mathbf{v}_1 + \mathbf{v}$ and $\mathbf{v}_0' = \mathbf{v}_0 + \mathbf{v}$. Therefore, the new acceleration $\mathbf{a}'$ is

$$\mathbf{a}' = \frac{\mathbf{v}_1' - \mathbf{v}_0'}{t_1 - t_0} = \frac{(\mathbf{v}_1 + \mathbf{v}) - (\mathbf{v}_0 + \mathbf{v})}{t_1 - t_0}$$

$$= \frac{\mathbf{v}_1 - \mathbf{v}_0}{t_1 - t_0} = \mathbf{a}$$

Thus, the acceleration measured in the moving reference frame is exactly equal to that measured in the reference frame at rest.

We may therefore draw the following important conclusion: *If Newton's laws are valid in a certain reference frame, then they will also be valid in any other reference frame that moves with constant velocity with respect to the first frame* (see Fig. 3.11). Thus, there is no *one* reference frame that is preferred over all others.

3.7 Momentum

Force and Change of Momentum

In addition to the concepts of force, mass, and acceleration, there are other physical quantities that are of importance in dynamical situations. One of these quantities is *momentum*, which, for a material object, is defined to be the product of the object's *mass* and its *velocity*. Because mass is a scalar and velocity is a vector, the product $m\mathbf{v}$ is a vector; we give to the momentum vector the symbol $\mathbf{p}$:

$$\text{Momentum} = \mathbf{p} = m\mathbf{v} \tag{3.7}$$

Frequently, we use the term *linear momentum* for $\mathbf{p}$ in order to distinguish it from *angular momentum*, which we introduce in Section 4.1. The dimensions of linear momentum are mass times velocity and the SI unit of $\mathbf{p}$ is kg·m/s.

Thus far, we have always considered a body's mass to remain constant. (But there is nothing in Newton's laws that *requires* mass to be constant with time.) In fact, there are important situations in which mass does change with time. For example, a rocket is propelled by the ejection of mass (usually in the form of gases at high velocity) and therefore the mass of the rocket-plus-fuel system decreases with time.[1] The rocket system is discussed later in this section. Newton allowed for the possibility that mass could change with time and actually originally stated his second law, not in terms of (mass × acceleration) as we have indicated, but in terms of the time rate of change of momentum. That is,

$$\overline{\mathbf{F}} = \frac{\Delta \mathbf{p}}{\Delta t} \quad \text{and} \quad \mathbf{F} = \lim_{\Delta t \to 0} \frac{\Delta \mathbf{p}}{\Delta t} \tag{3.8}$$

Using the definition of $\mathbf{p}$ and considering mass to remain constant in time, we have

$$F = \lim_{\Delta t \to 0} \frac{\Delta (m\mathbf{v})}{\Delta t}$$

$$= m \lim_{\Delta t \to 0} \frac{\Delta \mathbf{v}}{\Delta t}$$

or, using the definition of acceleration,

$$\mathbf{F} = m\mathbf{a} \tag{3.9}$$

so that we recover the form of the second law that we used in Eq. 3.2. *Equation 3.8 is the most general statement regarding force*. In the special case that the mass remains constant, Eq. 3.8 is equivalent to Eq. 3.9.

●*Example 3.6*

A 100-kg man jumps into a swimming pool from a height of 5 m. It takes 0.4 s for the water to reduce his velocity to zero. What average force did the water exert on the man?

The man's velocity on striking the water was (see Eq. 2.16)

$$v = \sqrt{2gh}$$
$$= \sqrt{2 \times (9.80 \text{ m/s}^2) \times (5 \text{ m})}$$
$$= 9.90 \text{ m/s}$$

---

[1] This does not violate the law of mass conservation. Here, it is only the mass of the accelerating body that is changing; the sum of the masses of the body and the ejected gases is constant.

Therefore, the man's momentum on striking the water was

$p_1 = mv$
$= (100 \text{ kg}) \times (9.90 \text{ m/s})$
$= 990 \text{ kg-m/s}$

The final momentum was $p_2 = 0$, so that the average force was

$\overline{F} = \dfrac{\Delta p}{\Delta t} = \dfrac{p_2 - p_1}{\Delta t}$

$= \dfrac{0 - 990 \text{ kg-m/s}}{0.4 \text{ s}}$

$= -2450 \text{ N}$

The negative sign means that the retarding force was directed opposite to the downward velocity of the man. ∎

Conservation of Linear Momentum

Next, we consider the importance of the concept of linear momentum. Suppose we have a system that is *isolated;* that is, the constituents of the system interact with one another but there is no outside agency that acts on them in any way. Truly isolated objects are not possible in the real physical world, but a group of objects whose mutual interaction is much greater than their interaction with other objects can frequently be treated as if they are isolated. For example, consider two hockey pucks that slide (almost) without friction over ice and collide with one another. The individual motions are influenced in a small way by the gravitational forces due to surrounding objects and to an even greater extent by friction, but the collisional interaction between the two pucks so overwhelms these other interactions that for most purposes the latter can be neglected.

If a group of objects constitutes an isolated system, there is not net force on this system. With $\mathbf{F} = 0$, Eq. 3.7 requires that there be no change of linear momentum with time; in other words, $\mathbf{p} = $ const. This is an extremely important result:

If there is no external force applied to a system, then the total linear momentum of that system remains constant in time.

This is the statement of the *principle* (or *law*) *of the conservation of linear momentum.*

The momentum to which the conservation principle applies is the *total linear momentum* of the sys-

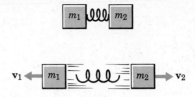

Figure 3.12 The motion of the masses after release is determined by the conservation of linear momentum.

tem. If the system consists of two objects, m_1 and m_2, then

$\mathbf{p} = \mathbf{p}_1 + \mathbf{p}_2$
$= m_1\mathbf{v}_1 + m_2\mathbf{v}_2 = \text{const.}$ (3.10)

Suppose we have two masses, at rest, that are connected by a compressed spring, as in Fig. 3.12. (As in all such cases, we neglect the mass of the spring.) In this condition the total linear momentum of the system is zero and the conservation law tells us that it must always be zero. After the release of the spring, the objects move away from each other with velocities $\mathbf{v}_1$ and $\mathbf{v}_2$. Thus,

$\underbrace{0}_{\substack{\text{before}\\\text{release}}} = \underbrace{\mathbf{p}_1 + \mathbf{p}_2}_{\substack{\text{after}\\\text{release}}}$

so that

$m_1\mathbf{v}_1 + m_1\mathbf{v}_2 = 0$ (3.11)

or, considering only the magnitudes of the velocities,

$m_1 v_1 = -m_2 v_2$ (3.12)

where the negative sign indicates that the velocities are oppositely directed.

● *Example 3.7*

Momentum conservation can be used to measure dynamic quantities in a straightforward way that otherwise might be rather difficult to measure. For example, if we wished to measure the muzzle velocity of a rifle bullet, we could employ sophisticated high-speed photography. However, a very simple arrangement that might be employed in an elementary laboratory will suffice.

Suppose we fire a 15-g bullet into a 10-kg wooden block that is mounted on wheels and measure the time required for the block to travel a distance of 0.45 m. This can easily be accomplished with a pair of photocells and an electronic clock. If

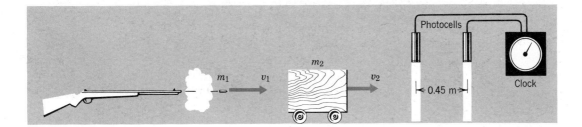

the measured time is 1 s, what is the velocity of the bullet?

The recoil velocity of the block is 0.45 m/s, and from momentum conservation we have

$$m_1 v_1 = m_2 v_2$$

If we take m_2 to be 10 kg, that is, we neglect the mass of the bullet embedded in the block, then,

$$v_1 = \frac{m_2 v_2}{m_1}$$

$$= \frac{(10 \text{ kg}) \times (0.45 \text{ m/s})}{0.015 \text{ kg}}$$

$$= 300 \text{ m/s} \blacksquare$$

In this example the bullet comes to rest in the block and imparts its momentum to the block. The process by which the bullet stops is a complicated one, but we need to know none of the details in order to calculate the velocity by using momentum conservation. This is indeed a powerful physical principle!

Momentum Conservation and the Third Law

Previously, we used Newton's third law to express the ratio of two isolated and interacting masses in terms of the ratio of their accelerations. Thus, the third law permits a definition of mass in terms of kinematic quantities. Equation 3.12, written in the form

$$\frac{m_1}{m_2} = -\frac{v_2}{v_1}$$

states that we can use momentum conservation to determine the ratio of two masses by measuring the ratio of their velocities. But we obtained the momentum conservation law from Eq. 3.8 which is a statement of Newton's *second* law. Can we therefore avoid the necessity of introducing the third law, which apparently serves no function other than to define mass, and obtain all of the machinery of dynamics from the second law?

The answer to this question is an emphatic "no." The reason is that we have really used the third law in obtaining the momentum conservation law. Consider the statement that led directly to momentum conservation: *If a group of objects constitutes an isolated system, there is no net force on this system.* This statement can be true only if the third law is obeyed, for if object 1 exerts a force on object 2 there will be a net force on the system unless object 2 exerts an equal and opposite force on object 1. An isolated system cannot accelerate itself; this is guaranteed by the third law, in the guise of momentum conservation.

Conservation of the Momentum Vector

Momentum is a *vector* quantity, so the principle of momentum conservation can be applied component by component. That is, if **p** is conserved, then p_x and p_y (and also p_x if the problem is three-dimensional) are *individually* conserved.

A situation may also arise for which one component of the momentum of a system is conserved while another component is not conserved. The following example illustrates such a case.

● *Example 3.8*

A railway gun whose mass is 70,000 kg fires a 500-kg artillery shell at an angle of 45° and with a velocity of 200 m/s. Calculate the recoil velocity of the gun if friction along the track may be neglected.

Let us take for our system the artillery shell and the railway gun. First, consider the horizontal or x-component of the shell's momentum:

$$p_{1x} = m_1 v_1 \cos 45°$$
$$= (500 \text{ kg}) \times (200 \text{ m/s}) \times 0.707$$
$$= 7.07 \times 10^4 \text{ kg-m/s}$$

There being no horizontal force acting on the system, this must equal (except for the sign) the recoil

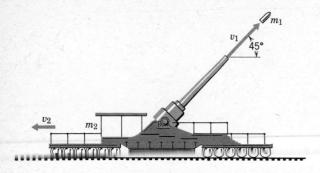

momentum of the gun which moves only horizontally:

$p_2 = m_2 v_2 = -7.07 \times 10^4$ kg-m/s

Therefore,

$$v_2 = \frac{7.07 \times 10^4 \text{ kg-m/s}}{7 \times 10^4 \text{ kg}}$$

$\cong -1$ m/s

or, approximately 2 mi/h.

What has happened to the *vertical* component of the momentum, $P_{1y} = 7.07 \times 10^4$ kg-m/s? Neglecting the deflection of the track, the vertical component of the momentum of the railway gun is zero, i.e. $p_{2y} = 0$. The acquired vertical component of the momentum of the system, $p_{1y} + p_{2y}$ is supplied by the vertical impulsive force exerted by the track against the wheels of the railway gun. If the firing time is, for example, 0.1 s then use of Eq. 3.8 permits us to estimate this average impulsive force $\overline{F}_y$:

$$\overline{F}_y = \frac{\Delta p_y}{\Delta t} = \frac{p_{1y} + p_{2y}}{\Delta t} = \frac{p_{1y} + 0}{\Delta t}$$

$$= \frac{7.07 \times 10^4 \text{ kg-m/s}}{0.1 \text{ s}}$$

$= 7 \times 10^5$ N

From a point of view that includes the Earth, which of course supports the track rails, as part of the system, the Earth absorbs the vertical momentum. The Earth does in fact recoil, but because of the extremely large value of the Earth's mass compared to that of the railway gun, this recoil velocity cannot be measured. ∎

Rocket Propulsion

Rocket propulsion is a striking example of momentum conservation in action. The onboard fuel is burned in the rocket engine and ejected out of a rear nozzle as a hot, high-velocity gas. The rocket, in re-

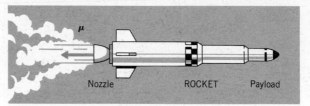

Figure 3.13 A rocket ejecting an exhaust gas with relative velocity μ.

coiling from the ejected gas, is thrust forward. In Example 3.8, we had the case of a gun recoiling from its fired projectile. The relationship of the rocket to its ejected exhaust gas is quite similar except, of course, that rocket propulsion is a continuous process. It is important to realize that unlike ordinary aircraft propulsion, a rocket does not make use of the air of the atmosphere. When traveling through the atmosphere the presence of the air produces, in fact, an undesirable drag on the rocket that impedes its motion. Rockets operate best in outer space.

Suppose the rocket shown in Fig. 3.13 ejects a mass Δm of exhaust gas in a time interval Δt. If the exhaust velocity relative to the rocket is μ, then the exhausted gas has acquired a momentum $\Delta \mathbf{p} = \mu \Delta m$. The associated force $\mathbf{F}_e$ is provided by the rocket engine nozzle chamber and using Eq. 3.8 is

$$\mathbf{F}_e = \frac{\Delta \mathbf{p}}{\Delta t} = \mu \frac{\Delta m}{\Delta t}$$

The quantity $\Delta m / \Delta t$ is the rate at which the hot exhausted gas is ejected. The thrust pushing the rocket forward is the reaction force to $\mathbf{F}_e$, that is the force the ejected gas exerts on the rocket,

$$\mathbf{F} = -\mathbf{F}_e$$

thus,

$$\mathbf{F} = -\mu \frac{\Delta m}{\Delta t} \qquad (3.13)$$

A modern large rocket might have an exhaust gas ejection rate of 150 kg/s at a speed of 3000 m/s. The developed thrust is, therefore,

$$F = \mu \frac{\Delta m}{\Delta t} = (3000 \text{ m/s})(150 \text{ kg/s})$$

$= 4.5 \times 10^5$ N

If the rocket (with the remaining fuel) at the instant under discussion has a mass of 2500 kg, then, its acceleration would be

$$a = \frac{F}{M} = \frac{4.5 \times 10^5 \text{ N}}{2.5 \times 10^3 \text{ kg}} = 180 \text{ m/s}^2$$

or about 18 g.

The mass of the rocket, with its onboard fuel, it continually decreasing as it ejects the burned fuel out the engine nozzle. The result is to produce a continually increasing acceleration until all the fuel is spent. The expression relating the final velocity v achieved by the rocket having an initial mass (with fuel included) M_i and a final mass (with all fuel spent) M_f is [2]

$$v = \mu \ln (M_i/M_f) \qquad (3.14)$$

where again μ is the exhaust-gas velocity relative to the rocket.

Summary of Important Ideas

Except for objects moving with velocities approaching the velocity of light and for effects that take place on an atomic scale, *Newton's laws* are a correct description of the dynamics of physical systems.

If Newton's laws are valid in a certain frame of reference, they are also valid in *any* frame that moves with *constant linear velocity* relative to the first frame.

A body can be accelerated only if a *force* is applied to it. Force is the time rate of change of linear momentum.

The motion of a body is determined by the net force acting *on* the body and does not depend on the forces that the body exerts on other bodies.

Friction always acts in a direction to retard the motion of a body.

Mass is an intrinsic property of matter; *weight* is a result of the *gravitational force* on an object.

Force and linear momentum are *vectors*.

If no external force is applied to a system, the linear momentum of the system remains *constant* in time.

◆ Questions

3.1 Suppose that you are pushing a block with constant velocity across a horizontal surface. On top of the block being pushed there rests a second block. If you stop the lower block suddenly, the upper block will slide forward and fall from the lower block. But if you gradually slow the lower block to a stop, the upper block remains in place. Explain carefully the difference between these two situations.

3.2 A heavy weight is suspended from the ceiling by a length of string. Hanging downward from the weight is another length of the same string. Which string will break when (a) a steady pull is exerted on the lower string and (b) a sudden pull is exerted on the lower string? Explain.

3.3 List the controls in an automobile which, when activated, tend to make a passenger alter her position in her seat. What is the nature of the acceleration produced by each?

3.4 Two students are attempting to stretch a spring between two posts. Which of the following methods is better? (Explain carefully.) (a) Each student grasps an end of the spring and each pulls his end toward one of the posts. (b) The students attach one end of the spring to one of the posts and then both pull the other end of the spring toward the other post.

3.5 In Section 3.4, reference was made to the measurement of weight by means of a bathroom-type scale. Explain how such a scale determines weight. Describe another method of determining weight.

3.6 A vendor who sells items according to the weight registered on a spring scale (i.e., "by the newton") establishes his stand in the elevator of a large building. During which portion of an elevator trip would it be advantageous to purchase some of the vendor's wares? When does the vendor wish to sell?

3.7 When an object is dropped it accelerates toward the Earth. This acceleration is due to the gravitational force exerted on the object by the Earth. According to Newton's third law, an equal and opposite force is exerted on the Earth by the object. Therefore, the Earth should also move toward the object. Does it?

3.8 Do you impart momentum to the Earth when you walk? Explain.

3.9 You have probably seen photographs of astronauts taking "space walks." If an astronaut were not connected to the space vehicle (he is actually tethered at all times), how could he propel himself about and return to the vehicle?

---

[2]The symbol l*n* stands for the natural logarithm; see Appendix A7.

3.10 A rocket is coasting at constant velocity in free space. When the rocket engine is turned on, the rocket begins to accelerate. Explain how this is possible. What is the nature of the force exerted on the rocket?

3.11 When an inflated balloon is released so that the air may escape from it, the balloon shoots off in a crazy zig-zag path. Explain why this happens.

★ Problems

Sections 3.2 and 3.3

3.1 Two forces, one twice as large as the other, both pull in the same direction on an object ($m = 10$ kg) and impart to it an acceleration $a = 3$ m/s^2. What are the magnitudes of the forces? If the smaller force is removed, what is the new acceleration?

3.2 A 20-kg object is at rest at the origin of an *x-y* coordinate system. At a certain instant two forces are applied to the object: 25 N in the $+y$ direction and 50 N in the $+x$ direction. Describe the subsequent motion of the object.

3.3 A 20-kg object rests on a horizontal frictionless surface and is subject to two forces: **F**$_1$ has a magnitude of 40 N and is directed *east*; **F**$_2$ has a magnitude of 60 N and is directed *northwest*. Describe the motion of the object.

3.4 A 3-kg block and 5-kg block are in contact as they rest on a horizontal frictionless surface. A constant force of 12 N is applied. Describe the motion of the blocks. What force does the 3-kg block exert on the 5-kg block? What is the net force on the 3-kg block? Explain why the two blocks don't separate.

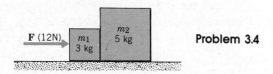

Problem 3.4

3.5 A 3-kg object is observed to be accelerating at a rate of 12 m/s^2 in the $+y$ direction while a 3-N force is acting on it in the $+x$-direction. What other force must also be acting? Give both its magnitude and direction.

Section 3.4

3.6 What is the mass in kilograms of a $\frac{3}{4}$-pound steak? A housewife buys a 2-"kilo" chicken in a Parisian market. What is its weight in pounds?

3.7 A 100-kg person stands on a set of scales in an elevator. The elevator begins to descend with an acceleration $a = 2$ m/s^2. What is the reading on the scales in newtons? What is the reading on the scales when the elevator is ascending uniformly with $v = 2.5$ m/s?

3.8 A ball is swung in a vertical circle with constant speed at the end of a 1-m rope. What must be the period of the motion for the ball to be "weightless" at the top of the swing? (Here, "weightless" means that there would be zero tension in the rope and yet it would remain straight.)

3.9* A small bolt loosens and drops from the ceiling of an elevator that is accelerating upwards at a rate of 2 m/s^2. If the floor to ceiling distance of the elevator is 2.5 m, how long does it take for the bolt to hit the elevator floor? What net gravitational acceleration would produce the same result in a stationary elevator?

3.10 A 2000-kg automobile is traveling with a speed of 25 m/s. What braking force is required to bring the automobile uniformly to a stop in 4 s? How far did the automobile travel while the brakes were being applied?

3.11 A block ($M = 15$ kg) rests on a frictionless horizontal plane and is acted upon by two forces. **F**$_1$ acts horizontally in the $+x$-direction with a magnitude of 12 N; **F**$_2$ acts upward at an angle of 30° above the horizontal in the $-x$-direction with a magnitude of 18 N. Describe the motion of the block.

3.12 A 10-kg object is pulled over a rough surface by a 20-N force. The object accelerates at a rate of 1.5 m/s^2. What is the frictional force between the object and the surface?

3.13 A 1-kg block is pulled across a horizontal surface by a force of 2 N that is directed at an angle of 45° above the horizontal. If there is a frictional retarding force of 0.5 N, what is the acceleration of the block?

3.14 A 40-kg block is sliding over a rough surface. At a certain instant the velocity of the block is 2 m/s. If the block is slowed to rest by friction in a uniform manner after the block has traveled 40 m, what was the average frictional force?

3.15 A pendulum consists of a 0.1-kg ball suspended from a 2-m string. The ball is drawn to one side until the string makes an angle of 30° with the vertical and is then released. Sketch the force diagram at the instant of release. What is the initial acceleration of the ball? (Find the magnitude and the direction.)

3.16 A block with a mass M rests on a plane that is inclined at an angle θ. A string attached to the block is directed up the plane where it is supported by a frictionless pulley; a block with a mass m hangs from the lower end of the string. Describe the motion of the blocks when $M = 12$ kg, $m = 4$ kg, and $\theta = 30°$.

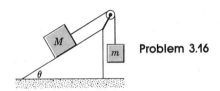

Problem 3.16

3.17 Rework Problem 3.16 for the case $M = 12$ kg, $m = 6$ kg, and $\theta = 20°$.

3.18 Describe the initial acceleration of the blocks in the diagram if $M = 20$ kg, $m = 5$ kg, $\theta = 30°$, $F = 60$ N. (Assume that friction is negligible.)

Problem 3.18

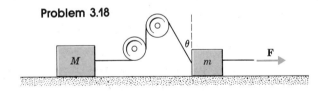

3.19 A block sent sliding across a horizontal surface is observed to decelerate at a rate of 3.5 m/s^2 because of the presence of friction. What is the coefficient of kinetic friction between the block and the surface?

3.20* A force $F = 60$ N is observed to pull an 8-kg block up the inclined plane at a constant speed. If the surface of the plane makes an angle of 30° with the horizontal find—the coefficient of kinetic friction between the block and the plane surface—what force F would allow the block to slide down the inclined plane with a constant speed?

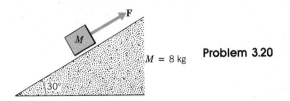

Problem 3.20

3.21 A 10-kg block slides down an inclined plane making an angle of 20° with the horizontal. Find the acceleration of the block if the coefficient of kinetic friction between the block and the plane is $\mu_k = \frac{1}{5}$.

3.22* A block is pushed along a horizontal surface by a force $\mathbf{F}$ that makes an angle θ with the horizontal. Taking the coefficient of kinetic friction to be μ_k, show that as θ is increased a critical angle will be reached at which forward motion is impossible, regardless of the magnitude of $\mathbf{F}$. Find the expression for this critical angle.

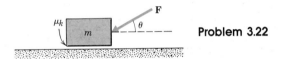

Problem 3.22

3.23* A 50-kg block is pulled upward by a rope and pulley combination (assumed to have negligible mass and friction). A 30-kg block is used as a counterweight. What force $\mathbf{F}$ must be applied to raise the block to a height of 20 m in 5 s, starting from rest?

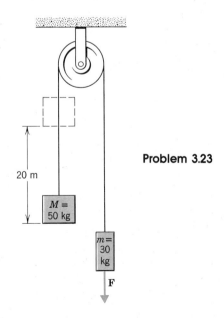

Problem 3.23

3.24 A force of 800 N acts upward on a pair of blocks held together by a 10-kg cable. Find the acceleration of the system. What is the tension in the cable at point **A**? At point **B**? At point **C** the center of the cable?

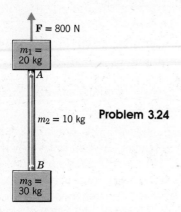

Problem 3.24

3.25* Find the accelerations of the blocks in the diagram. The pulleys are assumed to be frictionless. (What is the velocity, at any instant, of m_1 compared to that of m_2?)

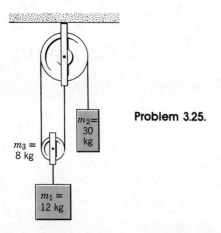

Problem 3.25.

3.26* A rope that is 10 m long has a mass of 20 kg. One end of this rope is attached to a beam that protrudes from the side of a building. An 80-kg individual slides down this rope at a constant speed of 0.8 m/s. What is the tension in the rope 5 m from the top when the individual has slid to this point?

3.27 An *accelerometer* is a device for measuring acceleration. A simple accelerometer can be made by suspending a small mass within a box by means of a length of string. A scale is provided for measuring the angle θ that the suspension string makes with the sides of the box. The box is placed on the flat floor of an object that undergoes a horizontal acceleration. Find the expression that relates the angle θ to the acceleration of the object in terms of g. When $\theta = 8°$, what is the acceleration? How would you construct an accelerometer to measure acceleration in *any* direction?

Section 3.7

3.28 When a 10-g bullet is fired from a rifle, a force of 3000 N is exerted for a millisecond. What is the muzzle velocity of the bullet?

3.29 A fire hose ejects 50 kg of water per second at a speed of 40 m/s. What force must be exerted by the firemen holding the hose in order to keep it stationary?

3.30 A ball ($m = 0.1$ kg) is thrown directly against a wall with a velocity of 10 m/s and it rebounds with a velocity of the same magnitude. If the ball was in contact with the wall for 1 ms (10^{-3}) s, what was the average force exerted on the ball by the wall? (Find the magnitude and the direction of the force.)

3.31* In Example 3.7 it is found that the bullet penetrates 10 cm of wood. Assume that there is a uniform decrease of the velocity to zero. How long did it take for the bullet to come to rest? What average force did the wood exert on the bullet?

3.32 In "shoot-'em-up" movies, the villain is often knocked down by a single bullet from the hero's gun. A .357-magnum bullet has a mass of approximately 10 g and a muzzle velocity of 400 m/s. What would be the recoil velocity of an average-size man struck by such a bullet? Is it likely that he would be knocked down?

3.33 An object of mass 1 kg is moving in the x-direction with a velocity of 45 m/s and another object of mass 2 kg is moving in the y-direction with a velocity of 15 m/s. The two objects collide and stick together. Describe the subsequent motion.

3.34 A circus cannon of mass 1000 kg is mounted on wheels and rests on a flat surface. If the "Great Zucchini" stunt performer of mass 70 kg is fired at an angle of 30° to the horizontal with a muzzle velocity of 20 m/s, what will be the recoil velocity of the cannon?

3.35 Two children ($m = 30$ kg each) are sliding on ice in "saucers" ($m = 1$ kg each). They are moving side by side in parallel paths with the same velocity, $v = 3$ m/s. Each child has a 2-kg block which he throws with a velocity of 10 m/s to the other child who catches it. Describe the subsequent motion of each child. (Assume that the saucer-ice friction is negligible.)

3.36 A 10-kg cart is moving over a horizontal surface at a constant velocity of 2 m/s. A 2-kg lump of clay is dropped into the cart from a height of 4 m and it sticks to the bed. Describe the subsequent motion. What happened to the vertical momentum?

3.37 A 600-kg block of wood is balanced on top of a vertical post 2-m high. a 10-g bullet is fired horizontally into the block. If the block and embedded bullet land at a point 4-m from the base of the post, find the initial velocity of the bullet.

3.38 A 60-g rubber ball is dropped from a height of $1\frac{1}{2}$ m and rebounds to a height of 1 m. If the ball was in contact with the floor for 2 ms $(2 \times 10^{-3}$ s$)$ what average force did the floor exert on the ball?

3.39 What thrust is developed by a rocket engine that can eject gas at a rate of 80 kg/s at an exhaust speed of 1800 m/s?

3.40 A rocket with a total mass of $1.6 = 10^5$ kg sits vertically on a launch pad. At what rate must exhaust gas be ejected on ignition if the thrust is to just overcome the gravitational force on the rocket? Assume a velocity for the ejected gas of 3000 m/s.

3.41 A 50-g plastic toy rocket is loaded with water and compressed air to a total mass of 200 g. Set to fly straight up into the air it discharges all the water in a negligibly short time at a discharge velocity of 15 m/s. How high will the rocket fly?

3.42* A spaceship is stationary in deep space when its rocket engine is ignited for a 100-s "burn." The exhaust ejection rate is 100 kg/s at a speed of 3500 m/s. The initial mass of the spaceship is 20,000 kg. Determine the thrust of the rocket engine and the initial acceleration. What is the final velocity of the spaceship?

Torque and Angular Momentum

The preceding chapter examined the dynamics of translational motion, motion not involving rotation. We introduced the important concept of linear momentum and the conservation principle associated with it. The application of these ideas greatly simplified our understanding of many complicated situations.

In this chapter we consider another type of motion, motion that involves the rotation of objects. We introduce a new dynamical quantity called *angular momentum*, a quantity that obeys a conservation principle similar to that for linear momentum. We also inquire into the question of how applied forces generate or modify rotational motion. These new ideas will also prove to offer useful insights.

4-1 Torque and Angular Momentum

A Torque Can Produce Rotation

If a force is applied to the handle of a door, the door can be opened or closed; that is, a rotation around the hinge line can take place (Fig. 4.1a). But if the same force is applied in such a way that the force vector passes *through* the hinge line, there can be no rotation (Fig. 4.1b). A force that is applied "off center," and can therefore produce rotation, constitutes a *torque*.

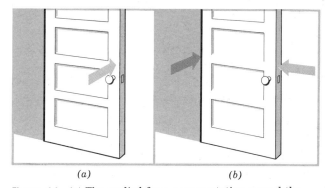

Figure 4.1 (a) The applied force causes rotation around the hinge line. (b) The applied forces each pass through the hinge line and no rotation results.

The quantitative measure of a torque is the distance from the point of application of the force to the center of rotation multiplied by the component of the force perpendicular to this distance. In Fig. 4.2a, the force **F** acts at the point A which is a distance r from the center of rotation. The force vector is perpendicular to the line AB and therefore the entire force is effective in producing the torque[1]:

$$\text{torque} = \mathcal{T} = r \times F \tag{4.1}$$

---

[1] $r \times F$ is read "r cross F."

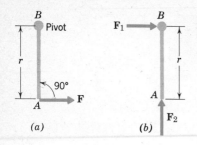

Figure 4.2 (a) The torque is $\mathcal{T} = r \times F$. (b) The force $\mathbf{F}_1$ is directed at the pivot point so that $r = 0$ and there is no torque. The force $\mathbf{F}_2$ has no component perpendicular to the line AB so there is no torque.

In Fig. 4.2b, neither F_1 nor F_2 can produce a torque around the point B.

If the force is applied at an angle to the line AB, as in Fig. 4.3, then only the *perpendicular* component of the force $F_\perp$ can contribute to the torque:

$$\mathcal{T} = r \times F_\perp$$

or, using $F_\perp = F \sin \theta$,

$$\mathcal{T} = rF \sin \theta \tag{4.2}$$

(Clearly, the other component of the force, $F_\perp$, can produce no rotation, because its direction is in line with the pivot point.)

It is not necessary to have a pivoted rod in order to define torque. The force F applied to the free particle of mass m in Fig. 4.4 constitutes a torque, $\mathcal{T} = r \times F$, referred to the arbitrary point O. If the constant force F acts for a certain time, the particle will accelerate, but because it will move in the direction of the force, the *torque* around O will remain the same. (Check this by noting that the product of $F_\perp$ and the distance from m to O remains constant.) The dimensions of torque are force times length and the SI unit of torque is the newton-meter.

Relationship between Torque and Angular Momentum

From Newton's laws we have obtained the important result that if the net force on a body is zero, the linear momentum will remain constant. It is also

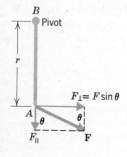

Figure 4.3 The torque is $\mathcal{T} = rF \sin \theta$. Note that θ is the angle between **F** and the line that connects the pivot point to the point at which the force is applied.

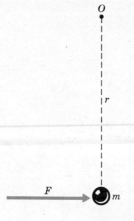

Figure 4.4 The torque on m around the arbitrary point O is $\mathcal{T} = r \times F$.

true that if the net *torque* on a body is zero, we obtain another important conservation law, on a level equal to the conservation of linear momentum. To state this new conservation principle, we define a quantity called the *angular momentum*, which is the product of the distance from the object to the axis of rotation and the perpendicular component of the linear momentum, as in Fig. 4.5. Angular momentum is denoted by L:

$$\begin{aligned} L &= r \times p_\perp \\ &= rmv_\perp \\ &= rmv \sin \theta \end{aligned} \tag{4.3}$$

where θ is the angle between **p** (or **v**) and the line connecting m and the axis point O; then, $v_\perp = v \sin \theta$, as in Fig. 3.15c.

Because the *linear* velocity v is equal to the product of r and the *angular* velocity ω (Eq. 2.33), we can use $v = r\omega$ to express the angular momentum as

$$L = mr^2 \omega \sin \theta \tag{4.4}$$

In terms of the angular momentum thus defined, we can state: *If no external torque is applied to a body or system of bodies, the angular momentum remains constant.* This is the statement of the principle (or law) of *conservation of angular momentum*. This is a well-established law of physics; no exceptions or contradictions are known.

Just as the *force* on a body is equal to the time rate of change of *linear* momentum (Eq. 3.8), the *torque* is equal to the time rate of change of *angular momentum*:

$$\overline{\mathcal{T}} = \frac{\Delta L}{\Delta t} \quad \text{or} \quad \mathcal{T} = \lim_{\Delta t \to 0} \frac{\Delta L}{\Delta t} \tag{4.5}$$

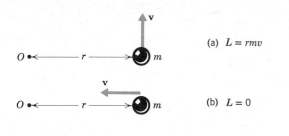

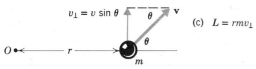

(a) $L = rmv$

(b) $L = 0$

(c) $L = rmv_\perp$

$v_\perp = v \sin \theta$

Figure 4.5 Three different cases for the angular momentum of m around an axis through the point O.

The angular momentum conservation principle follows directly from this statement because, if $\mathcal{T} = 0$, then L does not change with time.

It should be noted that angular momentum is a well-defined quantity even if the object is not moving in a curved path. For example, in Fig. 4.5a, if the mass m moves in a straight line with the velocity vector as shown, the perpendicular distance from its line of motion to the point O is always r, therefore, the angular momentum about this point is always $L = mvr$ (in the absence of any external torque). In Fig. 4.4, on the other hand, the torque on m around O is constant and therefore the angular momentum around O increases uniformly with time (that is, at a constant rate).

Naturally, there are many situations in which torques are present. What effect does an applied torque have on the motion of a particle? Let us answer this question by considering a special yet important case of a particle confined to move in a circular path and subjected to a torque. Figure 4.6 shows a particle of mass m rotating about a central point O at a fixed radius r. A force **F** is acting in the plane of motion producing a torque $\mathcal{T} = rF \sin \theta$.

If the (instantaneous) tangential velocity of the particle is v then its angular momentum is

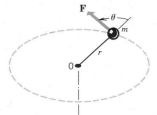

Figure 4.6 A particle of mass m executing circular motion under the action of a torque $\mathcal{T} = Fr \sin \theta$ produced by a coplanar force **F**.

$$L = mrv = mr^2\omega \quad (4.6)$$

where we have noted that $v = r\omega$. Making use of Eq. 4.5 we have

$$\mathcal{T} = \frac{\Delta L}{\Delta t} = \frac{\Delta(mr^2\omega)}{\Delta t}$$

$$= mr^2 \frac{\Delta \omega}{\Delta t}$$

Thus, in the limit of $\Delta t \to 0$ we have the final result

$$\mathcal{T} = mr^2 \alpha \quad (4.7)$$

In obtaining this result we have noted that m and r are constant and that the angular acceleration α is $\Delta\omega/\Delta t$ in the limit $\Delta t \to 0$. This important result gives the relationship between torques and the angular accelerations they produce.

● **Example 4.1**

A ball of mass 2 kg is attached to the end of a string and is swung in a circle of radius 50 cm in a vertical plane. At the instant shown $\theta = 20°$ and the ball is descending with a tangential velocity of 2.5 m/s. We wish to describe the acceleration of the ball.

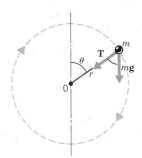

The only forces acting on the ball are: the string tension **T** and the gravitational force $m\mathbf{g}$. The torque acting about the center O is just $mgr \sin \theta$, since the string tension does not produce a torque. Thus, the total torque acting when $\theta = 20°$ is

$$\mathcal{T} = mgr \sin \theta$$
$$= (2 \text{ kg}) \times (9.80 \text{ m/s}^2) \times (0.5 \text{ m}) \sin 20°$$
$$= 3.35 \text{ N} \cdot \text{m}$$

The corresponding angular acceleration is

$$\alpha = \frac{\mathcal{T}}{mr^2} = \frac{3.35 \text{ N m}}{(2 \text{ kg}) \times (0.5 \text{ m})^2}$$
$$= 6.70 \text{ rad/s}^2$$

82 Torque and Angular Momentum

This results in a tangential acceleration

$a_t = r\alpha = (0.5 \text{ m}) \times (6.70 \text{ rad/s}^2)$
$= 3.35 \text{ m/s}^2$

Finally, the centripetal acceleration is

$a_c = v^2/r$
$= \dfrac{(2.5 \text{ m/s})^2}{(0.5 \text{ m})} = 12.5 \text{ m/s}^2$

We must emphasize that our results only apply to the instant shown. In this example α is not a constant. As θ increases the torque and hence α will increase, reaching a maximum value when $\theta = 90°$. Since the torque is clockwise throughout the angular range $0 < \theta < 180°$ and changes sign to counterclockwise for $\theta > 180°$, the angular velocity evidently has its maximum value for $\theta = 180°$, that is, at the bottom of the swing.

● *Example 4.2*

A ball of mass 10 kg is attached to the end of a string and is swung in a circle of radius 4 m with a constant tangential velocity of 2 m/s. While the ball is in motion, the string is shortened to 2 m. What is the change in the velocity and in the period of the motion?

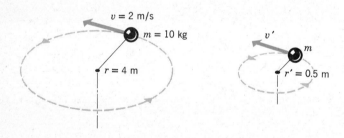

The initial angular momentum is

$L = mvr$
$= (10 \text{ kg}) \times (2 \text{ m/s}) \times (4 \text{ m})$
$= 80 \text{ kg-m}^2/\text{s}$

The initial period is

$\tau = \dfrac{2\pi}{\omega} = \dfrac{2\pi}{(v/r)} = \dfrac{2\pi r}{v}$
$= \dfrac{2\pi \times (4 \text{ m})}{2 \text{ m/s}}$
$= 12.56 \text{ s}$

Shortening the string does not apply any torque to the ball because the applied force lies along the line connecting the ball with the center of rotation. Therefore, the final angular momentum is equal to the initial angular momentum:

$L' = mv'r' = L$

Thus, the final velocity is

$v' = \dfrac{L}{mr'}$
$= \dfrac{80 \text{ kg-m}^2/\text{s}}{(10 \text{ kg}) \times (2 \text{ m})}$
$= 4 \text{ m/s}$

The new period is

$\tau' = \dfrac{2\pi r'}{v'}$
$= \dfrac{2\pi \times (2 \text{ m})}{4 \text{ m/s}}$
$= 3.14 \text{ s}$

Therefore, decreasing the radius by a factor of 2 has increased the linear velocity by the same factor and has decreased the period by a factor of 4.

● *Example 4.3*

A satellite of mass m moves around the Earth as shown (actually, the path is an *ellipse*). Which instantaneous velocity is greater, v_1 (at point P) or v_2 (at point A)?

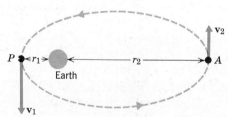

The gravitational force exerted on the satellite by the Earth is directed along the line connecting the two bodies. Therefore, the Earth exerts no torque on the satellite. Consequently, if we consider the Earth as a fixed object and neglect the influence of the Sun and other planets, the angular momentum of the satellite around the Earth is constant. Thus,

$mv_1 r_1 = mv_2 r_2$

Because $r_1 < r_2$, we must then have

$v_1 > v_2$

The velocity is greatest when the satellite is nearest the Earth; this point is called the *perigee* (labeled P in the diagram). The velocity is least at the farthest point from the Earth—the *apogee* (A) of the orbit. ∎

The Direction of Angular Momentum

Angular momentum is actually a *vector* quantity. The *magnitude* of the angular momentum vector has already been defined, but because *angular* motion is involved, the *direction* of the vector must be specified separately. If a particle is moving in a circular path around a certain point, the angular momentum vector is defined to have the direction in which a right-hand screw would advance if moved in the same sense (see Fig. 4.7). Thus, for motion in a plane, the angular momentum vector is *perpendicular* to that plane. In an x-, y-, z-coordinate system, a particle that moves in the x-y plane around the origin has an angular momentum vector that lies along the z axis. Alternatively, if the fingers of the right hand are curled in the direction of the motion of the particle, the thumb will point in the direction of **L**.

The conservation of angular momentum is important in the theory of the motion of planets around the Sun (and in all similar gravitational orbit problems). Consider the motion of the Earth. The gravitational force exerted on the Earth by the Sun acts along the line connecting the Earth and the Sun. Consequently, there is no torque acting on the Earth due to the Sun's gravitational pull, and the angular momentum of the Earth in its orbit must remain constant. That is, the Earth's angular momentum vector points in a fixed direction in space and cannot change in either magnitude or direction. This fact has two important implications: (1) Because **L** is constant in magnitude, the product rv for the Earth is constant (m is constant), as in Example 4.3; as the Earth moves in its orbit, its speed is greatest when nearest the Sun and least when farthest away. (The maximum and minimum distances differ by only about 3 percent, so the Earth moves with r and v essentially constant. (2) Because **L** is constant in direction, the Earth always moves in the same *plane*. We will discuss further the motions of planets in Section 6.2.

4.2 Center of Mass

The Significant Point for Extended Objects

We have stated the principles of conservation of linear momentum and angular momentum as applied to *particles*. But what is the significance of these principles for aggregates of particles (*systems*) or for bulk matter? In order to apply the conservation principles to a system we must find the single point within the system as a whole that always moves in accordance with the conservation laws. This point is called the *center of mass* of the system. When considering the implications of the conservation laws on the motion of a system of bodies, we usually find it most convenient to refer to the motion of the center of mass of that system.

For a body with a regular shape that consists of a uniform distribution of matter (i.e., if the density is everywhere the same), the center of mass is located at the geometrical center of the body. Therefore, when considering such objects as spheres, cubes, or rectangular bars, the center of mass is located at the obvious point. For collections of particles or for irregular objects, we must make a calculation in order to locate the center of mass. Once identified, the center of mass is the point at which the entire mass of the object may be considered to be located for the purposes of calculating the translational motion.

Calculation of the Position of the Center of Mass

For a simple system consisting of two particles we can calculate the position of the center of mass as follows. Imagine that the two masses are connected by a rigid, massless rod, as shown in Fig. 4.8. This system can be balanced on a pivot at a certain point O that corresponds to the center of mass. Because there is no rotation of the system around O, we know that the torque around O produced by the gravita-

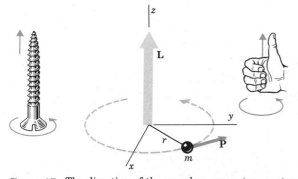

Figure 4.7 The direction of the angular momentum vector is the same as the direction of advance of a right-hand screw moving the same sense; or, if the fingers of the right hand are curled in the direction of motion; the thumb points in the direction of the vector **L**.

84 Torque and Angular Momentum

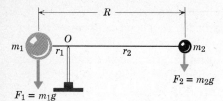

Figure 4.8 Two unequal masses are balanced on a pivot.

tional force on m_1 equals the torque around O produced by the gravitational force on m_2. That is,

$$m_1 g r_1 = m_2 g r_2$$

or,

$$m_1 r_1 = m_2 r_2 = m_2(R - r_1) \quad (4.8)$$

where $R = r_1 + r_2$ is the distance between m_1 and m_2.

We can always calculate the position of the center of mass of two objects by using Eq. 4.8. For systems consisting of three or more particles or for rigid bodies the expressions for the coordinates of the center of mass are easy to derive but they are more complicated and we have no need for them here.

● *Example 4.4*

What is the position of the center of mass of the bar in the diagram below? Each of the two sections has a uniform distribution of mass. The left-hand section has a mass of 4 kg and the right-hand section has a mass of 6 kg.

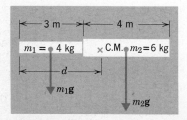

Because each section of the bar is uniform, the center of mass of each section will be located at its geometrical center. These points are indicated by the dots in the diagram. The center of mass of the entire bar is located at the position marked ×, which is a distance d from the left-hand end of the bar. Now, we imagine that the bar is balanced on a pivot located at the ×. Then, the torques due to the two sections must be equal. The distance from the × to the center of the left-hand section is $(d - 1.5 \text{ m})$, and the distance from the × to the center of the right-hand section is $5 \text{ m} - d$. Then,

$$(d - 1.5 \text{ m}) \times m_1 g = (5 \text{ m} - d) \times m_2 g$$

Substituting the values for m_1 and m_2, we solve for d and find

$$d = 3.6 \text{ m} \quad \blacksquare$$

Center of Mass and the Laws of Dynamics

Although we have not proved them here, we can summarize the importance of the center-of-mass concept with the following statements:

1. In the absence of external forces, the center of mass of a system moves with constant velocity.

2. If a force is applied to a system as a whole, the center of mass undergoes an acceleration $\mathbf{a} = \mathbf{F}/M$, where M is the total mass of the system.

This statement is true regardless of the point within the body or the system at which the force is applied.

In Section 4.3 we take up the question of the rotation of bulk matter in the form of rigid bodies. This study will permit us to add to the two statements above a third principle giving a complete description of motion for extended objects.

Figure 4.9 Time-lapse (stroboscopic) photograph of a wrench sliding across a smooth surface while rotating about its center of mass. The position of the C.M. is marked with black tape as a cross. Use a ruler and verify that the C.M. moves in a straight line, as required by Newton's laws, even though the wrench as a whole undergoes a complicated motion.

Distribution of Mass in the Human Body

Several studies have been made of the distribution of mass in the average human body. The following diagram and the table show the results obtained by the National Aeronautics and Space Administration (NASA) using data for male U.S. Air Force personnel. These figures will therefore differ slightly from those for females and for males from the population at large. (The solid circles labeled with letters represent the hinge points; the open squares labeled with numbers represent the centers of mass of the various body parts; the solid squares and the letter S show the position of the center of mass of the entire body.)

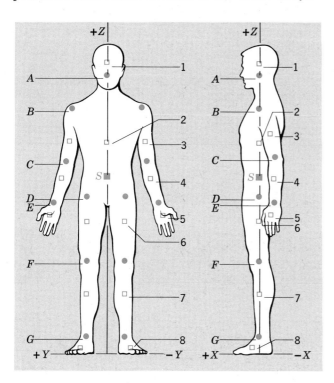

Symbol		Coordinates of point (percentage of height)			Mass (percentage of total body mass)
		x	y	z	
	Hinge Points				
A	Base of skull on spine	0.0	0.0	91.23	
B	Shoulder joints	0.0	± 10.66	81.16	
C	Elbows	0.0	± 10.66	62.20	
D	Hips	0.0	± 5.04	52.13	
E	Wrists	0.0	± 10.66	46.21	
F	Knees	0.0	± 5.04	28.44	
G	Ankles	0.0	± 5.04	3.85	
	Centers of Mass				
1	Head	0.0	0.0	93.48	6.9
2	Trunk-neck	0.0	0.0	71.09	46.1
3	Upper arms	0.0	± 10.66	71.74	6.6
4	Lower arms	0.0	± 10.66	55.33	4.2
5	Hands	0.0	± 10.66	43.13	1.7
6	Upper legs	0.0	± 5.04	42.48	21.5
7	Lower legs	0.0	± 5.04	18.19	9.6
8	Feet	3.85	± 6.16	1.78	3.4
S	Total (whole body)			57.95	100.00

Reference

Bioastronautics Data Book, National Aeronautics and Space Administration, Washington, D.C.

■ Exercises

1. A man has a mass of 80 kg (176 lb) and his height is 1.83 m (72 in.). What is the position of the center of mass of this man (above floor level) when he is in the normal standing position?

2. Suppose that the man in Exercise 1 raises both arms straight up? What is the new position of his center of mass?

3. Will a woman's center of mass be higher or lower (relative to her height) than a man's? Explain your reasoning.

4. Estimate the position of your center of mass when you bend over and touch your toes.

4.3 Rotation of Rigid Bodies

Another physical property associated with extended objects is *rotational inertia*. As the name suggests it is a property that moderates the effect torques have in changing the angular momentum. It is the rotational analog of mass, the property that moderates the acceleration produced by an applied force.

A practical type of extended object to study is the so called *rigid body*. In a rigid body all the constituent portions remain in a fixed relative position with respect to each other. Consider such an object that is constrained to rotating about a fixed axis as shown in Fig. 4.10.

Imagine dividing the body into a large number of mass segments such as m_1 illustrated in Fig. 4.10. Let us sum the relationship expressed by Eq. 4.7 over all the segments that make up the entire body.

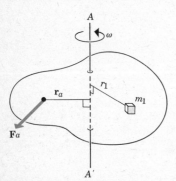

Figure 4.10 A rigid body rotating about a fixed axis AA'.

Then if $\mathcal{T}_a$ represents the torque about the axis AA' produced by one of the acting external forces $\mathbf{F}_a$, and supposing other external torques $\mathcal{T}_b$, $\mathcal{T}_c$, etc. to be present we have

$$\mathcal{T}_a + \mathcal{T}_b + \mathcal{T}_c + \cdots$$
$$= m_1 r_1^2 \alpha + m_2 r_2^2 \alpha + m_3 r_3^2 \alpha + \cdots$$
$$= (m_1 r_1^2 + m_2 r_2^2 + m_3 r_3^2 + \cdots)\alpha.$$

The fact that each segment has the *same* angular acceleration follows from the rigid body assumption. (Can you explain?) Finally, we write this result in short-hand form

$$\mathcal{T}_{\text{ext}} = I\alpha \qquad (4.9)$$

where $\mathcal{T}_{\text{ext}}$ is the *algebraic* sum of all the external torques acting on the rigid body and

$$I = m_1 r_1^2 + m_2 r_2^2 + m_3 r_3^2 + \cdots \qquad (4.10)$$

is the rotational inertia of the rigid body about the axis AA'.

The analogy between Eq. 4.9 and Newton's second law, Eq. 3.2, $\mathbf{F} = m\mathbf{a}$ is very striking. The torque $\mathcal{T}$, like the force $\mathbf{F}$, is the motive agent; the angular acceleration α, like linear acceleration $\mathbf{a}$, represents the resulting change in motion; and the rotational inertia I, like the mass m, is the inertial factor impeding the change in motion.

Note that unlike the center of mass of a rigid body that only depends on the mass distribution; the rotational inertia depends *both* on the mass distribution and the location of the axis of rotation. Equation 4.10 shows an important characteristic of the rotational inertia of objects; mass distributed further from the axis contributes relatively more than mass situated closer to the axis. Thus, in contrasting a disc and a hoop of the same radius and mass, the hoop has the larger rotational inertia. The evaluation of the rotational inertia for objects generally requires using the methods of calculus. For our purposes we simply give the values for typical cases in Fig. 4.11.

Object	Moment of inertia
Single particle	MR^2 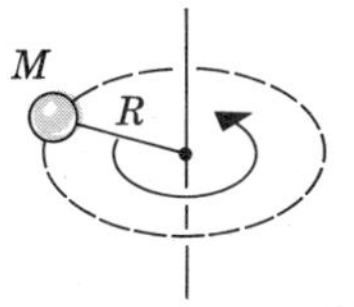
Thin hoop	MR^2 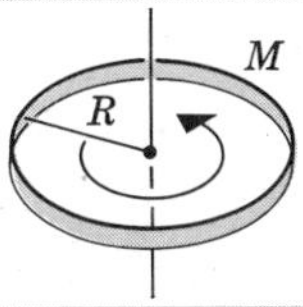
Disk (or cylinder)	$\frac{1}{2}MR^2$ 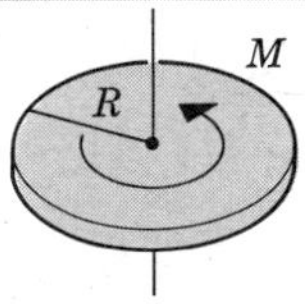
Sphere: (a) solid (b) hollow (thin shell)	$\frac{2}{5}MR^2$ $\frac{2}{3}MR^2$ 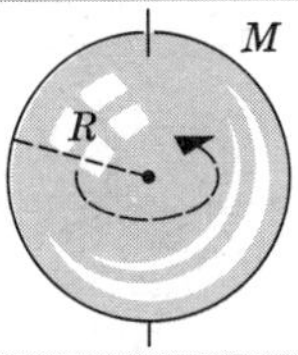
Thin rod (a) axis through center	$\frac{1}{12}ML^2$ 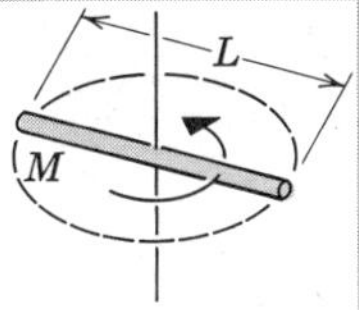
(b) axis through one end	$\frac{1}{3}ML^2$ 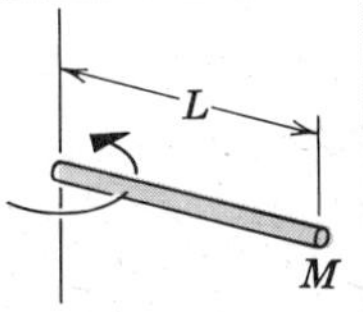

Figure 4.11 Rotational inertia for objects with various simple shapes.

Angular Momentum of Rigid Bodies

The rotational inertia is also useful in expressing the angular momentum of a rigid body. The angular momentum of a rigid body is defined to be the sum of the angular momenta of its composite pieces. Thus, using Eq. 4.3 and noting that for pure rotational motion v is perpendicular to r (i.e. $v = v_\perp$) we have for the sum of the mass fragments

$$L = m_1 r_1 v_1 + m_2 r_2 v_2 + m_3 r_3 v_3 + \cdots$$
$$= (m_1 r_1^2 + m_2 r_2^2 + m_3 r_3^2 + \cdots)\omega$$

or

$$L = I\omega \qquad (4.11)$$

In obtaining this result we have also repeatedly made use of the condition that $r_i \omega = v_i$ ($i = 1, 2, 3, \ldots$) is appropriate for rigid bodies.

Evidently, we also have

$$\mathcal{T}_{\text{ext}} = \frac{\Delta L}{\Delta t} = I\frac{\Delta \omega}{\Delta t} = I\alpha \qquad (4.12)$$

since I is a constant in time. This is just the result expressed by Eq. 4.9, as it should be. The conservation of angular momentum principle is also applicable to rigid bodies. In the absence of external torques or if the algebraic sum of the external torques acting on a rigid body is zero, then Eq. 4.12 implies that

$$\Delta L = 0 \qquad (\mathcal{T}_{\text{ext}} = 0) \qquad (4.13)$$

or that the angular momentum is a constant in time.

We may on occasion deal with an object that changes from one (prior) constant geometric shape to another new (afterwards) constant geometry because of the application of entirely *internal* forces. If I is the rotational inertia up to the time of the change and I_2 is the value after the change, we must have in the absence of external torques, a readjustment of the angular velocity from ω_1 to ω_2 such that

$$L_1 = I_1 \omega_1 = L_2 = I_2 \omega_2 \qquad (4.14)$$

Figure 4.12 shows a "rigid" body held in one fixed configuration changing to another configuration by using (internal) muscular forces.

Figure 4.12 A student is set to rotating on a stool while holding two massive dumbbells at arm's length. When he draws the dumbbells to his sides, r decreases so that the angular velocity ω must *increase* in order to conserve angular momentum (see Eq. 4.14).

●*Example 4.5*

A grinding wheel in the shape of a thick solid disc having a radius of 50 cm and a mass of 200 kg is used to sharpen a tool held against the wheel offering a constant tangential frictional force of 100 N. (See diagram.) Initially the grinding wheel was rotating at 100 rpm and the driving power was disconnected. What decelerating torque is applied by the tool friction? What is the resulting decelerating angular acceleration α? What is the angular velocity 2 seconds after the drive motor is disconnected?

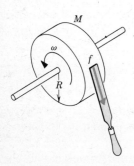

The decelerating torque produced by the (tangential) frictional force is

$$\mathcal{T}_{ext} = (0.5 \text{ m}) \times (100 \text{ N}) = 50 \text{ N-m}$$

The rotational inertia about the central axis is (see Fig. 4.11)

$$I = \tfrac{1}{2} MR^2 = \tfrac{1}{2} \times (200 \text{ kg})(0.5 \text{ m})^2$$
$$= 25 \text{ kg-m}^2$$

Thus,

$$\alpha = -\frac{\mathcal{T}_{ext}}{I} = -\frac{50 \text{ N-m}}{25 \text{ kg-m}^2} = -2 \text{ rad/s}^2$$

Now originally $\omega_0 = 100 \times 2\pi/60 = 10\pi/3$ rad/s. Also we have, using Eq. 2.40,

$$\omega = \omega_0 + \alpha t = \frac{10\pi}{3} - 2 \times 2 = 6.47 \text{ rad/s}$$

$$= 61.8 \text{ rpm}$$

●*Example 4.6*

A 3-kg mass hangs from a string wound on a 10-kg wheel in the shape of a solid disc with a radius of 10 cm. (See diagram.) Find the linear acceleration of the hanging mass and the angular acceleration of the wheel.

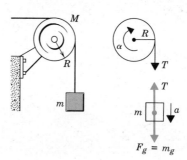

The diagram also shows the free-body diagrams for both the wheel and the hanging mass. Applying Newton's second law to the hanging mass, taking downward as positive, we have

$$m\mathbf{g} - T = m\mathbf{a} \quad (1)$$

Assuming frictionless bearings for the wheel the only external torque is produced by the string tension T giving $\mathcal{T}_{ext} = RT$ (positive clockwise). Consistent with the downward motion of the mass m, let us take α positive when clockwise. Using Eq. 4.9 and Fig. 4.11 we have, with $I = \tfrac{1}{2} MR^2$ for the wheel,

$$RT = \tfrac{1}{2} MR^2 \alpha \quad (2)$$

As the string unwinds and the mass m descends, an inelastic string requires that $R\Delta\theta = \Delta y$. From this it follows that $R\omega = v$ and $R\alpha = a$. (Prove these assertions.) Eliminating the string tension T between Eqs. 1 and 2 gives

$$mg = (m + M/2)a$$

or,

$$a = g/(1 + M/2m)$$

$$= \frac{9.80 \text{ m/s}^2}{1 + \frac{10 \text{ kg}}{2 \times 3 \text{ kg}}} = 3.68 \text{ m/s}^2$$

The angular acceleration of the wheel is therefore,

$$\alpha = a/R = \frac{3.68 \text{ m/s}^2}{0.10 \text{ m}} = 36.8 \text{ rad/s}^2$$

Summary of Important Ideas

Changes in the rotational motion of objects are produced by torques.

If no external torque is applied to a system, the angular momentum of the system around any point remains *constant* in time.

The center of mass of an object is the point at which all of the mass may be considered to be located for the purpose of making dynamical calculations.

The rotational inertia of an object depends on both its mass distribution and the location of the axis of rotation.

◆ Questions

4.1 *Torque* is actually a vector quantity. Use the way that the direction of the angular momentum vector was defined as a guide and devise a consistent definition of the direction of the torque vector.

4.2 A mass hanging from the end of a string is executing the simple back and forth oscillatory motion of a pendulum. What forces act on the mass and what torques do they produce about the point of suspension of the string as the mass swings through its lowest point? At the highest points reached by the mass? What is the angular momentum of the mass about the point of suspension of the string at these locations in its motion? Discuss the angular acceleration of the mass in terms of Eq. 4.7.

4.3 Is it always true that the center of mass of a body lies *within* the body? Where is the center of mass of a block letter C that is cut from a piece of wood?

4.4 Why do most flywheels have most of their mass concentrated around their rims?

4.5 What would be the effect on the Earth's rotation if the polar ice caps were to melt?

4.6 A wheel can turn freely in a horizontal plane on a vertical axle that is stuck in the ground. A cat is sitting on the rim of the wheel.
(a) The wheel is not in motion. Suddenly the cat begins to walk around the rim. What happens to the wheel and why?
(b) The wheel is spinning at a constant rate with the cat stationary at one point on the rim. Suddenly the cat begins to move toward the center of the wheel. What happens to the wheel and why?

4.7 Consider a stool that can move freely about a pivot at the center. A boy sits on the stool and holds a spinning top with its axis vertical; the stool is at rest. Suddenly, the boy grasps the top and stops the spinning. What is the result? Would it be possible to design a practical helicopter with only one set of rotating blades? Explain.

4.8 If released simultaneously at the top of an inclined plane which would reach the bottom first, a hollow cylinder or a solid cylinder?

4.9 Apparently contradicting the principle of the conservation of angular momentum all rotating objects slow down and eventually stop. Why? How does your explanation apply to the rotation of the Earth?

★ Problems

Section 4.1

4.1 A pipefitter's wrench is 50 cm long. If a worker can exert a maximum force of 450 N, what is the maximum torque he or she can apply with the wrench?

4.2 A 2-kg rock is dropped from a 1-m overhang on the roof of a building; see diagram. What is the torque exerted by the Earth's gravitational force on the rock relative to the point 0 at the base of the overhang? What is the angular momentum of the rock relative to the point 0 after 10 seconds?

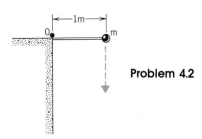

Problem 4.2

4.3 A 750-g stone is whirled in a horizontal circular path at the end of a string 80-cm long. The angular momentum of the stone about the fixed end of the string is 2.5 kg-m²/s. What is the angular velocity of the stone? If the string were to suddenly break, describe quantitatively the motion of the stone.

4.4 A 5-kg block moves uniformly in a horizontal circle at the end of a 2-m rope; the linear speed of the block is 4 m/s. At some point, a 1-kg piece of the block breaks off. Describe the subsequent motion of each part of the block.

4.5 The Earth executes an elliptic orbit about the Sun. The furthest distance from the Sun (at the aphelion) is 1.521×10^8 km, while the closest distance (at the perihelion) is 1.471×10^8 km. What is the ratio of the orbital velocities at these two points in the Earth's orbit?

4.6 At perigee an Earth orbiting satellite is 420 km above sea level and has an orbital speed of 8.39 km/s. What is the orbital speed at apogee 3820 km above sea level?

Section 4.2

4.7 Find the location of the center of mass for a uniform rod with a mass of 3 kg and length of 1.5 m with a small ball having a mass of 1 kg attached to one end.

4.8 When we say that the Earth revolves "around the Sun," we imply that the rotation takes place around the *center* of the Sun. Is this strictly true? Calculate the location of the point around which the rotation actually takes place. (The necessary data are listed in one of the tables inside the front cover.)

4.9 Four objects are located as follows: $m_1 = 1$ kg at $x = 0$, $y = 0$; $m_2 = 2$ kg at $x = 0$, $y = 6$ m; $m_3 = 6$ kg at $x = 4$ m, $y = 6$ m; and $m_4 = 3$ kg at $x = 4$ m, $y = 0$. Find the position of the center of mass of this system of objects. Proceed in the following way. Select one pair of objects, for example, m_1 and m_2. Calculate the position of the C.M. of this pair. Consider the total mass, $m_1 + m_2$, to be located at this position. Next, do the same with the remaining pair, m_3 and m_4. Now, the problem is reduced to one pair of objects and the C.M. of this pair (i.e., the C.M. of the *system*) is located in the usual way.

4.10* A uniform solid disc of radius 10 cm has a circular hole with a radius of 5 cm cut out of it to form a crescent. Referring to the diagram, locate the center of mass.

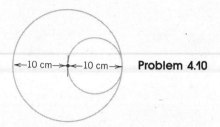

Problem 4.10

Section 4.3

4.11 A wheel with a radius of 15 cm and a rotational inertia of 0.085 kg-m² is rotating at 75 rpm. A brake applies a constant tangential frictional force of 10 N. Through how many rotations will the wheel turn before stopping?

4.12 A wheel with a radius of 15 cm is mounted on a horizontal frictionless axle. A mass of 100 g hangs from the free end of a string which is wound around the rim of the wheel. Ten seconds after the system is released the mass descends 1.5 m. What is the rotational inertia of the wheel about the axle?

4.13 A 150-cm length of a string is wrapped around the rim of a wheel free to rotate about a frictionless axle through its center. A constant force of 3 N is applied to the free end of the string unwinding it completely from the wheel. If the radius of the wheel is 10 cm and its rotational inertia is 0.015 kg-m², what is the angular acceleration of the wheel? What is the final angular velocity of the wheel? What is the linear velocity of the string as it leaves the wheel?

4.14 Two masses, $m_1 = 3$ kg and $m_2 = 2$ kg are connected by a massless string that passes

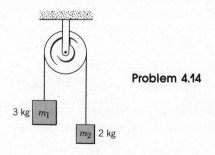

Problem 4.14

over a 500-g pulley as shown in the diagram. The pulley is in the form of a 10-cm diameter solid disc. Determine the acceleration of the masses and the angular acceleration of the pulley.

4.15 A spool in the form of a solid disc is attached to the ceiling by a string wrapped around its rim. Show that the spool descends with an acceleration of 2/3 g. (Hint: Apply Eq. 4.9 for rotation about the spool axis.)

4.16 A thin rod 1 m in length and mass 500 g rotates in a horizontal plane about a fixed axle through one end. A 200-g blob of putty is dropped onto its rotating end from directly above and sticks fast. What is the ratio of the original angular velocity to the new value?

4.17 Two 50-kg girls are standing on the opposite ends of a 6-m plank (which has negligible mass). The plank and the girls are set to spinning around an axis located at the midpoint of the plank; the uniform angular velocity is $\omega = 1.5$ rad/s. Suppose that one of the girls walks along the plank to a point halfway toward the axis of rotation. What is the new angular velocity of the system?

4.18 Rework Problem 4.17 taking the mass of the plank to be 30 kg. Refer to Fig. 4.11 and assume that the entry for the thin rod is applicable to the plank.

5
Forces in Equilibrium

In the two previous chapters we introduced and discussed Newton's laws of motion. There we emphasized the accelerations that are produced when *unbalanced* forces act on objects. In this chapter we consider the class of problems in which objects are subjected to *balanced* systems of forces. That is, we now specialize to the *equilibrium* situation in which the net force acting upon the object under study is *zero*. The absence of a net force means that there is no acceleration, that the velocity remains constant. Therefore, within the category of equilibrium situations we recognize the case in which the object is at *rest* (this is *static* equilibrium) as well as the case in which the object is in translational motion with *constant velocity* (this is *dynamic* equilibrium).

Large objects, in situations allowing the possibility of rotation for the objects, will be in *rotational equilibrium* if the net torque acting is zero.

Every object that we see around us has *some* force acting on it. Indeed, for an object to have *no* force acting on it, the object would have to be located in deep space where the gravitational forces due to all other objects and astronomical bodies are negligibly small. Most everyday objects are in equilibrium—they are either at rest or are moving uniformly (or approximately so). Several, perhaps many, forces act on these objects, but the *net* force in each case is *zero*. We can think of any number of such cases—for example, a sign held in place by beams and wires, or a cart or sled being pulled across the ground at constant speed, or a diver standing on the end of a diving board. In each of these situations we can analyze the forces that are acting by using our knowledge about the equilibrium conditions of the system.

5.1 Static Equilibrium of Translation

The Equilibrium Conditions

The basic requirement for an equilibrium situation to exist is that the net force on an object be zero:

$$\mathbf{F}_{net} = 0 \quad \text{(equilibrium)}$$

This simple equation means that if a number of forces, $\mathbf{F}_1$, $\mathbf{F}_2$, $\mathbf{F}_3$, and so on, act on the object, then the *vector sum* of all these forces is zero:

$$\mathbf{F}_1 + \mathbf{F}_2 + \mathbf{F}_3 + \ldots = 0$$

We usually abbreviate equations such as this by using the symbol Σ (Greek capital sigma), which means "sum all quantities of the type . . . . " Thus, we write

$$\Sigma \mathbf{F} = 0$$

This equilibrium equation has a further significance. It is a *vector* equation in three-dimensional space and therefore represents *three* separate equations, one for each space direction. That is,

$$\Sigma F_x = 0 \qquad \Sigma F_y = 0 \qquad \Sigma F_z = 0 \qquad (5.1)$$

If the net force $\mathbf{F}_{net}$ on an object is zero, then the sum of the force components in each of the three space directions is separately equal to zero. We can also have cases in which the sums of the forces in the x- and y-directions are zero, but the sum in the z-direction is not zero. Then, we will have equilibrium in the x- and y-directions, but we will have acceleration in the z-direction. (Can you think of a common example of this type of motion?) We will usually consider cases in which only two space directions are important, and we will represent these problems in terms of x- and y-coordinates.

There is one restriction we must place on the forces in an equilibrium situation. Consider the object illustrated in Fig. 5.1. Two forces act on this object. The magnitudes of these forces are equal and their directions are opposite so that the vector sum is zero. Nevertheless, the object is not in equilibrium because each force produces a torque around the center of mass of the object and the result is a rotation (see Section 4.1). (Such a pair of forces is called a *couple*.) If we are to ensure that any set of forces whose vector sum is zero actually produces an equilibrium situation, then we must restrict our attention to *particles*, objects that have no significant size. All forces that act on a particle necessarily are directed through the same point and no torque can be developed. Or, if we do consider an object with size, then we must specify that all forces are directed along the same line. In this way we again avoid the application of a torque to the object. (In Section 5.3 we treat cases in which we use the idea of torque to solve equilibrium problems.)

By considering only forces that act on the center

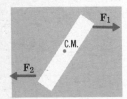

Figure 5.1 Two forces whose vector sum is zero, but that act on an object "off center," constitute a *couple*, and tend to cause a rotation. The forces illustrated here do not produce an equilibrium situation.

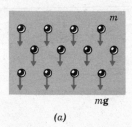
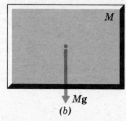

Figure 5.2 *(a)* The gravitational force mg acts on each of the particles of mass m that make up the block. *(b)* It is more convenient to analyze the forces acting on the block by representing the sum of all the individual forces mg as a single force Mg that acts on the center of mass of the block.

of mass of an object, we are doing nothing unusual or extraordinary. Indeed, it is natural to do this. For example, Fig. 3.4 shows the two gravitational force vectors acting on the centers of a pair of objects. The gravitational force on an object is actually the sum of the individual forces acting on each particle in the object (Fig. 5.2*a*). This sum is entirely equivalent to a single force acting on the center of mass of the object (Fig. 5.2*b*). It is always more convenient to use the single equivalent force than to deal with a large number of individual forces.

Graphical Analysis

In order to determine whether a system of forces is in equilibrium or not, we can perform a graphical summation of the individual force vectors. In Fig. 2.18 we showed how to make such a summation by placing the vector arrows foot-to-head. We can follow that procedure to sum the four force vectors acting on the object in Fig. 5.3*a*. We begin by drawing the vector $\mathbf{F}_1$ with its foot at the point A in Fig. 5.3*b* and with its magnitude and direction the same as in Fig. 5.3*a*. Next, we transfer another of the vectors (e.g., $\mathbf{F}_2$) in the same way, placing its foot at the head of $\mathbf{F}_1$. Then, we add $\mathbf{F}_3$ and $\mathbf{F}_4$ following the same procedure. We find that the head of the last vector added is at point B (Fig. 5.3*b*). Therefore, the sum of these four force vectors is equivalent to a vec-

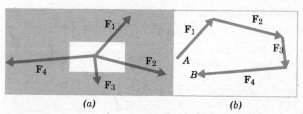

Figure 5.3 *(a)* Four force vectors act on an object. *(b)* Graphical summation of the four vectors shows that the sum is *not* zero and therefore that the object is *not* in equilibrium. The vector from point A to point B represents the net force acting on the object.

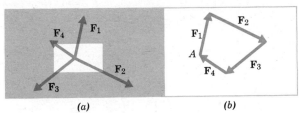

Figure 5.4 (a) Four force vectors act on an object. (b) Graphical summation of the four vectors shows that the sum is *zero* and therefore that the object is in equilibrium.

tor that extends from A to B. That is, the sum is *not* zero and the object is *not* in equilibrium.

Figure 5.4a shows another case of an object acted on by four forces. We proceed to add the individual force vectors in the same way, with the foot of F_1 placed at the point A (Fig. 5.4b). When all four vectors are properly added to the diagram, we find that the head of F_4 is exactly at the point A. Thus, the vector that represents the sum of the four vectors begins and ends at the point A—the sum vector is a *null* vector (that is, a vector with zero magnitude). The net force on the object is therefore zero, and the object is in equilibrium. Whenever the summation of the force vectors acting on an object produces a *closed* loop, as in Fig. 5.4b, the forces represent an equilibrium situation.

We can improve the accuracy of the graphical method of analysis by using trigonometric calculations. The following example illustrates this technique.

● **Example 5.1**

Suppose that an object is subjected to the forces shown in part (a) of the following diagram. What additional single force must be supplied to place the object in equilibrium?

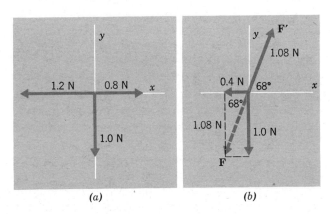

First, we note that the two forces acting in the x-direction can be combined and represented by a single force of 0.4 N acting to the left, as shown in

(b). Now, the vector sum of the 0.4-N and 1.0-N forces is **F**. We find the magnitude of **F** from

$$F = \sqrt{(0.4\ N)^2 + (1.0\ N)^2} = \sqrt{1.16}\ N = 1.08\ N$$

The angle that **F** makes with the negative x-axis is

$$\theta = \tan^{-1}\left(\frac{1.0\ N}{0.4\ N}\right) = \tan^{-1} 2.5 = 68°$$

Finally, the force **F**' that must be supplied to place the object in equilibrium is just the negative of **F**, that is, a force of 1.08 N acting above the x-axis at an angle of 68°, as shown in (b). ∎

Analysis by Vector Components

The trigonometric method used in Example 5.1 was easy to carry out because the forces acting on the object were directed along the coordinate axes. However, we know that *any* vector can always be resolved into components that are directed along the coordinate axes (see Section 2.7). Therefore, if we are confronted with a system of forces that act in various directions, the problem can always be attacked by the method of components. The procedure is first to resolve each force vector into its x- and y-components, then to apply the equilibrium conditions,

$$\Sigma F_x = 0 \quad \text{and} \quad \Sigma F_y = 0$$

The following examples show how problems of different types are solved by using vector components.

● **Example 5.2**

A ball with a mass $m = 5$ kg is suspended from a ceiling by means of a cord. When a horizontal force **F** is applied to the ball, as in part (a) of the following diagram, the ball hangs motionless with the cord making an angle of 35° with respect to the vertical. What is the magnitude of the force **F**, and what is the tension T in the cord?

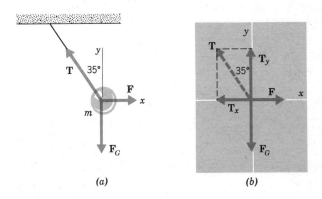

To solve this problem, we first express all three force vectors in terms of x- and y-components. The applied force **F** and the force of gravity $\mathbf{F}_G$ are already expressed in this form and it is necessary to decompose only the vector **T**. Part (b) of the diagram shows how this is done. Because the ball is in equilibrium we know that

$$\Sigma F_x = 0: \quad F - T_x = 0$$

(T_x enters the sum with a negative sign because $\mathbf{T}_x$ is directed to the left.) Thus,

$$F = T_x \tag{1}$$

Furthermore,

$$\Sigma F_y = 0: \quad T_y - F_G = 0$$
$$T_y - mg = 0$$

from which

$$T_y = mg \tag{2}$$

We also know that

$$T_x = T_y \tan 35° \tag{3}$$

and

$$T = \frac{T_y}{\cos 35°} \tag{4}$$

Using (1) and (2) in (3),

$$F = mg \tan 35° = (5 \text{ kg}) \times (9.8 \text{ m/s}^2) \times \tan 35°$$
$$= 34.4 \text{ N}$$

And using (2) in (4),

$$T = \frac{mg}{\cos 35°} = \frac{(5 \text{ kg}) \times (9.8 \text{ m/s}^2)}{\cos 35°}$$
$$= 59.8 \text{ N}$$

● *Example 5.3*

A block with a mass $m = 10$ kg rests on an inclined plane, as shown in part (a) of the following diagram. What mass M must be attached to the restraining cord in order to prevent the block from sliding down the (frictionless) plane?

When we resolve vectors into their components, our x-y coordinate system can have any orientation we choose. Naturally, we try to choose an orientation that will simplify the problem. In this case the vector **N**, representing the normal force of the plane on the block, and the vector **T**, representing the tension in the restraining cord, are perpendicular to one another. Therefore, if we choose our x- and y-axes to correspond with the directions of these two vectors, we will need to resolve only one vector, namely, $\mathbf{F}_G$,

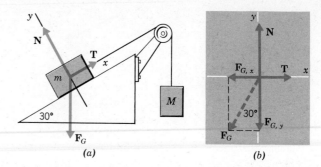

(a)   (b)

into components. The various vectors are shown in relation to this coordinate system in part (b) of the diagram.

To find the tension T, we need to solve only the x-component equation:

$$\Sigma F_x = 0: \quad T - F_{G,x} = 0$$

so that

$$T = F_{G,x} = F_G \sin 30° = mg \sin 30°$$
$$= (10 \text{ kg}) \times (9.8 \text{ m/s}^2) \times \sin 30°$$
$$= 49 \text{ N}$$

Because the system is in equilibrium, the tension in the cord is simply the gravitational force Mg acting on the mass M; then,

$$M = \frac{T}{g} = \frac{49 \text{ N}}{9.8 \text{ m/s}^2} = 5 \text{ kg}$$

● *Example 5.4*

A block with a mass $m = 10$ kg rests on an inclined plane, as shown in part (a) of the diagram at the top of the following page. What mass M must be attached to the restraining cord in order to prevent the block from sliding down the (frictionless) plane?

Although the wording of this problem is the same as that for Example 5.3, this case is somewhat more complicated because the restraining cord now makes an angle of 40° with respect to the vertical. This changes the problem in two ways. First, we now have only one vector (**N**) that is perpendicular or parallel to the plane so that there is no particular reason to choose the x-axis to be along the plane. Accordingly, we choose our axes to be vertical and horizontal, as shown in the diagram. The second change is that the upward force on the block now consists of two parts—one part due to the normal force **N** and one part due to the tension **T**. Resolving **N** and **T** along the coordinate axes, we obtain the components shown in part (b) of the diagram.

The equilibrium equations are

$$\Sigma F_x = 0: \quad T_x - N_x = 0$$

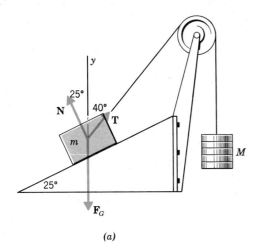
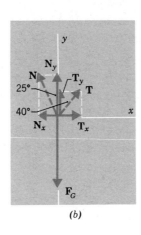

(a)   (b)

so that

$$T_x = N_x \quad (1)$$

and

$$\Sigma F_y = 0: \quad N_y + T_y - F_G = 0$$

from which (using $F_G = mg$)

$$T_y = mg - N_y \quad (2)$$

From the component equations, we have

$$T_y = \frac{T_x}{\tan 40°} \quad (3)$$

$$N_y = \frac{N_x}{\tan 25°} = \frac{T_x}{\tan 25°} \quad (4)$$

where we have used (1) to substitute for N_x in (4). Now, introducing (3) and (4) into (2), we can write

$$\frac{T_x}{\tan 40°} = mg - \frac{T_x}{\tan 25°}$$

Solving for T_x,

$$T_x = mg \left(\frac{1}{\tan 40°} + \frac{1}{\tan 25°} \right)^{-1}$$

$$T_x = (10 \text{ kg}) \times (9.8 \text{ m/s}^2) \times \left(\frac{1}{\tan 40°} + \frac{1}{\tan 25°} \right)^{-1}$$

$$= 29.4 \text{ N}$$

Next, from (3),

$$T_y = \frac{29.4 \text{ N}}{\tan 40°} = 35.0 \text{ N}$$

Then, the tension T is

$$T = \sqrt{T_x^2 + T_y^2} = \sqrt{(29.4 \text{ N})^2 + (35.0 \text{ N})^2}$$
$$= 45.7 \text{ N}$$

Finally, the mass M is

$$M = \frac{T}{g} = \frac{45.7 \text{ N}}{9.8 \text{ m/s}^2} = 4.66 \text{ kg} \quad \blacksquare$$

Traction Systems

In order to treat certain broken bones and other injuries, it is sometimes necessary to immobilize the affected regions and to remove the usual forces from the fracture while it heals. Various systems of traction using weights, ropes, and pulleys are found in current practice. In the design of all such systems, advantage is taken of the fact that the tension in a taut rope is everywhere the same. (In making this statement we assume that the mass of the rope is sufficiently small that the gravitational force acting on it is small compared to all other forces in the system; see, however, Problem 3.24). Thus, in Fig. 1 the tension T in the rope is the same in all three situations and is equal to Mg (assuming negligible friction in the pulleys). The tension in a rope is the result of the force that is exerted by some outside agency. According to Newton's third law, the rope therefore exerts a force equal to the tension on whatever is attached to this rope. Figure 1 shows that by using a pulley,

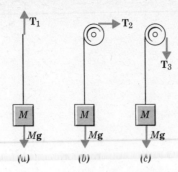

Figure 1 A pulley can be used to change the direction of the force that is exerted by a suspended weight without altering the magnitude: $T_1 = T_2 = T_3$. The tension in each rope is Mg.

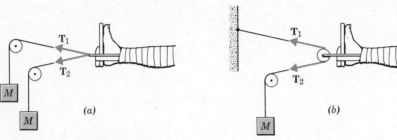

Figure 2 Two equivalent methods of exerting a force on a patient's leg. In each case, $T_1 = T_2$.

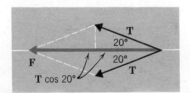

Figure 3 Net result of the traction system shown in Figure 2.

the force that is exerted on some other object can be changed in direction without affecting the magnitude.

Figure 2a shows a way to exert a certain force on a patient's leg by using two weight-and-pulley combinations. The two tensions have the same magnitude: $T_1 = T_2$. The same force can be produced by the simpler arrangement shown in Fig. 2b. In either case, the net force acting on the leg is equal to the vector sum of the two tensions. If the weight has a mass $M = 5$ kg, and if each rope makes an angle of 20° with respect to the horizontal, the force exerted on the patient's leg is represented by the vector **F** in Fig. 3. The magnitude of **F** is

$$F = 2\,T\cos 20° = 2\,Mg\cos 20°$$
$$= 2 \times (5\text{ kg}) \times (9.8\text{ m/s}^2) \times \cos 20° = 92\text{ N}$$

Figure 4a shows the more elaborate Russell traction system for immobilizing a fractured femur. This system extends that of Fig. 2b by adding two pulleys and a connection to the knee. If the thigh is to be maintained at an angle θ of approximately 20° with respect to the horizontal, the pulley positions and rope angles are adjusted until the vector sum of the three tensions is a force **F** that has the desired direction, as shown in Fig. 4b.

Reference

M. Williams and H. R. Lissner, *Biomechanics of Human Motion*, W. B. Saunders, Philadelphia, 1962, Chapters 6 and 7.

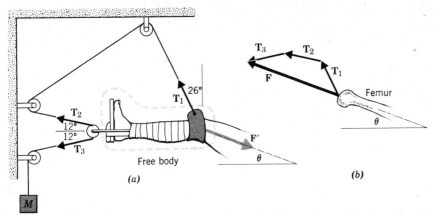

Figure 4 The Russell traction system used to immobilize a fractured femur. The system is adjusted so that the angle of elevation θ of the thigh is approximately 20°. (Adapted from Williams and Lissner.) The dashed curve isolates the lower leg as a free body. The force $\mathbf{F}'$ is the force exerted *by* the femur on the lower leg. The reaction force $\mathbf{F} = -\mathbf{F}'$ is exerted *on* the femur.

■ *Exercises*

1 In Fig. 2b each tension makes an angle of 20° with respect to the horizontal, the weight has a mass $M = 5$ kg, and the force exerted on the leg has a magnitude of 92 N and is directed horizontally. Now, suppose that the patient's leg is elevated until $\mathbf{T}_1$ is horizontal and $\mathbf{T}_2$ makes an angle of 36° with respect to the horizontal. What is the magnitude and direction of the force on the leg?

2 Find the magnitude and the direction of the force exerted on the femur by the system illustrated in Fig. 4. Let $M = 4$ kg.

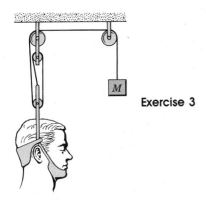

Exercise 3

3 Find the magnitude and direction of the force exerted on the patient's head by the cervical traction system illustrated above, if $M = 1.5$ kg.

5.2 Friction

In Section 3.5 we discussed the action of kinetic friction and postponed the discussion of static friction until this point. Although both forms of friction arise from the action of the intermolecular forces between the two surfaces in contact, the behavior is markedly different. When there is no relative motion between the surfaces, the intermolecular forces act in a manner similar to a spring in resisting the effect of any externally applied forces tending to produce motion. When the external forces become large enough, the static intermolecular bonds are ruptured and motion ensues. During such sliding motion, intermolecular bonds are in a continual state of

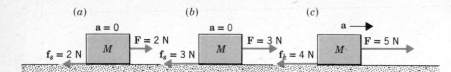

Figure 5.5 As the pulling force **F** is increased, (a) and (b), the frictional force also increases in order to maintain the block in static equilibrium. (c) Eventually, the applied force becomes sufficient to set the block into motion. The friction coefficient then appropriate is μ_k.

being formed and ruptured as the molecules of the two surfaces pass each other.

Static Friction

The kinetic frictional force that exists between a pair of surfaces is given by $f_k = \mu_k N$. That is, f_k always has a definite value for a particular pair of surfaces. This is not true, however, for *static* situations. For example, suppose that you pull horizontally on a block at rest with a force of 2 N and that the block does not move (Fig. 5.5a). Because the block is in equilibrium, there must be another 2-N force acting in the opposite direction—this is the static frictional force. Now, suppose that you increase the pulling force to 3 N and that the block still does not move (Fig. 5.5b). The equilibrium condition requires that the frictional force also increase to 3 N. As long as the pulling force is insufficient to set the block into motion, the force of static friction is always exactly equal to the applied force.

Because the frictional force varies with the applied force in static situations, it makes sense to talk about the coefficient of static friction only for the *maximum* force that still allows static equilibrium. That is, for the coefficient of static friction, μ_s, we use the value appropriate for the frictional force that exists just before the block breaks loose from its static condition and begins to move. Then, we define the coefficient in the usual way:

$$f_s = \mu_s N \qquad (5.2)$$

Some values of μ_s are shown along with those of μ_k in Table 5.1. The values of μ_k are repeated from Table 3.1 to provide a convenient comparison. Notice that μ_s for a particular pair of surfaces is always greater than μ_k. (For teflon, this difference appears in the next decimal of the values.) Can you see why this must be so?

The value of μ_s for a pair of surfaces can be determined in a simple way by using gravity as the pulling force. Suppose that we place a block on a plank and slowly raise one end of the plank. At a certain angle of inclination, the block will begin to slide down the plank. Figure 5.6 represents the situation immediately *before* the sliding motion begins.

Table 5.1 Coefficients of Kinetic and Static Friction

Materials	μ_k	μ_s
Steel on steel, dry	0.4	0.8
Steel on steel, lubricated	0.08	0.15
Steel on ice	0.06	0.10
Copper on cast iron	0.3	1.0
Oak on oak (grains parallel)	0.5	0.6
Rubber on concrete, dry	0.7	1.0
Rubber on concrete, damp	0.5	0.7
Teflon on teflon	0.04	0.04

That is, θ is the maximum angle of inclination for which the static condition exists. From the force diagram (Fig. 5.6b), we see that $f_s = mg \sin \theta$ and that $N = mg \cos \theta$. The coefficient of static friction is given by the ratio of these quantities:

$$\mu_s = \frac{f_s}{N} = \frac{mg \sin \theta}{mg \cos \theta} = \frac{\sin \theta}{\cos \theta} = \tan \theta \qquad (5.3)$$

Thus, if a block remains at rest on a plank until the inclination reaches 33°, the value of μ_s for this pair of surfaces is $\mu_s = \tan 33° = 0.65$.

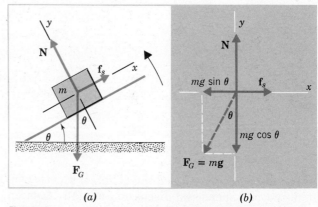

Figure 5.6 As the angle θ is increased, the block will eventually begin to slide. The coefficient of static friction for the pair of surfaces is given by the tangent of this maximum angle of inclination for the static condition.

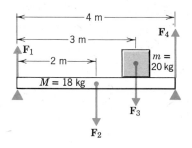

Figure 5.7 The method of torques must be used to determine the forces F_1 and F_4.

5.3 Static Equilibrium of Rotation

A New Equilibrium Condition

In the types of problems we have been discussing, the force vectors applied to an object were all considered to pass through the center of mass of the object. We restricted our attention to such forces, it will be recalled, so that there would be no tendency for the objects to rotate. But now we come to cases that involve "off-center" forces and the resulting torques. Figure 5.7 shows a plank that is subject to several forces and is in a condition of static equilibrium. Notice that only one of the forces—the gravitational force on the plank, F_2—acts on the center of mass of the plank. How do we analyze the forces in such a case? In particular, if the masses, M and m, are given, how do we find the forces, F_1 and F_4, that are exerted on the plank by the supports? First, we can use the equilibrium conditions for the x- and y-components of the forces. But here there are no forces acting in the x-direction, so we have only

$$\Sigma F_y = 0: \quad F_1 + F_4 - F_2 - F_3 = 0$$

or,

$$\begin{aligned} F_1 + F_4 &= (M + m)g \\ &= (18 \text{ kg} + 20 \text{ kg}) \times (9.8 \text{ m/s}^2) = 372.4 \text{ N} \end{aligned}$$

Thus, the usual equation enables us to determine the *sum* of the upward forces, $F_1 + F_4$, but not the individual values. We need some new condition to apply.

Look again at the plank in Fig. 5. Each of the forces, F_1, F_3, and F_4, produces a torque around the center of mass of the plank: F_1 and F_3 produce clockwise torques, and F_4 produces a counterclockwise torque. If an object is subject to a net torque, it will rotate. The plank here is in a condition of static equilibrium so we must conclude that the sum of the clockwise torques due to F_1 and F_3 exactly equals the counterclockwise torque due to F_4. We can summarize this statement by writing another equilibrium equation.[1]

$$\Sigma \mathcal{T} = 0 \qquad (5.4)$$

where we assign a positive value to each torque that tends to rotate the object in one direction and a negative value to each torque that tends to produce a rotation in the other direction.

We now have two conditions to apply in solving all types of equilibrium problems:

1 The net force is zero.

2 The net torque is zero.

These conditions are expressed by the equations,

$$\begin{aligned} \Sigma F_x &= 0 \\ \Sigma F_y &= 0 \\ \Sigma \mathcal{T} &= 0 \end{aligned} \qquad (5.5)$$

Let us return now to the case illustrated in Fig. 5.7 and use the torque equation to complete the solution. Recall that the torque is equal to the product of the force and the perpendicular distance to the pivot point: $\mathcal{T} = rF$. In this problem we choose the pivot point to be the center of mass of the plank, and we designate as *positive* those torques that tend to rotate the plank in the clockwise sense. Then, $F_3 = mg = 196$ N, and we can write

$$(2 \text{ m}) \times F_1 + (1 \text{ m}) \times F_3 - (2 \text{ m}) \times F_4 = 0$$

so that,

$$F_1 = F_4 - 98 \text{ N}$$

Substituting this result into our previous equation for $F_1 + F_4$, we find

$$(F_4 - 98 \text{ N}) + F_4 = 372.4 \text{ N}$$
$$F_4 = 235.2 \text{ N}$$

Finally,

$$F_1 = 372.4 \text{ N} - F_4 = 137.2 \text{ N}$$

---

[1] This is not a new dynamical postulate. The fact that the net torque is zero in an equilibrium situation can be obtained from Newton's laws. If we analyze the internal forces in an object in equilibrium, realizing that the net force on every particle must be zero, then we arrive at the same result as using Eq. 5.4. In most cases it is simpler to use the torque equation.

In solving this problem, we calculated the torques around a pivot point that coincided with the center of mass of the plank. How did we know to choose this particular point? Actually, it does not matter what point we choose! The plank is in static equilibrium and so there is zero rotation (and, therefore, zero net torque) around any point we may choose. For example, suppose that we choose a pivot point 1 m to the left of the left end of the plank in Fig. 5.7. Then, the torque equation becomes

$$(3 \text{ m}) \times F_2 + (4 \text{ m}) \times F_3 - (1 \text{ m}) \times F_1 - (5 \text{ m}) \times F_4 = 0$$

Solving for F_1, and using $F_2 = Mg = 176.4$ N, we have

$$F_1 = (3 \text{ m}) \times (176.4 \text{ N}) + (4 \text{ m}) \times (196 \text{ N}) - (5 \text{ m}) \times F_4$$
$$= 1313.2 \text{ N} - (5 \text{ m}) \times F_4$$

Substituting this expression for F_1 into the previous equation for $F_1 + F_4$, we find

$$(1313.2 \text{ N} - 5F_4) + F_4 = 372.4 \text{ N}$$

$$F_4 = \frac{940.8}{4} = 235.2 \text{ N}$$

which is the same as the result we obtained earlier.

Although it is possible to choose the pivot point anywhere, one should always make a choice that will simplify the problem. Usually, the pivot point should correspond to the point of application of one of the forces. In this way, one of the torques is eliminated from the calculation.

● *Example 5.5*

What horizontal force **F** is required to pull the wheel ($M = 20$ kg) over the curb?

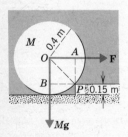

Although we are asked to find the force necessary to *move* the wheel over the curb, we can, in fact, treat this problem as if it were a problem in static equilibrium. Look at the diagram. As the force **F** is increased, the downward push of the wheel on the ground is diminished, as is the reaction force of the ground on the wheel. The force **F** will eventually reach a value that just barely removes the wheel from contact with the ground. In this condition the wheel is in equilibrium, but any tiny increase in **F** will cause the wheel to move over the curb. This is the value of **F** that we want. In this situation, there are three forces acting on the wheel: (1) the known downward force due to gravity, (2) the unknown applied horizontal force, and (3) the unknown force exerted by the top of the curb at P. We are required only to find F. Therefore, we need only one equation. Of the three equilibrium equations that we can use, the only one that involves F but not the force at P is the torque equation with P as the pivot point. From the diagram, we see that the force F acts through the perpendicular distance $\overline{AP} = 0.4$ m $- 0.15$ m $= 0.25$ m. The gravitational force acts through the perpendicular distance $\overline{BP}$:

$$\overline{BP} = \sqrt{(0.4 \text{ m})^2 - (0.25 \text{ m})^2} = 0.31 \text{ m}$$

Then,

$$\Sigma \mathcal{T} = 0: \quad (0.25 \text{ m}) \times F - (0.31 \text{ m}) \times Mg = 0$$

from which

$$F = \frac{(0.31 \text{ m}) \times (20 \text{ kg}) \times (9.8 \text{ m/s}^2)}{0.25 \text{ m}}$$
$$= 243 \text{ N}$$

● *Example 5.6*

Two boys ($m_1 = 60$ kg and $m_2 = 50$ kg) stand on a diving board in the positions shown in the following diagram. Find the forces, $\mathbf{F}_1$ and $\mathbf{F}_2$, that the support posts exert on the board.

There are no forces acting in the x-direction in this problem: consequently, we have only two of the three equilibrium equations to use. But there are only two unknowns, $\mathbf{F}_1$ and $\mathbf{F}_2$, so two equations are sufficient.

$$\Sigma F_y = 0: \quad F_1 - F_2 - Mg - m_1 g - m_2 g = 0$$

from which

$$F_1 = F_2 + (20 \text{ kg} + 60 \text{ kg} + 50 \text{ kg})g$$
$$= F_2 + 1274 \text{ N}$$

For the torque equation, we choose the center of mass of the board as the pivot point, and we let clockwise torques be positive:

$$\Sigma \mathcal{T} = 0: \quad (2 \text{ m}) \times F_1 + (0.4 \text{ m}) \times m_1 g + (2 \text{ m}) \times m_2 g - (0.6 \text{ m}) \times F_2 = 0$$

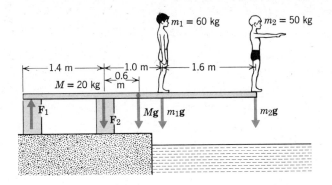
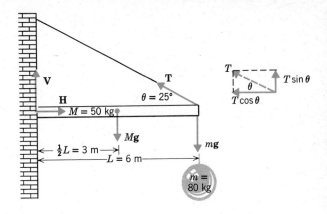

Then, using $m_1g = 588$ N and $m_2g = 490$ N,

$2F_1 + 235.2$ N $+ 980$ N $- 0.6F_2 = 0$

Substituting for F_1 from the force equation,

$2(F_2 + 1274$ N$) + 1215.2 - 0.6F_2 = 0$

$F_2 = -2688$ N

Then,

$F_1 = -2688$ N $+ 1274$ N $= -1414$ N

Notice that both F_1 and F_2 are *negative*. The reason is simply that in drawing the arrows to represent $\mathbf{F}_1$ and $\mathbf{F}_2$ in the diagram, the directions were reversed—$\mathbf{F}_1$ should point downward and $\mathbf{F}_2$ should point upward. But this error is automatically corrected in the calculation. If a direction for a force is ever chosen incorrectly in setting up a problem, the calculation will always produce the value for that force with a negative sign, thereby indicating that the force actually has the opposite direction.

● *Example 5.7*

A horizontal boom has one end attached to a wall and from the other end supports an 80-kg mass. Find the tension T in the guy wire and the vertical and horizontal components, V and H, of the force that the wall exerts on the boom.

There are three unknown quantities, T, V, and H, so the three equilibrium conditions are sufficient to yield a complete solution. We have

$\Sigma F_x = 0$: $H - T \cos \theta = 0$ (1)

$\Sigma F_y = 0$: $V + T \sin \theta - Mg - mg = 0$ (2)

Choosing the pivot point at the left end of the boom, we have

$\Sigma \mathcal{T} = 0$: $(\tfrac{1}{2}L) \times Mg + L \times mg$
$\qquad\qquad\qquad - L \times (T \sin \theta) = 0$ (3)

The torque equation (because of the judicious choice of pivot point) involves only one of the unknowns. Solving this equation for T:

$T = \dfrac{1}{L \sin \theta}(\tfrac{1}{2}LMg + Lmg) = \dfrac{g}{\sin \theta}(\tfrac{1}{2}M + m)$

$T = \dfrac{9.8 \text{ m/s}^2}{\sin 25°}(25 \text{ kg} + 80 \text{ kg})$

$= 2435$ N

Then, from (1),

$H = T \cos \theta = (2435$ N$) \times \cos 25°$

$= 2207$ N

Finally, from (3),

$V = Mg + mg - T \sin \theta$

$= (50 \text{ kg}) \times (9.8 \text{ m/s}^2) + (80 \text{ kg}) \times (9.8 \text{ m/s}^2)$
$\qquad\qquad\qquad - (2435 \text{ N}) \times \sin 25°$

$= 245$ N ∎

Forces in Muscles and Bones

We can use the ideas concerning the equilibrium of forces and torques to find the forces exerted by muscles and the forces exerted on bones in the body. Figure 1a shows the bones and muscles of the arm that are involved in holding an object in the hand. Figure 1b shows the various forces and distances that are important:

F_1 = force exerted by the humerus on the elbow joint
F_2 = force exerted by the biceps on the radius

Forces in Equilibrium

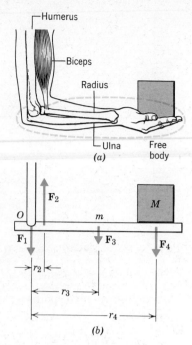

Figure 1 Forces involved in holding a block in the hand.

F_3 = gravitational force on the arm-hand combination

F_4 = weight of the block M

r_2 = distance from the elbow joint (point O) to the point of attachment of the biceps (typically, 4 cm)

r_3 = distance from O to the center of mass of the arm-hand combination

r_4 = distance from O to the center of mass of the block (assumed to coincide with the center of mass of the hand)

If we assume that the subject holding the block has a height of 1.83 m (6 ft) and a mass of 80 kg (176 lb), then we can use the information in the essay on page XX to compute the values of r_3, r_4, and F_3. We find

$$r_4 = (0.6220 - 0.4313) \times 1.83 \text{ m} = 0.35 \text{ m}$$

The value of r_3 is found by combining the data for the lower arm and the hand (verify this value):

$$r_3 = 0.19 \text{ m}$$

The value of F_3 is the weight of the arm-hand combination:

$$F_3 = mg = [\tfrac{1}{2}(0.042 + 0.017) \times 80 \text{ kg}] \times g = 23 \text{ N}$$

(The factor $\tfrac{1}{2}$ is necessary because the tabulated data refer to the masses of *both* arms and *both* hands.) Also, if the block has a mass of 10 kg,

$$F_4 = Mg = 98 \text{ N}$$

Now, we can evaluate F_1 and F_2 by solving the force and torque equations:

$\Sigma F_y = 0$: $\quad F_2 - (F_1 + F_3 + F_4) = 0$

$F_2 = F_1 + 23 \text{ N} + 98 \text{ N}$

$ = F_1 + 121 \text{ N}$

Using the point O as the pivot point, and letting clockwise torques be positive,

$\Sigma \mathcal{T} = 0$: $r_3 F_3 + r_4 F_4 - r_2 F_2 = 0$

$r_2 F_2 = r_3 F_3 + r_4 F_4$

$(0.04 \text{ m}) \times F_2 = (0.19 \text{ m}) \times (23 \text{ N}) + (0.35 \text{ m}) \times (98 \text{ N})$

$= 4.4 \text{ N-m} + 34.3 \text{ N-m}$

$= 38.7 \text{ N-m}$

From which

$$F_2 = \frac{38.7 \text{ N-m}}{0.04 \text{ m}} = 968 \text{ N}$$

Then,

$F_1 = F_2 - 121 \text{ N} = 847 \text{ N}$

These forces are considerably greater than the weight of the block ($w = F_4 = 98$ N):

$F_1 = 8.6w$
$F_1 = 9.9w$

Next, let us examine the forces in the foot that are involved in standing tiptoe. Suppose that the subject has a mass of 80 kg and stands on one foot with the heel barely raised from the floor. Figure 2a shows the various bones and muscles, and Fig. 2b shows the important forces and distances. For simplicity, we assume that the force $\mathbf{F}_3$ acts vertically although it actually makes a small angle with the vertical.

F_1 = normal force of the floor on the metatarsal (equal to the weight of the subject)
F_2 = force exerted by the tibia on the upper surface of the talus
F_3 = force exerted by the Achilles tendon on the calcaneus (or heel)
r_1 = distance from the ankle joint (point O) to the point of contact between the metatarsal and the floor (typically, 12 cm)
r_3 = distance from O to the point of attachment of the Achilles tendon (typically, 6 cm)

Using the force and torque equations (for the pivot at O), we find

$F_2 = F_1 + F_3$
$= (80 \text{ kg}) \times (9.8 \text{ m/s}^2) + F_3$
$= 784 \text{ N} + F_3$
$r_3 F_3 = r_1 F_1$
$(0.06 \text{ m}) \times F_3 = (0.12 \text{ m}) \times (784 \text{ N})$
$= 94 \text{ N-m}$

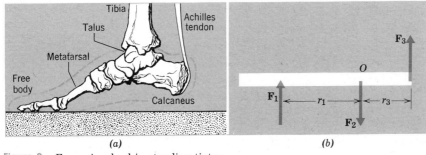

Figure 2   Forces involved in standing tiptoe.

From which

$$F_3 = \frac{94 \text{ N-m}}{0.06 \text{ m}} = 1567 \text{ N}$$

Then,

$$F_2 = 784 \text{ N} + 1567 \text{ N} = 2351 \text{ N}$$

These forces are considerably greater than the weight of the subject ($w = F_1 = 784$ N):

$F_2 = 3.0w$
$F_3 = 2.0w$

It is easy to see why standing tiptoe (or holding a heavy block) is tiring!

Reference
M. Williams and H. R. Lissner, *Biomechanics of Human Motion,* W. B. Saunders, Philadelphia, 1962, Chapters 5 and 7.

■ Exercises

1. The diagram below shows a sling attached to a subject's wrist. The subject pulls downward until the spring scale reads a force of 150 N. What force does the triceps exert on the ulna (F_1)? What force does the humerus exert on the ulna at the elbow joint (F_2)? The distance r_1 is typically 2.5 cm. Notice that F_3 and r_3 are the same as in the example above. The distance from the elbow to the wrist (where the sling is attached) can be found by using the data in the essay on page 85. (Again, assume that the height of the person is 1.83 m.)

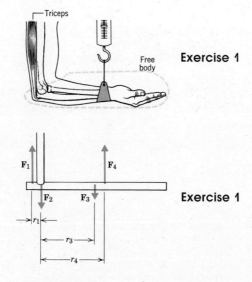

Exercise 1

Exercise 1

2. In Fig. 1a notice that the point of attachment of the biceps to the radius lies about 2 cm above a horizontal line through the elbow joint. In drawing the force diagram (Fig. 1b), this displacement does not matter because we are concerned with the perpendicular distance from $\mathbf{F}_2$ to O. But if the arm is

raised (or lowered) away from the horizontal, the displacement of the attachment point is important. Rework the example for the cases in which the arm is 40° above the horizontal and 40° below the horizontal.

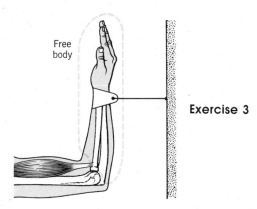

Exercise 3

3. The diagram above shows a subject's arm held vertical and pulling against a fixed support. If the tension in the string is 400 N, find the force exerted by the biceps on the radius and the force exerted by the humerus on the ulna. As in Exercise 1, assume that the sling is positioned on the wrist and that the subject has the same height and mass.

4. The following diagram shows in a schematic way the forces that are involved in holding a mass m in the hand when the arm is outstretched. The force $\mathbf{F}_1$ is that exerted by the deltoid muscle on the upper arm. Assume a "standard" man (1.83 m, 80 kg) and calculate the force exerted by the deltoid muscle when the mass m is 10 kg. The distance r_1 can be assumed to be 0.2 r_3. (M, r_2, and r_3) can be computed using the data in the essay on page 85.)

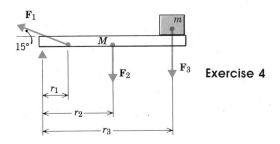

Exercise 4

Summary of Important Ideas

An object at rest is said to be in a condition of *static equilibrium;* an object in motion with constant velocity is said to be in a condition of *dynamic equilibrium*.

For an object to be in an equilibrium condition, there must be no net force and no net torque acting on the body. These two equilibrium conditions are expressed by the equations, $\Sigma \mathbf{F} = 0$ and $\Sigma \mathcal{T} = 0$. The force equation can be written in terms of components as (for two dimensions) $\Sigma F_x = 0$ and $\Sigma F_y = 0$.

The frictional force that exists between an object and a surface is directly proportional to the normal force that the surface exerts on the object. We identify two types of friction—*static* friction and *kinetic* (or *sliding*) friction. For a particular pair of surfaces, the force of static friction is always greater than that of kinetic friction.

◆ Questions

5.1 If a ball is thrown against a wall, the ball is momentarily at rest before it rebounds. Is the

108 Forces in Equilibrium

ball in a condition of static equilibrium during this instant? Explain.

5.2 A condition of *stable* equilibrium is one in which an object, if given a small displacement, will return to its original position. If an object is in a condition of *unstable* equilibrium, however, it will continue to move in the direction of any small displacement. Give some examples of both types of equilibrium situations.

5.3 An object in equilibrium is acted on by three forces, all of which have the same magnitude. What is the orientation of the forces? Is there more than one possible orientation?

5.4 Is there really any difference between *static* equilibrium and *dynamic* equilibrium? Include in your answer the idea of inertial reference frames.

5.5 Why do lubricants decrease friction?

5.6 Friction is often a nuisance and we take care to reduce it to a minimum. On the other hand, friction is often very useful. Give some examples of situations in which this is the case.

5.7 An automobile is moving down a gently sloping road that is covered uniformly with ice. The driver applies the brakes in order to slow the automobile. Is the automobile more likely to enter a skid if the front wheels lock (i.e., cease to rotate) or if the rear wheels lock? Explain. (Hints: Which pair of wheels experiences sliding friction and which pair experiences static friction? Which coefficient of friction is always greater for two particular surfaces?)

5.8 A cylinder has a height that is twice its diameter and rests on its base. If the cylinder is slowly tipped to one side, at what point will it topple over? Use a sketch to explain what happens. What is the angle at which toppling occurs?

★Problems

Section 5.1

5.1 Consider the following forces, all of which act in the x-y plane on an object:

$\mathbf{F}_1 = 4$ N, 30° above $+x$-axis

$\mathbf{F}_2 = 6$ N, 75° above $-x$-axis

$\mathbf{F}_3 = 3$ N, 60° below $-x$-axis

$\mathbf{F}_4 = ?$

Find $\mathbf{F}_4$ in the event that the object is in equilibrium.

5.2 A 15-kg crate is suspended by two ropes, each of which makes an angle of 35° with respect to the vertical. What is the tension in each rope?

5.3 In Example 5.3, find the normal force N.

5.4 In Example 5.4, find the normal force N.

5.5 If $T_2 = 2T_1$, find the angle θ and the tension in each rope.

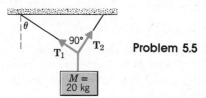

Problem 5.5

5.6 Bows that are used for archery or hunting are usually rated according to the force that is necessary to produce full extension. The British system of units is used in rating bows, and we commonly find 30-lb, 40-lb, and 50-lb bows. (Actually, these should be 648-N, 863-N, and 1079-N bows.) What is the tension T in the bowstring when a 40-lb bow is at full extension with the bowstring at an angle of 65° with respect to the arrow? Draw the force diagram expressing all forces in newtons.

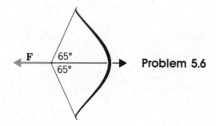

Problem 5.6

5.7 Find the tension in each of the cords that support the block.

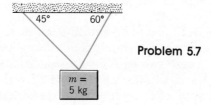

Problem 5.7

5.8 What is the largest mass boulder M that can be hung from the rope shown in the diagram before the rope breaks? Assume the maximum tension in the rope is 600 N and that relevant angles are as shown.

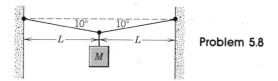

Problem 5.8

5.9 If the tension in the horizontal rope is 500 N, what is the magnitude of the mass M?

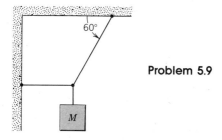

Problem 5.9

5.10 Find the mass m that is necessary to produce equilibrium in the system and compute the tension in each of the support cables.

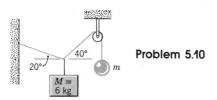

Problem 5.10

Section 5.2

5.11 A 6-kg block is pulled at constant velocity up a 30° plane by a force of 45 N that is directed parallel to the plane. What is the coefficient of kinetic friction between the block and the plane?

5.12 What force is necessary to pull a 10-kg block up a 20° plane at constant velocity if $\mu_k = 0.2$ for the block-plane combination?

5.13 A 5-kg block is on a horizontal surface. The coefficient of static friction between the block and the surface is 0.6. A force of 20 N is applied to the block and acts at an angle of 30° above the horizontal. Is the block in equilibrium?

5.14 A 3-kg block shown in Fig. 5.6 just begins to slide when the angle θ is increased to 28°. What is the coefficient of static friction? What is the value of the normal force N?

5.15 A block will remain at rest on a particular plank when the plank is inclined at an angle of 34° with respect to the horizontal, but the block will slide when the angle is 36°. How closely can the coefficient of static friction for these materials be determined from the information given?

5.16 A force **F** acts parallel to an inclined plane and on a block with a mass m. The angle of inclination θ of the plane is adjusted until the block moves up the plane with constant velocity. Express the coefficient of friction μ_k in terms of the angle θ.

5.17* What force **F** is necessary to pull the system in the diagram at constant velocity?

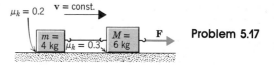

Problem 5.17

5.18* What force **F** will maintain the system moving with constant velocity? (See the diagram.)

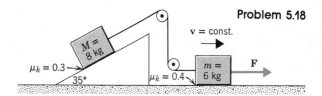

Problem 5.18

5.19* A man pulls a sled out from under a box that is held stationary by a child pulling on a rope attached to the box, as shown in the diagram. (The sled moves with constant velocity.) What pulling force **F** is necessary? If the child has a mass of 40 kg, what is the minimum value of the coefficient of static friction between the child's shoes and the snowy surface that will allow the child to maintain the box stationary as the sled is pulled?

110 Forces in Equilibrium

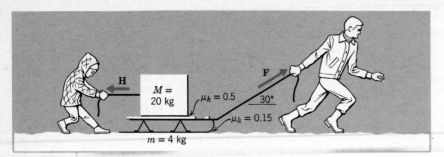

Problem 5.19

Section 5.3

5.20 What force **F** is necessary to maintain the plank ($M = 12$ kg) in equilibrium? What are the components of the force exerted on the plank by the pivot?

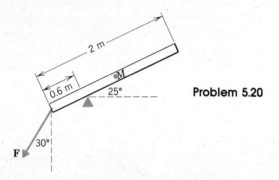

Problem 5.20

5.21 What is the magnitude of the force **F** that supports the slab? What are the components of the force that the pivot P exerts on the slab?

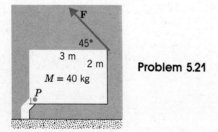

Problem 5.21

5.22 What force **F** must be applied to the disk in order to maintain equilibrium? What are the components of the force exerted on the disk by the pivot P?

5.23 The boom in the diagram had negligible mass and pivots freely about its lower end. Find the tension in the horizontal support rope and the components of the force exerted on the lower end of the boom by the wall and floor.

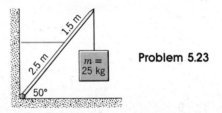

Problem 5.23

5.24 If a force $F = 150$ N is applied to the handle of a hammer as shown in the diagram, what force is applied to the nail by the claw of the hammer? (Assume a static situation.)

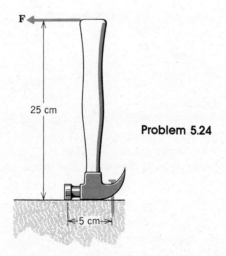

Problem 5.24

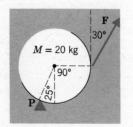

Problem 5.22

5.25 In the diagram, the man holds the safe of mass M in static equilibrium. ($M = 30$ kg, $m = 10$ kg.) What is the tension in each of the cables?

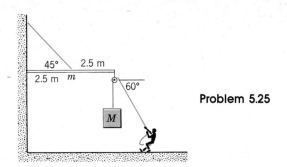

Problem 5.25

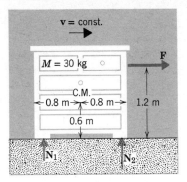

Problem 5.28

5.26 A sign has dimensions 2.4 m × 1.8 m and is suspended by a pivot pin at one corner and by a guy wire, as shown in the diagram. Find the tension in the wire and the force exerted on the pin.

5.29* Refer to Problem 5.28. How large can the coefficient of kinetic friction μ_k become before the chest will actually tip forward?

5.30 Two planks, each with a length of 2.40 m, are joined end-to-end by a hinge of point A, as shown in the diagram. The planks stand on a frictionless horizontal surface, forming an isosceles triangle. A rope with a length of 1.20 m is attached to the midpoints of the two planks. Each plank has a mass of 10 kg. Find the tension in the rope. Find the horizontal and vertical hinge forces.

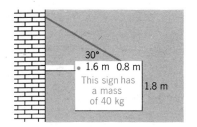

Problem 5.26

5.27 A stump with a mass $m = 5$ kg is attached to a plank whose mass is $M = 3$ kg. A force **F** is applied to one end of the plank, as shown in the diagram, and the plank is elevated to an angle of 35°. What is the force **F**, and what are the components of the force that the ground exerts on the plank at O?

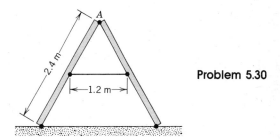

Problem 5.30

5.31* A uniform ladder with a mass m rests against a frictionless vertical wall at an angle of 45°. The lower end rests on a horizontal surface with $\mu_s = \frac{2}{3}$. A student with a mass $M = 2\,m$ attempts to climb the ladder. How far up the ladder, x/L, will he reach when the ladder begins to slip?

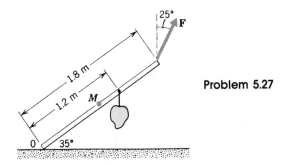

Problem 5.27

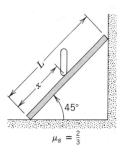

Problem 5.31

5.28* What force **F** is necessary to pull the chest across the floor at constant velocity if the coefficient of kinetic friction between the legs of the chest and the floor is 0.3? Find the normal forces, N_1 and N_2, that are exerted on the legs. (There are *four* legs!)

Gravitation

In our everyday experience we encounter a variety of forces—the muscular force exerted to open a door, the frictional force in the door hinges, and the elastic force in the door spring; the force that the atmosphere exerts on a barometer and the force that the Earth exerts on the Moon; the electrical force that starts an automobile engine, the hydraulic force that operates the brakes, or the mechanical force that stops the car if we are unfortunate enough to collide with a lamp post. In spite of the large number of names that we have given to forces that we use or must overcome, there are only *two* basic forces that govern the behavior of all everyday objects—the *gravitational* force and the *electric* force. All of the various forces mentioned above are actually only different manifestations of these two fundamental forces.

In this chapter we examine in detail the gravitational force. It is the gravitational force that holds the planets in orbit around the Sun and our Moon in orbit around the Earth. Evidently, the gravitational force can act between objects that are not in physical contact. Gravity is therefore sometimes described as an action-at-a-distance force. Although the gravitational force between two ordinary small objects is quite weak, it may assume enormous strength when acting between massive astronomical objects.

6.1 The Gravitational Force

Newton's Calculation

In 1666, young Isaac Newton, at his family's home in Woolsthorpe, made a simple observation, prompting an inquiry that led to important and far-reaching conclusions. Newton was well aware of the fact that the Moon describes an orbit around the Earth, completing a circuit in 27.3 days. But Newton did not know the reason for this regular behavior of the Moon. In the orchard at Woolsthorpe, Newton one day observed an apple fall to the ground. He wondered, "If a free and unattached apple falls toward the Earth, should not the Moon, which is also free and unattached, also fall toward the Earth?"

The Moon does not, of course, move directly toward the Earth. But the Moon does move in a (nearly) circular path around the Earth, and the continual deviation of the direction of motion from a straight line constitutes the "falling" of the Moon. That is, the Moon "falls" *around* the Earth (see

Problem 6.6). Newton wondered whether the falling motions of the apple and the Moon were due to the same cause, namely, some force of attraction originating in the Earth.

Near the surface of the Earth, an object falling freely experiences an acceleration of 9.8 m/s^2. What is the acceleration of the Moon? In order to maintain its almost circular orbit, the Moon must experience an almost constant centripetal acceleration, $a_c = v^2/R$. This centripetal acceleration is the *only* acceleration that the Moon experiences. To calculate a_c we need to know R and v. The Earth-Moon distance is $R = 3.84 \times 10^8$ m. (Newton knew this distance—in miles—less well than we do today, but he did know with reasonable accuracy that R is approximately 60 times the Earth's radius.) The speed v of the Moon in its orbit (see Fig. 6.1) is

$$v = \frac{\text{circumference of orbit}}{\text{orbit period}}$$

$$v = \frac{2\pi R}{\tau} = \frac{2\pi \times (3.84 \times 10^8 \text{ m})}{(27.3 \text{ days}) \times (86{,}400 \text{ s/day})}$$

$$= 1.02 \times 10^3 \text{ m/s}$$

Therefore, the centripetal acceleration of the Moon is

$$a_c = \frac{v^2}{R} = \frac{(1.02 \times 10^3 \text{ m/s})^2}{3.84 \times 10^8 \text{ m}} = 2.72 \times 10^{-3} \text{ m/s}^2$$

The acceleration experienced by the Moon is much smaller than that experienced by an object falling freely near the Earth's surface. To account for this difference, Newton drew on his knowledge of optics. When light radiates uniformly outward from a source, the amount of light that falls on a surface of unit area each second—that is, the *intensity* of the light—is inversely proportional to the *square* of the distance from the source to the surface. This is the famous *inverse square law* for light intensity. Newton reasoned that the force exerted by the Earth on objects near its surface and on the Moon—that is, *gravity*—must somehow spread out uniformly into space as does light. Thus, we should have

gravitational force $\propto \dfrac{1}{r^2}$

How do we choose the value to use for r? First, we know that the acceleration due to gravity is 9.8 m/s^2 at 1 m above the Earth's surface and also at 10 m and at 100 m. Therefore, the distance r is certainly not the height above the Earth's surface. Newton realized that it is the *entire* mass of the Earth that exerts the gravitational force on the *entire* mass of an object or the Moon. He therefore assumed that the distance in the force equation should be measured between the *centers* of the two bodies. The distance from the center of the Earth to the center of the Moon (3.84×10^8 m) is approximately 60 times the distance from the center of the Earth to its surface. Therefore, the falling Moon should experience a force (and, hence, an acceleration) $1/(60)^2$ times the force it would experience if it were located at the surface of the Earth. Thus,

$$a = \frac{1}{(60)^2} \times g = \frac{9.80 \text{ m/s}^2}{3600} = 2.72 \times 10^{-3} \text{ m/s}^2$$

The value of the Moon's acceleration agrees precisely with that calculated from the orbit parameters.

This result, obtained by Newton in 1666, confirmed the hypothesis that gravity obeys the inverse square law. But it was all based on an assumption, namely, that the calculations can be carried out by considering the entire mass of the Earth or other spherical object to be concentrated at its center. Newton struggled with a proof of this assumption. The problem was finally solved when Newton invented the calculus! With this new mathematical tool he was able to prove his original assertion.

Newton went on to complete his analysis by noting that the gravitational force exerted by, for example, the Earth on another object is proportional to the mass m of the object: $F_G \propto m$. But according to the third law of dynamics, the object must exert an equal force on the Earth and this force is proportional to the mass M of the Earth: $F_G \propto M$. Then, writing G for the proportionality constant, we have altogether

$$F_G = G\frac{mM}{r^2} \tag{6.1}$$

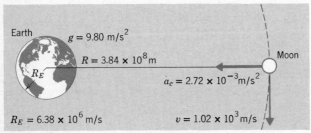

Figure 6.1 The centripetal acceleration of the Moon in its orbit is due to the gravitational attraction of the Earth.

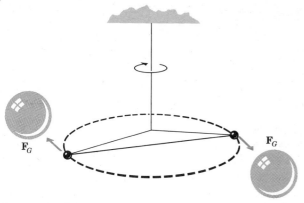

Figure 6.2 Schematic diagram of Cavendish's experiment to determine G.

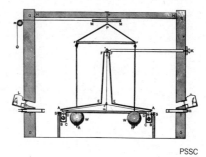

Figure 6.3 Sketch of Cavendish's apparatus as it appeared in this original paper. Notice that all of the manipulations, including the movement of the large balls W, were performed from outside the enclosure G. The measurements of the deflection angles were made with the telescopes T. Candles provided the illumination.

This is Newton's *universal law of gravitation*. It is truly a *universal* law, for it describes the gravitational interaction between pairs of objects wherever we observe them throughout the Universe.

The Gravitational Constant

The constant G that appears in the gravitational force equation is called the *gravitational constant*. The value of this constant must be determined by experiment. Gravity is actually a very weak force (compared to the other basic forces) and this means that the value of G is very small. Consequently, any measurement of G must be performed with extreme care. Such a measurement was first made in 1798 by the English chemist, Henry Cavendish (1731–1810), using an instrument now known as *Cavendish torsion balance*. The operation of the Cavendish balance is shown schematically in Fig. 6.2 and a sketch closely following that in Cavendish's original paper is shown in Fig. 6.3.

The idea of the Cavendish experiment was actually to measure the force of attraction between objects with known masses. As shown in Fig. 6.3, Cavendish mounted two small lead balls (2 in. in diameter, $m = 0.775$ kg) on opposite ends of a 2-m rod. The rod was supported by a fine wire. Two larger lead balls (8 in. in diameter, $m = 49.5$ kg) could be brought close to the small balls. The attractive force between the two sets of balls caused the smaller balls to move toward the larger balls, thus twisting the suspension wire by a small amount. The force constant for twisting the wire was obtained in a separate, calibration experiment using known forces. The degree of twist was then a measure of the force between the balls.

The result that Cavendish achieved in this experiment was a value for G only 1 percent different from that now accepted, that is,

$$G = 6.67 \times 10^{-11} \text{ N-m}^2/\text{kg}^2 \qquad (6.2)$$

(The units of G can also be expressed as m³/kg-s²; verify this.)

The Mass of the Earth

We can use Newton's law of gravitation to determine the mass of the Earth by a simple calculation. First, we consider an object with a mass m that is located near the surface of the Earth. Then, the gravitational force on that object is

$$F_G = G \frac{mM}{R_E^2}$$

where M is the mass of the Earth and R_E is the Earth radius. We also know that the gravitational force on m is given by

$$F_G = mg$$

Equating these two expressions for F_G and solving for M, we find

$$M = \frac{gR_E^2}{G} \qquad (6.3)$$

Using the known values of g, R_E, and G, we obtain

$$M = \frac{(9.80 \text{ m/s}^2) \times (6.38 \times 10^6 \text{ m})^2}{6.67 \times 10^{-11} \text{ m}^3/\text{kg-s}^2}$$

$$= 5.98 \times 10^{24} \text{ kg}$$

When Cavendish published the results of his

Figure 6.4 A globular cluster of stars in the constellation Hercules. Hundreds of thousands of stars are held together in this cluster by gravitation. Richard Feynman, winner of a share of the 1965 Nobel Prize in physics, has said that "if one cannot see gravitation acting here, he has no soul."

Mount Wilson and Palomar Observatories

experiment, he did not explicitly state a value for G. Instead, he used a calculation similar to this and gave his result as a value for the *density* of the Earth:

$$\rho = \frac{M}{V} = \frac{\frac{gR_E^2}{G}}{\frac{4}{3}\pi R_E^3} = \frac{3g}{4\pi G R_E} \quad (6.4)$$

Supply the values of the constants and show that $\rho = 5.5 \times 10^3$ kg/m^3.

● *Example 6.1*

With what force does the Earth attract the Moon?
We have

R = Earth-Moon distance = 3.84×10^8 m
m = mass of the Moon = 7.35×10^{22} kg
M = mass of the Earth = 5.98×10^{24} kg

Using Eq. 6.1, we find

$$F_G = G\frac{mM}{R^2}$$

$$= (6.67 \times 10^{-11} \text{ N-m}^2/\text{kg}^2) \times$$

$$\frac{(7.35 \times 10^{22} \text{ kg}) \times (5.98 \times 10^{24} \text{ kg})}{(3.84 \times 10^8 \text{ m})^2}$$

$$= 1.99 \times 10^{20} \text{ N}$$

We can obtain this result in another way by using the Moon's centripetal acceleration calculated above. The force on the Moon is equal to its acceleration multiplied by its mass:

$$F_G = ma_c = (7.35 \times 10^{22} \text{ kg}) \times (2.72 \times 10^{-3} \text{ m/s}^2)$$
$$= 1.99 \times 10^{20} \quad ■$$

Artificial Gravity

When a space vehicle orbits the Earth, the vehicle and all objects within it "fall" around the Earth at the same rate. Therefore, an astronaut within such a spacecraft will exert no force on any part of the vehicle; that is, the astronaut will have zero weight. If the vehicle is a space station or laboratory designed to remain in orbit for a long period of time, the weightlessness of the occupants may be a definite disadvantage. The human body, after all, is adapted to the Earth environment and is conditioned to work under the influence of gravity.

In order to avoid any problems arising from weightlessness, an "artificial gravity" can be produced in a space station so that the occupants can function in a near-normal manner. To accomplish this it is only necessary to set the space station into rotation. Figure 1 shows a donut-shaped space station that is rotating with an angular velocity ω around its central axis. The annular region of the donut is hollow and represents the working area occupied by the astronauts. The outer ring of the donut is at a distance r from the rotation axis. Any point on the outer rim experiences a centripetal acceleration,

$$a_c = \omega^2 r$$

The force on an astronaut in this position is the *centripetal force*,

$$F_c = ma_c = m\omega^2 r$$

This inward force is the *only* force acting upon the astronaut; it is the force required to maintain the circular motion.

What does the astronaut experience? The rim exerts a centripetal force $F_c = m\omega^2 r$ on the astronaut (Fig. 1a), and the astronaut exerts a reaction force of the same magnitude on the rim. This outward force is the *centripetal reaction* or the *centrifugal force*. In the astronaut's reference frame the rim is at rest and he is exerting a force upon it. This is exactly the situation that would exist if gravity were acting on the astronaut, and his body sensations are exactly as they would be in the presence of gravity. That is, the weight of the astronaut is $m\omega^2 r$. Moveover, the astronaut feels the centripetal force acting "upward" just as the reaction force of a floor would act in a normal situation. The astronaut therefore feels "up" to be toward the center of the space station, as indicated in Fig. 1b. Notice that "up" is

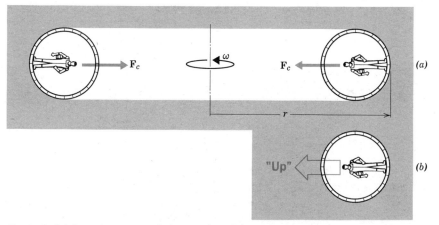

Figure 1 (a) An astronaut on the outer rim of a rotating space station is acted on by the centripetal force $\mathbf{F}_c$. (b) The astronaut interprets the effect of the centripetal force as indicating that "up" is toward the center of the space station.

always in the direction opposite to that in which a person exerts a force on other objects; that is, "up" is opposite to the direction of a person's weight. (This effect was imitated with remarkable realism in Stanley Kubrick's movie, *2001—A Space Odyssey*.)

Suppose that a space station such as that in Fig. 1a has a radius of 200 m. What angular velocity is necessary to duplicate the effect of Earth gravity on the outer rim? In this case we want

$$a_c = \omega^2 r = g = 9.8 \text{ m/s}^2$$

Then,

$$\omega = \sqrt{\frac{g}{r}} = \sqrt{\frac{9.8 \text{ m/s}^2}{200 \text{ m}}} = 0.22 \text{ rad/s}$$

The period of this motion is

$$\tau = \frac{2\pi}{\omega} = \frac{2\pi}{0.22 \text{ s}^{-1}} = 28.6 \text{ s}$$

Notice that the centripetal acceleration is directly proportional to r and therefore decreases as we move from the outer rim toward the rotation axis. In a rotating space station the magnitude of the artificial "gravity" will depend on position within the station. This effect may be very useful in an actual space laboratory. The astronauts could spend the largest fraction of their time living normally where "g" is greatest, and yet they could still conduct experiments requiring small values of "g" in the central part of the space station.

If a platform is rotating on or close to the Earth's surface, any object will exert on the platform an outward centrifugal force and a downward force $m\mathbf{g}$ due to the effect of gravity. These two forces combine to produce the net force exerted by the object, that is, the object's "weight." In Fig. 2 we see three identical objects at different distances from the rotation axis. The downward force $m\mathbf{g}$ is the same in each case, but the centrifugal force increases with distance away from the axis. Therefore, the direction of $\mathbf{F}_{net}$ (the "weight" of the object) is different in each case and a different direction corresponds to "up" at each position. The deviation of "up" from the direction defined by $\mathbf{g}$ is greatest at the outermost position.

It is easy to demonstrate that plants (as well as astronauts) sense "weight" due to rotation. If a number of plants grow in a container that rotates on a platform, the individual sprouts will grow in the direction that is locally "up." That is, plants near

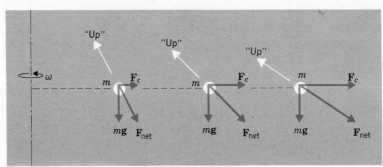

Figure 2 An object on a platform that is rotating near the Earth exerts a net force that depends on the distance of the object from the rotation axis. The direction sensed as "up" also changes with distance.

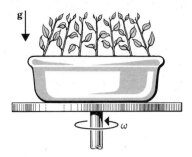

Figure 3 Plants growing in a rotating container sense different directions as "up."

the center of rotation will grow vertically, but those farther and farther from the rotation axis will grow with increasing inward slants, as illustrated in Fig. 3.

■ Exercises

1. A rotating space station has a radius of 100 m. An astronaut at the outer rim (where "g" = g) throws a ball to a "height" of 80 m. Describe qualitatively the motion of the ball. Will the astronaut notice any difference in the motion compared to that he or she would observe on Earth?

2. Refer to Fig. 2. Express the angle θ between the direction "up" and the true vertical in terms of ω, r, and g. Suppose that a number of plants grow in a container that has a radius $r = 30$ cm and is rotating with an angular velocity $\omega = 6$ rad/s. Make a sketch of the directions of the sprouts, showing individual sprouts at intervals of 5 cm from the center outward.

3. How do you suppose that plants sense the direction "up"?

6.2 Planetary Motion

Kepler's Laws

Between 1609 and 1611, Johannes Kepler (1571–1630) enunciated his famous three laws of planetary motion. Kepler's conclusions were based on his analysis of the extensive data relating to planetary positions (particularly pertaining to the planet Mars) that had been acquired by Tycho Brahe (1546–1601) during many years of observation.

The statements of Kepler's laws are:

I   The motion of a planet is an ellipse with the Sun at one focus.

II The line connecting the planet with the Sun sweeps out equal areas in equal times.

III The period of a planet's motion and its distance from the Sun are related by R^3/τ^2 = constant, where the constant is the same for all planets.

Kepler's First Law

By geometrical construction from his position data, Kepler showed that planetary orbits are elliptical (but nearly circular). Newton used more sophisticated mathematics to prove the more general result that all orbits of objects interacting via a $1/r^2$ force are *conic sections*. The four possible types of curves that are in this category are obtained from the intersections of a plane with a cone (see Fig. 6.5). If the plane sections the cone at right angles to the axis of the cone, the result is a *circle*. By sectioning the cone at an angle, an *ellipse* is produced. If the sectioning angle is increased until it coincides with the cone angle (that is, until the plane is parallel to one of the straight lines that runs the length of the cone on its surface), a *parabola* results. A further increase in sectioning angle yields an *hyperbola*. Because specific sectioning angles are required to produce the circle and the parabola, these curves are actually only special cases of the ellipse and the hyperbola, respectively. Thus, in Nature we do not find orbits that are *exactly* circular or parabolic, although in certain cases these shapes are approached closely. The orbits of Venus and Neptune, for example, are the most nearly circular of all planetary orbits. For Venus, for example, the difference between the furthest distance from the Sun (called the *aphelion*) and

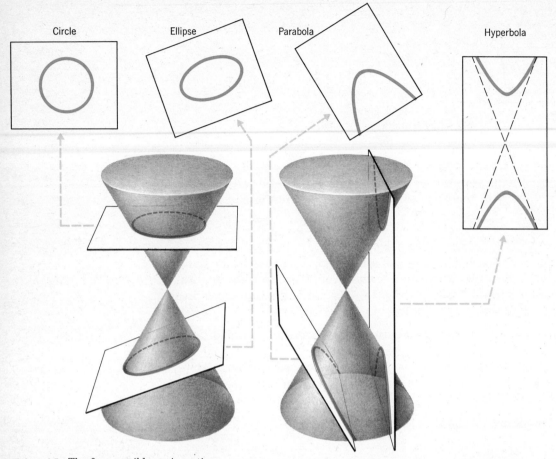

Figure 6.5 The four possible conic sections.

the closest distance (called the *perihelion*) is only 1.4% of the mean distance from the Sun.

There is a simple method for constructing an ellipse that follows from an alternate definition. First, select the two points, F_1 and F_2, that are the *foci* of the ellipse (Fig. 6.6). Next, attach the ends of a length of string (which is longer than the distance from F_1 to F_2) to pins at F_1 and F_2, then, with the tip of a sharp pencil, extend the string until it is taut. The pencil tip will be at a distance r_1 from F_1 and a distance r_2 from F_2 (Fig. 6.6). Finally, the ellipse is constructed by drawing the curve that the pencil follows as it is moved in such a way that maintains the tautness of the string. The corresponding definition is: an ellipse is the curve that is formed by connecting all points of equal values of the sum $r_1 + r_2$, where r_1 and r_2 are the distances from a given point on the curve to the foci.

Kepler described planetary orbits as ellipses with the Sun at one focus. (The other focal point has no physical significance for the case of planetary motion.) A more detailed analysis shows that it is not the center of the Sun that lies at one focus, but

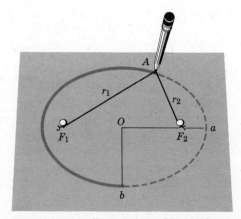

Figure 6.6 An ellipse is formed by the line connecting all points (such as A) for which $r_1 + r_2$ = constant. Oa is the semimajor axis and Ob is the semiminor axis.

rather the *center of mass* of the Sun-planet system. Because the Sun is so much more massive than any planet, the center of mass of any Sun-planet combination lies very near the Sun's center.

The ellipse is the general form of the orbit for *bound motion*. (*Bound motion* is the case in which two objects are bound together by their mutual

Table 6.1 Planetary Data

Planet	Semimajor axis (A.U.)*	Period (years)	Mass (Earth masses)	Diameter (Earth diameters)
Mercury	0.387	0.241	0.055	0.382
Venus	0.723	0.615	0.815	0.949
Earth	1.000	1.000	1.000	1.000
Mars	1.524	1.881	0.107	0.553
Jupiter	5.203	11.862	317.89	11.19
Saturn	9.539	29.458	95.17	9.49
Uranus	19.182	84.013	14.56	3.69
Neptune	30.058	164.793	17.24	3.50
Pluto	39.518	248.43	0.11(?)	<0.47

*One astronomical unit (1 A.U.) = Earth-Sun distance = 1.496×10^{11} m.

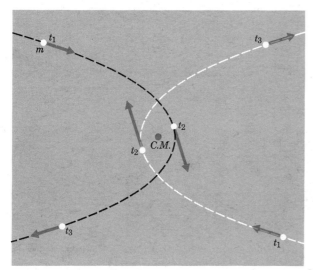

Figure 6.7 The hyperbolic orbits of two "colliding" stars that interact via a gravitational force. The positions and the velocity vectors of the two stars are shown for the times t_1, t_2, and t_3. The center of mass of the system remains fixed.

gravitational attraction and cannot escape from each other.) The hyperbola is the general form of the orbit for *unbound* motion. For example, if two isolated stars move toward each other, as in Fig. 6.7, they will execute hyperbolic orbits relative to their common center of mass. The position of the center of mass will, of course, move with constant velocity. If, at any instant, the center of mass is at rest with respect to some coordinate system, it will remain at rest as long as the two interacting stars are not influenced by any outside force.

Kepler's Second Law

The second law of Kepler simply expresses the fact that the angular momentum of a planet around the Sun remains constant, although Kepler did not realize this. Consider the motion of a planet about the

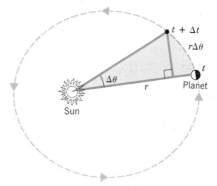

Figure 6.8 A planet in elliptic orbit around the Sun.

Sun shown in Fig. 6.8. In a short time interval Δt it moves from the point labeled t to a nearby point along its orbit labeled $t + \Delta t$ having swept out an angle $\Delta\theta$ with the line connecting it with the Sun. The corresponding area (shown shaded) that this line swept out may be approximated by the area of the right triangle with base $r\Delta\theta$ and altitude r. Thus, the area ΔA swept out in the time interval Δt is

$$\Delta A = \tfrac{1}{2}(r\Delta\theta)(r) = \tfrac{1}{2}r^2\Delta\theta.$$

The rate at which the radial line sweeps out orbital area is therefore,

$$\frac{\Delta A}{\Delta t} = \tfrac{1}{2} r^2 \frac{\Delta\theta}{\Delta t} = \tfrac{1}{2}r^2\omega \qquad (6.5)$$

where ω is the instantaneous orbital angular velocity.

Let us now recall that the angular momentum of the planet orbiting the Sun (see Eq. 4.6) is

$$L = mr^2\omega.$$

Substituting L/m for $r^2\omega$ in Eq. 6.5 gives

$$\frac{\Delta A}{\Delta t} = L/2m. \tag{6.6}$$

The gravitational force the Sun exerts on the planet is radial and hence produces no torque about the Sun. The planet's angular momentum about the Sun is therefore a constant. Because L must be constant Eq. 6.6 tells us that $\Delta A/\Delta t$ must also be constant. This is just the content of Kepler's second law. Note that our derivation was for the general case of an elliptic orbit.

Kepler's Third Law

The third law of Kepler is most easily derived for the case of a hypothetical planet executing a circular orbit about the Sun. Figure 6.9 shows a planet, mass m, circling the Sun, mass M_s, at a distance R.

The tangential velocity of the planet in terms of the radius R and the rotational period τ (the planetary year), is

$$v = \frac{2\pi R}{\tau}. \tag{6.7}$$

The planet's centripetal acceleration $a_c = v^2/R$, according to Newton's second law, must be provided by the gravitation force the Sun exerts on the planet or

$$ma_c = \frac{GM_s m}{R^2}$$

hence,

$$v^2 = \frac{GM_s}{R}. \tag{6.8}$$

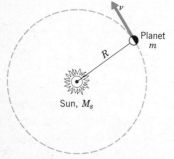

Figure 6.9 A planet of mass m in circular orbit around the Sun, mass M_s.

Combining Eqs. 6.7 and 6.8 gives the desired result

$$R^3/\tau^2 = G\frac{M_s}{4\pi^2}. \tag{6.9}$$

We note that the quantities appearing on the right-hand side of Eq. 6.9 do not include the planet's mass, and thus this is the statement of Kepler's third law.

Although we obtained the result for the special case of circular motion, the law is also valid for elliptical motion where R refers to the length of the semimajor axis (see Fig. 6.6).

Does the Gravitational Force Depend Exactly on $1/r^2$?

We have been using Newton's universal gravitational force law with a dependence on distance of the form $1/r^2$. But how do we know that the exponent is not 2.000 001 or 1.999 999? Is the exponent *exactly* 2? A sensitive test of a possible deviation of the exponent from 2 can actually be made by observations of planetary orbits.

Newton showed that if the gravitational force varies exactly as $1/r^2$, then the elliptical orbits described by the planets must remain in *fixed* positions. In particular, the point of the ellipse that is closest to the Sun (the perihelion) must remain fixed in its relation to the distant "fixed" stars. Of course, there are small deviations from exact elliptical orbits (perturbations) due to the influence of the other planets, but these deviations are small because of the dominant gravitational force of the Sun; furthermore, mathematical methods exist for the precise calculation of these perturbations. Therefore, if any motion of the perihelion (apart from that expected due to other planets) is observed, this would indicate that the exponent in the force law expression is not *exactly* 2.

About a hundred years ago, a small unexplained motion in the perihelion of Mercury was observed. The perihelion moves forward *(precesses)* at a very slow rate so that the orbit has the appearance of a slowly rotating ellipse (Fig. 6.10). After subtraction of all of the effects due to other planets, there is a net precession that amounts to 43 sec of arc per century. That is, it would require more than 4000 y (or about 16,000 revolutions) for the perihelion to move by an amount equal to the diameter of the Moon (0.5 degree = 1800 arc sec).

At first, the precession of Mercury's perihelion was thought to indicate the presence of an unobserved planet near the Sun that would influence the motion of Mercury and whose effects would not have

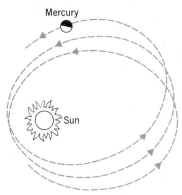

Figure 6.10 The perihelion of Mercury's orbit is observed to precess slowly. (The elongation of the orbit and the amount of precession have been greatly exaggerated.)

been included in the previous perturbation calculations. This planet (prematurely named *Vulcan*) was sought for many years without success, and so it was believed that Newton's gravitational force law was slightly in error. Early in this century, Einstein showed that there was indeed a small correction needed in the gravitational force law due to relativistic effects. This correction depends on the planet's velocity and therefore is important only for Mercury, which has the highest velocity of any planet. With this correction, the precession of Mercury's perihelion is entirely accounted for.

Although a small relativistic correction in Newton's form of the gravitational force law is required for objects in motion with high velocities, the Newtonian equation is valid for essentially all practical purposes and it appears to be entirely correct for *static* gravitational forces.

6.3 Satellite Orbits

A large number of satellites are in various orbits around the Earth, and their number is increasing with each passing year. The orbital characteristics of Earth-bound satellites (well above the Earth's atmosphere) are completely analogous to the motion of the planets about the Sun. Such satellite orbits are also ellipses (see Fig. 6.11). A number of points concerning these orbits should be noted. The center of the Earth (assumed to be spherical in shape) is at the focus, F, of the elliptic orbit and *not* at the center of the orbit, C. If the altitude of the satellite at perigee is h_p and at apogee h_a then the semimajor axis a is given by

$$a = R_E + \frac{h_p + h_a}{2} \tag{6.10}$$

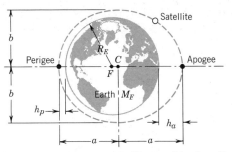

Figure 6.11 The general elliptic orbit of an Earth-bound satellite. The closest point to the Earth's surface is called the *perigee*, and the furthest point is the *apogee*. The length a is referred to as the *semimajor* axis and the length b is called the *semiminor* axis.

Kepler's third law, Eq. 6.9, is also applicable to satellite motion. Applying it to the general elliptic satellite orbit by replacing R with the semimajor axis a and the mass of the Sun, M_s, by the mass of the Earth, M_E, we have

$$a^3/\tau^2 = G\frac{M_E}{4\pi^2}, \tag{6.11}$$

where τ is now the period of the satellite.

Synchronous Satellites

One of the most significant recent advances in the field of communications has been the establishment of a system of artificial satellites that remain in fixed positions at a certain height over the Earth's equator. These *synchronous satellites* are equipped to provide relay facilities for many channels in the intercontinental communications network (Fig. 6.12).

The height at which these satellites must be placed into orbit is determined by the fact that the period of rotation must exactly equal the period of rotation of the Earth. If the periods are equal, the satellite and the Earth will rotate together *(synchro-*

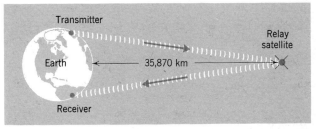

Figure 6.12 Synchronous satellites relay telephone, radio, and TV transmissions between points on the Earth that would not otherwise be able to communicate via radio because of the straight-line nature of the propagation of high-frequency radio waves.

nously) and the satellite will remain in a fixed position with respect to the Earth if it is in a circular orbit.

It is useful to rewrite Eq. 6.11 in a slightly different form. We know that at the surface of the Earth (refer to Eq. 6.3),

$$g = G\frac{M_E}{R_E^2}$$

Replacing GM_E in Eq. 6.11 with gR_E^2 the semimajor axis with the orbit radius R, we obtain

$$R^3 = \frac{\tau^2 g R_E^2}{4\pi^2} \qquad (6.12)$$

Using $\tau = 1$ day $= 86{,}400$ s and the values for g and R_E, we have, by taking the cube root,

$$R = \left[\frac{(86{,}400 \text{ s})^2 \times (9.80 \text{ m/s}^2) \times (6.38 \times 10^6 \text{ m})^2}{4\pi^2}\right]^{1/3}$$

$$= 42{,}250 \text{ km}$$

Thus, the height of the satellite above the surface of the Earth is

$$h = R - R_E$$
$$= 42{,}250 \text{ km} - 6{,}380 \text{ km} = 35{,}870 \text{ km}$$

or a little more than 22,000 miles.

If the satellite is placed into orbit at this height above the Earth's equator, it will be a *synchronous* satellite. (The behavior of the satellite if it is at this height, but not above the equator, is the subject of Question 6.1).

Summary of Important Ideas

The gravitational force depends on $1/r^2$ and is a *long-range* force.

All motions of two isolated objects that interact via a $1/r^2$ force are *conic sections*. Planets move in *elliptical* orbits around the Sun.

For a planet circling the sun in an orbit with radius R and with a period τ we have

$$\frac{R^3}{\tau^2} = \text{constant}$$

For purposes of gravitational calculations, the entire mass of a uniform spherical object can be considered to be concentrated at its *center*.

◆ Questions

6.1 An errant synchronous satellite is placed in orbit, not at the equator, but at a certain latitude λ. What is the motion of the satellite relative to the surface of the Earth?

6.2 The mass of the Moon is approximately $\frac{1}{81}$ of the mass of the Earth but the gravitational attraction on an object at its surface is only about $\frac{1}{6}$ of the gravitational attraction on the same object at the surface of the Earth. Why?

6.3 Comets are members of the solar system and travel in elliptical orbits. Explain why some comets (such as Halley's comet) appear remarkably bright for a brief period and then become unobservable, even with powerful telescopes, for a number of years.

6.4 In the Cavendish experiment, can the force on one of the small balls due to the farther large ball be neglected in comparison to the force due to the nearer large ball?

6.5 The Earth's surface has many irregularities, ranging from deep ocean trenches to high mountains. How is it possible to make the assumption, as Newton did, that the entire mass of the Earth can be considered to be concentrated at its center for purposes of gravitational calculations?

★ Problems

Section 6.1

6.1 Near the surface of the Earth, the acceleration due to gravity is 9.80 m/s². What value for g would be found at heights above the Earth's surface of 10 km and 6400 km?

6.2 At what height above the Earth's surface will the acceleration due to gravity be 7.5 m/s²?

6.3 The mass of the Moon is about $\frac{1}{81}$ of the mass of the Earth. At what point on a line between the Earth and the Moon will an object experience zero net gravitational force?

6.4 Use the data in Table 6.1 and compute the value of g on Mars. What is the gravitational force on a 0.1-kg mass on the Martian surface?

6.5 What is the value of g at the surface of the Sun?

6.6* The Moon's orbital velocity is 1.02 km/s. In traveling at this velocity for 3 s, how far has the Moon "fallen" toward the Earth? Compare this with the distance that an object will fall near the surface of the Earth in 3 s. Calculate the ratio of the two distances and account for the value.

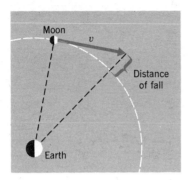

6.7 The centers of two identical spherical objects are separated by a distance of 1 m. If the gravitational force between them is 6.7×10^{-9} N, what is the mass of each object?

6.8* If the mass of the Earth were 4 times its actual value, what would be the length of the lunar month? (Assume the Earth-Moon distance to be the actual distance.)

6.9 What was the attractive force between one of the large balls and one of the small balls in the Cavendish experiment when the centers of the balls were separated by 0.22 m?

6.10 Compare the gravitational forces exerted on the Earth by the Moon and by the Sun.

6.11* Two blocks, each with mass $m = 2$ kg, are on a flat, frictionless, horizontal plane. There is no extraneous matter in the vicinity. If the blocks are initially separated by a distance of 2 m, how long will be required for the gravitational attraction between the blocks to reduce the separation by 1 cm?

6.12 Refer to the previous problem. What is the maximum value of the coefficient of friction between the blocks and the plane that will permit the blocks to move as a result of the mutual gravitational attraction?

Section 6.2

6.13 The speed of the Earth in its orbit around the Sun is approximately 30 km/s. What are the corresponding speeds for Mars, Venus, and Saturn?

6.14 If the Moon were suddenly placed in a circular orbit around the Earth at a distance twice that at present, what would be the new period? What would be the new orbital velocity?

6.15 Use the data in Table 6.1 and verify Kepler's third law by comparing R^3/τ^2 for 4 or 5 of the planets.

6.16 An astronomer claims to have found a new planet whose orbit is midway between the Earth and the Sun whose period is 240 days. What is your opinion of his claim and why?

6.17* Two spherical objects, interacting only via their mutual gravitational forces, execute circular orbits in space around their common center of mass. If one object is three times as massive as the other, sketch the orbits of each. Compare the periods of the orbits of the two objects.

6.18 Assume that the Earth is the only planet in the solar system. What is the radius of the orbit that the Sun executes around the center of mass of the Earth-Sun system? What is the period of this motion of the Sun?

6.19 One of the moons of Jupiter revolves around the planet in a nearly circular orbit with a radius of 1.07×10^6 km and a period of 7.16 days. What is the mass of Jupiter? Compare your result with the entry in Table 6.1.

Section 6.3

6.20 A satellite is placed in a circular orbit just skimming the surface of the Moon. (The radius of the Moon is 1738 km and the mass is 7.35×10^{22} kg.) What is the acceleration and speed of the satellite? What is the period of the satellite?

6.21 A 100-kg satellite is in a circular orbit at an altitude of 1500 km above the Earth's surface. What is the acceleration and speed of the satellite? What is the period of the satellite?

6.22 What is the gravitational force on a 100-kg synchronous satellite in orbit around the Earth?

6.23* A satellite is placed in an elliptic orbit

around the Earth. Its altitude at the perigee is 500 km and 4500 km at apogee. What is the semimajor axis? What is the period of the satellite? What is the ratio of the orbital velocities at the perigee to that at the apogee?

6.24* An Earth satellite is in an elliptic orbit with a period of 140 minutes and a closest approach to the Earth's surface of 400 km. What is its highest altitude? If an on-board rocket burst is fired at this point in the orbit, what speed must be reached to go into a circular orbit? In what direction must this rocket burst be aimed?

6.25* An Earth satellite at perigee has an altitude of 300 km. At apogee its orbital velocity is only $\frac{1}{3}$ that at perigee. What is the satellite's period?

7 Energy

Without a doubt, *energy* is the single most important physical concept in all of science. A clear understanding of energy and an appreciation of its importance was not fully realized until 1847 when the German physicist Hermann von Helmholtz (1821–1894) enunciated the general law regarding energy. Since that time the consideration of energy in physical (and biological) processes has been a crucial ingredient in the efforts to understand natural phenomena.

Closely associated with energy is the concept of *work*. Historically, there grew up intuitive conceptions regarding *energy* and *work* just as there did for length, time, and mass. Some of these ideas of the layman are, in fact, quite closely related to the precisely formulated ideas of the physicist. It could be said, for example, "Eat a good meal and you will have a lot of energy," and "A person who has a great deal of energy can do a large amount of work." These statements correspond closely to those that the physicist would make: "The stored chemical energy in foodstuffs can be transferred to biological systems," and "Energy is the capacity to do work."

In this chapter we first define *work* and then discuss the connection between work and *energy*. The importance of the concept of energy lies in the fact that various forms of energy within an isolated system can be transformed into one another *without a change in the total amount of energy*. That is, in any physical process, *energy is conserved*. The law of energy conservation, which is obeyed in every known process, is the most fruitful law in physics for the analysis of phenomena of every sort.

7.1 Work

Motion against a Force

If we exert a force on an object and move it a certain distance, we say that we have done *work*. Lifting a weight, for example, requires exerting a force sufficient to overcome the downward gravitational force. If we apply such a force sufficiently long to raise the weight to a height h, we have done a certain amount of work. Or, if we push an object across a rough surface with a force sufficiently large to overcome friction and move it a distance s, we have again done a certain amount of work. The amount of work done is proportional to both the applied force and the distance through which the force acts.

Suppose that we apply a constant force to an object and thereby cause it to move with *constant velocity* over a rough surface. Because the acceleration of the object is zero, we know that there is no *net* force being applied. The externally applied force in

Figure 7.1 When the displacement **s** is in the same direction as the applied force **F**, the work done is $W = Fs$.

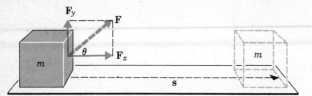

Figure 7.2 Only the component of the force **F** in the direction of the displacement **s** does any work. Therefore, $W = \mathbf{F} \cdot \mathbf{s} = Fs \cos \theta$.

this case just balances the retarding frictional force so that $F_{\text{net}} = 0$. Nevertheless, work *is* being done by the external force because the external force is working against the frictional force. The act of displacing an object against a retarding force (such as friction or gravity) by the application of an external force constitutes work, regardless of whether the object is accelerated or moves with constant velocity.

Work = Force × Distance

Work is the product of the force applied to an object by an outside agent and the distance through which the force acts on the object. In Fig. 7.1, the work done by the force **F** in producing the displacement **s** is $W = Fs$. In this case the direction of the displacement is the same as the direction of the applied force. In Fig. 7.2, however, the force is applied at an angle θ relative to the direction of motion. This force can be considered to be the vector sum of two independent forces, the *x*- and *y*-components of **F**, where

$F_x = F \cos \theta$
$F_y = F \sin \theta$

The component F_x has the direction of the displacement; therefore the work done by this component of the force is $W = F_x s = Fs \cos \theta$. There is no displacement in the *y*-direction; therefore, the component F_y does no work. In general, then, our definition of work must be modified:

Work is the product of the component of force in the direction of the displacement and the magnitude of the displacement produced by the force, or

$$W = Fs \cos \theta \tag{7.1}$$

where θ is the angle between the force vector **F** and the vector **s** that represents the displacement of the object moved. Although *work* is equal to the product of two quantities, **F** and **s**, that are vectors, work is a *scalar* quantity. In fact, the particular way shown in Eq. 7.1 of multiplying two vectors is called the *scalar product* of the vectors and is usually written as $\mathbf{F} \cdot \mathbf{s} = Fs \cos \theta$; that is,[1]

$$W = \mathbf{F} \cdot \mathbf{s} = Fs \cos \theta \tag{7.2}$$

To form the scalar product (or *dot product*) of two vectors, take the product of the two magnitudes multiplied by the cosine of the angle between the two vectors. This amounts to taking the product of one vector with the component of the other vector that is in the direction of the first vector. Thus, in Fig. 6.2 $\mathbf{F} \cdot \mathbf{s}$ means the product of s and the component of **F** that lies in the direction of **s**, that is, $F_x = F \cos \theta$. In the metric system of units, the dimensions of *work* are the dimensions of *force* times the dimensions of *distance*. For the unit of work we use the special name *joule* (J):

$1 \text{ N-m} = 1 \text{ kg-m}^2/\text{s}^2$
$= 1 \text{ joule (J)}$

The designation *joule* for the unit of work is in honor of James Prescott Joule (1818–1889), the English physicist whose efforts greatly clarified the concepts of work and energy.

● *Example 7.1*

A horizontal force of 5 N is required to maintain a velocity of 2 m/s for a box of mass 10 kg sliding over a certain rough surface. How much work is done by the force in 1 min?

First, we must calculate the distance traveled:

$s = vt$
$= (2 \text{m/s}) \times (60 \text{ s})$
$= 120 \text{ m}$

Then,

$W = Fs$
$= (5 \text{ N}) \times (120 \text{ m})$
$= 600 \text{ N-m} = 600 \text{ J}$

● *Example 7.2*

In Fig. 7.2, all of the work done in moving the block is due to F_x, and the component F_y does no work at

---

[1] $\mathbf{F} \cdot \mathbf{s}$ is read "F dot s."

all. Could we, therefore, eliminate F_y without affecting the work done? Not at all! The reason is that F_x does work against the frictional force which depends directly on the normal force (Eq. 3.6). The normal force, in turn, depends on the force component F_y. If F_y is made larger, the normal force and, hence, the frictional force will decrease; then, the value of F_x required to move the block at constant velocity will be smaller and the work done will be smaller. If F_y is made equal to mg, the normal force will become zero; then, there will be no frictional force at all to work against.

The free-body-force diagram below shows how to analyze the situation. We assume that the block is being pulled at constant velocity. Then, we can immediately write two force equations:

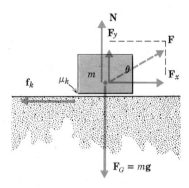

$$f_k = F_x \tag{1}$$

$$N + F_y = mg \tag{2}$$

We can also use Eq. 3.6 to write

$$f_k = \mu_k N = \mu_k(mg - F_y) = F_x \tag{3}$$

Also,

$$F_x = F \cos\theta \qquad F_y = F \sin\theta \tag{4}$$

Using (4) in (3), we can write

$$F \cos\theta = \mu_k(mg - F \sin\theta) \tag{5}$$

Solving (5) for F,

$$F = \frac{\mu_k mg}{\cos\theta + \mu_k \sin\theta} \tag{6}$$

Let us suppose $m = 4$ kg, $\theta = 30°$, $\mu_k = 0.4$. Then,

$$F = \frac{(0.4) \times (4 \text{ kg}) \times (9.8 \text{ m/s}^2)}{\cos 30° + (0.4) \times (\sin 30°)}$$

$$14.7 \text{ N} \tag{7}$$

Now, if the distance through which this force acts is $s = 15$ m,

$$W = \mathbf{F} \cdot \mathbf{s} = Fs \cos\theta$$
$$= (14.7 \text{ N}) \times (15 \text{ m}) \times (\cos 30°)$$
$$= 191 \text{ J}$$

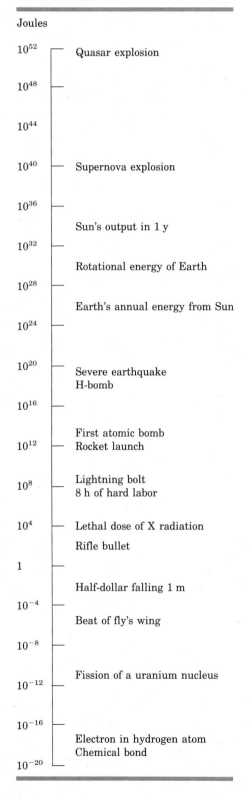

Table 7.1 Range of Energies found in the Universe

Suppose that we eliminate the vertical component of the force and apply a horizontal force just sufficient to balance the frictional force. Then, the work done in moving the block 15 m is

$$W = Fs = \mu_k mgs$$
$$= (0.4) \times (4 \text{ kg}) \times (9.8 \text{ m/s}^2) \times (15 \text{ m})$$
$$= 235 \text{ J}$$

which shows that the effect of the vertical component of the applied force is to reduce the amount of work done in pulling the block at constant velocity. (From your own experience, is it easier to drag a box or carton across a floor by applying a horizontal force or by lifting on the box while pulling?)

● *Example 7.3*

To a good approximation, the force *required to stretch a* spring is proportional to the distance the spring is extended. That is,

$$F = kx$$

where k is the so-called *spring constant* or *force constant* and depends on the dimensions and material of the spring. Many elastic materials, if not stretched too far, obey this simple relationship—called *Hooke's law* after Robert Hooke (1635–1703), a contemporary of Newton.[2] (We will consider Hooke's law further in Chapter 11.)

Suppose that a force of 0.2 N is required to extend a certain spring 5 cm. What force is required to stretch the spring from its natural length to a length 20 cm greater?

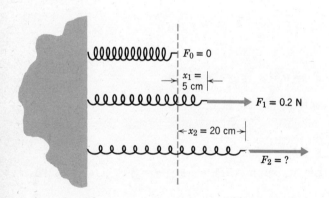

The force constant is

$$k = \frac{F_1}{x_1}$$

$$= \frac{0.2 \text{ N}}{0.05 \text{ m}}$$
$$= 4 \text{ N/m}$$

Therefore, the force required to extend the spring to 20 cm is

$$F_2 = kx_2$$
$$= (4 \text{ N/m}) \times (0.2 \text{ m})$$
$$= 0.8 \text{ N}$$

How much work is expended in stretching the spring to 20 cm?

To answer this question, we must realize that we are dealing here not with a *constant* force (as in the previous cases), but with a force that varies with the extension of the spring. We can still obtain the correct value for the work done if we are careful to use the *average* force that is exerted in stretching the spring. Now, a Hooke's law force varies linearly with the extension (as shown in the diagram), so the average force is simply the arithmetic average (compare Fig. 2.11):

$$\overline{F} = \tfrac{1}{2}(F_2 + F_0) = \tfrac{1}{2}(F_2 + 0) = \tfrac{1}{2} kx_2$$

and the work expended is the average force multiplied by the distance:

$$W = \overline{F} x_2 = \tfrac{1}{2} kx_2^2$$
$$= \tfrac{1}{2} \times (4 \text{ N/m}) \times (0.2 \text{ m})^2$$
$$= 0.08 \text{ J} \quad \blacksquare$$

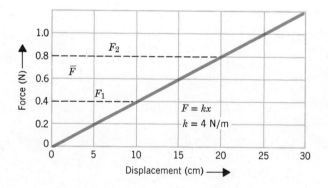

7.2 Kinetic and Potential Energy

Kinetic Energy—The Energy of Motion

Suppose we have a *free* object—for example, an object in space initially at rest (with respect to some coordinate system) and subject to no forces. Then, if

---
[2]Hooke's law is often expressed in terms of the force the *spring* exerts on an object that has stretched it by an amount x, namely $F_s = -kx$.

we apply a certain constant force F, the object will be accelerated. After moving a distance $s = \frac{1}{2}at^2$, the object will have a velocity $v = at$. The work done on the object is

$$F \times s = (ma) \times (\tfrac{1}{2}at^2)$$
$$= \tfrac{1}{2}m \times (at)^2$$
$$= \tfrac{1}{2}m \times v^2$$

or,

$$Fs = \tfrac{1}{2}mv^2$$

Thus, an amount of work, $W = Fs$, has been performed on the object and we say that the object has acquired an amount of *energy* equal to $\frac{1}{2}mv^2$. This energy, which the object possesses *by virtue of its motion*, is called *kinetic energy*:

Kinetic energy: $KE = \tfrac{1}{2}mv^2$ (7.3)

Clearly, the unit of kinetic energy is the same as that for work, namely, *joules*.

• *Example 7.4*

A free particle, which has a mass of 2 kg, is initially at rest. If a force of 10 N is applied for a period of 10 s, what kinetic energy is acquired by the particle?

In order to calculate the kinetic energy we must first compute the final velocity acquired by the particle. Using $a = F/m$, we have

$$v = at = \left(\frac{F}{m}\right)t = \left(\frac{10 \text{ N}}{2 \text{ kg}}\right) \times (10 \text{ s}) = 50 \text{ m/s}$$

Then,

$$KE = \tfrac{1}{2}mv^2 = \tfrac{1}{2} \times (2 \text{ kg}) \times (50 \text{ m/s})^2 = 2500 \text{ J}$$

How much work was done by the applied force? The distance moved is

$$s = \tfrac{1}{2}at^2 = \tfrac{1}{2}\left(\frac{F}{m}\right)t^2 = \tfrac{1}{2} \times \left(\frac{10 \text{ N}}{2 \text{ kg}}\right) \times (10 \text{ s})^2 = 250 \text{ m}$$

so that the work done is

$$W = F \times s$$
$$= (10 \text{ N}) \times (250 \text{ m}) = 2500 \text{ J}$$

Thus, the work done is transformed entirely into the kinetic energy of the particle. ∎

The Energetics of Running

Consider an object that is being moved at constant velocity across a horizontal surface. The only work that must be done to maintain this motion is that necessary to overcome friction and air resistance. In the case of a runner, the frictional effects are not large, and yet a considerable expenditure of energy is involved in running at constant velocity. Energy is consumed in the up-and-down bobbing motion of running and in the collisions of the feet with the ground. Also, energy is converted into heat by the runner's body. An additional reason for energy loss is that the runner's legs, which constitute approximately 30 percent of the body mass (see the essay on page 85), are continually being accelerated and decelerated as the runner strides. Therefore, a considerable amount of work is being done by the leg muscles just to propel the remainder of the body forward at constant velocity.

The amount of work done by a runner's leg muscles is

$$W = Fd = \tfrac{1}{2}mv^2$$

where F is the muscular force, d is the distance through which the muscles act in each striding motion, and m is the leg mass. From the essay on scaling (page 8), we know that the muscular force F is proportional to the square of the characteristic length L. Furthermore, the distance d is proportional to L and the mass m is proportional to L^3. Therefore,

$$v^2 = \frac{2Fd}{m} \propto \frac{L^2 \times L}{L^3} = \text{constant}$$

Thus, the speed that a runner can maintain does not depend on size. In fact, this analysis is not limited to human running. All animals that have similar shapes will have similar running speeds regardless of their sizes. The list of animal running speeds below indicates that this conclusion is approximately correct. With the exception of the entry for humans, there is only a difference of a factor of 2 between the fastest and the slowest speeds.

Animal Running Speeds

Animal	Speed (m/s)
Cheetah	30
Gazelle	28
Ostrich	23
Fox	20
Horse	19
Rabbit	18
Wolf	18
Racing dog	16
Human	11

Humans are not very efficient runners because their propulsion is provided by muscles located entirely in the moving legs, thereby increasing the mass that must be accelerated and decelerated. The fastest animals are those with slender legs and with most of the muscle mass located within the body.

If animal running speeds were not approximately independent of size, the natural balance between preying animals and their prey would be upset. Notice that the wolf and the rabbit have essentially the same running speed; the same is true of the cheetah and the gazelle. The large cats do not have the slender legs of the cheetah and have somewhat lower running speeds. The large leg muscles of these cats are adapted for leaping (i.e., high acceleration) instead of high-speed running and they catch their prey by pouncing on it after a stealthy approach. If a lioness (the chief hunter) cannot down her prey after a sudden leap and a chase of a few strides, she will give up and await a better opportunity.

As humans evolved, presumably at some stage they ceased using all four limbs for locomotion and began to stand and move erect. Thus, their hands became free for other purposes. It has been suggested that humans have paid a price in running efficiency for this change in function of their upper limbs. Recent experiments, however, have demonstrated that the change from a four-legged to a two-legged stance does not significantly affect running efficiency. A group of chimpanzees was trained to run using both four and two limbs. Measurements of oxygen consumption showed that running at constant speed with either style required the same amount of muscular work. Therefore, it appears that humans' lower running speed compared to most other animals is not a result of the fact that they use only two legs.

Reference

C. R. Taylor and V. J. Rowntree, "Running on Two or on Four Legs: Which Consumes More Energy?" *Science* **179**, 186 (12 January 1973).

■*Exercise*

1 Estimate the average muscular force in each leg that is necessary to propel a human at top running speed. Compare this figure with the person's total weight.

Potential Energy—
The Energy of Position

Suppose we lift an object of mass m, originally at rest, to a position that is a height h above the initial position and leave the object again at rest. Clearly, we have done work against the gravitational force, but there is no net change in velocity and therefore we have imparted no kinetic energy. However, the object *does* possess energy by virtue of its *position*. We can easily see this by allowing the object to fall toward its original position. After falling through a distance h, it will have acquired a velocity $v = \sqrt{2gh}$ (see Eq. 2.16) and its kinetic energy will be

$$KE = \tfrac{1}{2}mv^2 = \tfrac{1}{2}m(2gh) = mgh$$

This amount of energy, mgh, may be accounted for as follows. In order to balance the gravitational force on an object of mass m, the application of a force mg is required. If a force infinitesimally greater[3] than mg acts through a distance h (the height to which the object is raised), an amount of work $W = Fs = mgh$ is done. The object then possesses an energy mgh that has the potential of being transformed into kinetic energy (by falling through the height h). We call this energy, which the body possesses *by virtue of its position*, the *potential energy* of the body. That is, the work done against the gravitational force is stored by the object raised and is retained as potential energy (Fig. 7.3).

Potential energy: $PE = mgh$ (7.4)

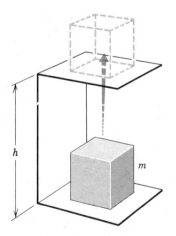

Figure 7.3 An object acquires a potential energy mgh by being raised to a height h.

---

[3] By choosing a force infinitesimally greater than mg, we are assured that we can actually *move* the object (not just balance the gravitational force); but, at the same time, we make no appreciable error by equating the work done to mgh.

Figure 7.4 The horizontal movement of the box at constant velocity involves no work being done against any force acting on the box.

Conservative Forces

Work is done only when a force acts to displace an object along the direction of the force. The gravitational force always acts *downward*. Therefore, if a force acts to displace an object *upward*, then work will be done by that force against the gravitational force. What happens when an object is moved *horizontally* (in the absence of friction)? Consider the situation in Fig. 7.4. The person is carrying a box and is moving at constant velocity. The box is neither gaining nor losing speed, so there is no change in its kinetic energy. Furthermore, the box is neither gaining nor losing height, so there is no change in its potential energy. There are only two forces acting on the box: the downward force due to gravity and the upward force exerted by the person's hand. There is no motion of the box in the vertical direction, so neither force acting on the box does any work. (Work *is* being done by the person's muscles during this motion. After all, the person will feel tired after carrying the box. But the internal work being done to trigger the muscles into action does not alter the fact that no work is done by either force that acts on the *box*.)

Suppose that we raise a block with mass m from the floor straight up to a height h (Fig. 7.5a). We know that the work required is mgh. Next, suppose that we raise the same block from the floor to the height h but this time we follow the path shown in Fig. 7.5b. We have just argued that horizontal motions do not involve work done. Because the vertical portions of the path are the same in (b) as in (a), we conclude that the total amount of work done in the second case is also mgh. Finally, we note that any arbitrary displacement can always be represented by a combination of horizontal and vertical displacements. Only the vertical portions require that work be done. Consequently, if we move the block from the floor to the height h along *any* path, even that shown in Fig. 7.5c, the work done is still mgh.

If the amount of work done against a force depends only on the initial and final positions of the object moved (and not on the path followed), we call

134 Energy

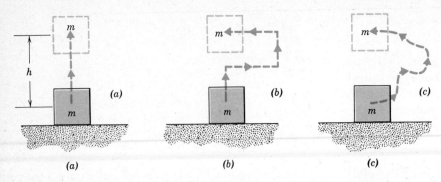

Figure 7.5 The work done in lifting the block from the floor to the height h is mgh regardless of the path taken.

this force a *conservative force*. We have just seen that the constant gravitational force that acts near the surface of the Earth is a conservative force. In fact, the gravitational force is conservative even in its long-range actions. The electrostatic force is another conservative force.

In general, any force whose direction is always along the line connecting the two objects that are interacting is a conservative force. Such forces are called *central forces* and all central forces are conservative.[4]

Frictional forces are *nonconservative* because the amount of work done against friction depends on the total path length, not merely on the initial and final positions. It *does* require work, for example, to slide a box across the floor and back to the starting point.

Energy and Work

As we have mentioned earlier, *energy is the capacity to do work*. The potential energy acquired by an object in being raised to a certain height can be converted into work in a number of ways. One way of accomplishing this is by means of a *pile-driver* as illustrated schematically in Fig. 7.6. By falling

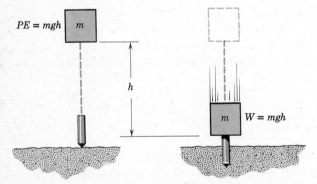

Figure 7.6 The potential energy of the block is transformed into kinetic energy by falling and does work by driving the stake into the ground.

---

[4]Except for those special forces that depend on the relative velocity of the objects.

through a height h a block of mass m transforms its potential energy mgh into kinetic energy $\frac{1}{2}mv^2$, where $v = \sqrt{2gh}$. The moving block, in being stopped, can do an amount of work $mgh = \frac{1}{2}mv^2$. In Fig. 7.6, this amount of work is expended in driving the stake into the ground and is equal to $\overline{F} \times \Delta s$, where $\overline{F}$ is a very large average force and Δs is the small distance through which the stake moves. (We assume that no work was done in deforming the stake nor, as we will discuss later, was any work converted into heat.)

● *Example 7.5*

How much work is required to raise a 0.1-kg block to a height of 2 m and simultaneously give it a velocity of 3 m/s?

The work done is the sum of the potential energy, $PE = mgh$, and the kinetic energy, $KE = \frac{1}{2}mv^2$:

$PE = mgh = (0.1 \text{ kg}) \times (9.8 \text{ m/s}^2) \times (2 \text{ m})$
$\quad = 1.96 \text{ J}$

$KE = \frac{1}{2}mv^2 = \frac{1}{2} \times (0.1 \text{ kg}) \times (3 \text{ m/s})^2$
$\quad = 0.45 \text{ J}$

$W = PE + KE$
$\quad = 1.96 \text{ J} + 0.45 \text{ J} = 2.41 \text{ J}$ ∎

Power

Power is the *rate* at which work is done or energy is expended. That is,

$$\text{power} = \frac{\text{work}}{\text{time}}$$

or,

$$P = \frac{W}{t} \tag{7.5}$$

If a machine can lift an object to a certain height in 20 s and if another machine can perform the same task in 10 s, the second machine is working with twice the power of the first machine.

Work is measured in *joules* and time is measured in *seconds,* and so the unit of power is the *joule per second* (J/s). To this unit we give the special name *watt* (W):

1 J/s = 1 W

Also, 10^3 W = 1 kilowatt (kW) and 10^6 W = 1 megawatt (MW).

The unit of power is named in honor of the Scottish engineer, James Watt (1736–1819), whose improvements of existing designs resulted in the first commercially practical steam engine.

Power is often measured in terms of another unit called the *horsepower* (h.p.) which is now defined in terms of the watt:

1 h.p. = 746 W
$\cong \frac{3}{4}$ kW

● *Example 7.6*

During a 5-min period, a man lifts 75 boxes, each with mass m = 20 kg, from floor level to a height of 1.5 m. At what average power does the man work?

The total amount of work done in lifting n boxes (each with mass m) through a height h is $W = nmgh$. Therefore, the rate at which the person does work is

$$P = \frac{W}{t} = \frac{nmgh}{t}$$

$$= \frac{75 \times (20 \text{ kg}) \times (9.8 \text{ m/s}^2) \times (1.5 \text{ m})}{(5 \text{ min}) \times (60 \text{ s/min})}$$

= 73.5 W

$\cong \frac{1}{10}$ h.p. ∎

7.3 Conservation of Energy

Our Most Useful Physical Principle

In the simple examples given in the preceding sections it was implicit that potential energy could be transformed into kinetic energy, and *vice versa,* without any loss of energy. That is, a mass m in falling through a height h acquires a kinetic energy $\frac{1}{2} mv^2$, where $mgh = \frac{1}{2} mv^2$, or $(PE)_{\text{initial}} = (KE)_{\text{final}}$. Thus, energy (in one form or another) has been *conserved* during the process. This is, in fact, a general result embodied in the principle (or *law*) of *the conservation of energy.*

Energy conservation is a far-reaching principle, and, just as for linear momentum and angular momentum, there is no known exception to the rule that energy is conserved in every physical process. Indeed, we have the attitude that if we do not find a balance of energy that exactly makes up the deficit! This is not really a trick or a dishonest attempt to cover up our ignorance about Nature, for once we invent a new form of energy we must therefore use the same definition and always incorporate this new form into our calculations in the same way. If we have made a poor choice, we will rapidly come upon a contradiction. Following this attitude, we have invented, for example, thermal energy, electromagnetic energy, and nuclear energy. Using these ideas consistently, we find no contradictions.

Many problems in which complicated forces are involved and therefore the solutions to which are extremely difficult to construct by using Newton's laws can, nevertheless, be solved in a simple way by using the conservation laws, particularly energy conservation. We will use energy conservation to solve many of the problems in the remainder of this book. We will discuss thermal energy in Chapters 9 and 10, and nuclear energy will be treated in Chapters 17 and 21. Electric energy is the subject of Chapter 12.

● *Example 7.7*

A 3-kg block starts from rest and slides 4 m along a plane that is inclined to the horizontal at an angle of 35°. At the 4-m position the speed of the block is 5 m/s. What is the coefficient of friction between the block and the plane?

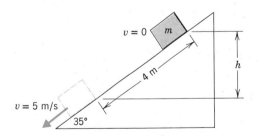

First, we need to know the amount of energy dissipated by friction. The initial potential energy relative to the final position of the block is

$PE = mgh = (3 \text{ kg}) \times (9.8 \text{ m/s}^2) \times (4 \text{ m}) \times (\sin 35°)$
     = 67.5 J

The kinetic energy at the final position is

$KE = \frac{1}{2} mv^2 = \frac{1}{2} \times (3 \text{ kg}) \times (5 \text{ m/s}^2)$
$= 37.5 \text{ J}$

The total amount of energy in the system must remain constant. This total energy is the initial potential energy, 67.5 J. The final energy of the block is the kinetic energy, 37.5 J, so the difference in these energies, or 30 J, must have been dissipated by friction (and appears as thermal energy or heat, as we will discuss in Chapter 9). That is, 30 J represents the work done against the frictional force f_k:

$W = f_k s = (\mu_k N) \times s = (\mu_k mg \cos \theta) \times s$

So that

$\mu_k = \dfrac{W}{mgs \cos \theta}$

$= \dfrac{30 \text{ J}}{(3 \text{ kg}) \times (9.8 \text{ m/s}^2) \times (4 \text{ m}) \times (\cos 35°)}$

$= 0.31$ ∎

Only Energy Differences Are Meaningful

Unlike length and mass, *energy* has no absolute value. Suppose we ask the question "How much potential energy does a certain body have?" The answer is not simply "*mgh*," because then one can ask "*h* above what?" One could always drop the object into a hole and release some extra potential energy as kinetic energy. Therefore, the potential energy of the object at the surface of the Earth surely is not zero. The absolute value of the potential energy at any particular point has, in fact, no physical significance; it is only the *difference* in potential energy between two points that is important. Figures 7.7 and 7.8 emphasize this property of potential energy. In Fig. 7.7 we have two identical blocks that are at different heights above the Earth's surface (level *B*). If the blocks are released, the block on the left will fall through a height of 1 m and strike surface *A*; the right-hand block will also fall through a height of 1 m and strike surface *B*. Each block will therefore convert the same amount of potential energy ($mgh = 9.8$ J) into work on striking the surface to which it falls. Alternatively, we could say that the left-hand block begins at a height of 4 m above the surface of the Earth and therefore has $PE_{\text{initial}} = mgh = (1 \text{ kg}) \times (9.8 \text{ m/s}^2) \times (4 \text{ m}) = 39.2$ J. The final position is 3 m above the surface of the Earth, so $PE_{\text{final}} = (1 \text{ kg}) \times (9.8 \text{ m/s}^2) \times (3 \text{ m}) = 29.4$ J. The *change* in potential energy is converted into work:

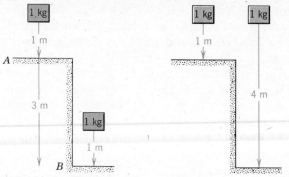

Figure 7.7 Although the two blocks are initially at different heights, each converts the same amount of potential energy to work in falling to the surface below it.

Figure 7.8 The two blocks are initially at the same height, but the right-hand block releases four times as much potential energy to work when the blocks fall to the surfaces.

$W = PE_{\text{initial}} - PE_{\text{final}} = 39.2 \text{ J} - 29.4 \text{ J} = 9.8 \text{ J}$

which is the same as we calculated for a fall through a height of 1 m.

Figure 7.8 shows the two blocks initially at the same height. However, when released they fall through different heights and therefore convert different amounts of potential energy into work.

In moving an object between two points, only the difference in potential energy can be converted into kinetic energy. Because we can always add a constant amount to the value of the potential energy at each of the two positions without altering the *difference* in potential energy between the positions, the absolute value of the potential energy is arbitrary. We have seen that it is often convenient to choose a position at which the potential energy is set equal to zero. But this is always an arbitrary choice; we specify the position of zero potential energy for convenience, not because of any physical requirement.

Kinetic energy is also a relative concept. The kinetic energy of a moving automobile, for example, appears to have different values for an observer standing on the road and for an observer in a train traveling on a track alongside the road. It is the *relative* velocity that determines the kinetic energy through the relation $\frac{1}{2} mv^2$. The kinetic energy of an object has different values with respect to different moving coordinate systems. Again, it is only the *change* in kinetic energy that is important because it is just this change in energy that appears as work in any frame of reference.

• *Example 7.8*

A roller-coaster car ($M = 200$ kg) is at the top of a 30-m hill and is moving with a speed of 4 m/s. The car dives down into a valley whose deepest point is at a height of 10 m and then climbs to the top of a 20-m hill. Assume that the car moves frictionlessly and find the speed at the top of the 20-m hill.

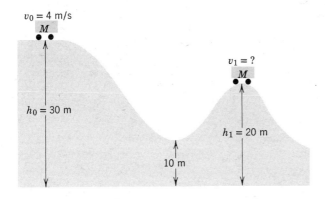

At every position, the car possesses both potential energy and kinetic energy. Because no energy is lost to friction, the sum of these two energies must remain constant:

$PE + KE =$ constant

That is,

$$Mgh_0 + \tfrac{1}{2} Mv_0^2 = Mgh_1 + \tfrac{1}{2} Mv_1^2$$

Rearranging,

$$\tfrac{1}{2} Mv_1^2 = (Mgh_0 - Mgh_1) + \tfrac{1}{2} Mv_0^2$$

The term in parentheses is just the *change* in potential energy. If we use

$\Delta h = h_0 - h_1$, we see that

$$\tfrac{1}{2} Mv_1^2 = Mg\Delta h + \tfrac{1}{2} Mv_0^2$$

Notice that the mass M appears in each term and can be canceled. The final speed v_1 therefore does not depend on the mass of the car. We can now write

$$\begin{aligned} v_1^2 &= 2g\Delta h + v_0^2 \\ &= 2 \times (9.8 \text{ m/s}^2) \times (10 \text{ m}) + (4 \text{ m/s})^2 \\ &= 212 \text{ m}^2/\text{s}^2 \end{aligned}$$

from which

$v_1 = \sqrt{212 \text{ m}^2/\text{s}^2} = 14.6$ m/s ∎

The Energetics of Jumping

The Standing High Jump

If a human or an animal crouches and then uses the leg muscles to jump vertically, the center of mass of the body can be raised to a certain height. If the crouch lowers the center of mass a distance d, this is the distance through which the muscles work during the jump, and the amount of work done is

$W = Fd = mg(d + h)$

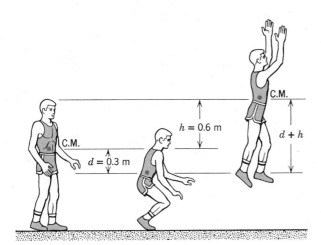

Figure 1 The standing high jump.

where F is the force exerted by the muscles, m is the total body mass, and $d + h$ is the height through which the body center of mass is raised from the crouch position to maximum height (see Fig. 1). In the essay on scaling (page 8), we found that $F \propto L^2$ and $m \propto L^3$; also, d is proportional to L, the characteristic length. Then, the total distance through which the center of mass is raised is

$$h + d = \frac{Fd}{mg} \propto \frac{L^2 \times L}{L^3} = \text{constant}$$

Most jumping animals (a human is an exception) can jump much higher than the crouch distance. That is, h is rather large compared to d. Thus, we can conclude that the height of jump by animals of similar shape should be approximately independent of the size. Indeed, the tiny rat kangaroo (which is about the size of a rabbit) can jump about the same height as the giant kangaroo (about 2.5 m).

The best vertical jump that an average athletic male can make raises his center of mass about 0.6 m (Fig. 1). (Some talented "leapers" can raise the center-of-mass point by nearly 1 m.) During this process the leg muscles act through a distance of about 0.3 m. Therefore, the muscular force required for the jump is

$$F = \frac{mg(d + h)}{d} = mg + mg \times \frac{h}{d}$$

$$= w\left(1 + \frac{h}{d}\right) = w\left(1 + \frac{0.6 \text{ m}}{0.3 \text{ m}}\right)$$

$$= 3w$$

where $w = mg$ is the man's weight. Thus, the force exerted by the leg muscles in the jump is about 3 times the total body weight.

The Running High Jump

In the running high jump, the jumper must raise his body to clear a horizontal bar. The world record for this type of jump is about 2.3 m. This means that the jumper must raise his center of mass from its normal height of about 1.0 m (see the essay on page 85) through a distance of 1.3 m to reach the bar height. But because the center of mass of the body lies *within* the body, the center of mass must be raised about 0.10 m higher in order to clear the bar using the *straddle* technique (Fig. 2). (Some high jumpers can so distort their bodies in clearing the bar that the center of mass does not rise above the level of the bar. See also Exercise 4.) Therefore, the total height through which the jumper must raise his center of mass is

$$h = 2.3 \text{ m} + 0.10 \text{ m} - 1.0 \text{ m} = 1.40 \text{ m}$$

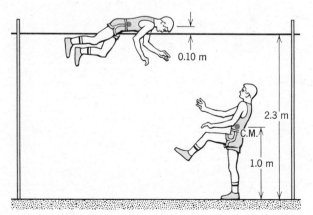

Figure 2 The running high jump.

We have seen that a jumper can raise his center of mass in a vertical jump by about 0.6 m. The remaining 0.80 m required to clear the horizontal bar must be derived from the running motion. That is, some of the kinetic energy of the horizontal running motion must be converted into jumping energy. A high jumper does not approach the bar with full sprinter's speed; many jumpers approach rather slowly. Let us assume $v = 6$ m/s. Then, the kinetic energy of a 70-kg jumper is

$KE = \frac{1}{2} mv^2 = \frac{1}{2} (70 \text{ kg}) \times (6 \text{ m/s})^2 = 1260$ J

The energy necessary to negotiate the remaining 0.80 m is

$\mathscr{E} = mgh = (70 \text{ kg}) \times (9.8 \text{ m/s}^2) \times (0.80 \text{ m}) = 549$ J

Thus, the jumper actually needs to convert less than half of his running energy into jumping energy. If it were possible to make this conversion more efficiently, the jumper would be able to clear a considerably greater height.

The Pole Vault

Although a jumper cannot convert a large fraction of his running energy into jumping energy using his legs alone, he can make this conversion with relatively high efficiency by using a vaulting pole. The vaulter runs at top speed carrying a long flexible pole (modern poles are made of fiberglass). He jams the tip of the pole into a cup at the base of the bar structure and his forward motion bends the bar almost double (Fig. 3). In this process, his running kinetic energy is converted into elastic energy in the pole. As the pole unbends, the elastic energy does work on the vaulter by raising him to clear the bar. If all of the vaulter's kinetic energy were used to raise him, we would have

$\frac{1}{2} mv^2 = mgh$

Then, the bar height H he could clear would be h plus the original height of his center of mass (1.0 m) plus the height of a standing vertical jump (0.6 m). (We cannot add the height of a *running* high jump because we assume that all of the running kinetic energy is converted into elastic energy in the pole.) Then,

$H = h + 1.0 \text{ m} + 0.6 \text{ m} = \dfrac{v^2}{2g} + 1.6 \text{ m}$

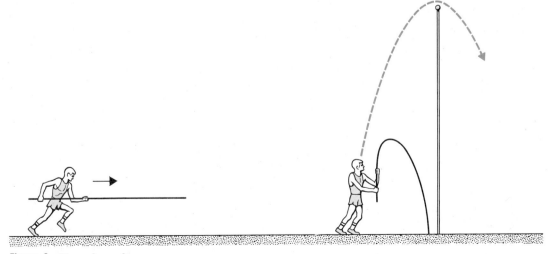

Figure 3 The pole vault.

If we assume a top speed of 9.5 m/s (we should not use the maximum speed of 10.5 m/s because the vaulter is carrying the pole),

$$H = \frac{(9.5 \text{ m/s})^2}{2 \times (9.8 \text{ m/s}^2)} + 1.6 \text{ m} = 4.6 \; m + 1.6 \; m = 6.2 \text{ m}$$

This estimate for H is slightly too large because not all of the vaulter's kinetic energy can be used to bend the pole; he must have some residual horizontal velocity to move across the bar. The current world record for the pole vault is about 5.5 m. It is apparent that the elastic vaulting pole is quite efficient in using the vaulter's kinetic energy to raise him over the bar. (One reason for the apparently high efficiency is that we have neglected the push that the vaulter exerts with his arms on the pole; this push increases the height.)

References

R. McNeill Alexander, *Animal Mechanics,* University of Washington Press, Seattle, 1968, Chapter 1.

P. Davidovits, *Physics in Biology and Medicine,* Prentice-Hall, Inc., Englewood Cliffs, N.J., 1975, Chapter 3.

G. Dyson, *The Mechanics of Athletics,* University of London Press, London, 1968, Chapters 8 and 9.

J. G. Hay, "Straddle or Flop?" *Athletic Journal* **55,** 8 (April 1975).

■ Exercises

1. We have argued here that the height of jump by animals of similar construction should be approximately the same. The body structure of a flea, for example, is quite different from that of a human or any other mammal. Furthermore, the jumping ability of a flea is inhibited by the relatively large effect of air resistance. (The ratio of surface area to mass is much greater for a flea than for mammals.) It is therefore not surprising that the jumping height of a flea is rather small compared to that for a human or a kangaroo (but it is large compared to its own size!). From its crouched position, an oriental rat flea can accelerate to a vertical velocity of 2 m/s in a time of 1.5 ms and then rise to a height of 0.090 m. Express the acceleration in units of g. Express the average frictional force due to air resistance in terms of the weight of the flea. (M. Rothschild, Y. Schlein, K. Parker, C. Neville, and S. Sternberg, "The Flying Leap of the Flea," *Scientific American,* November 1973).

2. Calculate the amount of work done by the jumper ($m = 70$ kg) in Fig. 1 in moving from the crouch position to maximum height. If the muscles act for a time of 0.25 s, at what power are they working?

3. With what vertical velocity must a high jumper leave the ground to clear a bar at 2.3 m? (See Fig. 2.)

4. In the discussion of the pole vault, the additional height requirement of 0.10 m included in the running high jump, was not mentioned. Can you think of a reason why it was omitted? (Consider the form of the high jumper executing a straddle jump, as in Fig. 2, compared to that of a pole vaulter. What can you say about the motion of the vaulter's center of mass?) Using this same reasoning, explain why a high jumper using the technique called the "Fosbury Flop" is able to clear a greater height than if he were to make a straddle jump.

5 *The Energetics of Throwing.* The world's throwing records for the shotput, the discus, and the javelin are approximately 22 m, 71 m, and 95 m, respectively. The corresponding masses of these objects are 7.3, 2.0, and 0.80 kg. In addition, although no official records are kept, a good player can throw a baseball ($m = 0.14$ kg) about 120 m. Assume that each throw is made at the optimum angle (for no air resistance) of 45°. (Eq. 2.30) and compute the initial kinetic energy of each object. What is the striking feature of these numbers? Can the difference be explained, entirely or in part, by the neglect of air resistance? What other features of the problem have been neglected? Comment on the efficiency of the muscles in throwing objects with different masses. Is there any obvious reason for this?

7.4 Elastic and Inelastic Collisions

The Importance of Internal Energy

In addition to the energy that a body possesses by virtue of its motion or position, there is also the so-called *internal energy* of a body. An increase in the internal energy can take place, for example, if some of the kinetic energy of the body *as a whole* is converted into increased relative motion of the atoms or molecules that make up the body. This motional energy of the atoms or molecules does not manifest itself on a scale that can be detected by observing the motion or position of the body as a whole; instead, the result is an increase in the *temperature* of the body. In Chapter 10 we discuss in detail a theory of heat based on the motion of atoms and molecules. For the present purpose it suffices to realize that there exists the possibility of altering the internal energy of a body.

When two objects collide, we know from our previous discussion that momentum and energy are conserved. We do not require, however, that kinetic energy be conserved; it is necessary only that the *total* energy be conserved. If some of the kinetic energy is converted into thermal energy, we say that the collision is *inelastic*. In general, an inelastic collision is one in which there is a change of the internal energy of one or both of the colliding objects. If there is no change of the internal energy, the collision is *elastic*. If the amount of kinetic energy converted into internal energy is the maximum allowed by the conservation of momentum, the collision is said to be *completely inelastic*.

Figures 7.9 and 7.10 show two elastic and two inelastic collisions. If a ball dropped onto a fixed surface rebounds to its original height, an *elastic* collision has occurred (Fig. 7.9). Of course, this is an idealized process and does not occur in Nature. During the collision there is always some loss of kinetic energy to the motion of the molecules in the ball. Therefore, all such *real*, macroscopic collisions are inelastic and the ball will never rebound to its original height. Collisions between macroscopic objects are always inelastic to some degree although in favorable cases the fraction of the kinetic energy that is converted into internal energy may be quite small.

If a body of mass m collides with an object of mass M (where $n < M$), as in Fig. 7.10a, the smaller mass will rebound, thereby imparting some energy and momentum to the larger mass, which will recoil along the original direction of motion. Elastic collisions of this type can actually occur between elementary particles (such as an electron incident on a proton) because there is no way to change the "internal energy" of such particles.

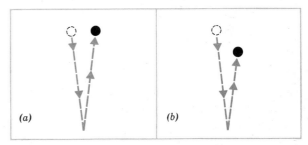

Figure 7.9 (a) The elastic and (b) the inelastic bouncing of a ball from a fixed surface.

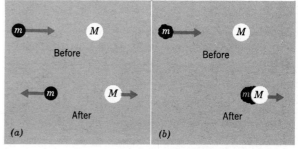

Figure 7.10 (a) An elastic and (b) an inelastic collision.

As a final example (see Fig. 7.10b), suppose that a blob of putty is projected toward an object at rest. On collision the putty sticks to the object instead of rebounding as in the previous case. In this situation conservation of momentum alone determines the final velocity of the combination. A maximum amount of kinetic energy has been converted into internal energy and the collision is said to be completely inelastic.

● *Example 7.9*

A billiard ball is projected with a velocity of 100 m/s. The ball strikes another identical billiard ball and is deflected through an angle of 60°. In what direction does the struck ball recoil and what is its velocity, assuming the collision to be elastic?

We proceed by using the equations that represent momentum and energy conservation.

Before collision the x- and y-components of the momentum are

$$p_x = m_1 v_1$$
$$p_y = 0$$

These quantities must be individually conserved. *After* collision we have for the x- and y-components of the momentum,

$$m_1 v_1' \cos\theta_1 + m_2 v_2' \cos\theta_2 = m_1 v_1 \quad \text{(x-components)}$$
$$m_1 v_1' \sin\theta_1 - m_2 v_2' \sin\theta_2 = 0 \quad \text{(y-components)}$$

where a negative sign is required in the equation for the y-components because these momenta are oppositely directed.

Because the collision is *elastic*, we also have an energy equation:

$$\tfrac{1}{2} m_1 v_1^2 = \tfrac{1}{2} m_1 v_1'^2 + \tfrac{1}{2} m_2 v_2'^2$$

Because $m_1 = m_2$, the momentum and energy equations become

$$v_1' \cos\theta_1 + v_2' \cos\theta_2 = v_1 \quad (1)$$
$$v_1' \sin\theta_1 - v_2' \sin\theta_2 = 0 \quad (2)$$
$$v_1^2 = v_1'^2 + v_2'^2 \quad (3)$$

These equations are all that we require to solve the problem because v_1 and θ_1 are known; that is, we have three equations for the three unknowns, v_1', v_2', and θ_2. We proceed as follows. First, we solve (2) for v_1'. Then, we substitute this result into (1) and solve for v_2'. Thus, we have equations for both v_1' and v_2' that can be substituted into (3). This equation then contains the single unknown, θ_2. The value of θ_2 is substituted back into the previous equations in order to solve for v_1' and v_2'. The actual calculation requires considerable algebraic manipulation and is left as an exercise for the interested reader. The results are:

$$v_1' = 50 \text{ m/s}$$
$$v_2' = 86.6 \text{ m/s}$$
$$\theta_2 = 30°$$

The important point to notice is that the angle between the deflected ball and the recoiling object ball is 60° + 30° = 90°; that is, the two velocity vectors, $\mathbf{v}_1'$ and $\mathbf{v}_2'$, are at *right angles*. This is, in fact, a general result for the collision of a particle or ball in motion with another particle or ball or *equal mass* which is at rest; the two objects always leave the point of collision at right angles. Thus, when a moving proton collides with another proton at rest the result is always[5] that the two protons take perpendicular paths away from the point collision. An example of such an event is shown in Fig. 7.11. ■

---

[5]"Always" if the velocities involved are sufficiently small so that relativistic effects can be neglected. (See also Section 17.3.)

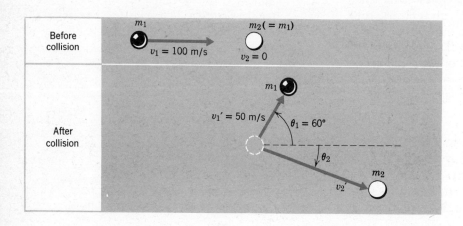

7.4 Elastic and Inelastic Collisions

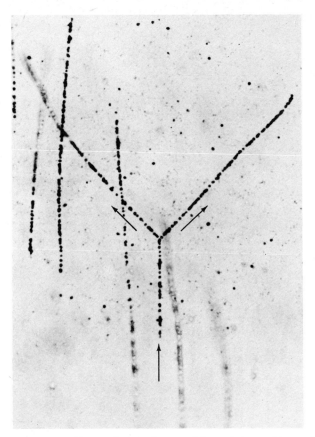

Figure 7.11 The tracks are the record in a photographic emulsion of the collision of a moving proton with a proton at rest in the emulsion. The incident proton moves upward from the bottom of the photograph. The angle between the tracks after collision is 90°.

Driving a Golf Ball—An Example of an Inelastic Collision

Figure 7.12 shows the essential quantities involved in driving a golf ball. The clubhead with mass M and velocity $\mathbf{V}_0$ strikes the golf ball, which has a mass m. After the ball has separated from the club face, the ball moves with a velocity $\mathbf{v}$ and the clubhead moves with a velocity $\mathbf{V}$.

We consider the clubhead and the ball to represent an isolated system. Then, conservation of momentum allows us to write

$$MV_0 = MV + mv$$

We can obtain a connection between the various velocities in this equation by examining the degree of elasticity in the collision between the clubhead and the ball. This is expressed in terms of a quantity called the *coefficient of restitution e*. If an object makes a collision with another object, the ratio of the relative velocity between the objects *after* collision to the relative velocity *before* collision is termed the coefficient of restitution for the particular pair of objects. For example, if a steel ball is projected against a hard steel plate, the ball will rebound with a velocity essentially equal to the incident velocity. Then, we have, approximately, $e = 1$, and the collision is almost perfectly elastic. On the other hand, if a blob of clay is thrown against the plate, the clay will stick and not rebound at all. Then, $e = 0$, and the collision is completely inelastic. For a golf ball and the head of a golf club, the value of e depends somewhat on the impact velocity. If a full driving swing is made, then e is approximately 0.7.

We can now write the ratio of the relative velocities as

$$e = \frac{\text{relative velocity after collision}}{\text{relative velocity before collision}}$$
$$= \frac{v - V}{V_0} \qquad (7.6)$$

Solving for V, we have

$$V = v - eV_0$$

Substituting this expression for V into the momentum equation, we find

$$MV_0 = M(v - eV_0) + mv$$

and solving for v,

$$v = \frac{M(1 + e)}{M + m} V_0 \qquad (7.7)$$

The mass of a regulation golf ball is 1.62 ounces, or $m = 0.046$ kg. The mass of a clubhead is, approximately, $M = 0.2$ kg. High-speed photographs show

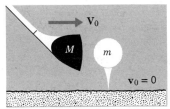 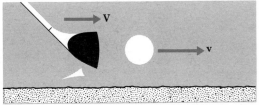

Figure 7.12 The impact of the clubhead with the golf ball sets the ball into motion with a velocity v.

that the clubhead velocity at impact is about $V_0 = 60$ m/s for long-hitting professional golfers. Thus,

$$v = \frac{(0.2 \text{ kg})(1 + 0.7)}{(0.2 \text{ kg}) + (0.046 \text{ kg})} \times (60 \text{ m/s})$$
$$= 83 \text{ m/s}$$

or about 185 mi/h.

The velocity of the clubhead after driving the ball can be obtained from the momentum equation:

$$V = V_0 - \frac{m}{M} v = 60 \text{ m/s} - \frac{0.046 \text{ kg}}{0.2 \text{ kg}} \times (83 \text{ m/s})$$
$$= 41 \text{ m/s}$$

We can now compute the kinetic energies before and after the collision:

KE of clubhead before collision:

$$= \tfrac{1}{2} M V_0^2 = \tfrac{1}{2} \times (0.2 \text{ kg}) \times (60 \text{ m/s})^2$$
$$= 360 \text{ J}$$

KE of clubhead after collision:

$$= \tfrac{1}{2} M V^2 = \tfrac{1}{2} \times (0.2 \text{ kg}) \times (41 \text{ m/s})^2$$
$$= 168 \text{ J}$$

KE of golf ball after collision:

$$= \tfrac{1}{2} m v^2 = \tfrac{1}{2} \times (0.046 \text{ kg}) \times (83 \text{ m/s})^2$$
$$= 158 \text{ J}$$

Thus, we see that

(Total *KE* before collision)
$$- \text{(Total } KE \text{ after collision)}$$
$$= 360 \text{ J} - (168 \text{ J} + 158 \text{ J})$$
$$= 34 \text{ J}$$

We therefore conclude that 34 J of the original kinetic energy has been converted into other forms of energy, particularly thermal energy and sound energy (after all, we can *hear* the club strike the ball!).

Collisions without Contact

When one billiard ball strikes another billiard ball, we say that a *contact* collision has taken place. That is, the molecules on the surface of one ball have come into close proximity with those on the surface of the other ball. The term "contact" as used here is somewhat vague, however, because it is not clear what is meant by saying that one molecule "touches" another molecule. What actually happens in a so-called "contact" collision is that the molecules interact by means of electric forces to the extent that there occurs a mutual repulsion. By extending this reasoning, we can state that, in general, the effect of a collision between two objects is due to the electric and/or gravitational forces that act between them. (Of course, for collisions involving nuclear particles, the nuclear force must be included.) Thus, it is possible to have a "collision" in which the objects never really come very close to one another. For example, a comet is swept around the Sun by virtue of the mutual gravitational attraction. This is, in fact, a form of "collision" even though no physical "contact" is involved.

7.5 Rotational Kinetic Energy

Rotational Inertia

We have discussed the kinetic energy of an object in terms of its translational motion. Now, consider an object whose center of mass is stationary but which is rotating around an axis through the center of mass. The object has no translational motion, so the kinetic energy associated with this type of motion is zero. Nevertheless, every particle within the object is moving and therefore does possess kinetic energy. We call this *rotational kinetic energy* to distinguish it from the *translational kinetic energy* that an object possesses by virtue of its translational motion.

Figure 7.13*a* shows an object in the shape of a sphere with a radius R and a mass M that is rotating with an angular velocity ω around an axis through its center. (We assume in all such cases that the distribution of mass within the object is uniform.) How

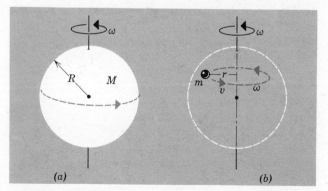

Figure 7.13 Calculation of the rotational kinetic energy of a solid object.

do we calculate the rotational kinetic energy of this object? First, look at Fig. 7.13b where we see one of the large number of particles that make up the object. This particle has a mass m and is situated at a distance r from the axis of rotation. The particle—and all other particles within the object—rotates around the axis with the angular velocity ω.

The kinetic energy of the particle is $\frac{1}{2}mv^2$, and we can express the velocity v as $r\omega$ (Eq. 2.33). Hence,

$$KE = \tfrac{1}{2}mv^2 = \tfrac{1}{2}m(r\omega)^2 = \tfrac{1}{2}mr^2\omega^2$$

To obtain the rotational kinetic energy of the entire object, we must sum the contributions from all of the particles that comprise the object:

$$KE_{rot} = \tfrac{1}{2}(\Sigma mr^2)\omega^2 = \tfrac{1}{2}I\omega^2 \qquad (7.8)$$

The quantity, $I = \Sigma mr^2$, appearing in Eq. 7.8 is the rotational inertia introduced in Section 4.3. Values for some simple-shaped objects are given in Fig. 4.11.

The expression for the rotational kinetic energy of an object has the same general appearance as that for the translational kinetic energy:

$$\text{translational kinetic energy} = KE_{trans} \qquad (7.9a)$$
$$= \tfrac{1}{2}Mv^2$$
$$\text{rotational kinetic energy} = KE_{rot} \qquad (7.9b)$$
$$= \tfrac{1}{2}I\omega^2$$

and the total kinetic energy when both forms of motion are present, is the sum of these two terms:

$$KE_{total} = KE_{trans} + KE_{rot} \qquad (7.10)$$

● *Example 7.10*

A solid sphere ($M = 12$ kg, $R = 8$ cm) rolls without slipping down an inclined plane, starting from rest, from a height of 2 m to ground level. Assuming no frictional losses, what is the translational velocity v of the sphere when it is rolling across the ground?

Using energy conservation, we can write

$$PE = KE_{trans} + KE_{rot}$$
$$Mgh = \tfrac{1}{2}Mv^2 + \tfrac{1}{2}I\omega^2$$

For a solid sphere, $I = \tfrac{2}{5}MR^2$ (Fig. 4.11), and using $\omega = v/R$, we have

$$Mgh = \tfrac{1}{2}Mv^2 + \tfrac{1}{2} \times (\tfrac{2}{5}MR^2) \times (v/R)^2$$
$$= \tfrac{1}{2}Mv^2 + \tfrac{1}{5}Mv^2$$
$$= \tfrac{7}{10}Mv^2$$

Solving for the velocity v,

$$v = \sqrt{\tfrac{10}{7}gh}$$
$$= \sqrt{\frac{10 \times (9.8 \text{ m/s}^2) \times (2 \text{ m})}{7}}$$
$$= 5.29 \text{ m/s}$$

Notice that the final velocity is *less* than the velocity we would find for a block sliding down the plane (for which $v = \sqrt{2gh}$). The reason is that some of the final kinetic energy appears in the form of rotational energy.

In obtaining the solution to this problem, we used the expression, $\omega = v/R$, where ω is the angular velocity of the rotational motion around an axis through the center of the sphere and where v is the translational velocity of the center of mass. Can you see why this expression for ω is valid? The sphere is rotating instantaneously around the point of contact with the inclined plane; if there is no slipping, the angular velocity of the center of mass of the sphere around this point is the same as that of the sphere around its center, as shown in the diagram below.

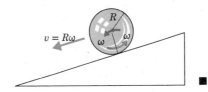

7.6 Gravitational Potential Energy

Work Required To Lift an Object

Near the surface of the Earth, where we can consider the gravitational force to be constant, we know that to lift an object of mass m through a height h requires an amount of work $W = mgh$. After lifting, we say that the object has a gravitational potential energy $PE_G = mgh$. If we consider distances that are no longer negligible compared to the radius of the Earth, we must take account of the fact that the gravitational force varies as the square of the distance from the center of the Earth.

We now calculate the amount of work required to raise an object of mass m from an initial position that is a distance r_1 from the center of the Earth, to a final position that is a distance r_2 from the center of

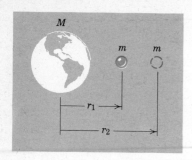

Figure 7.14 Work done by an outside agency is required to move m from r_1 to r_2.

the Earth, without any change in the kinetic energy of the object (see Fig. 7.14). The force on m at r_1 is

$$F_1 = G\frac{Mm}{r_1^2} \quad \text{at } r_1$$

and at r_2 the force is

$$F_2 = G\frac{MM}{r_2^2} \quad \text{at } r_2$$

We know that the amount of work done on an object is equal to the average force exerted multiplied by the distance through which the object is moved:

$$W = \overline{F} \times s \qquad (7.11)$$

For the situation in Fig. 7.14, the distance moved is clearly $s = r_2 - r_1$, but what is the average gravitational force on m in the interval from r_1 to r_2? In Example 7.3 we calculated the average force by taking the *arithmetic* average of the forces at the initial and final positions. But in that case we were dealing with a force that varies *linearly* with distance. In the present case, we have a $1/r^2$ force and the arithmetic average is no longer appropriate. We need a procedure that yields an average force that is closer to F_1 than to F_2 because the force falls off rapidly in going from r_1 to r_2. Such an average (and the one appropriate for a $1/r^2$ force) is the *geometric* average, namely,

$$\overline{F} = G\frac{Mm}{r_1 r_2}$$

(We have not proved that this average is correct; we have only argued that it is reasonable. In spite of this arbitrariness, the result is nevertheless *exact*.) Substituting this expression for $\overline{F}$ and $s = r_2 - r_1$ into Eq. 7.11, we find

$$W_{12} = G\frac{Mm}{r_1 r_2} \times (r_2 - r_1)$$

$$= GMm\left(\frac{1}{r_1} - \frac{1}{r_2}\right) \qquad (7.12)$$

The increase in gravitational potential energy of m in moving from r_1 to r_2 is just the work required to effect this change of position; that is,

$$PE_G = GMm\left(\frac{1}{r_1} - \frac{1}{r_2}\right) \qquad (7.13)$$

Notice that the gravitational potential energy depends only on the initial and final positions of the object, r_1 and r_2. Although we considered here a particularly simple path by which the object was moved from r_1 to r_2, in fact, PE_G depends only on r_1 and r_2 for *any* path that connects the initial and final positions because the gravitational force is a conservative force.

Potential Energy Near the Surface of the Earth

Let us use the general (and exact) result for PE_G that we have just derived and compute the potential energy gained by an object in raising it to a height h above the surface of the Earth. If h is small compared with R_E, the radius of the Earth, then we expect that we will obtain our previous result, that is, $PE_G = mgh$. From Eq. 7.13 the gain in potential energy is raising m from R_E to $R_E + h$ is

$$PE_G = GMm\left(\frac{1}{R_E} - \frac{1}{R_E + h}\right)$$

$$= GMm\frac{h}{R_E(R_E + h)} \qquad (7.14)$$

If h is small compared to R_E ($h \ll R_E$), we can neglect h in the denominator and write

$$PE_G \cong \frac{GMmh}{R_E^2} \qquad (7.15)$$

Now, the force on m at the surface of the Earth is

$$F_G = G\frac{Mm}{R_E^2}$$

But this force is just mg, so

$$mg = G\frac{Mm}{R_E^2}$$

Therefore,

$$GMm = mgR_E^2$$

Substituting this value for GMm into Eq. 7.15 the potential energy becomes

$$PE_G \cong \frac{(mgR_E^2)h}{R_E^2}$$

or,

$$PE_G \cong mgh \quad \text{(for } h \ll R_E\text{)} \tag{7.16}$$

which is our previous result.

Establishing the "Zero" of Potential Energy

According to our general expression (Eq. 7.13), the increase in gravitational potential energy in raising an object of mass m from the surface of the Earth (radius R_E) to a position that is a distance r from the Earth's center is

$$PE_G = GMm\left(\frac{1}{R_E} - \frac{1}{r}\right) \tag{7.17}$$

The term GMm/R_E is a constant and so does not affect *differences* in PE_G for various values of r. Because the zero level of PE_G is arbitrary, we can subtract this term from the expression for PE_G and write

$$PE_G = -\frac{GMm}{r} \tag{7.18}$$

We see from this equation that $PE_G = 0$ as r becomes indefinitely large, $r \to \infty$. That is, we arbitrarily select the zero level of PE_G to be the situation in which the two objects are separated by an infinite distance. The use of an infinite separation distance as the reference level for PE_G simply means that we measure PE_G relative to a configuration (infinite separation) for which there is no interaction between the objects.

According to our choice of position for zero gravitational potential energy, the value of PE_G is always *negative* and increases (that is, becomes less negative) as r is increased, becoming zero as $r \to \infty$ (seeing Fig. 7.15). Work is always required to increase the potential energy (that is, to separate the bodies by acting against the attractive gravitational force).

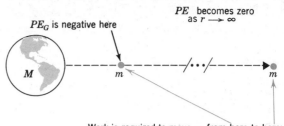

Figure 7.15 Work is required to *increase* the gravitational potential energy from a negative value to a less negative value or to zero (as $r \to \infty$).

Potential Energy Differences

It must be remembered that only *changes* in potential energy are physically meaningful. Therefore, we are usually concerned with calculating $\Delta(PE_G)$ in moving an object of mass m from r_1 to r_2 as measured from M. According to Eq. 7.12,

$$\Delta(PE_G) = W_{12} = -GMm\left(\frac{1}{r_2} - \frac{1}{r_1}\right) \tag{7.19}$$

If $W_{12} > 0$, then work was done by some outside agent *against* the attractive gravitational force. If $W_{12} > 0$, then work was done *by* the gravitational force and this amount of energy can be used, for example, to increase the kinetic energy of the bodies. If two masses with a certain separation are released, the attractive gravitational force will cause the masses to accelerate toward one another and the separation will decrease. Consequently, the gravitational potential energy will decrease, and there will be an equivalent gain in kinetic energy.

● *Example 7.11*

The *escape velocity* v_e is the minimum velocity with which an object must be propelled from the surface of the Earth if it is to move an infinite distance away, thereby "escaping" from the gravitational attraction of the Earth. (We neglect here the fact that a real object launched from the surface of the Earth would encounter air resistance before leaving the atmosphere.) In order to calculate v_e, we note that the initial kinetic energy at the Earth's surface,

$\frac{1}{2}mv_e^2$, will be completely expended in raising the initial gravitational potential energy, $-GMm/R_E$, to zero as $r \to \infty$.

Using energy conservation, we can write

$(KE + PE_G)_{\text{surface of Earth}} = (KE + PE_G)_{r \to \infty}$

or,

$\frac{1}{2}mv_e^2 - \frac{GMm}{R_E} = 0$

from which

$v_e = \sqrt{2GM/R_E}$

Substituting $G = 6.67 \times 10^{-11}$ N-m^2/kg^2, $M = 5.98 \times 10^{24}$ kg, and $R_E = 6.37 \times 10^6$ m, we find

$v_E = 1.12 \times 10^4$ m/s

= 11.2 km/s

= 25,040 mi/h   (for escape from Earth)

Actually, such a velocity is not sufficient to allow an object to escape from the solar system because the gravitational influence of the Sun is so much greater than that of the Earth. In order to calculate the velocity of escape *from the Sun,* starting from the position of the Earth, we use the same expression for v_e but substitute the mass of the Sun for $M(1.99 \times 10^{30}$ kg) and the Earth-Sun distance for $R_E(1.50 \times 10^{11}$ m). Then, we find

$v_e = 42.1$ km/s   (for escape from the Sun)

The result we have just obtained allows us to make a good estimate for the impact velocity of meteoroids when they strike the Earth. A particle falling toward the Sun from infinitely far away[6] (starting from rest) will acquire, by the time it has reached the position of the Earth's orbit around the Sun, a velocity just equal to that calculated above, namely, 42.1 km/s. The velocity of a meteoroid relative to the Earth is the vector sum of this velocity and the orbital velocity of the Earth around the Sun (29.9 km/s). If the meteoroid collides "head on" with the moving Earth, the impact velocity will be approximately $42.1 + 29.9 = 72.0$ km/s, whereas an "overtaking" meteoroid will have an impact velocity of approximately $42.1 - 29.9 = 12.2$ km/s. Most meteoroids have impact velocities in the range 10–70 km/s. ∎

Summary of Important Ideas

The *work* done by a force **F** in moving an object through a displacement **s** is equal to the *scalar product* of **F** and **s**; that is, $W = \mathbf{F} \cdot \mathbf{s} = Fs \cos \theta$.

An object possesses *kinetic energy* by virtue of its motion. An object possesses *potential energy* as the result of work done against an opposing force such as the gravitational or electric force.

The work done in moving an object from one point to another against a *conservative force* depends only on the initial and final positions and not on the path followed. All *central forces* (such as the gravitational and electrostatic forces) are conservative forces. Frictional forces are not conservative.

The absolute value of energy has no physical meaning; only *changes* in energy are significant.

The total energy of an isolated system remains constant—energy is *conserved*. Energy may be changed from one form to another (for example, from potential energy to kinetic energy) without loss.

The *internal energy* of a body is that energy due to the motion of the constituent atoms or molecules.

In an *elastic* collision, the internal energies of the colliding bodies remain constant; if the internal energies change, as is the most general case, the collision is *inelastic*. When two colliding objects stick together, the collision is *completely inelastic*.

A moving object can possess both *translational* and *rotational* kinetic energy. The rotational kinetic energy of an object is equal to $\frac{1}{2}I\omega^2$, where I is the *rotational inertia*.

◆ Questions

7.1 The Earth moves around the Sun in an almost circular orbit. Does the gravitational force do work to maintain the Earth in its circular orbit? Explain.

7.2 When an object is pushed across a rough surface, work is done against the frictional force. Does the potential energy increase in this case? Explain.

7.3 Does it make any sense to specify a value for the potential energy of a book on a shelf if the book remains permanently on the shelf? Under what circumstances does it become meaningful to assign a value to the book's potential energy?

---

[6] Actually, meteoroids arise from material in the solar system and so do not originate infinitely far away. But the velocity acquired by an object falling from far outside the Earth's atmosphere is not too different from that of an object originating infinitely far away. Therefore, the numerical result here is not seriously in error. A meteoroid that is visible in the sky (due to heating) is called a *meteor;* one that reaches the surface of the Earth is called a *meteorite.*

7.4 In Example 7.8 we considered a frictionless case and found that the final speed did not depend on the mass of the car. If friction is included, will the final speed still be independent of the mass? Explain.

7.5 One often hears the statement that engines or machines are inefficient, that "they waste energy." Does this mean that the energy actually disappears? Explain.

7.6 Some of the energy we use in our homes is electric energy that is generated in hydroelectric facilities. What is the ultimate source of this energy? Trace the flow of this energy from the source to your use.

7.7 One sometimes reads in a newspaper a statement such as, "The power plant generated 1000 MW of energy during the year." What is wrong with such a statement?

7.8 When a baseball player hits a ball with his bat, is this an elastic or inelastic collision? Explain. Do you have evidence for your conclusions?

7.9 When a meteoroid enters the Earth's atmosphere, it collides with the molecules in the air. Are these collisions elastic or inelastic? How do you know?

7.10 An object of mass m_1 collides "head on" with an object of mass m_2 that is initially at rest. If $m_1 < m_2$, will m_1 always reverse its direction of motion, independent of the magnitude of its initial velocity? What will happen if $m_1 > m_2$? What will happen if $m_1 = m_2$? (Assume elastic collisions.)

★Problems

Section 7.1

7.1 A person whose mass is 100 kg climbs stairs to a height of 10 m. How much work did the person do? Is there a difference between the work required to climb stairs (which are slanted) to a given height and that required to climb a ladder (which is vertical) to the same height? Explain.

7.2 A pile driver is used to drive a stake into the ground. The mass of the pile driver is 2500 kg and is dropped through a height of 10 m on each stroke. The resisting force of the ground for this stake is 4×10^6 N. How far into the ground is the stake driven on each stroke?

7.3 A force of 50 N that acts parallel to an inclined plane is required to push a 4-kg object up the plane. The total distance moved along the plane is 10 m and the final position is 4 m above the initial position. How much work was done by the force?

7.4 Refer to the previous problem. How much work was done against gravity? How much work was done against friction?

7.5* The coefficient of kinetic friction between a certain block ($m = 6$ kg) and a surface is 0.5. Find the force that must be applied to the block at an angle of 35° upward with respect to the horizontal to accelerate the block from rest to a velocity of 8 m/s in a time of 4 s.

7.6* Refer to Example 7.2. Assume that the force **F** is directed *downward* at an angle of 30° with respect to the horizontal. Take μ_k, m, and s to be the same as in the example. Find the magnitude of **F** and the amount of work done. Comment on the work done in this case compared to the value found in Example 7.2.

Section 7.2

7.7 What is the kinetic energy of a 1600-kg automobile moving with a speed of 80 km/h?

7.8 What is the kinetic energy of an athlete ($m = 75$ kg) while running a 10-s 100-m dash? (Assume constant speed.)

7.9 A baseball pitcher can throw a ball ($w = 5$ oz or 0.14 kg) at a speed of 140 km/h. How much energy is absorbed by the catcher when he catches the ball?

7.10 An object is mass 0.1 kg rests on a frictionless horizontal plane. A certain constant force is applied to the object and the object is accelerated to a velocity of 2 m/s, moving a distance of 10 m in the process. At this point an additional, decelerating force is applied to the object and after moving an additional 10 m, the object is again at rest. How much work was done on the object in moving it the first 10 m? What was the net amount of work done on the object during the entire process?

7.11 A block of mass 1 kg slides down an inclined plane from a height of 2 m. The length of the plane is 4 m. When the block reaches the bottom it is found to be traveling with a speed of 2 m/s. How much work was done against the frictional force? What was the average frictional force on the block while it was sliding?

7.12* A 3-kg block rests on a plank. The coefficients of friction for the block-plank combination are $\mu_s = 0.6$ and $\mu_k = 0.4$. One end of the plank is raised slowly until, at a certain angle, the block slides down the plane. After the block has moved 2.5 m, how much work has been done against friction? What is the kinetic energy of the block?

7.13* A spring has a force constant of 20 N/m. One end of the spring is attached to an overhead support. On the other end is attached a 0.2 kg mass. How far does the spring stretch? If the spring and mass are extended an additional 0.1 m and then released, the mass will be set into motion. What maximum velocity will the mass attain?

7.14 How much work is required to increase the speed of an automobile ($m = 1500$ kg) from 10 m/s to 20 m/s? (Neglect friction.)

Section 7.3

7.15 A water storage tank contains 2000 m³ of water and is at an average height of 40 m above ground level. How much work was required to fill the tank from a reservoir at ground level? How much work can be done by the water if it is piped to a place which is 20 m lower than ground level at the tank site? Is energy conservation violated here? Explain.

7.16 Four identical boxes of uniform composition have a mass of 10 kg and a height of 0.1 m; all rest on a floor. How much work is required to arrange the boxes in a vertical stack?

7.17 A 500-kg roller-coaster car starts from rest at a point of 35-m above ground level. The car dives down into a valley 4 m above ground level and then climbs to the top of a hill that is 24 m above ground level. What is the velocity of the car in the valley and at the top of the hill? (Neglect friction.)

7.18 In the preceding problem, the length of track from the starting position to the top of the hill is 100 m. If the car just reaches the top of the hill ($v = 0$), what is the average frictional force between the car and the track?

7.19 An automobile ($m = 2000$ kg) is traveling on a level road at a speed of 20 m/s. What is the automobile's kinetic energy? If friction were negligible, could the automobile *coast* to the top of a hill that is 15 m higher than the road? If so, what would be the speed at the top?

7.20 A person ($m = 80$ kg) runs up two flights of stairs (3 m per flight) in 4.5 s. At what power level is the person working? Do you suppose that the person could maintain this pace very long?

7.21* Look up the specifications of some new automobile and compare the rated horsepower of the engine with the power necessary to achieve a certain speed in an acceleration test. (For example, a 1200-kg automobile with a 140-h.p. engine might accelerate from rest to 60 mi/h in 9 s.

7.22 A constant force **F** moves an object with a constant velocity **v**. Show that the power expended is $\mathbf{F} \cdot \mathbf{v}$.

7.23 A common unit of energy is the kilowatt-hour (kWh), particularly on bills from the local power company. How many joules are there in 1 kWh?

Section 7.4

7.24 A 600-g glider strikes and sticks to a stationary 300-g glider resting on an "air track." If the initial velocity of the moving glider was 5 m/s, what is the final velocity of the two gliders? How much energy was lost in the collision?

7.25 A 1500-kg automobile is traveling due north with a velocity of 85 km/h when it collides with a 4000-kg truck traveling due east with a velocity of 60 km/h. The two tangled wrecks move off stuck together. What is the velocity (magnitude and direction) of the wreck just after collision? How much energy was lost in the collision?

7.26 Refer to the discussion of driving a golf ball. High-speed photographs show that the clubhead remains in contact with the ball for about 5×10^{-4} s. What is the average force on the ball? Express the result in terms of the weight of an average man. Over what distance is the clubhead in contact with the ball?

7.27 Two golf balls, one with $e = 0.60$ and one with $e = 0.80$, are each dropped from a height of

2 m onto the same hard surface. How high will each rebound?

7.28* A ball is dropped onto a steel plate from a certain height h. Each time the ball strikes the plate it loses 10 percent of its motional energy. How many collisions will be required before the ball fails to rise to a height greater than $\frac{1}{2}h$? What is the coefficient of restitution for the ball?

Section 7.5

7.29 It is estimated that in 10^8 years the Earth's day has increased by approximately 4 hours. If we estimate the Earth's rotational inertia to be

$$I = \tfrac{1}{5} M_E R_E^2$$

how much rotational energy has been lost during this time? If this energy were lost at a uniform rate, how many kW does it correspond to?

7.30 Two spherical balls roll, starting from rest, down the same inclined plane. Both balls have the same radius and mass, but one ball is a solid sphere, whereas the other is a hollow thin spherical shell. What is the ratio of their velocities as they roll down the plane?

7.31 A solid cylinder is sent rolling up an inclined plane making an angle of 30° with the horizontal. If the initial velocity is 2.5 m/s along the plane, how far up the plane will it go before it stops?

7.32 The yo-yo shown in the diagram descends by unwinding the string wrapped around its central spool. What is its velocity, starting from rest, when 60 cm of string has unwound?

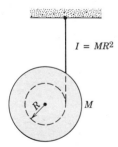

Problem 7.32

7.33* A meter stick is sent sliding along a frictionless horizontal surface. In a time interval of 5 s the 50 cm mark has traveled a distance of 16 m and the stick has completed 20 full rotations. What is the kinetic energy of the meter stick if its mass is 200 g? Consider the meter stick to be a slender rod.

7.34* A solid disk rolls down a plane that is inclined to the horizontal at an angle of 37°. The disk starts from rest at a height of 3 m above ground level. How much time is required for the disk to reach ground level? What is the acceleration of the disk down the plane expressed in units of g?

Section 7.6

7.35* How much work is required to raise a 100-kg object from the surface of the Earth to a height of 1000 km? If the object were dropped from such a height, what would be its impact velocity (neglecting air resistance)?

7.36 The Andromeda galaxy has a mass of approximately 4×10^{11} solar masses and a diameter of about 10^5 L.Y. What velocity is necessary to escape from Andromeda from a position near its outer boundary?

7.37 The Sun is but one star in the local (Milky Way) galaxy and is located near the outer edge about 30,000 L.Y. from the center. The Sun's orbital velocity about the galactic center is approximately 250 km/s. What is the period with which the Sun circles the galactic center? What is the approximate mass of the Milky Way galaxy? Assuming the Sun to be a typical star in the galaxy, estimate the number of stars in our local galaxy. What is the kinetic energy of the Sun in its galactic orbit? What is its potential energy?

7.38* A 750-kg Earth satellite is in a circular orbit at an altitude of 400 km. Because of the effect of air friction it eventually falls to the Earth's surface with an impact velocity of 3 km/s. How much energy was absorbed by the atmosphere during the fall?

7.39 Find the gravitational potential energy of a cluster of 8 stars, each of solar mass, located at the corners of a cube whose sides are 1 L.Y. in length. How much energy would it take to remove one star to infinity if the others remained fixed in their original positions?

Fluids

In our previous discussions we have been concerned exclusively with solid objects. We now come to the study of substances that do not have rigid structure or form. In this category we identify two different states of matter, that is, liquids and gases, which together we call *fluids*. The important characteristic of fluids that distinguishes this type of matter from solid matter is that fluids *flow*. In solid matter the atoms and molecules are locked in place by strong intermolecular forces and do not move relative to one another (except when the object is deformed, as in the stretching of a spring). The intermolecular forces in fluids, however, are much weaker than in solids so that fluid molecules are able to move about. The molecules in a liquid slip easily past one another, but the forces are strong enough to hold the substance loosely together. In gases, the intermolecular forces are almost nonexistent and the molecules move nearly independent of one another, filling any volume in which they are confined.

In this chapter we concentrate on the properties of *liquids,* although many of the terms and ideas that are developed and used here apply equally well to the discussion of *gases.* Unlike gases, however, liquids are essentially incompressible. To simplify matters, we shall assume throughout this chapter that when we are dealing with liquids they are incompressible. In Chapter 10 we study the unique properties of gases.

8.1 Pressure

Force per Unit Area

Both liquids and gases can exert forces. The atmosphere exerts a force on the surface of the Earth, and the ocean waters exert a force on the sea floor. Because fluids flow and adjust themselves to the shape of any container, contact exists between a fluid and the entire container surface. The force that is exerted by the fluid on the container is distributed over the entire surface. The most convenient way to describe this situation is in terms of the force *per unit area* of the surface. This quantity we call the *pressure:*

$$\text{pressure} = \frac{\text{force}}{\text{area}}$$

$$P = \frac{F}{A} \tag{8.1}$$

The units of pressure are *newtons per square meter* (N/m^2). We also use the unit 1 *pascal* (Pa) to stand for 1 N/m^2. The *pascal* is named in honor of the French scientist, Blaise Pascal (1623–1662), whose principal of hydrostatics we discuss in the next section. Later we will see some additional units that are used to measure pressure.

•*Example 8.1*

A thumbtack is pushed into a piece of wood by applying a force of 15 N to the head of the tack. The radius r of the point is 0.1 mm and the radius R of the head is 5 mm. (a) What is the pressure applied to the head of the tack and (b) what is the pressure exerted by the point on the wood?

(a) $P_{\text{head}} = \dfrac{F}{A_{\text{head}}} = \dfrac{F}{\pi R^2} = \dfrac{15 \text{ N}}{\pi \times (5 \times 10^{-3} \text{m})^2}$

$= 1.91 \times 10^5$ N/m^2 or 1.91×10^5 Pa

(b) $P_{\text{point}} = \dfrac{F}{A_{\text{point}}} = \dfrac{F}{\pi r^2} = \dfrac{15 \text{ N}}{\pi \times (0.1 \times 10^{-3})^2}$

$= 4.77 \times 10^8$ Pa

For comparison, the weight of a 75-kg human when standing upright is distributed over about 250 cm^2. The pressure that the person's feet exert on the floor is

$$P = \frac{F}{A} = \frac{Mg}{A} = \frac{(75 \text{ kg}) \times (9.80 \text{ m/s}^2)}{(0.025 \text{ m}^2)}$$

$= 2.94 \times 10^4$ Pa

which is less than 1/10,000 of the pressure exerted by the point of the tack. It is clear why pushing on the wrong end of a thumbtack is uncomfortable! ∎

When a fluid exerts a force on a surface, in what direction does this force act? If we examine the entire surface that contains a fluid, we find that both the magnitude and the direction of the force vary from place to place over the surface. Therefore, let us consider only a tiny part of the container surface. We let the area of this portion of the surface be sufficiently small so that a single vector **F** describes the force that is acting there. Next, we ask whether **F** can have a component parallel to the surface. The answer is *no*, because a fluid not in motion (a *static fluid*) cannot exert a force *along* the surface of a container. Why is this so? Suppose that we place a solid block on a surface and then exert on the block a force at some angle. We know in this case that the force exerted by the block on the surface will have a component parallel to the surface. This force is just the reaction to the force of static friction exerted on the block by the surface, and we know that frictional forces always act *along* surfaces. A block can exert a force parallel to a surface because a block has *rigidity*. A fluid, on the other hand, has no rigidity and so there is no way that a fluid at rest can exert a parallel force on a surface. We conclude, therefore, that a static fluid always exerts a force *perpendicular* to a surface (Fig. 8.1).

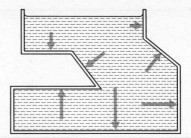

Figure 8.1 The force exerted by a static fluid on the walls of a container is always perpendicular to the surface of the container.

Pascal's Principle

Suppose that we fill a container, such as the one in Fig. 8.2, with a fluid. The fluid stands at the same level in each of the three pipes. Next, tight-fitting pistons with negligible mass are inserted into the pipes and rest on the fluid surface. If a force F_1 is applied to the piston with area A_1, the pressure exerted on the fluid in this pipe (and also the pressure exerted by the fluid on the piston) will be $P = F_1/A_1$. If there is no external force exerted on the pistons A_2 and A_3, the application of the force F_1 will cause the other pistons to rise in the respective pipes. In order to maintain the two pistons in fixed positions, the forces F_2 and F_3 must be applied (see Fig. 8.2). The magnitudes of these two forces must be such that the pressure in each pipe is the same; otherwise, with a

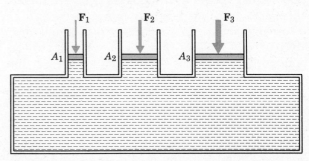

Figure 8.2 If the fluid in the container is static, the pressure exerted by the fluid on each piston must be the same: $P = F_1/A_1 = F_2/A_2 = F_3/A_3$.

pressure difference between two points in the fluid, the fluid would tend to flow toward the low-pressure point. Because we are considering a static fluid, we conclude that the pressure in each pipe is $P = F_1/A_1 = F_2/A_2 = F_3/A_3$. The areas A_2 and A_3 are greater than A_1; therefore, F_2 and F_3 are correspondingly greater than F_1.

In Fig. 8.2 the pistons are all at the same level and we found that the pressure exerted by the fluid on each piston is the same. We can make an even broader statement: *any pressure that is applied to the surface of a confined fluid is transmitted undiminished to all points within the fluid.* This idea is called *Pascal's principle,* after the discoverer, Blaise Pascal. According to this principle, the pressure applied to the first piston, $F = F_1/A_1$, is transmitted to the side walls and the bottom of the container, as well as to the other pistons. In fact, every point *within* the fluid experiences a pressure change of F_1/A_1. (It is necessary to make this statement in terms of the *change* in pressure because, as we will see in the next section, the weight of the fluid causes the pressure to increase with depth.)

● *Example 8.2*

Design a rock crusher that will exert a force of 2×10^5 N on a rock when an external force of 100 N is applied.

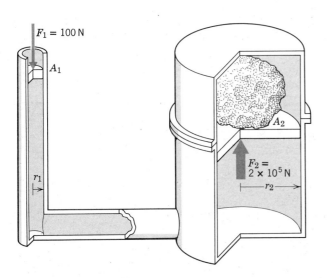

From Pascal's principle, we know that the pressure in each arm of the rock crusher is the same, so we can write

$$\frac{F_1}{A_1} = \frac{F_2}{A_2} \tag{1}$$

Let us suppose that the pipe through which the external force is applied has a radius $r_1 = 2$ cm. If the rock-crusher arm has a radius r_2, then

$$\frac{F_1}{\pi r_1^2} = \frac{F_2}{\pi r_2^2}$$

from which

$$r_2^2 = \frac{F_2}{F_1} \times r_1^2 = \frac{2 \times 10^5 \text{ N}}{10^2 \text{ N}} \times (0.02 \text{ m})^2 = 0.800 \text{ m}^2$$

Then,

$r_2 = 0.894$ m

Notice in the diagram that the area of contact between the rock and the chamber is small compared to the area of the piston. Consequently, the pressure experienced by the rock is very large. If the contact area is 2 cm^2, the pressure is

$$P = \frac{2 \times 10^5 \text{ N}}{2 \times 10^{-4} \text{ m}^2} = 10^9 \text{ Pa}$$

This example also affords a good opportunity to apply the principle of the conservation of energy. Suppose that in applying the force F_1 the small piston moves down a distance ℓ_1 and the rock-crusher piston moves up a distance ℓ_2 (while exerting the force F_2). The energy input to the device is the work done by the force F_1, namely $W_1 = F_1\ell_1$. The energy delivered by the device is $W_2 = F_2\ell_2$. If the liquid is incompressible, then the volume $A_1\ell_1$ must be the same as the volume $A_2\ell_2$, thus

$$\ell_1/\ell_2 = A_2/A_1 \tag{2}$$

Also, an incompressible liquid cannot store any internal energy. (If the liquid were compressible it would act like a spring and could store energy, see Example 7.3.) It then follows from the conservation of energy that

$$W_1 = W_2$$

or,

$$F_1\ell_1 = F_2\ell_2 \tag{3}$$

Combining Eqs. 2 and 3 results in Pascal's principle stated in Eq. 1. ∎

8.2 Pressure within Static Fluids

The Variation of Pressure with Depth

What pressure does a static fluid exert on the bottom of a container in which the fluid stands at a height h (Fig. 8.3)? The *force* that the fluid exerts on the bot-

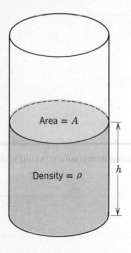

Figure 8.3 The pressure in a fluid at a depth h is due to the weight of the fluid above that depth: $P = \rho g h$.

tom of the container is just the *weight* of the fluid, $F = w = mg$. The volume of the fluid is $V = Ah$, and is the fluid density is ρ, we can express the mass as $m = \rho V = \rho A h$. Then,

$$F = mg = (\rho A h) \times g = \rho g h \times A$$

and the pressure is

$$P = \frac{F}{A} = \rho g h \qquad (8.2)$$

The distance h in Eq. 8.2 is not necessarily the total depth of the fluid. At *any* depth h, whether or not this corresponds to the bottom of the fluid, the fluid pressure is $\rho g h$. This pressure is due to the presence of the fluid alone. If the top of the container is open to air, the *absolute pressure* at the bottom of the fluid must also include the atmospheric pressure, discussed in the next section.

● *Example 8.3*

What is the fluid pressure (a) at a depth of 50 m in a lake and (b) at the bottom of the ocean where the depth is 8000 m?

(a) $P = \rho g h$
$= (10^3 \text{ kg/m}^3) \times (9.80 \text{ m/s}^2) \times (50 \text{ m})$
$= 4.90 \times 10^5 \text{ Pa}$

(b) The density of sea water is $1.025 \times 10^3 \text{ kg/m}^3$, so

$P = \rho g h$
$= (1.025 \times 10^3 \text{ kg/m}^3) \times (9.80 \text{ m/s}^2) \times (8000 \text{ m})$
$= 8.04 \times 10^7 \text{ Pa}$

Notice that the density must be expressed in units of kg/m^3 in order to obtain the pressure directly in N/m^2 or Pa. ∎

Atmospheric Pressure

The Earth is surrounded by a blanket of air and this air has weight. The force exerted on each square meter of the surface of the Earth (at sea level) amounts to just over 10^5 N, under normal atmospheric conditions. That is, normal atmospheric pressure (called 1 *atmosphere* or 1 *atm*) is

$$1 \text{ atm} = 1.013 \times 10^5 \text{ Pa} \qquad (8.3)$$

The *atm* is another unit we often use for the measurement of pressure. In English units one atmosphere corresponds to 14.70 lbs/in^2. The combination of units lbs/in^2 (pounds per square inch) is usually abbreviated psi.

At a great depth in the ocean, even at the bottom of the deepest ocean trench, the density of the water differs by only a small amount from the density at the surface. The reason is that water has a very low *compressibility* (all liquids have this property). That is, when a volume of water is subjected to a large pressure, the volume does not decrease very much. Thus, the density of water increases only slightly as the pressure is increased. On the other hand, gases compress very easily. In a column of air that stands above the Earth's surface, any portion of the air at the bottom is compressed by the weight of all the air above it. As a result, the density of the Earth's atmosphere decreases rapidly with increasing altitude. Figure 8.4 shows the effect of this density decrease. Notice that only 1/100,000 (10^{-5}) of the atmosphere lies above an altitude of 80 km and that all but 10 percent of the atmosphere is concentrated in the first 16 km above the surface.

At altitudes of a few tens of kilometers the value of g is still essentially equal to the surface

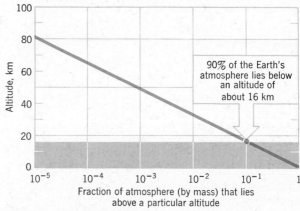

Figure 8.4 The fact that air is easily compressed by its own weight means that most of the Earth's atmosphere is concentrated near the surface.

value of 9.8 m/s². Therefore, we can compute the mass of the column of air above 1 m² of the Earth's surface from

$$m = \frac{W}{g} = \frac{F}{g} = \frac{10^5 \text{ N}}{9.8 \text{ m/s}^2} \cong 10^4 \text{ kg}$$

If a 1 m × 1 m slab of rock with a mass of 10^4 kg were placed on your body, you would be crushed! But your body is not crushed by atmospheric pressure because every tissue surface has the same (or essentially the same) pressure acting on both sides. However, if you dive to a depth of 5 m in water, the external pressure on your body is approximately 0.5 atm due to the water plus 1 atm due to atmospheric pressure on the water surface. Thus, your body is subjected to an inward pressure of 1.5 atm and an outward pressure of 1 atm due to trapped air in your body. You readily feel the effect of the net inward pressure of 0.5 atm. If you wish to dive to great depths, then it is necessary to use a breathing apparatus that will supply high-pressure air to the lungs in order to counteract the effect of the high external pressure.

Measuring Pressure

Pressure can be measured in a variety of ways. One method makes use of the fact that the pressure at a depth h in a fluid with a density ρ is $\rho g h$. Suppose that we take a long glass tube, seal it at one end, and then fill it with a dense liquid such as mercury ($\rho = 13.6$ g/cm³). The tube is then inverted and the open end is placed in a reservoir of mercury, as shown in Fig. 8.5. A vacuum space develops above the mercury column at the sealed end of the tube (if we have been careful to make the tube sufficiently long). Because the mercury that fills the column and the reservoir is static, the pressure is the same along any horizontal plane through the mercury. In particular, along the plane that is the upper surface of the reservoir, the pressure is 1 atm. That is, a point on the exposed surface of the mercury reservoir is subjected to a downward force due to an atmospheric pressure of 1 atm. Moreover, a point within the tube at the same horizontal level is subjected to a downward force due to the weight of the mercury column. (Remember, there is only vacuum above the mercury column, so there is no contribution due to air pressure.) We conclude that the weight of the mercury column per unit area is equal to 1 atm. Therefore, to find the height of the column, we solve Eq. 8.2 for h and supply the appropriate values of the quantities:

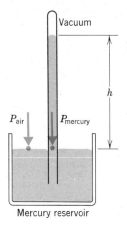

Figure 8.5 In a mercury barometer, normal atmospheric pressure will support a 0.76-m column of the liquid.

$$h = \frac{P}{\rho g} = \frac{1013 \times 10^5 \text{ PA}}{(13.6 \times 10^3 \text{ kg/m}^3) \times (9.80 \text{ m/s}^2)}$$
$$= 0.76 \text{ m} \quad \text{or} \quad 760 \text{ mm}$$

That is, normal atmospheric pressure will support a column of mercury that is 0.76 m or 760 mm high. Local weather conditions can cause variations in atmospheric pressure and these changes are observable as changes in the height of the mercury column in the tube. That is, the measurement of h provides a determination of the local atmospheric pressure. The pressure-measuring device shown in Fig. 8.5 is called a *mercury barometer*.

Sometimes pressure is expressed in terms of the height of the mercury column that the pressure will support. A pressure of 0.5 atm, for example, could be expressed as 380 mm Hg (Hg is the chemical symbol for mercury). This is an awkward way to designate pressure, so in some scientific fields the practice is to define a pressure that will support 1 mm of mercury as 1 *torr*, in honor of the Italian physicist, Evangelista Torricelli (1608–1647), who invented the mercury barometer. Thus,

1 atm = 760 mm Hg = 760 torr

● *Example 8.4*

By how much will the height of the column in a mercury barometer change if the barometer is moved from ground level (sea level) to the top of a 10-story building? (Assume each story corresponds to 3 m).

The change in atmospheric pressure ΔP due to a change in altitude $\Delta h = 30$ m is obtained from

$$P_1 - P_2 = \rho g(h_2 - h_1)$$

or,

$$\Delta P = \rho g \, \Delta h$$

Neither the value of g nor the density of air will change for the relatively small altitude change in this example. According to Table 1.9, ρ(air) = 1.29 kg/m³. Therefore,

$$\Delta P = (1.29 \text{ kg/m}^3) \times (9.80 \text{ ms}/^2) \times (30 \text{ m})$$
$$= 379 \text{ Pa}$$

so that the fractional change in the pressure is

$$\frac{\Delta P}{P} = \frac{379 \text{ Pa}}{1.013 \times 10^5 \text{ Pa}} = 0.37 \times 10^{-2} \text{ or } 0.37\%$$

Because the height ℓ of the mercury column is directly proportional to the pressure that supports it, the fractional change in ℓ will be the same as the fractional change in P:

$$\frac{\Delta \ell}{\ell} = 0.37 \times 10^{-2}$$

so that

$$\Delta \ell = (0.37 \times 10^{-2}) \times (760 \text{ mm})$$
$$= 2.8 \text{ mm}$$

which is a change that is easily observed. ∎

Another type of pressure-measuring device is the *aneroid barometer* (Fig. 8.6). The sensing of pressure changes in this instrument is by a flexible can that is evacuated and sealed. If the atmospheric pressure decreases, the can expands and the pointer moves downward. Conversely, if the pressure increases, the can contracts and the pointer moves upward. Because it only senses pressure changes and does not measure pressure directly, an aneroid barometer must be calibrated in terms of a mercury barometer. A sensitive aneroid barometer will give an indication of the pressure change corresponding to an altitude change of only a few meters. Indeed, aircraft altimeters are actually aneroid barometers that are calibrated to read altitude in feet or meters.

Two other types of pressure-measuring instruments are discussed in Question 8.3 at the end of the chapter.

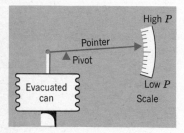

Figure 8.6 In an aneroid barometer, pressure changes are sensed by the expansion or contraction of a flexible evacuated can.

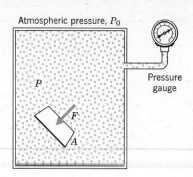

Figure 8.7 The pressure indicated by the gauge (the *gauge pressure*) is the difference between the *absolute pressure P* and the atmospheric pressure P_0.

Absolute Pressure and Gauge Pressure

Suppose that we have a quantity of gas confined in a container, as in Fig. 8.7. We measure the pressure of the gas with some type of pressure gauge. What does the reading of this gauge actually mean? Almost all mechanical pressure gauges indicate the *difference* in pressure between the volume to which the gauge is attached and the ambient atmosphere. This pressure is called the *gauge pressure*. The actual pressure P within the container is given by the force per unit area exerted on any surface within the container; this pressure is called the *absolute pressure*:

gauge pressure = absolute pressure
  − atmospheric pressure

or,

$$P_g = P - P_o \qquad (8.4)$$

In Example 8.3a we calculated the pressure at a depth of 50 m in a lake to be 4.9×10^5 Pa or 4.8 atm. This is actually just the *water* pressure and is equal to the increase of pressure at a depth of 50 m compared to that at the surface where the atmosphere exerts a pressure of 1 atm. Therefore, the pressure of 4.8 atm corresponds to the *gauge pressure;* the *absolute pressure* is 5.8 atm.

Notice that a pressure gauge constructed in the same way as a mercury barometer (with a vacuum space above the mercury column) will indicate *absolute pressure*. (Why?)

8.3 Buoyancy

Archimedes' Principle

Why does a block of wood *float* in water? If the same block were released in air, it would *fall*. Clearly, there is some extra force acting on the block when in water that buoys it up and prevents it from sinking.

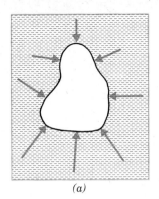
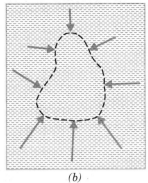

Figure 8.8 *(a)* An object immersed in a fluid. *(b)* The object replaced by an equal volume of fluid. The pressures exerted by the surrounding fluid are the same in both cases.

This force is called the *buoyant force* and is due to the displacement of the water by the block.

We can obtain an expression for the buoyant force on an object immersed in a fluid by the following reasoning. In Fig. 8.8a we have an irregularly shaped object of mass m, immersed in a fluid. The net force acting on the object, considered a free body, is the sum of the gravitational force exerted on it by the Earth $F_g = mg$ and the pressure-developed surface forces due to the fluid. We note that the fluid-pressure-related forces do not depend on the material composition of the object. Obeying Pascal's principle, the forces exerted by the fluid only depend on the orientation of the surface of the object at a particular point and, according to Eq. 8.2, the depth in the fluid at that point. Suppose we remove the object and replace it with exactly an equal volume of the fluid—refer to Fig. 8.8b. The surface forces acting on this volume of fluid, considered a free body, is exactly the same as on the immersed object. This volume of fluid, however, is at rest and hence the vector sum of the external fluid forces must just cancel the gravitational force acting on the fluid within the volume. Thus, the direction of the total external surface forces exerted on either the object or its fluid replacement is upward (i.e., a buoyant force) and has a magnitude equal to the weight of the fluid displaced by the object, w_f. Thus,

$$F_{\text{net}} = mg - w_f \tag{8.5}$$

The quantity w_f is the amount by which the normal downward force of gravity mg is reduced because of the buoyant effect of the fluid. That is, w_f is the buoyant force:

$$\begin{aligned}\text{buoyant force} = w_f &= \rho_f g V \\ &= \text{``weight'' of fluid displaced} \\ &\quad \text{by the object}\end{aligned} \tag{8.6}$$

The idea that an object immersed in a fluid experiences a buoyant force equal to the weight of the fluid displaced by the object was originally conceived by Archimedes (c. 287 B.C.–212 B.C.), the greatest scientist of ancient Greek times, and is known as *Archimedes' principle*.

Equation 8.5 tells us that if the buoyant force w_f exceeds the normal weight of the object mg, then F_{net} becomes negative or *upward*. This will happen whenever the density ρ of the object is smaller than the *density* ρ_f of the fluid. The net upward force causes the object to rise in the fluid. In the case of a liquid, when the object begins to emerge from the surface, the volume of liquid displaced by the object decreases and this causes the buoyant force to decrease. The object will continue to rise until the buoyant force decreases to a value exactly equal to mg. Then, in this condition of equilibrium ($F_{\text{net}} = 0$), the object rests with part of its volume submerged and part exposed; that is, the object *floats*. If the density of the object is only slightly less than that of the liquid, only a small fraction of the volume of the object will protrude above the surface.

● *Example 8.5*

Although we usually think of Archimedes' principle in terms of objects immersed in liquids, the idea is also valid for objects in *gases*. For example, when a balloon is filled with a low-density gas such as hydrogen or helium, the weight of the displaced air can be greater than the total weight of the balloon and the filling gas. Then, the balloon will rise in the air.

One of the most famous of the rigid lighter-than-air craft was the Zeppelin *Hindenburg*, built in Germany in 1936. The *Hindenburg* was buoyed up by filling its gas bags with hydrogen, a gas with a density much smaller than that of air ($\rho_H = 0.09$ kg/m^3). The *Hindenburg*'s gas sections were filled with 2×10^5 m^3 of hydrogen gas, and the mass M of the airship (not including the hydrogen gas or the useful load) was approximately 1.8×10^5 kg.

What useful load m_ℓ could be carried by the *Hindenburg*?

We will assume that the buoyant force was exactly equal to the total weight of the airship when loaded to capacity. Then, we can write

total weight = buoyant force

and the total weight is equal to the combined weights of the gas, the frame, and the useful load:

$$\rho_H g V_H + Mg + m_\ell g = \rho_f g V_{\text{frame}}$$

where V_H is the volume occupied by the hydrogen gas ($V_H = 2 \times 10^5$ m^3) and where V_{frame} is the total volume of displaced air, which we assume is 10 percent greater than the hydrogen volume ($V_{\text{frame}} = 2.2 \times 10^5$ m^3). The density of the air is $\rho_f = 1.29$ kg/m^3. Solving for the useful load,

$$\begin{aligned} m_\ell &= \rho_f V_{\text{frame}} - \rho_H V_H - M \\ &= (1.29 \text{ kg/m}^3) \times (2.2 \times 10^5 \text{ m}^3) - (0.09 \text{ kg/m}^3) \\ &\quad \times (2 \times 10^5 \text{ m}^3) - 1.8 \times 10^5 \text{ kg} \\ &= 8.6 \times 10^4 \text{ kg} \end{aligned}$$

or approximately 95 tons.

The *Hindenburg* was a luxurious way to travel, but the fact that it was filled with flammable hydrogen gas (Germany had no helium) made it potentially hazardous. The worst was realized on May 6, 1937, when the *Hindenburg* exploded and burned while landing at Lakehurst, New Jersey, killing 36 persons. The whole concept of lighter-than-air commercial airlines died with the *Hindenberg*. ∎

Measuring the Density of Irregular Objects

The density of an object is the ratio of its mass to its volume. If the object has an irregular shape, this will not affect the measurement of its *mass* but the *volume* may be very difficult to determine. However, by using Archimedes' principle, we can obtain the density of an irregular object without ever determining the volume. The procedure is as follows. We make two measurements of the weight of the object. One measurement is made while the object is immersed in a fluid of density ρf (Fig. 8.9a); we call this the *immersed weight* w_i. The second measurement is made while the object is suspended in air (Fig. 8.9b); this is the normal weight and here we call it the *weight in air* w_a.

According to Archimedes' principle, the immersed weight w_i is equal to the normal weight, $w_a = mg$, less the buoyant force, which is the weight of the displaced fluid, w_f:

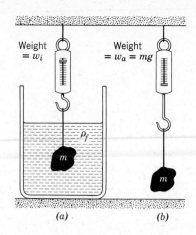

Figure 8.9 The density of an irregular object can be determined by measuring (a) the immersed weight w_i and (b) the weight in air w_a.

$$w_i = w_a - w_f \tag{8.7}$$

If the volume of the object is V, the weight of the displaced fluid is $\rho_f g V$. Thus, we can write

$$\rho_f g V = w_a - w_i$$

from which

$$V = \frac{w_a - w_i}{\rho_f g} \tag{8.8}$$

Now, the density ρ of the object is $\rho = m/V = (w_a/g)/V$, and substituting Eq. 8.8 for V,

$$\rho = \frac{w_a/g}{(w_a - w_i)/\rho_f g}$$

so that we have, finally,

$$\rho = \frac{w_a}{w_a - w_i} \rho_f \tag{8.9}$$

The density of the object is given directly in terms of the two weight measurements and the density of the fluid. Notice that in using this method, neither the mass nor the volume needs to be determined explicitly.

Gold is a very soft metal and is usually alloyed with copper or silver in order to increase its strength and enhance its usefulness. The density of gold is 19.3 g/cm^3; the densities of copper and silver are considerably smaller—8.9 g/cm^3 and 10.5 g/cm^3, respectively. Consequently, when gold is combined with one of these metals, the density of the alloy is less than that of pure gold. Therefore, the measurement of the density of a sample is sufficient to determine whether it is 100 percent gold. Moreover, if the alloying material is known, a density measurement can be used to determine the fraction of gold present.

This was the problem presented to Archimedes by the king of Syracuse, Hieron II. Hieron had just received a new gold crown from his goldsmith and he wished to know (without destroying the crown) whether the gold content had been excessively diluted with silver. Archimedes considered the problem, and, according to the legend (which is probably true), he discovered the solution while in his bath. Archimedes noted that as he stepped into the bath water, the water level increased; he quickly deduced that the amount of rise was equivalent to the volume of the immersed part of his body. He saw that by comparing in this way the volume of the crown with the volume of an equal weight of known pure gold, he could determine the purity of the crown. (Archimedes is supposed to have leaped from his bath and to have run naked to the palace shouting "Eureka!"—"I've got it!")

In modern times we have rephrased Archimedes' discovery in terms of "buoyant forces," but this does not detract from the importance of his original idea.

● *Example 8.6*

Suppose that a gold crown has a mass $m = 2.1$ kg. It is supposed to contain 10 percent silver (by mass). How can this be checked?

In order to determine the purity of the crown, it is weighed while immersed in water with the result

$w_i = 19.20$ N

We also have

$w_a = mg = (2.1 \text{ kg}) \times (9.80 \text{ m/s}^2) = 20.58$ N

The density of the crown is (Eq. 8.9)

$$\rho = \frac{w_a}{w_a - w_i} \rho_f = \frac{20.58 \text{ N}}{20.58 \text{ N} - 19.20 \text{ N}} \times (10^3 \text{ kg/m}^3)$$
$$= 14.9 \times 10^3 \text{ kg/m}^3$$

If the crown consisted of 10 percent silver, the volumes occupied by the two metals would be (gold = Au; silver = Ag):

$$V_{Au} = \frac{m_{Au}}{\rho_{Au}} = \frac{0.9 \times 2.1 \text{ kg}}{19.3 \times 10^3 \text{ kg/m}^2} = 9.79 \times 10^{-5} \text{ m}^3$$

$$V_{Ag} = \frac{m_{Ag}}{\rho_{Ag}} = \frac{0.1 \times 2.1 \text{ kg}}{10.5 \times 10^3 \text{ kg/m}^3} = 2.00 \times 10^{-5} \text{ m}^3$$

The total volume should be

$V_{Au} + V_{Ag} = 1.18 \times 10^{-4}$ m^3

The actual volume is (Eq. 8.8)

$$V = \frac{w_a - w_i}{\rho_f g} = \frac{20.58 \text{ N} - 19.20 \text{ N}}{(10^3 \text{ kg/m}^3) \times (9.80 \text{ m/s}^2)}$$
$$= 1.41 \times 10^{-4} \text{ m}^3$$

The actual volume is considerably larger than it should be, indicating that somewhat more than 10 percent silver was added to the gold used in making the crown. (A calculation of the fraction of silver actually in the crown is requested in Problem 8.20.)

In the case of Hieron's crown, Archimedes found that it too had been degraded. The goldsmith was subsequently executed.

● *Example 8.7*

Determine the fraction of an iceberg that lies beneath the surface of the sea.

Although the exact density of sea water and sea ice depends on the salinity of the water, let us use typical values $\rho_w = 1.028 \times 10^3$ kg/m^3 and $\rho_i = 0.917 \times 10^3$ kg/m^3 for water and ice respectively.

For equilibrium the buoyant force is equal to the gravitational force acting on the entire iceberg, thus, referring to the diagram,

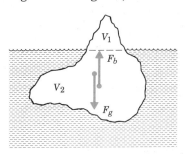

$\rho_i(V_1 + V_2)g = \rho_w V_2 g$

or

$$\frac{V_2}{V_1 + V_2} = \frac{\rho_i}{\rho_w} = \frac{0.917 \times 10^3 \text{ kg/m}^3}{1.028 \times 10^3 \text{ kg/m}^3} = 0.892$$

Thus, 89.2% of the iceberg lies beneath the surface of the sea. ■

8.4 Surface Tension

The Effects of Cohesive Forces

In a liquid the attractive forces between molecules, although not as strong as in solids, are sufficient to maintain the substance in a condensed state displaying a *surface*. The attractive forces that exist between nearby molecules of the same type are called *cohesive forces*, and they are particularly important in phenomena associated with liquid surfaces.

To increase the surface area of a quantity of liquid requires that work be done. For example, suppose that we dip a tube into soapy water so that a thin film of liquid soap closes one end of the tube. We can increase the surface area of the film and form a soap bubble by blowing through the tube. In order to do so, we exert a force on the film and move it through a certain distance; that is, we must do work against the cohesive forces in the film to increase the surface area. In its extended condition, the soap bubble has a certain amount of potential energy. If the blowing pressure is removed, the film will seek the condition of minimum potential energy, namely, the original condition before blowing began.

Another manifestation of the potential energy associated with a liquid surface is found in the spherical nature of raindrops. The surface of a raindrop seeks a condition of minimum potential energy and this means a condition of minimum surface area. For a particular amount of liquid, the surface area will be least when the shape is that of a sphere. Because the raindrop has weight and because of air resistance effects, the shape of a falling raindrop is not exactly spherical. If all of the external forces could be eliminated, the shape would be precisely spherical. Suppose that you introduce some olive oil into a liquid (such as vinegar or a water-alcohol mixture) that has the same density as the oil but with which the oil does not mix. If you then shake the mixture to break up and distribute the olive oil, you will see tiny spherical droplets of oil suspended in the main liquid. The droplets are indeed spherical because the external forces have been eliminated by placing the oil in an equal-density medium. This observation was made by the Belgian physicist, Joseph Plateau, in 1873. Perhaps you have observed this effect with an oil-and-vinegar salad dressing.

The amount of work required per unit area to increase the surface area of a liquid is called the liquid's *surface tension* σ:

$$\sigma = \frac{W}{A} \tag{8.10}$$

The quantity σ is also a measure of the amount of potential energy per unit area associated with a surface. Surface tension is measured in units of J/m^2 or, equivalently, N/m.

The reason for using the term *tension* to describe work per unit area becomes evident when we look at the effect of surface tension in the following

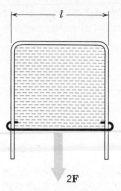

Figure 8.10 Work must be done against the surface tension in order to increase the area of the film.

way. Suppose that we have a U-shaped piece of wire that is fitted with a straight runner that slides smoothly along the outer sections of the wire, as in Fig. 8.10. If we dip the wire into a soap solution, the enclosed rectangular area will be covered with a soap film. We can enlarge the surface area of the film by pulling on the runner. There are *two* surfaces of the film attached to the runner and we pull with a force **F** on each surface so that the total force is 2**F**. If the runner is moved a distance s, the work done is

$$W = 2Fs$$

The increase in surface area of the film is $2\ell s$ (remember, the film has *two* sides), so that the work per unit area is

$$\frac{W}{A} = \frac{2Fs}{2\ell s} = \frac{F}{\ell}$$

Therefore, we can express the surface tension of the film as the *force per unit length* necessary to stretch the film:

$$\sigma = \frac{F}{\ell} \tag{8.11}$$

The surface tension of a liquid depends on the particular gas with which it forms an interface. For example, the surface tension of a water-air interface is slightly different from that when the gas is pure water vapor. We will consider here only liquid-air interfaces. Some representative values of surface tension are given in Table 8.1.

Capillarity

One of the most familiar effects of surface tension is the ability of a liquid to rise inside narrow tubes in apparent violation of the normal response to grav-

Table 8.1 Values of the Surface Tension for Various Liquids in Contact with Air

Liquid	Temperature (°C)	σ (J/m² or N/m)
Acetone	20	0.0237
Alcohol, methyl	20	0.0226
Benzene	20	0.0288
Chloroform	20	0.0271
Glycerol	20	0.0634
Mercury	15	0.487
Soap solution	20	0.025
Sodium bromide	melting point	0.103
Water	0	0.0756
Water	20	0.0728
Water	30	0.0712
Water	100	0.0589

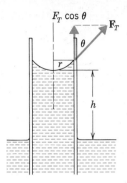

Figure 8.12 The upward force due to surface tension depends on the angle of contact θ.

ity. This effect is called *capillary action* or *capillarity*. Suppose that a glass tube with a small bore is inserted into a reservoir of water. The water will rise to a certain height in the tube and the surface will exhibit a concave shape with the liquid in the center of the tube at a lower height than the liquid in contact with the tube wall (Fig. 8.11a). The water literally climbs the tube wall until the upward pull of the surface tension is balanced by the weight of the raised column of water.

The height to which a liquid will rise in a tube depends on the relative strengths of the *cohesive* forces between molecules in the liquid and the *adhesive* forces between the liquid molecules and the molecules in the tube material. If the adhesive forces are large, the liquid is said to "wet" the tube material and the liquid will rise. The top surface of the liquid (called the *meniscus*) will contact the wall at a certain angle θ, as shown in Fig. 8.11a. The more effectively that the liquid "wets" the tube, the smaller will be the angle θ and the higher will be the rise of the liquid in the tube. If the cohesive forces are large compared to the adhesive forces, the liquid will not "wet" the tube, the contact angle will be greater than 90°, and the level of the liquid in the tube will be depressed (Fig. 8.11b).

Water and ethyl alcohol will completely "wet" a clean glass tube and the contact angle for these combinations is $\theta = 0°$. For a glass tube in kerosene, $\theta = 26°$. For a silver tube in water, $\theta = 90°$. And for a glass tube in mercury, $\theta = 140°$.

We can obtain an expression for the height to which a liquid will rise (or will be depressed) in a tube in the following way. Figure 8.12 shows a liquid that "wets" a tube and rises to a height h; the contact angle is θ. In this condition of equilibrium, the downward gravitational force acting on the water column must be exactly balanced by the upward force due to the surface tension. The surface tension force $\mathbf{F}_T$ on any small part of the edge of the meniscus acts *along* the surface of the liquid. Consequently, $\mathbf{F}_T$ makes an angle with respect to the tube that is equal to the contact angle θ. The vertical component of $\mathbf{F}_T$ is $F_T \cos \theta$. The total upward force due to surface tension is then equal to the upward component of the force per unit length (i.e., the surface tension) multiplied by the length of the line of contact (i.e., the circumference of the meniscus):

upward force = $(\sigma \cos \theta) \times (2\pi r)$

The downward force is the gravitational force on the water column:

downward force = $mg = V\rho g = (\pi r^2 h) \times (\rho g)$

Equating these two forces and solving for h, we find

$$h = \frac{2\sigma \cos \theta}{\rho g r} \qquad (8.12)$$

● *Example* 8.8

(a) To what height will water (at 20°C) rise in a glass tube with a bore radius of 0.1 mm?

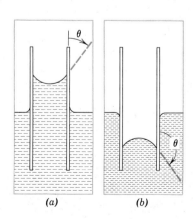

Figure 8.11 Capillary rise and depression of a liquid in a narrow tube. (a) If the liquid "wets" the tube material, the angle of contact θ between the meniscus and the tube will be less than 90° and the liquid will rise in the tube. (b) If the liquid does not "wet" the tube, $\theta > 90°$, and the liquid will be depressed.

The contact angle for water in a glass tube is 0° so the cos θ factor in Eq. 8.12 is equal to unity. Then,

$$h = \frac{2\sigma}{\rho g r} = \frac{2 \times (0.0728 \text{ N/m})}{(10^3 \text{ kg/m}^3) \times (9.8 \text{ m/s}^2) \times (10^{-4} \text{ m})}$$

$$= 0.15 \text{ m} = 15 \text{ cm}$$

(b) To what depth will mercury (at 15°C) be depressed in the same tube? In this case the contact angle is 140°; therefore,

$$h = \frac{2\sigma \cos\theta}{\rho g r}$$

$$= \frac{2 \times (0.487 \text{ N/m}) \times (-0.766)}{(13.6 \times 10^3 \text{ kg/m}^3) \times (9.8 \text{ m/s}^2) \times (10^{-4} \text{ m})}$$

$$= -0.056 \text{ m} = -5.6 \text{ cm}$$

Notice that cos 140° is negative, thereby making h negative and automatically indicating a depression of the surface instead of a rise. Notice also that the surface tension force in this case is equal to the weight of the mercury that originally filled the cavity left by the downward pull of surface tension. ∎

Capillarity is responsible for a number of familiar phenomena—the soaking up of liquids in blotters and towels, the rise of fuel in a wick, and the rise of ground water in soil. There are some materials that water will not "wet" effectively—for example, some types of fibers—and these materials will not soak up water by capillarity. However, if a *detergent* (or *wetting agent*) is added to the water, the contact angle can be reduced considerably, thereby allowing water to "wet" and be absorbed by the material. This process is particularly useful in the washing of many fabrics.

8.5 Fluids in Motion

The Energy Equation

Thus far we have been considering fluids only in *static* situations. We now turn to the study of some of the *dynamic* properties of fluids.

One of the important equations used to analyze fluids in motion was first obtained by the Swiss mathematician and physicist, Daniel Bernoulli (1700–1782). Bernoulli's treatment applied to the internal streamline flow patterns that exist within quietly flowing ideal fluids. Figure 8.13 shows the defining of streamlines in a quietly flowing fluid. The small dye particle is carried along with the fluid and identifies the particular local small segment of the fluid with which it moves.

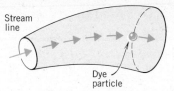

Figure 8.13 A streamline is the line traced out by a small particle (the dye particle) as it moves with a flowing fluid.

We proceed to derive a more general energy equation applicable to the bulk flow of fluids through pipes. We, however, also assume rather ideal properties for the fluid. That is, we ignore frictional or viscous forces and the temperature changes that such forces would produce. We also assume that the flow proceeds with very little turbulence or irregular motion. We consider only incompressible fluids so that a particular mass of the fluid will occupy the same volume regardless of its position or condition of flow. These conditions exist closely enough in many cases involving liquids. The resulting energy equation may also be used for gases if the pressure differences are sufficiently small that there is no appreciable compression.

We can derive the energy equation by considering energy conservation in the flow of a liquid through a pipe. As shown in Fig. 8.14, we allow the pipe to have a changing cross-sectional area and a changing height above some reference level. If we do a certain net amount of work W on the liquid, this will cause a change in the potential and kinetic energies of the liquid:

$$W = \Delta PE + \Delta KE \qquad (8.13)$$

We do work on the liquid by pushing with a force F_1 on a quantity of the liquid at position 1 and moving each small element of this volume of liquid through a net displacement s_1 at a constant velocity v_1. Thus,

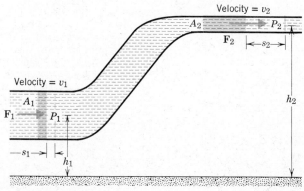

Figure 8.14 Geometry for the derivation of the energy equation for a liquid.

a volume $V_1 = A_1 s_1$ has been displaced a distance s_1 and the amount of work done *on* the liquid is $F_1 s_1$.

What is the effect of this displacement on the liquid at position 2? Because the liquid is incompressible, the displacement of a volume $A_1 s_1$ through a distance s_1 at position 1 means that there must be an equal volume displacement at position 2. Therefore, a quantity of liquid with a cross-sectional area A_2 and a length s_2 is displaced a distance s_2 at a constant velocity v_2, where $A_1 s_1 = A_2 s_2$. The volume of liquid at position 2 moves along the pipe only by pushing on the next element of liquid. This push F_2 acts through a distance s_2, so an amount of work $F_2 s_2$ is done *by* this volume of liquid. Thus, the *net* work done *on* the liquid is

$$W = F_1 s_1 - F_2 s_2$$

Writing each force as the product of the pressure and the cross-sectional area, we have

$$W = P_1 A_1 s_1 - P_2 A_2 s_2 = (P_1 - P_2)V \tag{8.14}$$

where $V = V_1 = A_1 s_1 = V_2 = A_2 s_2$.

The net effect of the change caused by pushing on the liquid with the force F_1 is exactly the same as the transfer of a mass ρV of liquid from position 1 to position 2. That is, a mass ρV of liquid at a pressure P_1, moving with a velocity v_1 at a height h_1, is transformed into an equal mass of liquid at a pressure P_2, moving with a velocity v_2 at a height h_2. Then, the changes in the potential and kinetic energies between positions 1 and 2 are

$$\Delta PE = \Delta(mgh) = mg\,\Delta h = \rho V g (h_2 - h_1) \tag{8.15a}$$
$$\Delta KE = \Delta(\tfrac{1}{2} mv^2) = \tfrac{1}{2} m\,\Delta v^2 = \tfrac{1}{2}\rho V(v_2^2 - v_1^2) \tag{8.15b}$$

where we have expressed the mass of each volume of liquid as $m = \rho V$.

Substituting Eqs. 8.14 and 8.15 into Eq. 8.13, we have

$$(P_1 - P_2)V = \rho V g(h_2 - h_1) + \tfrac{1}{2}\rho V(v_2^2 - v_1^2)$$

Dividing by V and rearranging, we obtain

$$P_1 + \rho g h_1 + \tfrac{1}{2}\rho v_1^2 = P_2 + \rho g h_2 + \tfrac{1}{2}\rho v_2^2 \tag{8.16}$$

This is the energy equation we are seeking. Because the positions 1 and 2 are completely arbitrary, Eq. 8.16 really states that the quantity $P + \rho g h + \tfrac{1}{2}\rho v^2$ evaluated at any point in the pipe is always the same.

The simple form of the energy equation, Eq. 8.16, can be modified as necessary, to include other energy input and output terms if circumstances warrant. The pipe system might, for example, include a pump that would increase the flow velocity and pressure. Heat energy might be added or removed, and so forth.

When the fluid flow is very quiet (no turbulence at all), we may imagine the internal flow pattern to be the equivalent of a large number of imaginary parallel tubes of flow within the fluid even though no physical partitions exist to define these tubes. Rather, the tube walls are definable in terms of the streamlines in the fluid. When Eq. 8.16 is applied to anyone of these tubes of flow, it is referred to as *Bernoulli's Equation*.

Applications of the Energy Equation

There are several important and interesting situations to which the energy equation can be easily applied.

Hydrostatics. If a fluid is not in motion, the velocity terms in the energy equation do not appear and we can write

$$P_1 - P_2 = \rho g(h_2 - h_1) \tag{8.17}$$

which is another way of expressing the content of Eq. 8.2.

Flow at constant height. Figure 8.15 shows the flow of a liquid through a horizontal pipe that has a varying cross-sectional area. With $h_1 = h_2$, the energy equation becomes

$$P_1 + \tfrac{1}{2}\rho v_1^2 = P_2 - \tfrac{1}{2}\rho v_2^2 \tag{8.18}$$

Because the pipe is constricted in position 2, we have $A_2 < A_1$ and $v_2 > v_1$. Equation 8.19 then tells us

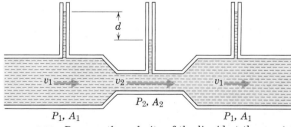

Figure 8.15 Because the velocity of the liquid at the constriction is large $(v_2 > v_1)$, the pressure at this point is small $(P_2 < P_1)$.

that $P_2 < P_1$. The difference d in the heights to which the liquid rises in the side tubes serves to measure the pressure difference:

$$P_1 - P_2 = \rho g d \qquad (8.19)$$

● **Example 8.9**

In Fig. 8.15 suppose that $v_1 = 15$ cm/s and that $A_1/A_2 = 4$. If the liquid is water, what is the difference in height d of the water in the side tubes?

Combining Eqs. 8.18 and 8.19 we can write

$$P_1 - P_2 = \rho g d = \tfrac{1}{2} \rho (v_2{}^2 - v_1{}^2) \qquad (1)$$

so that

$$d = \frac{1}{2g}(v_2{}^2 - v_1{}^2) \qquad (2)$$

Now, the liquid is incompressible, so the flow rate through the small section of the pipe is the same as that through the large section. We can express this by

$$A_1 v_1 = A_2 v_2 \qquad (3)$$

Solving this equation for v_2 and substituting into (2), we find

$$d = \frac{1}{2g}\left(\frac{A_1{}^2}{A_2{}^2} v_1{}^2 - v_1{}^2\right) = \frac{v_1{}^2}{2g}\left(\frac{A_1{}^2}{A_2{}^2} - 1\right)$$

$$= \frac{(0.15 \text{ m/s})^2}{2 \times (9.8 \text{ m/s}^2)}(16 - 1) = 0.017 \text{ m} = 1.7 \text{ cm}$$

Notice that d does not depend on the nature of the fluid that flows in the pipe (if the fluid is an *ideal* fluid).

As a practical matter, water flowing at 20°C with pipe radii r_1 greater than a few centimeters would lead to nonquiet flow for which streamlines and tubes of flow could not be defined. (See Problem 8.39.) However, the energy equation would, nonetheless, apply with very little error.

Torricelli's equation. Figure 8.16 shows a tank from which liquid flows through three openings. What are the upward, downward, and horizontal velocities, v_u, v_d, and v_h, of the liquid just as it leaves these three holes?

In this case we may think in terms of flow lines, that is, path lines followed by any particular small segment of the liquid, to define the walls of Bernoulli's tubes of flow. The tubes of flow end on one or the other of the three openings and confine the flow to occur from the upper surface all the way down to the particular opening.

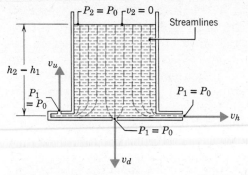

Figure 8.16 The flow velocities v_u, v_d, and v_h, are obtained from Bernoulli's equation.

The upper surface of the liquid corresponds to position 2 in Bernoulli's equation for any one of the tubes of flow. Here, the pressure is atmospheric pressure $P_2 = P_0$. The liquid is flowing out of the tank so that the upper surface is actually falling. But if the holes are sufficiently small, the rate of fall of the surface is very slow and we can set $v_2 = 0$. Position 1 corresponds to the bottom of the tank opposite any one of the openings. Just outside the holes, at the positions where we measure the flow velocities, the pressure is also atmospheric pressure, $P_1 = P_0$. Therefore, Bernoulli's equation becomes

$$P_0 + \rho g h_1 + \tfrac{1}{2} \rho v_1{}^2 = P_0 + \rho g h_2$$

Solving for v_1, we find

$$v_1 = \sqrt{2g(h_2 - h_1)} \qquad (8.20)$$

which is *Torricelli's equation*. Notice that the velocity of flow is exactly the same as that of an object falling from rest through a height $h = h_2 - h_1$. (Compare Eq. 2.16.) Notice also that the result v_1 does not depend on the direction of flow; that is,

$$v_1 = v_u = v_d = v_h$$

In fact, the velocity of flow v_u from the upward-directed opening is just sufficient for the stream of liquid to reach the level of the upper surface. In a real case, however, frictional effects would prevent the stream from rising all the way to this level. ∎

Another application of this effect is in the flow of air around an airfoil. (The pressure differences are sufficiently small in this case that we can consider the air to be incompressible and use Bernoulli's equation.) Figure 8.17 shows the cross-section of an aircraft wing and two flow lines of the air passing over and under the wing. The air that passes over the wing must travel a path that is longer than that traveled by the air passing under the wing. (Can you see why? The two streamlines of air indicated by the

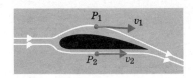

Figure 8.17 An aircraft wing experiences a lift because $v_1 > v_2$ and $P_2 > P_1$.

arrows that are adjacent when in *front* of the wing must also be adjacent when they are *behind* the wing.) The air that moves through the longer path must have a greater velocity; that is, $v_1 > v_2$. Then, Eq. 8.16 with $h_1 = h_2$, requires that $P_2 > P_1$. Because the air beneath the wing has a higher pressure than the air above the wing, a net upward force or *lift* results.

The Magnus effect. The curving of a pitched baseball and the hooking or slicing of a golf ball can be explained in terms of Bernoulli's equation if we allow for the occurrence of frictional drag, an effect we have thus far ignored. Figure 8.18 shows a ball moving through air. The ball actually moves from left to right through still air, but this is entirely equivalent to the air moving from right to left past a stationary ball. As the ball rotates, some air near the surface is dragged along because of friction between the air and the surface of the ball. Near the ball, the air velocity due to the drag is u. In addition, the ball moves through the air with velocity v, and in our stationary reference frame the air moves past the ball with velocity v. In position 1 (above the ball in the diagram) the velocity u of the dragged air adds to the velocity v of the air rushing past the ball so that the net air velocity is $v_1 = v + u$. In position 2 the velocities are opposed, so $v_2 = v - u$. Therefore, $v_1 > v_2$, and according to Eq. 8.16 with $h_1 = h_2$ we have $P_2 > P_1$. As in the case of the airfoil, this pressure difference means that there is a force **F** acting on the ball in the direction from position 2 to position 1. Thus, the ball curves to the left as it moves through the air. This is called the *Magnus effect* after the German physicist, H. G. Magnus (1802–1870), who first demonstrated the presence of the transverse force in 1852.

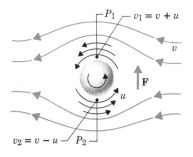

Figure 8.18 The velocity of the air dragged along by a rotating ball *adds* to the air velocity at position 1 but *subtracts* in position 2. Therefore, $v_1 > v_2$ and $P_2 > P_1$. The pressure difference causes a force **F** to act on the ball.

We have not shown that smooth quiet flow is possible when viscous drag is present. However, advanced analysis shows this to be possible particularly when the fluid is air, thus permitting streamlines to be defined and Bernoulli's equation to be applied. The next section gives some additional details for smooth quiet flow with viscosity present.

8.6 The Viscous Flow of Fluids

The Coefficient of Viscosity

Real fluids are not friction free. We know that water flows more easily than heavy oil and that oil flows more easily than molasses. The reason is that, as they flow, the molecules in oil or molasses drag against one another much more strongly than do the molecules in water. Substances that have a high degree of internal friction are said to be highly *viscous*. Because the interaction between gas molecules is much weaker than that between liquid molecules, the viscosity of gases is much smaller than that of liquids.

We can obtain a measure of the viscosity of a fluid in the following way. Suppose that we enclose a layer of the fluid between two large parallel plates, as shown in Fig. 8.19. The fluid is in contact with both plates and has a uniform thickness d. Consider the lower plate to be fixed. What happens when we pull on the upper plate with a horizontal force **F**? Because the fluid is viscous, it tends to stick to both plates. That is, the fluid layer immediately adjacent to a plate does not move relative to that plate. When an equilibrium condition is established, the force **F** pulls the upper plate with a constant velocity **v**. This means that the uppermost layer of fluid moves with the velocity **v**. Because of the viscous drag, this layer pulls on the layer immediately below causing it to move, but with a velocity slightly less than **v**. Each layer in turn imparts motion with a smaller velocity to the layer below until, finally, the lowest layer does not move at all. If the velocity **v** is sufficiently small, this transfer of motion causes each layer to move smoothly, without turbulence or irregularity. This type of fluid motion we call *laminar flow*.

How does the force **F** vary with the quantities

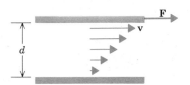

Figure 8.19 The laminar flow of a viscous fluid.

that characterize the system? First, if we make the plates larger, the force necessary to maintain a constant velocity **v** will increase in direct proportion to the plate area; that is, $F \propto A$. Also, if the thickness d of the fluid is decreased, the change in velocity across the fluid is much sharper and a correspondingly larger force is required; that is, $F \propto 1/d$. Finally, a larger force is required to maintain a higher velocity; that is, $F \propto v$. Combining these statements, we have

$$F \propto \frac{vA}{d}$$

We call the constant of proportionality the *coefficient of viscosity* and write

$$F = \eta \frac{vA}{d} \qquad (8.21)$$

We expect a small value of η for air and a large value for oil or molasses. Because the motion of molecules within all types of matter depends on the temperature, we expect that the coefficient of viscosity for a particular fluid will be a function of temperature. In fact, the coefficient of viscosity for liquids *decreases* with increasing temperature. (Can you guess why?)

According to Eq. 8.21, the units of η are

$$[\eta] = \frac{\text{N} \times \text{m}}{\frac{\text{m}}{\text{s}} \times \text{m}^2} = \frac{\text{N}}{\text{m}^2} \times \text{s} = \text{Pa-s}$$

Sometimes η is given in units called *poise* (P) or *centipoise* (cP):

1 Pa-s = 10 P = 10^3 cP.

Table 8.2 gives some typical values of η.

The Reynolds Number

The simplest type of fluid motion is *laminar flow* in which the layers of fluid slide smoothly past one another. At low velocities, the flow of a fluid through a pipe will be laminar. But if the velocity is increased beyond a certain critical value (which depends on the properties of the fluid and the radius of the pipe), laminar flow cannot be maintained. Instead, the flow becomes highly irregular as random circular currents (*vortices*) develop in the fluid and the resistance to flow increases sharply. This type of fluid motion is called *turbulent flow*.

Table 8.2 Some Typical Values of the Coefficient of Viscosity

Substance	Temperature (°C)	η (Pa-s)
Air	0	1.71×10^{-5}
	20	1.84×10^{-5}
	40	1.96×10^{-5}
Blood	37	4.0×10^{-3}
Castor oil	20	0.99
Ethyl alcohol	20	1.20×10^{-3}
Glycerine	-42	6710
	20	1.49
Mercury	20	1.55×10^{-3}
Water	0	1.79×10^{-3}
	20	1.00×10^{-3}
	40	6.56×10^{-4}

Experiments show that turbulence sets in when a certain combination of quantities exceeds a critical value. This combination is called the *Reynolds number*:

$$\mathcal{R} = \frac{2\rho_f av}{\eta} \qquad (8.22)$$

where ρ_f and η are the density and viscosity coefficient, respectively, of the fluid, and a is the radius of the pipe; v is average forward flow velocity. It is easy to verify that $\mathcal{R}$ has no units—it is a dimensionless number.

The flow of a fluid through a pipe will be laminar if $\mathcal{R}$ is less than about 2000. For $\mathcal{R} > 3000$, the flow is turbulent. The regime of $\mathcal{R}$ between 2000 and 3000 is a transition region in which the flow is unstable and may change erratically between laminar and turbulent flow.

● *Example 8.10*

(a) Suppose that water (at 20°C) flows with a velocity of 20 cm/s through a pipe whose radius is 1 cm. Is the flow laminar or turbulent?

Using the value of the viscosity coefficient for water given in Table 8.2, the Reynolds number for this situation is

$$\mathcal{R} = \frac{2\rho_f av}{\eta}$$

$$= \frac{2 \times (10^3 \text{ kg/m}^3) \times (0.01 \text{ m}) \times (0.20 \text{ m/s})}{1.00 \times 10^{-3} \text{ Pa-s}}$$

$$= 4000$$

and the flow is turbulent.

(b) If we substitute air (also at 20°C) for water in (a), will the flow be laminar or turbulent?

We now have

$$\mathcal{R} = \frac{2 \times (1.29 \text{ kg/m}^3) \times (0.01 \text{ m}) \times (0.20 \text{ m/s})}{1.84 \times 10^{-5} \text{ Pa-s}}$$
$$= 280$$

and the flow is laminar. ∎

The Flow of Blood in the Circulatory System

Blood is a viscous fluid that is pumped through a complicated system of arteries and veins by the muscular action of the heart. The rate of flow of blood through the body is sufficiently small that the flow is generally laminar instead of turbulent. Therefore, we can treat the flow of blood through an artery in the same way that we would treat the laminar flow of a fluid through a smooth pipe. Because of the attractive molecular forces between the blood and the inner wall of an artery, there is no flow of the blood that is in contact with the artery. (This is true for any fluid in a pipe.) Consequently, the velocity of the blood flow is zero at the arterial wall and flows most rapidly at the center of the artery. It is possible to derive (using the methods of calculus) an expression for the velocity of flow of a fluid as a function of the distance r from the center of a pipe. As shown in Fig. 1a, we consider a pipe of inner radius a and a length ℓ across which a pressure difference $P_1 - P_2$ exists. If the viscosity of the fluid is η, the velocity is given by

$$v = \frac{1}{4\eta\ell}(P_1 - P_2)(a^2 - r^2) \tag{1}$$

Figure 1b shows how the velocity varies across the diameter of the pipe. Notice that the shape of the velocity curve is *parabolic*.

Because the flow velocity changes with the radial distance in the arterial tube, Bernoulli's equation indicates that there must be an accompanying pressure change within the tube. The low velocity near the wall means that the pressure in this region is relatively high. At the center of the tube, where the velocity is greatest, the pressure is the least. That is, the pressure increases radially outward. Any small object, such as a blood cell, that is flowing through the tube will therefore experience a radial pressure difference. This pressure difference produces a force that

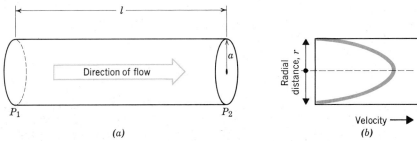

Figure 1 A viscous fluid that flows through a smooth pipe has a parabolic velocity distribution.

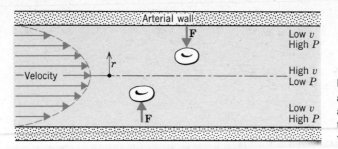

Figure 2 The variation in velocity across the diameter of an artery is accompanied by a pressure variation that pushes the blood cells toward the center of the artery.

tends to push the cell toward the center of the tube, as shown in Fig. 2. We know from other evidence that blood cells are concentrated in the central portions of the arteries.

We can use the expression above for the velocity to calculate (again by using the methods of calculus) the rate of flow of a fluid through a pipe. The flow rate Q is measured in m³/s and is given by

$$Q = \frac{\pi a^4}{8\eta \ell}(P_1 - P_2) \tag{2}$$

This is known as *Poiseuille's equation,* after the French physiologist Jean Marie Poiseuille (1799–1869) who first derived and tested the expression. Notice how sensitively Q depends on the radius a; when a is doubled, Q increases by a factor of 16. Similarly, when a is made smaller, the flow rate decreases drastically. If some condition results in the thickening of the arterial walls (which has the effect of making a smaller), the lowered flow rate of the blood can cause *angina pectoris,* characterized by pains in the chest following physical exertion. The most common cause of angina pectoris is *artereosclerosis,* hardening of the arteries. Relief can be effected by administering of some substance (e.g., nitroglycerine) that relaxes the muscle in the arterial walls and permits a to become larger, thereby increasing the blood flow rate and decreasing the load on the heart.

In the expressions for v and Q, we see that both the velocity and the flow rate depend on the pressure difference between the ends of the section of tube considered. How can the pressure of the blood in arteries be measured? We have already remarked that the flow of blood through the body is generally laminar. The laminar flow of blood through an artery is a "quiet" process; turbulent flow, on the other hand, is "noisy." If the blood in an artery is made to flow in a turbulent manner, the characteristic sound that is produced can be detected by a stethoscope placed on the artery. This way of sensing turbulent flow is used in the most common method of measuring blood pressure. Actually, there are two values of the pressure to be measured. The maximum (or *systolic*) pressure occurs when the heart muscles contract and blood is forced from the left ventricle into the *aorta* and, from there, into the arteries. Between contractions, the pressure drops to its lowest (or *diastolic*) pressure. When using the sound technique to measure blood pressure, it is assumed that any artificially forced constriction of an artery will result in turbulent flow. Such a constriction can be produced by applying to the exterior of an artery a pressure in excess of the blood pressure within the artery.

Usually, the main artery in the upper arm is chosen for the measurement, and an inflatable strap is placed around the arm. The strap is then pumped up with air to a pressure sufficient to collapse the artery and stop all flow. The valve on the strap is then opened and the air pressure is allowed to decrease slowly. The operator listens with the stethoscope placed on the artery downstream from the strap (usually, near the inside of the elbow). No sound is heard until some blood is forced through the constricted artery. This occurs when the air pressure in the strap has dropped just below the systolic pressure. The pressure in the strap at the first sound is recorded as the systolic pressure. As the strap pressure decreases further, the flow continues to be turbulent until the strap pressure is sufficiently low that there is no constriction of the artery. The pressure in the strap when the sound of turbulent flow ceases is recorded as the diastolic pressure. (Do these pressures correspond to the pressure in the *heart*? See Exercise 3.)

In a healthy adult, the systolic pressure is approximately 120 torr and the diastolic pressure is approximately 80 torr. (These are *gauge* pressures.) Thus, the average pressure of the blood as it leaves the heart (that is, at the entrance to the *arch of the aorta*, which connects directly to the heart) is about 100 torr. Because of losses of energy to friction, the pressure of the blood decreases along the complex system of arteries, capillaries, and veins. By using Poiseuille's equation, we can see that the pressure drop in the aorta is not large. The total flow rate of blood in the human body (when inactive) is about 80 cm³/s or 8×10^{-5} m³/s. The radius of the aorta is approximately 1 cm or 10^{-2} m. Therefore, solving Poiseuille's equation for $(P_1 - P_2)/\ell$, we find

$$\frac{P_1 - P_2}{\ell} = \frac{8\eta Q}{\pi a^4} = \frac{8 \times (4.0 \times 10^{-3} \text{ Pa-s}) \times (8 \times 10^{-5} \text{ m}^3/\text{s})}{\pi \times (10^{-2} \text{ m})^4}$$

$$= 80 \text{ Pa/m} = 0.6 \text{ torr/m}$$

This is the pressure drop per meter of tube length. Therefore, in the aorta, before it begins to branch into the large arteries (a distance of at most 0.4 m), the pressure drop is negligible compared with the average pressure of the blood leaving the heart. Even during strenuous activity, when the flow rate may be increased by a factor of 5, the pressure drop in the main aorta tube is still small.

Farther along the system, the pressure decrease becomes evident. After passing through the large arteries, the pressure is about 90 torr, and after passing through the small arteries (the arterioles), the pressure is only about 25 torr. When the veins are finally reached, the pressure is 10 torr or less. Figure 3 shows, in a schematic way, the pressure decrease along the system. This graph shows only the *average* pressure. There are, of course, pulsations in pressure caused by the beating heart. These pulsations cause the pressure in the aorta and the large arteries to oscillate between about 120 torr and about 80 torr. In the small arteries the pulsations are partially damped, and when the capillaries are reached, there are almost no pressure oscillations and the flow is uniform.

Let us now return to the question of the flow rate of blood in the circulatory system. The flow rate Q can be expressed as the product of the cross-sectional area of the tube and the average flow velocity $\bar{v}$: $Q = A\bar{v}$. Therefore, we can find $\bar{v}$ from

$$\bar{v} = \frac{Q}{A} = \frac{Q}{\pi a^2}$$

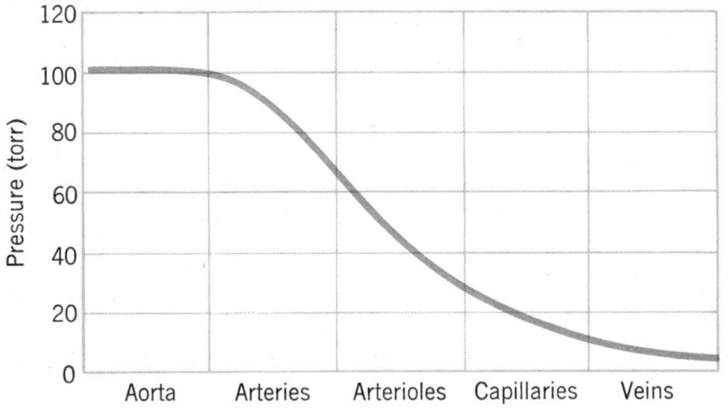

Figure 3 Variation of average blood pressure through the system of arteries, capillaries, and veins (schematic).

For the aorta, $a = 0.01$ m and $Q = 8 \times 10^{-5}$ m³/s (inactive condition). Then,

$$\bar{v} = \frac{8 \times 10^{-5} \text{ m}^3/\text{s}}{\pi \times (0.01 \text{ m})^2} = 0.25 \text{ m/s}$$

The Reynold's number for this flow is

$$\mathscr{R} = \frac{\rho \bar{v} d}{\eta} = \frac{(1.05 \times 10^3 \text{ kg/m}^3) \times (0.25 \text{ m/s}) \times (0.02 \text{ m})}{4.0 \times 10^{-3} \text{ Pa-s}}$$

$$= 1300$$

This value is smaller than the critical value for turbulent flow ($\mathscr{R} \cong 2000$), so the flow in the aorta in the inactive condition is laminar. During strenuous activity, however, the flow rate increases, causing $\bar{v}$ to increase. This results in a Reynold's number exceeding 2000 and the flow in the aorta becomes turbulent. In the smaller tubes, the velocity remains sufficiently small that the flow is still laminar.

References

A. C. Burton, *Physiology and Biophysics of the Circulation,* Year Book Medical Publishers, New York, 1965.

F. R. Hallett, P. A. Speight, and R. H. Stinson, *Introductory Biophysics,* Halsted (Wiley), New York, 1977, Chapter 11.

D. A. McDonald, *Blood Flow in Arteries,* Edward Arnold, London, 2nd ed., 1974, Chapters 1–4.

T. C. Ruch and H. D. Patton, eds., *Physiology and Biophysics,* 19th ed., W. B. Saunders, Philadelphia, 1965.

■ Exercises

1. Compare the value of the *average* velocity of the blood in the aorta (0.25 m/s) with the velocity in the *center* of the aorta by using Eq. 1. [Use the value of $(P_1 - P_2)/\ell$ computed for the aorta.] Finally, use Bernoulli's equation and find the pressure difference (in torr) between the center of the aorta and the wall.

2. If an artery is constricted by deposits of *plaque* that reduce the effective radius by 20 percent, by what factor must the pressure increase to maintain the same flow rate?

3. Blood pressure measurements are made using the *brachial* artery in the upper arm. This artery has a radius of about 0.005 m and the distance from the arch of the aorta to the point of measurement is approximately 0.3 m. The flow rate through the brachial artery is about 10^{-5} m³/s. Calculate the pressure drop between the aorta and the point of measurement. Express the result in torr. Does the measurement of the pressure in the brachial artery give an accurate representation of the pressure of the blood leaving the heart?

4. While a subject is lying in a horizontal position, one of the subject's main arteries is punctured. (The puncture is in the upper surface.) To what maximum height will the blood spurt?

5. In the measurement of blood pressure, it is the pressure in the *heart* that is important. Why is the upper arm chosen for the measurement? Would an artery in the leg serve equally well?

6 From a knowledge of the properties of the aorta and the blood contained in the aorta ($P = 100$ torr, $a = 0.01$ m, $\bar{v} = 0.25$ m/s), calculate the average power expended by the heart in pumping blood. The heart, like most muscles, is about 20 to 25 percent efficient. How much power is expended by the heart in producing heat in the inactive condition (when the efficiency is about 20 percent)?

Terminal Velocity

If an object is allowed to fall through vacuum near the surface of the Earth, we know that it will experience a constant acceleration equal to g. However, if the object falls through a fluid—even air—the rate of fall will be influenced by viscous effects. The falling object is subject to a constant downward force mg due to gravity and an upward retarding force F_{vis} due to viscous drag (air friction). Thus,

$$F_{net} = mg - F_{vis} \qquad (8.23)$$

(If the object falls through a dense medium such as water or oil, then there is an additional force—the buoyant force—that acts upward.) The viscous force increases with increasing velocity. Therefore, as the object falls, gravity tends to increase the downward velocity and this, in turn, tends to increase the viscous force. Eventually, an equilibrium condition is reached in which the upward force F_{vis} is exactly equal to the downward force mg. Then, F_{net} is zero and the object moves thereafter with constant velocity. This maximum velocity of fall is called the *terminal velocity*.

The expression for the viscous force has two different forms, depending on whether the motion is laminar or turbulent. For the case of a sphere moving through a fluid, laminar flow results when the Reynolds number is less than about 10. (In the expression for $\mathcal{R}$, Eq. 8.22, we now interpret a as the radius of the falling sphere. Note that the critical values of the Reynolds number separating the regions of laminar and turbulent conditions depend on the geometry. The critical value for spheres and cylinders moving through a fluid is much lower than for flow through pipes.) The result of a theoretical calculation is that

$$F_{vis} = 6\pi a \eta v \qquad (\mathcal{R} \lesssim 10;\ \text{laminar}) \qquad (8.24)$$

For turbulent flow ($\mathcal{R}$ greater than a few hundred), theory tells us that

$$F_{vis} = \tfrac{1}{4}\pi \rho_f a^2 v^2 \qquad (\mathcal{R} \gtrsim 200;\ \text{turbulent}) \qquad (8.25)$$

The terminal velocity v_t can be found for each case by setting F_{net} equal to zero (Eq. 8.23), so that

$$F_{vis} = mg$$

Then, we have

$$v_t = \frac{mg}{6\pi a \eta} \qquad (\mathcal{R} \lesssim 10) \qquad (8.26)$$

$$v_t = \frac{2}{a}\sqrt{\frac{mg}{\pi \rho_f}} \qquad (\mathcal{R} \gtrsim 200) \qquad (8.27)$$

For the case of large $\mathcal{R}$ notice that the *density* is the only property of the fluid that influences the terminal velocity.

● *Example 8.11*

What is the terminal velocity of a raindrop with a radius of 1 mm falling through air at 20°C?

First, we need to know whether the motion is laminar or turbulent. We might guess that this is a case of laminar flow, but let us obtain an approximate value for $\mathcal{R}$. By observing raindrops, we can guess that they fall with a velocity of at least 3 m/s. Using this value for v, we find

$$\mathcal{R} = \frac{2\rho_f a v}{\eta} = \frac{2 \times (1.29\ \text{kg/m}^3) \times (10^{-3}\ \text{m}) \times (3\ \text{m/s})}{1.84 \times 10^{-5}\ \text{Pa-s}}$$
$$= 420$$

So our intuition was wrong; the flow is in fact turbulent. Therefore, we must use Eq. 8.27 for the terminal velocity. First, we calculate the mass of the raindrop:

$$m = \rho V = \frac{4\pi}{3}a^3\rho = \frac{4\pi}{3} \times (10^{-3}\ \text{m})^3 \times (10^3\ \text{kg/m}^3)$$
$$= 4.2 \times 10^{-6}\ \text{kg}$$

Then,

$$v_t = \frac{2}{10^{-3}\ \text{m}}\sqrt{\frac{(4.2 \times 10^{-6}\ \text{kg}) \times (9.8\ \text{m/s}^2)}{\pi \times (1.29\ \text{kg/m}^3)}}$$
$$= 6.4\ \text{m/s}$$

The value of v_t is about twice as large as we assumed to estimate the Reynolds number; consequently, $\mathcal{R}$ is actually about 900 for this case, indicating a high degree of turbulence.

● *Example 8.12*

What is the size of a fog droplet that falls through air at 0°C with $\mathcal{R} = 0.2$?

Substituting Eq. 8.26 for v_t into the expression for $\mathcal{R}$, we have

$$\mathcal{R} = \frac{2\rho_f a v}{\eta} = \frac{2\rho_f a}{\eta} \times \frac{mg}{6\pi a \eta}$$

Writing $m = 4\pi a^3 \rho/3$, the expression for $\mathcal{R}$ becomes

$$\mathcal{R} = \frac{2\rho_f g}{6\pi \eta^2} \times \frac{4\pi a^3 \rho}{3} = \frac{4\rho \rho_f g}{9\eta^2}$$

Substituting $\mathcal{R} = \frac{1}{5}$ and solving for a, we find

$$a = \left(\frac{9}{20} \frac{\eta^2}{\rho \rho_f g}\right)^{1/3}$$

$$= \left[\frac{9 \times (1.71 \times 10^{-5} \text{ Pa-s})^2}{20 \times (10^3 \text{ kg/m}^3) \times (1.29 \text{ kg/m}^3) \times (9.8 \text{ m/s}^2)}\right]^{1/3}$$

$$= 2.2 \times 10^{-5} \text{ m} \cong \frac{1}{50} \text{ mm}$$

This is a typical size for a fog droplet. ■

Summary of Important Ideas

Pressure is force per unit area.

The forces associated with fluid pressure in static systems always act *perpendicular* to the container surface.

Pascal's principle states that any pressure applied to the surface of a confined fluid is transmitted undiminished to all points within the fluid.

The *fluid pressure* at a depth h in a fluid with density ρ is $\rho g h$.

Normal *atmospheric pressure* is sufficient to balance a column of mercury 760 mm high: 1 atm = 760 mm Hg = 760 torr = 1.013×10^5 Pa.

According to *Archimedes' principle*, the buoyant force on an object immersed in a fluid is equal to the weight of the displaced fluid.

Surface tension is due to the cohesive forces between the molecules in a liquid. Surface tension is responsible for *capillary action*.

Bernoulli's equation describes the *dynamic* properties of ideal fluids.

The *viscosity* of a fluid is the result of internal friction among the constituent molecules. One of the effects of viscosity is that an object falling in a viscous medium will reach a *terminal velocity*.

The *Reynolds number* is useful in determining whether a dynamic system will exhibit *laminar* (smooth) flow or *turbulent* flow.

◆ Questions

8.1 Why are dams thicker at the bottom than at the top?

8.2 How will temperature changes affect the reading indicated by a mercury barometer?

8.3 The diagrams below indicate the measurement of pressures by (a) an *open-tube manometer* and (b) a *closed-tube manometer*. Which type of manometer is used to measure pressures near atmospheric pressure and which is used for pressures near zero? Explain. Should a high-density liquid (such as mercury) or a low-density liquid (such as oil) be used in the manometers if maximum precision is to be obtained? Explain. Write down the expressions for the pressures P_a and P_b.

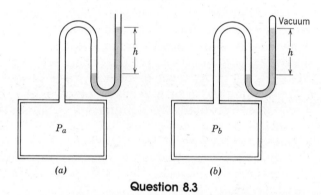

Question 8.3

8.4 In the Question 8.3, which manometer will indicate the *gauge pressure* in the container and which will indicate the *absolute pressure*?

8.5 Explain how you could use a manometer to measure the pressure in your lungs during exhalation.

8.6 One of the ideas concerning energy and the stability of systems is that any system will spontaneously attempt to achieve a condition in which the potential energy is minimum. (Falling objects, for example, obey this rule.) If the parts of a system can move relative to one another, they will tend to adjust themselves so that the center of mass of the system is at the lowest possible point. Use this rule to explain the concept of buoyancy for floating objects.

8.7 Will a ship sink or rise with respect to the water level as it moves from the Houston Ship

Channel (fresh water, more or less) into the Gulf of Mexico? Explain.

8.8 Solid pieces of gold, lead, and copper are placed in a beaker of mercury. What will happen to each of the pieces of metal? (You will need the information in Table 1.9.)

8.9 How does a waterproofing substance applied to a fabric affect the contact angle between water and the fibers of the fabric material?

8.10 In the energy equation (8.16) are the pressures P_1 and P_2 *absolute* or *gauge* pressures? Under what circumstances is the distinction important? When is it unimportant?

8.11 Perhaps you have seen a beach ball being supported on a column of air being blown upward. (This eye-catching demonstration is frequently used in department stores by vacuum-cleaner salesmen who direct the air sucked through a cleaner into a vertical tube.) Explain how this is possible by using Bernoulli's equation. Consider carefully what happens when the ball starts to move horizontally and away from the air stream.

8.12 Cut two pieces of paper to a size of about 5 cm × 20 cm. Hold the strips vertical about one finger width apart. Now, gently blow down between the strips. What happens? Why?

8.13 A *siphon* is a device that transfers a liquid from one level to a lower level by first raising the liquid in a tube above the level in the higher reservoir. Make a sketch of a siphon and explain how it works.

8.14 Water does not "wet" paraffin. Suppose that you sprinkle some water on a horizontal paraffin surface. Make a sketch of the way the water rests on the surface. What would happen if a drop of liquid detergent were added to one of the blobs of water? (Try it.) Show the effect in another sketch.

★Problems

Section 8.1
(For problems in this section, assume 1 atm = 1.013 N/m^2)

8.1 Until the mid-1960s a popular style of women's shoes had "spike" heels. The area of one of these heels was about 1 cm^2. What pressure did a 50-kg woman exert on the floor when walking in this type of shoe? (These shoes were not too popular with owners of soft floors!)

8.2 At a distance of 6 km from a 1-megaton nuclear weapon exploded at ground level, the peak overpressure will be 0.2 atm. What force will such an explosion cause to be exerted on the side of a house whose dimensions are 4 m × 20 m?

8.3 Two hemispherical steel shells with diameters of 0.6 m are placed together and the air is pumped out of the enclosed volume. What force is necessary to pull the shells apart? (In 1654 Otto von Guericke of Magdeburg demonstrated the effect of air pressure by having two teams of eight horses attempt—unsuccessfully—to pull apart two such shells.)

8.4 A cylindrical can 8 cm in diameter has a loosely fitting top cover. What is the lowest pressure inside the can that would permit a man, capable of exerting a force of 500 N, to lift off the cover?

8.5 A suction cup 10 cm in diameter is attached to the ceiling and supports a hanging mass. If the pressure inside the cup is $\frac{1}{10}$ atmospheric, what is the largest mass the suction cup can support?

8.6 The automobile lift in a repair garage consists of a hydraulic piston that has a diameter of 30 cm and a total mass of 400 kg. The lift is capable of handling automobiles with mass up to 3000 kg. The pump is a cylinder with a diameter of 2.5 cm. What force must be applied to the piston in the control arm in order to raise an automobile with the rated mass?

Section 8.2

8.7 A water tank has the shape of a truncated cone. The bottom surface has a diameter of 0.5 m and the top rim has a diameter of 3 m. The top rim is 5 m above the bottom surface. If the tank is filled to within 0.4 m of the top rim, what is the water pressure at the bottom of the tank?

8.8 At what depth in the ocean is the pressure due to the water equal to 100 atm?

8.9 A swimming pool has a length of 15 m and a width of 5 m. The bottom slopes smoothly from a depth of 1 m at one end to 3 m at the other end. What is the maximum pressure in the pool? If the pool is completely filled with water, what is the total force exerted on the bottom?

8.10* A cylinder with a diameter of 4 m is filled with glycerine ($\rho = 1.26$ g/cm^3). The fluid pressure at the bottom of the cylinder is 55.6 torr. What is the volume of glycerine in the cylinder?

8.11 We sometimes see pressure expressed in *pounds per square inch* (lb/in.^{2}). In this context, *1 pound* means the gravitational force on a mass of one pound. Express normal atmospheric pressure in lb/in.^{2}

8.12 Normal atmospheric pressure is 760 mm Hg. Express this pressure also in *inches* of mercury and in *feet* of water.

8.13* What is the atmospheric pressure at an altitude of 16 km? At 80 km? (Use Fig. 8.4.)

8.14 Suppose that air is completely incompressible and that the density is everywhere equal to the sea-level value (Table 1.9). At the surface of the Earth the pressure is 1 atm. At what altitude would we then find the top of the atmosphere? How does this altitude compare with that of the highest mountains and that of the highest-flying aircraft?

8.15 Suppose that pressure were expressed in centimeters of oil ($\rho = 0.85$ g/cm^3). What is 1 atm in these units?

8.16 A sealed room has a single, tight-fitting door with dimensions 2 m × 0.8 m. If the pressure in the room is greater than atmospheric pressure by 0.3 percent, what force would be required to open the door? (The door opens *into* the room!)

8.17 A pump is used at the Earth's surface to raise water from a well. What is the maximum depth of the water surface from the level of the pump for which this method will work? How could water be raised from a 100-m well by using a pump?

8.18 The bottom of a water tank is 15 m above the ground and is filled to a depth of 6 m. A pipe leads from the side of the tank (1.5 m above the bottom) to ground level. What is the absolute pressure in the pipe at ground level? (The bottom end of the pipe is closed.)

8.19 A fish fills the air spaces in its porous bone with air at a pressure of 1 atm. The fish then dives to a depth of 100 m in the ocean ($\rho = 1.025$ g/cm^3). What pressure difference must the porous bone be able to withstand?

Section 8.3

8.20 Refer to Example 8.6. Calculate the mass and the fraction by mass of silver in the crown.

8.21* A fish can maintain its depth in water without the expenditure of energy by adjusting the air content of porous bone or air sacs to make its density the same as that of the water. A certain fish has a density of 1.07 g/cm^3 when its air sacs are collapsed. To what fraction of the body volume must the air sacs be inflated to reduce the density of the fish to that of water?

8.22 A raft made from styrofoam ($\rho = 0.3$ g/cm^3) has dimensions 2 m × 2 m × 0.2 m. How much further will the raft sink when two 60-kg swimmers climb aboard?

8.23 A wooden stick ($\rho = 0.7$ g/cm^3) has a diameter of 6 cm and a length of 0.8 m. A spike with a mass of 300 g is driven into one end and the stick is then thrown into a pool of water. Make a sketch showing how the stick floats.

8.24 You are determining the mass of an aluminum object by using a beam balance which balances when two standard brass 1-kg masses are placed in the pan opposite the object. The mass of the object is not *exactly* equal to 2 kg. Why? By how much does the mass differ from 2 kg? (The density of brass is 8.44 g/cm^3.)

8.25 A certain balloon has a volume of 10 m^3 when it contains helium at normal atmospheric pressure. What is the maximum mass of the balloon proper (not including the gas) and any payload if the balloon is to just lift off the ground? (Refer to Table 1.9.)

8.26* A 0.5 m × 0.2 m block of wood ($\rho_w = 0.65$ g/cm^3) has a thickness of 6 cm. (a) How much of the block will be exposed when it floats in water? (b) Next, a 0.5-cm thick piece of aluminum, also 0.5 m × 0.2 m, is placed on top of the wood. How will the combination now float in the water? (Use Table 1.9.)

8.27* A certain object is weighed in air with the

result $v_a = 0.980$ N. Next, it is weighed while immersed in water: $w_i = 0.855$ N. Finally, the object is weighed while immersed in oil: $w_i' = 0.880$ N. (a) What is the density of the oil? (b) What is the density of the object? Identify the material.

Section 8.4

8.28 Figure 8.10 shows a film of soap solution on a wire frame. If $\ell = 4$ cm, how much work is required to extend the runner by 2 cm? Is a greater amount of work required to extend the runner by an *additional* 2 cm?

8.29 What is the minimum diameter that a mercury barometer tube can have if the correction for the height of the mercury column due to capillarity is not to exceed 0.2 torr?

8.30 Two glass plates, each with a width ℓ, are held parallel to one another with a spacing of 0.2 mm. If the plates are dipped vertically into water (at 20°C), to what height above the surface will the water rise between the plates?

8.31 A glass tube with a diameter of 0.3 mm is lowered vertically into a pan of water (at 20°C) until the top of the tube is 8 cm above the surface of the water. Describe the action of the water. Will water squirt from the top of the tube? Explain.

8.32 A capillary tube with an inside diameter 0.30 mm can support a 10-cm column of a liquid that has a density of 1.15×10^3. If the observed contact angle is $\theta = 20°$, what is the surface tension of the liquid?

8.33 One of the mechanisms by which fluids could rise in plants is capillary action. To see if this is reasonable for tall plants, such as trees, calculate the capillary rise of water in a tube that has a diameter of 0.015 mm. Actually, water is moved upward in plants by *negative pressure* developed at the top of the plant. What negative pressure would be necessary to raise water to the top of a 100-m sequoia?

Section 8.5

8.34 A cylindrical water tank rests on the ground and is filled to a height of 4 m. At a point 1.5 m below the water surface, a small hole is punched in the side. How far away from the tank will the stream of water strike the ground? (The stream of water follows the same path as a ball thrown horizontally.)

8.35* An enclosed cylindrical water tank is filled with water to a height of 5 m and the space above the water is pressurized to a gauge pressure of 5 atm. At the level of the bottom of the tank a hole is drilled and the water is allowed to flow out through a small tube that is directed upward. To what height will the water rise?

8.36 A large water tank has a hole in its side at a point 2 m below the surface. If the mass discharge rate through the hole is 0.5 kg/s, what is the area of the hole?

8.37 The arrangement shown in Fig. 8.15 is used to determine the flow rate of a liquid. The pipe section radii are $r_1 = 5$ cm, $r_2 = 2$ cm, and d stands at 12 cm. What is the flow rate in liters per second?

8.38 The device shown in the diagram (called a *Prandtl tube* or sometimes a *Pitot tube*) is a type of manometer that is used to measure fluid flow velocities. Assume that the Prandtl tube is stationary and that the fluid moves past the tube with a velocity v_1. Obtain an expression for v_1 in terms of the difference d in the heights of the liquid (density ρ_m) in the two manometer arms. Could this device be used in an aircraft as an airspeed indicator? (In the diagram, notice carefully how the device is constructed from a single piece of tubing, one end of which has a section with a large diameter, at point 1.)

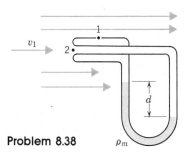

Problem 8.38

Section 8.6

8.39 In Example 8.9 we used an expression for d that was obtained from Bernoulli's equation. Were we justified in using the expression in this case? (Calculate the Reynolds number for the flow in both sections of the pipe.)

8.40 How rapidly can mercury by pumped

through a tube with a diameter of 8 mm without having the flow become turbulent?

8.41 What is the terminal velocity of the fog droplet discussed in Example 8.12?

8.42* An aluminum sphere with a radius of 3 mm falls through a column of glycerine ($\rho_f = 1.26$ g/cm^3). What is the terminal velocity for a temperature of 20°C? For a temperature of −42°C? (The buoyant force is important.)

8.43* A thin-walled glass sphere has a radius of 12 cm and a mass of 2 g. If the sphere is filled with helium to a pressure of 1 atm (absolute), what will be the terminal velocity if the sphere falls through air at 20°C?

8.44* What is the terminal velocity of an air bubble with a diameter of 1 mm rising through water at 20°C? (Neglect the density of air in the bubble compared to that of water.)

9
Temperature and Heat

From time to time in our previous discussions we have mentioned the existence of *thermal energy* or *heat* and the fact that energy dissipated through friction appears in this particular form. We now turn our attention to the thermal properties of matter and consider how heat plays a role in the behavior of objects and substances. As in the last chapter, here we study the ways in which *bulk matter* acts. We will find, however, that the explanation for the thermal properties of bulk matter lies ultimately in the actions of the individual molecules that make up the substance. In this chapter and the next, we draw the connection between the motions and interactions of molecules and the effects we see in the large-scale matter that they comprise.

9.1 Temperature

The Fahrenheit and Celsius Scales

In discussing thermal phenomena we will introduce several new terms and concepts. The easiest to understand of the thermal quantities is *temperature*. We all appreciate that temperature is a measure of the degree of "hotness" or "coldness" of an object. Before we can make real use of this idea, however, we must find some convenient way to specify temperature with *numbers*. In the next section we will discuss the effect that temperature has on the *size* of an object. Usually, when the temperature of an object is increased, there is a corresponding increase in the size of the object. For example, if a quantity of mercury has a certain volume when it is "cold," it will have a greater volume when it is "hot." This property of mercury (and most other liquids, as well) can be utilized in making an instrument to specify temperature in a quantitative way.

Suppose that a small amount of mercury is sealed in a glass tube that has a narrow bore and a reservoir bulb at one end. If we cause the temperature of the mercury to increase, then its volume will increase and the level to which the mercury stands in the tube will rise. Similarly, if the temperature of the mercury is lowered, the level will fall (Fig. 9.1). This is the operating principle of the mercury *thermometer*. Most inexpensive household thermometers use colored alcohol as the expensive liquid. (The working liquid in the first thermometer of this type was *wine!*)

In order to establish a temperature *scale,* we make use of the properties of water—in particular, the melting point of ice and the boiling point of liq-

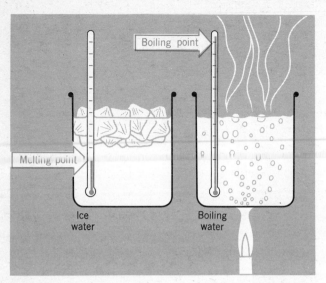

Figure 9.1 Because the volume of a sample of mercury (or other liquid) increases when the temperature is increased, the level at which mercury stands in a thermometer tube can be used to specify temperature.

uid water (Fig. 9.1). In the temperature system commonly used in the United States (but almost nowhere else in the world!), we designate the melting point of ice as 32 degrees and the boiling point (or steam point) of water as 212 degrees. This system is known as the *Fahrenheit* temperature scale, after the German physicist Gabriel Daniel Fahrenheit (1686–1736) who constructed the first practical mercury thermometer. In the Fahrenheit system we write the temperature of the ice point as 32°F and that of the steam point as 212°F. Between these two points, the scale is divided into 180 equal parts (Fig. 9.2).

In most of the world, temperature is measured on the *Celsius* scale (previously called the *centigrade* scale). In this temperature system, the interval between the melting point of ice and the boiling point of water is divided into 100 parts, with the melting point called 0°C and the boiling point called 100°C. This temperature scale was first used by the Swedish astronomer Anders Celsius (1701–1744), although Celsius at first called the steam point 0° and the ice point 100°!

We can obtain the relationship between temperatures on the Fahrenheit and Celsius scales in the following way. As indicated in Fig. 9.2, 100 degrees on the Celsius scale corresponds to the same temperature interval as 180 degrees on the Fahrenheit scale. That is, the Fahrenheit degree is a smaller unit than is the Celsius degree, smaller by the factor $100/180 = 5/9$. Thus, a change of 72 degrees on the Fahrenheit scale corresponds to a change of $72 \times \frac{5}{9} = 40$ degrees on the Celsius scale. Because the zero of the Celsius scale corresponds to 32° on the Fahrenheit scale, we must include this factor in the relationship connecting the two scales. We find

$$\left. \begin{array}{l} T_C = \frac{5}{9}(T_F - 32°) \\ T_F = \frac{9}{5} T_C + 32° \end{array} \right\} \quad (9.1)$$

where T_C stands for the Celsius temperature and T_F stands for the Fahrenheit temperature.

● *Example 9.1*

(a) We often refer to a temperature of 20°C as *room temperature*. What is this temperature on the Fahrenheit scale?

We use

$$T_F = \tfrac{9}{5} T_C + 32° = \tfrac{9}{5} \times 20° + 32° = 68°F$$

(b) On a warm summer day the temperature is 86°F. What is the temperature on the Celsius scale?

In this case we use

$$T_C = \tfrac{5}{9}(T_F - 32°) = \tfrac{5}{9}(86° - 32°) = 30°C$$

● *Example 9.2*

At what temperature are the values of the Fahrenheit and Celsius scales *equal*?

If we set $T_F = T_C = T$ in either of the equations in 9.1, we can write

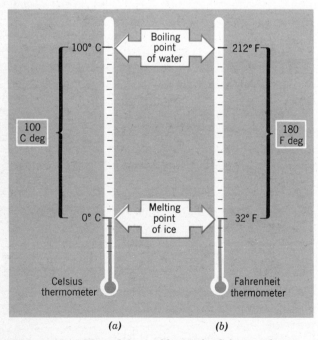

Figure 9.2 (a) Most of the world uses the Celsius scale to measure temperature. (b) In the United States the Fahrenheit scale is commonly used.

$T = \tfrac{9}{5}T + 32°$

from which

$$T = \frac{32°}{1-\tfrac{9}{5}} = -\tfrac{5}{4} \times 32° = -40°$$

Thus, $-40°F = -40°C$. ∎

A word concerning the distinction between *temperature* and *temperature interval* is in order here. When we state that "the *temperature* is 20 degrees on the Celsius scale," we indicate this temperature as 20°C. But when we state that "the *temperature interval* between 15°C and 35°C is 20 degrees," we cannot express this interval as 20°C because the designation 20°C means a *particular temperature*. If we mean a *temperature interval,* as between 15°C and 35°C, we write this as 20 deg. That is, 35°C − 15°C = 20 deg. To be more precise, we could write the temperature interval as 20 C deg to indicate that the Celsius scale is used. But we will use the Fahrenheit scale so seldom that there will be no confusion. (To solve the problem, we *will* write 20 F deg).

When we wish to indicate that some physical quantity changes with temperature, we can use the *reciprocal degree* deg^{-1}. For example, if there is an energy change of 15 joules for every temperature change of 1 deg, we could express this as 15 joules per degree or as 15 J/deg = 15 J deg^{-1}. Here, we use the reciprocal unit in the same way that we would write a velocity as 25 m/s = 25 m s^{-1}. The exponent −1, when applied to a unit, always mean *per*.

The Absolute Temperature Scale

By the middle of the nineteenth century, physicists realized that there existed a temperature below which the concept of temperature has no physical significance. This minimum temperature is called *absolute zero* and on the Celsius scale it corresponds to −273°C. (Actually, the precise value is −273.15°C.) It is useful to have a temperature scale whose zero is the absolute zero of temperature. (Indeed, for many purposes *only* this scale is meaningful, as we will see in the next chapter.) This temperature system was first used by William Thomson, Lord Kelvin (1824–1907), the great Scottish mathematical physicist. Kelvin set the zero of his scale at absolute zero and he used a temperature unit equal to 1 deg. That is, the basic interval on the *absolute* (or *Kelvin*) temperature scale is exactly the same as that on the Celsius scale. The relationship between the absolute temperature T and the Celsius temperature T_C is

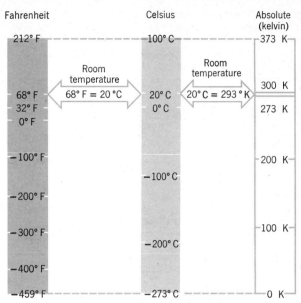

Figure 9.3 Comparison of the three temperature scales.

$$T = T_C + 273° \qquad (9.2)$$

When we write an absolute temperature, we indicate this by using the symbol K without any degree sign; for example, room temperature is expressed as 20°C = 293 K.

The symbol K is used to indicate both the *temperature* and a *temperature interval*. That is, 300 K − 250 K = 50 K. The unit 1 K is called a *kelvin*. Therefore, to indicate a certain amount of energy per Kelvin degree we would write 25 joules per kelvin or 25 J/K or 25 JK^{-1}. Although 1 K = 1 deg, we customarily use the *deg* notation when the situation is described in Celsius temperatures and we use the *K* notation along with absolute temperatures.

A comparison of the three temperature scales is given in Fig. 9.3. Table 9.1 shows the range of temperatures found in the Universe.

9.2 Thermal Expansion

The Effect of Temperature on Length

A mercury thermometer is a useful device for indicating temperature because mercury expands when its temperature is increased. In fact, almost all substances—solids, liquids, and gases—have the property that they increase in size as the temperature is raised. For solid objects we can describe these thermal expansion effects in terms of length changes or volume changes. For liquids and gases only volume changes are meaningful.

182 Temperature and Heat

Table 9.1 Range of Temperatures in the Universe

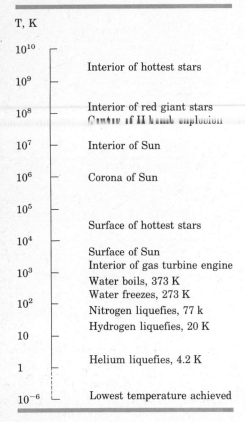

The change in length $\Delta\ell$ of the rod increases linearly with the temperature change ΔT. Moreover, $\Delta\ell$ increases in direct proportion to the original length ℓ of the rod. (That is, if a 5-m rod increases in length by 3.6 mm for a temperature increase of 40 deg, a 10-m rod will increase in length by 7.2 mm.) Altogether, we can write

$$\Delta\ell = \alpha\ell\,\Delta T \text{ or } \Delta\ell/\ell = \alpha\Delta T \tag{9.3}$$

where ℓ is the original length and where the proportionality constant α is called the coefficient of linear thermal expansion. The length ℓ' of the rod after a temperature change ΔT is

$$\ell' = \ell + \Delta\ell = \ell(1 + \alpha\Delta T) \tag{9.4}$$

Notice that if the temperature is *decreased*, then ΔT is negative and ℓ' is less than ℓ. That is, the rod shrinks when the temperature is lowered.

From Eqs. 9.3 and 9.4 we see that the units of α are *per degree Celsius* or deg^{-1}. Note carefully that α gives the *fractional change* in length ($\Delta\ell/\ell$) per unit temperature change. Values of α for several substances are given in Table 9.2. (Notice that *all* of the values are multiplied by the factor 10^{-6}.) There is some variation of α with temperature.

●*Example 9.3*

Use the information in Fig. 9.4 to obtain the value of α for copper.

Solving for Eq. 9.4 for α, we have

$$\alpha = \frac{\ell' - \ell}{\ell\Delta T} = \frac{5.0036 \text{ m} - 5.0000 \text{ m}}{(5.0000 \text{ m}) \times (40 \text{ deg})}$$

$$= 18 \times 10^{-6} \text{ deg}^{-1}$$

Table 9.2 Values of the Coefficient of Linear Thermal Expansion for Various Solids

Substance	α (deg^{-1}) (for $T = 0°C - 100°C$)
Aluminum	24×10^{-6}
Brass	19
Concrete	≈12
Copper	18
Glass (ordinary)	≈ 8
Gold	14
Invar (36% Ni steel)	1
Iron	12
Lead	29
Quartz (fused)	0.5
Silver	19

Let us begin by examining the effect of temperature on a copper rod. Suppose that we cut a section of the rod to a length of exactly 5 m when the temperature is 20°C. Next, we raise the temperature of the rod and plot the length versus the temperature. We then obtain the graph shown in Fig. 9.4. At a temperature of 60°C, the length of the rod is 5.0036 m, an increase of 3.6 mm over the original length.

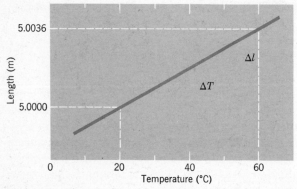

Figure 9.4 Increase in length of a copper rod with increasing temperature.

● *Example 9.4*

What will be the length at 95°C of an iron bar that has a length of 6.2 m at 15°C?

In Table 9.2 we see that the value of α for iron is 12×10^{-6} deg^{-1}. Then, using Eq. 9.4,

$\ell' = \ell(1 + \alpha \Delta T)$
$= (6.2 \text{ m}) \times [1 + (12 \times 10^{-6} \text{ deg}^{-1}) \times (80 \text{ deg})]$
$= 6.206 \text{ m}$

so that the length increases by 6 mm or about $\frac{1}{4}$ in.

● *Example 9.5*

In certain parts of the United States (e.g., in the Midwest), the temperature extremes between summer and winter can be quite large. The temperature of highway materials in summer can reach 60°C (140°F). (Because of the absorption of the Sun's direct rays, highway temperatures can exceed the air temperature.) In winter, the temperature can drop to −40°C, producing an extreme range of about 100 deg. What will be the expansion of a 10-m section of concrete highway between the lowest and highest temperatures encountered?

In Table 9.2 we see that α for concrete is approximately 12×10^{-6} deg^{-1}. Using this value in Eq. 9.3, we find

$\Delta \ell = \alpha \ell \Delta T = (12 \times 10^{-6} \text{ deg}^{-1}) \times (10 \text{ m})$
$\times (100 \text{ deg})$
$= 12 \times 10^{-3} \text{ m} = 12 \text{ mm}$

or about $\frac{1}{2}$ in. It is to accommodate thermal expansion effects that concrete highways are built in sections with tar-filled gaps between sections. ■

The coefficient of linear thermal expansion for a particular material usually has some variation with temperature. For example, iron has $\alpha = 11.7 \times 10^{-6}$ deg^{-1} at 27°C, whereas $\alpha = 12.9 \times 10^{-6}$ deg^{-1} at 127°C. If the temperature range is not too great, α can be considered to remain constant for most materials. It is interesting to note that a few substances, in particular temperature ranges, actually have linear expansion coefficients that are *negative*. That is, these substances shrink as the temperature increases and expand when the temperature decreases. For example, at $T = 75$ K, silicon has $\alpha = -0.6 \times 10^{-6}$ K^{-1}. But for temperatures above about 125 K, silicon has the normal behavior and expands with increasing temperature. Quartz has a negative expansion coefficient at *high* temperatures.

In Table 9.2, notice that the values of α for Invar (a special alloy of nickel and iron) and fused quartz are quite small. These materials are often used in applications that require the absolute minimum of length change with temperature.

The Volume Expansion of Liquids and Solids

When there is a rise in temperature, an object expands in all three dimensions, so there is an increase not only in length but also in the *volume*. In analogy with the case of linear equations, which we have just discussed, we can define a *coefficient of volume thermal expansion* β according to the relation

$$\Delta V = \beta \Delta T \quad \text{or} \quad \Delta V/V = \beta \Delta T \tag{9.5}$$

where ΔV is the change in the original volume V caused by the temperature change ΔT. The new volume V' is (compare Eq. 9.4).

$$V' = V + \Delta V = V(1 + \beta \Delta T) \tag{9.6}$$

Values of the coefficient β for some liquids are given in Table 9.3. (Notice that *all* of the values are multiplied by the factor 10^{-3}.)

The volume expansion coefficient for a solid is related in a simple way to its linear expansion coefficient. Consider a solid that has dimensions $\ell_1 \times \ell_2 \times \ell_3$. The original volume is $V = \ell_1 \ell_2 \ell_3$ and the volume after a temperature change ΔT will be $V' = \ell'_1 \ell'_2 \ell'_3$, where

$\ell'_1 = \ell_1(1 + \alpha \Delta T)$
$\ell'_2 = \ell_2(1 + \alpha \Delta T)$
$\ell'_3 = \ell_3(1 + \alpha \Delta T)$

Multiplying these three expressions together and using the definitions of V and V', we find

Table 9.3 Values of the Coefficient of Volume Thermal Expansion for Various Liquids

Substance	β (deg^{-1}) (for $T = 20$°C)
Alcohol, ethyl	1.12×10^{-3}
Alcohol, methyl	1.20
Carbon tetrachloride	1.24
Gasoline	0.95
Glycerine	0.50
Mercury	0.18
Water	0.21

184 Temperature and Heat

$$V' = V(1 + \alpha \, \Delta T)^3$$
$$= V(1 + 3\alpha \, \Delta T + 3\alpha^2 \Delta T^2 + \alpha^3 \Delta T^3)$$

Now, α is a small quantity; as we can see in Table 9.2, α is about 10^{-5} deg^{-1} for most solids. Therefore, if ΔT is not too large, the terms involving $\alpha^2 \Delta T^2$ and $\alpha^3 \Delta T^3$ will be small compared with the term involving $\alpha \, \Delta T$. Consequently, we neglect these higher-order terms and write

$$V' = V(1 + 3\alpha \, \Delta T) \qquad (9.7)$$

If we compare this expression with Eq. 9.6, we see that

$$\beta = 3\alpha \qquad (9.8)$$

Using this relation, we find that the volume expansion coefficient of aluminum is $3 \times (24 \times 10^{-6}$ deg$^{-1}) = 0.072 \times 10^{-3}$ deg^{-1} and that for iron is 0.036×10^{-3} deg^{-1}. Notice that these volume expansion coefficients (which are rather typical for solids) are much smaller than the values for liquids (see Table 9.3).

● *Example 9.6*

A 20-gallon steel tank is filled with gasoline when the temperature is 15°C. The tank and its contents are then heated by the Sun's rays to a temperature of 45°C. How much of the gasoline will spill from the tank?

(Notice, in Eq. 9.5, that the volume change ΔV is automatically given in the same units as those used to express the original volume V. Therefore, if we use V in *gallons*, we will obtain ΔV in *gallons*. Alternatively, we could convert gallons to metric units—cubic meters or liters—by using $1 \, \ell = 10^3$ cm$^3 = 10^{-3}$ m$^3 = 0.264$ gal.)

In this problem, both the tank and the gasoline will expand because of the temperature increase. First, we calculate the expansion of the gasoline. In Table 9.3, we find $\beta = 0.95 \times 10^{-3}$ deg^{-1} for gasoline. Therefore,

$$\Delta V_{\text{gas}} = \beta V \, \Delta T$$
$$= (0.95 \times 10^{-3} \text{ deg}^{-1}) \times (20 \text{ gal}) \times (30 \text{ deg})$$
$$= 0.57 \text{ gal}$$

Next, we determine the volume expansion of the tank. We use the value of α for iron:

$$\Delta V_{\text{tank}} = 3\alpha V \, \Delta T$$
$$= 3 \times (12 \times 10^{-6} \text{ deg}^{-1}) \times (20 \text{ gal}) \times (30 \text{ deg})$$
$$= 0.02 \text{ gal}$$

Therefore, the spillage is

$$V_{\text{spill}} = \Delta V_{\text{gas}} - \Delta V_{\text{tank}} = 0.55 \text{ gal}$$

If an automobile gasoline tank is filled during the cool hours of early morning and is then allowed to sit in the sun all day (without having used any appreciable amount of the gasoline), by afternoon a half-gallon or so of gasoline will have been spilled.

(In this example we calculated the volume expansion of the *hollow* tank by using the expression for a *solid* material. Can you see why this is correct? Consider a solid block to consist of a series of hollow shells.) ∎

The Effect of Temperature on Density

A change in temperature will not affect the *mass* of an object but it will affect the *volume*. Because the density of an object is the ratio of its mass to its volume, $\rho = M/V$, an *increase* in the volume means a *decrease* in the density. Similarly, a decrease in the volume means an increase in the density. The fractional change in the density, $\Delta \rho / \rho$, is the same as the fractional change in the volume, $\Delta V / V$, except that the sense in opposite. That is, a 1 percent increase in volume results in a 1 percent decrease in density. We can express this relationship in the following way:

$$\frac{\Delta \rho}{\rho} = -\frac{\Delta V}{V} = -\beta \, \Delta T \qquad (9.9)$$

Notice that the negative sign automatically gives an increase in $\Delta \rho$ when ΔV decreases (i.e., when ΔV is negative).

Most substances undergo a uniform increase in volume (and, therefore, a uniform decrease in density) as the temperature is raised. Water, however, exhibits a very unusual behavior in the vicinity of its freezing point. Instead of the density decreasing uniformly with increasing temperature, the density of water actually exhibits an *increase* as the temperature is raised from 0°C to 4°C. Further increases in temperature cause the density to decrease in the normal way. Thus, the density of water is *maximum* at 4°C. Figure 9.5 shows the variation in the density of water with temperature. Notice that both scales have been changed at $T = 10$°C in order to show more clearly the peak of the density curve at 4°C.

Although the density of water at 4°C is only about $\frac{1}{100}$ of 1 percent greater than the density at 0°C, this tiny difference has extremely important consequences. When there is a decrease in the air

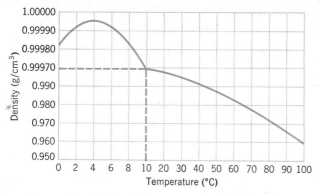

Figure 9.5 The density of pure water as a function of temperature. Both scales have been changed at 10°C in order to show clearly that the density of water is maximum at 4°C. Notice that the maximum density is slightly less than 1 g/cm³.

temperature over a body of water, such as a pond or lake, the surface water is cooled. This cool water has a density greater than the warmer water immediately below it, so the cool water sinks. As further cooling takes place, the coldest water accumulates at the bottom. When the surface temperature is reduced to 4°C, however, further cooling actually *decreases* the density. Therefore, the water with temperature nearest to 0°C remains on the surface. Consequently, it is the *surface* of a body that freezes first. The layer of ice that is formed tends to insulate the water beneath from the cold air so that much of the water remains liquid unless the air temperature drops to unusually low values. If the density of water increased uniformly with decreasing temperature, ponds and lakes would freeze from the bottom to the top. The consequences for marine life and most other life would be severe! Thus, the Earth's biosystem is based on a very unusual fluid, water.

The process of freezing decreases the density of water still further. The density of ice is only 0.917 g/cm³. Thus, ice always floats on water.

9.3 Heat

The Microscopic View

In every piece of bulk matter—in solids, liquids, and gases—the constituent molecules are continually in motion. In gases the molecules move freely between collisions with other molecules or with the walls of the container. In liquids the molecules slide easily past their neighbors although collisions between molecules are frequent. In solids, on the other hand, the molecules or atoms maintain their positions (at least, their *average* positions) within the material but they vibrate rapidly back and forth around these positions. In all types of matter we find molecular movement, and this movement means that the molecules possess kinetic energy. Therefore, in addition to the kinetic energy that an object possesses due to its motion as a whole, there is an "extra" amount of kinetic energy associated with the internal motions of the atoms and molecules that comprise the object.

The degree to which the particles move about in an object is related to the *temperature* of the object: the higher the temperature, the more rapid are the movements of the particles. The kinetic energy that an object possesses by virtue of the movement of its molecules and atoms is called the *internal energy* or *thermal energy* of the object. In Chapter 10 we will make the connection between the average speed of the molecules or atoms and the temperature of the material.

In Chapter 7 we solved a number of different types of problems by applying the principle of energy conservation. Are the solutions invalid because we did not include the internal energy in treating those situations? Not at all. Remember, only energy *differences* are physically meaningful. If an object participates in an interaction with another object in such a way that its internal energy does not change in the process, then we may eliminate the internal energy from our considerations. That is, the internal energy is constant and therefore does not contribute to any energy change in the process. This is the case for all of the ideal, frictionless, or perfectly elastic processes we have discussed.

When the internal energy of an object *does* change, then we must be careful to include this change in our calculations. There are a number of ways that such changes can be brought about. We can increase the internal energy of an object (i.e, the atoms and molecules can be made to move more rapidly) by doing *work* on the object. For example, it is easy to verify that the temperature of a small metal bar is increased when struck a series of blows with a hammer. This increase in temperature indicates an increase in the internal energy of the bar. Or a frictional force can do work on an object and change its internal energy. We have treated such cases and have explicitly taken into account the work done against friction. Another way to increase the temperature (and the internal energy) of an object is to place it over a flame. In this case we add energy to the object, in the form of *heat*, even though we do no mechanical work on the object. Indeed, we can define heat as energy that is transferred as the result of a temperature difference.

The Measurement of Heat

Heat is simply one form of energy and so the physical unit appropriate to measure heat is the same as that for mechanical energy, namely, the *joule*. Before the theories of heat and mechanical energy became unified, a different unit was used to measure heat. As in the case of the temperature scale, water was used as the substance for the definition of the heat unit. The amount of heat required to raise the temperature of 1 kg of water by 1 deg (actually, from 14.5°C to 15.5°C) was defined to be 1 *Calorie* (Cal). You may also see references to another heat unit called the *calorie* (cal) with a lower-case "c" instead of a capital "C." The calorie is the amount of heat required to raise 1 g of water by 1 deg. Therefore,

$$1 \text{ Cal} = 10^3 \text{ cal} = 1 \text{ kcal (or kilocalorie)} \qquad (9.10)$$

The Calorie is the unit used by dieticians to specify the energy content of foods (see Table 9.4). We will usually express heat quantity in terms of joules.

In the 1840s Julius Robert Mayer (1814–1878), a German physician and physicist, first realized that heat is simply another form of energy and that the total amount of energy in a system remains constant. Mayer also determined the heat unit in terms of the unit used to measure mechanical energy; this relationship is known as the *mechanical equivalent of heat*. Mayer has not received proper credit for either of these accomplishments. The German physiologist and physicist Hermann von Helmholtz (1821–1894) is usually cited as the originator of the principle of energy conservation, due largely to the careful and convincing way in which he presented his ideas. The first determination of the mechanical equivalent of heat is usually credited to James Prescott Joule (1818–1889), an English physicist. Joule devoted a decade to careful measurements of heat production by every conceivable means. He found that a particular amount of work always produced a particular amount of heat. The value for the mechanical equivalent of heat measured by Joule is very close to the value accepted today:

$$1 \text{ Cal} = 4186 \text{ J} \qquad (9.11)$$

● *Example 9.7*

In one of the experiments performed by Joule, he allowed a falling block M to turn a paddle wheel immersed in water, as shown in the diagram. In this way the gravitational potential energy Mgh was converted into heat through frictional dissipation in the water. If the amount of water in the container is 1 kg and if the 5-kg block falls through a distance of 2 m, what will be the temperature rise of the water?

The amount of work done on the water by the falling block is

$$W = Mgh = (5 \text{ kg}) \times (9.80 \text{ m/s}^2) \times (2 \text{ m})$$
$$= 98 \text{ J}$$

An amount of work equal to 4186 J (i.e., 1 Cal) is required to raise the temperature of 1 kg of water by 1 deg. Therefore, the temperature rise in this case is

$$\Delta T = \frac{98}{4186} = 0.0234 \text{ deg}$$

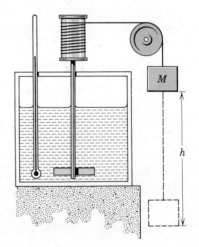

Table 9.4 Energy Content in Calories of Some Foods

Food	Calories	Food	Calories
Bacon, 1 slice	35	Egg	75
Banana, medium	85	Fish, 4 oz	140
Beef, lean, 4 oz	190	Ham, 4 oz	250
Bread, 1 slice	70	Jello, cup	145
Cake, 1 piece	200	Milk, 1 glass	165
Diet pop	1	Potato, medium	90
Doughnut	240	Yogurt, cup	120

From the smallness of this temperature rise, it can be appreciated how carefully Joule had to conduct his experiments in order to determine the mechanical equivalent of heat. He was eventually able to construct a thermometer that could be used to measure temperature changes as small as 0.005 deg!

Heat expressed in English units is the *British thermal unit* (BTU). The conversion between BTU and Calories is given by

1 Cal = 3.968 BTU. ∎

Energy and Metabolic Rates of Animals

In the essay on page 7, we argued that the rate of heat loss by an animal is proportional to the animal's surface area and therefore to the square of its characteristic length L. Because the mass M of the animal is proportional to L^3, we can write $M \propto L^3$ or $M^{1/3} \propto L$. Therefore, we expect the rate of heat loss to be

(rate of heat loss) $\propto L^2 = (M^{1/3})^2 = M^{2/3}$

Thus, for any animal we expect the rate of heat loss (or any other metabolically related variable, such as oxygen consumption) to be proportional to the mass of the animal to the $\frac{2}{3}$ power.

In 1932, Kleiber studied the rates of heat production for several different animals. These results are shown in Fig. 1. (Notice that the data are presented in a log-log plot. If you are unfamiliar with such plots, refer to the Appendix.) The dashed line is the expected variation of the rate of heat production with mass ($M^{2/3}$). Clearly, the data are more accurately represented by a line with a slightly greater slope. In fact, the points fall extremely close to a line with a slope of $\frac{3}{4}$. That is, the actual rate of heat production varies as $M^{3/4}$.

In order to understand the variation of the rate of heat production with mass, Thomas McMahon (see the reference below) has developed the following model. First, McMahon assumes that all animals can be represented as consisting of body parts that are *cylinders*. That is, the trunk, head, and limbs of an animal are all considered to be approximately cylindrical in shape. Then, the mass of any body part is proportional to the cylinder length ℓ multiplied by the square of the cylinder diameter d:

$M \propto \ell d^2$

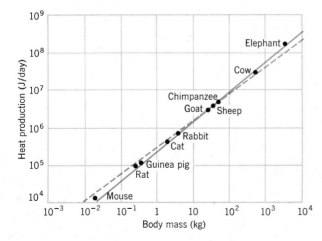

Figure 1 Variation of the rate of heat production with mass for several different animals. The dashed line is the result expected if the rate of heat production varies as $M^{2/3}$. The data actually follow a line that represents $M^{3/4}$.

(Notice that this analysis is more sophisticated than the simple model we have used previously. Instead of a single characteristic length L, McMahon uses two lengths, ℓ and d.)

Next, it is necessary to obtain a relationship connecting ℓ and d. McMahon points out that the limiting length of a cylindrical column that can support itself is proportional to the $\frac{2}{3}$ power of its diameter: $\ell \propto d^{2/3}$. He cites evidence from the sizes of tree trunks and limbs to demonstrate the validity of this relationship. He then argues that animal trunks and limbs should follow the same rule. Then, the mass of an animal should be

$$M \propto \ell d^2 \propto d^{2/3} \cdot d^2 = d^{8/3}$$

or,

$$d \propto M^{3/8}$$

Measurements made by comparative zoologists of the trunk and limb diameters of various animals show that they are indeed related to the mass in this way.

Finally, McMahon uses the fact that the power output of a muscle depends on its cross-sectional area, that is, on d^2. This is known as Hill's law of muscular power. We can understand this law in the following way. The power developed by a muscle is equal to the work done divided by the time:

$$P = \frac{\text{work}}{\text{time}} = \frac{F \times \Delta x}{\Delta t}$$

where F is the muscular force exerted and where Δx is the distance through which the force acts. The muscular force is proportional to the number of fibers in the muscle, that is, to the cross-sectional area of the muscle of d^2. Hill found that the speed of shortening of a muscle, $\Delta x/\Delta t$, is constant in any particular muscle from species to species. Therefore, $P \propto d^2$.

The power output of a muscle is directly related to the rate of heat production. That is,

$$(\text{rate of heat production}) \propto d^2 \propto (M^{3/8})^2 = M^{3/4}$$

which is the observed relationship.

As further evidence that this relationship is correct, notice that the metabolic rate of an animal is governed by the rate at which oxygen is supplied to the blood. The oxygen supply rate, in turn, is proportional to the lung volume V and to the respiratory rate (or respiratory frequency). Therefore,

$$(\text{rate of heat production}) \propto (\text{lung volume}) \times (\text{respiratory frequency})$$

Now, the lung volume is proportional to the size (or the mass) of the animal: $V \propto M$. Thus,

$$(\text{respiratory frequency}) \propto \frac{(\text{rate of heat production})}{(\text{lung volume})}$$

$$\propto \frac{M^{3/4}}{M} = M^{-1/4}$$

Several studies have, in fact, demonstrated that respiratory frequency (as well as the heart rate) depends on the mass in this way.

References

T. McMahon, "Size and Shape in Biology," *Science* **179**, 1201, 23 March 1973.

D'A. W. Thompson, *On Growth and Form*, Cambridge University Press, London, 1917. (Reprint of essay, "On Magnitude" in *The World of Mathematics*, J. R. Newman, ed., Simon and Schuster, New York, 1956, p. 1001.)

■ Exercises

1. From the solid line in Fig. 1, estimate the rate of heat production for a 70-kg person. Express this value in watts. Compare the result with the rate of energy production by a sprinter who accelerates from rest to a speed of 10.5 m/s in 2 s (see the essay on page 30).

2. If the $M^{2/3}$ scaling law were used to estimate the heat production of a 5000-kg elephant based on the value for a 20-g mouse, how large an error, compared to the correct $M^{3/4}$ scaling law, would be introduced?

3. Use your own mass and respiratory rate to estimate the respiratory rates of a 2-kg cat and a 5000-kg elephant. Can you see why there might be a connection between an animal's size and its lifespan?

Specific Heat

When 1 Cal of heat is supplied to 1 kg of water, the temperature of the water will increase by 1 deg. But if the same amount of heat is supplied to the same mass of methyl alcohol, the temperature increase will be approximately 1.5 deg. Or, if 1 Cal of heat is added to 1 kg of aluminum, the temperature of the metal will increase by almost 5 deg. In fact, every substance responds to an input of heat in a characteristic way. The amount of heat required to raise 1 kg of a substance by 1 deg is called the *specific heat* of that substance. Thus, the specific heat of water is 1 Cal/kg-deg (which we read as 1 Calorie per kilogram per degree). In SI units we would express the specific heat of water as 4186 J/kg-deg. The specific heat values for some solids and liquids are given in Table 9.5. Generally, the specific heat of a substance depends to some extent on the temperature. However, for a range of temperatures around room temperature we can consider the specific heat of a substance to be constant. (Specific heat values for gases require special attention; we will devote the following subsection to this topic.)

The amount of heat Q that must be supplied to a mass m of a substance whose specific heat is c in order to produce a temperature change ΔT is given by

$$Q = mc\,\Delta T \qquad (9.12)$$

● Example 9.8

How much heat is required to raise the temperature of 3ℓ of methyl alcohol from 20°C to 45°C?

First, we need to find the mass of the alcohol. One liter is equal to 10^3 cm^3 and, according to Table

Table 9.5 Specific Heat Values for Some Solids and Liquids near Room Temperature

Substance	Specific Heat Cal/kg-deg	Specific Heat J/kg-deg
Solids		
Aluminum	0.214	876
Copper	0.0921	386
Gold	0.0316	132
Iron	0.107	448
Lead	0.0306	128
Silver	0.0558	234
Liquids		
Acetone	0.506	2118
Alcohol, ethyl	0.572	2394
Alcohol, methyl	0.600	2512
Benzene	0.406	1700
Ether	0.529	2214
Mercury	0.0332	139
Turpentine	0.411	1720
Water	1	4186

1.9, the density of methyl alcohol is 0.81 g/cm³. Therefore,

$$m = \rho V = (0.81 \text{ g/cm}^3) \times (3 \times 10^3 \text{ cm}^3)$$
$$= 2.43 \times 10^3 \text{ g} = 2.43 \text{ kg}$$

In Table 9.5 we find that the specific heat of methyl alcohol is 2512 J/kg-deg. Then,

$$Q = mc\,\Delta T = (2.43 \text{ kg}) \times (2512 \text{ J/kg-deg}) \times (25 \text{ deg})$$
$$= 1.52 \times 10^5 \text{ J} \quad (\text{or } 36.5 \text{ Cal})$$

●*Example 9.9*

A container holds 2.4 kg of water at 22°C. Into this water is placed a 0.60-kg piece of iron that has a temperature of 95°C. Assuming that no heat is lost by the water-iron system, what will be the equilibrium temperature of the mixture?

It is clear that the final temperature will be between 22°C and 95°C and that, in the process of reaching this temperature, heat will be lost by the iron and gained by the water. This problem can be solved by using the fact that energy is conserved, so we write

(heat lost by iron) = (heat gained by water)

or,

$$(mc\,\Delta T)_{\text{iron}} = (mc\,T)_{\text{water}}$$

If we let the final temperature be T, then

$$m_{\text{iron}} c_{\text{iron}} (95°C - T) = m_{\text{water}} c_{\text{water}} (T - 22°C)$$

that is,

$$(0.60 \text{ kg}) \times (448 \text{ J/kg-deg}) \times (95°C - T)$$
$$= (2.4 \text{ kg}) \times (4186 \text{ J/kg-deg}) \times (T - 22°C)$$

Solving this equation for the temperature T, we find $T = 23.9°C$

Because the specific heat of water is much larger than that of iron (and, indeed, is much larger than that of most other substances—see Table 9.5), the final temperature of the mixture is only slightly higher than the original water temperature.

●*Example 9.10*

The measurement of the specific heat of a substance can be made by using the *method of mixtures*. (This is essentially the same as the previous example except that we measure the final temperature and use this information to determine the specific heat). Suppose we add 500 g of copper shot that has been heated in boiling water to 0.8 kg of water at 19°C. We allow the mixture to reach equilibrium and find that the temperature is then 23.41°C. Assuming no heat losses, what is the specific heat of copper?

Again, we equate the heat gained to the heat lost:

$$(mc\,\Delta T)_{\text{copper}} = (mc\,\Delta T)_{\text{water}}$$

that is,

$$(0.5 \text{ kg}) \times c_{\text{Cu}} \times (100°C - 23.41°C)$$
$$= (0.8 \text{ kg}) \times (4186 \text{ J/kg-deg}) \times (23.41°C - 19°C)$$

or,

$$38.3 \text{ kg-deg} \times c_{\text{Cu}} = 14{,}768 \text{ J}$$

from which

$$c_{\text{Cu}} = 386 \text{ J/kg-deg}$$

●*Example 9.11*

(In this example we express the specific heat values in Cal/kg-deg.) A glass beaker has a mass of 150 g and a heat capacity of 0.2 Cal/kg-deg. After 400 g of water is poured into the beaker, the system is allowed to reach room temperature (20°C). Next, 300 g of ethyl alcohol at 70°C is poured into the beaker. If there are no heat losses from the beaker-water-alcohol system, what will be the final temperature?

In this problem we have an additional part of the system (the beaker) that changes temperature. We now have

(heat lost by alcohol) = (heat gained by water)
  + (heat gained by beaker)

$$(mc\,\Delta T)_{\text{alcohol}} = (mc\,\Delta T)_{\text{water}} + (mc\,\Delta T)_{\text{beaker}}$$

that is,

$$(0.3 \text{ kg}) \times (0.572 \text{ Cal/kg-deg}) \times (70°C - T)$$
$$= (0.4 \text{ kg}) \times (1 \text{ Cal/kg-deg}) \times (T - 20°C)$$
$$+ (0.15 \text{ kg}) \times (0.2 \text{ Cal/kg-deg}) \times (T - 20°C)$$

Solving for T, we find

$T = 34.24°C$ ∎

The Specific Heats of Gases

Since gases are readily compressible, we must specify the precise method used when heat is added. Two common methods are: first, to hold the gas volume a constant while heat is added, and second, to maintain a constant pressure in the gas as heat is added.

If an amount of heat Q is added to a mass m of a gas that is maintained at a constant volume, then a change in temperature ΔT means that the specific heat of the sample is

$$c_v = \frac{Q}{m\,\Delta T}$$

where we have added a subscript v to the symbol for the specific heat to indicate that the value is appropriate for the case in which the *volume* is held constant; c_v is the *specific heat at constant volume*.

It is also possible to measure the specific heat of a gas at constant *pressure*. Suppose the gas is confined to a cylinder with a movable piston. As heat is added, the pressure would tend to rise. Thus, the piston must be moved outward to keep the pressure the same. As the piston moves, it does work on the outside force keeping the piston in place and this represents a loss of energy by the confined gas. In the constant-volume case, all of the added heat contributes to the internal energy of the gas; that is, all of the added heat is utilized in increasing the temperature of the gas. In the constant-pressure case, however, some of the added heat is used in doing work by the piston, and therefore the temperature rise is less than in the constant-volume case. A measurement made in this way yields the *specific heat at constant pressure*:

$$c_p = \frac{Q}{m\,\Delta T'}$$

where, as we have argued, $\Delta T' < \Delta T$, and, consequently, $c_p > c_v$. Table 9.6 lists the values of c_v and c_p and the ratio $\gamma = c_p/c_v$ for some gases. Notice that the values of γ for the monatomic gases (He and Ar) are approximately 1.67, whereas those for the diatomic gases (H_2, N_2, O_2, and CO) are all near 1.40; the values for the more complex gases (NH_3 and CO_2) are slightly smaller. Gases with very complex molecular structures have values of γ near 1. In Section 10.1 we will see how the quantity γ enters into the explanation of certain types of dynamical processes in gases.

Although we did not specify it, solids and liquids, when heated under ordinary conditions, involve a specific heat at constant pressure (atmospheric pressure). However, since the expansion of such substances under ordinary circumstances is very small, the work done during expansion is negligible and, therefore, there is no substantial difference between c_v and c_p.

●*Example 9.12*

How much heat is required to raise the temperature of 1 mol of nitrogen gas by 1 deg at constant pressure?

According to Table 1.10, the molecular mass of nitrogen (N_2) is 28 u. Therefore 1 mol of N_2 contains 28 g or 0.028 kg. The heat required to increase the temperature of this mass of N_2 by 1 deg is

$$Q = mc_p\Delta T = (0.028 \text{ kg}) \times (1037 \text{ J/kg-deg}) \times (1 \text{ deg})$$
$$= 29.0 \text{ J}$$

This amount of heat is called the *molar heat capacity* (at constant pressure) for nitrogen and is usually denoted by a capital letter:

$$C_p \text{ (for } N_2\text{)} = 29.0 \text{ J/mol-deg} \quad ■$$

9.4 Heat Transfer

Conduction

There are three distinct processes by which heat energy can be transferred from one place to another. The most familiar of these is *conduction*, a process in which heat flows from a hotter object to a cooler object in contact. For example, heat is conducted from the heating element of an electric stove to the bottom of a metal pan and from there to the water contained in the pan. Or, if you hold one end of a metal rod and place the other end in a block of ice, heat will flow from your hand, through the rod, to the ice; in this process the ice will be warmed and your hand will be cooled by the transfer of heat.

Although we use such expressions as "flow of heat" to describe the transfer of internal energy from one object to another, we must remember that there is no substance that *flows* during these processes. By "flow of heat" we mean that energy is transferred by collisions in which the rapidly moving atoms and molecules (in the hotter object) pass on some of their energy to the more slowly moving atoms and molecules (in the cooler object). Conduc-

Table 9.6 Specific Heat Values for Some Gases near Room Temperature

Gas	c_v (J/kg-deg)	c_p (J/kg-deg)	$\gamma = c_p/c_v$
He	3134	5202	1.66
Ar	314	524	1.67
H_2	10,061	14,186	1.41
N_2	739	1037	1.40
O_2	651	912	1.40
CO	739	1037	1.40
NH_3	1672	2190	1.31
CO_2	639	833	1.30

tion is a process in which energy is transferred by means of collisions at the microscopic level but completely without any gross movement of matter.

The ability to conduct heat varies greatly from one substance to another. Generally, metals are good conductors of heat, whereas such materials as wood, brick, asbestos, and glass are poor conductors. Indeed, these latter materials (called *insulators*) are used in the construction of buildings to minimize the outward flow of heat during winter as well as the inward flow heat during summer.

Let us consider the flow of heat through some material shaped in the form of a bar with a uniform cross-sectional area A and a length L. Suppose that we arrange to maintain each end of the bar at a particular constant temperature. For example, we might immerse one end in a water-ice mixture and the other end in boiling water. Let the temperature of the cooler end be T_1 and that of the warmer end be T_2. Then, heat will flow along the rod from the end at T_2 to the end at T_1, as shown in Fig. 9.6.

If there is a large temperature difference, $T_2 - T_1 = \Delta T$, between the ends of a bar with a length L, the heat flow will be more rapid than for a small temperature difference. Similarly, if ΔT is constant, the heat flow rate (i.e., the amount of heat conducted per unit time) will be greater if L is small. We can combine these statements by writing, (rate of heat flow) $\propto \Delta T/L$. This quantity, $\Delta T/L$, is called the *temperature gradient* in the bar. Moreover, if the cross-sectional area A of the bar is doubled (e.g., by placing a second, identical bar alongside the original bar), the rate of heat flow will also double. Therefore, the amount of heat Q conducted between the two ends of the bar will be directly proportional to the temperature gradient $\Delta T/L$, the cross-sectional area A, and the time t. The proportionality constant κ is called the *thermal* (or *heat*) *conductivity* of the substance. Altogether, we have

$$Q = \kappa A t\, \Delta T/L \qquad (9.13)$$

The units of κ are J/m-s-deg. Clearly, the thermal

Table 9.7 Values of the Thermal Conductivity for Some Materials

Material	κ (J/m-s-deg)
Conductors	
Aluminum	205
Copper	381
Gold	397
Iron	67
Silver	414
Tungsten	159
Insulators	
Air	0.024
Asbestos	≈ 0.2
Brick, insulating	≈ 0.15
Brick, ordinary	≈ 0.6
Glass	≈ 0.8
Wood	≈ 0.08

conductivity κ will be large for materials that are good heat conductors and small for materials that are poor conductors or good insulators. Values of the thermal conductivity for some materials are given in Table 9.7.

● *Example 9.13*

A copper rod has a diameter of 4 cm and a length of 0.8 m. One end of the rod is placed in boiling water and the other end is placed in an ice bath. How much heat is conducted through the bar in 1 min?

The cross-sectional area of the bar is

$$A = \pi r^2 = \pi \times (0.02 \text{ m})^2 = 1.257 \times 10^{-3} \text{ m}^2$$

The temperature difference is $\Delta T = 100°C - 0°C = 100$ deg, and the time is $t = 1$ min $= 60$ s. From Table 9.7 we find $\kappa = 381$ J/m-s-deg. Then, using Eq. 9.13

$$Q = \kappa A t\, \Delta T/L$$

$$= \frac{(381 \text{ J/m-s-deg}) \times (1.257 \times 10^{-3} \text{ m}^2) \times (60 \text{ s}) \times (100 \text{ deg})}{(0.8 \text{ m})}$$

$$= 3592 \text{ J}$$

or approximately 0.86 Cal.

● *Example 9.14*

What is the heat flow per hour and per square meter through a wall consisting of ordinary brick that has a thickness of 8 in. (0.20 m) if the inner and outer temperatures are 19°C and 36°C, respectively?

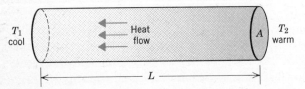

Figure 9.6 The rate of heat flow through the rod is proportional to $\Delta T/L$.

Using $\kappa = 0.6$ J/m-s-deg, we have

$$\frac{Q}{\text{h-m}^2} = \frac{\kappa \Delta T}{L}$$

$$= \frac{(0.6 \text{ J/m-s-deg}) \times (36°C - 19°C) \times (3600 \text{ s/h})}{(0.20 \text{ m})}$$

$$= 1.84 \times 10^5 \text{ J/h-m}^2$$

● *Example 9.15*

What temperature gradient must be applied to an aluminum bar (2 cm × 4 cm) if the heat flow is to be 100 J/s?

Solving Eq. 9.13 for the temperature gradient $\Delta T/L$,

$$\frac{\Delta T}{L} = \frac{1}{\kappa A} \times \frac{Q}{t}$$

From Table 9.7 we find $\kappa = 205$ J/m-s-deg. Therefore,

$$\frac{\Delta T}{L} = \frac{(100 \text{ J/s})}{(205 \text{ J/m-s-deg}) \times (0.02 \text{ m} \times 0.04 \text{ m})}$$

$$= 610 \text{ deg/m}$$

This temperature gradient could be achieved by placing a 100-deg temperature difference across a bar that has a length of 16.4 cm.

In the construction industry insulating materials of fiber glass and styrofoam are available in sheets and slabs of standard thicknesses. Their resistance to heat transfer is in terms of their $\mathcal{R}$-value.

$\mathcal{R}$-value = L/κ

Thus, Eq. 9.13 may be written

$Q = At\,\Delta T/\mathcal{R}$.

The $\mathcal{R}$-value is quoted in the English system of units and corresponds to the heat flow in BTU/h through a 1-ft² area of the particular insulating slab or sheet per °F of temperature difference. In the northern United States, insulation with an $\mathcal{R}$-value of 30 ft²·F·h/BTU (expressed as $\mathcal{R}$30) is recommended in outside walls. ■

Convection

When a substance is heated, it expands. That is, heating causes the volume of a substance to increase and its density to decrease. If the substance is a liquid or a gas and if the heating is localized, then the heated fluid will experience a buoyant force and will rise through the cooler surrounding fluid. Figure 9.7

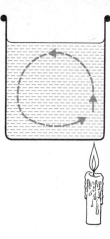

Figure 9.7 Convective circulation in a beaker of water heated at one point. Convective circulation is an important heat transfer mechanism in the atmosphere and in the oceans.

shows how heating a beaker of water at one point causes a *convective* circulation of the liquid. As the warm water rises above the spot heated by the flame, cooler water moves in to replace the rising water. This cooler water, in turn, is heated and rises, making way for the inflow of more cool water. The warmed water is gradually cooled as it rises and eventually makes its way back to the bottom of the beaker, thereby completing the circuit. In this way, the heat added to the water at one point is distributed throughout the container and the water reaches essentially the same temperature everywhere in the container.

If, instead of heating a quantity of water at the *bottom*, we cool the water at the *top*, we will find the same type of convective circulation with the cool water sinking to the bottom. This pattern will continue until the water reaches a temperature of 4°C. Further cooling brings the water into the temperature regime of reversed density variation (Fig. 9.5). Convection continues in this temperature regime, but with the *warmest* water sinking to the bottom.

Convection is one method of household heating. If hot water is piped from a boiler to a radiator, the air immediately adjacent to the radiator is heated and convective circulation begins, carrying heat throughout the room. In modern construction, a fan or blower is used to distribute the heat more quickly and more uniformly than would result from natural convection. This type of heating is called *forced convection*.

Notice that in heat transfer by *conduction* (and by the method we will discuss next, namely, *radiation*), there is no net movement of matter. In the convection process, however, the transfer of heat is due entirely to the gross movement of the heated fluid.

Radiation

If you bring your hand near a glowing electric light bulb or near a glowing log in a fire, you will experience a definite sensation of heat even though you do not touch either the bulb or the log. This heating effect is due only slightly to conduction and convection. Instead, it is due primarily to *radiation*, the third process by which heat energy can be transferred from one place to another.

We know that the atoms in any substance are continually in motion and that the degree of agitation increases with the temperatures. At very high temperatures, collisions occur in which some of the kinetic energy of the colliding atoms can be absorbed by the atomic electrons. The atoms do not retain this extra energy for more than a tiny fraction of a second. When the electrons return to their normal condition in the atom, the energy that had been absorbed is emitted in the form of *electromagnetic radiation*. That is, some of the kinetic energy of the rapidly moving atoms is converted into radiant energy by this process.

The most familiar type of electromagnetic radiation is *light*, radiation that we can detect with our eyes. But the only feature of light that distinguishes it from other forms of electromagnetic radiation (such as radio and television signals, infrared radiation, and X rays) is the *frequency* (or *wavelength*) of the radiation. Thus, X rays and ultraviolet radiation have higher frequencies than visible light, whereas radio signals and microwaves have lower frequencies. Although we assign various names to these radiations, we must remember that they all have the same basic properties and differ only in frequency. (We will discuss electromagnetic radiation more thoroughly in later chapters.)

When a nonflammable object is heated, the radiation that it emits increases in frequency as the temperature rises. At temperatures below about 1000°C, the radiated energy consists of long wavelength radiation (infrared radiation); at these temperatures we readily feel the heat radiation, but very little visible radiation is emitted and we do not perceive any "glow" from the object. At 2000°C, a sufficient amount of the emitted energy consists of visible light that the object can be seen to glow with a dull red color. As the temperature is increased further, the predominant color of the emitted light changes to bright red, then yellow, and finally blue-white. For a temperature of 6000°C (the temperature of the Sun's surface), more yellow light is emitted than any other color (and we see the Sun's disc as yellow). At 20,000°C (a temperature that can be achieved only for brief periods in the laboratory but which is the temperature of some particularly hot stars), the predominant color is blue-white. The fact that the color of the light emitted by an object depends on its temperature is the basis of an instrument (called a *pyrometer*) for measuring very high temperatures.

When an object receives heat energy, its temperature increases. The Earth continually receives vast amounts of radiant energy from the Sun, and yet the temperature of the Earth remains essentially constant (on the average). Why is this so? The reason is that the amount of energy radiated by an object depends strongly on its temperature. The radiant solar energy *absorbed* by the Earth has increased the temperature to the point that the energy *radiated* by the Earth is nearly equal to the absorbed energy. (In fact, the energy radiated by the Earth is slightly greater than the amount absorbed because heat is generated within the Earth by the decay of natural radioactive substances.)

The ability of an object to emit radiation is characterized by a quantity called the *emissivity* and represented by the symbol e. The emissivity is a pure number between 0 and 1. A perfect radiator would have $e = 1$. Dark objects usually have emissivities of 0.9 or greater, whereas light-colored or shiny objects have values of $e = 0.2$ or smaller. For example, carbon black (soot) has $e = 0.95$, and polished aluminum has $e = 0.05$.

Any object that is a good emitter of radiation is also a good absorber. Suppose that we have two identical sheets of aluminum; we polish one sheet and we coat the other with carbon black. The two sheets are now exposed to direct sunlight and allowed to come to temperature equilibrium. If we measure the temperature of each sheet, we find that the black sheet has a much higher temperature than the shiny sheet. This temperature difference is a result of the difference in emissivity between the two sheets. The blackened sheet will absorb energy and increase in temperature until the rate of absorption of energy is just balanced by the rate of emission. The polished sheet, on the other hand, reflects most of the incident energy and absorbs relatively little. Therefore, this sheet must emit energy at a much lower rate than the blackened sheet in order to maintain an equilibrium condition; this lower rate of emission is reached at a much lower temperature.

Experiments have shown that the amount of energy radiated per second by an object (i.e., the radiated *power*) is proportional to the *fourth power* of the absolute temperature, T^4. The radiated power is also proportional to the surface area A and the emissivity e. Altogether, we have

$$P = \sigma A e T^4 \quad (9.14)$$

where the proportionality constant σ is called the *Stefan-Boltzmann constant* and the radiation law expressed by Eq. 9.14 is called the *Stefan-Boltzmann law*. Numerically,

$$\sigma = 5.67 \times 10^{-8} \text{ J/s-m}^2\text{-K}^4$$
$$= 5.67 \times 10^{-8} \text{ W/m}^2\text{-K}^4$$

That is, when the area is expressed in square meters and the temperature in kelvins, the radiated power is expressed in *watts*.

● *Example 9.16*

An ordinary incandescent electric light bulb contains a fine tungsten wire that is heated to a high temperature by passing an electric current through it. A typical filament has a surface area of about 1 cm² (10^{-4} m²) and is heated to a temperature of approximately 3000 K. At this temperature, the emissivity of tungsten is 0.34. What is the radiated power of such a light bulb?

Using Eq. 9.14, we find

$$P = \sigma A e T^4$$
$$= (5.67 \times 10^{-8} \text{ W/m}^2\text{-K}^4) \times (10^{-4} \text{ m}^2) \times (0.34)$$
$$\times (3000 \text{ K})^4$$
$$= 156 \text{ W}$$

This power is close to that of most household light bulbs. But it must be emphasized that only a small fraction (several percent) of the radiated power is in the form of visible light. Most of the power radiated by an ordinary light bulb is in the form of infrared (heat) radiation. It is for this reason that light bulbs make rather efficient warming elements. ■

If a heated object is in an enclosure, the object will be emitting radiation as well as absorbing radiation emitted by the enclosure walls. The radiation emitted by the object will depend on its absolute temperature T_2, and the radiation absorbed will depend on the amount emitted by the walls at the temperature T_1. The *net* amount of power being radiated away from the object is then

$$P_{net} = \sigma A e T_2^4 - \sigma A e T_1^4 = \sigma A e (T_2^4 - T_1^4) \quad (9.15)$$

Because the temperature terms appear raised to the fourth power, the effect of the term T_1^4 is important only if T_1 and T_2 are approximately equal. (See Problem 9.50 for a discussion of this case.) Thus, in Example 9.16 we were justified in neglecting the effect of the room in which the light bulb was located because the walls of that room would have a temperature of about 300 K. Comparing,

$$(3000 \text{ K})^4 - (300 \text{ K})^4 = (8.1 \times 10^{13} - 8.1 \times 10^9) \text{K}^4$$

and the second term is only 0.1 percent of the first term.

The present-day Universe is uniformly bathed in black body radiation corresponding to 2.7 K. Much of this radiation is in the radio-wave portion of the electromagnetic spectrum and was first detected by radio telescopes. The source of this radiation is the relativistically red-shifted remains of the big-bang fireball start of the Universe some 10 to 20 billion years ago. At the tender age of one million years after creation the fireball temperature dropped to about 3000 K and became transparent enough to allow radiation to escape. It is this radiation we see today. See also Chapters 17 and 21.

Metabolic Rates of Humans

(In the life sciences, heat values are commonly expressed in Calories instead of joules; we will follow that practice in this essay.)

In order to sustain the activities of an average man, his daily food intake must amount to about 3000 Cal. How this energy is used depends on the degree of strenuousness of the activity. Some typical situations are:

	Metabolic rate	
	Cal/h	Watts
Sleeping, resting, attending lectures	70	80
Light activity (walking, housework)	250	290
Moderate activity (slow cycling, slow swimming, playing a video game)	450	520
Heavy activity (playing basketball, chopping wood, jogging)	600	700
Extreme activity (limit of physical exertion that can be sustained for 1 min)	700	810
(limit of physical exertion that can be sustained for 10 s)	1200	1400

Only a small fraction of the energy generated by metabolic processes in the body is converted into useful work. For a specific activity, such as riding a bicycle, the conversion efficiency can be as high as 20 percent. But averaged over a day's activities, the efficiency figure drops to only a few percent. Almost all of the food energy taken into the body is converted into heat.

The energy that maintains the body's metabolic functions is derived from the oxidation of such foodstuffs as carbohydrates, fats, and proteins. The conversion of the carbohydrate *glucose* ($C_6H_{12}O_6$) into carbon dioxide and water is typical of the biochemical reactions that supply energy to the body. The actual chain of reactions that constitutes this oxidation process is quite complex, but the net result is

$$C_6H_{12}O_6 + 6\,O_2 \longrightarrow 6\,CO_2 + 6\,H_2O$$

Every gram of glucose that is oxidized releases 3.8 Cal of energy. Other foodstuffs release different amounts of energy per gram consumed. (Some fats, for example, release about 10 Cal/g). In spite of these differences, there is one aspect of the various reactions that is the same; that is, the energy released in the oxidation of all foodstuffs is very closely 5 Cal for every liter of oxygen that is consumed.

Because oxygen enters the body primarily through the lungs (only about 2 percent is absorbed through the skin), the rate of oxygen consumption and, hence, the metabolic rate is controlled by the breathing rate. While resting or sleeping, the breathing is slow and shallow, but during violent exercise, the breathing is rapid and deep in order to meet the increased demand for oxygen.

The volume of an average pair of lungs is about 6 ℓ, but this entire volume is not involved in the ordinary breathing process. In fact, the volume of air taken into the lungs during a breath while the body is resting amounts to only about 0.5 ℓ. Air consists of about 80 percent nitrogen and 20 percent oxygen. Therefore, each breath contains about 0.1 ℓ of oxygen. Not all of this oxygen is absorbed by the body. Measurements have shown that approximately 22 percent of inhaled oxygen is absorbed and used by the body. In the resting condition, the breathing rate is about 11 per minute. Therefore, the rate of energy release by oxidation reactions in the resting condition is

Rate of energy release = (0.1 ℓ/breath) × (0.22) × (11 breaths/min) × (60 min/h) × (5 Cal/ℓ)

= 72.6 Cal/h

which is quite close to the metabolic rate for the resting condition listed above.

It is also interesting to enquire how the body rids itself of the heat generated in the metabolic processes. In the resting condition, most of the heat can be radiated away. To see this, we can use the Stefan-Boltzmann law (Eq. 9.15), and we express the constant σ as 1.36×10^{-11} Cal/m²-s-K⁴. The skin temperature actually depends on internal and external conditions; let us assume a value $T_2 = 34°C = 307$ K. The temperature T_1 is the temperature of the surroundings, for which we use $27°C = 300$ K. The area of an average man is about 1.8 m². Finally, we need the emissivity of the skin. The radiation emitted by the skin is in the infrared region and the corresponding emissitivity is approximately unity. Then,

$P = \sigma e A(T_2^4 - T_1^4)$
$ = (1.36 \times 10^{-11} \text{ Cal/m}^2\text{-s-K}^4) \times (1) \times (1.8 \text{ m}^2) \times [(307 \text{ K})^4 - (300 \text{ K})^4]$
$ = 69$ Cal/h

This value is also quite close to the metabolic rate for the resting condition. Of course, significant radiation will be emitted only by the exposed area of the skin. Moreover, if the surroundings are at a temperature above skin temperature, the body will receive energy more rapidly than it can be radiated away. In this case, the body begins to perspire and evaporation removes the excess heat. For activity above the resting level, the radiative emission remains essentially the same and the heat generated by the activity is carried away by convection and by the evaporation of perspiration. Since many of the body's blood vessels are at one's head, one should wear a head cover to minimize heat loss (see, Mother was right!).

Reference
G. B. Benedek and F. M. H. Villars, *Physics with Illustrative Examples from Medicine and Biology*, Addison-Wesley, Reading, Mass., 1973, Vol. 1, Chapter 5.

■Exercises

1. What must be the respiration rate of a person walking at a brisk pace so that his or her metabolic rate is 350 Cal/h? (Assume that the effective volume of each breath is 0.7 ℓ.)

2. The loss of 1 g of body fat by oxidation requires the expenditure of 10 Cal by muscular work. Suppose that the person in the previous exercise wishes to lose 0.5 kg (about 1 lb) during the hike. What must be the duration of the hike? Does this suggest that exercise is an effective method of weight control?

3. A man is working at a moderate level (metabolic rate = 400 Cal/h) in a room in which the temperature is 20°C. (He wears only shorts and the area of exposed skin is 1.5 cm².) How much energy does the man radiate per hour? How much perspiration must be evaporated per hour to account for the remainder of the heat loss? (An amount of energy equal to 580 Cal/kg is required to vaporize water at normal skin temperature.)

4. Some of the energy used by the body on a cold day is expended in heating the air that is breathed. Assume the respiration conditions for minimal activity (11 breaths per minute and 0.5 liter per breath), and calculate the heat loss due to breathing when the ambient temperature is $-20°C$. (The specific heat at constant pressure for air is 0.24 Cal/kg-deg.)

Summary of Important Ideas

Temperature is a measure of the degree of "hotness" or "coldness" of an object. On the *Celsius* temperature scale, we choose the melting point of ice to be 0°C and the boiling point of water to be 100°C. On the *absolute* or *Kelvin* scale, temperature is measured from the absolute zero of temperature at $-273°C$. Thus, $0°C = 273$ K.

When most substances are heated, they expand. The change in length $\Delta \ell$ of a solid is proportional to the original length ℓ, to the temperature change ΔT, and to a quantity α (the *coefficient of linear thermal expansion*) that is characteristic of the particular material: $\Delta \ell = \alpha \ell \, \Delta T$. Similarly, the volume of a substance changes with temperature according to $\Delta V = \beta V \, \Delta T$, where β is the *coefficient of volume thermal expansion*. For a solid, $\beta = 3\alpha$.

The change in volume with temperature of *water* in the range 0°C to 4°C is peculiar in that the expansion coefficient β is negative. That is, water decreases in volume (and increases in density) as the temperature is raised from 0°C to 4°C.

Heat is one form of energy. In addition to the normal units used to measure energy (i.e., joules), heat is often specified in terms of *Calories*. One Calorie is the amount of heat required to raise the temperature of 1 kg of water by 1 deg. The *mechanical equivalent of heat* is 1 Cal = 4186 J.

The *specific heat* of a substance is the amount of heat energy required to raise the temperature of 1 kg of the substance by 1 deg.

For *gases*, the specific heat at constant pressure is always greater than the specific heat at constant volume: $c_p > c_v$.

Heat can be transferred from one object or location to another by three different processes: *conduction*, *convection*, and *radiation*.

The amount of heat per unit time and per unit area that is conducted through a slab of material is proportional to the *temperature gradient* in the material and to its *thermal conductivity* κ: $Q/\text{s-m}^2 = \kappa \, \Delta T/L$, where L is the thickness of the slab.

Convective circulation in a fluid depends on the change in density that results from heating or cooling a substance.

A heated object *radiates* energy in the form of electromagnetic radiation at a rate that is proportional to the *fourth* power of the absolute temperature of the object.

◆ Questions

9.1 As pointed out in Section 9.1, Celsius originally chose the ice point to be 100° and the steam point to be 0°. Why was this choice not particularly clever?

9.2 Both the Celsuis scale and the absolute scale are used in scientific matters. When an astronomer refers to the temperature of the Sun's core, he might say that the temperature is "about 15 million degrees" without specifying the scale being used. Does it matter?

9.3 A clock is regulated by the back-and-forth movement of a pendulum. The time required for one complete swing is called the *period* of the pendulum and is proportional to the square root of the length of the pendulum. If the pendulum rod is made from a material such as iron, will the clock "run fast" or "run slow" when the temperature rises? Explain.

9.4 Iron reinforcing rods are often used in concrete structures such as buildings and highways. Examine Table 9.2 and explain why this is a reasonable procedure. Could aluminum rods be substituted?

9.5 When a wheelwright places an iron rim on a wagon wheel, he first heats the rim. Why?

9.6 A circular hole is drilled in a steel plate. When the temperature of the plate is increased, what happens to the size of the hole?

9.7 A bucket is made with a brass bottom and steel (iron) sides. Do you suppose this bucket will resist leaking for very long? Explain.

9.8 Some thermometers consist of *bimetallic strips*—strips of two different metals that are welded or riveted together along their largest dimension. Strips of iron and aluminum are often used in these bimetallic thermometers. What will happen when an iron-aluminum strip is heated? Show how a bimetallic strip could be used as a *thermostat* to turn on an air conditioner at a predetermined temperature.

9.9 Many solid substances (e.g., metals and minerals) are *crystalline* in form—that is, the atoms are arranged in a regular, repeating array. (We will discuss crystals in Chapter 19.) In crystals, the forces that hold the atoms together can be slightly different in different directions because of the way the atoms are arranged. Is it likely that a *single* coefficient of linear thermal expansion could be defined for a crystalline substance? Explain.

9.10 A mercury thermometer has a tube with a very small bore. Explain the reason in detail. (Refer to Table 9.3.)

9.11 Why is it important to prevent the water in pipes from freezing in winter?

9.12 Will you gain weight if you drink only *hot* water? Explain.

9.13 When a gas at high pressure is released into the air from a container the expanding gas does work against the force due to atmospheric pressure. What happens to the temperature of the gas? Can you give some examples of this effect?

9.14 Why are the handles of cooking utensils made of wood or plastic instead of metal as is the rest of the utensil?

9.15 On a day when the temperature is considerably below freezing, it is possible to handle a piece of wood with your bare hand. But if you pick up a piece of metal with a bare hand, it will feel much colder and your skin may even freeze to the surface! Explain the difference between these two situations.

9.16 On a cold day, outdoorspersons frequently wear a loose-knit sweater beneath a tightly knit outer coat. Explain the reason for dressing in this way.

9.17 Why are above-ground fuel storage tanks usually painted with aluminum or another light-colored paint?

★ Problems

Section 9.1

9.1 What is normal body temperature (98.6°F) on the Celsius scale?

9.2 When a person has a "high fever," the body temperature may be 6 F deg above normal. To what interval on the Celsius scale does this increase correspond?

9.3 The melting point of tin is 182°C. Express this temperature on the Fahrenheit scale.

9.4 The lowest recorded temperature on Earth was —127°F, at Vostok Station, Antarctica, and the highest was 136°F, at Azizia, Libya. What are the corresponding temperatures on the Celsius and Absolute scales?

9.5 Express the temperature interval from −30°F to 150°F in kelvins.

Section 9.2

9.6 A bridge over a river has a total length of 1.8 km. The bridge is constructed from steel (iron) and the temperature of the steel ranges from −20°C in winter to 50°C in summer. If the bridge is built when the temperature is 15°C, what is the total length of gaps that must be left between the steel sections in order to allow for the maximum expansion? What will be the total length of gaps during the coldest part of winter?

9.7 When a 100-m length of chromium wire is cooled from 20°C to 0°C, it is found that the length decreased in length by 1 cm. What is the coefficient of linear thermal expansion for chromium?

9.8 At a temperature of 15°C, an iron rod and an aluminum rod each have a length of 2.5 m. When the rods are heated to a temperature of 150°C, which rod will be longer and by how much?

9.9 A meter "stick" is constructed from steel (iron) and is found to be accurate at 20°C. By how much will a full-scale measurement be in error at 80°C.

9.10 A weight is hung from a 30-m copper wire that is suspended from the ceiling over a stairwell. A pointer is attached to the weight. Marks are made on the wall to indicate the position of the pointer at various temperatures. What is the distance between the marks for 8°C and 45°C? (Do you need to know that the building is constructed with steel beams?)

9.11 Atmospheric pressure is determined with a mercury barometer by measuring the height of the mercury column above the surface of the mercury in the reservoir (see Fig. 8.5). If an aluminum scale (accurate at 9°C) is used to make this measurement, what amount of correction must be applied when the scale reading is 758 mm at a temperature of 40°C. (Does the expansion of the glass tube enter into this problem?)

9.12 A rod has a length of 2.1578 m at a temperature of 18°C. When the temperature is raised to 34°C, the length is 2.1588 m. From what material (among those listed in Table 9.2) was the rod constructed?

9.13 A brass rod is to increase in length by 5 mm when the temperature is raised from 0°C to 20°C. What must be the length of the rod at 0°C?

9.14 An aluminum plate is cut to a size of 3 m × 4 m when the temperature is 15°C. What will be the area of the plate if the temperature is increased to 40°C?

9.15 When a material in the form of a plate or

sheet experiences a temperature rise, its area increases. In analogy with the method used to obtain Eqs. 9.7 and 9.8, derive an expression for the area of a plate and express the *coefficient of area expansion* in terms of α.

9.16 A block of copper (4 cm × 6 cm × 12 cm) is heated from room temperature (20°C) to 125°C. What is ΔV for the block?

9.17 What is the minimum volume (measured at 20°C) that a glass beaker must have so that 500 cm^3 of methyl alcohol (at 20°C) will not overflow when the temperature is raised to 60°C? (Neglect the effects of surface tension at the rim.)

9.18 A 3-ℓ quantity of carbon tetrachloride is placed in a glass container that has a narrow neck (1 cm diameter). The liquid extends into the neck when the temperature is 18°C. How far will the carbon tet rise if the temperature is increased to 40°C?

9.19 Methyl alcohol at 18°C exactly fills two 1.5-ℓ cups, one of which is made of ordinary glass and one of which is made of aluminum. When the temperature is raised to 50°C, how much alcohol will spill from each cup?

9.20 If you purchase 15 gal of gasoline when the temperature is 10°F, how much gasoline will you have when the temperature becomes 65°F?

9.21 An aluminum cup with a volume of 150 cm^3 is filled to the brim with a liquid when both are at a temperature of 5°C. When their temperature is raised to 35°C, 1.93 cm^3 of the liquid spills over. What is the volume expansion of the liquid? Identify the liquid using Table 9.3.

9.22 A mercury thermometer is made from ordinary glass and has a reservoir volume of 0.4 cm^3. The diameter of the column is 0.01 cm. How far apart will be the scale graduations if each interval represents 0.1 deg?

9.23 What will be the fractional change in density of a quantity of glycerin if the temperature is increased by 60 deg?

9.24 By how much will the density of gold change between a temperature of 0°C and a temperature of 100°C?

9.25 Carbon tetrachloride has a density of 1.595 g/cm^3 at a temperature of 20°C. What will be the density at 0°C?

9.26 What is the average fractional change in density per degree for water between 0°C and 4°C. (Use Fig. 9.5.)

9.27* A block of copper (8 cm × 8 cm × 8 cm) is placed in a container that is filled with methyl alcohol. What is the buoyant force on the copper block when the temperature of the system is 0°C? (The densities listed in Table 1.9 are appropriate for 0°C.) If the alcohol and copper are heated to 50°C, what is the new buoyant force on the copper block?

9.28 A cylinder contains 10 g of helium at a temperature of 20°C. How much heat must be added at constant volume to raise the temperature to 60°C?

9.29 How much heat must be added if the gas in Problem 9.28 were heated at constant pressure?

Section 9.3

9.30 In the paper that describes his experiments concerning the mechanical equivalent of heat (1845), Joule remarked on the expected rise in temperature of water as it goes over Niagara Falls. What should be the water temperature at the bottom of the Falls if the temperature at the top ($h = 160$ ft) is 14°C? (Assume that there are no heat losses.)

9.31 A man takes in 3000 Cal of food energy per day. During the course of a day's work, the man lifts 20-kg boxes from ground level to a truck bed that is 1.5 m above ground level. He lifts such a box every 10 s throughout his 8-h work day. What fraction of his energy intake has been expended in this work? Comment on the efficiency of the human body as a machine.

9.32 An amount of work, 10^4 J, is done by hammering on each of two 1-kg bars, one made of copper and the other of gold. Assuming no heat losses, which bar will become hotter and by how much?

9.33 A kilogram of copper shot is placed in a bag. The bag is raised to a height of 10 m and dropped. The process is repeated until 10 drops have been made. Assuming that all of the heat is produced in the copper and that there is no heat loss from the copper, what is the temperature rise of the copper?

9.34* A copper bar has a mass of 70 kg and a cross-section that is 4 cm × 6 cm. If the bar absorbs 6 Cal of heat, by how much will the length change? Did you need to know the mass of the bar in order to calculate $\Delta\ell$? (Express $\Delta\ell$ in terms of known quantities before substituting numerical values.)

9.35 A block of lead at 20°C is allowed to fall through a height of 20 m and strike a hard floor. If all of the kinetic energy at the instant of impact is converted into internal energy of the lead, what is the final temperature of the block?

9.36 One of the units that can be used to measure the quantity of heat is the *British thermal unit* (Btu). The specific heat of water can be expressed as 1 Btu (lb)$^{-1}$ (F deg)$^{-1}$. Use this information to find the relationships between the Btu and the Calorie and between the Btu and the joule.

9.37 Use the specific heat values for the metals in Table 9.5 and construct a table of the *molar heat capacities* for these elements. (Refer to Example 9.12.) Comment on the results.

9.38 A lead bullet ($m = 10$ g) enters a wood block with a velocity of 300 m/s. When the bullet comes to rest in the block, by how much will its temperature have risen, assuming that one-half of the heat generated in stopping remains in the bullet?

9.39 A beaker contains 1.2 ℓ of water at 18°C. Into this beaker is poured 0.9 ℓ of methyl alcohol at 27°C. Assuming no heat losses, what is the final temperature of the mixture?

9.40 If 3 kg of mercury at 80°C is poured into 2 ℓ of water at 20°C, what will be the final temperature of the mixture? (Assume no heat losses.)

9.41 The container in Example 9.9 is aluminum with a mass of 0.30 kg. Repeat the calculation for the equilibrium temperature of the aluminum-water-iron system assuming no heat is lost by the system. Explain why the inclusion of the container material in the calculation makes such a small difference even though the aluminum and iron masses require approximately the same added heat to raise their temperatures by one degree.

9.42 One way to determine the temperature of liquid nitrogen (or other low-temperature liquid) is the following. A piece of tin ($m = 0.2$ kg) is placed in the liquid nitrogen and allowed to reach temperature equilibrium. The tin is then removed and placed in a mercury bath (M = 2 kg) which is at a temperature of 90°C. When the tin and mercury reach equilibrium, the temperature is 53.2°C. The specific heat of tin is in the appropriate temperature range is 205 J/kg-deg. Find the temperature of liquid nitrogen.

9.43* An aluminum can has a mass of 0.3 kg and contains 1.6 kg of water; the combination is at a temperature of 18°C. A lead bar ($m = 0.8$ kg) at a temperature of 85°C is placed in the water. If no heat is lost from the can-water-lead system, what will be the final temperature?

Section 9.4

9.44 A house has an 8-in. brick wall that is 10 ft × 45 ft. How much heat flows through this wall per hour if the outer and inner temperatures are 85°F and 70°F? (Convert to metric units.)

9.45 One end of a 2-m copper rod ($r = 3$ cm) is maintained at a temperature of 22°C. What must be the temperature of the other end if the heat flow through the bar is 800 J/min?

9.46 The effective surface area of a typical refrigerator is about 4 m^2. The walls are the equivalent of 10 cm of asbestos. At what rate must heat be removed from the interior in order to maintain the inside temperature at -2°C when the outside temperature is 20°C?

9.47 A sphere that has a surface area of 20 cm^2 is heated to a temperature of 2400 K. At this temperature it is found that the sphere radiates at a rate of 1600 W. What is the emissivity of the sphere material?

9.48* What is the rate of heat flow in Cal/s through a 1 m^2 slab of $\mathcal{R} = 30$ insulating material when the temperature difference between the two slab surfaces is 20°C?

9.49 The temperature of the Sun's surface is approximately 6000 K and its emissivity is essentially unity. How many watts per square meter does the Sun radiate? The Sun's radius is 7×10^8 m. What is the total power output of the Sun? (Remember, we consider a 1000-MW power plant to be a large plant!)

9.50 What is the temperature of a perfect emitter that radiates at a rate of 0.907 MW/s-m^2?

9.51 Newton found that the rate of cooling of an

isolated object is directly proportional to the temperature difference between the object and its surroundings. That is, $Q/t \propto (T_2 - T_1)$, as long as T_1 is relatively close to T_2. This is called *Newton's law of cooling*. In such cases, the loss of energy by the object is due almost entirely to radiation. Show that Newton's law of cooling follows from Eq. 9.15 when $T_1 \cong T_2$. [Proceed by factoring $(T_2^4 - T_1^4)$ and then setting $(T_2^2 + T_1^2) \cong 2 T_2^2$; repeat for the next step.]

9.52 To demonstrate the concept of using an *intermediate heat shield* consider the arrangement shown in the diagram. A vacuum dewar holds liquid helium at $T_0 = 4.2$ K. The outer wall is at room temperature $T_2 = 20°C$. Let the the heat entering the liquid helium through the two equal areas A facing each other be P. When an intermediate heat shield with area A is placed in between, it will arrive at some equilibrium temperature T_1 (i.e. when the heat flow to the shield is equal to the heat flow from the shield). The new rate of heat flow to the liquid helium is now reduced to P'. Assuming all surface emissivities to be the same, determine the temperature T_1. What is the ratio of heat flows P'/P into the helium?

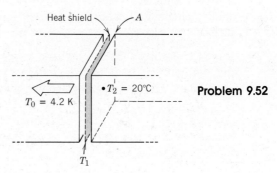

Problem 9.52

Gas Dynamics

In this chapter we continue our study of thermal phenomena by concentrating on the properties of gases subject to varying conditions of pressure and temperature. We will find that there is a very simple way to relate the pressure, the temperature, and the volume of a particular sample of gas. Moreover, we will discover that this relationship can be easily derived from an elementary view of the way that gas molecules move inside a container. The theory that results from this approach—the *kinetic theory* of gases—is one of the most satisfactory theories we have for describing the properties of bulk matter in gaseous form from the molecular viewpoint.

10.1 The Gas Laws

Boyle's Law

In 1662, the British chemist Robert Boyle (1627–1691) studied the way that pressure influences the volume of a confined gas. Boyle discovered that the volume is inversely proportional to the pressure, if the temperature of the gas remains constant (Fig. 10.1). This rule is obeyed approximately by all gases and is called *Boyle's law*. We can express Boyle's result as $P \propto 1/V$, or as

$$PV = \text{constant} \quad (T \text{ constant}) \tag{10.1}$$

Isothermal and Adiabatic Processes

Boyle's law will be valid if the initial temperature of the gas T_1 is equal to the final temperature T_2; that is,

$$P_1 V_1 = P_2 V_2 \quad \text{if} \quad T_1 = T_2$$

However, while we are changing the conditions from P_1 and V_1 to P_2 and V_2, Boyle's law may not be an

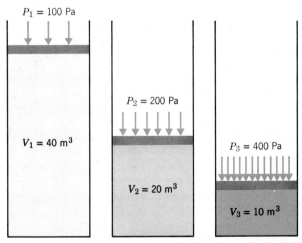

Figure 10.1 Robert Boyle discovered the inverse relationship between the volume of a confined gas and the applied pressure for constant temperature.

appropriate description of the gas. In Fig. 10.1 we see that increasing the pressure exerted on the piston from 100 Pa to 200 Pa results in the compression of the gas from a volume of 40 m³ to a volume of 20 m³. A force has been exerted on the gas and the gas molecules have been moved; clearly, work has been done on the gas. If we allow the gas to retain as internal energy the work done on it, the temperature of the gas will rise. Therefore, to make the final temperature equal to the initial temperature (as is required for Boyle's law to be valid), heat must be removed from the gas.

Suppose that the gas we are studying is confined in a cylinder (as in Fig. 10.1) and that the cylinder is immersed in a water-ice bath. Then, if we compress the gas very slowly, the heat generated by the work done is immediately absorbed by the water-ice bath and the temperature of the gas remains at 0°C throughout the compression. This kind of process is called an *isothermal* (equal temperature) process and Boyle's law is valid throughout the change from the initial pressure and volume to the final pressure and volume.

Now, instead of immersing the cylinder in a constant-temperature bath, suppose that we wrap the cylinder with some kind of insulating material and perform a *rapid* compression from V_1 to V_2. In this case, heat is prevented from leaving the gas during the compression, and so all of the work done on the gas is transformed into internal energy. Consequently, the temperature of the gas rises and Boyle's law is not valid. A process in which heat does not flow into or out of the system is called an *adiabatic* process.

Because the temperature change in an adiabatic process is related to the amount of energy introduced into the gas, we can imagine that the description of such a process involves the specific heat of the gas. In fact, instead of Boyle's law, we have, for an adiabatic process,

$$PV^\gamma = \text{constant} \quad \text{(adiabatic process)} \tag{10.2}$$

where $\gamma = c_p/c_v$ is the ratio of the specific heat of the gas at constant pressure to the specific heat at constant volume (see Section 9.3).

It is very instructive to show the isothermal and adiabatic compression processes on a graph of pressure versus volume, the so called *PV diagram*. The isothermal compression from V_1 to V_2 following Eq. 10.1 is seen to be a portion of a hyperbola. The pressure increases from P_1 to P_2. The adiabatic compression follows Eq. 10.2 and since $\gamma > 1$ arrives at a

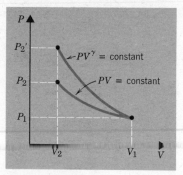

Figure 10.2 The *PV*-diagrams for an isothermal and an adiabatic compression.

higher pressure P_2', starting at pressure P_1 for the same compression.

●*Example 10.1*

A quantity of nitrogen gas has an initial pressure $P_1 = 2 \times 10^5$ Pa and an initial volume $V_1 = 4$ m³. The gas is compressed rapidly to a volume $V_2 = 2.5$ m³. (a) What is the final pressure? (b) If the nitrogen is now cooled to the original temperature, what is the pressure?

(a) For an adiabatic compression, we have

$$P_1 V_1^\gamma = P_2 V_2^\gamma$$

For nitrogen, $\gamma = 1.40$ (see Table 9.6). Therefore,

$$P_2 = \frac{V_1^\gamma}{V_2^\gamma} \times P_1 = \left(\frac{V_1}{V_2}\right)^\gamma \times P_1$$

$$= \left(\frac{4 \text{ m}^3}{2.5 \text{ m}^3}\right)^{1.40} \times (2 \times 10^5 \text{ Pa})$$

$$= 1.93 \times (2 \times 10^5 \text{ Pa})$$

$$= 3.86 \times 10^5 \text{ Pa}$$

(b) After the adiabatic compression, the temperature of the gas is higher than the original temperature. If the gas is now allowed to cool to the original temperature, it will be in exactly the same condition as if the compression had been carried out isothermally. Therefore, we can use Boyle's law and write

$$P_2 = \frac{V_1}{V_2} \times P_1 = \frac{4 \text{ m}^3}{2.5 \text{ m}^3} \times (2 \times 10^5 \text{ Pa})$$

$$= 3.2 \times 10^5 \text{ Pa}$$

Thus, the pressure of the nitrogen gas is 3.86×10^5 Pa immediately after the compression and gradually decreases to 3.2×10^5 Pa as the gas cools to its initial temperature.

In a diesel engine, the compression of the air-fuel mixture in the cylinder is essentially adiabatic, and the temperature increase is sufficient to ignite the fuel. Such engines do not require spark plugs. ∎

The Law of Charles and Gay-Lussac

The French physicists Jacques Charles (1746–1823) and Joseph Louis Gay-Lussac (1778–1850) independently discovered the relationship connecting the temperature of a gas and its volume (at constant pressure) and so extended Boyle's law. This relation states that the volume of a gas is proportional to its temperature, $V \propto T$, or,

$$\frac{V}{T} = \text{constant} \quad (P \text{ constant}) \tag{10.3}$$

A graph of V versus T is a straight line and can be extrapolated to the point at which the volume would become *zero*, as in Fig. 10.3. The temperature at this point is the temperature we have called "absolute zero" (see Section 9.1), and corresponds to $-273.15°C$. (We will usually refer to absolute zero as $-273°C$.) Absolute zero is the zero point for the Kelvin temperature scale; both Kelvin and Celsius temperatures are shown in Fig. 10.3.

Notice that the temperature T in the Charles–Gay-Lussac law (Eq. 10.3) must be expressed on the Kelvin or absolute scale. It makes no sense to discuss the relationship $V = (\text{constant}) \times T$ if T can be a *negative* temperature. (What would a negative volume mean?) When the temperature is expressed on the Kelvin scale, the temperature is always *positive*.

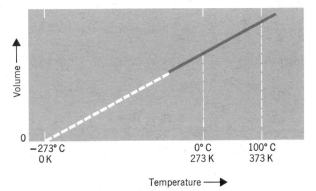

Figure 10.3 Measurements of volume and temperature can be extrapolated to define the absolute zero of temperature at $-273°C = 0$ K.

The Ideal Gas Law

The laws of Boyle and of Charles and Gay-Lussac can be combined into a single statement:

$$\frac{PV}{T} = \text{constant} \tag{10.4}$$

When the temperature is held constant, this expression reduces to $PV = \text{constant}$, which is Boyle's law. When the pressure is held constant, Eq. 10.4 reduces to $V/T = \text{constant}$, which is the law of Charles and Gay-Lussac.

What can we say about the constant in Eq. 10.4? Suppose that we hold the pressure and the temperature constant. Then, the volume clearly depends on the quantity of gas in the sample. That is, the constant is proportional to the number of molecules N that make up the sample. Accordingly, we can write

$$PV = NkT \tag{10.5}$$

where the proportionality constant k is called the *Boltzmann constant* in honor of Ludwig Boltzmann (1844–1906), an Austrian physicist who was one of the leading contributors to the development of the modern theory of heat (kinetic theory). The experimentally determined value of k is

$$k = 1.380 \times 10^{-23} \text{ J/K} \tag{10.6}$$

It is often convenient to express the number of molecules N in a sample in terms of the number of moles n. In Section 1.3 we defined the mole to be that quantity of a substance having a mass M in grams equal to the molecular mass m of the substance expressed in atomic mass units. That is,

$$n(\text{mol}) = \frac{M(\text{in g})}{m(\text{in u})} \tag{10.7}$$

Remember, 1 mol of a substance contains Avogadro's number of molecules:

$$N_0 = 6.022 \times 10^{23} \text{ molecules/mol}$$

We can now rewrite Eq. 10.5 in terms of the number of moles n in the sample. When we do this we have a new proportionality constant:

$$PV = nRT \tag{10.8}$$

where R is called the *universal gas constant* and has the value,

$$R = 8.314 \text{ J/mol-K} \tag{10.9}$$

Equating the constants in Eqs. 10.5 and 10.9 we have

$$Nk = nR$$

Now, the ratio N/n is the number of molecules per mole, which is Avogadro's number:

$$N_0 = \frac{N}{n} \tag{10.10}$$

so that the gas constant can also be expressed as

$$R = N_0 k \tag{10.11}$$

Equation 10.8 (or Eq. 10.5) is a remarkably good representation of real gases at low densities. We cannot expect the relation to be perfect at all temperatures (after all, real gases liquefy at low temperatures), but the departures from the predictions of Eq. 10.8 for a particular real gas are generally small. Because only a perfect gas will follow Eq. 10.8 exactly, this expression is called the *ideal gas law*. Usually, we will consider ideal gases and will assume that Eq. 10.8 is valid.

When we use the ideal gas law to solve problems, we will often find it convenient to write the equation in the form

$$\frac{P_1 V_1}{n_1 T_1} = \frac{P_2 V_2}{n_2 T_2} \tag{10.12a}$$

or,

$$\frac{P_1}{P_2} \times \frac{V_1}{V_2} \times \frac{n_2}{n_1} = \frac{T_1}{T_2} \tag{10.12b}$$

Because we have an equation that involves *ratios* of pressure, volume, temperature, and the number of moles, we can express these quantities in whatever units are convenient. For example, we could express the pressures in Pa, torr, atm, or even lb/in.^{2}. And we could express the volumes in m^3, cm^3, ℓ, or ft^3. Furthermore, if the type of gas does not change, the ratio n_2/n_1 could be expressed in kg, g, mol, or lb. The temperatures must be expressed in terms of a scale whose zero point is absolute zero; the only such scale we have mentioned is the Kelvin scale, but others are sometimes used.

● *Example 10.2*

What is the pressure of 7 kg of nitrogen gas confined to a volume of 0.4 m^3 at 20°C?

Solving Eq. 10.8 for the pressure, we have

$$P = \frac{nRT}{V}$$

The number of moles in n of nitrogen is

$$n = \frac{M}{m} = \frac{7 \times 10^3 \text{ g}}{28 \text{ g/mol}} = 250 \text{ mol}$$

Then,

$$P = \frac{(250 \text{ mol}) \times (8.314 \text{ J/mol-K}) \times (293 \text{ K})}{(0.4 \text{ m}^3)}$$

$$= 15.2 \times 10^5 \text{ Pa}$$

or about 15 atm.

● *Example 10.3*

A 500-ℓ tank contains air at a pressure of 3 atm when the temperature is 0°C. What will be the pressure if the tank is heated to a temperature of 120°C?

In this case, the volume of the gas and the number of moles remain the same. Therefore, the ratios V_1/V_2 and n_2/n_1 are unity, and Eq. 10.12b becomes

$$\frac{P_1}{P_2} = \frac{T_1}{T_2}$$

so that, converting the temperatures to kelvins,

$$P_2 = \frac{T_2}{T_1} \times P_1 = \left(\frac{393 \text{ K}}{273 \text{ K}}\right) \times (3 \text{ atm})$$

$$= 4.3 \text{ atm}$$

● *Example 10.4*

A high-altitude research balloon that is designed to carry a 500-kg payload to an altitude of 30 km (about 100,000 ft) is filled with 400 m^3 of helium at sea level when the temperature is 15°C. What will be the volume of the balloon when it reaches maximum altitude?

At an altitude of 30 km, the pressure is about 0.01 of sea-level pressure P_0 and the temperature is about −50°C. Therefore, we have

$$P_1 = P_0 \qquad T_1 = 288 \text{ K} \qquad V_1 = 400 \text{ m}^3$$

$$P_2 = 0.01 P_0 \qquad T_2 = 223 \text{ K} \qquad V_2 = ?$$

Again, the factor n_2/n_1 is unity, so solving Eq. 10.9b for V_2, we have

$$V_2 = \frac{P_1}{P_2} \times \frac{T_2}{T_1} \times V_1$$

$$= \left(\frac{P_0}{0.01\,P_0}\right) \times \left(\frac{223\text{ K}}{288\text{ K}}\right) \times (400\text{ m}^3)$$

$$= 3.1 \times 10^4 \text{ m}^3$$

or about 77 times the original volume. If you have ever witnessed or seen photographs of the launching of a high-altitude balloon, you probably noticed that the gas bag was only partially filled. This is necessary in order to allow for the large amount of expansion that takes place in the upper atmosphere where the pressure is very low.

● *Example 10.5*

What volume is occupied by 1 mol of a gas under *standard conditions* ($P = 1$ atm, $T = 0°C = 273$ K)?
Solving Eq. 10.8 for V,

$$V = \frac{nRT}{P} = \frac{(1\text{ mol}) \times (8.314\text{ J/mol-K}) \times (273\text{ K})}{(1.013 \times 10^5\text{ Pa})}$$

$$= 2.24 \times 10^{-2} \text{ m}^3$$

$$= 22.4 \text{ }\ell$$

This useful number can sometimes simplify the solutions to problems.

● *Example 10.6*

What is the volume of 8 g of methane gas under standard conditions?
According to Table 1.10, the molecular mass of methane (CH_4) is 16 u. Therefore, an 8-g sample contains $\frac{1}{2}$ mol. The result of Example 10.5 is that the volume of 1 mol of *any* gas under standard conditions is 22.4 ℓ. Thus, the volume occupied by 8 g of methane gas is $\frac{1}{2} \times 22.4\text{ }\ell = 11.2\text{ }\ell$. (The same result can, of course, be obtained by using Eq. 10.8, as was done previously.) ■

10.2 Kinetic Theory

A Microscopic Description of Gas Dynamics

Experiments have demonstrated that the behavior of a gas is closely described by a simple relationship that connects the pressure, the volume, and the temperature of the sample. This rule—the ideal gas law—was deduced by examining the *bulk* properties of gases. We now change our viewpoint and ask what conclusions can be drawn concerning the properties of gases by studying the behavior of the microscopic

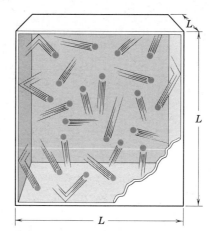

Figure 10.4 Molecules move at random in a gas.

constituents of the gases, namely, molecules. We will base our discussion on the application of Newtonian dynamics to the motion of gas molecules.

We begin by considering a certain large number N of ideal gas molecules that are confined within a cubical box whose sides have a length L (Fig. 10.4). Within this box the molecules move at *random* with velocities v. These velocities are not all the same but this need not concern us at the moment because we will eventually require only the *average* velocity.

Consider first a single molecule of mass m that has an x component of velocity equal to v_x, as in Fig. 10.5. The initial momentum is $p_x = mv_x$. If the molecule collides elastically with the wall, the momentum after collision will be $p'_x = -mv_x$. Thus, the *change* in momentum of the molecule during the collision is

$$\Delta p_x = p'_x - p_x = -2\,mv_x$$

and the negative of this amount of momentum is delivered to the right-hand wall of the cube (Fig. 10.5).

A collision with the right-hand wall occurs once every round trip of the molecule; hence, the time interval between successive collisions is

$$\Delta t = \frac{2L}{v_x}$$

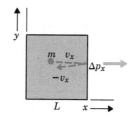

Figure 10.5 A molecule collides elastically with the wall of the container and alters the sign (but not the magnitude) of its velocity.

The *average* force (averaged over the time Δt) that is exerted on the right-hand wall by the collision of this single molecule is

$$\overline{F}_x = \frac{-\Delta p_x}{\Delta t} = \frac{2mv_x}{2L/v_x} = \frac{mv_x^2}{L} \qquad (10.13)$$

We must now take into account the fact that N molecules contribute to the *total* force on the right-hand wall. We can compute the total force by using the *average* of the x component of the velocity for the N molecules. However, this is not quite correct because in Eq. 10.13 it is the *square* of v_x that is related to the force. Therefore, in computing the total force we must use the average of v_x^2 (that is, $\overline{v_x^2}$) instead of the square of the average of v_x (that is, $\overline{v}_x^2$).

● *Example 10.7*

Show that $\overline{v}^2$ does *not* equal $\overline{v^2}$.

Consider four particles with the following velocities: 1, 2, 3, and 4 m/s. The square of the average of v is

$$\overline{v}^2 = \left(\frac{1+2+3+4}{4}\right)^2 = \left(\frac{10}{4}\right)^2$$

$$= (2.5)^2 = 6.25 \ (m/s)^2$$

whereas the average of v^2 is

$$\overline{v^2} = \frac{(1)^2 + (2)^2 + (3)^2 + (4)^2}{4}$$

$$= \frac{1+4+9+16}{4} = \frac{30}{4} = 7.5 \ (m/s)^2$$

so that there is a substantial difference between the two methods of averaging.

If the individual velocities are $+1$, -2, -3, and -4 m/s, $\overline{v^2}$ will still be 7.5 $(m/s)^2$, but the average velocity $\overline{v}^2$ will be *zero*. ∎

If we multiply the right-hand side of Eq. 10.13 by N and take the average value of v_x^2, we obtain the total force exerted on the wall of the cube:

$$F_x = \frac{Nm\overline{v_x^2}}{L} \qquad (10.14)$$

The average of v_x^2 is very simply related to the average of the square of v:

$$\overline{v^2} = \overline{v_x^2} + \overline{v_y^2} + \overline{v_z^2}$$

(Think of v_x, v_y, and v_z as the sides of a cube and v as the diagonal. Then, use the Pythagorean Theorem twice to obtain v^2). Now, the molecules in the sample move at random. That is, there is no distinction between v_x and v_y or v_z. In fact,

$$\overline{v_x^2} = \overline{v_y^2} = \overline{v_z^2}$$

Therefore,

$$\overline{v^2} = 3\overline{v_x^2}$$

Substituting $\overline{v_x^2}$ from this expression into Eq. 10.14, we find for the total force on any wall,

$$F = \frac{Nm\overline{v^2}}{3L}$$

The area of each wall is L^2, so the pressure is

$$P = \frac{F}{A} = \frac{F}{L^2} = \frac{Nm\overline{v^2}}{3L^3}$$

Because the volume of the box is $V = L^3$, we can rewrite this expression as

$$PV = \tfrac{1}{3} Nm\overline{v^2}$$

Now, the average kinetic energy of one of the molecules is

$$\overline{KE} = \tfrac{1}{2} m\overline{v^2}$$

so that the equation for PV becomes

$$PV = \tfrac{2}{3} N \, \overline{KE} \qquad (10.15)$$

The ideal gas law, $PV = nRT$, is a description of the *macroscopic* or bulk properties of a gas. Equation 10.15, on the other hand, is based on a *microscopic* view of the interaction of gas molecules with the walls of a container. The left-hand sides of these two equations (PV) are identical. Therefore, we can equate the right-hand sides of these equations and obtain the vital link between the macroscopic concept of temperature and the microscopic description of molecular motion. That is,

$$\overline{KE} = \tfrac{3}{2} kT \quad \text{(average } KE \text{ per molecule)} \qquad (10.16)$$

We see in Eq. 10.16 that temperature is basically a mechanical quantity that expresses *average kinetic energy*; temperature is therefore not a basic physical quantity as are length, mass, and time. But notice that temperature is statistical in character because it is a measure of the *average* kinetic energy of the molecules in a sample. Thus, there is no meaning to the "temperature" of a single molecule, or even a dozen; we need a large sample before the temperature concept is useful.

Having obtained the connection between the macroscopic temperature and the microscopic kinetic energy by equating our experimental and theoretical expressions for PV, we must now ask how we can verify our connecting equation. One of the most direct ways to establish Eq. 10.16 is by the measurement of molecular velocities.

Molecular Velocities

Equation 10.16 can be expressed as

$$\overline{KE} = \tfrac{1}{2} m \overline{v^2} = \tfrac{3}{2} kT$$

from which

$$\overline{v^2} = \frac{3kT}{m}$$

If we take the square root of $\overline{v^2}$, we do not obtain $\overline{v}$ (compare Example 10.7), but a different quantity, which we call the *root-mean-square* (or *rms*) velocity:

$$v_{rms} = \sqrt{\frac{3kT}{m}} \qquad (10.17)$$

The velocities of the molecules in a gas are actually distributed according to a function that can be obtained from kinetic theory by using more powerful mathematical techniques than we have introduced. This distribution of velocities is called the *Maxwellian distribution*, in honor of James Clerk Maxwell (1831–1879), the Scottish theoretical physicist who contributed greatly to kinetic theory and developed the theory of electromagnetism. Figure 10.6 shows the Maxwellian distribution functions for two different temperatures. As the temperature is increased, the peak of the distribution moves toward higher velocities, in accordance with Eq. 10.17 which states that v_{rms} increases as $\sqrt{T}$.

The accuracy of the Maxwellian formula for the distribution of molecular velocities has been directly tested and verified in numerous experiments, thus establishing that Eq. 10.17 for v_{rms} is correct.

● *Example 10.8*

Compute the *rms* velocity of oxygen molecules at room temperature.

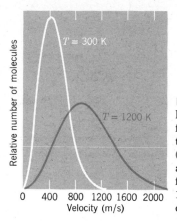

Figure 10.6 Maxwellian velocity distribution functions for nitrogen molecules at two temperatures. For T = 300 K (approximately room temperature), v_{rms} = 517 m/s, and for T = 1200 K, v_{rms} = 1033 m/s. The peaks of the curves occur at 0.82 v_{rms}.

Room temperature is 20°C, so we use T = 293 K. The molecular mass of oxygen is 32 u (Table 1.10); thus, $m = 32 \times (1.66 \times 10^{-27} \text{ kg}) = 5.3 \times 10^{-26}$ kg. Then,

$$v_{rms} = \sqrt{\frac{3kT}{m}}$$

$$= \sqrt{\frac{3 \times (1.38 \times 10^{-23} \text{ J/K}) \times (293 \text{ K})}{5.3 \times 10^{-26} \text{ kg}}}$$

$$= 478 \text{ m/s}$$

which is approximately 1000 mi/h! Of course, molecules do not travel very far at these speeds because of frequent collisions that change the direction of motion (see Table 10.1). Therefore, it takes a certain amount of time for the scent of perfume from a bottle uncorked in one part of a room to penetrate (or *diffuse*) to the other side of the room.

● *Example 10.9*

Compute the average kinetic energy of the gas molecules in a gas at room temperature.

According to the result expressed in Eq. 10.16 average kinetic energy does not depend on the mass

Table 10.1 Approximate Values for Various Quantities in the Atmosphere at Sea Level and 0°C

No. of molecules per unit volume	2.7×10^{25} m^{-3}
Pressure	1.01×10^5 Pa
Density	1.29 kg/m^3
Molecular velocity	$\begin{cases} 460 \text{ m/s (oxygen, } O_2) \\ 490 \text{ m/s (nitrogen, } N_2) \end{cases}$
Mean distance between molecular collisions	8×10^{-8} m
Collision frequency	6×10^9 s^{-1}
Typical distance between molecules	3.5×10^{-9} m
Mean molecular mass	4.8×10^{-26} kg

of the molecule; therefore, the molecules of *all* gases have the same average kinetic energy (but *different* values of v_{rms}) at a given temperature. Thus,

$$\overline{KE} = \tfrac{3}{2} kT$$
$$= \tfrac{3}{2} \times (1.380 \times 10^{-23} \text{ J/K}) \times (293 \text{ K})$$
$$= 6.065 \times 10^{-21} \text{ J}$$

Since one mole of any gas contains the same number of molecules, namely Avogadro's number N_0, we also have that the total translational kinetic energy of one mole of gas is

$$KE_{\text{TOTAL}} = \tfrac{3}{2} k N_0 T = \tfrac{3}{2} RT.$$

If we equate the total kinetic energy of a monatomic gas with its internal energy, then n moles of any monatomic gas will have an internal energy U given by [1]

$$U = \tfrac{3}{2} nRT \text{ (monatomic gas)}.$$

At room temperature

$$U = \tfrac{3}{2} n (8.31 \text{ J/mol} \cdot \text{K})(293 \text{ K})$$
$$= 3650 \, n \text{ J}$$

● *Example 10.10*

How many molecules are there in 1 m^3 of air under standard conditions?

Solving Eq. 10.5 for N, we have

$$N = \frac{PV}{kT} = \frac{(1.013 \times 10^5 \text{ Pa}) \times (1 \text{ m}^3)}{(1.380 \times 10^{-23} \text{ J/K}) \times (273 \text{ K})}$$
$$= 2.7 \times 10^{25} \text{ molecules (in } 1 \text{ m}^3) \quad \blacksquare$$

Notice the following important conclusions we have reached concerning the motion of gas molecules:

1 Different types of gas molecules at the same temperature will have different average velocities that depend on the molecular masses.

2 The average kinetic energy of the molecules in a gas sample (regardless of type) is directly proportional to the absolute temperature.

In any sample of mixed gases, the lighter molecules will have the higher velocities and the heavier molecules will have the lower velocities. Each individual type of gas in the sample will maintain its own particular average velocity. Although the average velocity of the molecules of one type depends on the molecular mass, the average kinetic energy does not. Each species of molecules in a sample will have the same average kinetic energy at the same temperature. In a mixture of hydrogen and oxygen gases, for example, the average velocity of the hydrogen molecules will be four times greater than the average velocity of the oxygen molecules, but the average kinetic energy of each molecular species will be exactly the same.

The Assumptions in Kinetic Theory

In deriving the microscopic version of the ideal gas law equation, we have made explicit or implicit use of several assumptions:

1 A gas consists of a large number of molecules that have no appreciable size compared to the average distance between molecules.

2 No forces act on the molecules except during collisions with the walls; that is, there is no interaction between molecules.

3 The molecules in the sample are in random motion $(\overline{v_x^2} = \overline{v_y^2} = \overline{v_z^2})$.

4 The molecules obey Newton's laws.

5 Collisions of the molecules with the walls are elastic.

It is indeed remarkable that such a simple, classical model of an ideal gas is capable of yielding a gas law equation that is so closely obeyed by real gases. This is especially true when it is realized that the assumptions in the theory are only approximations to the real physical case. The molecules in a gas *do* interact with one another and the molecules *do* have finite sizes. It has been possible to take account of these facts in the more complete theory. And, indeed, the theory has provided us with an independent determination of molecular sizes (about 10^{-10} m) and a description of the way in which molecules interact. Modifications having to do with quantum effects have been made in the theory to improve its accuracy. As a result, the *kinetic theory of gases* is now one of the most useful (and most highly developed) theories in physics.

Absolute Zero

According to the kinetic theory as we have developed it, it would appear that all molecular motion should cease at the absolute zero of temperature. Such a conclusion is not really justified. We have

---

[1] The restriction to monatomic gases (such as He, Ne, Ar, Kr) is required if the internal energy is to consist of only translational kinetic energy. Diatomic molecules (such as H_2, N_2, O_2) may also possess rotational kinetic energy contributions to the internal energy.

based the development of kinetic theory strictly on classical or Newtonian ideas. At very low temperatures the molecular energies become very small and the classical explanation of the behavior of matter is no longer adequate. Atoms and molecules are not tiny balls of matter whose motions and interactions can be described in the same way that we describe ordinary balls. The deviations from classical behavior are particularly evident at low temperatures. In this regime, we find a number of phenomena that can be explained only in terms of *quantum theory*. We will describe some of these phenomena and will outline the basic ideas of quantum theory in Chapters 18 through 20. According to this theory, molecular energies are not exactly zero at absolute zero. Instead, every molecule possesses a small amount of energy (the so-called *zero-point energy*) even at 0 K. This energy can be thought of in terms of a residual vibratory motion, although the molecules do not, in fact, actually vibrate back and forth as would a ball on a spring. The absolute zero of temperature has never been reached in the laboratory, although bulk matter has been cooled to 0.001 K.

Diffusion

According to kinetic theory, the molecules in a gas (or a liquid) are continually moving about and are colliding with each other. Because of these collisions, no particular molecule or group of molecules will move very rapidly from one region of the container to another. If a tiny crystal of iodine is introduced into a container of air, the purple I_2 vapor will spread only slowly through the volume of the container. We say that the iodine vapor *diffuses* through the air molecules and eventually becomes distributed uniformly throughout the container.

The diffusion process is similar to the conduction of heat through a rod (Section 9.4). The conduction of heat takes place in response to a *temperature gradient;* diffusion in a gas (or in a liquid) takes place in response to a *concentration gradient*. If the concentration of iodine molecules in air is equal to C_1 molecules per unit volume at position 1 and is equal to some larger value C_2 at position 2 a small distance Δx away, the molecules will tend to diffuse from the region of high concentration toward the region of low concentration at a rate proportional to the concentration gradient $(C_2 - C_1)/\Delta x = \Delta C/\Delta x$. The rate, $\Delta N/\Delta t$, at which the diffusing molecules are transported across any surface in response to a concentration gradient perpendicular to that surface must be proportional to the surface area A. Therefore, we have for the number of molecules ΔN that diffuse across the surface in a time Δt,

$$\frac{\Delta N}{\Delta t} = -DA \frac{\Delta C}{\Delta x} \tag{1}$$

where the negative sign is introduced to remind us that the direction of diffusion is opposite to that of increasing concentration, and where D is the *diffusion coefficient* for the particular substances involved. Equation 1 is known as *Fick's law*, first formulated in 1855 by the physiologist, Adolf Fick (1829–1901). Notice that this equation has exactly the same form as Eq. 9.13 for the rate of heat transfer by conduction.

It seems clear that the rate of diffusion must depend, through D, on the speed of the diffusing molecules. One of the results of kinetic theory (Eq. 10.17) is that the rms velocity of molecules with mass m in a gas at a temperature T is proportional to $\sqrt{T/m}$. Thus, we expect that D depends on temperature and molecular mass in the same way; experimentally, we find that this is indeed the case, to a good approximation (see Exercise 1).

The diffusion coefficients for some small molecules in air (at normal atmospheric pressure) are given in the table at the top of the next page. We see in these data primarily the effect of the molecular mass on the diffusion coefficient. For example, the square root of the ratio of the mass of CS_2 to that of H_2 is $\sqrt{76/2} = 6.2$, and the value of D for CS_2 is smaller than that for H_2 by the same factor.

Molecule	m(u)	T(°C)	D(m²/s)
Hydrogen, H_2	2	0°	6.34×10^{-5}
Water, H_2O	18	8°	2.39×10^{-5}
Oxygen, O_2	32	0°	1.78×10^{-5}
Carbon dioxide, CO_2	44	0°	1.39×10^{-5}
Methyl alcohol, CH_3OH	32	40°	1.37×10^{-5}
Carbon disulfide, CS_2	76	20°	1.02×10^{-5}

We have framed the discussion so far in terms of *gases,* but diffusion also takes place in *liquids.* Indeed, the diffusion of substances in water is extremely important in biological systems. Fick's law is valid for molecules diffusing through water, but the diffusion coefficient has a different form in this case. A complete analysis shows that D depends on the temperature, the size of diffusing molecules, and on the viscosity of the medium:

$$D = \frac{kT}{6\pi a \eta} \tag{2}$$

where a is the radius of the diffusing molecules (assumed spherical), and where the other symbols have their usual meanings. Because the mass m of the molecule is proportional to a^3, we have the interesting result that $D \propto m^{-1/3}$. That is, the diffusion coefficient is relatively insensitive to the mass of the diffusing molecules. This effect is seen in the diffusion coefficients for some molecules in water (at 20°C) that are given in the table below.

Molecule	m(u)	D(m²/s)
Water, H_2O	18	2.0×10^{-9}
Oxygen, O_2	32	1.0×10^{-9}
Urea, $CO(NH_2)_2$	60	1.1×10^{-9}
Glucose, $C_6H_{12}O_6$	180	6.7×10^{-10}
Sucrose, $C_{12}H_{22}O_{11}$	342	5.2×10^{-10}
Ribonuclease	13,683	1.2×10^{-10}
Hemoglobin	68,000	6.9×10^{-11}
Urease	480,000	3.5×10^{-11}

According to Eq. 2, a measurement of the diffusion coefficient for a particular molecular species in water (at a temperature T) permits a calculation of the radius a of the molecule. This is one method for determining molecular sizes. (Strictly, Eq. 2 is valid only for spherical molecules, but we can still use the expression to obtain an estimate of the "average size" for molecules with other shapes.) With a knowledge of a, we can proceed to calculate the molecular mass according to

$$m = \tfrac{4}{3}\pi a^3 \rho \tag{3}$$

where ρ is the density. The *dry* density of most large biological molecules is about 1.27 g/cm³. However, when these molecules are placed in water, water molecules adhere to the surfaces and increase the effective volume of the molecules. This has the result that the molecular mass given by the simple formula is too large by a factor of about 1.5. Consequently, the corrected molecular mass m_c is

$$m_c = \frac{m}{1.5} \tag{4}$$

In this way we can obtain an estimate of the molecular mass of a substance from a knowledge of its diffusion coefficient. (See Exercise 2). (Diffusion coefficients for substances in water can be determined by measuring the rate of sedimentation in an ultracentrifuge.)

The diffusion of molecular ions through the membranes that form the walls of cells is responsible for maintaining the proper electric potential within cells. Moreover, chemical processes that take place in cell membranes alter the diffusion rates for different ions in such a way that electric signals representing nerve impulses are propagated along nerve fibers (see the essay in Chapter 12).

About 98 percent of the oxygen required by the body is absorbed through the lungs by a diffusion process. The tiny cavities (called *alveoli*) within the lung at which diffusion occurs have a total area of about 70 m² in the normal adult. The alveolar walls are extremely thin (average thickness about 0.5 μm) and are tightly packed with a network of interconnecting capillaries. Therefore, the blood circulates through the lung membranes almost as a thin sheet. (At any instant the total amount of blood in the lung capillaries is about 100 mℓ or 10^{-4} m³.) Because the area through which oxygen diffuses into the blood (and carbon dioxide diffuses out) is so large and because the membrane thickness is so small, it is easy to see why the diffusion process is so efficient.

References

R. B. Setlow and E C. Pollard, *Molecular Biophysics,* Addison-Wesley, Reading, Mass., 1962, Chapter 4.

A. J. Vander, J. H. Sherman, and D. S. Luciano, *Human Physiology: The Mechanisms of Body Function,* McGraw-Hill, New York, 1970, Chapter 2.

F. M. H. Villars and G. B. Benedek, *Physics with Illustrative Examples from Medicine and Biology,* Vol. 2, Addison-Wesley, Reading, Mass., 1974, Chapter 2.

■Exercises

1. Check the validity of the proposition that the diffusion coefficient for a gas in another gas should vary as $\sqrt{T/m}$ by calculating the product $D\sqrt{m/T}$ for the gases listed in the table. What is the average percentage deviation from the mean for this sample?

2. The diffusion coefficient for *pepsin* in water at 20°C is 9.0×10^{-11} m²/s. Estimate the size and the mass of the pepsin molecule. (The viscosity of water at 20°C is 1.0×10^{-3} Pa-s; see Table 8.1.)

3. Comment on the following statement: The diffusion of gas molecules through a gas is limited by molecular collisions, whereas the diffusion of molecules through a liquid is limited by viscous frictional effects. (Compare the factors in Eq. 2 with Eq. 8.24 for the viscous retarding force in laminar flow.)

4. Estimate the average thickness of the "sheet" of blood that flows through the lung membranes. Why is this number important in determining the rate of transfer of alveolar oxygen to the blood?

Osmosis

In the preceding essay we learned about the diffusion of one substance through another when the diffusion process is unimpeded—that is, we studied the case of diffusion in the absence of any barriers. Now, we wish to consider the selective diffusion through porous barriers, or *semipermeable membranes*.

Many biological membranes are permeable to water but not to materials dissolved in the water. This is the situation when the pores in the membrane are sufficiently large to permit the free passage of water molecules (diameter = 3.8 Å) but are too small to permit the passage of solute molecules such as glucose (diameter = 8.8 Å). The effect of this selective diffusion (called *osmosis*) is illustrated in Fig. 1.

In Fig. 1a we see the initial situation. The two parts of a cell are separated by a semipermeable membrane; the right-hand part is filled with pure water and the left-hand part is filled with a sugar solution. (The dots represent glucose molecules; the water molecules are not shown.) At first, the liquid level in each part of the cell is the same. Because the molecules are in thermal motion, the membrane is bombarded from both sides by water molecules and from the left side by glucose molecules. The pores in the membrane allow the water molecules to pass through, but the glucose molecules are blocked and remain entirely in the left-hand section of the cell. The rate at which water molecules reach the membrane surface is smaller in the left-hand side than in the right-hand side because the presence of the glucose molecules near the membrane prevents some of the water molecules from reaching the surface. Consequently, there is a greater rate of transport of water molecules through the membrane from right to left than from left to right, as shown by the arrows in Fig. 1a. This net flow of water into the glucose solution continues until the pressure in the left-hand part of the cell reaches a point at which the rate of bombardment of the membrane by water molecules is the same on both sides. This is the situation in Fig. 1b. If h is the difference between the liquid levels in the two sections, the pressure differential is $\rho g h$—this is called the *osmotic pressure* of the sugar solution.

The total liquid pressure in a liquid is equal to the sum of the partial pressures of the various constituents. (This is the same as for gases.) In Fig. 1a the *total* liquid pressure is the same in both sections (because the liquid levels are the same). But the *water* pressure is greater in the right-hand section. The solute pressure does not affect the rate of diffusion, so the water flows from the section with the higher water pressure (the right-hand side) to the section with the lower water pressure (the left-hand side). This continues until the *water* pressure is the same on both sides of the membrane (Fig. 1b).

In 1887 the chemist J. H. van't Hoff (1852–1911) discovered that the osmotic pressure Π varies directly with the concentration C of the solute and with the absolute temperature T. Thus, using $\mathcal{R}$ to denote the proportionality constant,

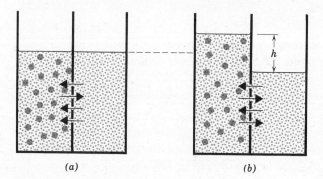

Figure 1 (a) Initial condition. A sugar solution (left) is separated from pure water (right) by a semipermeable membrane. (b) Final condition. Diffusion of water molecules through the membrane takes place preferentially from right to left until the osmotic pressure increases to the point that the diffusion is in equilibrium.

$$\Pi = \mathcal{R}CT \tag{1}$$

where $C = n/V$ is equal to the number of moles of solute per unit volume of solution. Therefore, we can write

$$\frac{\Pi V}{T} = n\mathcal{R} \tag{2}$$

This expression has the same form as the ideal gas law equation (Eq. 10.8). In fact, when the constant $\mathcal{R}$ is determined from osmotic pressure measurements, it is found that

$$\mathcal{R} = 0.0827 \text{ atm-}\ell/\text{mol-K} \tag{3}$$

The unfamiliar units of $\mathcal{R}$ are those generally used in discussing osmotic effects. If we transform the conventional units of the gas constant R into those in Eq. 3 (see Exercise 1), we find

$$R = 0.0821 \text{ atm-}\ell/\text{mol-K} \tag{4}$$

That is, Eq. 2 represents (to a good approximation) an ideal gas law equation for the solute molecules, a rather remarkable result. It should be pointed out, however, that Eq. 2 is valid only for concentrations of solute in the solvent that are not too large (a few tenths mol/ℓ or less). For large solute concentrations, the osmotic pressure is actually *greater* than that predicted by Eq. 2 and a more complex theory is needed to explain the results.

The osmotic pressure of a solution depends on the concentration of particles that will not pass through the membrane. Thus, ionizing substances will exert greater osmotic pressures than will nonionizing substances. Sodium chloride, for example, is ionized in solution to Na^+ and Cl^- ions, both of which are effective osmotic particles. We can measure the concentration of nonpassing particles in the solution in terms of the *osmolality*. *One osmole* of a nondissociating substance is equal to 1 mol, but 1 mol of sodium chloride would be equivalent to 2 osmol because each NaCl unit would produce two particles.

An osmolality of 1 means that the solution consists of 1 osmol per kilogram of water. Another measure of concentration is in terms of the *osmolarity*. An osmolarity of 1 means that the solution consists of 1 osmole per liter of solution. If we confine our attention to low solute concentrations, there is very little distinction between osmolality and osmolarity. The usual physiological studies is to express concentrations in terms of osmolarity.

The molecular mass of glucose ($C_6H_{12}O_6$) is 180 u. Therefore, if 0.1 mol (18 g) of glucose is dissolved in 1 ℓ of water, the osmolarity will be 0.1 osmol/ℓ. On the other hand, if 0.1 mol of sodium chloride is dissolved in water, the osmolarity will be 0.2 osmol/ℓ because sodium chloride in water dissociates into Na^+ and Cl^- ions.

The normal osmolarity of cellular fluids is about 0.3 osmol/ℓ. The corresponding osmotic pressure at body temperature (37°C) is

$$\Pi = \mathcal{R}CT = (0.0827 \text{ atm-}\ell/\text{mol-K}) \times (0.3 \text{ osmol}/\ell) \times (310 \text{ K})$$
$$= 7.7 \text{ atm}$$

This is an enormous pressure. However, remember that it corresponds to the pressure that would exist if the extracellular fluid were pure water. This fluid actually has a similar concentration of large molecules and therefore also has a high osmotic pressure. The *difference* in osmotic pressure between a cell and the surrounding fluid is a measure of the tendency of water to pass into or out of the cell. If the cell were placed in pure water, water would readily pass into the cell and the internal pressure would increase until the cell ruptures.

References

R. W. Stacy, D. T. Williams, R. E. Worden, and R. O. McMorris, *Essentials of Biological and Medical Physics,* McGraw-Hill, New York, 1955, Chapter 4.

O. Stuhlman, Jr., *An Introduction to Biophysics,* Wiley, New York, 1943, Chapter V.

■ Exercises

1. Show that the gas constant R can be expressed as in Eq. 4.

2. When a patient is fed intravenously with a sugar solution, the osmolarity of the solution must be the same as that of the intracellular fluid, that is, 0.3 osmol/ℓ. (Why?) How many grams of glucose must be used to produce 2 ℓ of appropriate solution?

3. When 0.2 mol of sodium chloride (molecular mass = 58.5 u) is dissolved in 1 ℓ of water at 18°C, 82 percent of the NaCl units are dissociated into Na^+ and Cl^- ions. What is the osmotic pressure of such a solution?

4. In low concentrations (0.001 mol/ℓ or less) substances such as NaCl or KCl are essentially completely ionized. What is the osmotic pressure of a solution that consists of 50 mg of KCl (molecular mass = 74.5 u) and 200 mg of glucose at a temperature of 20°C?

10.3 The First Law of Thermodynamics

In our previous discussions of dynamics we were concerned only with the mechanical properties of objects and systems. Now, we will include *thermal* properties as we consider the subject of *thermo*dynamics. First, we must reexamine the statement of the energy conservation principle and frame it in such a way that changes in internal energy are included.

Let us denote by U the internal energy of a body and by Q the amount of thermal energy or heat supplied to the body by any means. If a body takes in an amount of heat Q, the internal energy is increased by exactly that amount: $\Delta U = Q$. Alternatively, the body could do a certain amount of work W when supplied with the heat Q; for example, if the body is a gas, it could expand against a restraining force as in Fig. 10.7. The increase in internal energy is diminished if the body does work when supplied with heat. The principle of energy conservation states that the change in internal energy of a body is equal to the heat supplied less the work done *by* the body:

$$\Delta U = Q - W \quad (10.18)$$

This equation is called the *first law of thermodynamics,* but there is no new physical idea contained in this law—it is a restatement of the principle of energy conservation when thermal energy is included.

As indicated in Fig. 10.7, the work done by the gas expansion is evidently $W = Mgh$ (explain). It is, however, more useful to determine the work done during an expansion in terms of the thermodynamic parameters describing the process. Figure 10.8

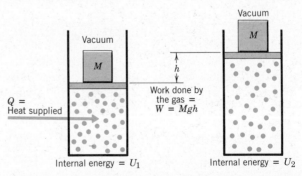

Figure 10.7 An amount of heat Q is supplied to a gas and the expanding gas does an amount of work W. The increase in the internal energy of the gas is $\Delta U = Q - W$. Notice that part of the thermal energy is transformed into gravitational potential energy: the work done by the expanding gas is $W = mgh$.

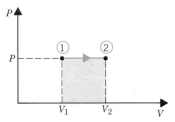

Figure 10.8 A constant pressure expansion from V_1 to V_2.

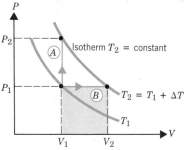

Figure 10.9 Achieving the temperature change $\Delta T = T_2 - T_1$ by a constant volume process A and by a constant pressure process B.

shows the constant pressure expansion on a PV-diagram.

If the piston area is A and the pressure is P, then the force on the piston is $F = PA$. If while maintaining the pressure a constant, the piston moves (outward) through a distance h, then the work done is

$$W = Fh = (PA)h = P(Ah)$$

or,

$$W = P(V_2 - V_1) \qquad (10.19)$$

Note that on the PV-diagram (Fig. 10.8) the quantity appearing in Eq. 10.19 is just the area under the thermodynamic path from state 1 to state 2 of the gas. This is *always true for any thermodynamic process* shown on a PV diagram.

The result expressed by Eq. 10.19 is, of course, in agreement with the earlier stated result, that $W = Mgh$, for

$$W = Mgh = (Mg/A) \times Ah = P(Ah)$$
$$= P(V_2 - V_1)$$

In Section 9.3 we discovered that a certain amount of heat added to a gas at constant pressure raised the temperature by a smaller amount than the same heat added at a constant volume. Or equivalently, the added heat required to achieve the same temperature increase is larger for the constant-pressure process than for the constant-volume process. We may illustrate this difference on a PV-diagram; see Fig. 10.9. The constant-volume process, labeled A, does no work, hence $Q_A = \Delta U$ relates the heat required Q_A to the change in internal energy ΔU accompanying the rise in temperature $\Delta T = T_2 - T_1$. During the constant-pressure process, labeled B, an amount of work $W_B = P_1(V_2 - V_1)$ is done by the expanding gas. Because the internal energy only depends on the temperature the change in internal energy, ΔU, is the same. Thus, from Eq. 10.18

$$Q_B = \Delta U + W_B > Q_A$$

10.4 The Second Law of Thermodynamics

The Direction of Heat Flow

We have established the fact that every piece of bulk matter has a certain amount of internal energy associated with the motion of its constituent molecules. We can now ask the question "Is it possible to extract this internal energy and use it to do work on another substance?" Suppose we have a certain mass m_1 of a material that is at a temperature T_1. If we place this object in contact with another mass m_2 of a material that is at a *lower* temperature T_2, as in Fig. 10.10, then we know from experience that the pair of objects will eventually come to a common temperature between T_1 and T_2. In other words, some of the internal energy of m_1 has been transferred to m_2 and has increased the internal energy and temperature of m_2. This transfer of internal energy is what is meant by heat.

Heat can flow from a hotter body to a colder body. But the reverse is not true. We cannot use the internal energy of m_2 to increase the temperature of m_1 while the temperature of m_2 is *decreased*. Thus, *unless work is done by an outside agent, heat always flows from objects at higher temperatures to objects at lower temperatures*. This is the substance of the *second law of thermodynamics*.

The second law of thermodynamics is a new physical idea, not connected with the first law (en-

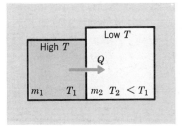

Figure 10.10 The spontaneous flow of heat is always from a hotter body to a colder body.

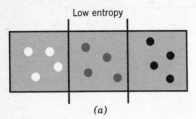 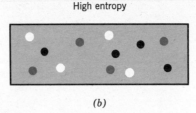

Figure 10.11 (a) An ordered system with a low entropy. (b) Removal of the barriers in the box in (a) allows the particles to mix: the degree of order is lowered and the entropy is increased.

ergy conservation). Nothing in our previous development of physical theory tells us that heat must always flow from a hotter to a colder body. The second law represents a new fact about the way in which Nature behaves.

There *are* situations in which heat flows from a colder body to a hotter body. But in these situations, we always find that the contrary flow of heat is due to work done by some outside agent. For example, heat can be made to flow out of the freezer compartment of a refrigerator and into the warmer room. In this process, water can be made to freeze while the temperature of the room increases. We know, however, that this is all possible only because work is being done by the refrigerator motor. The point of the second law is that heat will never flow *spontaneously* from a colder body to a hotter body.

Order and Disorder

The second law of thermodynamics can be expressed in terms of degree of *order* in a system. Imagine that we can line up a number of nitrogen molecules along one edge of an otherwise empty box. This condition is one with a high degree of *order*. But we know that the molecules will not remain in this condition. Instead, they will naturally tend to move about until they are distributed uniformly throughout the box. This latter condition is one with a high degree of *disorder*. Any ordered system in Nature if left to itself always tends to proceed spontaneously to a configuration with a lesser degree of order; that is, the trend in natural occurrences is always toward *disorder*.[2] The situation shown in Fig. 10.10 is an example of such a process: the initial condition is one in which the molecules in m_1 have a high average velocity while those in m_2 have a lower average velocity (that is, $T_1 > T_2$)—thus, there is a certain degree of order in the system. But, when the objects are placed in thermal contact, this order is decreased as the excess energy of the molecules in m_1 is shared with the molecules in m_2. Heat always flows in the direction that corresponds to a decrease in the order of a system; in other words, molecular motions always tend to produce a condition that is maximally random, a condition in which the internal energy is shared as equitably as possible among the constituents.

Entropy

The degree of order in a system can be expressed quantitatively by using the concept of *entropy*. An *ordered* system has a *low* entropy; a *disordered* system has a *high* entropy (Fig. 10.11). The second law tells us that an isolated, ordered system tends spontaneously toward disorder; that is, a system tends naturally to alter its condition in such a way that its entropy increases. Because the system is *isolated*, the internal energy remains constant throughout the process—no heat flows into or out of the system and no work is done on or by the system. The tendency toward disorder continues until the entropy of the system is maximum, and we say that the system has reached *thermodynamic equilibrium*.

A process that carries a system from one condition of equilibrium to another can take place in an (idealized) *reversible* way or in an *irreversible* way. If a cup of red dye is poured into a container of water, the dye molecules distribute themselves throughout the water, and in a relatively short time the water is a uniform pink color. This is clearly an *irreversible* process: the dye molecules will not reassemble so that they can be dipped out with the cup. A *reversible* process is one in which a system proceeds from a condition A to a condition B through a series of intermediate equilibrium conditions; by slightly altering the environment, the system can be made to retrace its path and proceed from B to A. No real process is truly reversible, but the concept is a useful one in the same way that the study of idealized frictionless processes is useful.

---

[2]Any housekeeper will be able to provide ample evidence in support of this statement.

Example 10.11

An example of a reversible thermodynamic process is the following. Consider a piece of ice that is in contact with a heat reservoir at a temperature of 0°C. (We consider the heat reservoir to be sufficiently large so that any changes brought about by the presence of the ice do not affect its temperature.) We will see in the next section (Fig. 10.16) that the melting point of ice (which is 0°C at normal atmospheric pressure) is *lowered* if the pressure is *increased*. Therefore, suppose that we slowly increase the pressure to which the piece of ice is subjected while maintaining its temperature at 0°C by contact with the heat reservoir. Because the ice is now at a temperature *above* the melting temperature for the increased pressure, the ice changes to liquid water. By increasing the pressure by only a tiny amount, we have caused the ice to change to a new condition, namely, liquid water. Now, by slowly removing the applied pressure, we can cause the liquid water to return to its frozen condition. At each step in the ice-to-water transition and in the water-to-ice transition, we imagine the system to be in an equilibrium condition. During the first transition, heat is absorbed by the ice; during the second transition, the same amount of heat is returned to and is reabsorbed by the reservoir. ∎

When we consider thermodynamic processes in which we focus upon the entropy, it is convenient to identify the *system* and its *environment*. Indeed, the environment of the system is everything in the Universe *except* the system: system + environment = Universe. Depending on the way that energy is transferred between the system and the environment during a particular process, the degree of order (i.e., the entropy) of the system may increase or decrease. Thus, in Example 10.11, the entropy of the system *increased* during the ice-to-water transition (the degree of order of the water molecules is greater in the ice phase than in the liquid phase), but the entropy *decreased* during the water-to-ice transition.

We denote entropy by S and entropy changes by ΔS. In any thermodynamic process, the change in entropy of the Universe ΔS_u is equal to the sum of the entropy changes of the system and the environment:

$$\Delta S_U = \Delta S_S + \Delta S_E \qquad (10.20)$$

If the process is *reversible*, then there is no net effect on the entropy of the Universe; thus,

reversible: $\Delta S_U = 0 \qquad \Delta S_E = -\Delta S_S \qquad (10.21)$

In any real (and, consequently, *irreversible*) process, the degree of order in the Universe as a whole decreases; thus, there is a net increase in the entropy of the Universe:

irreversible: $\Delta S_U > 0 \qquad \Delta S_E > -\Delta S_S \qquad (10.22)$

(Remember, a *negative* value of ΔS means an *increase* in the degree of order; such a change is contrary to the spontaneous tendency.)

If an amount of heat Q is absorbed in a reversible way by a system at the temperature T (an isothermal process), the entropy of the system increases, so that $\Delta S_S > 0$; in such a situation, the entropy change is defined to be

reversible, isothermal; $\Delta S_S = \dfrac{Q}{T} \qquad (10.23)$

Because the process is reversible, we also have $\Delta S_E = -Q/T$.

The units of entropy are, by referring to Eq. 10.23, equal to J/K (or Cal/K).

Biological systems are not exempt from the second law of thermodynamics. The metabolic processes in a cell increase the order in the cell by forming large molecules from small molecules as, for example, in photosynthesis. Although the entropy of the cell is decreased by these processes, the entropy of the surroundings is increased by an even greater amount. Any isolated system, and indeed the entire Universe, follows a course that continually increases its entropy. Therefore, we can look forward to the time (albeit in the distant future!) when the Universe will have reached a state of maximum entropy. Then, all objects will be at a common temperature so that no more work can be done by thermal energy and the Universe will die a "heat death."

Time's Arrow

The second law of thermodynamics, which can be characterized by the statement that the entropy of the Universe (or of any isolated system) cannot decrease, makes a unique contribution to our understanding of the nature of the physical Universe. The fact that entropy is ever-increasing shows the *direction* of thermodynamic changes—that is, *time flows only in one direction*. The conclusion that time is unidirectional is consistent with our experiences (birth, growth, death, and so forth), but most of the fundamental laws of physics are not altered by the substitution of $-t$ for t. For these laws, the idea of the unidirectional flow of time is of no consequence. It is only the fact that entropy forever increases that provides us with a proper concept of the unidirectional flow of time.

10.5 Heat Engines

A very important application of thermodynamics is to cyclic processes. During such processes the gas behavior maybe represented by a closed path on the PV-diagram. The gas arrives back at its original state at the end of the cycle. Figure 10.12 illustrates one such possible cycle. It might be convenient to think of a quantity of gas confined to a cylinder with a movable piston such as illustrated in Figure 10.7. Starting at state A the first step carries the gas to state B by adding heat Q_{AB} at a constant volume. The step from state B to C adds heat Q_{BC} at constant pressure. The two return steps reject heat Q_{CD} at constant volume and reject heat Q_{DA} at constant pressure. The net result is that heat energy in the amount of $Q_{in} = Q_{AB} + Q_{BC}$ is supplied to the gas by external sources of heat and an amount of heat energy $Q_{out} = Q_{CD} + Q_{DA}$ is rejected by the gas to the environment. Finally, the first law of thermodynamics, Eq. 10.18, states that an amount of work $W = Q_{in} - Q_{out}$, corresponding to the shaded area in Figure 10.12, is delivered to some external agency connected to the movable piston. Note that in applying Eq. 10.18 to the entire cyclic process, there is no net change in the internal energy because the gas returns to its original state A.

A heat engine is a device for converting heat energy into mechanical energy by performing some cyclic thermodynamic process over and over. We define the efficiency ϵ of such an engine by

$$\epsilon = \frac{\text{work output/cycle}}{\text{heat input/cycle}} = \frac{W}{Q_{in}} = \frac{Q_{in} - Q_{out}}{Q_{in}} \quad (10.24)$$

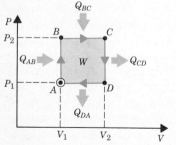

Figure 10.12 A simple cyclic process, starting and ending with state A.

● *Example 10.12*

A heat engine consisting of a cylinder with a movable piston, see Fig. 10.7, uses 10 g of He gas as its working substance. The cycle followed is that illustrated in Fig. 10.12. The strokes of the cycle are such that $V_2/V_1 = 2$ and $P_2/P_1 = 2$. If the temperature of state A is 200 K find Q_{in}, Q_{out}, W, and ϵ.

Using Table 9.6 we note that $c_v = 3134$ J/kg·deg and $c_p = 5202$ J/kg·deg for helium. Applying the gas law, Eq. 10.8, and given $T_A = 200$ K, we calculate that $T_B = 400$ K, $T_C = 800$ K, and $T_D = 400$ K. Thus,

$$Q_{AB} = mc_v \Delta T = (0.010 \text{ kg}) \times (3134 \text{ J/kg-deg}) \times (200 \text{ deg})$$
$$= 6{,}270 \text{ J/cycle}.$$

$$Q_{BC} = mc_p \Delta T = (0.010 \text{ kg}) \times (5202 \text{ J/kg-deg}) \times (400 \text{ deg})$$
$$= 20{,}810 \text{ J/cycle}.$$

Similarly we find that

$$Q_{CD} = 12{,}540 \text{ J/cycle}$$
$$Q_{DA} = 10{,}400 \text{ J/cycle}.$$

Therefore, we have

$$Q_{in} = Q_{AB} + Q_{BC} = 27{,}080 \text{ J/cycle}$$
$$Q_{out} = Q_{CD} + Q_{DA} = 22{,}940 \text{ J/cycle}$$
$$W = Q_{in} - Q_{out} = 4{,}140 \text{ J/cycle}$$
$$\epsilon = W/Q_{in} = 4{,}140/27{,}080 = 0.153 \quad \text{or} \quad 15.3\%$$

If the engine executed 10 cycles per second, it would deliver mechanical power at a rate of 41.4 kW. In practice there would be considerable energy losses present that we have not considered in this example, and as a result the efficiency would be lower and the delivered mechanical power would also be lower. ■

A very important cycle is the Carnot cycle illustrated in Figure 10.13 on a PV diagram.[3] It consists of two reversible adiabatic steps and two reversible isothermal steps. The first step, starting at state A at temperature T_1, is an adiabatic compression to state B, during which there is no heat exchange. From state B to C we have an isothermal expansion at the constant temperature T_2, during which an amount of heat energy Q_{BC} is supplied by an external heat source. Step C to D is an adiabatic further expansion. Finally, we return to state A from D by

---
[3] Named after Nicholas Léonard Sadi Carnot (1796–1832), French military engineer and physicist.

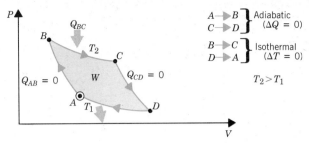

Figure 10.13 The Carnot cycle.

10.6 Changes of Phase

Vaporization

If a dish of water is exposed to the air, we know that the water will eventually disappear. This disappearance or *evaporation* of the water is a result of molecular agitation in the liquid. We have learned that the molecules in a gas are continually in motion and that the average velocity depends on the temperature. The same is true in a liquid. The molecules in a liquid are continually moving and colliding with their neighbors. In these collisions, energy can be transferred, with one molecule gaining energy and the other molecule losing energy. Just as in a gas, energy exchanged during molecular collisions in liquids produce a distribution of velocities for the molecules.

Because the attractive intermolecular forces are stronger in liquids than in gases, a liquid is a condensed state of matter and exhibits a *surface*. If a molecule in the surface layer of a liquid suddenly acquires sufficient energy in a collision, the molecule can overcome the forces that hold it in the surface and can escape into the surrounding air. This process, by which liquid molecules are freed from the liquid and become gaseous molecules, is called *evaporation* or *vaporization*. It is important to realize that only the molecules with exceptionally high energies actually escape from the liquid. Each vaporized molecule carries away an energy that is greater than the average, and consequently, the vaporization process causes a lowering of the average energy per molecule in the liquid and results in the *cooling* of the liquid.

When a dish that contains a liquid is exposed to the open air, the vaporized molecules move through the air in erratic paths due to the collisions they make with air molecules. As a result, the vapor molecules have very little chance of wandering back to the liquid surface. However, suppose that the space above the liquid is limited by the volume of a container. The vapor molecules will now bounce off the container walls and will eventually strike the liquid surface where they will be absorbed. In this situation, the liquid and vapor quickly come to an equilibrium condition in which equal numbers of molecules are being vaporized and absorbed each second. That is, the air above the liquid surface contains a constant number of vapor molecules per unit volume and we say that the air is *saturated* with vapor.

The vapor that exists in a confined volume above a liquid surface has a definite pressure. That is, there is a definite number of molecules per unit

an isothermal compression at the constant temperature T_1 during which an amount of heat Q_{DA} is rejected. The work done during the cycle is again $W = Q_{in} - Q_{out} = Q_{BC} - Q_{DA}$ represented by the shaded area in Fig. 10.13. The efficiency is

$$\epsilon = \frac{W}{Q_{in}} = \frac{Q_{BC} - Q_{DA}}{Q_{BC}}. \qquad (10.25)$$

Because the cycle consists of only reversible steps, we arrive at the end with no net change in entropy. Thus, the increase in entropy $\Delta S = Q_{BC}/T_2$ during the isothermal expansion must be equal to the decrease in entropy $\Delta S = Q_{DA}/T_1$ during the isothermal compression. (There is no change in entropy during the adiabatic steps.) Thus,

$$\frac{Q_{BC}}{T_2} = \frac{Q_{DA}}{T_1} \qquad \text{Carnot cycle.} \qquad (10.26)$$

Using this result in Eq. 10.25 gives

$$\epsilon = \frac{T_2 - T_1}{T_2} \qquad \text{Carnot cycle.} \qquad (10.27)$$

Carnot proved that no engine operating between a heat source at temperature T_2 and a heat sink at temperature T_1 can have an efficiency greater than that for a reversible Carnot cycle operating between the same temperatures. Equation 10.27 also explains why power plants operate with high-temperature furnaces and dump rejected heat at the lowest temperature available in the environment such as a cool river or lake. For example, a furnace operating as a heat source at 500°C and using a cool lake at 15°C would give the maximum possible efficiency of

$$\epsilon = \frac{500 - 15}{500 + 273} = 0.627 \quad \text{or} \quad 62.7\%$$

volume at a definite temperature; with n/V and T specified, the pressure is determined according to the ideal gas law. Notice that the pressure of the vapor does not depend on the nature of the gas in the confined volume. The space above the liquid can contain the vapor alone, vapor plus air, or vapor plus a mixture of gases. Moreover, the gas or gases in the volume can be at any pressure. The molecules in a gas or vapor move independently. This means that the molecules leave the liquid surface and return to it without regard to the presence of other types of molecules. If dishes of a number of different liquids are exposed to the same confined volume, each liquid will contribute its own particular vapor pressure (or *partial* pressure) to the total pressure in the container.

We can also argue that the vapor pressure of a liquid cannot depend on the volume of the space above the liquid. Suppose that we have a liquid and its vapor in equilibrium within a certain confined volume. Now, suppose that we suddenly double the volume. The vapor expands into the new volume and the concentration of vapor molecules becomes one-half of the former value. As a result, there are fewer vapor molecules that strike the liquid surface each second. On the other hand, the number of molecules leaving the surface per second is unchanged. (We assume that the temperature remains constant.) The liquid and vapor are no longer in equilibrium and the concentration of vapor molecules builds up until a balance is reestablished. In the new equilibrium condition, the value of n/V is the same as before the change in volume. With the temperature constant, this means that the vapor pressure has returned to its original value.

We must conclude from this analysis that the only quantity on which the vapor pressure of a particular liquid can depend is the *temperature*. It is easy to see why the vapor pressure of a liquid does depend on the temperature. The greater the number of molecules that leave a liquid surface per second (at equilibrium), the higher will be the vapor pressure. The rate at which molecules leave a surface can be increased only by increasing the average molecular energy and this means increasing the temperature.

Vapor Pressure Values

We can measure the vapor pressure of a liquid in the following way. We place the liquid in a container with a tight lid and we connect to the container a vacuum pump, as in Fig. 10.14. First, we evacuate the space above the liquid using the pump, and then

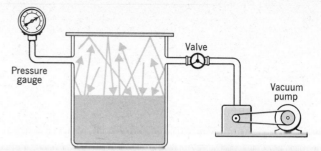

Figure 10.14 The saturated vapor pressure of a liquid can be determined by allowing the liquid to vaporize into a vacuum space created by a pump.

we close the valve between the pump and the container. The liquid vaporizes until the liquid and the vapor are in equilibrium. (Actually, equilibrium is reached only a small fraction of a second after the valve is closed.) Then, the vapor pressure can be read directly by means of a gauge or manometer attached to the upper part of the container. This pressure is appropriate for the particular temperature of the liquid. We call this maximum vapor pressure the *saturated vapor pressure* even though, in the situation illustrated in Fig. 10.14, there is no air or other gas to be "saturated." Remember, the vapor pressure is the same whether or not the air is present.

In different liquids the strengths of the intermolecular forces are different. Therefore, we can expect that the vapor pressure values for different liquids at the same temperature will be different. Molecules break away from the surface of methyl alcohol much more readily than they do from the surface of water. (Notice, in Table 8.1, that the surface tension of methyl alcohol is considerably smaller than that of water.) Accordingly, we expect the vapor pressure of methyl alcohol to be greater than that of water at any temperature. (We say that alcohol is more *volatile* than water.) Figure 10.15 shows the saturated

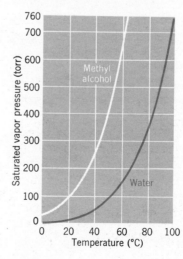

Figure 10.15 Values of the saturated vapor pressure for methyl alcohol and water.

Table 10.2 Saturated Vapor Pressure and Vapor Density for Water at Various Temperatures

Temperature (°C)	Vapor Pressure (torr)	Vapor Density (kg/m³)
0	4.58	0.00485
5	6.54	0.00680
10	9.21	0.00941
15	12.8	0.0128
20	17.5	0.0173
25	23.8	0.0230
30	31.8	0.0304
40	55.3	0.0511
50	92.5	0.0832
60	149	0.130
70	234	0.198
80	355	0.294
90	526	0.424
100	760	0.598

vapor pressure values for methyl alcohol and water; the values for alcohol are always about four times greater than the corresponding values for water. Table 10.2 lists the values of the saturated vapor pressure for water along with the values of the density of the water vapor.

Molecules can also break away from the surfaces of solids. But because the binding forces are much stronger than those in liquids, the vapor pressure values for solids are always quite low. For example, arsenic is a very volatile metal, and yet its vapor pressure becomes as large as 1 torr only at a temperature of 372°C.

Relative Humidity

When a dish of water is exposed to a confined volume of air, the water will evaporate until the saturated vapor pressure is reached for the particular temperature. The equilibrium situation will be reached, however, only if there is a sufficient quantity of water in the dish. If the water has been completely evaporated before equilibrium is attained, the air will not be saturated with water vapor. The ratio of the actual vapor pressure of water in air to the saturated vapor pressure at the same temperature is called the *relative humidity*. That is, the relative humidity is the amount of water vapor in air expressed as a percentage of the maximum amount that could be vaporized at that temperature.

When the relative humidity is low, we say that the air is *dry*. In this situation, water evaporates quickly because the rate at which molecules are vaporized far exceeds the rate at which they strike and are absorbed by the liquid. Body perspiration readily evaporates in a low-humidity condition and we feel cooled by this process. However, if the relative humidity is high, we say that the air is *muggy* or *sticky*. The reason that high-humidity conditions feel so oppressive is that evaporation is very slow. The air contains such a high concentration of water vapor, that molecules return to the liquid almost as rapidly as they vaporize. If body perspiration is slow to evaporate, there is very little cooling effect and we feel uncomfortable. If the relative humidity is 100 percent, there can be no net evaporation—wet clothing or damp skin will never dry under such conditions.

●*Example 10.13*

Air that is saturated with water vapor at 20°C is heated to 30°C. What is the relative humidity in the final condition?

The original vapor pressure is the saturated value for 20°C; according to Table 10.2, this value is 17.5 torr. This vapor pressure does not change when the temperature is increased (no molecules are lost in the process). At 30°C, the saturated vapor pressure is 31.8 torr. Therefore, the relative humidity at 30°C is

$$\text{Relative humidity} = \frac{17.5 \text{ torr}}{31.8 \text{ torr}} = 0.55 = 55\%$$

●*Example 10.14*

A house has a volume of 500 m³. At a temperature of 25°C the relative humidity in the house is 80 percent. How much water must be removed from the air to reduce the relative humidity to 30 percent?

According to Table 10.2, the saturated density of water vapor at 25°C is 0.0230 kg/m³. Therefore, the total mass of water vapor in the air for 80-percent relative humidity is

$$M(80\%) = 0.8 \times (0.0230 \text{ kg/m}^3) \times (500 \text{ m}^3) = 9.20 \text{ kg}$$

and for the lower relative humidity, the mass of water vapor is

$$M(30\%) = 0.3 \times (0.0230 \text{ kg/m}^3) \times (500 \text{ m}^3) = 3.45 \text{ kg}$$

Thus, 9.20 − 3.45 kg = 5.75 kg of water must be removed from the air to lower the relative humidity to 30 percent. This is equivalent to about 1.5 gal. ■

A volume of air at a temperature of 15°C contains sufficient water vapor to make the relative humidity equal to 72 percent. If this air is now cooled to 10°C, the air will become saturated. (Verify this statement by using the values in Table 10.2.) If the air, or any surface to which the air is exposed, is cooled to a lower temperature, there will be an excess amount of water vapor in the air. This excess vapor will return to the liquid state by *condensation* to form water droplets on a cooled surface (*dew*) or water droplets in the air (*fog* or *clouds*). The temperature at which the water vapor in a particular sample of air becomes saturated is called the *dew point*. If the temperature must be lowered below 0°C in order to achieve saturation, the excess water vapor will form *frost*.

● *Example 10.15*

When the air temperature is 25°C, a piece of metal is cooled until, at 15°C, water droplets appear. What is the relative humidity of the air?

The air in the immediate vicinity of the metal is cooled by conduction to the temperature of the metal. The fact that water vapor condenses on the metal when the temperature is 15°C means that the air is saturated at this temperature. Therefore, the partial pressure at 25°C is equal to the saturated pressure at 15°C, that is, 12.8 torr. Hence, the relative humidity is

$$\text{Relative humidity} = \frac{12.8 \text{ torr}}{23.8 \text{ torr}} = 0.54 = 54\%$$

This technique of determining the *dew point* is often used to measure relative humidity. ■

Boiling

When a flame is applied to a beaker of liquid, the temperature of the liquid increases. As a result, the vapor pressure increases and vaporization from the surface takes place at an increasing rate. Actually, vaporization takes place *within* the liquid as well. This vapor tends to form small bubbles, but these are quickly collapsed by the liquid because the liquid pressure is greater than the vapor pressure in these microscopic bubbles. Now, the vapor pressure of the liquid in proximity to the heat source will be higher than that of the surrounding cooler liquid. The result of this difference is not apparent until the vapor pressure of the hottest part rises to equal the pressure at the bottom of the liquid. (In a shallow container, this pressure will be essentially atmospheric pressure, because the liquid pressure due to the depth of the liquid is small.) Then, the vapor pressure overcomes the collapsing effect of the liquid and the vapor pushes the liquid aside and bubbles can actually form within the liquid; the liquid *boils*. The temperature cannot increase further because the vapor pressure cannot exceed the pressure in the liquid (essentially atmospheric pressure). In fact, we define the *boiling point* of a liquid to be the temperature at which the vapor pressure is equal to atmospheric pressure. Thus, boiling always takes place at constant temperature.

At normal atmospheric pressure (760 torr), water boils at 100°C. If the atmospheric pressure is lower than 760 torr, as it would be on a mountain, the boiling temperature will be less than 100°C. For example, at the top of a 20,000-ft (6-km) mountain, the atmospheric pressure is approximately 355 torr. This means (see Table 10.2) that water will boil at 80°C at this altitude.

In order to maintain a liquid at a constant temperature during vaporization, heat must be added to the liquid from an outside source. The quantity of heat that is required to vaporize a liquid at a particular temperature is called the *heat of vaporization*, H_v. For example, the conversion of water at 100°C to vapor (steam) at the same temperature requires 2.26×10^6 J/kg (540 Cal/kg). At any lower temperature, the degree of molecular agitation is less, and, consequently, the heat of vaporization is larger than the value for 100°C. Table 10.3 gives the heat of vaporization of water at various temperatures.

● *Example 10.16*

How much heat is required to convert 4 kg of water at 20°C into vapor at 60°C?

First, the water must be raised from 20°C to 60°C. This requires

$$Q_1 = mc\Delta T = (4 \text{ kg}) \times (4186 \text{ J/kg-deg}) \times (40 \text{ deg})$$
$$= 6.7 \times 10^5 \text{ J} = 0.67 \times 10^6 \text{ J}$$

Next, the water at 60°C must be converted to vapor at the same temperature. Using the heat of vaporization H_v from Table 10.3.

Table 10.3 Heat of Vaporization of Water

	Heat of Vaporization, H_v	
Temperature (°C)	(J/kg)	(Cal/kg)
0	2.49×10^6	596
20	2.45	585
40	2.40	574
60	2.36	563
80	2.31	552
100	2.26	540

$$Q_2 = mH_v = (4 \text{ kg}) \times (2.36 \times 10^6 \text{ J/kg})$$
$$= 9.44 \times 10^6 \text{ J}$$

The total amount of heat required is

$$Q = Q_1 + Q_2 = 0.67 \times 10^6 \text{ J} + 9.44 \times 10^6 \text{ J}$$
$$= 1.01 \times 10^7 \text{ J}$$

or about 2400 Cal. ∎

Melting

In order to convert a substance from the liquid state to the gaseous state, the intermolecular bonds that hold the liquid together must be completely broken apart. (Remember, there are almost no intermolecular forces in gases.) Consequently, the *change of phase* from liquid to gas requires the input of energy. Similarly, the conversion of a substance from the solid state to the liquid state also involves the input of energy and the breaking of intermolecular bonds. However, in this change of phase the bonds are not completely broken apart and we expect that a smaller amount of energy will effect the phase change. In fact, 1 kg of ice at 0°C will be converted into water at the same temperature with the input of 3.35×10^5 J (80 Cal); this amount of energy is considerably smaller than that necessary to convert 1 kg of water at 100°C into vapor at the same temperature (2.26×10^5 J or 540 Cal). The energy required for the solid-to-liquid phase change is called the *heat of fusion*, H_f. Some relevant thermodynamic properties for various substances are given in Table 10.4.

Many of the substances found in Nature exist in the form of *crystals*. As we pointed out in Section 1.3, crystals exhibit a well-defined melting point because they have the same composition and structure throughout. Noncrystalline or *amorphous* materials, such as glass or plastics, have no regular structure pattern and so the binding forces are not all the same. When these substances are heated, some of the bonds break before the others. As a result, the material gradually softens and does not exhibit a sharp melting point.

Phase Diagrams

Many of the interesting and important thermodynamic properties of a substance can be summarized in a diagram that shows the various phases of the substance in a pressure-versus-temperature plot. The *phase diagram* of water is shown in Fig. 10.16. The central feature of this diagram is the *triple point*, which occurs at a temperature of 0.01°C and a pressure of 4.58 torr. The triple point represents the only conditions under which the three phases of water are in equilibrium. That is, at the triple point, liquid water, ice, and water vapor can all exist together.

The curves that radiate away from the triple point divide the diagram into three regions that correspond to the three phases. The *vaporization curve* separates the liquid phase from the vapor phase and corresponds to the saturated vapor pressure values listed in Table 10.2. The vapor and the liquid phases remain in equilibrium until a temperature of 374°C is reached. At this temperature the vapor pressure is 218 atm and, of course, this condition can exist only within a container of high strength—a kind of super pressure cooker. At temperatures above 374°C, *no* pressure, however great, can maintain the liquid phase. The point in the diagram at $T = 374°C$, $P = 218$ atm is called the *critical point*. Notice that for temperatures below the critical temperature we call the gaseous phase *vapor* because it can be in equilibrium with the liquid phase. For temperatures above the critical temperature, the liquid phase cannot exist and we call this phase *gas*.

Table 10.4 Thermodynamic Properties of Some Substances

Substance	Normal Melting Point (°C)	Heat of fusion, H_f (J/kg)	Normal Boiling Point (°C)	Heat of Vaporization, H_v (J/kg)
Helium			−268.6	2.5×10^4
Lead	327	2.45×10^4		
Mercury	−39	1.18×10^4	356.6	2.95×10^5
Nickel	1435	3.09×10^5		
Nitrogen	−210	2.55×10^4	−195.6	1.99×10^5
Silver	961	8.83×10^4		
Tin	232	5.86×10^4		
Water	0	3.35×10^5	100	2.26×10^6
Methyl alcohol	−97.8	9.21×10^4	64.7	1.10×10^6

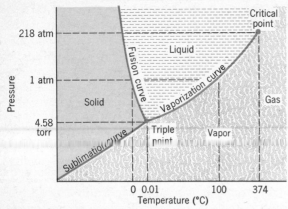

Figure 10.16 Phase diagram for water. Both the pressure and the temperature scales have been grossly distorted in order to show all of the important features of the diagram in a limited space.

The *fusion curve* separates the solid and liquid phases, and the *sublimation curve* separates the solid and vapor phases. Notice that transitions across this latter curve represent direct phase changes from vapor to solid or solid to vapor without the formation of the liquid phase. For water, the sublimation process that converts ice into water vapor takes place only very slowly because the vapor pressure of ice is so low (it is always less than the triple-point pressure of 4.58 torr). Carbon dioxide, on the other hand, has a very high triple-point pressure (5.1 atm). That is, at normal atmospheric pressure, liquid carbon dioxide cannot exist and sublimation is the usual phase change. For this reason, solid carbon dioxide is known as *dry ice*.

Summary of Important Ideas

The laws concerning the behavior of gases that were discovered by Boyle and by Charles and Gay-Lussac can be combined into a single statement which is called the *ideal gas law*: $PV = nRT$. In using this equation, the temperature must always be given in kelvins (degrees absolute).

The *kinetic theory of gases* is a description of the behavior of gases which results from the application of Newtonian dynamics to the motion of gas molecules. This theory relates the *microscopic properties* of a gas (molecular velocities and kinetic energies) to the *bulk properties* of the gas (pressure and temperature).

Different types of gas molecules at the same temperature will have different *average velocities* that depend on the molecular masses. The average *kinetic energy* of the molecules in a gas sample (regardless of type) is directly proportional to the absolute temperature.

The *first law of thermodynamics* is an expression of the conservation of energy to processes involving heat. It states that the heat added to a system less the work done by the system results in an increase in the internal energy.

The *second law of thermodynamics* expresses the fact that heat will flow from a cooler body to a warmer body only if work is done by some outside agent. Alternately, we can state that the natural tendency of an isolated system is to alter itself spontaneously toward a condition of a lower degree of orderliness. That is, the *entropy* of an isolated system—and of the Universe as a whole—never decreases.

Heat engines employ cyclic thermodynamic processes that convert heat energy input into mechanical work and to unavoidably accompanying heat-energy loss. The reversible Carnot cycle is the most efficient cycle to operate between a heat source and a heat sink.

In the process of *vaporization*, the faster moving molecules escape from the surface of a liquid, thereby resulting in a cooling of the liquid. The *saturated vapor pressure* that exists in the confined space above a liquid surface depends only on the *temperature*.

The *relative humidity* is the amount of water vapor in the air at a particular temperature expressed as a percentage of the maximum amount that the air could contain at the same temperature.

At the *boiling point*, the vapor pressure of a liquid is equal to the atmospheric pressure.

The *heat of vaporization* for a substance is the energy per kilogram that is necessary to convert the substance from the liquid phase to the vapor phase at a particular temperature.

The *heat of fusion* for a substance is the energy per kilogram that is necessary to convert the substance from the solid phase to the liquid phase. The heat of fusion is always smaller than the heat of vaporization for a particular substance because stronger intermolecular forces must be overcome to convert the liquid to a vapor than to convert the solid to a liquid.

Only crystalline materials exhibit sharp, well-defined *melting points*.

The *triple point* refers to the pressure-temperature combination that allows a substance to exist simultaneously in all three phases—solid, liquid, and vapor. The triple point of water occurs for a pressure of 4.58 torr and a temperature of 0.01°C. The *critical point* refers to the pressure-temperature combination that includes the highest temperature at which the vapor phase of a substance can be in equilibrium with its liquid phase, regardless of the pressure. The critical point for water occurs for a temperature of 374°C and a pressure of 218 atm.

◆ **Questions**

10.1 How would you expect the ideal gas law to be modified if the mean spacing between molecules in a gas could not be considered to be very large compared to molecular dimensions?

10.2 How would you expect the ideal gas law to be modified if the effect of having attractive forces between molecules were also included?

10.3 Why are the conservation laws of momentum and energy inadequate to predict the direction in which processes will tend with time? In what ways does the second law of thermodynamics, involving the concept of entropy, resolve this problem?

10.4 Two identical springs, one compressed and the other having its relaxed length, are dropped into identical beakers of acid and completely dissolved. Is there any difference in the two processes? Explain.

10.5 Make a graph of the Carnot cycle, Figure 10.13, where the coordinates are: the temperature T as abscissa and entropy S as ordinate. Identify the points (states) A, B, C, and D and the path direction of the cycle. What is the significance of the area enclosed by the closed path of the cycle?

10.6 A liter of water is placed in a 10-ℓ can (a lightweight "tin" can) and the can is heated until most of the water has boiled away. The top of the can is then screwed tightly into place. When the can cools to room temperature, it collapses. Explain in detail what has happened.

10.7 A quantity of nitrogen gas is sealed in an insulated container at a pressure of several atmospheres. A leak develops in the tank and the gas escapes slowly through a small hole. If the gas in the tank does not exchange heat with the tank or the surroundings, what happens to the temperature of the gas as the leaking continues? (Hint: Which molecules are most likely to find their way to the hole and leak out?)

10.8 Why does the Earth's atmosphere consist primarily of "heavy" gases (nitrogen, oxygen, carbon dioxide, water vapor), with only traces of "light" gases (hydrogen, helium)? (Refer to Example 7.10.)

10.9 A salt-water solution stands in a glass. As the water evaporates, the salt organizes itself into crystals which collect on the bottom of the glass. The salt crystals represent a highly ordered state compared to that of salt ions wandering randomly through the solution. Has the entropy law been violated here? Explain.

10.10 A pressure of 10 torr is maintained within a certain volume that contains a dish of water. The entire system is at a constant temperature of 20°C. Describe what happens to the water.

10.11 At the peak of Mount Everest, the atmospheric pressure is about 240 torr. If an open container of water is heated, its vapor pressure can never be increased to 760 torr. (Why?) What is meant by "boiling" in this situation?

10.12 How can water be made to boil at 10°C?

10.13 During the winter, frost can form on the inner surface of window panes even though the room is warm. Explain why this happens.

10.14 Energy is required to convert water into steam even though the temperature is not changed. Explain why this is so.

10.15 Why does the use of a *pressure cooker* reduce the cooking times for foods? Should any precautions be exercised in using a pressure cooker? Explain.

10.16 Describe the process of boiling in a deep container that is heated at the bottom.

10.17 Suppose that you have a quantity of water at room temperature. You wish to vaporize as much of this water as possible, but you have only a certain amount of energy to "spend." Is it better to use this energy to vaporize the water at room temperature or should you raise the temperature of the water to the boiling point? Explain the physical reasoning behind your answer.

10.18 A quantity of a solid material is mixed with its liquid phase. Heat can be added to this mixture and yet the temperature of the substance does not change. Explain why this is possible.

10.19 Heat is added at a constant rate to a quantity of water in the solid phase (ice). Make a sketch of the temperature of the water as a function of time. Extend the diagram to the time that the water becomes steam. (Assume constant pressure throughout.)

10.20 Snow and ice will eventually disappear from the ground or a street even if the temperature remains continually below the freezing point. Explain how this can happen.

★ **Problems**

Section 10.1

10.1 Construct a graph of pressure (in atm) versus volume (in m^3) at constant temperature for a gas sample that has a volume of 4 m^3 when the pressure is 4 atm.

10.2 A quantity of nitrogen gas, initially at a temperature of 20°C and at atmospheric pressure, is adiabatically compressed to $\frac{1}{5}$ of its original volume. What is the final pressure of the gas?

10.3* A hollow cylinder has an inside diameter of 2 cm and a length of 2.5 m. One end of the cylinder is sealed and the cylinder is stood upright with the sealed end on the ground. A piston whose mass is 10 kg is fitted snugly into the open end of the cylinder. Assuming that no air escapes past the piston, to what height above the ground will the piston fall when released? What will be the final position of the piston when thermal equilibrium is reached?

10.4 A 6-ℓ quantity of gas is at a pressure of 15 atm and a temperature of 27°C. If the temperature is increased by 200 deg while the volume is increased to 45 ℓ, what will be the final pressure?

10.5 If the gas in the preceding problem is oxygen, what is its mass?

10.6 Gas is stored in a steel tank at a pressure of 300 atm and a temperature of 20°C. If the bursting pressure of the tank is 450 atm, at what temperature will the tank explode?

10.7 A certain quantity of gas is confined in a cylinder with a movable piston. Heat is added to the gas and the temperature rises from its initial value to 27°C to a final value of 177°C. At the same time the volume is changed to $\frac{1}{2}$ the original value. What is the change in pressure?

10.8 A container has a volume of 1 m^3. The gas in the container has a temperature of 150°C and is at a pressure of 1 atm. If the temperature is lowered to $-20°C$, what is the new pressure if the volume remains constant? What change in volume will be necessary to maintain the pressure at 1 atm?

10.9 A gas, confined to a volume V, is heated from a temperature of 10°C (at which temperature the pressure is 2 atm) to a temperature of 400°C. What is the new pressure? What temperature decrease would be required to reduce the pressure to 1 atm?

10.10 The best laboratory vacuum apparatus can produce pressures of approximately 10^{-15} atm. How many molecules per cubic centimeter are there in a gas at this pressure and at a temperature of 20°C?

10.11 A 1.2-m^3 tank contains 100 mol of sulfur dioxide (SO_2) gas at a temperature of 25°C. What is the pressure in the tank?

10.12 Refer to Fig. 10.2 and draw a similar graph that relates the pressure and the temperature of an ideal gas at constant volume. This graph can be represented by the equation, $P = P_0(1 + aT_C)$, where P_0 is the pressure of the gas at 0°C and where P is the pressure at any Celsius temperature T_C. What is the value of the constant a? What is the significance of $1/a$?

10.13 Show that $TV^{\gamma-1} = $ constant for an adiabatic process in an ideal gas.

10.14 Use the result of the preceding problem to find the final temperature of the compressed gas in Problem 10.2.

10.15 Show that for gases the coefficient of volume thermal expansion β in the relationship $\Delta V = \beta V \Delta T$ at constant pressure is $\beta = 1/T$ and hence for all gases near 0°C, $\beta = 3.66 \times 10^{-3}$ deg^{-1}.

Section 10.2

10.16 What is the total kinetic energy of 5 mol of nitrogen gas at a temperature of 300 K?

10.17 Show that $P = \frac{1}{3}\rho \overline{v^2}$ for a gas with density ρ.

10.18 A tank whose volume is 4 m^3 contains 3 kg of a certain gas at a pressure of 2 atm. What is the rms velocity of the molecules in the gas? (Use Eq. 10.12; what is the significance of the product NM that appears in this equation?)

10.19 A gas is confined to a volume V. Initially, the pressure is 1 atm and the temperature is

100°C. What temperature is required to double the rms velocity of the gas molecules? At this new temperature, what is the pressure?

10.20 A certain volume of helium gas is at a temperature of 800 K. What is the average kinetic energy of the helium atoms? (Express the result in electronvolts.) What is the rms velocity of the atoms in the sample?

10.21 The interiors of most stars are at temperatures of approximately 10^8 K. What is the rms velocity of a hydrogen nucleus (i.e., a proton; there are no *atoms* at these temperatures) in such a stellar interior? What is the average kinetic energy of such a proton in keV?

10.22 At what temperature will the rms velocity of nitrogen molecules be equal to that of hydrogen molecules at 0°C?

Section 10.3

10.23 The movable piston shown in Fig. 10.7 is mechanically linked to a hoist. Heat supplied to the gas at a constant temperature enables the coupled hoist to raise a 100-kg mass through a vertical distance of 2 m. How many Cal of heat was supplied to the gas? (Assume no loss in energy in the mechanical linkage.)

10.24 The cylinder shown in Fig. 10.7 has a piston of mass 5 kg and contains $\frac{1}{10}$ g of helium gas. How many Cal of heat must be supplied to raise the piston $h = 10$ cm? Use Table 9.6.

10.25* The cylinder shown in Figure 10.7 contains $\frac{1}{300}$ mol of helium gas. The cylinder walls are well insulated against heat transfer between the gas inside and the exterior environment. The piston mass is slowly doubled to 10 kg and the piston is observed to subsequently sink to a new equilibrium position 5 cm lower. What are the initial and final temperatures of the gas? What is the increase in the internal energy of the gas? How much work was done during the compression? Note, the internal energy of helium may be expressed as $U = (3/2)nRT$, see Example 10.9.

Section 10.4

10.26 When 1 kg of ice completely melts at 0°C, an amount of heat input of 3.35×10^5 J must be provided. What is the increase in entropy? What would be the increase in entropy if only enough heat were provided to melt $\frac{1}{2}$ of the ice?

10.27 In an isothermal expansion at 100°C, 0.01 Cal of heat is supplied. What is the change in entropy?

10.28 A 1-kg block of aluminum is heated from 0°C to 100°C. The entropy change under these circumstances is given by

$$\Delta S = S_2 - S_1 = mc \, \ell n(T_2/T_1)$$

where c is the specific heat and ℓn stands for the natural logarithm. What is the change in entropy? How much error is made if we approximate ΔS by

$$\Delta S = \frac{\Delta Q}{T} = \frac{mc(T_2 - T_1)}{(T_2 + T_1)/2}$$

Section 10.5

10.29 It is suggested that an ocean temperature gradient of the order of 10°C might be made use of to provide usable energy. If the surface temperature is 20°C, what is the highest-efficiency engine that could operate between the temperatures corresponding to these conditions?

10.30 A high-pressure steam turbine operates at 60% of its maximum possible efficiency. What is its realized efficiency if it uses steam heated to 650°C and exhausts at 80°C?

10.31 An ideal Carnot cycle engine has a heat input of 150 Cal at 400°C and exhausts 100 Cal. What is the exhaust temperature?

10.32 A very efficient engine operates between 1800 K and 700 K with a realized efficiency of 35%. What percent of its maximum possible efficiency is attained?

10.33 An ideal Carnot cycle engine takes in 10^5 W of heat at 400°C and exhaust at 80°C. How much mechanical power is delivered?

10.34* In a two-stage Carnot-cycle engine the heat exhausted from the first stage constitutes the input heat to the second stage. The first stage operates between 2000 K and 700 K. The second stage operates between 600 K and 300 K after the transferred heat lost 100 K in temperature and 20% of the heat energy. At what efficiency is the combined work output generated?

Section 10.6

10.35 The cover is removed from a dish of water in a sealed room in which the perfectly dry air is at a temperature of 20°C and a pressure of 760 torr. If the temperature does not change, what will be the eventual pressure in the room? (Assume that sufficient water is available to saturate the air in the room.)

10.36 Sealed containers of water and of methyl alcohol are placed in a confined volume that contains dry air at normal atmospheric pressure. When the entire system comes to temperature equilibrium at 50°C, the covers are removed from the containers of alcohol and water. What is the new equilibrium pressure in the confined volume? (Use Fig. 10.15.)

10.37* There is a tiny bubble in a quantity of water at 20°C. What are the partial pressures of air and water vapor in the bubble? The water is now heated to 100°C. What happens to the bubble? What are the new partial pressures?

10.38 What is the mass of water vapor in a room with dimensions 6 m × 4 m × 3 m when the temperature is 25°C and the relative humidity is 60 percent?

10.39* A 1-m^3 sample of air at 25°C is isolated in a container and cooled to 10°C. During this process, 8 cm^3 of water condenses on the bottom of the container. What was the original relative humidity of the air?

10.40* At a temperature of 25°C the relative humidity of a sample of air is 60 percent. What is the dew point? (Devise some method of interpolating in Table 10.2 or use Fig. 10.15.)

10.41 How much heat (in Cal) is required to vaporize 6 mol of helium at the normal boiling point?

10.42* How much energy (on the average) is required to vaporize one molecule from the surface of water at 100°C and normal atmospheric pressure? Express the result in J and in eV. Compare the result with $\frac{3}{2}kT$ at the same temperature (Example 10.9).

10.43 How much heat (in Cal) is required to convert 20 kg of ice at 0°C into steam at 100°C (at constant pressure)?

10.44* An insulated box contains 6 ℓ of water at 20°C. Into this box is introduced 3 kg of ice at 0°C and 2 kg of steam at 100°C. What is the final temperature of the system, assuming no heat losses to the box or to the surroundings?

10.45 How much heat (in Cal) is required to melt 2.5 mol of nitrogen at the normal melting point?

10.46 What is the least amount of water at 20°C that could be used to convert molten lead at 327°C to solid lead at the same temperature? (The water is vaporized in the process.)

10.47 What is the maximum amount of ice that can be melted by the condensation of 4 kg of methyl alcohol vapor at 64.7°C? (See Table 10.4).

10.48 How much heat is required to convert 8 kg of ice at −15°C to water at 15°C? (The specific heat of ice is approximately one-half that of water.)

10.49 An ordinary ice cube has a mass of approximately 25 g. If one ice cube (at 0°C) is placed in 200 g of water at 20°C, what will be the temperature of the water when the ice cube has melted completely? (Assume no heat exchange with the surroundings.)

10.50* According to Fig. 10.16, the solid-to-liquid transition temperature (i.e., the melting point) decreases with increasing pressure. Thus, water remains a liquid at temperatures below 0°C if the pressure is sufficiently high. The melting temperature is lowered by 1 deg for each 1.1×10^7 Pa (approximately) of applied pressure. Use this information to explain why ice skaters move on a film of *water* even though the ice temperature is maintained at the normal freezing point. Proceed by estimating the pressure exerted on the ice by the skate runners for an average skater and then calculating the resulting depression in the melting temperature.

Elasticity and Vibrations

If an attempt is made to deform any solid object by changing its size or shape, the object resists this attempt. Moreover, when the force that has produced a deformation is removed, the object tends to return to its initial condition. *All* solids exhibit this property of *elasticity,* and the recovery of an object to its original size and shape will be essentially complete if the distortion has not been too severe. Of course, any object can be permanently bent or broken if too great a deformation is produced.

The reason that any solid object has an intrinsic elasticity is that the object is held together by electric forces acting among the constituent atoms and molecules. As in every natural situation, the atoms in a solid spontaneously adjust themselves into the position of minimum potential energy. If a force is applied to the solid and alters the arrangement of the atoms, work has been done and the potential energy of the system is increased. When the disturbing force is removed, the system again seeks the condition of minimum potential energy, and in a deformed solid the electric forces produce the necessary readjustment. For example, when a rubber band is stretched, some of the electrons in the rubber molecules are pulled farther from the positively charged nuclei than they are in the normal condition. This produces an extra attractive force which acts to return the rubber band to its original length when the stretching force is removed. Gases and liquids also exhibit elastic properties, but in this chapter we are concerned mainly with the elasticity of solid materials.

The phenomenon of *elasticity* is closely linked with that of *vibration* (or *oscillation*). Everyone has observed that when some elastic material (e.g., a spring suspended from a rigid support or a metal bar that is fixed at one end) is displaced from its normal or equilibrium position and then released, it will move through the equilibrium position and become extended in the opposite direction. That is, the material "overshoots" the mark in its attempt to return to its original condition. Having extended itself too far, the motion is reversed and is directed back toward the equilibrium position where another overshoot occurs. The material continues this back-and-forth vibratory motion until frictional effects cause the motion to cease. All vibratory motions of this type depend directly on the elastic properties of the vibrating material.

In this chapter we first study some of the basic ideas and principles concerning elasticity and the elastic properties of materials. Then, we continue by concentrating on vibratory motion because this topic is vital to the understanding of many of the discus-

11.1 Basic Elastic Properties of Materials

Hooke's Law

One of the early interesting and useful observations concerning the elastic properties of matter was made by Robert Hooke (see Example 7.3). *Hooke's law*[1] states that the force required to stretch (or compress) a particular spring is proportional to the distance the spring is stretched (or compressed). That is,

$$F = kx \qquad (11.1)$$

where x is the distance that the spring is displaced from its original equilibrium position. The factor k is called the *force constant* of the spring and depends on the type and size of the material from which the spring is made and on the way it is wound; the units of k are N/m.

Not only do springs obey Hooke's law, but so does matter in other forms and matter subject to different types of deformation. For example, if a metal bar is fixed at one end and is stretched by applying a force to the other end, the amount of stretch will be proportional to the force. Or, if the bar is given a lateral push at the free end, the displacement will again be proportional to the force.

Stress and Strain

The force that is applied to an elastic material divided by the cross-sectional area of the material is called the *stress*:

$$\text{stress} = \frac{F}{A} \qquad (11.2)$$

The simplest type of stress involves two forces acting on an object along the same straight line, tending either to elongate the object (Fig. 11.1a) or to com-

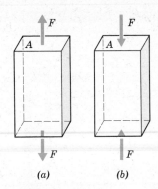

Figure 11.1 *(a)* A *tensile stress* tends to elongate an object. *(b)* A *compressive stress* tends to decrease the length of an object.

press it (Fig. 11.1b). In the first case we refer to the stress as *tensile stress* and in the second case we use the term *compressive stress*.

The effect produced on an object by a stress is called the *strain*. For a tensile or compressive stress, the measure of the strain is the fractional change in length:

$$\text{strain} = \frac{\Delta \ell}{\ell} \qquad (11.3)$$

where ℓ is the original unstressed length of the object and where $\Delta \ell$ is the change in length produced by the application of the stress (see Fig. 11.2). Notice that when the length of the sample is increased, the cross-sectional area decreases slightly to compensate. If the elongation is relatively small, the area change can be neglected and we do not consider the stress to be altered by any variation in the area.

For any object or material, the stress and the strain are related in a simple way only over a restricted range of stress values. When the strain is directly proportional to the stress, Hooke's law is obeyed, but if too large a stress is applied to any particular sample, this linear relationship ceases to be valid. Figure 11.3 shows a stress-strain graph that is similar to those for some real materials. An ideal material would follow the linear Hooke's law

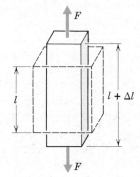

Figure 11.2 The strain produced by the application of a stress to an object is $\Delta \ell / \ell$. The deformation is exaggerated.

---

[1] This is not a *law* in the same sense that Newton's statements are laws. Hooke's relation is merely a convenient approximate expression that is useful but does not constitute a fundamental rule of Nature. Unfortunately, historically many such expressions have been labeled "laws."

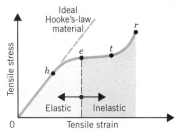

Figure 11.3 A schematic stress-strain graph for a hypothetical material. The point h denotes the limit of validity for Hooke's law, and r represents the point at which the material ruptures.

line (dashed line) for all stress values, but for real materials, deviations occur for large stress values.

Let us examine the details of the stress-strain graph in Fig. 11.3. Between the origin O and the point h, the strain is directly proportional to the stress—this is the region of validity of Hooke's law. Between points h and e, the stress-strain relationship is no longer linear, but it is reversible; that is, when the stress is removed, the material returns to its original length. For any stress up to the value corresponding to point e, the material is said to be *elastic*. For any greater stress, the material is permanently deformed and it will not return to its original condition when the stress is removed. Accordingly, the point e is called the *elastic limit* of the material. Notice that between points e and t the strain increases (i.e., the material elongates) even though there is almost no change in the applied stress. This means that the material is flowing plastically in a way that is not reversible. The point t is called the *tensile yield point* and is usually defined to be the point at which the permanent deformation is not sufficiently large (perhaps 0.2 percent) to have seriously degraded the performance of the material. (Thus, the distance between the points e and t in Fig. 11.3 has been greatly exaggerated.) Finally, point r corresponds to the stress at which the substance ruptures and is called the *tensile strength* of the material.

Figure 11.3 shows only the tensile region of the stress-strain relationship. Generally, a metallic substance, either in bulk form or wound as a spring, will have the same Hooke's-law force constant under compressive stress as under tensile stress. That is, the slope of the stress-strain line between points O and h in Fig. 11.3 is the same for compression as for tension. On the other hand, composite or heterogeneous materials, such as wood, concrete, bone, or plastic, will usually exhibit quite different properties under compression and tension. These materials will show different slopes for the stress-strain line

(as we will see, this means different values of Young's modulus), different elastic limits, and different rupture points for the two modes of stress application.

Other Types of Stresses

In addition to tensile and compressive stresses, materials are often subjected to more complex combinations of forces that produce stresses with different characteristics. Figure 11.4a shows a force F applied to the top of a block whose base is fixed. (Of course, this means that the base experiences a force F directed to the left.) This type of stress is called *shear* and tends to distort the object in the manner shown by the dashed lines.

In Fig. 11.4b we see a twisting force that produces a *torsional stress* on the rod. This torsional stress is very similar to shear, as can be realized by examining the forces on any small segment of the rod. We will not have occasion to deal further with shear or torsional stress.

Elastic Moduli

If a material obeys Hooke's law, the stress-strain relationship is linear and the ratio of the stress, F/A, to the strain, $\Delta \ell/\ell$, is a constant:

$$\frac{\text{stress}}{\text{strain}} = \frac{F/A}{\Delta \ell/\ell} = \text{constant} = Y \qquad (11.4)$$

The quantity Y is called *Young's modulus*, the dimensions of which are Nm^2. The slope of the linear portion of the stress-strain curve, Fig. 11.3, is equal to the Young's modulus. Some typical values of Young's modulus are listed in Table 11.1. It must be emphasized, however, that the elastic properties of a particular sample depend on the precise nature of the molecular or crystalline structure. For example, Table 11.2 gives the Young's modulus and some ad-

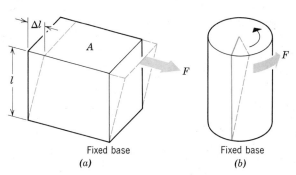

Figure 11.4 (a) The block is subjected to a *shearing stress*. (b) The rod is subjected to a *torsional stress*.

Table 11.1 Elastic Moduli (Approximate) for Some Materials

Substance	Young's Modulus, Y (N/m^2)	Bulk Modulus, B (N/m^2)
Aluminum	6.9×10^{10}	7.7×10^{10}
Copper	12	12
Iron, wrought	19	16
Lead	1.6	5.2
Mercury	—	2.5
Rubber	0.05	—
Steel	20	16
Water	—	0.22
Wood (pine)	0.5	—

ditional elastic information for four different types of aluminum, two that are alloys and two that are pure aluminum but with different preparation. Although the values of Y are not different, the other elastic parameters exhibit considerable variation. Notice that alloys of aluminum can have tensile strengths more than 10 times that of pure annealed aluminum and that they remain elastic over a much greater stress range.

● *Example 11.1*

A steel rod, 8 m in length and 12 mm in diameter, has one end attached to a rigid support. A 1200-kg block (about the mass of a small automobile) is suspended from the lower end. By how much will the rod stretch?

Solving Eq. 11.4 for $\Delta \ell$ and substituting mg for F, we have

$$\Delta \ell = \frac{F\ell}{AY} = \frac{mg\ell}{AY} = \frac{mg\ell}{\pi r^2 Y}$$

In Table 11.1 we find $Y = 20 \times 10^{10}$ N/m^2 for steel; therefore,

$$\Delta \ell = \frac{(1200 \text{ kg}) \times (9.8 \text{ m/s}^2) \times (8 \text{ m})}{\pi (6 \times 10^{-3} \text{ m})^2 \times (20 \times 10^{10} \text{ N/m}^2)}$$

$$= 4.16 \times 10^{-3} = 4.16 \text{ mm} \quad \blacksquare$$

A quantity closely related to Young's modulus is the *bulk modulus* B, which relates the *volume stress* on a substance to the *volume strain*. The volume stress is simply the pressure change ΔP, and the volume strain is the fractional change in volume. Thus,

$$\text{bulk modulus} = \frac{\text{pressure change}}{\text{volume strain}}$$

$$= -\frac{\Delta P}{\Delta V/V} = B \quad (11.5)$$

where the negative sign is necessary because the volume *decreases* when pressure is applied.

Values of the bulk modulus are listed in Table 11.1 for some substances.

● *Example 11.2*

A block of lead is placed in a container filled with liquid and is subjected to a (gauge) pressure of 800 atm. What is the fractional change in density of the lead?

First, from Table 1.9, we find $\rho(\text{lead}) = 11.3$ g/cm^3; and, from Table 11.1, we find $B(\text{lead}) = 5.2 \times 10^{10}$ N/m^2. Solving Eq. 11.5 for $\Delta V/V$, we have

$$-\frac{\Delta V}{V} = \frac{\Delta P}{B}$$

Also, using Eq. 9.9, we can write for the density change,

$$\frac{\Delta \rho}{\rho} = -\frac{\Delta V}{V}$$

Table 11.2 Some Elastic Properties of Types of Aluminum

Substance	Young's Modulus, Y (N/m^2)	Stress at Elastic Limit (N/m^2)	Tensile Strength (N/m^2)	Elongation at Rupture (percent)
Aluminum, pure annealed	6.9×10^{10}	1.2×10^7	4.7×10^7	49
Aluminum, pure, cold rolled	6.9	10.6	11.2	5
Typical aluminum alloy (Cu, Mg)	7.1	15	20	3
Aluminum alloy (5.5% Zn, 2.5% Mg, 1.5% Cu, 0.3% Cr, 0.2% Mn)	7.2	50	56	11

Combining these equations,

$$\frac{\Delta\rho}{\rho} = \frac{\Delta P}{B} = \frac{800 \times (1.013 \times 10^5 \text{ N/m}^2)}{5.2 \times 10^{10} \text{ N/m}^2}$$

$$= 0.0016 \quad \text{or} \quad 0.16 \text{ percent}$$

This represents a density *increase* of 1 percent.

● *Example 11.3*

Find the value of the bulk modulus for an ideal gas at normal atmospheric pressure and constant temperature.

From Eq. 11.5, we have

$$B = -\frac{\Delta P}{\Delta V/V}$$

For constant temperature, the ideal gas law becomes Boyle's law:

$$PV = \text{const.}$$

This relationship means that if P is *increased* by a small amount, then V is *decreased* by the same fractional amount; that is,

$$\frac{\Delta P}{P} = -\frac{\Delta V}{V}$$

Using this relation in the expression for B, we find

$$B = -\frac{\Delta P}{(-\Delta P/P)} = P$$

When P is equal to 1 atm,

$$B = 1.013 \times 10^5 \text{ N/m}^2$$

This analysis shows that the bulk modulus of an ideal gas at any pressure is always equal to that pressure. ∎

Elastic Properties of Biological Materials

Bone is a composite material consisting primarily of organic fibers (chiefly *collagen*), inorganic crystals (*hydroxyapatite*), cementing substances, and water. The viscoelastic properties of compact bone are due mainly to the protein (collagen) and mineral (hydroxyapatite or, usually, apatite) components. The mineral component accounts for about 70 percent of the bone mass and the protein component accounts for about 20 percent. The responses of these materials to applied stresses are quite different. Strength tests have been made, for example, on specimens of cortical bone from the tibias of steers. It was found that the mineral component has a compressive strength that is about 30 percent and a tensile strength that is about 5 percent of that of compact bone. On the other hand, the protein component has a compressive strength that is less than 0.1 percent and a tensile strength that is about 7 percent of that of compact bone. That is, the collagen contributes little to the compressive strength of bone but it is slightly more important than apatite in determining the tensile strength. This is plausible, because a protein fiber is a stringy structure and therefore has almost no compressive strength. A significant and interesting feature of the results of the strength tests is that the two bone constituents, which are individually weak, combine to give strengths that are comparable with those of metals.

Not only are the strengths of the protein and apatite components of bone quite different, but so are (as we might expect) the elastic moduli. The table at the top of p. 236 lists Young's modulus and the strength for compact bone and its components in compression and in tension (data from steer tibia). Notice that the tensile strength of compact bone is about the same as that for aluminum (Table 11.2).

	Young's Modulus (10^{10} N/m²)	Strength (10^7 N/m²)
Compression:		
Compact bone	1.02	14.7
Mineral component	0.64	4.4
Protein component	< 0.001	0.01
Tension:		
Compact bone	2.24	9.8
Mineral component	1.66	0.5
Protein component	0.02	0.7

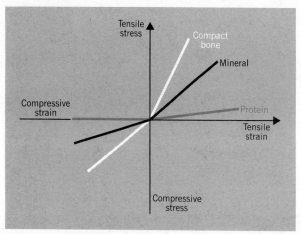

Figure 1 Schematic representation of the stress-strain relationships for compact bone and its components in tension and in compression.

Type of Femur	Tensile Strength (10^7 N/m²)	Compressive Strength (10^7 N/m²)
Human	12.4	17.0
Horse	12.1	14.5
Cow	11.3	14.7
Wild boar	10.0	11.8
Pig	8.8	10.0
Deer	10.3	13.3
Ostrich	7.1	12.0

Figure 1 shows a stress-strain graph for typical bone and its components. The differences between tensile and compressive effects are reflected in the different slopes of the lines in the two regions. (Remember, Young's modulus is equal to the inverse of the slope of the line.)

The tensile and compressive strengths (as well as Young's modulus) of bone vary only slightly from species to species, even though the structures of the animals are quite different. The table above lists the tensile and compressive strengths of the femur of man and of various animals. Notice that the maximum variation is less than a factor of 2. (These data refer to bones selected and processed in ways different from those in the previous table. Consequently, the values for human bone in the two cases are slightly different.)

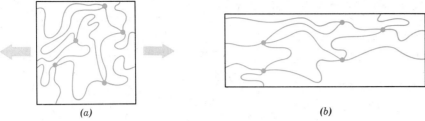

Figure 2 An elastomer consists of long, cross-linked molecules. When such a material is stressed *(a)*, the molecules extend until, at maximum extension, they are nearly parallel *(b)*.

Soft biological materials, such as tissue and muscle, have elastic properties that are distinctly different from those of bone and other hard materials. These soft materials resemble rubber much more than they resemble solid substances such as bone or near-solid materials such as cartilage. Soft elastic materials are often referred to as *elastomers*. The basic characteristic shared by these materials is that they are composed of long cross-linked molecules that can be stretched until they are nearly parallel (Fig. 2).

Elastomers differ from bony materials in three important respects:

1. Whereas bony materials exhibit stress-strain graphs that are linear over a considerable range of stresses (Fig. 1), elastomers tend to have S-shaped stress-strain graphs (Fig. 3).

2. Bony materials have values of Young's modulus near 10^{10} N/m². Elastomers, because of the effect in (1), have values of Y that vary with stress and are in the range 10^5 to 10^6 N/m².

3. The maximum strain that bony materials can withstand is about 0.01, that is, a change in length of about 1%. On the other hand, elastomers can be extended to two or three times their unstressed lengths without rupture.

References

R. McNeill Alexander, *Animal Mechanics,* University of Washington, Seattle, 1968, Chapters 3 and 4.

F. G. Evans, *Mechanical Properties of Bone,* Charles C. Thomas, Springfield, Ill., 1973, Chapter 9.

R. W. Stacy, D. T. Williams, R. E. Worden, and R. O. McMorris, *Essentials of Biological and Medical Physics,* McGraw-Hill, New York, 1955, Chapters 6 and 7.

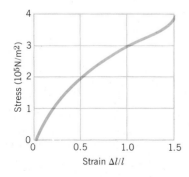

Figure 3 Stress-strain graph for contracted muscle fibers. Notice that the value of Young's modulus depends on the applied stress.

Exercises

1. When subjected to a compressive stress, the bone in the leg most likely to fracture is the tibia. At its narrowest point, the tibia has a cross-sectional area of about 3 cm². What is the maximum compressive force that both of these bones in the legs can withstand without fracture? (The compressive strength of the tibia is essentially the same as that of the femur.)

2. It has been determined that a human skull will be penetrated by a moving object with a surface area of a few cm² when the stress is about 5×10^7 N/m². Suppose that a hammer (mass = 2 kg, head diameter = 2.5 cm) is dropped from a height h and lands on a person's head with the face of the hammer head parallel to the skull surface. If the contact lasts 1 ms, what is the maximum value of h that will result in no skull penetration?

3. Estimate the variation in Young's modulus for muscle fiber by obtaining values of Y from Fig. 3 for stresses of 1, 2.5, and 3.5×10^5 N/m².

4. One of the results of the theory of cross-linked elastomers is that Young's modulus is given approximately by

$$Y = 3\rho RT/M$$

where ρ is the density, R is the universal gas constant, T is the absolute temperature, and M is the average mass (in kg/mol) of the piece of molecule between one cross-link position and the next. *Resilin* is a rubbery protein substance found in anthropods, serving various functions involving movement. The value of Young's modulus for resilin is approximately 1.8×10^6 N/m²; the dry density of the protein component (the effective component) of resilin is about 0.5 g/cm³. Use this information to estimate the molecular mass between cross links in resilin. The average molecular mass of an amino acid unit is about 90 u. How many such units are there between cross links? Does your result seem reasonable?

11.2 Simple Harmonic Motion

The Equation of Motion

The vibratory motion of a mass attached to a spring is characteristic of a large and important class of oscillatory phenomena called *simple harmonic motion* (SHM). Figure 11.5a shows a block at rest in its equilibrium position on a frictionless surface. If we apply an external force to displace the block to the *right*, there will be a restoring force **F** exerted on the *block by the spring* and this force is directed to the *left*, as shown in Fig. 11.5b. Note that this force F is the reaction force to the one appearing in Eq. 11.1. We assume that the displacement of the block by the amount x_0 has not carried the spring beyond the range of validity of Hooke's law. Then, we can write the connection between the force and the displacement in the standard form of Hooke's law. But in the discussion of SHM, we are usually interested in the *restoring* forces, not in the *applied* forces. Therefore,

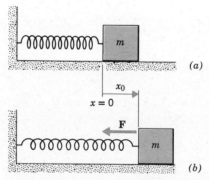

Figure 11.5 (a) A mass m rests on a frictionless surface and is attached to a spring. The mass and spring combination is in its normal (equilibrium) condition. (b) If m is displaced to the right an amount x_0 (by an external force), there will be a restoring force to the left given by $F = -kx_0$, where k is the force constant characteristic of the particular spring.

to emphasize that the direction of the displacement is opposite to that of the restoring force, we write Hooke's law with a negative sign:

$$F = -kx \qquad (11.6)$$

When the extension is to the left of the equilibrium position, the restoring force acts toward the right; that is, F has a positive value, whereas x is negative, and Eq. 11.6 is still valid.

We know from Newton's equation of motion that the net force on an object must equal the product of its mass and its acceleration. Therefore, at any extension x,

$$F = ma = -kx$$

so that

$$a = -\frac{k}{m}x \qquad (11.7)$$

This is the basic *equation of motion* for an object undergoing *simple harmonic motion*.

A Graphical Solution

If the block in Fig. 11.5b is released from its position of initial extension ($x = x_0$), it will be accelerated to the left by the restoring force. At $x = x_0$ the acceleration is a maximum: as the block moves toward $x = 0$ the velocity increases while the acceleration decreases. When the block reaches $x = 0$, the restoring force (and, hence, the acceleration) will have decreased to zero, but the velocity of the block will have been increased to its maximum value at this point and the inertia of the block will carry it into the region of negative x (to the *left* of $x = 0$). In this region the spring is compressed. The restoring force (and, hence, the acceleration) is directed to the *right* and the block will be slowed down. At $x = -x_0$, the motion will stop and the acceleration (which is still toward the right) will cause the block to reverse its motion and move toward $x = x_0$ again. The entire process is one of *cyclic* (or *oscillatory* or *periodic*) motion, with the block vibrating back and forth between $x = x_0$ and $x = -x_0$.

We can obtain a record of the motion of the block as a function of time in the following simple way. As shown in Fig. 11.6, we attach a pen to the block and allow it to touch a roll of paper that is moved uniformly in a direction perpendicular to the direction of motion of the block. In this way we obtain a displacement-time graph of the motion. Examination of the graph shows that it is a cosine curve of the form[2]

---

[2]$x(t)$ means "the displacement x as a function of time t."

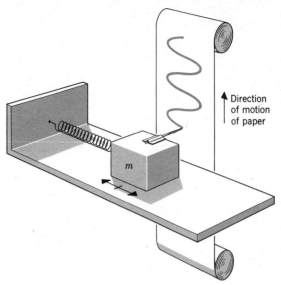

Figure 11.6 A simple method for recording the motion of an oscillating block as a function of time.

$$x(t) = x_0 \cos \frac{2\pi}{\tau} t \qquad (11.8)$$

where τ is the *period* of the motion, that is, after every interval of time τ, the motion repeats itself. The quantity x_0 is the *amplitude* of the motion, the maximum excursion from the equilibrium position experienced by the block. A sinusoidal function (sine or cosine) varies in a simple and regular way (i.e., the variation is *harmonic*), and therefore motion described by such functions is termed *simple harmonic motion*.

Equation 11.8 describes an oscillatory motion that continues indefinitely without any change in amplitude. This is the ideal case in which frictional losses are ignored. Of course, in a real situation, friction will be present and the motion will eventually cease. We will briefly discuss this case of *damped* motion at the end of this chapter.

The Connection between Circular Motion and SHM

Suppose that we have a particle executing uniform circular motion. For example, consider a ball attached to one end of a string and moving in a circular path, as in Fig. 11.7. The motion of the ball is described by the angle θ between some fixed line and the line connecting the ball and the center of rotation. The motion is *uniform* when θ increases linearly with the time, that is, when

240 Elasticity and Vibrations

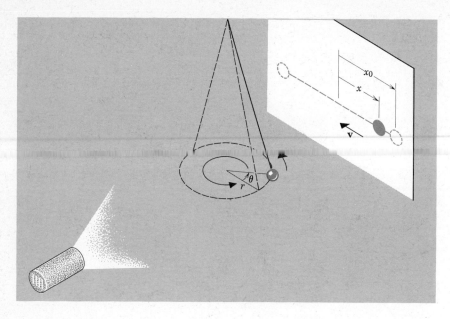

Figure 11.7 The projection of *circular motion* is *simple harmonic motion*.

$$\theta = \omega t \qquad (11.9)$$

where ω is the angular velocity (Eq. 2.31).

Now, suppose we examine the *projection* of this circular motion on a surface that is perpendicular to the plane of motion of the ball. One simple way to visualize this projection is shown in Fig. 11.7. A light source is directed toward the revolving ball and a shadow of the ball is cast on the screen behind. As the ball moves in its circular path, the shadow moves back and forth along a straight line. The shadow actually executes SHM. We can see that this is the case by analyzing the projection of the ball's motion on a diameter of its path. This situation is illustrated in Fig. 11.8a. Here, the position of the particle is described by the vector **r**, and the projection of **r** on the diameter is x. The radius of the circular path is r and this corresponds to the maximum projected displacement x_0 (the *amplitude* of the motion). Then, we can write

$$x = x_0 \cos \theta \qquad (11.10)$$

Notice that we take x to be *positive* when the displacement is to the right; when the displacement is to the left, $\cos \theta < 0$ and the necessary negative sign is automatically supplied. Using Eq. 11.9, we can write Eq. 11.10 as

$$x(t) = x_0 \cos \omega t \qquad (11.11)$$

Note that x_0 is both the value of the displacement at $t = 0$ and the amplitude of the motion as well. From Eq. 2.32 we know that the period τ of the motion is $\tau = 2\pi/\omega$. Therefore, Eq. 11.11 is exactly the same as Eq. 11.8. That is, the projection of circular motion is simple harmonic motion.

We can extend this type of analysis and obtain expressions for the velocity and the acceleration of the projected motion. Figure 11.8b shows the geometrical construction to find the velocity. The veloc-

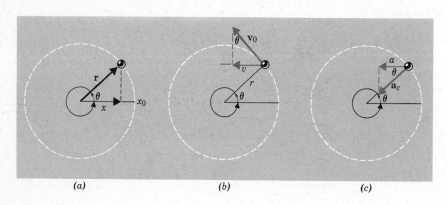

Figure 11.8 The projected displacement (a), velocity (b), and acceleration (c) of a particle undergoing uniform circular motion.

ity vector of the particle moving in the circular path is indicated by $\mathbf{v}_0$ and the magnitude of the projection is v. The maximum value of v is equal to v_0 and occurs when $\theta = 90°$. Thus, v_0 is the velocity amplitude and we have

$$v = -v_0 \sin \theta \qquad (11.12)$$

where the negative sign is included to give the correct direction of the velocity with respect to the displacement. From Eq. 2.33 we know that $v_0 = r\omega = x_0\omega$. Therefore, we can express v as

$$v(t) = -x_0 \omega \sin \omega t \qquad (11.13)$$

Finally, we can obtain an expression for the projection of the acceleration by referring to Fig. 11.8c. The only acceleration that the particle experiences is the centripetal acceleration $\mathbf{a}_c$. The projection is

$$a = -a_c \cos \theta \qquad (11.14)$$

where the negative sign again is included to give the proper relationship between the directions of the acceleration and the displacement. The centripetal acceleration is related to the velocity and the radius according to (Eq. 2.34).

$$a_c = \frac{v_0^2}{r} = \frac{(x_0\omega)^2}{x_0} = x_0\omega^2 \qquad (11.15)$$

Using this result in Eq. 11.14, we have

$$a(t) = -x_0 \omega^2 \cos \omega t \qquad (11.16)$$

The expressions for the displacement, velocity, and acceleration in SHM are summarized in Table 11.3.

Table 11.3 Important Quantities in SHM

Displacement	$x(t) = x_0 \cos \omega t$
Velocity	$v(t) = -v_0 \sin \omega t$
	$\quad\ = -x_0 \omega \sin \omega t$
Acceleration	$a(t) = a_c \cos \omega t$
	$\quad\ = -x_0 \omega^2 \cos \omega t$
Angular velocity (angular frequency)	$\omega = \sqrt{\dfrac{k}{m}}$
Period	$\tau = \dfrac{2\pi}{\omega} = 2\pi\sqrt{\dfrac{m}{k}}$
Frequency	$\nu = \dfrac{1}{\tau} = \dfrac{\omega}{2\pi} = \dfrac{1}{2\pi}\sqrt{\dfrac{k}{m}}$

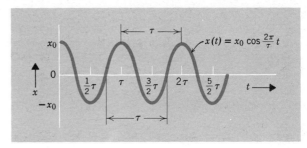

Figure 11.9 Displacement as a function of time for a particle undergoing simple harmonic motion. There is a time interval τ (the period) between any two successive corresponding points on the curve.

Period and Frequency

In SHM the quantity τ is the *period* of the motion; that is, τ is the period of time required for the system to proceed through one complete cycle of its repetitive motion. Figure 11.9 shows graphically the displacement as a function of time for a particle undergoing SHM. Notice that τ can be defined as the time interval between *any* two successive corresponding points on the sinusoidal curve.

We also find it convenient to define and use the term *frequency* when describing SHM of various types. The frequency of the motion indicates how frequently the oscillations repeat themselves. If there are 100 complete cycles of the motion each second, we say that the frequency is $\nu = 100$ cycles per second. But "cycles" is not a physical unit, so in terms of physical units only, we would write $\nu = 100 \text{ s}^{-1}$. The unit of frequency is given the special name *hertz* (Hz) in honor of the German physicist Heinrich Hertz (1857–1894) who made great contributions to the study of electrical oscillations; thus, 1 cycle per second = 1 Hz.

If the frequency of a particular vibration is 100 Hz, this means that each complete cycle requires 1/100 s. That is, the frequency ν and the period τ are related according to

$$\nu = \frac{1}{\tau} \qquad (11.17)$$

Frequency is also measured in terms of the quantity ω which we call *angular velocity* or *angular frequency*. Whereas ν gives the number of complete vibrations per second, ω gives the number of radians through which the angle θ (Fig. 11.8) progresses each second. Because each complete vibration corresponds to an increase in θ of 360° or 2π radians, we see that

$$\omega = 2\pi\nu \qquad (11.18)$$

The angular frequency ω is measured in radians per second or s^{-1}; the unit Hz is usually reserved for measuring ν and is not used for ω.

Now, let us look again at the acceleration of a particle undergoing SHM. We have two expressions for the acceleration—Eqs. 11.7 and 11.16. If we equate these expressions and use Eq. 11.11 for x, we find

$$a = -x_0 \omega^2 \cos \omega t = -\frac{k}{m}x = -\frac{k}{m}x_0 \cos \omega t \quad (11.19)$$

From the second and fourth parts of this equation, we obtain the connection between the frequency ω and the mechanical properties of the system described by k and m, that is, $\omega^2 = k/m$. Thus,

$$\omega = \sqrt{\frac{k}{m}} \quad (11.20)$$

We can now also write

$$\tau = \frac{2\pi}{\omega} = 2\pi \sqrt{\frac{m}{k}} \quad (11.21)$$

The expressions for ω and τ are also summarized in Table 11.3.

● *Example 11.4*

A spring hangs vertically from a fixed support. When a 2.5-kg block is attached to the lower end, the spring extends by 15 cm. The block and spring are then placed on a frictionless table, as in Fig. 11.5. The spring is now stretched a distance of 8 cm from its equilibrium position and released from rest. Find k, ω, τ, ν, and v_0.

First, we find the spring constant by using the information relating to the block hanging from the spring:

$$k = \frac{F}{x} = \frac{mg}{x} = \frac{(2.5 \text{ kg}) \times (9.80 \text{ m/s}^2)}{0.15 \text{ m}} = 163 \text{ N/m}$$

Next,

$$\omega = \sqrt{\frac{k}{m}} = \sqrt{\frac{163 \text{ N/m}}{2.5 \text{ kg}}} = 8.07 \text{ rad/s} = 8.07 \text{ s}^{-1}$$

$$\tau = \frac{2\pi}{\omega} = 2\pi \sqrt{\frac{m}{k}} = 2\pi \sqrt{\frac{2.5 \text{ kg}}{163 \text{ N/m}}} = 0.778 \text{ s}$$

$$\nu = \frac{1}{\tau} = \frac{1}{0.778 \text{ s}} = 1.29 \text{ Hz}$$

$$v_0 = x_0 \omega = (0.08 \text{ m}) \times (8.07 \text{ s}^{-1}) = 0.646 \text{ m/s} \quad \blacksquare$$

Energy in SHM

We often refer to a system that undergoes SHM as a *simple harmonic oscillator*. In an ideal oscillator there are no frictional losses, so energy is conserved. At any instant of time, we have

$$\text{total energy} = \mathcal{E} = PE + KE = \text{const.} \quad (11.22)$$

From the result of Example 7.3, we know that an amount of work $\frac{1}{2}kx^2$ (supplied by an outside agency) is required to extend the spring by an amount x from its equilibrium position. The potential energy of the simple harmonic oscillator is just the energy stored in the spring, so that

$$PE = \tfrac{1}{2} kx^2 \quad (11.23)$$

The kinetic energy of the oscillator is the motional energy, so

$$KE = \tfrac{1}{2} mv^2 \quad (11.24)$$

Therefore,

$$\mathcal{E} = \tfrac{1}{2} kx^2 + \tfrac{1}{2} mv^2 = \text{const.} \quad (11.25)$$

As the oscillator moves, the values of x and v change. That is, energy shifts back and forth between PE and KE. At the extremes of the motion, $x = \pm x_0$ and the velocity is zero; thus, the energy is entirely PE. Similarly, when the oscillator passes through its equilibrium position, $x = 0$, the PE is zero and the velocity is maximum, so the energy is entirely KE. Thus, when $x = \pm x_0$, $\mathcal{E} = \frac{1}{2} kx_0^2$, and when $x = 0$, $\mathcal{E} = \frac{1}{2} mv_0^2$. That is,

$$\mathcal{E} = \tfrac{1}{2} kx_0^2 = \tfrac{1}{2} mv_0^2 \quad (11.26)$$

Solving for the relationship between x_0 and v_0, we find

$$v_0 = \pm x_0 \sqrt{\frac{k}{m}} \quad (11.27)$$

where a positive or a negative sign must be chosen for the square root in order to indicate the direction of the velocity relative to the direction of the initial displacement. (The mass can pass through $x = 0$ moving in *either* direction; the magnitude of v_0 will be the same for each case, but the *signs* will be different.)

Notice that we obtain the same relationship from our previous analysis where we found $v_0 = x_0 \omega$ and $\omega = \sqrt{k/m}$.

The Simple Pendulum

The results we have obtained for the oscillations of a block attached to a spring are not unique to this particular case. There are many other situations in which the motion is simple harmonic. In fact, almost all oscillatory motions are approximately simple harmonic, *if the amplitude of the motion is sufficiently small.*

The oscillatory motion of a simple pendulum will be SHM if we can find a force constant for the system that relates the restoring force and the displacement. According to Eq. 11.6, the force constant is given by

$$k = -\frac{\text{restoring force}}{\text{displacement}} \quad (11.28)$$

As shown in Fig. 11.10, there are two forces acting on the pendulum bob—the downward gravitational force, $\mathbf{F}_G = m\mathbf{g}$, and the tension in the suspension string, $\mathbf{T}$. The net force acting on the bob is $\mathbf{F}_{net} = \mathbf{F}_G + \mathbf{T}$, the magnitude of which is $mg \sin\theta$; this is the *restoring force*. The *displacement* x is the distance through which the bob moves along the arc starting at A; that is, $x = \ell\theta$, where θ is measured in radians. Then, Eq. 11.28 becomes

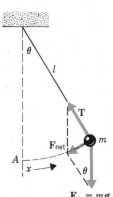

Figure 11.10 A sample pendulum of length l. The net force on the pendulum bob is the vector sum of the gravitational force $\mathbf{F}_G$ and the tension in the string $\mathbf{T}$.

$$k = \frac{mg \sin\theta}{\ell\theta} = \frac{mg}{\ell}\left(\frac{\sin\theta}{\theta}\right) \quad (11.29)$$

where we have cancelled the negative sign by taking into account the fact that the displacement and the restoring force have *opposite* directions.

Now, if θ is small, $\sin\theta \cong \theta$ (see Fig. 11.11). Then, $(\sin\theta)/\theta \cong 1$ so that

$$k \cong \frac{mg}{\ell} \quad (11.30)$$

With this expression for the force constant k, we can proceed to obtain expressions for τ and ω:

$$\tau = 2\pi\sqrt{\frac{m}{k}} = 2\pi\sqrt{\frac{m}{mg/\ell}}$$

so that

$$\tau = 2\pi\sqrt{\frac{\ell}{g}} \quad (11.31)$$

Also,

$$\omega = \sqrt{\frac{k}{m}} = \sqrt{\frac{mg/\ell}{m}} = \sqrt{\frac{g}{\ell}} \quad (11.32)$$

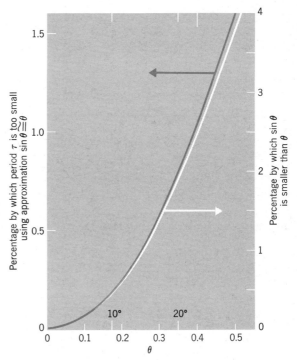

Figure 11.11 Percentage errors that result from the approximation, $\sin\theta \cong \theta$.

In the approximation that the angular amplitude of the motion is small so that $\sin\theta \cong \theta$, the period τ and the angular frequency ω are both *independent of the amplitude of the motion*. Figure 11.11 shows the percentage error in the period of a simple pendulum introduced by using the approximation $\sin\theta \cong \theta$. Notice that the percentage error in τ is only about $\frac{1}{4}$ the percentage error in $\sin\theta$ itself. The error in the period is less than 0.1 percent if θ is less than $7°$.

• *Example 11.5*

What must be the length of a simple pendulum for which each swing requires 1 s at the surface of the Earth? (Such a pendulum is called a "seconds pendulum.")

If each swing requires 1 s, the period of the pendulum is 2 s. Then, squaring Eq. 11.31 and solving for the length ℓ,

$$\ell = \frac{\tau^2 g}{4\pi^2} = \frac{(2\text{ s})^2 \times (9.80\text{ m/s}^2)}{4\pi^2} = 0.9929\text{ m}$$

Thus, the length is approximately 1 m. ∎

Damped Oscillations

We have been discussing ideal, friction-free oscillators. In such oscillators, there are no energy losses and the total energy, $\mathscr{E} = PE + KE$, remains constant. Real oscillators, on the other hand, *do* have friction and the initial energy of the oscillator is eventually dissipated (primarily as heat) and the motion finally ceases. The graph of displacement versus time for frictionless oscillations is sinusoidal, as shown in Fig. 11.9. When friction is introduced, energy is continually removed from the system. For example, if a block vibrates up and down at the end of a spring while immersed in some viscous fluid (see Fig. 11.12), energy is transferred from the oscillating block to the fluid. Consequently, the maximum amplitude of the motion attained during each cycle

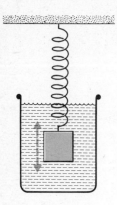

Figure 11.12 If a block vibrates up and down while immersed in a viscous fluid, energy is transferred from the block to the fluid and the motion is damped.

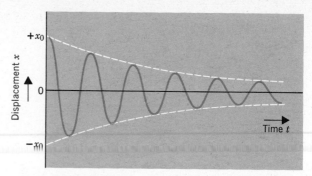

Figure 11.13 Displacement versus time for a moderately damped oscillator.

decreases with time; the maximum velocity (i.e., the velocity as the system passes through its equilibrium point) likewise decreases with time. This type of motion is called a *damped* oscillation.

Figure 11.13 shows the displacement or amplitude of a damped oscillator as a function of time. If the amplitude decreases by only a small amount during each cycle, the motion is said to be *lightly damped*. In such a case, the time interval between successive passages of the motion through the equilibrium position is almost exactly the same as for the same oscillator in the absence of damping. For oscillators that are more heavily damped, the frequency departs more severely from the undamped frequency. If the damping is sufficiently heavy, the system does not execute even a single oscillation; instead, it simply decreases from its original displacement to zero displacement without ever reversing its direction of motion. This case is called *overdamping*, and no frequency is defined.

Resonance

Every pendulum or block on a spring or any other type of vibratory system has a natural frequency at which it will oscillate freely. We call this the *characteristic frequency* of the system. A simple pendulum has a single characteristic frequency, but a more complex system, such as a metal bar or plate, may vibrate at several characteristic frequencies.

Consider a simple oscillator that consists of a block attached to a spring. This system has a single characteristic frequency, $\omega_0 = \sqrt{k/m}$. Now, suppose we apply to this oscillator a sinusoidal driving force with a frequency ω. If ω is quite different from ω_0, the driving force will seldom be in the same direction as the natural motion of the oscillator. A driving force with such a frequency will never be in step with the oscillator for a time long enough to cause the oscillator to vibrate with a very large amplitude. On the other hand, suppose that ω is close to ω_0.

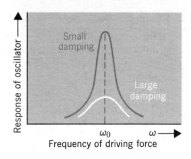

Figure 11.14 The amplitude response of an oscillator to a driving force with frequency ω. The characteristic frequency of the oscillation is ω_0.

Now, the driving force tends to act in concert with the natural motion of the system. When the oscillator block moves toward the right, it receives a push in this same direction from the driving force. Then, when the block begins to move toward the left, the driving force has also reversed direction and again pushes the block along in its direction of motion. When $\omega = \omega_0$, the driving force imparts the maximum energy to the oscillator; we call this the *resonance* condition.

If an oscillator were completely friction-free, a resonant driving force would cause an increase in amplitude with each vibration. The result would be that the amplitude of the oscillator (and, hence, its energy) would increase without limit. But, of course, every real oscillator is subject to a certain amount of frictional damping. Generally, the rate at which the damping removes energy from the system increases with the amplitude. Therefore, as the amplitude grows due to the input of energy from the driving force, the rate at which energy is dissipated as friction also increases. Eventually, a condition will be reached in which the rate of energy delivered by the driving force is just equal to the rate of frictional energy loss. At this point, the oscillator will have its maximum amplitude. If the oscillator is damped more severely, the magnitude of the maximum amplitude will be correspondingly reduced. Also, if the frequency of the driving force is not exactly equal to ω_0, the vibration amplitude will be less than the maximum. These characteristics of resonance phenomena are illustrated in the schematic response curves in Fig. 11.14.

We will return to the discussion of resonance phenomena when we study electrical oscillations (Chapter 14) and sound waves (Chapter 15.)

Summary of Important Ideas

The *elasticity* of a solid is due to the electric forces among the constituent atoms and molecules that act to restore the object to its original shape when it is deformed.

If the deformation of an object (e.g., the amount by which a spring is stretched) is directly proportional to the deforming force, the object is said to obey *Hooke's law*: $F = kx$.

The *stress* applied to a material is equal to the force per unit cross-sectional area, F/A. If the force tends to stretch the material, the stress is a *tensile* stress; if the force tends to compress the material, the stress is a *compressive* stress.

The *strain* that a stressed material experiences is equal to its frictional change in length $\Delta \ell/\ell$.

If a material obeys Hooke's law, the ratio of stress to strain, which is called *Young's modulus Y*, is constant.

The *bulk modulus* of a material is equal to the ratio of the applied pressure to the fractional change in volume: $B = -\Delta P/(\Delta V/V)$.

When a material or system that obeys Hooke's law (e.g., a block attached to a spring) is displaced from its equilibrium position and released, the displacement varies sinusoidally with time, and the system is said to execute *simple harmonic motion* (SHM).

In SHM, the displacement, the velocity, and the acceleration all vary sinusoidally with time.

The back-and-forth motion of a *simple harmonic oscillator* is exactly the same as the *projection* on a diameter of the motion of a particle traveling uniformly in a *circular* path.

The *period* (and also the *frequency*) of a simple harmonic oscillator is independent of the amplitude of the motion.

The *energy* of a simple harmonic oscillator shifts between being entirely *PE* to being entirely *KE*, with $PE + KE$ = constant.

A *simple pendulum* is approximately a simple harmonic oscillator. The approximation becomes better as the angular amplitude of the motion is made smaller.

When friction is present in an oscillating system, the motion is *damped*. The energy is dissipated (usually as heat), and the motion eventually ceases.

When an oscillator with a characteristic frequency ω_0 is acted on by a sinusoidal driving force whose frequency is ω, the amplitude of the oscillations will become large if ω is near ω_0. When $\omega = \omega_0$, the oscillator is in *resonance* with the driving force.

◆ Questions

11.1 A beam is supported at both ends and is loaded at the middle. Show with a sketch that the upper part of the beam is under compression, whereas the lower part is under tension.

11.2 Speculate on the reason that the stress-strain graph of a material might have different shapes in the tensile and compressive regions of the diagram. (The compressive region, which is not shown in Fig. 11.3, lies below and to the left of the origin O.)

11.3 How do you suppose the tensile strength of concrete compares with its compressive strength? Why do you suppose reinforcing rods are incorporated into concrete structures?

11.4 A single crystal of a substance consists of a large number of atoms arranged in a particular repeating geometrical pattern. Do you suppose it would be possible to specify the tensile properties of such a substance by means of a single number (Young's modulus)? Explain. (Do you expect the stress-strain properties of a crystal to be the same in all directions?)

11.5 Give some examples (other than those mentioned in the text) of oscillations that are simple harmonic oscillations (or are approximately so).

11.6 When the science of mechanics was first being developed, the acceleration due to gravity g was most easily determined by measurements of the periods of simple pendula instead of measurements on freely falling objects. Why do you suppose that this was the case? Is it the case *now*?

11.7 A clock is regulated by the swinging of a pendulum. At the bottom of the pendulum there is an adjusting screw (often a nut on a fixed, threaded screw). In which direction (up or down) must the nut be moved in order to correct the clock if it runs too slowly? Explain.

11.8 Make a sketch of *PE* and *KE* versus θ for a simple pendulum.

11.9 Two identical simple pendula are suspended from the same point. One is displaced 2° to the left and the other is displaced 4° to the right. The pendula are released at the same instant. At what point do they collide? Explain.

11.10 A pendulum clock is adjusted to keep correct time in Atlantic City, N.J. The clock is then moved to Denver, Col. How will the accuracy of the time keeping be affected?

11.11 The springs in an automobile suspension have a certain characteristic oscillation frequency. Suppose that the automobile is driven over a "washboard" road. Is there any particular speed that the driver should avoid? Explain.

★ Problems

Section 11.1

11.1 What force is required to stretch a steel rod 3 mm in diameter) by 0.15 percent of its original length?

11.2 A copper tube 5 m in length has an outside diameter of 1.4 cm and an inside diameter of 1.2 cm. By how much will this tube stretch if a force of 1200 N is applied along its axis?

11.3 What is the maximum mass that can be suspended from a 2-cm diameter vertical steel rod without exceeding the elastic limit of the material (8.3×10^8 N/m^2)?

11.4 A rod of cold rolled aluminum with a length of 3.5 m and a diameter of 12 mm is stretched until it ruptures. What is the length just before rupture occurs?

11.5 Express the Hooke's-law force constant k, defined by Eq. 11.6, in terms of Young's modulus for the material.

11.6 A copper wire has a diameter of 1 mm and a length of 40 m. If one end of this wire is fixed and a 60-kg block is suspended from the other end, by how much will the wire stretch?

11.7 A pure, cold rolled aluminum rod has a diameter of 1 cm. What is the maximum mass that could be suspended from this rod without causing it to break?

11.8 A wire has a length ℓ, a cross-sectional area A, and a Young's modulus Y. Derive an expression for the work done in stretching this wire by an amount $\Delta \ell$.

11.9 The *compressibility* κ of a substance is the reciprocal of the bulk modulus, $\kappa = 1/B$. Show that $\kappa = (1/\rho)(\Delta\rho/\Delta P)$, where ρ is the density of the substance.

11.10 Use the result of the previous problem and the information in Table 11.1 to determine the change in density of water when it is subjected to a pressure of 1000 atm (the pressure at the bottom of a deep ocean trench).

11.11* When a cylindrical rod is bent into a circle, the length of the rod measured along its axis does not change. The outer circumference, however, is greater than the original length, so this region is subject to a tensile stress. Suppose that a rod made from an aluminum alloy (last entry in Table 11.2) has a diameter of 2 cm and is to be bent into a circle with a radius R. How small can R be if the elastic limit of the material is not to be exceeded? (Hint: Find the strain on the outer edge and set this equal to the allowed stress divided by Young's modulus.)

11.12 A copper block has the form of a cube with sides 10 cm. Two opposite faces are subjected to equal compressive forces $F = 3 \times 10^4$ N. Assuming Young's modulus to be the same in compression as in tension, find both the stress and the strain. What is $\Delta \ell$?

11.13 A lead cube with sides 10 cm has one face attached to a fixed inelastic surface while the opposite face is subjected to a shear force of $F = 4 \times 10^4$ N; see Fig. 11.4a. If the shear modulus (the equivalent of Young's modulus in Eq. 11.4 when this equation is applied to shear) is 0.54×10^{10} N/m², find both the stress and the strain. What is $\Delta \ell$?

11.14 A solid copper cylinder with a radius of 1 cm and a length of 10 cm is tightly fitted lengthwise between two parallel flat fixed surfaces. The cylinder originally at a temperature of 20°C is heated to a temperature of 150°C. Find the stress and strain in the cylinder when hot. Assume Young's modulus to be the same in compression as in tension. Use Table 9.2.

Section 11.2

11.15 Refer to Fig. 11.8. The projections here are made on a diameter that corresponds to the *x*-axis. Repeat the analysis by making the projections on a diameter that corresponds to the *y*-axis. (The angle θ remains the same as in the previous analysis.) Show that the *y*-projected motion is also SHM.

11.16 The velocity of a simple harmonic oscillator is expressed by (in metric units) $v(t) = 9 \sin(3\pi t)$. What is the displacement of the oscillator as a function of time?

11.17 A particle moves according to the expression $a(t) = -32 \cos(4\pi t)$. Write expressions for $x(t)$ and $v(t)$.

11.18 A force of 4 N applied to a 0.1-kg mass that is attached to a spring displaces the mass 0.1 m from its equilibrium position. What maximum velocity will the mass attain after release? What maximum acceleration will the mass experience? What will be the period of the motion?

11.19 A mass attached to a spring vibrates with a period of 0.5 s. When the mass passes through its equilibrium position, it is moving with a velocity of 0.2 m/s. What is the amplitude of the motion?

11.20 A 10-g mass attached to a spring vibrates with a period of 2 s. How much force is required to stretch the spring by 10 cm starting from its equilibrium position?

11.21 A 0.2-kg mass vibrates at the end of a spring. When the mass passes through the equilibrium position, the velocity is 0.4 m/s. What amount of work was required initially to stretch the spring?

11.22 The motion of a particle is described by the expression $x(t) = 4 \cos(6\pi t)$. (The displacement is measured in meters and the time is measured in seconds.) Find the following quantities: x_0, v_0, ω, v, and τ.

11.23 The period of a certain simple harmonic oscillator is 4 s. At $t = 0$, the oscillator is at rest at a position 10 cm from its equilibrium position. Write expressions for $x(t)$, $v(t)$, and $a(t)$.

11.24* A particle moves in a circular path in the *xy*-plane. The radius of the path is 2 m and the speed is 6 m/s. Find the following quantities: ω, v, τ, $x(t)$, and $y(t)$.

11.25 Write Eq. 11.25 for the total energy of a simple harmonic oscillator and substitute the expressions for $x(t)$ and $v(t)$. Show that $\mathscr{E} = $ const.

11.26 A force of 6 N stretches a spring from its equilibrium position ($x = 0$) to $x = 15$ cm. How much work is required to stretch this spring from $x = 15$ cm to $x = 40$ cm?

11.27 A simple harmonic oscillator consists of a 2-kg block attached to a spring with a negligible mass. The period of the oscillations is 0.8 s. The oscillator is set into motion in such a way that it has an energy $\mathcal{E} = 15$ J. Find the amplitude x_0 and the maximum speed v_0.

11.28 What must be the force constant of a spring if a compression of 6 cm stores 20 J of potential energy?

11.29 What difference in length is there between a "seconds pendulum" constructed for use at the North Pole and one constructed for use at the Equator? (See Fig. 2.15.)

11.30 What is the length of a "seconds pendulum" that would be suitable for use on the surface of the Moon? (Refer to Example 11.5.)

11.31* A "seconds pendulum" is given an initial displacement of 25°. Assuming no energy loss, by how much will the time-keeping of the pendulum be in error after 1 h (1800 vibrations). (Use Fig. 11.11.)

11.32 A simple pendulum is set up in the stairwell of a building by suspending a weight from the end of an 80-ft wire. What is the period of the oscillations?

11.33* Some future astronauts take a 2-m simple pendulum to Mars. What do they find for the period? (The diameter of Mars is 6788 km and its density is 3.9 g/cm³.)

11.34 An experiment is performed to determine the value of g by measuring the period of a simple pendulum. For a pendulum with a length of 1.023 m, a period of 2.029 s is found. What is the value of g?

11.35 The amplitude of a lightly damped oscillator decreases by 3 percent during each oscillation cycle. What fraction of the energy of the oscillator is lost each cycle?

11.36 A 4-cm diameter hollow thin-walled tube is sealed at the bottom and filled with lead shot to give a total mass of 0.10 kg. The tube is floated in water; see diagram. Find the depth z. The tube is now pushed 1 cm further into the water and then released to bob up and down. Neglecting any damping, what is the nature of the motion? What is the period of the motion?

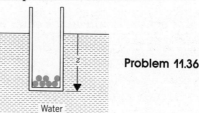

Problem 11.36

11.37* A pan with a mass of 0.50 kg hangs from a spring elongating it by 2 cm, see diagram. A blob of putty with a mass of 0.20 kg is dropped onto the pan from a height of 15 cm. The putty, on striking the pan, sticks to it. Describe the motion that ensues. What is the period of the motion? What is the amplitude of the motion and the maximum velocity of the motion? [Hint: Be sure to determine the pan-plus-putty equilibrium position and make use of the conservation of linear momentum.]

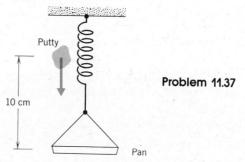

Problem 11.37

12
Electric Fields and Currents

We now begin a series of three chapters devoted to electric and electromagnetic phenomena. In this chapter, we first study the effects produced by electric charges at rest—this is the subject of *electrostatics*. We will then turn to a discussion of charges in motion, that is, electric *currents*. In the succeeding chapters we consider electromagnetism and electric circuits.

Basic to the study of electrostatics is the idea of the *field*, and this is the first subject of our discussion. Newton realized the fact that the gravitational force can act over vast regions of empty space without the benefit of any contact between material bodies to propagate the force. The same is true of the electric force. These forces were accordingly called *action-at-a-distance forces*. There was, however, no insight into the reason for this extraordinary character of the forces. Indeed, we still cannot explain *why* the forces of Nature act the way they do. But we do have a means, in field theory, of conveniently picturing the action-at-a-distance forces and of making calculations in the clearest and easiest way. We cannot explain *why* the forces act as they do, but we can describe *how* they act to a high degree of precision.

The motion of electric charges is a fundamental aspect of much that takes place in our world and, indeed, in the Universe as a whole. Moving charges in the form of electric currents perform countless tasks in our present-day electrically oriented society by operating all manner of electric motors, electronic circuits, and other devices. There is almost no area of modern society that does not depend in a crucial way on the effects of moving charges. In addition to these practical benefits, moving charges are also responsible for large-scale physical effects such as the Earth's magnetism and the surface dynamics of the Sun and other stars.

12.1 The Electrostatic Force

Positive and Negative Electricity

It has been known for about 200 years that in Nature there are two basic types of electricity which we designate *positive* and *negative*. The basic units of negative electricity are *electrons,* which constitute the outer portions of all atoms. The atomic cores, the nuclei, are the seats of positive electricity, the basic units of which are *protons*. All macroscopic matter is basically electrically *neutral*, because the magnitude of the negative electric charge carried by an electron is equal to that of the positive electric charge carried by a proton and all atoms in their natural states contain equal numbers of protons and electrons.

The distribution of electric charge is almost always accomplished by the movement of *electrons*; the more massive positively charged nuclei remain essentially stationary in almost all electrical processes. That is, a material is given a negative charge by the addition of excess electrons or a positive charge by the removal of electrons; the number of *atoms* is not changed in either case.

A basic property of electricity is that *like charges repel* and *opposite charges attract*. These facts can easily be demonstrated by some simple experiments. When a glass rod is rubbed with a silk cloth, the contact between the materials causes electrons to be transferred from the glass to the silk. Thus, the glass rod becomes *positively* charged. When a rubber rod is rubbed with a piece of fur, electrons are transferred to the rubber rod and it becomes *negatively* charged. If we suspend a pair of light-weight balls (such as styrofoam balls) on strings, as in Fig. 12.1, and touch each of the balls with a glass rod that has been rubbed with a silk cloth, the balls will both acquire positive charges and will repel each other. On the other hand, if one of the balls has been touched instead with a rubber rod that has been rubbed with fur, this ball would acquire a negative charge and would therefore be attracted to the other ball.

These experiments clearly establish the rules for attraction and repulsion. They also demonstrate that the glass rod carried a charge opposite to that carried by the rubber rod, but other experiments are necessary to show which was positively charged and which was negatively charged (according to the convention that electrons carry negative charge).

Certain types of materials (such as metals) have an interesting property—a fraction of the atomic electrons are not bound to any particular atom but are free to move about in the material. Such materials are termed *conductors*. If an electric charge is placed on such a material, it will quickly distribute itself. On the other hand, if an electric charge is placed on an *insulating* material, the charge will remain localized. Insulators (for example, glass, plastics) do not have free electrons and therefore electricity does not readily flow in such materials.

The Electroscope

The *electroscope* is a simple instrument for the detection of the presence of electric charge (but it cannot distinguish between positive and negative). An electroscope (Fig. 12.2) is simply a more sophisticated version of the charge "detector" that consists of a pair of suspended balls. Instead of styrofoam balls, two thin metallic sheets or leaves are used. If the electroscope is to be a sensitive detector of charge, the leaves must have very little mass so that, when charged, their displacement will be large. A common material for the leaves is gold, which can be pounded or rolled into extremely thin sheets. In order to eliminate the effects of air movements on the lightweight leaves, it is customary to enclose the leaves in a transparent container made of an insulating material (such as glass or clear plastic). The connection to the leaves is made by a thin conduction rod at the top (see Fig. 12.2). At the top of the rod there is usually mounted a large metal ball or plate. If an electric charge (of either sign) is placed on the leaves via the outside ball, the electrostatic repulsion will cause the leaves to separate; the greater the amount of charge, the greater will be the separation.

The easiest way to charge an electroscope is simply to touch the ball with a charged rod or other object. This is called charging by *conduction*. It is also possible to charge an electroscope by *induction* in which the charged rod does not make contact with the electroscope. The sequence of operations for charging by induction is shown in Fig. 12.2. We begin with an uncharged electroscope to which is attached a wire that is grounded (i.e., is connected to the Earth), as shown in Fig. 12.2a. Next, we bring a positively charged rod near the electroscope. As we see in Fig. 12.2b, electrons are drawn from the Earth, which acts as a giant reservoir, toward the positive charge. There is no charge on the leaves and they hang undeflected. In the third step, Fig. 12.2c, the grounding wire is removed while the charged rod remains near the electroscope. The electrons that had been drawn toward the positive charge remain on the electroscope ball. The electroscope

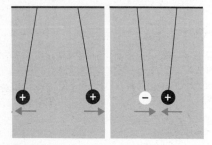

Figure 12.1 Like charges repel; opposite charges attract.

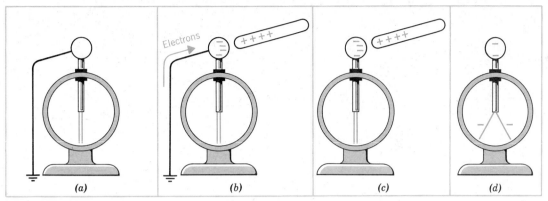

Figure 12.2 Sequence of steps in charging an electroscope by induction. The wire in (a) and (b) is connected to ground.

now carries a net charge, but the leaves are still uncharged and remain undeflected. Finally, as shown in Fig. 12.2d, the charged rod is removed, and the charge distributes itself between the ball and the leaves. The leaves are now charged (by *induction*) and stand apart.

The Conservation of Charge

An object can be given an electric charge by the transferral of electrons to or away from the object. Charge may be *lost* by the object but it is then *gained* by some other object. It is one of the fundamental conservation laws of Nature that *the total amount of electric charge in an isolated system remains constant.*

We shall see later that it is possible, under certain conditions and subject to certain restrictions, to *create* a pair of electric charges—a *negative* electron (i.e., an ordinary electron) and a *positive* electron (i.e., a *positron,* a particle identical in every respect to an electron except that it carries a *positive* charge). Although processes such as these are possible, they do not violate the law of charge conservation because the total charge of the created pair is *zero* so that the net charge of the system remains constant.

No electron or proton has ever been observed to be created or destroyed *by itself*. Creation and destruction processes *always* occur with positive-negative pairs of particles. No deviation from the law of charge conservation has ever been discovered.

Coulomb's Law of Electrostatic Force

In 1785, the French physicist Charles Augustin de Coulomb (1736–1806), extensively studied electrostatic forces using a sensitive torsion balance (Fig. 12.3) similar to that used by Cavendish in his investigation of the gravitational force constant (Fig. 6.3). From his measurements, Coulomb concluded that the electrostatic force between two charged objects varies as the inverse square of the distance between them.[1] That is, the electrostatic force has the same dependence on distance as does the gravitational

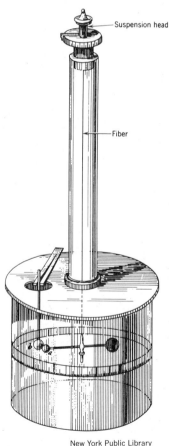

Figure 12.3 Coulomb's torsion balance for the study of electrostatic forces as it appeared in his 1785 paper. The force is measured between the charged balls *a* and *b*.

New York Public Library

---

[1]Coulomb concluded that the power of the distance was 2 with an accuracy of about 1 percent. Measurements made recently at Princeton University imply that the exponent is 2 to within 1 part in 10^{12}.

force. The electrostatic force is proportional to the product of the charges involved (again, of the same general form as the gravitational case with, of course, *charge* substituted for *mass*). Therefore, the expression for the electrostatic force law (also called *Coulomb's law*) is

$$F_E = K \frac{q_1 q_2}{r^2} \qquad (12.1)$$

where q_1 and q_2 are the charges on the two objects and r is their separation.

The electric force is a vector quantity and therefore has direction as well as magnitude. The direction of the force on one charge due to another charge can always be determined by examining the signs of the charges. If the signs are *different*, the force is *attractive* and the direction of the force on one charge is *toward* the other charge. If the signs are the *same*, the force is *repulsive* and the direction of the force on one charge is *away from* the other charge. Usually, we calculate only the magnitude of $\mathbf{F}_E$ (i.e., we do not carry the signs of q_1 and q_2); we determine the direction separately by comparing the signs of the charges.

Earlier, we learned that when using Newton's universal gravitational law, the entire mass of a uniform sphere could be considered to be concentrated at its center. This is a result of the $1/r^2$ nature of the gravitational force. Because Coulomb's law also depends on $1/r^2$, we can conclude that when using this law we can consider the entire charge of a uniformly charged sphere to be concentrated at its center.

Notice also in Eq. 5.11 that we have an "electrostatic force constant" analogous to G for the gravitational case. The units for the force and distance have already been specified, but those of charge and the constant K have not. Therefore, we have some flexibility in choosing the units for q and for K. We can elect to specify the size of the unit for charge in terms of the charge on the basic carrier of negative electric charge, the *electron*.

We call the unit of charge the *coulomb* (C) and in terms of this unit the magnitude of the electron charge is

$$e = 1.602 \times 10^{-19} \text{ C} \qquad (12.2)$$

(The electron charge is *negative*, and the symbol e represents only the *size* of this basic charge unit.)

The coulomb is a very large unit since it corresponds to an excess (or deficit) of approximately 6×10^{18} electrons. In practice, we usually deal with microcoulombs (μC) or millicoulombs (mC) of charge.

With the unit of charge defined in this way, the value of K is then determined. The value is

$$K = 8.99 \times 10^9 \text{ N-m}^2/\text{C}^2 \qquad (12.3)$$

That is, to calculate the electrostatic force in newtons between two charges by using Eq. 12.1, the charges must be given in coulombs, the separation must be given in meters, and K must be set numerically equal to 8.99×10^9 or approximately 9.0×10^9.

● *Example 12.1*

Calculate the electrostatic force *on q_1 due to q_2* (F_{12}) and the force *on q_2 due to q_1* (F_{21}) for the case illustrated in the figure.

$q_1 = +3 \times 10^{-3}$ C $q_2 = +5 \times 10^{-3}$ C

Using Coulomb's law (Eq. 12.1),

$$F_{12} = (8.99 \times 10^9) \times \frac{(3 \times 10^{-3}) \times (5 \times 10^{-3})}{(3)^2}$$

$$= 1.50 \times 10^4 \text{ N (to the } left\text{)}$$

$$F_{21} = (8.99 \times 10^9) \times \frac{(5 \times 10^{-3}) \times (3 \times 10^{-3})}{(3)^2}$$

$$= 1.50 \times 10^4 \text{ N (to the } right\text{)}$$

Both forces have the same magnitude (Newton's third law applied to electrostatic forces) and the force on each charge is *repulsive*.

● *Example 12.2*

Suppose that all of the electrons in a gram of copper could be moved to a position 0.3 m away from the copper nuclei. What would be the force of attraction between these two groups of particles?

The atomic mass of copper is 63.5 u. Therefore, 1 g of copper contains a number of atoms given by Avogadro's number divided by the mass of 1 mol (i.e., 63.5 g):

$$\text{no. atoms} = \frac{6.02 \times 10^{23} \text{ mol}^{-1}}{63.5 \text{ g mol}^{-1}}$$

$$= 0.92 \times 10^{22} \text{ atoms/g}$$

The atomic number of copper is 29; thus, each neutral copper atom contains 29 electrons. Therefore, the number of electrons in 1 g is

no. electrons in 1 g = 29 electrons/atom × $(0.92 \times 10^{22}$ atoms/g$) = 2.7 \times 10^{23}$ electrons/g

Thus, the total charge on the group of electrons is

$q_e = 2.7 \times 10^{23} \times (-e)$
$= 2.7 \times 10^{23} \times (-1.6 \times 10^{-19}$ C$)$
$= -4.3 \times 10^4$ C

A similar positive charge resides on the group of copper nuclei. Hence, the attractive electrostatic force is

$$F_E = K\frac{q_e^2}{r^2} = \frac{(9.0 \times 10^9 \text{ N-m}^2/\text{C}^2) \times (4.3 \times 10^4 \text{ C})^2}{(0.3 \text{ m})^2}$$

$$= 1.85 \times 10^{20} \text{ N}$$

This force is as great as the gravitational force between the Earth and the Moon! (See Example 6.1).

● *Example 12.3*

Compare the electrostatic and gravitational forces that exist between an electron and a proton.

The electrostatic force law and the gravitational force law both depend on $1/r^2$:

$$F_E = K\frac{q_1 q_2}{r^2}; \quad F_G = G\frac{m_1 m_2}{r^2}$$

Therefore, the ratio F_E/F_G is independent of the distance of separation:

$$\frac{F_E}{F_G} = \frac{Kq_1q_2}{Gm_1m_2}$$

For the case of an electron and a proton this becomes

$$\frac{F_E}{F_G} = \frac{Ke^2}{Gm_e m_p}$$

Substituting the values of the quantities, we find

$$\frac{F_E}{F_G} = \frac{(9.0 \times 10^9 \text{ N-m}^2/\text{C}^2) \times (1.6 \times 10^{-19} \text{ C})^2}{(6.67 \times 10^{-11} \text{ N-m}^2/\text{kg}^2) \times (9.11 \times 10^{-31} \text{ kg}) \times (1.67 \times 10^{-27} \text{ kg})}$$

$$= 2.3 \times 10^{39}$$

Thus, the electrostatic force between elementary particles is enormously greater than the gravitational force. Therefore, only the electrostatic force is of importance in atomic systems. In nuclei, the strong nuclear force overpowers even the electrostatic force but not to the extent that electrostatic forces are completely negligible. Many important nuclear effects are the result of electrostatic forces, as we shall see in Chapter 21.

● *Example 12.4*

In the Bohr model of the hydrogen atom, the electron is considered to move in a circular orbit around the nuclear proton. The radius of the orbit is 0.53×10^{-10} m. What is the velocity of the electron in this orbit?

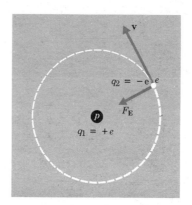

Even though the electron is moving around the proton, we can assume, without appreciable error, that the primary force between the two particles is the electrostatic force,

$$F_E = K\frac{q_1 q_2}{r^2} = \frac{Ke^2}{r^2}$$

$$= \frac{(9.0 \times 10^9 \text{ N-m}^2/\text{C}^2) \times (1.6 \times 10^{-19} \text{ C})^2}{(0.53 \times 10^{-10} \text{ m})^2}$$

$$= 8.2 \times 10^{-8} \text{ N}$$

In order to maintain a circular orbit, the electron must be subject to a centripetal acceleration:

$$a_c = \frac{v^2}{r}$$

The mass of the electron times this acceleration must equal the force on the electron:

$$m_e a_c = \frac{m_e v^2}{r} = F_E$$

or, solving for v,

$$v = \sqrt{\frac{rF_E}{m_e}}$$

$$= \sqrt{\frac{(0.53 \times 10^{-10} \text{ m}) \times (8.2 \times 10^{-8} \text{ N})}{(9.1 \times 10^{-31} \text{ kg})}}$$

$$= 2.18 \times 10^6 \text{ m/s} = 2180 \text{ km/s}$$

which is about 1 percent of the velocity of light.

• *Example 12.5*

Three changes are situated as shown in the diagram. What is the net force on q_3?

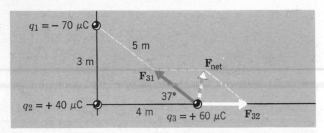

The electric force is a vector just as is every other force we have studied. Therefore, this type of problem can be solved in the standard way of separating the forces into components:

$(F_{net})_x = (F_{32})_x + (F_{31})_x$

$= K \dfrac{q_2 q_3}{(4 \text{ m})^2} - K \dfrac{q_1 q_3}{(5 \text{ m})^2} \cos 37°$

$= (9 \times 10^9 \text{ N-m}^2/\text{C}^2) \times$
$\left[\dfrac{(40 \times 10^{-6} \text{ C})(60 \times 10^{-6} \text{ C})}{(4 \text{ m})^2} \right]$

$- (9 \times 10^9 \text{ N-m}^2/\text{C}^2) \times$
$\left[\dfrac{(70 \times 10^{-6} \text{ C})(60 \times 10^{-6} \text{ C})}{(5 \text{ m})^2} \right] \times (0.8)$

$(F_{net})_x = 1.35 \text{ N} - 1.21 \text{ N}$
$= 0.14 \text{ N}$

$(F_{net})_y = (F_{31})_y$

$= K \dfrac{q_1 q_3}{(5 \text{ m})^2} \sin 37°$

$= (9 \times 10^9 \text{ N-m}^2/\text{C}^2) \times$
$\left[\dfrac{(70 \times 10^{-6} \text{ C})(60 \times 10^{-6} \text{ C})}{(5 \text{ m})^2} \right] \times (0.6)$

$= 0.91 \text{ N}$

Written as a vector
$\vec{F}_{net} = (0.14\vec{i} + 0.92\vec{j}) \text{ N}$

The magnitude of $\mathbf{F}_{net}$ is

$F_{net} = \sqrt{F_{net,x}^2 + F_{net,y}^2}$
$= \sqrt{(0.14 \text{ N})^2 + (0.91 \text{ N})^2}$
$= 0.92 \text{ N}$

The angle θ that $\mathbf{F}_{net}$ makes with the x-axis is

$\theta = \tan^{-1} \dfrac{(F_{net})_y}{(F_{net})_x} = \tan^{-1} \left(\dfrac{0.91 \text{ N}}{0.14 \text{ N}} \right)$
$= \tan^{-1} 6.5 = 81°$ ∎

Electric Forces in General

In this section we have discussed only the force that acts between electric charges at rest—the electrostatic force. When charges are in motion relative to one another we will find that an additional force—the *magnetic* force—comes into play. The term *electromagnetic force* is frequently used to indicate that both electrostatic and magnetic effects are present. However, magnetic forces have no existence independent of electric charges; these forces arise *exclusively* from charges in motion. Therefore, we will use the term *electric force* in a general sense to indicate an electrostatic force (if the charges are at rest) and to include the possibility of a magnetic force (if the charges can be in motion).

12.2 Electrostatic Potential Energy

Comparison of Gravitational and Electrostatic Potential Energies

We consider now only that portion of electromagnetic energy that is associated with charges at rest—*electrostatic* energy.

The gravitational force between two masses has exactly the same dependence on the separation distance (i.e., $1/r^2$) as does the electrostatic force between two charges. We expect, therefore, that the expression for the electrostatic potential energy PE_E will have exactly the same form as that for the gravitational potential energy,

$$PE_G = -G \dfrac{m_1 m_2}{r}$$

The only difference between the gravitational and electric cases is that mass values are always positive, whereas charge values can be either positive or negative. In order to take account of the fact that like charges repel and opposite charges attract, we write an expression for PE_E that has the same form as the expression for PE_G, except for the sign:

$$PE_E = K \dfrac{q_1 q_2}{r} \quad (12.4)$$

Now, when q_1 and q_2 have the same sign, $PE_E > 0$; then, the force is repulsive and the potential energy decreases with increasing separation. When q_1 and q_2 have opposite signs, $PE_E < 0$; then, the force is attractive and the potential energy increases (becomes less negative) with increasing separation.

The work done, or change in PE_E, in moving a charge q_1 from r_1 to r_2 as measured from q_2 is

$$W_{12} = \Delta(PE_E) = Kq_1q_2\left(\frac{1}{r_2} - \frac{1}{r_1}\right) \quad (12.5)$$

Compare this expression with Eq. 7.19 for $\Delta(PE_G)$.

The difference of sign between PE_G and PE_E or between $\Delta(PE_G)$ and $\Delta(PE_E)$ is easy to understand by referring to the concept of *work*. This is illustrated schematically in Fig. 12.4 for the case of electrostatic potential energy. Whenever work is done *against* a force such as F_E or F_G, the potential energy *increases*; $W > 0$ and $\Delta(PE) > 0$. If work is done *by* such a force, the potential energy *decreases*; $W < 0$ and $\Delta(PE) < 0$. The first two cases[2] in Fig. 12.4 apply also for the gravitational case.

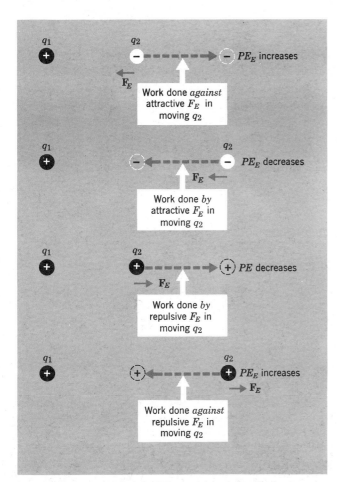

Figure 12.4 The changes in PE_E are shown for the four possible cases of a charge (q_2) moved in the presence of a positive charge (q_1).

---

[2]Only the first two cases because F_G is always *attractive*.

●*Example 12.6*

Suppose we have two charges, $q_1 = +4 \times 10^{-4}$ C and $q_2 = -8 \times 10^{-4}$ C, with an initial separation of $r_1 = 3$ m. What is the change in potential energy if we increase the separation to 8 m?

Using Eq. 12.5,

$$\Delta(PE_E) = Kq_1q_2\left(\frac{1}{r_1} - \frac{1}{r_2}\right)$$

$$= (9 \times 10^9)(+4 \times 10^{-4})(-8 \times 10^{-4}) \times (\tfrac{1}{8} - \tfrac{1}{3})$$

$$= (-2.88 \times 10^3) \times (-\tfrac{5}{24}) = +600 \text{ J}$$

In this case there is a net *increase* in the electrostatic potential energy (that is, $\Delta(PE_E) > 0$) because work was done by an outside agent against the attractive electrostatic force. ■

12.3 Potential Difference and the Electron Volt

Volts

A positive charge q that is in the vicinity of another positive charge and a negative charge will be attracted by the negative charge and repelled by the positive charge. Thus, in Fig. 12.5, an expenditure of work by an outside agency will be required to move the charge q from A to B. If $q = 1$ coulomb (C) and if the amount of work required is $W = 1$ joule (J), we say that the *potential difference* between points A and B is 1 *volt* (V). That is, the potential difference between two points is a measure of the *work per unit charge* required to move a charge from one point to the other. That is,

$$\text{Potential difference, } V = \frac{W}{q} \quad (12.6)$$

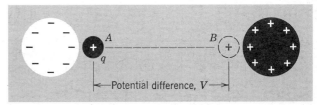

Figure 12.5 If 1 J of work is required to move 1 C of positive charge from A to B, the *potential difference* between the two points is 1 V.

and[3]

$$1 \text{ V} = 1 \text{ J/C} \tag{12.7}$$

We often measure potential differences with respect to the *ground,* which we elect to say is at zero potential.

The *volt* is the unit familiar in terms of household electricity. The potential difference between the two wires of common household electric circuits is 110 V; the potential difference between the terminals of a flashlight battery is 1.5 V.

The Electronvolt

A unit of energy that is quite useful in dealing with problems in atomic and nuclear physics is obtained in the following way. Suppose that a charge e, equal to the charge on an electron (which is the same as the charge on a proton, disregarding the sign), is moved from one position to another between which exists a potential difference of 1 V. How much work has been done *on* the charge? Using Eq. 12.6, we have

$$\begin{aligned} W = qV &= e \times V \\ &= (1.602 \times 10^{-19} \text{ C}) \times (1 \text{ V}) \\ &= 1.602 \times 10^{-19} \text{ J} \end{aligned} \tag{12.8}$$

This unit of *work* or energy is given the special name *electronvolt,* and is denoted by the symbol eV.

$$1 \text{ eV} = 1.602 \times 10^{-19} \text{ J} \tag{12.9}$$

Larger units of energy are convenient for many problems, especially in nuclear physics; those most frequently used are

1 kiloelectron (keV) $= 10^3$ eV $= 1.602 \times 10^{-16}$ J

1 megaelectronvolt (MeV) $= 10^6$ eV $= 1.602 \times 10^{-13}$ J

1 gigaelectronvolt (GeV) $= 10^9$ eV $= 1.602 \times 10^{-10}$ J

Typical energies in atomic and molecular processes are a few eV. X rays can have energies of a few keV. Nuclear reactions and transitions frequently involve energies of a few MeV. And elementary particle interactions often involve energies of a few GeV.

● *Example 12.7*

A proton, starting from rest, moves through a potential difference of 10^6 V. What is its final kinetic energy and final velocity?

For the kinetic energy we have, simply,

$$KE = 10^6 \text{ eV} = 1 \text{ MeV} = 1.601 \times 10^{-13} \text{ J}$$

In order to compute the final velocity, we use

$$KE = \tfrac{1}{2} m_p v^2$$

or

$$v = \sqrt{\frac{2KE}{m_p}} = \sqrt{\frac{2 \times (1.60 \times 10^{-13} \text{ J})}{1.67 \times 10^{-27} \text{ kg}}}$$

so that

$$v = 1.38 \times 10^7 \text{ m/s}$$

If we had considered an electron, instead of a proton, falling through a potential difference of 10^6 V, we could not have computed the final velocity in such a simple way because the velocity of a 1-MeV electron is close to the velocity of light. Relativity theory provides the correct method of calculation. The relativistic effect, which is manifest for a 1-MeV electron, is negligible for a proton of the same energy owing to the much larger mass of the proton. However, for protons with energies of about 100 MeV or more we must also use the relativistic expression for computing the velocity. ■

The *electronvolt* is commonly used as a unit of energy even in the event that the particle has not fallen through any potential difference. For example, a neutron (which has approximately the same mass as a proton but no electric charges and, therefore, cannot experience an electrostatic force) which is moving with a velocity of 1.38×10^7 m/s is said to have an energy of 1 MeV. That is, the unit "1 MeV" means that the particle has a definite amount of kinetic energy and it does not matter how the particle acquired this energy.

The eV notation is rarely used for objects larger than atomic size because the unit is too small to be convenient. The energy of a 0.1 g meteoroid which strikes the Earth with a velocity of 50 km/s, for example, has an energy of approximately 8×10^{14} GeV; we usually express such an energy in units of joules.

---

[3] Do not confuse the symbol for voltage V with that for the *unit* of voltage, the volt V.

12.4 The Electric Field

The Field Vector

According to Coulomb's law (Eq. 12.1), the magnitude of the electric force on a charge q due to a charge Q that is a distance r away is

$$F_E = K \frac{qQ}{r^2} \tag{12.10}$$

We also know that the electric force on q is a vector quantity with a direction *toward* Q if the charges have the opposite sign or *away from* Q if the charges have the same sign.

It proves convenient to describe electric forces in the following way. We say that Q sets up a certain condition in space to which q reacts by experiencing a force directed toward or away from Q (depending on the signs of the charges). This "condition in space" is called the *electric field* of the charge Q. (It is also true that q sets up a field and that Q reacts to this field, but we will concentrate on the effects of the field due to Q.) Because Q in some way produces the electric field that affects q, we say that Q is the *source* of the field and we refer to Q as the *source charge*. Any charged object that is placed at some point in this field will experience a force that depends on the magnitude and direction of the electric field at that point.

Instead of writing a *force* equation for the case of a particular charge q responding to the source charge Q, let us divide both sides of Eq. 12.10 by q:

$$\frac{F_E}{q} = K \frac{Q}{r^2} \tag{12.11}$$

The right-hand side of this expression is independent of q and depends only on the magnitude of Q and the distance from Q to q. That is, the right-hand side of Eq. 12.11 is a specification of the electric field at a distance r from the source charge. This electric field is characteristic only of Q and is the same for any charge q placed at the distance r. We use the label E to denote the quantity we call the *electric field strength*:

$$E = \frac{F_E}{q} = K \frac{Q}{r^2} \tag{12.12}$$

Notice that the units of E are those of force per unit charge or newtons per coulomb (N/C). As we will see later, we usually express electric field strength in the equivalent units of volts per meter (V/m).

In Section 6.1 we discussed the fact that for the purposes of making calculations of gravitational force, the entire mass of a uniform spherical object could be considered to be concentrated at its center. This simplification is due to the $1/r^2$ character of the gravitational force. Now, the electric force F_E also varies as $1/r^2$, so we expect that calculations of F_E can be carried out by considering the total charge on a uniformly charged sphere to be concentrated at its center. Because the electric force $\mathbf{F}^E$ is actually a vector, the electric field strength is also a vector quantity which we write as $\mathbf{E}$. The direction of $\mathbf{E}$ at a particular point is chosen to be the direction of the force on a *positive* charge placed at that point. Therefore, the electric field vector for a positive source charge always points *away from* the source and the field vector for a negative source charge always points *toward* the source (Fig. 12.6).

If a charge q is placed in an electric field described by the vector $\mathbf{E}$, the charge will experience a force given by

$$\mathbf{F}_E = q\mathbf{E} \tag{12.13}$$

This is the basic force equation for the electric field.

The Principle of Superposition

One of the facts that makes the field concept a useful one is that the electric force vector and the electric field vector obey the *principle of superposition*. This principle states that if we wish to find the total electric force on a particular charge q due to a number of other charges (Fig. 12.7), this total force is equal to the vector sum of all the individual forces:

$$\mathbf{F}_{\text{net}} = \mathbf{F}_1 + \mathbf{F}_2 + \mathbf{F}_3 + \mathbf{F}_4 + \ldots \tag{12.14}$$

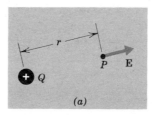 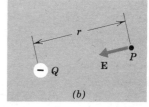

Figure 12.6 (a) The electric field vector for a positive source charge is directed *away* from the source. (b) The field vector for a negative source charge is directed *toward* the source. The direction of $\mathbf{E}$ is always the direction of the force exerted on a positive charge in the field.

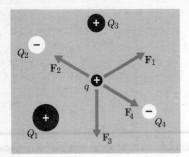

Figure 12.7 The net electric force on a charge q is the vector sum of the individual forces due to Q_1, Q_2, Q_3, and Q_4. Notice that two of these forces are attractive and two are repulsive.

Thus, each of the individual charges contributes a force just as if the other charges were not present.

Similarly, the total electric field at a point is expressed as the vector sum of the individual contributing fields at that point:

$$\mathbf{E}_{net} = \mathbf{E}_1 + \mathbf{E}_2 + \mathbf{E}_3 + \mathbf{E}_4 + \ldots \quad (12.15)$$

As we will see in the next chapter, magnetism is another phenomenon that can conveniently be described in terms of a field. And although we have not made use of the idea, gravitation also lends itself to a field description. All of these fields can exist in empty space. Furthermore, they can all exist together without interference. When two or more different kinds of fields are present at a particular point in space, it makes no sense to sum the *field* vectors (indeed, the field quantities have different physical units). Instead, we first determine the force on a particle due to each field and then sum the individual *force* vectors in the conventional way.

● **Example 12.8**

Two electric charges are located as follows: $Q_1 = +30 \times 10^{-6}$ C at the origin (0,0) and $Q_2 = -40 \times 10^{-6}$ C at the point (0,3). What is the electric field strength at point P (2,1)?

The following diagram shows that the distance from Q_1 to P is $\sqrt{5}$ m and that the distance from Q_2 to P is $\sqrt{8}$ m. Notice that Q_1 is positive, so that $\mathbf{E}_1$ is directed *away* from this source charge (see Fig. 12.6a); also, Q_2 is negative, so that $\mathbf{E}_2$ is directed *toward* this charge (see Fig. 12.6b). The magnitudes of $\mathbf{E}_1$ and $\mathbf{E}_2$ are found by using Eq. 12.12:

$$E_1 = K\frac{Q_1}{r_1^2} = \frac{(9 \times 10^9 \text{ N-m}^2/\text{C}^2) \times (30 \times 10^{-6} \text{ C})}{5 \text{ m}^2}$$
$$= 5.4 \times 10^4 \text{ V/m} = 54 \text{ kV/m}$$

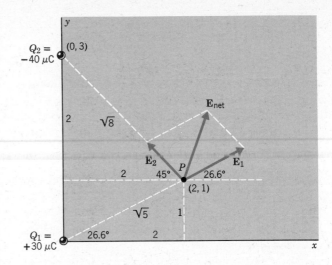

$$E_2 = K\frac{Q_2}{r_2^2} = \frac{(9 \times 10^9 \text{ N-m}^2/\text{C}^2) \times (40 \times 10^{-6} \text{ C})}{8 \text{ m}^2}$$
$$= 4.5 \times 10^4 \text{ V/m} = 45 \text{ kV/m}$$

Referring again to the diagram, we see that $\mathbf{E}_1$ makes an angle $\theta_1 = \tan^{-1}(1/2) = 26.6°$ with the x-axis and that $\mathbf{E}_2$ makes an angle of 45° with the negative x-axis. The net electric field at P is the vector sum of $\mathbf{E}_1$ and $\mathbf{E}_2$. By using the standard methods for combining vectors, we find $\mathbf{E}_{net}$:

$E_{net} = 58.3$ kV/m at an angle of 73.6° with respect to the x-axis. ∎

The Electric Potential

According to Eq. 12.4, the potential energy of a charge q that is a distance r from a source charge Q is

$$PE_E = K\frac{qQ}{r} \quad (12.16)$$

If we divide this expression by q, we obtain a quantity that is characteristic of the source charge Q. This new quantity is called the *electric potential* Φ:

$$\Phi = \frac{PE_E}{q} = K\frac{Q}{r} \quad (12.17)$$

The quantity Φ is potential energy per unit charge and has units of joules per coulomb (J/C). The electric potential at a point is equal to the work required to move a unit positive charge from infinitely far away to the particular point. Because Φ is *work* (per unit charge), it is a *scalar* quantity. Consequently, the electric potential also obeys the principle of superposition:

$$\Phi_{net} = \Phi_1 + \Phi_2 + \Phi_3 + \Phi_4 + \ldots \quad (12.18)$$

where Φ is the potential at a point P due to charge Q_1 at point P_1, etc.

In Section 12.2 we discussed the work required to move a charge from one point to another in the presence of a source charge. This work is just the potential energy difference ΔPE_E between the two points. The *work per unit charge* required to make this movement is the change in potential $\Delta \Phi$ between the points. This is the quantity we have given the symbol V (Eq. 12.6):

$$\frac{W}{q} = \frac{\Delta PE_E}{q} = \Delta \Phi = V \quad (12.19)$$

The quantity V is the *potential difference* or the *voltage* between the points. Thus, we see that the electric potential Φ can also be measured in volts (V). We also notice that electric field strength E (Eq. 12.12) has units of Φ divided by length; that is, E is measured in V/m. (Do not confuse *voltage V* with *volts* V.)

Table 12.1 summarizes the various electrical quantities, their units, and their defining equations.

● *Example 12.9*

Compute the electric field and the electric potential at point P midway between two charges, $Q_1 = Q_2 = +5 \times 10^{-6}$ C, separated by 1 m.

Because P is located midway between two identical charges, any charge placed at this point will experience equal but *oppositely directed* forces due to Q_1 and Q_2. The *net* force is therefore zero, so that $\mathbf{E} = 0$ at P.

Although the electric field at P is zero, this does *not* imply that the electric potential is also zero. The total potential Φ_{total} is the sum (the *algebraic* sum since potential is a *scalar*) of the potentials due to Q_1 and Q_2:

$$\Phi_1 = K\frac{Q_1}{r_1} = (9 \times 10^9) \times \frac{5 \times 10^{-6}}{0.5}$$
$$= 9 \times 10^4 \text{ V} = 90 \text{ kV}$$

$$\Phi_2 = K\frac{Q_2}{r_2} = (9 \times 10^9) \times \frac{5 \times 10^{-6}}{0.5}$$
$$= 9 \times 10^4 \text{ V} = 90 \text{ kV}$$

Therefore,

$$\Phi_{total} = \Phi_1 + \Phi_2 = 1.8 \times 10^5 \text{ V} = 180 \text{ kV}$$

Notice that if either Q_1 or Q_2 is changed from $+5 \times 10^{-6}$ C to -5×10^{-6} C, the electric *potential* will vanish but the electric *field* will not. Therefore, the fact that either the field or the potential is zero in any particular case does not necessarily mean that the other quantity will also be zero; each quantity must be calculated separately.

● *Example 12.10*

In the Bohr model of the hydrogen atom (Section 19.2 and our earlier discussion in Example 12.4) the electron is considered to move around the nuclear proton in a circular orbit that has a radius of

Table 12.1 Important Electrical Quantities

Quantity	Units	Equation[a]
Force, F_E	newton (N)	$F_E = K\frac{qQ}{r^2} = qE$
Charge, Q	coulomb (C)	
Electric force constant, K	newton-meter^2 per coulomb^2 (N-m^2/C^2)	
Electric field strength, E	newton per coulomb (N/C), volts per meter (V/m)	$E = \frac{F_E}{q} = K\frac{Q}{r^2}$
Electric potential, Φ	joule per coulomb (J/C), volt (V)	$\Phi = K\frac{Q}{r}$
Voltage, V	volt (V)	$V = \Delta\Phi = \frac{W}{q}$

[a]For spherical or point charges.

0.53×10^{-10} m. What is the electric field and what is the potential at the position of the electron?

$$E = K\frac{Q}{r^2}$$

$$= \frac{(9 \times 10^9 \text{ N-m}^2/\text{C}^2) \times (1.6 \times 10^{-19} \text{ C})}{(0.53 \times 10^{-10} \text{ m})^2}$$

$$= 5.1 \times 10^{11} \text{ V/m}$$

which is a very large field indeed. To give a comparison, sparking usually occurs in air when a field strength of 3×10^6 V/m is reached.

$$\Phi = K\frac{Q}{r}$$

$$= \frac{(9 \times 10^9 \text{ N-m}^2/\text{C}^2) \times (1.6 \times 10^{-19} \text{ C})}{0.53 \times 10^{-10} \text{ m}}$$

$$= 27 \text{ V}$$

which is a rather small potential. For example, the potential difference between the terminals of an automobile battery is 12 V.

The electric field E depends on $1/r^2$, whereas the potential Φ depends on $1/r$; because r is extremely small (0.5×10^{-10} m) in the case of the hydrogen atom, the field strength is large whereas the potential is small. ∎

Lines of Force and Electric Field Lines

How do we provide a visual picture of a field quantity? A *weather map* is actually just such a picture—the field quantity in this case is the *barometric pressure*. The map is constructed by drawing lines to connect together points that have the same measured pressure. The completed map shows the variation of barometric pressure across the country, with regions of high and low pressure encircled by closed lines of equal pressure. We can treat barometric pressure as a field quantity and can map it in this way because pressure varies smoothly from place to place; that is, there are no sudden and disjoint jumps in the pressure between neighboring spots.

The electric force and electric field strength also are quantities that are well defined and vary smoothly from place to place so that they can be treated with a field description. But there is an essential difference between a pressure field and the field of electric force. Pressure is a *scalar* quantity; that is, the pressure at a particular location is completely specified by a single number (and appropriate units). The electric force and the electric field

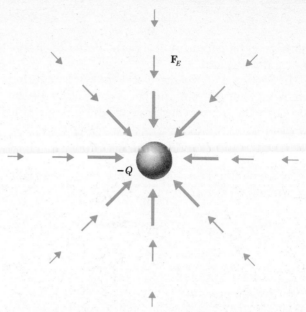

Figure 12.8 Mapping the electric force field surrounding the source charge $-Q$. Because we always consider the test charge to carry a positive charge, all of the force vectors are directed toward the source charge.

strength, however, are *vector* quantities; that is, both magnitude and direction must be stated to specify completely these quantities.

Suppose that we begin to map the electric force field around a certain negative source charge $-Q$ by measuring the force on a small test charge.[4] The results of such measurements can be represented by a series of arrows, as in Fig. 12.8. The length of each arrow is proportional to the electric force on the (positive) test charge when it is located at the end of the arrow opposite the point. This procedure provides a pictorial representation of the electric force field (or the electric field strength) with as much detail as we require.

The vector map, although it can be as complete as desired, tends to be awkward to use. Usually, we simplify the picture by constructing around the source charge a set of continuous lines such that, at any point, the direction of the line is the same as the direction of the force vector at that point. These lines are called *lines of force*. For a negative source charge located at a point, the lines of force are simply straight lines that are directed toward the source charge (Fig. 12.9).

Because the electric field strength is equal to

---

[4] A *test charge* is a hypothetical object whose charge (always taken to be positive) is so small that it does not distort the electric field of the source charge and whose size is sufficiently small that it samples the field essentially at a point.

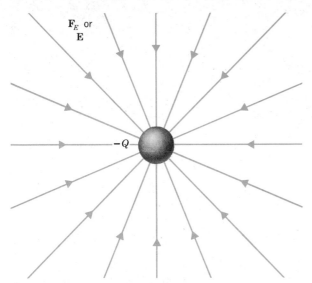

Figure 12.9 The lines of force or field lines **E** around a point source charge are all straight lines. The density of lines (in three dimensions) decreases with the square of the distance from the source charge because the force and the field strength vary as $1/r^2$.

the electric force divided by the magnitude of the test charge, $E = F_E/q$, the lines of electric force also describe the electric field **E**. When a diagram is designed to represent the electric field, the lines are called *field lines*. Notice that the direction of a field line at a point is always the same as the direction of the electric force on a *positive* test charge located at that point. Thus, the field lines are always directed *toward* a negative source charge and *away from* a positive source charge (Fig. 12.10).

Not only does a lines-of-force diagram show us the direction of the force and the electric field at any point, it also provides information about the magnitudes of these quantities. Suppose that we fill Fig. 12.9 with many more field lines, all drawn according to the same procedure. Now, count the number of lines that pass through a small element of area oriented perpendicular to the lines at some point. It is easy to see from the geometry (in three dimensions) that the number of lines is inversely proportional to the square of the distance from the source charge to the point at which the area element is located. We know from Coulomb's law and from Eq. 12.12 that F_E and E decrease with distance from the source charge as $1/r^2$. That is, the *density* of the field lines at any point is proportional to the field strength at that point. Where the lines bunch together, the field is strong; where the lines are sparse, the field is weak.

For a point charge or a simple uniform spherical distribution of charge, the field lines are all straight lines that proceed radially outward from or inward toward the source charge. Where do the field lines come from and where do they go? If we had an isolated charge, the field lines would proceed indefinitely far into space as straight lines. But, of course, it is not possible to have a truly isolated charge. All bulk matter (indeed, presumably, the entire Universe) is composed of equal numbers of elementary positive and negative charges and is therefore electrically neutral. (Objects can be charged, but this condition is brought about by separating the positive and negative charge of an originally neutral object.) Let us consider the case of two objects carrying equal and opposite charges (Fig. 12.11). As always, we can generate the field lines by measuring or calculating the magnitude and direction of the force on a positively charged test body. If we do this, we find that the field lines emerge from the object carrying positive charge and go smoothly in curves that terminate on the negatively charged object.

If we perform this kind of experiment or do the calculation for an arbitrary assembly of charges (but with zero *net* charge), we always find the same re-

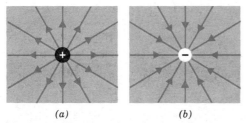

Figure 12.10 The electric field lines from a positive source charge *(a)* are directed radially outward whereas those from a negative source charge *(b)* are directed radially inward.

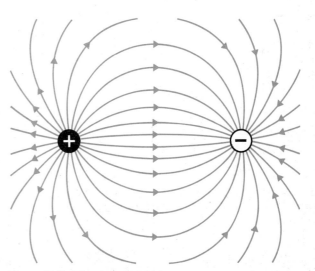

Figure 12.11 The electric field lines originate on the positive charge and terminate on the negative charge.

sult: *the electric field lines originate on positive charges and terminate on negative charges.* This is one of the important results in the theory of electrostatics.

Figure. 12.11 illustrates a simple case in which the field lines are curved instead of straight. But notice that in the close vicinity of either charge the field lines are approximately straight. That is, we can consider a charge to be approximately isolated (straight field lines) if we confine our attention to distances from the charge that are small compared to the distance to another charge of similar magnitude.

Fields for Distributions of Charge

The shapes of the field lines for different geometrical configurations of charges can be obtained in a variety of ways that present graphic pictures of the fields. One such way is to suspend grass seeds in an insulating liquid (such as oil or glycerine) that surrounds the configuration of charges under investigation. The electric field induces a separation of the charge in the seeds: one end of a seed becomes negatively charged and the other end becomes positively charged, but the seed remains electrically neutral. This process is called *polarization*. The polarized seeds then orient themselves along the field lines, thereby rendering "visible" the shape of the field lines. Some photographs of field configurations made in this manner are shown in Fig. 12.12. The first two photographs show the field lines for pairs of charges with the same and with opposite signs, respectively. Notice that Fig. 12.12*b* is the same as Fig. 12.11. The remaining photographs demonstrate the following points:

1 *The electric field inside a current-free conductor (solid or hollow) is zero.* (Figures 12.12*c* and 12.12*d* illustrate this point for the case of the hollow conductor.) First, consider a *solid* conductor. If an amount of charge is placed within such a conductor and if the charges are free to move, there will be mutual repulsion among all of the individual charges and they will rapidly migrate to the surface. Now, if this surface charge sets up an electric field *within* the conductor, then the conduction electrons will be induced to move and will then constitute a current, contrary to the stipulation that the conductor be current free. (Beginning in Section 12.6 we investigate the effects of electric current.)

Next, consider a *hollow* conductor. Let us take the simple case of a hollow sphere. If the sphere is charged, there will be a uniform distribution of charge on the surface.[5] A test charge placed at the center of the sphere will experience no net force; the field at the center must be zero. But what about other interior points? Figure 12.13 shows the geometry of the situation. We wish to determine the net force on a test charge at the point P. Two concentric cones, extending in opposite directions, are constructed with their vertices at P. These cones intercept two areas on opposite sides of the sphere. Because the charge is uniformly distributed on the surface of the sphere, the force on the test charge due to the charge on either intercepted segment of the shell is proportional to the area of that segment. The two forces are oppositely directed. But the larger area is farther away from P, and the increase in force due to the larger area ($\propto r^2$) is offset by the decrease in the force due to the greater distance ($\propto 1/r^2$). Therefore, the two forces are equal in magnitude and opposite in direction, giving no net effect. Because the same argument applies to the remaining area of the sphere, there is zero net force on the test charge. An identical result is found for any point within the sphere. Hence, the electric field *within* the spherical shell is everywhere *zero*. By using more complicated mathematics, the same result can be obtained for closed conducting surfaces of arbitrary shape.

2 *The electric field between two uniformly charged parallel plates of equal area is uniform* (see Fig. 12.12*f*). First, consider the situation if the two plates are of infinite extent, one with a uniform positive charge and the other with a uniform and equal negative charge. The lines of force must originate on positive charge and terminate on negative charge. Because of the uniformity of the charge, no region between the plates can be different from any other region. Therefore, the lines of force must be uniform and parallel. The absence of any bunching of the lines of force means that the electric field is also uniform. In the real case of plates with finite size (Fig. 11.9), the field lines tend to curve at the edges. However, if the spacing between the

---

[5]This is insured by the *symmetry* of the sphere; no point on the surface is different from any other point, and so, because the charges mutually repel each other, they distribute themselves uniformly over the surface.

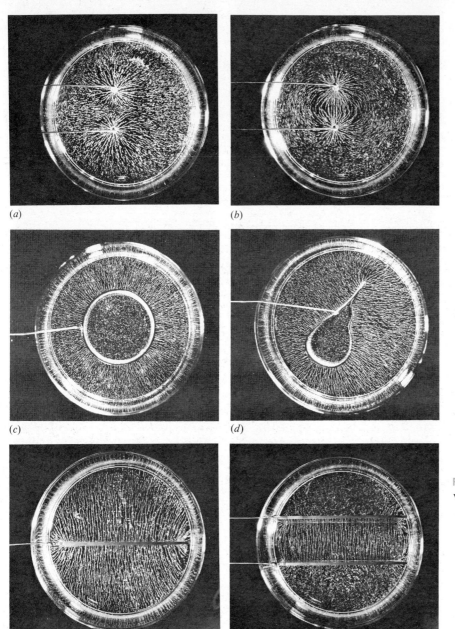

Figure 12.12 Electric field lines around various charged objects.
(a) Two charges of the same sign.
(b) Two charges with opposite signs.
(c) A charged ring (cross section of a charged sphere); the field inside is zero.
(d) A charged conductor of arbitrary shape; the field inside is zero.
(e) A charged plate.
(f) A pair of plates carrying equal and opposite charges, uniformly distributed.

plates is small compared to the linear dimensions of the plates, this so-called *edge effect* is unimportant for most calculations and is therefore neglected.

A pair of parallel plates separated by a gap is called a *capacitor* (or *condenser*). We will discuss these devices in detail in the following section.

3 *Electric field lines at the surface of a conductor are always perpendicular to the surface.* In Section 8.1 we argued that the force exerted by a static fluid on the walls of a container is always

Figure 12.13 The fact that the electrostatic force varies as $1/r^2$ insures that there is no force on a test charge at any point within a uniformly charged spherical shell; therefore, the electric field within the shell is zero.

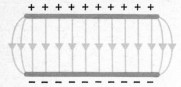

Figure 12.14 The actual curving of the field lines at the edges of a parallel-plate capacitor is usually neglected and only the ideal situation is treated. The field lines in an ideal parallel-plate capacitor are uniform both in magnitude and in direction.

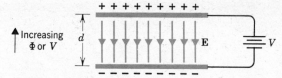

Figure 12.16 The magnitude of the electric field between a pair of parallel plates connected to a battery is V/d. Notice the symbol used to indicate a battery.

perpendicular to the container surface. We have a similar situation in the case of electric field lines at the surface of a conductor. Field lines originate and terminate at the surfaces of conductors whenever charges are located on the surfaces. If the electric field had any component along the surfaces, the charges would be subject to a force and they would move—that is, a current would flow. The situations we are considering are *static* (no current flow), so we conclude that the field lines are all perpendicular to the surface. The various photographs in Fig. 12.12 illustrate this feature of electric field lines. Figure 12.15 shows a more detailed view of one end of the parallel-plate capacitor in Fig. 12.14. Notice that the field lines emerge from the upper plate and terminate on the lower plate in directions that are perpendicular to the surfaces.

Voltage

The difference in electric potential between two points is called the *voltage* between those points: $\Delta \Phi = V$. How do we produce a voltage? One of the common ways is to take advantage of the separation of charge that occurs in certain chemical reactions. This is the principle of the common *battery*. In a battery, chemical reactions lead to the accumulation of positive charge on one electrode, called the *positive terminal* (or *anode*) and to the accumulation of negative charge on the other electrode, called the *negative terminal* (or *cathode*). For a particular choice of chemical reactants, there is a definite voltage that is developed across the terminals of a battery. For example, in a lead-acid battery of the type used in automobiles, the potential difference across the terminals of each battery cell is 2 V. Ordinary 6-V or 12-V batteries are constructed by connecting together either three or six such cells. Suppose we have a pair of parallel plates separated by a distance d and across which there is a voltage V supplied by a battery (Fig. 12.16). The electric field between the plates is uniform and the magnitude is given by the electric force on a test charge q in the field (Eq. 12.12): $E = F_E/q$. If we move the (positive) test charge from the negative plate to the positive plate, the work required is

$$W = F_E \times d = qE \times d \qquad (12.20)$$

or,

$$E = \frac{W}{qd} \qquad (12.21)$$

The work done per unit charge, W/q, is just the voltage (Eq. 12.19); therefore, the strength of the uniform electric field between a pair of parallel plates is

$$E = \frac{V}{d} \quad \text{(uniform field)} \qquad (12.22)$$

It must be emphasized that this expression for the electric field strength is valid only for the case of a uniform field. The equivalent expression for the case of Fig. 12.11, for example, is somewhat more complicated.

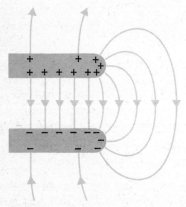

Figure 12.15 Electric field lines at the surfaces of conductors are always perpendicular to the surfaces.

The Determination of the Electron Charge e

In 1911 Robert A. Millikan performed a beautifully simple but highly significant experiment that established the fact that every electron carries a charge of the same magnitude; his experiment also obtained, for the first time, a precise value for this quantity. Millikan set up an electric field between a pair of parallel plates, as in Fig. 12.17. He sprayed a fine mist of oil droplets in this field. (Oil does not evaporate as readily as water.) Some of the droplets became negatively charged by friction in the process of spraying. Millikan viewed the droplets with a microscope and found that by adjusting the voltage V between the plates to a particular value he could suspend a given droplet in an equilibrium position. When suspended, the downward gravitational force, mg, equaled the upward electric force, qE, where the magnitude of E is just V/d. Therefore, at equilibrium,[6]

$$mg = qE = \frac{qV}{d}$$

so that

$$q = \frac{mgd}{V} \qquad (12.23)$$

The charge on the droplet q is given in terms of measurable quantities. The values of g, d, and V were measured directly but it was necessary to determine the mass m by an indirect method. First, the density of the oil was obtained by bulk measurements. Next, the rate of fall of the droplet in the absence of the electric field was measured. For the tiny droplet that Millikan used, the rate of fall was the *terminal velocity* (see Section 8.6) and it was known from the theory of motion of objects in a viscous medium, such as air, that the terminal velocity depends on the radius of the droplet (see Eq. 8.26). Millikan also took into account the effect of the buoyancy of the air. Therefore, a measurement of the terminal velocity determined the radius (and, hence, the volume) and, when combined with the density measurement, provided a value for the mass.

Millikan found that the values of the charge on various droplets, determined in this way, were not of arbitrary sizes. Instead, he found that every charge was an integer number times some basic unit of charge; that is, $q = Ne$, $N = 1, 2, 3, \ldots$. This basic unit of charge is the charge on the electron, $e = 1.602 \times 10^{-19}$ C. (Millikan's 1911 value of e was slightly in error due to the use of a faulty value for the viscosity of air. When the correction was made later, the resulting value was quite close to that accepted at present.)

12.5 Capacitance

The Farad

A pair of conducting surfaces separated by a gap constitutes a *capacitor*. We have learned that the electric field strength in the region between a pair of parallel plates is related to the voltage V across the plates and the distance d that separates the plates: $E = V/d$. Let us examine the way that a capacitor becomes charged.

Suppose that we have a battery which has a potential difference V across its terminals. We connect the positive terminal to one plate of a capacitor and the negative terminal to the other plate, as in Fig. 12.18. Each wire and plate now becomes an extension of the terminal to which it is attached. Thus, a negative charge accumulates on the plate connected to the negative terminal, and a positive charge accumulates on the plate connected to the positive terminal. The buildup of charge on the plates continues (for a tiny fraction of a second after the connection to the battery is made) until the voltage across the ca-

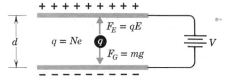

Figure 12.17 Schematic diagram of Millikan's oil drop experiment for the determination of e.

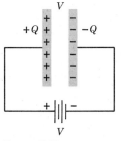

Figure 12.18 A capacitor is charged by connecting the plates to the terminals of a battery.

---

[6] Millikan actually used a *dynamic* instead of a *static* method for determining e in his oil-drop experiment, but the distinction is not important here.

pacitor plates is the same as the voltage across the terminals of the battery. This is the equilibrium (or static) condition. In this condition the amount of positive charge Q on the left-hand plate (Fig. 12.18, is exactly equal to the amount of negative charge on the right-hand plate. (Can you see why this must be so?)

The amount of charge Q that will accumulate on the plates of a capacitor is directly proportional to the applied voltage V. The greater the voltage, the greater will be the amount of charge that the plates can accommodate before equilibrium is reached. The proportionality constant connecting the charge and the voltage is called the *capacitance C* of the capacitor:

$$Q = CV \qquad (12.24)$$

The unit of measure for capacitance is the coulomb per volt, which we call the *farad* (F) in honor of Michael Faraday (1791–1867), the great English physicist who contributed greatly to our understanding of electrostatic phenomena. The capacitance values of the devices most commonly used are very small fractions of 1 F. Therefore, we usually see capacitance measured in units of $1 \, \mu F = 10^{-6}$ F or $1 \, pF = 10^{-12}$ F.

On what factors does the capacitance of a pair of parallel plates depend? The most obvious factor is the area A of the plates. The larger the plates, the greater will be the amount of charge that will accumulate for a particular voltage V; that is, $C \propto A$. Next, the amount of charge that will accumulate on a pair of plates of a certain size depends on the distance d between the plates. If d is large, the field strength for a particular voltage is low because $E = V/d$. This means that the amount of charge on the plates is small. If d is made smaller, the field strength and the amount of accumulated charge will increase. Thus, we expect $C \propto 1/d$. Finally, the capacitance value for a particular pair of plates depends on the medium that fills the space between the plates. Later in this section we will consider various materials (*dielectrics*) that can be used in capacitors; for the moment, we assume that the space between the plates is vacuum. Then, a complete derivation shows that the capacitance of a pair of plates with area A and separated by a distance d is given by

$$C = \frac{A}{4\pi K d} \qquad (12.25)$$

where K is the constant in the Coulomb force law (Eq. 12.1). The proportionality constant connecting C and A/d is sometimes written as ϵ_0 and is called the *permittivity of free space* (vacuum):

$$\epsilon_0 = \frac{1}{4\pi K} = 8.85 \times 10^{-12} \, C^2/N\text{-}m^2$$
$$= 8.85 \times 10^{-12} \, F/m \qquad (12.20)$$

(Verify the units of ϵ_0.) Then,

$$C = \frac{\epsilon_0 A}{d} \qquad (12.25a)$$

Although some capacitors are made in the form of rigid parallel plates separated by an air gap (the tuning capacitors in radios are usually constructed this way), most capacitors are more compact in form. A common method for manufacturing capacitors is to insulate two sheets of metallic foil with two sheets of some plastic material (a dielectric). The sandwich is then rolled into a tight cylindrical bundle. A wire that is attached to one of the plates (i.e., one of the metallic sheets) is brought out through one end of the cylinder and a wire that is attached to the other plate is brought out through the other end. Prepared in this way, the capacitor is a compact device that is easy to insert into electric and electronic circuits.

● *Example 12.11*

How many electrons accumulate on the negative plate of a 3-μF capacitor when it is connected across a 12-V battery?

Using Eq. 12.24,

$$Q = CV = (3 \times 10^{-6} \, F) \times (12 \, V) = 3.6 \times 10^{-5} \, C$$

This total charge is due to N electrons: $Q = Ne$. Therefore,

$$N = \frac{Q}{e} = \frac{3.6 \times 10^{-5} \, C}{1.60 \times 10^{-19} \, C} = 2.25 \times 10^{14}$$

● *Example 12.12*

It is desired to construct a 20-pF capacitor using two plates with dimensions 4 cm × 6 cm. What must be the separation of the plates? (The plates are separated by air, which is almost the same as a vacuum gap.)

Solving Eq. 12.25 for the separation d,

$$d = \frac{A}{4\pi KC} = \frac{\epsilon_0 A}{C}$$

$$= \frac{(8.85 \times 10^{-12} \text{ F/m}) \times (0.04 \text{ m} \times 0.06 \text{ m})}{20 \times 10^{-12} \text{ F}}$$

$$= 1.06 \times 10^{-3} \text{ m}$$

or slightly more than 1 mm. ∎

Combinations of Capacitors

In many types of electric circuits we find capacitors connected together in various combinations. The simplest type of capacitor combination is the *parallel* connection, shown in Fig. 12.19a. We would like to know the total capacitance C_t of this combination. That is, we would like to know the capacitance value of the *single* capacitor (Fig. 12.19b) that could be inserted into the circuit instead of C_1, C_2, and C_3, and would have exactly the same effect as the three capacitors connected in parallel.

First, we note that the same voltage V is across each of the three capacitors. (The wires have no effect, so the points a and b are common to all of the capacitors.) Then, we can write for the charges on the three capacitors,

$$Q_1 = C_1 V \qquad Q_2 = C_2 V \qquad Q_3 = C_3 V$$

The total amount of charge Q accumulated on the three capacitors is, therefore,

$$Q = Q_1 + Q_2 + Q_3 = (C_1 + C_2 + C_3)V \qquad (12.27)$$

Now, if a single capacitor with capacitance C_t is to have the same effect as the parallel combination, then it must carry the same total charge Q when the voltage V is applied. That is,

$$Q = C_t V \qquad (12.28)$$

If we compare Eqs. 12.27 and 12.28 for Q, we identify the total capacitance as

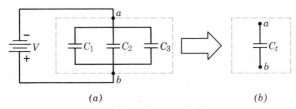

Figure 12.19 A parallel connection of three capacitors.

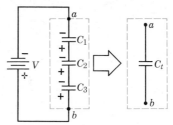

Figure 12.20 A series connection of three capacitors.

$$C_t = C_1 + C_2 + C_3$$
(parallel connection) $\qquad (12.29)$

We could have anticipated this result by noting that the capacitance of a capacitor is directly proportional to its area (Eq. 12.25). When a set of capacitors is connected in parallel, the connection in effect adds the areas together. If the areas are added, this means the capacitance values are also added; thus, $C_t = C_1 + C_2 + C_3$.

Figure 12.20a shows three capacitors in a series connection. Again, we would like to know the total capacitance of the combination. In this case, we do not have the same voltage applied to each capacitor. In fact, we must write

$$Q_1 = C_1 V_1 \qquad Q_2 = C_2 V_2 \qquad Q_3 = C_3 V_3 \qquad (12.30)$$

We do know, however, that the battery voltage V must equal the sum of the individual voltages (we discuss this idea in detail in Chapter 14); thus,

$$V = V_1 + V_2 + V_3 \qquad (12.31)$$

Next, we make use of the fact that the amount of positive charge on the positive plate of C_1 must exactly equal the amount of negative charge on the negative plate of C_2; and so forth. (How could it be otherwise? Charge cannot leap across the gap of a capacitor. This means that the part of the circuit containing, for example, the positive plate of C_1, the negative plate of C_2, and the connecting wire is an isolated section. The net charge in this section is zero before the battery connection is made and it must remain zero afterwards. Therefore, the positive charge on the C_1 plate must exactly balance the negative charge on the C_2 plate.) We conclude that all of the individual charges are equal and we label the common value Q:

$$Q_1 = Q_2 = Q_3 = Q \qquad (12.32)$$

In the series combination, the total amount of negative charge provided to the system by the battery is just the negative charge Q on the upper plate of C_1. (The negative charges on C_2 and C_3 are canceled by the positive charges in the respective isolated sections.) Similarly, the total amount of positive charge is that on the lower plate of C_3. Thus, the entire series combination has the same effect as a single capacitor C carrying the charge Q across which is applied a voltage V. That is,

$$Q = C_t V \quad \text{or} \quad V = \frac{Q}{C_t} \tag{12.33}$$

Combining Eqs. 12.30, 12.31, and 12.32, we have

$$V = V_1 + V_2 + V_3$$
$$= \frac{Q_1}{C_1} + \frac{Q_2}{C_2} + \frac{Q_3}{C_3}$$
$$= Q\left(\frac{1}{C_1} + \frac{1}{C_2} + \frac{1}{C_3}\right) \tag{12.34}$$

Comparing Eqs. 12.22 and 12.34 for V, we identify the total capacitance as

$$\frac{1}{C_t} = \frac{1}{C_1} + \frac{1}{C_2} + \frac{1}{C_3}$$
(series connection) $\tag{12.35}$

• *Example 12.13*

Three capacitors have capacitances $C_1 = 1\,\mu\text{F}$, $C_2 = 3\,\mu\text{F}$, and $C_3 = 6\,\mu\text{F}$. What is the total capacitance of the combination when connected (a) is parallel and (b) in series?

(a) For the parallel connection, we have
$$C_t = C_1 + C_2 + C_3 = 1 + 3 + 6 = 10\,\mu\text{F}$$

(b) For the series connection, we have
$$\frac{1}{C_t} = \frac{1}{C_1} + \frac{1}{C_2} + \frac{1}{C_3} = \frac{1}{1} + \frac{1}{3} + \frac{1}{6}$$
$$= \frac{6+2+1}{6} = \frac{9}{6}$$

so that
$$C_t = \frac{6}{9} = \frac{2}{3}\,\mu\text{F} \quad \blacksquare$$

We have been considering capacitors that consist of a pair of parallel plates separated by a vacuum space. If we fill the vacuum space of a particular capacitor with an insulating material (a *dielectric*), the capacitance will generally increase. We can characterize a dielectric material by measuring the capacitance of a pair of parallel plates when the space between the plates is filled with the material. If this capacitance value is C_d and if the value with a vacuum space is C_0, the ratio is called the *dielectric constant* k of the material:

$$k = \frac{C_d}{C_0} \tag{12.36}$$

The dielectric constants of several materials are listed in Table 12.2.

• *Example 12.14*

Suppose that Teflon is used as the dielectric in the capacitor in Example 12.12. What spacing d will give the same capacitance (20 pF)?

Solving Eq. 12.36 for C_0 and inserting this value for C in Eq. 12.25, we have

$$d = \frac{\epsilon_0 A}{C} = \frac{\epsilon_0 A}{(C_d/k)} = \frac{\epsilon_0 A}{C_d} \times k$$

The combination $\epsilon_0 A/C_d$ is just that calculated in Example 12.12 (because C_d is also 20 pF), that is, 1.06 mm. Therefore, using the Teflon dielectric increases the spacing by a factor equal to the dielectric constant. In Table 12.2, we find $k(\text{Teflon}) = 2.1$. Therefore, the new spacing is

$$d = (1.06\text{ mm}) \times 2.1 = 2.2\text{ mm} \quad \blacksquare$$

What is the physical process that causes the

Table 12.2 Dielectric Constants for Various Materials at Room Temperature

Material	k
Vacuum	1
Air, dry	1.0005
Benzene	2.28
Carbon dioxide	1.0009
Glasses	5–9
Glycerol	43.5
Methyl alcohol	33.6
Nylon 66	3.88
Plexiglas	3.40
Polyethylene	2.26
Porcelins	6–8
Rubber, natural	2.94
Teflon	2.1
Transformer oil	2.22
Water	80.4

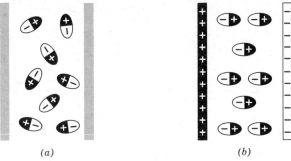

Figure 12.21 (a) Polar molecules in a field-free region have random orientations. (b) Polar molecules in the space between two charged plates tend to be aligned along the field direction.

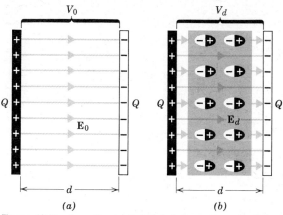

Figure 12.22 Inserting a dielectric between a pair of charged plates decreases the electric field and the potential difference.

capacitance of a capacitor to increase when a dielectric is inserted into the space between the plates? The answer lies in the fact that many materials consist of molecules in which the electric charge is naturally separated or becomes separated when the material is placed in a field. That is, one end of such a molecule carries a positive charge, whereas the other end carries a negative charge, but the molecule as a whole is electrically neutral. Such molecules are called *polar* molecules.

If a polar molecule is in a field-free region, it will move about at random and will have no particular orientation (Fig. 12.21a). On the other hand, if a polar molecule is in the space between a pair of charged plates, the negative end of the molecule will be attracted toward the positively charged plate and the positive end will be attracted toward the negatively charged plate (Fig. 12.21b). Thus, a polar molecule in an electric field will tend to be aligned with the direction of the field lines. (Notice that the molecule is electrically neutral as a whole, so there is no net force on the molecule. However, there is a *couple*, due to the forces on the two charged ends, and the resulting torque tends to align the molecule with the field direction.)

Let us now look more closely at the effect of placing a material consisting of polar molecules between a pair of charged plates. Figure 12.22a shows the initial condition: a pair of plates separated by a distance d carries a uniform charge of magnitude Q on each plate. The voltage across the plates is V_0 and the electric field strength is $E_0 = V_0/d$. Notice that each field line originates on a positive charge and terminates on a negative charge, as must always be the case. The capacitance of the plates is C_0, so that

$$Q = C_0 V_0 = C_0 E_0 d \qquad (12.37)$$

Next, we insert between the plates a dielectric material (Fig. 12.22b). The polar molecules align with the field and produce a negatively charged surface at the left-hand edge and a positively charged surface at the right-hand edge. Now there are negative charges on the left which terminate some of the field lines that originate on the positively charged plate. Furthermore, the same number of field lines originate on the surface charges at the right and terminate on the negatively charged plate. The net result is that there are fewer field lines in the dielectric than in the vacuum space of the initial situation (Fig. 12.22a). That is, the electric field strength E_d in the dielectric is weaker than the original field E_0. (Now that we have seen in detail how the termination of the field lines arises, we can imagine that the dielectric is enlarged to fill the entire space between the plates.)

Because no charge has been lost in the process of inserting the dielectric between the plates, the charge on each plate is still Q. Therefore, we can write for the case in Fig. 12.22b.

$$Q = C_d V_d = C_d E_d d \qquad (12.38)$$

Equating the corresponding parts of Eqs. 12.37 and 12.28, we find

$$C_0 V_0 = C_d V_d \qquad C_0 E_0 = C_d E_d$$
$$V_d = \frac{C_0}{C_d} V_0 = \frac{V_0}{k} \qquad E_d = \frac{C_0}{C_d} E_0 = \frac{E_0}{k} \qquad (12.39)$$

where we have used Eq. 12.36 to substitute $k = C_d/C_0$. We see that the potential difference between the plates is reduced from V_0 to V_0/k and the field strength is reduced from E_0 to E_0/k by introducing the dielectric.

Why are dielectrics used in capacitors? First, the use of a dielectric simplifies the mechanical construction of a capacitor; it is much easier to maintain a uniform spacing between two plates or metallic sheets when a solid material is used as a spacer than when an air gap is used. An additional reason is associated with the fact that a capacitor is a device for storing electric charge (and, as we will see in the next subsection, also for storing *energy*). Depending on the application, a capacitor may be called on to store charge for a tiny fraction of a second (as in high-speed electronic circuits) or for many minutes or hours (as in some types of electric discharge apparatus). The amount of charge that can be stored depends on the applied voltage and on the capacitance. The use of a dielectric medium in a capacitor not only increases the capacitance but it also permits the use of higher voltages before the capacitor breaks down because of sparking between the plates.

Energy Storage in Capacitor

Suppose that we begin with a pair of uncharged parallel plates. Imagine that we detach and move electrons from one plate to the other until each plate carries a charge Q. Work has been done to bring about this distribution of charge. We know that moving a charge q through a potential difference V requires an amount of work $W = qV$ (Eq. 12.19). The first electron that is moved between the uncharged plates requires (almost) zero work. The final electron is moved through a potential difference V and therefore the work required to effect this last transfer of charge is $e \times V$. Because the voltage between the plates builds up steadily as we transfer charge, we can say that the *average* potential difference through which *all* of the charge is moved is one-half the final voltage, or $\frac{1}{2}V$. Consequently, the work required to move the entire charge Q is

$$W = \frac{1}{2} QV \qquad (12.40)$$

Using Eq. 12.24, we can write

$$W = \frac{1}{2} QV = \frac{1}{2} CV^2 = \frac{Q^2}{2C} \qquad (12.41)$$

The work done in moving the charge Q from one plate to the other is stored energy—electric potential energy—that can be recovered by connecting the two plates together and allowing the excess charges to flow. These flowing charges constitute a current and, as we will see in Section 12.7, a current can do work. Where does this potential energy reside? Usually, we say that the energy is contained in the electric field between the plates. If we use Eq. 12.25a for C and remember that $V = Ed$ (Eq. 12.22), we have for the stored energy.

$$\mathscr{E} = \frac{1}{2} CV^2 = \frac{1}{2} \times \frac{\epsilon_0 A}{d} \times (Ed)^2 = \frac{1}{2} \epsilon_0 E^2 \times (Ad)$$

Now, Ad is just the *volume* of the space between the plates. Therefore, the energy stored per unit volume of the field is

$$\left.\begin{array}{l}\text{energy stored in the}\\\text{electric field per unit}\\\text{volume}\end{array}\right\} = \frac{\mathscr{E}}{Ad} = \frac{1}{2} \epsilon_0 E^2 \qquad (12.42)$$

That is, the energy stored is proportional to the *square* of the electric field strength.

● *Example 12.15*

How much energy is stored in a 10-μF capacitor that is charged to a voltage of 12 V?
We use Eq. 12.41 and write

$$\mathscr{E} = \tfrac{1}{2} CV^2 = \tfrac{1}{2} \times (10 \times 10^{-6} \text{ F}) \times (12 \text{ V})^2$$
$$= 7.2 \times 10^{-4} \text{ J}$$

For comparison, the energy stored in a typical 12-V automobile battery is about 2×10^6 J.

● *Example 12.16*

Two plates, 1 m × 1 m, are separated by a uniform air gap of 1 mm. If a voltage of 2.5 kV is applied to the plates, how much energy is stored in the field?
First, we need to calculate the capacitance of the plates; we use Eq. 12.25a:

$$C = \frac{\epsilon_0 A}{d} = \frac{(8.85 \times 10^{-12} \text{ F/m}) \times (1 \text{ m} \times 1 \text{ m})}{10^{-3} \text{ m}}$$
$$= 8.85 \times 10^{-9} \text{ F}$$

Then, using Eq. 12.41,

$$\mathscr{E} = \tfrac{1}{2} CV^2 = \tfrac{1}{2} \times (8.85 \times 10^{-9} \text{ F}) \times (2.5 \times 10^3 \text{ V})^2$$
$$= 0.0277 \text{ J}$$

This is actually a rather small amount of energy; it is approximately the kinetic energy that a 1-g mass acquires in falling through a distance of 3 m.
In this example, the electric field strength is $E = V/d = (2.5 \times 10^3 \text{ V})/(10^{-3} \text{ m}) = 2.5$ MV/m. This is close to the maximum field strength (about 3 MV/m) that an air gap can withstand without a breakdown of the insulating quality of the air, which results in a spark. ∎

Membrane Potentials and Nerve Impulses

The walls of animal cells are thin membranes that consist of two layers of protein separated by a layer of lipid (fat) molecules (Fig. 1). Each of these layers has a thickness of about 30 Å, so the thickness of the entire membrane is about 90 Å.

A cell membrane separates two regions that contain a variety of ions in solution. In the intracellular region we find a preponderance of Na^+ and Cl^- ions, whereas in the interior region of a cell we find that K^+ ions have the greatest concentration. These ions can diffuse through the porous protein structure of the membrane, but, as we will see, the diffusion rates for the various ions are quite different. In addition to the Na^+, K^+, and Cl^- ions, the extracellular and intracellular fluids contain a variety of negative ions (such as phosphate ions, carbonate ions, and larger organic ions). These ions are all larger than the pores in the protein layers through which the small ions diffuse, so we can essentially ignore diffusion effects for the large negative ions.

The table at the top of p. 272 lists the concentrations of the important ions in the extracellular and intracellular regions. (The concentrations are measured in 10^{-3} moles per liter, mmol/ℓ.) The concentration of the potassium ions, for example, is 30 times greater inside the cell than outside. If we had a passive membrane separating the two regions at the same electric potential, the K^+ ions would diffuse through the membrane with equal rates in both directions and no concentration gradient could exist. However, on the surfaces of a cell membrane there exists a double layer of charge that produces a potential difference of approximately 70 mV between the intracellular and extracellular regions (Fig. 2). This potential difference assists in maintaining the concentration gradients of the various ions. Notice that the excess charge in each region resides on the membrane surface; the fluid in each region is electrically neutral (see Exercise 1).

The potential difference V that can support an equilibrium concentration ratio C_i/C_e across a membrane at normal body temperature (310 K) is given by the *Nernst equation*:

$$V = V_i - V_e = -61 \log \frac{C_i}{C_e} \text{ mV} \tag{1}$$

This equation applies for positive ions; for negative ions, the sign changes. We interpret the Nernst equation as follows. The concentrations of Cl^- ions are $C_e = 125$ mmol/ℓ and $C_i = 9$ mmol/ℓ. If these concentrations are to be in equilibrium, the potential difference across the membrane must be

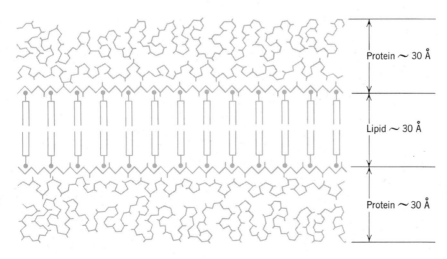

Figure 1 Schematic cross section of a cell membrane. Pores in the protein layers allow small ions to diffuse through the membrane. Membrane thicknesses are in the range from about 75 Å to 100 Å.

Ion	Intracellular Concentration, C_i (mmol/ℓ)	Extracellular Concentration, C_e (mmol/ℓ)	C_i/C_e	Nernst Potential (mV)
Na$^+$	15	145	0.103	+60
K$^+$	150	5	30	−90
Cl$^-$	9	125	0.072	−70

$$V(\text{Cl}^-) = +61 \log \frac{9}{125} = -70 \text{ mV}$$

The actual potential difference is indeed −70 mV, so the Cl$^-$ ions are in equilibrium; that is, the chlorine ions diffuse into and out of the cell with equal rates and the concentration ratio is maintained by the potential difference. The same calculation for the Na$^+$ ions gives $V(\text{Na}^+) = +60$ mV. This potential is far from the actual potential of −70 mV, so the sodium ions tend to diffuse from the region of high concentration to the region of low concentration (*into* the cell). Finally, the calculated potential for the K$^+$ ions is −90 mV, not far from the actual potential. But the potential of the cell interior is not sufficiently negative to cause an equilibrium situation. Consequently, potassium ions diffuse *out* of the cell in response to the concentration difference. In the case of the K$^+$ ions, the electric gradient tends to promote the diffusion in the opposite direction (*into* the cell). Thus, both directions of diffusion are indicated in Fig. 2. The net diffusion effect, however, is an outward drift of the potassium ions.

If diffusion were the only process governing the ions in and around a cell, there would be an inward movement of the sodium ions and an outward movement of the potassium ions; the concentration ratios would change until new equilibrium conditions were reached. This does not happen because, in addition to diffusion, there is

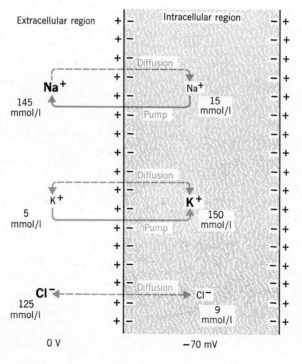

Figure 2 Schematic representation of the effects of diffusion and active transport in the presence of a potential difference across a cell membrane. It is customary to set the potential in the extracellular region equal to zero: $V_e = 0$.

an *active transport* of Na$^+$ and K$^+$ ions through the membrane. This active transport is caused by a mechanism (some sort of chemical reaction) called the *sodium-potassium pump*. The pump operates to inject K$^+$ ions into the cell and to drive Na$^+$ ions out of the cell, in opposition to the normal diffusion for both ion species. Energy must be expended to operate the pump and this energy is furnished by the metabolic processes within the cell. The details of how the pump actually functions are not known.

The resting membrane is 50 to 100 times as permeable to K$^+$ ions as it is to Na$^+$ ions. That is, as soon as potassium ions are pumped *into* the cell, they tend to diffuse *out* again. But, remember, the Nernst potential for potassium ions (-90 mV) is near the actual intracellular potential (-70 mV). This means that the potassium pump does not have to work very hard to maintain the proper concentration ratio. In fact, the potassium pump is considerably weaker than the sodium pump which must work to maintain a Nernst potential of $+60$ mV in the face of an actual potential of -70 mV.

Because a membrane is so thin, the fact that a potential difference of 70 mV exists between its external and internal surfaces means that the electric field strength E within the membrane is very high:

$$E = \frac{V}{d} = \frac{0.070 \text{ V}}{90 \times 10^{-10} \text{ m}} = 7.8 \times 10^6 \text{ V/m}$$

This is indeed a very high field strength: In air, sparking will occur between two points when the field strength reaches about 3×10^6 V/m. For comparison, special insulating oils have breakdown field strengths of 10^7 V/m or greater.

Even though the field strength within a membrane is high, relatively few of the cell's ions need to be deposited on the membrane surfaces to establish the field. We can see this in the following way. A typical cell in the human body has a volume of about 10^{-15} m^3 and a surface area of about 5×10^{-10} m^2. Moreover, it has been determined that cell membranes have capacitance values of about 10^{-2} F/m^2 (or 1 μF/cm^2). The total charge on a typical cell's membrane is, using Eq. 12.24,

$$Q = CV = (C \text{ per unit area}) \times (\text{area}) \times V$$
$$= (10^{-2} \text{ F/m}^2) \times (5 \times 10^{-10} \text{ m}^2) \times (0.070 \text{ V})$$
$$= 3.5 \times 10^{-13} \text{ C}$$

Then, the number n of (singly charged) ions is

$$n = (3.5 \times 10^{-13} \text{ C}) \times \frac{1}{1.6 \times 10^{-19} \text{ C/ion}} = 2 \times 10^6$$

We can compare n with the number N_K of potassium ions inside the cell:

$$N_K = (\text{concentration}) \times (\text{volume}) \times (\text{Avogadro's number})$$
$$= \left(\frac{150 \times 10^{-3} \text{ mol}}{10^{-3} \text{ m}^3}\right) \times (10^{-15} \text{ m}^3) \times (6 \times 10^{23} \text{ ions/mol})$$
$$= 9 \times 10^{10}$$

Thus, only about $(2 \times 10^6)/(9 \times 10^{10}) = 2 \times 10^{-5}$ or 1 part in 50,000 of the potassium ions inside the cell must be transported to the external surface to establish the field within the membrane.

The Action Potential

The nervous system of an animal consists of nerve fibers that extend throughout the body and are grouped into bundles as the spinal column and then the brain are approached. An individual nerve cell (or *neuron*) is composed of a cell body and a single long extension called an *axon*. These axons can be quite long; the axons that control finger movements extend the entire length of the arm, with the cell body located in the spinal column. Electrical signals are propagated along the axons to carry sensory information to the brain or to carry motor instructions from the brain to various parts of the body.

When carrying information to or from the brain, axons receive stimuli from other axons only at the ends (through *synapses*). However, a nerve cell can be stimulated into action at any point along the axon by electrical, chemical, thermal, or mechanical means. Figure 3 shows the result of such a stimulus. In the resting condition (Fig. 3a), the positive and negative charges are distributed uniformly along the axon; the interior potential is -70 mV. In Fig. 3b some sort of stimulus is applied, the effect of which is a sudden increase in the permeability of the membrane to sodium ions (by a factor of 5000 or so). The influx of Na^+ ions from the extracellular region overcomes the negative potential of the cell and brings about a local positive potential (Fig. 3b). This positive potential, which develops within a fraction of a millisecond and rises to about $+40$ mV, is called the *action potential* (see Fig. 4).

The reversal of the membrane charge is called *depolarization*. The effect of depolarization at one point is to induce depolarization on both sides of the stimulated point. Thus, the action potential propagates along the axon (Fig. 3c) until the entire region is depolarized (Fig. 3d).

Almost immediately after depolarization takes place, the membrane returns to its previous condition of low permeability to sodium ions. This allows a large quan-

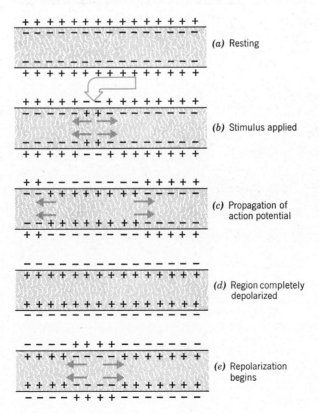

Figure 3 Schematic representation of the development and propagation of an action potential along an axon of a nerve cell. (See also Fig. 4.)

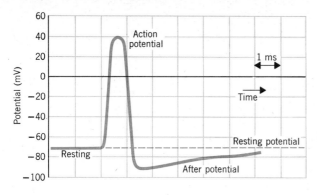

Figure 4 The action potential for a nerve axon. The potential as a function of time is shown for a particular point in the axon.

tity of K^+ ions to diffuse quickly *out* of the cell, thereby lowering the potential of the cell to negative values. The potassium "gate" remains open too long in this *repolarization* process and, consequently, the potential drops below the resting potential (see Fig. 4). Finally, active transport becomes effective and returns the cell to its original condition after an interval of 10 ms or so. The cell is incapable of "firing" again until the recovery process is completed. (The role of active transport in the recovery process is not clear; the cell can "fire" a number of times in succession even if the active transport mechanism is disabled.)

The action potential is an all-or-nothing response; either it is there in its entirety or it does not occur. If a stimulus has a magnitude sufficient to trigger an action potential, the potential has the same size and shape regardless of the type, magnitude, or duration of the stimulus. (A neuron is similar to a bipolar circuit element—it is either *on* or *off*.)

The velocity with which an action potential propagates along a nerve fiber depends on the diameter of the fiber and whether or not the axon is covered with a sheath of *myelin* (a lipid protein structure). For small unmyelinated fibers the velocity can be as low s 0.5 m/s, whereas for large myelinated fibers the velocity can be as high as 130 m/s. Clearly, nerve signals do not propagate as ordinary electric signals do. Nerve propagation is a complex electrochemical process, one that is not yet understood in detail.

References

E. Ackerman, *Biophysical Science,* Prentice-Hall, Inc., Englewood Cliffs, N.J., 1962, Chapter 4.

A. C. Guyton, *Textbook of Medical Physiology,* 5th ed., Saunders, Philadelphia, 1976, Chapter 10.

F. R. Hallett, P. A. Speight, and R. H. Stinson, *Introductory Biophysics,* Halsted (Wiley), New York, 1977, Chapter 13.

R. B. Setlow and E. C. Pollard, *Molecular Biophysics,* Addison-Wesley, Reading, Mass., 1962, Chapter 14.

A. J. Vander, J. H. Sherman, and D. S. Luciano, *Human Physiology: The Mechanisms of Body Function,* McGraw-Hill, New York, 1970, Chapters 5 and 9.

■Exercises

1. Current does not flow spontaneously through a cell's membrane. This means that both the extracellular fluid and the intracellular fluid are electrically neutral. What concentrations (in mmol/ℓ) of organic negative ions are needed in the two regions to maintain neutrality?

2 Use the information in the essay and obtain a value for the dielectric constant for a typical membrane. Is your value reasonable? Explain. (For lipid, $k \cong 3$.)

3 It has been estimated that between 10 and 40 percent of the body's energy intake (which totals about 3000 Cal/day) is expended in maintaining the electric gradient of the cellular membranes. Justify this estimate. Proceed in the following way. The energy stored in the electric field of a capacitor is (see Eq. 11.32) $\mathscr{E} = \frac{1}{2}CV^2$. The voltage V is known (0.070 V) and so is the capacitance per unit area (10^{-2} F/m^2); therefore, we need the total membrane area in the body. Assume a body mass of 75 kg, of which about 20 percent is in the form of extracellular fluid. The remainder of the body mass is divided about equally between large muscle cells and smaller cells. Assume that a typical small cell has a volume of 10^{-15} m^3 and an area of 5×10^{-10} m^2. Assume that a typical muscle cell is 20 μm in diameter and has a length of 1 cm. (Because we are concerned here with the total cell *area*, the length assumed for a typical muscle cell does not really matter. Can you see why?) Assume that both types of cells have $\rho = 1$ g/cm^3. With this information, calculate $\mathscr{E}$. Next, we need to know the rate at which this energy must be supplied. That is, at what rate would the membrane charge spontaneously leak away if it were not supported by an energy input from some source? When a nerve cell "fires," the action potential has a duration of about 1 ms; then, an interval of about 10 ms must pass before the cell can "fire" again. The 1-ms value is probably not the correct time to use in the calculation because the "firing" of a nerve cell is a special kind of process that is analogous to the discharge of a capacitor by a short circuit. The recovery process somehow seems to be a more "normal" event. Therefore, assume that the "normal" time during which the charge would leak from the membrane capacitor is equal to the recovery time (about 10 ms). Using this value, complete the calculation.

4 Estimate the number of sodium ions that enter a typical cell during the onset of the action potential. By what fraction does the intracellular concentration of Na$^+$ ions increase during this process?

12.6 Electric Current

The Transport of Electrons

The movement of electric charge from one position to another constitutes a *current*. When an electric current flows in a wire, the positively charged nuclei of atoms remain essentially stationary compared to the movement of the much less massive electrons. Thus, in almost all of the cases with which we shall be concerned, electric current is due to the motion of *electrons*.

In any material that conducts a current (whether that material is in solid, liquid, or gaseous form), a fraction of the electrons are not attached to specific atoms but are free to move about in the material. In the absence of an applied electric field, these *free electrons* (or *conduction electrons*) move at random with high speeds (about 10^6 m/s), colliding with the stationary atoms at frequent intervals and thereby changing their directions of motion. Across any given surface in the material, just as many electrons will move in one direction as in the other (Fig. 12.23). Therefore, there is no *net* transport of electrons across the surface and the electric current is *zero*. As we will see, when a current flows, the net speed at which the electrons move through the material is small compared with their random speeds.

Figure 12.23 If no electric field is present, the motion of the free electrons is random.

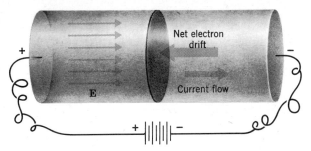

Figure 12.24 The application of an electric field to a conductor causes a net motion of the electrons in the direction opposite to the electric field lines. By convention, the current is said to flow in the same direction as the field lines.

If the ends of the conductor are attached to the terminals of a battery (Fig. 12.24), an electric field is established within the conductor. There will now be a net transport of electrons across any surface that is perpendicular to the field lines. That is, in Fig. 12.24, there will be a few more electrons crossing the surface from *right* to *left* than will be crossing from *left* to *right*.

The direction in which the electrons drift in an electric field is opposite to the direction of the field lines because electrons carry a negative charge. However, just as we arbitrarily elected to consider the direction of electric field lines to be the direction of the force on a *positive* charge, it is traditional to refer to the direction of *current* flow as the direction in which *positive* charges would move (even though it is only the electrons that actually move). A negative current (or an electron current) in one direction would produce exactly the same effects and is therefore equivalent to a positive current in the opposite direction. We shall always use the terms *current* or *current flow* to mean this so-called *conventional* current which flows in the same direction as the electric field lines. The terms *electron flow* or *electron current* will be used to refer explicitly to the motion of the electrons. The advantage in retaining this seemingly awkward convention will become apparent when we discuss electromagnetic phenomena in the next chapter.

Electromotive Force

What causes an electric current to flow? Actually, there are many ways—including mechanical, thermal, and chemical effects—by which an electric current can be produced. We give the name *electromotive force* or *EMF* to any agency that can cause a current to flow. Thus, the chemical reactions within a battery produce an EMF and so does the mechanical motion of the rotor in an electric generator (see Section 14.4). In every case the source of EMF transforms nonelectric energy into electric energy.

The measure of EMF is in units of *volts*. (Thus, we see that *electromotive force* is not really a proper term: *force* is not measured in *volts*. For this reason it is customary to use the abbreviation EMF instead of the original term.) We say that the EMF of a flashlight battery is 1.5 V or that the EMF of a generator is 400 V.

The Definition of Current

Electric current is defined quantitatively as the *rate* at which electric charge flows across a surface. When 1 C of charge crosses a given surface in 1 s, a current of 1 *ampere* (A) is flowing. In general, the current I is given by the net charge Q crossing a surface during a time t:

$$I = \frac{Q}{t} \qquad (12.43)$$

The unit of current is the ampere (A):

$$1 \text{ A} = 1 \text{C/s} \qquad (12.44)$$

Household circuits are usually capable of carrying currents of 15–20 A and household appliances usually require currents of a few amperes. A lightning stroke may involve a current of 100,000 A, whereas the current flowing through a camera's light meter may be only $1\mu\text{A} = 10^{-6}$ A.

Electron Drift Velocities

With what net velocity do electrons move along a current-carrying wire? A typical type of household copper wire has a cross-sectional area of approximately 1 mm². A 1-mm length of this wire has a volume of 1 mm³. Each cubic millimeter of copper contains approximately 8×10^{19} atoms and each atom contributes one free electron. If the electrons move with a net velocity (or *drift* velocity) of 1 mm/s, all of the free electrons in the segment will pass a point in the wire in 1 s. Hence, the current is $I = Q/t = (8 \times 10^{19} \text{ electrons}) \times 1.6 \times 10^{-19}$ C/electron/$1 \text{ s} \cong 13$ A, which is near the limit for current in this type of wire. Thus, we see that because conductors contain such enormous numbers of free electrons, only very small drift velocities are necessary to produce sizable currents. It should be remembered that this net drift velocity of a mm/s is superimposed on the large random electron velocities of about 10^6 m/s.

12.7 Electrical Resistance

Ohm's Law

When a wire is connected to a battery (as in Fig. 12.24), how much current will flow? Many materials have the property that the current flow through a particular sample is directly proportional to the voltage across the sample. That is, if a voltage of 6 V causes a current of 2 A to flow, a voltage of 12 V will cause a current of 4 A to flow, and so forth. The relationship between the voltage and the current is

$$I = \frac{V}{R} \quad \text{or} \quad V = IR \qquad (12.45)$$

where the proportionality constant R is called the *resistance* of the sample. The higher the resistance, the smaller will be the current for a given voltage. This relationship is called *Ohm's law*, in honor of its discoverer, George Simon Ohm (1787–1854). If a voltage of 1 V causes a current of 1 A to flow through a certain sample, the sample is said to have a resistance of 1 *ohm* (1 Ω). Ohm's law is an *approximate* description of the way many materials behave.

Copper is one of the best conductors known (at room temperature); only silver is better, and only slightly so. A 100-m length of household copper wire, with a cross-sectional area of 1 mm², has a resistance of about 1.8 Ω. The resistance of a similar piece of silver is 1.6 Ω, and for carbon the value is 3500 Ω. The resistance of any ordinary household light bulb (the filament of which consists of fine tungsten wire) is about 100 Ω.

● *Example 12.17*

An electrical automobile starter is jammed and cannot move, and yet it draws a current of 50 A. What is the resistance of the starter?

An automobile battery has a voltage of 12 V; therefore, solving Eq. 12.45 for R,

$$R = \frac{V}{I} = \frac{12 \text{ V}}{50 \text{ A}} = 0.24 \text{ Ω}$$

If the starter were not jammed, would you expect it to draw *more* or *less* current? ■

Series and Parallel Circuits

In all of the electric circuits that we will be considering, the connecting wires have such small resistance compared to the other parts of the circuit that the

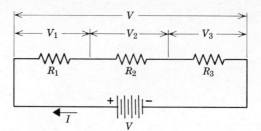

Figure 12.25 A *series* circuit consisting of three resistors connected to a battery.

resistive effects of the wires can be ignored. Suppose that we have three resistive circuit elements (called *resistors* and represented by sawtooth lines in diagrams) connected to a battery, as in Fig. 12.25. This type of connection is called a *series* circuit. What is the *total* resistance of the circuit?

In a series circuit, the same current I must flow through each of the resistors. The voltage across each resistor is given by Ohm's law:

$$V_1 = IR_1; \quad V_2 = IR_2; \quad V_3 = IR_3 \qquad (12.46)$$

The sum of these three voltages must equal the voltage across the batter (because the wires have negligible resistance):

$$\begin{aligned} V &= V_1 + V_2 + V_3 \\ &= IR_1 + IR_2 + IR_3 \\ &= I(R_1 + R_2 + R_3) \end{aligned} \qquad (12.47)$$

Now, the total resistance R_t is related to the battery voltage V and current I by

$$V = IR_t \qquad (12.48)$$

Combining Eqs. 12.47 and 12.48, we see that the total resistance is

$$R_t = R_1 + R_2 + R_3$$
(series connection) $\qquad (12.49)$

Next, suppose that the three resistors are connected together in a *parallel* arrangement, as shown in Fig. 12.26. In this case, the total current I flowing in the circuit is divided among the three resistors. However, the voltage across each resistor is now the same. Therefore, we can write for each resistor,

$$I_1 = \frac{V}{R_1}; \quad I_2 = \frac{V}{R_2}; \quad I_3 = \frac{V}{R_3} \qquad (12.50)$$

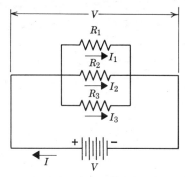

Figure 12.26 A *parallel* circuit consisting of three resistors connected to a battery.

The sum of these three currents is equal to the total current I:

$$I = I_1 + I_2 + I_3$$
$$= \frac{V}{R_1} + \frac{V}{R_2} + \frac{V}{R_3}$$
$$= V\left(\frac{1}{R_1} + \frac{1}{R_2} + \frac{1}{R_3}\right) \qquad (12.51)$$

The total resistance R_t is related to V and I by

$$I = V\left(\frac{1}{R_t}\right) \qquad (12.52)$$

Comparing Eqs. 12.51 and 12.52, we see that the total resistance is

$$\frac{1}{R_t} = \frac{1}{R_1} + \frac{1}{R_2} + \frac{1}{R_3}$$
(parallel connection) \qquad (12.53)

Notice that the expressions for the total resistance in series and parallel connections (Eqs. 12.49 and 12.53) have the *opposite* form compared to the equivalent expressions for the capacitor connections (Eqs. 12.29 and 12.35).

● *Example 12.18*

What current flows in the circuit illustrated below?
The total resistance in the circuit is obtained from

$$\frac{1}{R_t} = \frac{1}{R_1} + \frac{1}{R_2} + \frac{1}{R_3} = \frac{1}{4} + \frac{1}{6} + \frac{1}{12}$$
$$= \frac{6+4+2}{24} = \frac{12}{24} = \frac{1}{2}$$

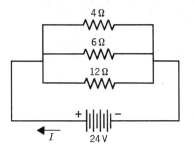

Therefore,

$$R_t = 2 \;\Omega$$

Then, the current is

$$I = \frac{V}{R_t} = \frac{24 \text{ V}}{2 \;\Omega} = 12 \text{ A} \quad \blacksquare$$

Resistivity

When a sample of some material is connected to the terminals of a battery, the conduction electrons move through the sample and a current flows. The rate at which the electrons move is determined by the number of collisions they make with the atoms of the sample. If the collision rate is high, the motion of the electrons is significantly impeded and we say that the electrical resistance of the sample is high. If the collision rate is low, the resistance is low. Whether a sample has a high or a low resistance to electron flow depends on the atomic character of the material; a tungsten wire, for example, will have a higher resistance than an identical wire made from copper or silver.

The resistance of a sample depends not only on the type of material and the temperature but also on the size and shape of the sample. Thus, in Fig. 12.27, we expect that the long, thin conductor will have a higher resistance than the short, fat sample (of the same material). Because the number of collision sites is directly proportional to the length of the sample, we expect $R \propto L$. Moreover, if the cross-sectional area A of the sample is increased, more conduction electrons can pass a given point per unit time; this means $R \propto 1/A$. If we let the proportionality constant be ρ, we have

$$R = \rho \frac{L}{A} \qquad (12.54)$$

The quantity ρ is called the *resistivity* of the material and is an intrinsic property of the material (as is the *density*), depending neither on the size nor the

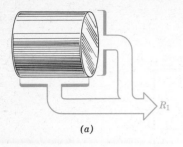
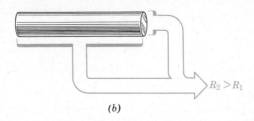

Figure 12.27 The electrical resistance of a sample of a particular material is proportional to the ratio of its length to its area, $R \propto L/A$.

shape of any particular sample. It is easy to see that the units of ρ are ohm-meters (Ω-m).

Table 12.3 lists the resistivity values for several common metals and, for contrast, the values for the very poor conductor, carbon, and for the good insulator, Teflon.

The electrical properties of materials are sometimes described in terms of a quantity called the *conductivity*, usually denoted by the symbol σ. The conductivity of a material is equal to the reciprocal of its resistivity: $\sigma = 1/\rho$.

When the temperature of a material is increased, the atoms vibrate more rapidly. Consequently, electrons that are moving through a metal collide more frequently with the agitated atoms. For this reason, the electrical resistance of metals increases with temperature. The resistivity values of the common metals increase about 0.4 percent per degree at ordinary temperatures. Therefore, if ρ_0 is the resistivity at some reference temperature (we use 20°C), then the resistivity at other temperatures not too different is given by

$$\rho = \rho_0(1 + \alpha \Delta T) \qquad (12.55)$$

where the *temperature coefficient of resistivity* is $\alpha = 0.004$ deg^{-1} (and is approximately the same for most metals) and ΔT is the temperature difference. Notice that the value of α that represents the sensitivity to change in electrical resistance with temperature is much greater than the equivalent coefficient that represents the sensitivity to change in physical length of a material with temperature (see Table 9.2).

● *Example 12.19*

A tungsten wire has a radius of 1 mm and a length of 100 m. What is the electrical resistance of this wire at 20°C?

We use Eq. 12.54 with the value of ρ for tungsten taken from Table 12.3

$$R = \rho \frac{L}{A} = \rho \frac{L}{\pi r^2}$$

$$= (5.6 \times 10^{-8} \, \Omega\text{-m}) \times \frac{100 \text{ m}}{\pi \times (10^{-3} \text{m})^2}$$

$$= 1.78 \, \Omega$$

● *Example 12.20*

What will be the resistance of the tungsten wire in the preceding example if it is heated to a temperature of 500°C?

Equation 12.55 gives the *resistivity* of a material as a function of temperature. Because the resistance R is directly proportional to ρ (Eq. 12.54), we can also express the *resistance* of a sample in the same form by Eq. 12.55:

$$R = R_0(1 + a\Delta T)$$

Then, we have

$$R = (1.78 \, \Omega) \times [1 + (0.004 \text{ deg}^{-1}) \times (500°C - 20°C)]$$

$$= (1.78 \, \Omega) \times (2.92)$$

$$= 5.2 \, \Omega$$

which is a substantial increase. Actually, the value of R will be even greater than 5.2 Ω because the coefficient α for tungsten increases from 0.004 deg^{-1} at 20°C to 0.0057 deg^{-1} at 500°C. At a temperature of 1000°C, the value of α has increased to 0.0089 deg^{-1}. ■

Table 12.3 Electrical Resistivities for Various Materials at Room Temperature (20°C)

Material	Resistivity, ρ (Ω-m)
Silver	1.59×10^{-8}
Copper, annealed	1.72×10^{-8}
Gold	2.44×10^{-8}
Aluminum	2.82×10^{-8}
Tungsten	5.6×10^{-8}
Iron	1.0×10^{-7}
Platinum	1.0×10^{-7}
Nichrome (an alloy of nickel, iron, and copper)	0.9–1.1×10^{-6}
Carbon	3.45×10^{-5}
Teflon	$> 10^{16}$

Equation 12.55 tells us that the resistivity of a material decreases as the temperature is lowered. If this relation were strictly valid at all temperatures, we should expect the resistivity to become *zero* at a temperature of −250°C (or 23 K). Actually, this does not happen for most materials. Instead, the resistivity levels off at some value as the temperature is decreased, even near absolute zero. There does exist a class of materials called *superconductors* which have the property that at very low temperatures (about 20 K or lower) the resistivity *does* become zero. These extraordinary materials offer no resistance at all to the flow of electric current! We will learn more about these exceptional substances and their applications in Chapter 20.

12.8 Electric Power

Power and Energy

Power is the rate at which work is done (or, equivalently, the rate at which energy is used). If an amount of work W is done in a time t, the average power expended during that time is (Eq. 7.5)

$$P = \frac{W}{t} \qquad (12.56)$$

The unit of work (or energy) is the *joule*, so the unit of power is the J/s, which we call the *watt* (W):

$$1 \text{ watt} = \frac{1 \text{ joule}}{\text{second}}; \quad 1 \text{ W} = 1 \text{ J/s} \qquad (12.57)$$

Suppose that a voltage of 1 V causes a current of 1 A to flow in a certain piece of wire. We can calculate the *power* expended in the wire by taking the product of voltage and current:

$$(1 \text{ V}) \times (1 \text{ A}) = \left(1 \frac{\text{J}}{\text{C}}\right) \times \left(\frac{1 \text{ C}}{\text{s}}\right) = 1 \frac{\text{J}}{\text{s}} = 1 \text{ W} \qquad (12.58)$$

where we have used the definitions of the volt (1 J/C) and the ampere (1 C/s). We see that this product is equal to 1 W; that is, the *power* or the rate at which energy (electrical energy) is being expended in the wire is 1 W. In general, for a voltage V and a current I, the power is

$$P = VI \qquad (12.59)$$

and using Ohm's law, $V = IR$, to substitute for V, we can express the power supplied to a resistance R by a current I as

$$P = I^2 R \quad \text{or} \quad P = \frac{V^2}{R} \qquad (12.60)$$

The total amount of energy consumed by an element of an electrical circuit is equal to the *rate* of energy use (that is, the *power*) multiplied by the total time:

$$W = Pt$$

If a current of 2 A flows for 1 h through a circuit whose resistance is 5 Ω, the energy used is

$$W = Pt = I^2 R \times t$$
$$= (2 \text{ A})^2 \times (5 \text{ Ω}) \times (3600 \text{ s}) = 72{,}000 \text{ J}$$

In electrical practice we usually express energy in units of watts multiplied by the time in hours. For the case above,

$$W = (2 \text{ A})^2 \times (5 \text{ Ω}) \times (1 \text{ h})$$
$$= 20 \text{ watt-hours (Wh)}$$
$$= 0.020 \text{ kilowatt-hours (kWh)}$$

The kWh is the usual unit by which electrical energy is sold by commercial power companies. A typical household use pays several cents per kWh. (The range is from about 2 cents per kWh near large hydroelectric plants to nearly 10 cents per kWh in high-demand areas where expensive fossil fuels must be used to generate electrical energy.)

• *Example 12.21*

A room heater has a resistance of 10 Ω and is used continuously for one day. What is the cost of this operation if the billing rate is 4 cents per kWh?

Ordinary household voltage is 110 V. Therefore, the current drawn by the heater is

$$I = \frac{V}{R} = \frac{110 \text{ V}}{10 \text{ Ω}} = 11 \text{ A}$$

The power is

$$P = I^2 R = (11 \text{ A})^2 \times (10 \text{ Ω}) = 1210 \text{ W}$$

The energy used in one day is

$$W = Pt = (1210 \text{ W}) \times (24 \text{ h})$$
$$= 2.9 \times 10^4 \text{ Wh} = 29 \text{ kWh}$$

and the cost is

cost = (29 kWh) × ($0.07/kWh) = $2.03 per day ∎

The resistance to current flow in a material is caused by the free electrons colliding with the atoms of the material and thereby causing them to be agitated more severely than in their normal state. The flow of current therefore raises the internal energy of the material, and electrical energy is converted into thermal energy. When current flows through the filament of a light bulb, some of the energy used appears as light, but the efficiency of a light bulb is very low and most of the energy—90 percent or more—appears as heat. Indeed, in almost all ordinary applications of electricity, most of the electrical energy is consumed in heating effects.

When James Prescott Joule was making measurements to determine the mechanical equivalent of heat (see Section 9.3), he performed experiments involving the heating effect of an electric current. Joule discovered in these experiments that the heat produced by an electric current is proportional to the square of the current and to the resistance of the sample, that is, $I^2 R$. For this reason, we sometimes refer to $I^2 R$ heating as *Joule heating*.

Summary of Important Ideas

The electrostatic force depends on $1/r^2$ and is a *long-range* force.

For purposes of electrical calculations, the entire charge on a uniform spherical distribution of charge can be considered to be concentrated at its center.

The net electric charge of an isolated system remains constant.

A convenient unit of energy called the electron volt (eV) is the energy acquired by a charge e (1.602×10^{-19} C) falling through a potential difference of one volt.

An electric charge of a distribution of charges sets up in space an *electric field* to which any other charge will react by experiencing a force. The electric field is characterized at any point by a *magnitude* and a *direction;* these quantities define the *field vector* **E**. The magnitude of E at a point is equal to the force per unit charge, F_E/q, exerted on a charge q located at that point. The direction of E is the same as the direction of the electric force F_E exerted on a *positive* test charge.

A convenient way to represent an electric field is in terms of *lines of force* or *electric field lines*. Electric field lines originate on positive charges and terminate on negative charges.

The *electric potential* at a point in space is equal to the electric potential energy per unit charge at the point. The electric potential energy, in turn, is equal to the work required to move a charge from infinitely far away to the particular point.

The difference in electrical potential between two points is equal to the *voltage* V (or *potential difference*) between those points.

The *electric field* inside a current-free conductor is *zero*.

The electric field between two uniformly charged parallel plates of equal area is *uniform* and the field strength is equal to the voltage between the plates divided by the spacing between the plates.

The electric field lines at the surface of a conductor are always *perpendicular* to the surface.

A *capacitor* consists of a pair of conducting plates that can accumulate and store electric charge. The *capacitance* of a capacitor is measured in *farads* and is equal to the charge on the plates divided by the voltage between the plates. The capacitance value can be increased by filling the space between the plates with a *dielectric* material.

The total capacitance of a group of capacitors connected in *parallel* is equal to the sum of the individual capacitances. The total capacitance of a group of capacitors connected in *series* is equal to the reciprocal of the sum of the reciprocals of the individual capacitances.

The amount of energy stored in the electric field in a capacitor C that is charged to a voltage V is equal to $\frac{1}{2} CV^2$.

The net flow of electric charge (usually electrons) past a particular point in a conductor constitutes an *electric current*. The magnitude of the current I is equal to the net charge Q that moves past the point per second: $I = Q/t$. Electric current is measured in *amperes*.

Any agency that can cause a current to flow by converting nonelectric energy into electric energy is called an *EMF*.

Ohm's law states that the current that flows through a particular sample is directly proportional to the voltage across that sample: $I = V/R$. The quantity R is the electrical *resistance* of the sample and is measured in ohms (Ω): 1 Ω = 1 V/A.

Resistors that are connected in *series* and *parallel* combinations have total resistance values given by expressions that apply for the *opposite* type of connection for capacitors.

The *power* dissipated in a resistor R is given by $P = I^2 R$. This type of power expenditures is called *Joule heating*.

◆ Questions

12.1 Two protons approach each other in space. Describe their relative orbits if they approach (a) head-on and (b) with initially parallel but displaced paths.

12.2 In the Cavendish experiment, can the force on one of the small balls due to the farther large ball be neglected in comparison to the force due to the nearer large ball?

12.3 Two objects carry small electric charges. Explain how an electroscope could be used to determine whether the charges are of the same or opposite signs.

12.4 A conducting rod A is supported on an insulating stand. Another conducting rod B is suspended near the end of rod A by strings. What will happen to B when an positively charged object C is brought close to the end of rod A that is farthest from rod B? What would happen if C were *negatively* charged?

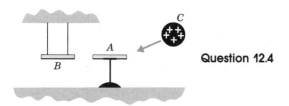

Question 12.4

12.5 Two styrofoam balls, suspended on threads as in Fig. 5.10, carry equal and opposite charges so they are attracted toward each other and the threads are deflected. If an *uncharged* copper sphere is placed midway between the styrofoam balls, in what way will the deflection of the threads be changed? Explain in detail.

12.6 If, on a dry day, you shuffle across a carpet made from some artificial fiber and then touch a door knob, what will happen? Explain.

12.7 You have two identical copper spheres that are suspended by insulating threads. In addition, you have a plastic rod that carries a certain negative charge. If you wish to place exactly equal negative charges on the copper spheres, explain how you would go about doing this.

12.8 A metal rod is supported on an insulating stand. You wish to place a *negative* charge on the rod, but you have only an object that carries a *positive* charge. Explain how you could accomplish your aim.

12.9 The wingspan of birds is a factor in determining the spacing between high-voltage power lines. Why?

12.10 If a charged object is released (from rest) in an electric field, its motion does not necessarily follow a field line. Explain why. Under what special conditions would the motion be exactly along a field line?

12.11 Can two electric field lines ever cross? Explain.

12.12 Under what conditions can the following be usefully considered to be fields? Specify whether the field is *scalar* or *vector*.

(a) The mass density distribution in the Earth.
(b) The population density in a country.
(c) The population density in a city block.
(d) The density of stars in a galaxy.
(e) The flow of air masses in the atmosphere.

12.13 Four equal positive charges are located at the corners of a square. Sketch the pattern of the resulting electric field.

12.14 Sketch the electric field lines in the vicinity of a positively charged conductor in the shape of a pear.

12.15 Two concentric, hollow conducting spheres carry equal and opposite charges. Argue whether an electric field exists in each of the regions A, B, and C. Reexamine the situation if both spheres carry equal charges of the same sign.

Question 12.15

12.16 At any point along an electric field line, the direction of the line represents the direction that the electric force would act on a positive test charge placed at that point. Now, we know that no work is required to move an object perpendicular to the direction of the net force acting on the object. (See the discussion of conservative forces in Section 7.2.) In an electric field, any path along which a test charge can be moved without the necessity of doing work is called an *equipotential path* (or, in three dimensions, an *equipotential surface*). That is, the potential energy of the test charge is the same anywhere along the path (or surface). Sketch some equipotential paths around an isolated negative charge. Do the same for the field distribution illustrated in Fig. 12.12a.

12.17 Electric current can be conducted through certain liquids (called *electrolytes*) by the movement of *ions* rather than electrons. Even water is normally ionized to a small degree into H^+ and OH^- ions. Sodium sulfate, Na_2SO_4, forms two Na^+ ions and an SO_4^{--} ion. Explain the qualitative differences between the conduction of electric current in an electrolyte and the conduction in a copper wire.

12.18 What are some common devices in which you expect that some part of the electric current is due to the motion of *ions*?

12.19 A wire is connected to the terminals of a battery. Toward what terminal do the electrons in the wire move? Toward what terminal do the negative charges (electrons and negative ions) move in the battery? Explain why the negative charges move toward opposite terminals in the two cases.

12.20 All ordinary conductors have some resistance to the flow of current and so there are always heating losses associated with current flow. When electric power is delivered from the generating plant to users, the transmission lines operate at high voltage and low current. Why is this done? Why not use low voltage and high current? (The product $P = VI$ would be the same. Think about this carefully. If you have difficulty, look ahead to the discussion following Example 14.9.)

12.21 An electron that moves in a wire suffers a collision with an atom of the wire after moving, on the average, a distance ℓ (called the *mean free path*). If ℓ is caused to become smaller, will the resistivity of the material change? What could cause the mean free path to decrease?

★ **Problems**

Section 12.1

12.1 What was the attractive force between one of the large balls and one of the small balls in the Cavendish experiment when the centers of the balls were separated by 0.2 m?

12.2 Two identical spherical copper balls carry charges of $+7 \times 10^{-4}$ C and -3×10^{-4} C and their centers are separated by 2 m. If the balls are brought together until they *touch* and then returned to their original positions, what will be the electrostatic force between them?

12.3 A certain object is given a positive electric charge by removing 10^{-15} kg of electrons. How many electrons were removed and what is the charge on the object?

12.4* A droplet of water has a radius of 0.1 mm. If there is 1 extra electron per 10^9 water molecules, what is the charge on the droplet?

12.5* A 1-kg mass, when allowed to fall freely near the surface of the Earth, experiences an acceleration of 9.8 m/s^2. If this mass carries a charge of 10^{-5} C, what would the charge on the Earth have to be in order to double the acceleration? (Consider the Earth to be uniformly charged.)

12.6 Three charges are located at the following positions along a straight line: $q_1 = +10^{-7}$ C at $x = 0$; $q_2 = -4 \times 10^{-7}$ C at $x = 0.2$ m; $q_3 = +4.5 \times 10^{-7}$ C at $x = 0.5$ m. What is the net force (magnitude and direction) on q_2?

12.7* An equilateral triangle has sides of 0.5 m. Equal charges, $q = +30 \mu$C, are located at each vertex. What is the electric force on any one of the charges?

12.8* A square has sides of 0.5 m. Equal charges, $q = +30 \mu$C, are located at each corner. What is the electric force on any one of the charges?

12.9* In the preceding problem, suppose that the charges alternate in sign around the square. What is the electric force on one of the positive charges?

12.10 A proton and an electron are at rest a distance 10^{-10} m apart. What is the electrostatic force exerted by one of the particles on the other? When they begin to move under the influence of this force, what is the initial acceleration of each particle?

12.11 Two α particles (helium nuclei) are separated by a distance of 1 Å. What electrostatic force does each α particle exert on the other?

Section 12.2

12.12 Two small spherical charges of $+3\,\mu C$ and $-8\,\mu C$ are initially separated by 4 cm. What is the change in electrostatic potential energy if the separation is increased by 2 cm? How much work had been done. What is the change in electrostatic potential energy if the separation is reduced to 2 cm? Is work done on the charges in the latter case?

12.13 How much work is required to move the charge $q = +2$ from A to B? (All charges are in units of mC = 10^{-3} C.)

12.14 What is the difference in electrostatic potential energy between configurations A and B? (All charges are in units of mC = 10^{-3} C.)

12.15 The potential at a point 3 m above ground level is $+10^5$ V compared to ground. How much work is required to raise a 1-kg object that carries a charge of $+2 \times 10^{-4}$ C from ground level to the 3-m point? How much work would be required if the charge on the object were $q = -2 \times 10^{-4}$ C?

12.16 Two protons are separated by 10^{-9} m. An electron is on the line connecting the protons and is at a distance of 10^{-10} m from one of the protons. How much work is required to move the electron to point B which is at a distance of 10^{-10} m from the other proton?

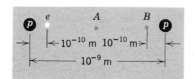

What is the change in potential energy if the electron is moved to point A, midway between the two protons?

Section 12.3

12.17* A 2-MeV neutron strikes a proton (initially at rest) and attaches to it to form a deuteron (the nucleus of "heavy hydrogen," consisting of a neutron and a proton bound together). Consider the collision to be perfectly inelastic and compute the velocity of the deuteron.

12.18 If a completely ionized carbon nucleus (six protons and six neutrons) falls through a potential difference of 3×10^6 V, what will be its final kinetic energy and final velocity?

12.19* A 5-MeV proton makes an elastic, "head-on" collision with a helium nucleus. What are the velocities of the particles after collision?

Section 12.4

12.20 What is the strength of an electric field that exerts a force of 4.8×10^{-16} N on an electron?

12.21* Three charges are placed on the vertices of an equilateral triangle. One charge is $+2$ C and the other two are each -1 C. Sketch the electric field lines.

12.22 A charge $Q = -50$ mC is located at the origin of a coordinate system. What is the electric field vector and the potential at the point $x = 4$ m, $y = 4$ m?

12.23 At a certain point P in space a source charge produces an electric field of 10^6 V/m in the $+x$ direction. At this same point another source charge produces a field of 2×10^6 V/m in the $+y$ direction. What force will a proton experience at P?

12.24 An electron is placed in a uniform field of 3 MV/m. What force does the electron experience?

12.25* At a distance of 100 km from a small, electrically charged asteroid it is found that there is an electric field of 3000 V/m. What charge does the asteroid carry? If the asteroid is spherical with a radius of 1 km, what is the charge per unit area on its surface. (The total charge is uniformly distributed.)

12.26 A completely ionized helium nucleus is in a uniform electric field of 1.5 MV/m. What acceleration does the helium nucleus experience?

12.27 Consider the proton to be a uniformly charged sphere with a radius of 10^{-15} m. (This is a dubious model of the proton.) What is the electric field at the surface of the proton?

12.28* Four charges are situated at the corners of a square with 1-m sides. The charges are, reading clockwise, $+3\,\mu C$, $-8\,\mu C$, $-5\,\mu C$, and $+10\,\mu C$. What is the field strength at the center of the square?

12.29* In the previous problem, what is the force on the $-8\text{-}\mu C$ charge?

12.30* Two charges are located as follows: $+Q$ at $x = -a/2$ and $-Q$ at $x = -a/2$. Find the electric field strength at a point $x = r$ where $r > |a|/2$. Show that when $r \gg a$, the field strength (to a good approximation) is given by $E = 2KQa/r^3$.

12.31 At a certain position in space the electric potential is $\Phi = +24{,}000$ V. What is the potential energy at this point of (a) an electron and (b) a proton?

12.32 Two points, A and B, have electric potentials of $+2000$ V and -3000 V, respectively. How much work is required to move an electron from A to B? Is the same amount of work required to move a proton from A to B?

12.33 Calculate the electric potential at a distance of 10^{-10} m from a proton.

12.34 Suppose that the surface of the Earth carries a uniform surplus of electrons amounting to 1 electron per cm². What is the charge on the Earth? What is the electric potential at the surface of the Earth?

12.35 What will be the velocity of an electron that is accelerated through a potential difference of 120 V, starting from rest?

12.36 A charge $Q_1 = +6\,\mu C$ is located at $x = 0$; a charge $Q_2 = -8\,\mu C$ is located at $x = 1$ m. What is the potential difference between the points $A(x = 0.2\text{ m})$ and $B(x = 0.7\text{ m})$?

12.37 A sphere of radius 1 m carries a uniform surface charge of 10^{-6} C/m². What is the electric field and the potential at the surface of the sphere? What are the values 1 m above the surface?

12.38 Two charges, $Q_1 = +5$ mC and $Q_2 = -3$ mC, are placed at opposite corners of a square whose sides are 0.5 m in length. What is the potential at each of the unoccupied corners?

12.39 Three charges are located in a straight line as follows: $+2$ mC at $x = -10$ m; -4 mC at $x = 0$ m; $+2$ mC at $z = +10$ m. Sketch the electric field lines. What is the field strength at $x = -5$ m?

12.40 Two parallel plates are separated by 2 cm. A battery is used to put a potential difference of 600 V across the plates. What electrical force will an oil droplet carrying a charge of $4e$ experience in the field between the plates?

12.41 Two parallel plates are separated by an air gap of 4 cm. What voltage must be placed across the plates to produce a field strength of 2 kV/m?

Section 12.5

12.42 What amount of charge accumulates on the plates of a $25\text{-}\mu F$ capacitor when it is connected to a 200-V power supply?

12.43 What is the charge density (i.e., the charge per unit area, Q/A) on the plates of parallel-plate capacitor when the field strength in the vacuum space between the plates is E?

12.44 It is desired to place a charge of $3\,\mu C$ on each of two parallel plates by using a 12-V battery. What must be the capacitance of the plates?

12.45 It is desired to make a 300-pF air-gap capacitor by using a single sheet of metal 0.2 m × 0.8 m. Explain how this can be done.

12.46 Four capacitors are connected in series: $C_1 = 4\,\mu F$, $C_2 = 6\,\mu F$, $C_3 = 8\,\mu F$, $C_4 = 12\,\mu F$. What is the total capacitance of the combination?

12.47 Two capacitors, $C_1 = 2\,\mu F$ and $C_2 = 6\,\mu F$, are connected in series. What capacitance C_3 must be connected in series with the first two capacitors to produce a combination that has a total capacitance of $1\,\mu F$?

12.48 Two capacitors, $C_1 = 4\,\mu F$ and $C_2 = 6\,\mu F$, are connected in series and a 24-V battery is connected across the combination. What is the potential difference across each capacitor? What is the charge on each capacitor?

12.49 If you have two metal plates 1.2 m × 1.6 m and a supply of 0.1-mm Nylon 66, what is the maximum capacitance of the capacitor you could make using these materials?

12.50 A capacitor that has an oil dielectric is charged to a voltage of 240 V and is then disconnected from the voltage source. The oil is then drained from the capacitor. What is the final potential difference between the plates?

12.51 Two plates, each with an area of 0.4 m², are separated by a uniform air gap of 2 mm. The plates are given equal and opposite charges of 1 μC. What is the capacitance of the plates and the field strength between the plates?

The charged plates are now placed in pure water. What is the new capacitance and the new field strength?

12.52 An air-gap parallel-plate capacitor has dimensions 0.4 m × 0.6 m. What amount of charge can be accumulated on the plates before the air gap will break down and sparking will occur? (Sparking in air occurs at a field strength of about 3 MV/m.)

Section 12.6

12.53* Plexiglas can withstand an electric field strength of about 100 MV/m. (Air breaks down at about 3 MV/m.) What is the minimum area of the plates of 0.2-μF capacitor with a Plexiglas dielectric if a potential difference of 600 V is to be placed across the capacitor?

12.54 How much energy can be stored by the capacitor in the previous problem?

12.55 If a capacitor is rated only for low-voltage applications, the gap separation can be made quite small. Suppose we use a very thin ("quarter mil" or 0.00025 in.) polyethylene dielectric. (a) What plate area will be required to produce a capacitance of 1 F? (b) What will be the total volume between the plates of this capacitor? (c) How much energy can be stored in this capacitor at a potential difference of 120 V? Do you expect that 1-F capacitors are common devices?

12.56 A capacitor consists of a pair of parallel plates separated by 1.2 mm of polyethylene. How large must the plates be if 0.05 J of energy are to be stored when the potential difference between the plates is 2400 V?

12.57 Two plates, 20 cm × 15 cm, are separated by a 1.5-mm piece of Plexiglas. If the plates are charged to a potential difference of 3 kV, how much electric energy is stored in the capacitor?

12.58 The number of conduction electrons in copper is 8.2×10^{22} per cm³. If these electrons drift with a net speed of 1.6 mm/s in a copper bar with a cross-sectional area of 1.5 cm², what current flows in the bar?

12.59 A 5-μC-charged ball ties to the end of a 25-cm length of insulating string is whirled at an angular speed of 500 RPM. What effective current may be associated with the motion of the charge?

Section 12.7

12.60 Two resistors, $R_1 = 12\ \Omega$ and $R_2 = 18\ \Omega$, are connected in parallel and the combination is connected to a 12-V battery. What is the current through each resistor?

12.61 What is the total resistance of n resistors R that are connected (a) in series and (b) in parallel?

12.62 Four identical resistors R are connected to form a square with one resistor on each side. What is the total resistance (a) between opposite corners of the square and (b) between adjacent corners of the square?

12.63 What is the total resistance between points A and B in the circuit below?

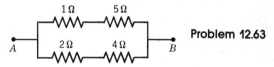

Problem 12.63

12.64 What is the resistance of a tungsten wire that has a length of 400 m and a diameter of 1.2 mm?

12.65 A copper wire (diameter = 1.5 mm) has a resistance of 8 Ω. What is the volume of the wire?

12.66 A piece of copper wire has the same diameter and the same mass as a piece of aluminum wire. What is the ratio of the resistance values of the wires?

12.67 A quantity of mercury (30 cm^3) is poured into a glass tub that has a diameter of 4 mm. The resistance of the mercury column is determined to be 0.01138 Ω. What is the resistivity of mercury?

12.68 A cube of aluminum has dimensions 1 cm $\times$ 1 cm $\times$ 1 cm. What is the resistance between opposite faces of the cube?
  The aluminum is now formed into a wire that has a cross-sectional area of 2 mm^2. What is the resistance of the wire?

12.69 A sample of copper with a mass of 45 g is drawn into a wire with a diameter of 1.5 mm. What is the total resistance of the wire?

Section 12.8

12.70 A current of 8 A is drawn from a 240-V power supply. What is the output power of the supply? How much energy (expressed in J and in kWh) is delivered by the supply if it operates continuously for a week?

12.71 A total current of 6 A flows through two resistors, 1 Ω and 2 Ω, that are connected in parallel. What is the voltage across the resistors? What is the total power expended?

12.72 Two resistors, 4 Ω and 8 Ω, are connected in series with a 24-V battery. How much power is extracted from the battery? What would be the power extracted if the resistors were connected in parallel?

12.73 A certain home swimming pool contains 10^5 liters (about 25,000 gallons) of water. An electric heater is used to raise the water temperature from 20°C to 28°C. If electric energy costs 3.5 cents per kWh and if half the input heat is lost by conduction and radiation, what is the total cost of heating the water? (This calculation will demonstrate why swimming pools are almost never heated electrically!)

12.74 The useful life of an automobile battery before recharging is necessary is usually given in units of ampere-hours (A-h). A typical 12-V battery will have a rating of 60 A-h. This means that a current of 60 A can be drawn from the battery for 1 h, or a current of 1 A for 60 h, or a current of 3 A for 20 h, and so on. Suppose that you forgot to turn off the headlamps of your automobile. Each lamp consumes 50 W of power. How long before your battery will be "dead"?

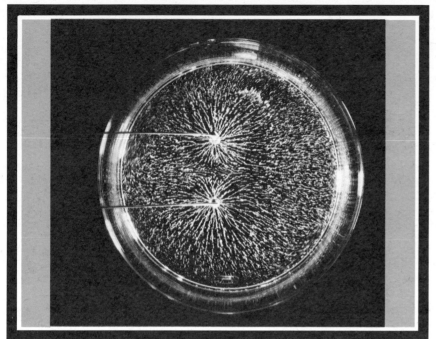

13
Electromagnetism

In this chapter we discuss the *magnetic force* and its effects. This force, as we will see, is not a new fundamental force; instead, the magnetic force is really a manifestation of the electric force when the charges have relative motion. We will find it instructive to describe magnetic effects by using the *field* concept just as we did for electric effects. Because electric and magnetic effects are so closely related, we use the term *electromagnetism* to describe this entire class of phenomena.

This chapter will be devoted to discussions of the fundamental aspects of electromagnetism. In the next chapter we consider several applications of the ideas we develop here.

13.1 Magnetism

Lodestones and the Compass

More than 2000 years ago in Asia Minor it was known that certain peculiar natural stones (called *lodestones*) would attract one another and would also attract and hold small bits of iron. Because these stones were found in the region called Magnesia, they became known as *magnets*. The property of attraction exhibited by the Magnesian stones (the *magnetism*) is somehow a permanent aspect of the material; magnetism cannot be "rubbed off" in the way that an electrostatic charge can be removed from an object.

Lodestones, when suspended freely or floated in still water on pieces of wood, were found to take up a definite direction with respect to the Earth—long, narrow lodestones orient themselves in a north-south direction. This observation led to the introduction of a practical magnetic compass for purposes of navigation.

The compass provides us with the standard by which we define directions in describing magnetic effects. The end of a compass magnet that is *north-seeking* is called the *N pole* of the magnet. Similarly, the south-seeking end is the *S pole* of the magnet.

Magnets have the familiar property (similar to that of electric charges) that *opposite poles attract* and *like poles repel* (Fig. 13.1). Therefore, the north-seeking N pole of a compass magnet actually is at-

Figure 13.1 Like magnetic poles repel and opposite magnetic poles attract.

tracted to and points toward the Earth's S magnetic pole which is located near (but not at) the geographic north pole (see Fig. 13.3).

Magnetic Fields

It is easy to verify that at every point in the vicinity of the Earth a compass will assume a definite orientation. That is, some property of the Earth influences the compass and causes it to point in a definite direction. This property is the *magnetic field* of the Earth. We can map the Earth's magnetic field, or the field of any magnetic object, by using a compass. (A compass will indicate the *direction* of the field at any point, but we must use other means to determine the *strength* of the field, as we will see.) Compass measurements of the field of a simple bar magnet show that the magnetic field lines are as indicated in Fig. 13.2. By convention, we take the direction of the field lines to be the direction in which the N pole of a compass magnet points; that is, the field lines in the region external to the bar magnet run *from the N pole to the S pole,* as in Fig. 13.2. Similar measurements of the Earth's magnetic field show that the Earth's magnetism is practically the same as that of a bar magnet (Fig. 13.3).[1]

Magnetic field lines can be made "visible" by a simple technique. If a sheet of paper is placed over a bar magnet and iron filings are sprinkled on the

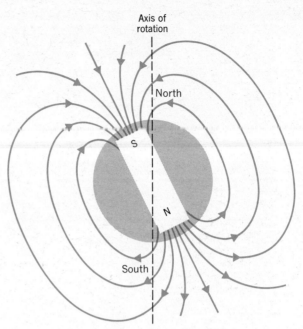

Figure 13.3 The Earth's magnetic field is like that of a bar magnet with the S pole near the north geographic pole.

paper, the tiny pieces of iron will take up positions with their long dimensions along the field lines. (That is, the pieces of iron are induced to become small compass magnets.) Figure 13.4 shows the magnetic field map of a bar magnet obtained in this way.

Elementary Magnets

If a bar magnet is cut into two pieces, as in Fig. 13.5, we find that the two halves are themselves complete magnets with N and S poles in the same orientation as the original magnet. Further division of the magnet produces additional magnets, again with N and

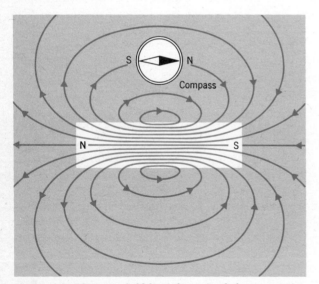

Figure 13.2 Magnetic field lines for a simple bar magnet.

---

[1]The Earth's magnetism is not static; instead, it changes slowly with time. Measurements made on archeologically dated samples of magnetic materials have shown that the intensity of the Earth's magnetic field fluctuates with a time interval of about 5000 years between successive maxima and minima. Furthermore, during the last 10–20 million years, the field has actually *reversed* its polarity every 300,000 years or so.

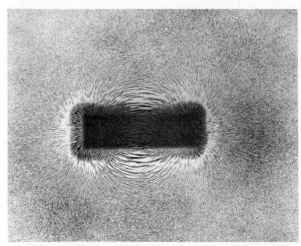

Figure 13.4 The magnetic field of a bar magnet obtained by sprinkling iron filings on a piece of paper covering the magnet.

Figure 13.5 Cutting a magnet produces two magnets with N and S poles in the same orientation as the original magnet.

S poles oriented in the same direction as the original magnet.

What will happen if we continue this process of division down to the atomic level? Can we ever separate the N pole from the S pole? As we will discuss in detail in later sections, even individual atoms can behave as microscopic but *complete* magnets with N and S poles. In no case has it ever been found possible to take an atom apart and separate the poles. We must conclude that the N and S poles of a magnet have no independent existence.

13.2 Electromagnetism

The Field of a Current-carrying Wire

The magnetism of a bar magnet appears to be a completely static affair. The magnetic material is electrically neutral and there does not appear to be any electric current flowing in the magnet. What, then, is the connection between electric currents and magnetism? Until early in the 19th century, electricity and magnetism were thought to be two independent phenomena. In 1820 Hans Christian Oersted (1777–1851), a Danish physicist, discovered (quite by accident) that a current-carrying wire influenced the orientation of a nearby compass magnet. The reason for the phenomenon that Oersted observed is that a current-carrying wire produces a magnetic field which can influence the orientation of a compass placed in the vicinity of the wire. Measurements with a compass show that the field lines near a current-carrying wire are *circles* centered on and perpendicular to the wire, as shown in Fig. 13.6.

The Right-hand Rule

We can establish the *direction* of the magnetic field lines due to a current-carrying wire by observing the orientation of a compass magnet when placed in the vicinity of the wire. The results of such an experiment are summarized in the so-called *right-hand rule*:

If a current-carrying wire is grasped with the right hand in such a way that the thumb is in the direction of conventional current flow, then the fingers encircle the wire in the same direction as the magnetic field lines (see Fig. 13.7).

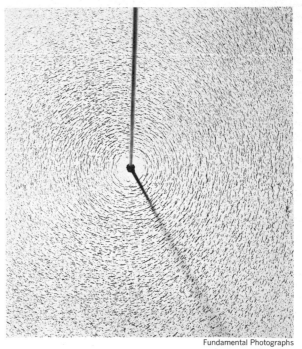

Figure 13.6 The circular magnetic field lines surrounding a current-carrying wire are revealed by using the iron-filing technique.

The magnetic field vector (analogous to the electric field vector **E**) is given the symbol **B**.

We see here the advantage of having adopted the convention of always using the term *current flow* to mean the (equivalent) flow of *positive* charge,

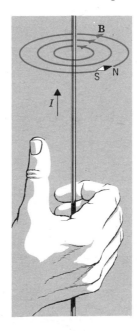

Figure 13.7 Illustration of the right-hand rule for determining the direction of the magnetic field lines due to a current flowing in a wire.

namely, that it permits us to use a *right*-hand rule for the direction of the field lines and the field vector **B**. We have already defined the direction of the angular momentum vector (Section 4.1) in terms of the direction of advance of a right-hand screw and we shall later have additional conventions regarding *right* hands. Therefore our choice allows us always to specify the directions of these vector quantities in terms of rules using the *right* hand.

The Field of a Current Loop

If a wire that carries a steady current is bent into the shape of a circular loop, what is the form of the resulting magnetic field? Imagine that the wire in Fig. 13.7 is bent into a loop and look at the way the right-hand rule specifies the field direction. On the inside of the loop the field lines due to all parts of the loop have the same direction, as shown in Fig. 13.8. Because the field lines spread out and produce a field very similar to that of a bar magnet (Fig. 13.2), we can define an N pole and an S pole for the field of the loop. Notice that another type of right-hand rule is in operation here. If the fingers of the right hand are curled in the direction of current flow in the loop, the thumb points in the same direction as the magnetic field lines that pass through the loop. Or, we can say that the thumb points in the direction from the S pole to the N pole of the field.

One of the types of magnets often found in electrical devices consists of a current-carrying wire that is wrapped around a piece of iron; this is an

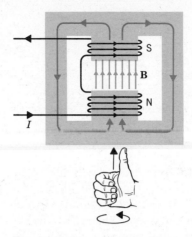

Figure 13.9 Schematic design of an electromagnet.

electromagnet. If the magnet has the shape shown in Fig. 13.9, the iron guides the field lines so that the field in the gap between the poles is strong and uniform. Furthermore, the field can be changed by varying the current that flows in the windings. Notice that the direction of the field lines follows the prescription discussed in connection with the field in Fig. 13.8.

Atomic Magnetism

The smallest units of magnetism are found at the atomic level. In a highly simplified model of atomic structure, electrons are considered to move around the atomic nucleus in definite orbits. A single electron that executes a circular orbit around a stationary positively charged nucleus is shown in Fig. 13.10. The motion of this electron is equivalent to a current loop (but with the current flowing in the direction opposite to that of the electron velocity). Therefore, a magnetic field is produced with the same configuration as that shown in Fig. 13.8.

The orbital motions of atomic electrons do play a role in the magnetic properties of matter, but, as we will see in Section 13.6, the effects of *spinning* electrons are much more important in most materials.

Where do Magnetic Field Lines Begin and End?

We have previously seen (Section 12.4), that electric field lines originate on positive charges and terminate on negative charges. The lines that specify the magnetic field are distinctly different in character: *magnetic field lines have no beginning and no end.* Thus, the field lines surrounding a straight current-carrying wire are circles; the field lines of a current loop are not circles but nevertheless the lines have no point of origination or termination. Even the field lines of a bar magnet or electromagnet do not begin at the N pole and end at the S pole; the lines extend

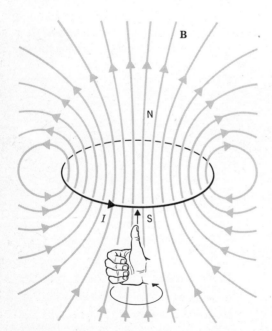

Figure 13.8 The field lines for a current-carrying loop of wire. Notice that the field resembles that of a (very short) bar magnet.

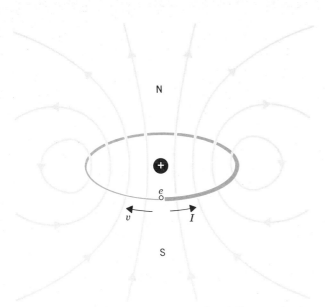

Figure 13.10 An electron moves around a stationary atomic nucleus and produces a magnetic field equivalent to that of a current loop. Notice that the direction of current flow is opposite to that of the electron's velocity because the electron carries a negative charge.

through the interior of the bar or core without termination.

The fact that magnetic field lines have neither a beginning nor an end is equivalent to the statement that single magnetic poles isolated from other magnetic poles (*monopoles*) do not exist. Electric field lines originate and terminate on electric monopoles (charges) but there are no magnetic monopoles to terminate the magnetic field lines. The continuity of magnetic field lines and the nonexistence of magnetic monopoles is one of the important facts of electromagnetism.[2]

13.3 Effects of Magnetic Fields on Moving Charges

The Strength of the Magnetic Field

Although we have completely specified the *direction* of the magnetic field lines (or, equivalently, the direction of the magnetic field vector **B**), we have as yet made no quantitative statement regarding the *strength* of the field (that is, the *magnitude* of **B**). In the case of the electric field, the magnitude of **E** was defined in terms of the *force* on a stationary test charge in the field. Similarly, we can define the magnitude of **B** in terms of the force exerted by the field on a test charge. But a test charge that is *stationary* in a magnetic field experiences no force. Only in the event that the test charge is in motion is there a magnetic force on the charge.

In a given magnetic field it is found that the magnetic force is directly proportional to both the charge and the velocity of the test particle. The factor that connects the magnetic force F_M with the charge and velocity of the test particle is the magnetic field strength B:

$$F_M = qvB \quad \text{(for } \mathbf{v} \perp \mathbf{B}\text{)} \tag{13.1}$$

The SI unit of magnetic field strength is the *tesla*. Another unit that is often used is the *gauss*:

$$1 \text{ tesla (T)} = 10^4 \text{ gauss (G)}$$

(The *gauss* is the unit of magnetic field strength in an earlier vision of the metric system and is still retained in many publications.)

Some of the magnetic field strengths found in various natural and man-made situations are shown in Table 13.1.

Table 13.1 Range of Magnetic Field Strengths in the Universe

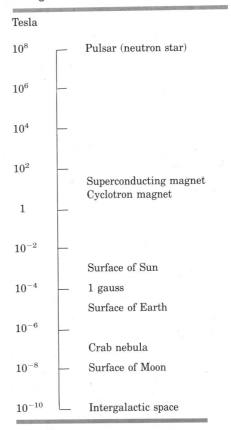

---

[2]There is a theoretical suggestion that magnetic monopoles might actually exist, and there has been an active search for these peculiar objects for several years. None has yet been found. But if monopoles do exist, the theory of electromagnetism will require fundamental modifications.

294 Electromagnetism

● *Example 13.1*

What is the force on a 1-MeV proton that moves perpendicular to a magnetic field whose strength is 2000 G?

From Example 12.6 we know that the velocity of a 1-MeV proton is 1.38×10^7 m/s. Therefore, using Eq. 13.1,

$$F_M = qvB = (1.60 \times 10^{-19} \text{ C}) \times (1.38 \times 10^7 \text{ m/s})$$
$$\times (0.2 \text{ T})$$
$$= 4.42 \times 10^{-13} \text{ N}$$

This is a very small force, but the proton also has a very small mass! Therefore, the acceleration experienced by the proton ($a = F/m$) is significant and is sufficient to make the proton move in a circular path of reasonable size, as we will see in Example 13.3. ■

The Direction of the Magnetic Force

The quantities v and B that appear on the right-hand side of Eq. 13.1 are the magnitudes of the vectors **v** and **B**. And, of course, F_M is the magnitude of the force vector $\mathbf{F}_M$. What is the relationship among the directions of these three vectors?

The case of the magnetic force is distinctly different from that of the electric force. As shown in Fig. 13.11, the electric force vector $\mathbf{F}_E$ that acts on a positive test charge has the *same* direction as the electric field vector **E** and is independent of the direction of the velocity vector **v**. On the other hand, it has been found experimentally that when a charged particle enters a magnetic field, the direction of the magnetic force is *perpendicular* to both **v** and **B** (Fig. 13.12).

The direction of $\mathbf{F}_M$ relative to **v** and **B** is given by another right-hand rule: The vector $\mathbf{F}_M$ has the same direction as that of the advance of a right-hand

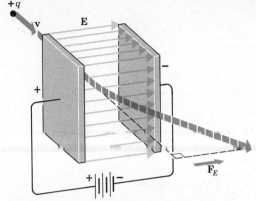

Figure 13.11 The electric force vector $\mathbf{F}_E$ has the same direction as the electric field vector **E**.

screw when rotated in the sense that moves the vector **v** toward the vector **B** (see Fig. 13.13a). Alternatively, we can state the rule in the following way: Point the fingers of your right hand in the direction of **v** and curl the fingers toward the direction of **B**; the thumb then points in the direction of $\mathbf{F}_M$ (see

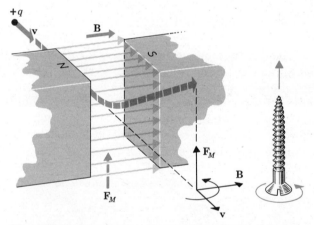

Figure 13.12 The magnetic force is perpendicular to both **v** and **B**. Application of the right-hand rule shows that the direction of $\mathbf{F}_M$ is up.

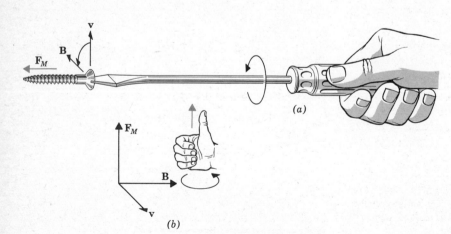

Figure 13.13 Illustration of the right-hand rules for determining the direction of the magnetic force $\mathbf{F}_M$ on a moving positive charge.

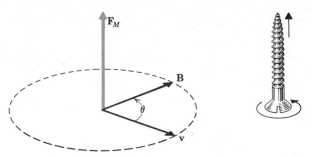

Figure 13.14 The magnitude of F_M depends on the sine of the angle θ between **v** and **B**. The force vector $\mathbf{F}_M$ is *always* perpendicular to the plane defined by **v** and **B**.

Fig. 13.13*b*). Note that the rules apply to the case of a particle carrying a *positive* charge; for a negatively charged particle the direction is *opposite* to that given by the rules.

The maximum force exerted on a moving charged particle by a magnetic field occurs when the velocity vector of the particle is perpendicular to the field vector **B**, as in Fig. 13.12. The magnitude of this maximum force is given by Eq. 13.1. If **v** is not perpendicular to **B**, the force is less and becomes zero when **v** is parallel to **B**. In general, if θ is the angle between **v** and **B**, the magnetic force is

$$F_M = qvB \sin \theta \qquad (13.2)$$

If the angle between **v** and **B** is not zero, the two vectors define a plane and the force vector $\mathbf{F}_M$ is always perpendicular to this plane, as in Fig. 13.14.

Another way of stating the result expressed by Eq. 13.2 is the following: The magnetic force on a moving charged particle depends only on the component of the velocity vector that is perpendicular to **B**. The perpendicular component is (Fig. 13.15)

$$v_\perp = v \sin \theta \qquad (13.3)$$

which is just the factor that appears in Eq. 13.2.

In Section 7.1 we discussed how to take the product of two vectors in such a way that the result is a *scalar* quantity. This type of product is called the *scalar product* or *dot product*. We now find a situation in which we take the product of two vectors (**v**

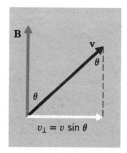

Figure 13.15 The component of **v** perpendicular to **B** is $v_\perp = v \sin \theta$.

and **B**) in such a way that the result is a *vector* quantity ($\mathbf{F}_M$). This type of product, defined by the right-hand rules we have described, is called the *vector product* and is written as

$$\mathbf{F}_M = q\mathbf{v} \times \mathbf{B} \qquad (13.4)$$

This product is also called the *cross product* because the multiplication cross distinguishes the method of multiplication from that of the dot product. When expressed in terms of the *magnitude* of the vectors, Eq. 13.4 becomes Eq. 13.2.

The Force on a Current-carrying Wire

Suppose that a certain straight wire carries a steady current I. Suppose further that a segment of this wire with a length L lies within a uniform magnetic field B and that the direction of the wire makes an angle θ with respect to **B**. Each electron that moves through the wire experiences a magnetic force given by Eq. 13.2. If there are n moving electrons per unit length in the wire, the total number of electrons acted on by the field is nL and the total force on the segment of wire with length L is

$$F_M = nLevB \sin \theta$$

Now, look at the combination nev. How much charge q will pass a particular point in the wire during a time t? We can write

$q =$ (number of electrons per unit length, n)
  × (charge of electron, e)
  × (distance ℓ each electron moves in time t)

The distance ℓ is equal to vt. Therefore,

$$q = nevt$$

The charge passing a particular point per unit time is the *current*, so we have

$$\frac{q}{t} = I = nev$$

Thus, the force on the wire becomes

$$F_M = BIL \sin \theta \qquad (13.5)$$

Now, when a current flows, at any point there is a net movement of the charges in a particular direction. Therefore, we can express the direction of the wire at that point as a vector **L** and can then write the magnetic force in terms of a vector product:

$$\mathbf{F}_M = I\mathbf{L} \times \mathbf{B} \tag{13.6}$$

● *Example 13.2*

Near the Earth's equator the geomagnetic field is horizontal and has a strength of approximately 0.25 G. A straight wire that has an east-west orientation has a length of 1 km and carries a current of 12 A. What is the total force exerted on the wire by the Earth's magnetic field?

Using Eq. 13.5 with $\theta = 90°$ and $\sin \theta = 1$, we have

$$F_M = BIL = (0.25 \times 10^{-4} \text{ T}) \times (12 \text{ A}) \times (10^3 \text{ m})$$
$$= 0.3 \text{ N} \quad \blacksquare$$

Electromagnetic Forces

In combined electric and magnetic fields, a moving charged particle will experience a separate force due to each of the individual fields. These forces are

$$\mathbf{F}_E = q\mathbf{E}$$
$$\mathbf{F}_M = q\mathbf{v} \times \mathbf{B} \tag{13.7}$$

The *total* force on the particle is the vector sum of these two forces:

$$\mathbf{F}_L = \mathbf{F}_E + \mathbf{F}_M = q(\mathbf{E} + \mathbf{v} \times \mathbf{B}) \tag{13.8}$$

This electromagnetic force $\mathbf{F}_L$ is called the *Lorentz force* after Hendrik Antoon Lorentz (1853–1928), the Dutch physicist whose studies of electromagnetism paved the way for Einstein's development of relativity theory.

It is important to realize that the magnetic force is not a basic force of Nature in the sense that the electric and gravitational forces are. Magnetic fields are produced only by the motion of electric charges, whether they are free charges flowing in a wire or bound charges that are circulating or spinning in an atom. Therefore, the magnetic force is just one manifestation of the fundamental electric force.

Magnetic Fields and Relative Motion

Suppose that we have a charge q and a meter that is sensitive to a magnetic field; suppose that both are at rest in some coordinate system, as in Fig. 13.16a. Clearly, the meter will show zero field, $B = 0$. (There is, of course, an *electric* field at the position of the meter.) However, if the charge is in motion, as in Fig. 13.16b, we know that this is equivalent to a current and that there is produced a magnetic field which will be registered by the meter. But it is only the *relative* motion of the charge and the meter that is important. (The laws of physics are the same in all inertial reference frames.) Therefore, if q remains at rest in the coordinate system and the *meter* moves, as in Fig. 13.16c, this is entirely equivalent to the situation in Fig. 13.16b, and the presence of a magnetic field will again be shown by the meter. *A magnetic field is produced only by a changing or moving electric field.*

13.4 Orbits of Charged Particles in Magnetic Fields

Circular Orbits

A charged particle can be started into motion by allowing it to be accelerated in an electric field. Once in motion, if we project the charged particle into a uniform magnetic field, we find that the magnetic force exerted on the particle by the field causes it to move in a *circular* path. (We must assume here that the particle moves in a vacuum; otherwise, the par-

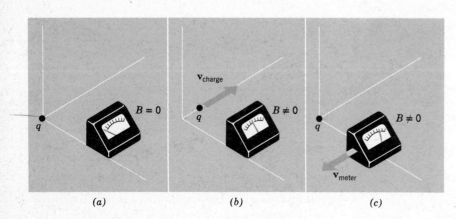

Figure 13.16 A magnetic field is produced solely as the result of relative motion between the charge q and the "**B** meter."

13.4 Orbits of Charged Particles in Magnetic Fields

Figure 13.17 A charge particle moves in a circular orbit in a uniform magnetic field if $\mathbf{v} \perp \mathbf{B}$.

ticle will collide with and lose energy to other particles and the orbit will not be circular—see Fig. 13.18.)

Consider a uniform magnetic field $\mathbf{B}$ and a charged particle of mass m moving in the field with a velocity $\mathbf{v}$, where $\mathbf{v} \perp \mathbf{B}$, as in Fig. 13.17. The magnetic field exerts on the particle a force of constant magnitude:

$$F_M = qvB$$

The direction of this magnetic force is always perpendicular to the instantaneous direction of motion of the charged particle; that is, $\mathbf{F}_M \perp \mathbf{v}$. Hence, there is no component of the force in the direction of motion and *no work is done on the particle by the magnetic field*. Although the direction of motion is continually changing as a result of the magnetic force, the *speed* of the particle (that is, v) is constant and the kinetic energy remains always the same.

The magnetic force produces a centripetal acceleration of constant magnitude that is always perpendicular to $\mathbf{v}$. This means (see Section 2.9) that the particle moves in a *circular* orbit. The centripetal acceleration is

$$a_c = \frac{F_M}{m} = \frac{qvB}{m}$$

In terms of the velocity and the orbit radius, the centripetal acceleration is given by (see Eq. 2.34)

$$a_c = \frac{v^2}{R}$$

Equating these two expressions for a_c and solving for R, we have

$$R = \frac{mv}{qB} \tag{13.9}$$

Therefore, a charged particle moving at right angles with respect to a uniform magnetic field executes a *circular* orbit in the field with a radius that is directly proportional to its momentum (mv) and inversely proportional to the field strength (see Fig. 13.18).

It is important to realize that a charged particle executing an orbit in a static magnetic field in vacuum will neither gain nor lose energy from its interaction with the magnetic field.[3]

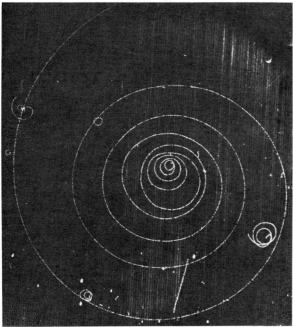

Lawrence Radiation Laboratories

Figure 13.18 The path of a charged particle in a *bubble chamber* (consisting of liquid hydrogen) is made visible by the many tiny bubbles that are formed in the wake of the particle. This photograph shows the orbit of a fast electron in a bubble chamber that is in a strong magnetic field. The electron loses energy through collisions with the hydrogen atoms and so the radius of the orbit decreases, causing the electron to move in a spiral path. The tracks of some secondary electrons released in encounters with hydrogen atoms can be seen near the main track.

● *Example 13.3*

What is the radius of the orbit of a 1-MeV proton in a 10^4-G field?

---

[3] In the event that the magnetic field changes with time, the energy of the particle will, in general, be altered; see Section 13.6. Even with a static field the moving electron represents an accelerating charge and consequently emits electromagnetic radiation referred to as *synchrotron radiation*.

298 Electromagnetism

We have

$m = 1.67 \times 10^{-27}$ kg

$q = e = 1.60 \times 10^{-19}$ C

Again, from Example 12.6 we know that the velocity of a 1-MeV proton is $v = 1.38 \times 10^7$ m/s. Therefore,

$$R = \frac{mv}{eB}$$

$$= \frac{(1.67 \times 10^{-27} \text{ kg}) \times (1.38 \times 10^{-7} \text{ m/s})}{(1.60 \times 10^{-19} \text{ C}) \times (1 \text{ T})}$$

$$= 0.14 \text{ m} \quad \blacksquare$$

The Cyclotron

One type of device that is often used in the acceleration of charged particles (protons, deuterons, α particles, etc.) to high velocities is the *cyclotron*. The basic idea of cyclotron operation is to accelerate charged particles by means of electric fields while confining the particles with a magnetic field.

A schematic representation of a cyclotron is shown in Fig. 13.19. The essential elements of a cyclotron are a hollow, cylindrical cavity which is split along a diameter to form two "dees" (so-called because of their shape), an electromagnet (not shown in the figure) which produces a magnetic field perpendicular to the plane of the "dee" structure, and high voltage apparatus which produces a potential difference V between the "dees." Neutral gas atoms (for example, hydrogen) are ionized at the source S, located near the center of the "dees," to produce charged particles that are injected into the left-hand "dee." These particles move in a circular orbit under the influence of the field $\mathbf{B}$ until they emerge from the left-hand "dee." The particles are then accelerated across the "dee" gap by the electric field and are increased in energy by an amount qV. The electric field exists only *between* the "dees;" the interiors of the conducting "dees" have no electric field. Therefore, when the particles enter the right-hand "dee" they again move in a circular orbit but now of increased radius corresponding to their greater velocity. By the time the particles reach the "dee" gap again, the oscillating high-voltage apparatus has reversed the signs of the voltage on the "dees" so that the particles experience another accelerating voltage at the gap. This process is continued for many passages through the gap; each passage increases the energy by the amount qV and the particles spiral outward to greater and greater radii. Near the outer wall of the "dee" structure the particles pass into an *extractor* (usually a pair of plates across which is placed a high voltage) and emerge as a beam of high energy particles at A.

It is essential for the operation of a cyclotron that the voltage across the "dee" gap always be of the correct sign to accelerate the particles rather than to retard them. This is relatively easy to accomplish because of the important fact that a charged particle moving in a given uniform magnetic field requires a *fixed* time to execute an orbit *independent of its velocity*[4] (see Problem 13.18). Therefore, the particles require the same time to complete each half revolution in the "dees" and arrive at the gap *in phase* with those particles executing orbits with the different radii, ready to accept the next accelerating voltage. For a given type of particle and for a given magnetic field there is a single frequency (called the *cyclotron frequency*) at which the polarity of the voltage must be alternated to provide continuing acceleration.

The first cyclotron (only 11 inches in diameter) was constructed by E. O. Lawrence and M. S. Livingston at the University of California in 1930. (Lawrence received the 1939 Nobel Prize in physics for this work.) One of Lawrence's early cyclotrons is shown in Fig. 13.20. Modern cyclotrons are used throughout the world in research programs designed to learn more about the structure of atomic nuclei. Some cyclotrons are used for the production of radioisotopes and as sources of particles for radiation therapy of certain diseases (see Chapter 21).

Modifications of the basic cyclotron principle have been made to permit the acceleration of particles to ultra-high (relativistic) energies. These ma-

---

[4]This statement is only true for velocities sufficiently low that relativistic effects can be ignored.

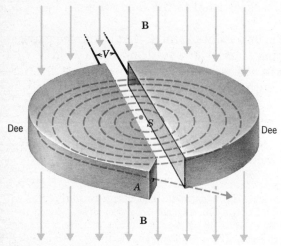

Figure 13.19 Schematic of a cyclotron.

Figure 13.20 One of E. O. Lawrence's early cyclotrons constructed at the University of California in the 1930s.

chines are known as *synchrocyclotrons* and *synchrotrons*.

Helical Orbits

If a charged particle moves in a uniform magnetic field with its velocity vector at an angle other than 90° with respect to the field direction, the orbit will

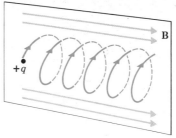

Figure 13.21 If the velocity vector is not perpendicular to **B**, the orbit is a helix.

not be circular. Recall that only that component of the velocity that is *perpendicular* to **B** contributes to the magnetic force (Eqs. 13.2 and 13.3); the parallel component is unaffected by the field. Therefore, if a particle moves in a field with velocity components both parallel and perpendicular to the field, the total motion will be a combination of circular motion (the field action on $v_\perp$) and a steady drift along the field direction ($v_\parallel$ unaffected by the field). The combination of circular motion and a steady drift produces a *helical orbit*, the axis of which coincides with the direction of the magnetic field, as shown in Fig. 13.21.

Electromagnetic Blood Flowmeters

Many different methods have been developed for measuring the flow rate of blood in the various vessels of the circulatory system. One widely used technique involves an *electromagnetic flowmeter*. The operating principal of this instrument is based on the movement of electric charges in a magnetic field. Within the blood there is a large concentration of electric charges in the form of ions. In fact, blood plasma is a typical extracellular fluid and contains about 145 mmol/ℓ of Na^+ ions and about 125 mmol/ℓ of Cl^- ions (see the essay beginning on page 271); all other ionic concentrations are negligible in comparison.

Suppose that a number of singly charged ions are moving with a velocity **v** within an artery. If we place the artery between the poles of a magnet, the ions now move within a magnetic field. For the directions of **v** and **B** indicated in Fig. 1, the magnetic force $\mathbf{F}_M^+$ on the positively charged ions is *up,* and the force $\mathbf{F}_M^-$ on the negatively charged ions is *down*. Under the influence of these forces, the ions will drift toward opposite sides of the artery. This polarization of the arterial ions produces an electric field **E** (Fig. 2) that we can approximate as the uniform field of a parallel plate capacitor. Then, the potential difference V across the artery (whose diameter is d) is related to E by Eq. 12.22.

$$E = \frac{V}{d} \tag{1}$$

This electric field acts on the ions to produce electric forces, $\mathbf{F}_E^+$ and $\mathbf{F}_E^-$, whose directions are opposite to the directions of $\mathbf{F}_M^+$ and $\mathbf{F}_M^-$, as shown in Fig. 2.

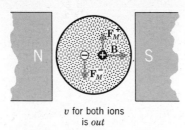

Figure 1 Positive and negative ions moving through an artery will each experience a force due to the applied magnetic field **B**.

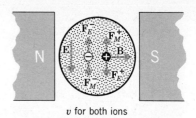

Figure 2 The magnetic forces acting on the moving ions results in a polarization of the charge and an electric field **E**.

The accumulation of charge on opposite sides of the artery will continue until the electric field has increased to the point that $F_E^\pm = F_M^\pm$. For this equilibrium condition we can write

$$F_E^\pm = F_M^\pm$$

$$eE = e\frac{V}{d} = evB$$

so that

$$v = \frac{V}{Bd} \tag{2}$$

Thus, the flow velocity is directly proportional to the voltage that builds up across the artery. This voltage is measured by placing electrodes on opposite sides of the artery, and the result, coupled with a knowledge of B and d, permits a determination of the flow velocity. It is a mathematical accident that the velocity v in this analysis is just the average velocity of the ions in the artery, regardless of the velocity distribution and regardless of whether the flow is laminar or turbulent (as long as the flow is not asymmetric within the artery).

The magnitudes of the potential differences that must be measured in an electromagnetic flowmeter are quite small. For example, consider an artery with a diameter of 1 cm in which the blood flow rate is 30 cm/s. If a magnet with a field strength of 1000 G is used, the potential difference that is developed will be

$$V = vBd = (0.30 \text{ m/s}) \times (0.1 \text{ T}) \times (0.01 \text{ m})$$
$$= 300 \, \mu V$$

This small value of V means that measurements must be made with great care.

One of the difficulties encountered in using an electromagnetic flowmeter in the simple way we have just outlined arises because the artery is immersed in an ionic fluid (or *electrolyte*). When the ions accumulate on the artery wall, they attract ions with the opposite charge from the outside fluid. It is difficult to discriminate between these surface polarization potentials (which represent unwanted *noise*) and the desired flow-induced potential. This problem can be overcome by using a system in which the magnetic polarity is reversed with a frequency sufficiently high that the surface potentials do not have an opportunity to develop but sufficiently low that the flow potential *does* have an opportunity to develop. It has been found that frequencies of a few hundred Hz are satisfactory.

Figure 3 illustrates the procedure using an alternating magnetic field in the flowmeter. The upper graph shows the *square-wave* form of the current through the magnet coils. The lower graph shows the potential signal due to the switching of the magnet current and the signal due to the ionic flow. Measurements of the desired

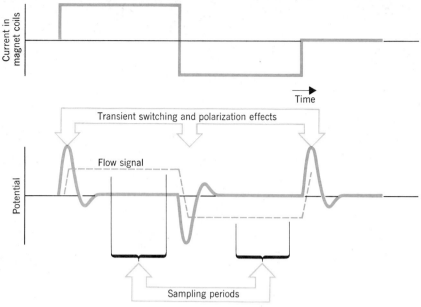

Figure 3 The use of an alternating magnetic field to eliminate the effects of undesired surface polarization.

flow signal are made only during a portion of each half-cycle, namely, after the transient effects have subsided and before the next current reversal. In this way the effects of the spurious surface polarization potentials are eliminated. (This is one example of a large class of electrical measurements in which an alternating or AC signal is generated artificially from an inherently DC source in order to improve the signal-to-noise ratio.)

In addition to surface polarization effects, electromagnetic flowmeters are subject to other difficulties arising from the various electric fields within the body. However, techniques have been devised to solve these problems, and electromagnetic flowmeters are now used routinely for blood flow measurements.

References

A. C. Burton, *Physiology and Biophysics of the Circulation,* Year Book Medical Publishers, Chicago, 1965.

D. A. McDonald, *Blood Flow in Arteries,* 2nd ed., Edward Arnold, London, 1974, Chapter 9.

C. J. Mills in *Cardiovascular Fluid Dynamics,* D. H. Bergel, ed., Vol. 1, Academic Press, New York, 1972.

■Exercises

1. A potential of 250 μV is measured across an artery ($d = 1.2$ cm) when a magnet with a field strength of 800 G is used in a flowmeter. What is the volume flow rate Q (in m^3/s) in the artery?

2. For the conditions of the example given in the essay, what is the electric current through the artery due to the sodium ions?

3. In an artificial kidney machine, the blood is pumped through a series of cleansing operations. If mechanical pumps were used, the blood cells could be damaged by the moving parts of the pump. Explain how an *electromagnetic pump* could be used in such a situation. What would be the advantage? (Hint: Use crossed electric and magnetic fields.)

13.5 Magnetic Fields Produced by Electric Currents

The Field Due to a Current Element

We know that a current flowing in a wire produces a magnetic field. The strength of the field produced in this way at a particular point in space must be determined from experiment. Refer to Fig. 13.22. Measurements of the strength of the magnetic field in the vicinity of a current-carrying wire can be interpreted in the following way. If the wire carries a steady current I, the contribution ΔB to the total magnetic field at the point P due to the small segment of wire with length ΔL is

$$\Delta B = K_M \frac{I \Delta L}{r^2} \sin \phi \quad \text{(current element)} \quad (13.10)$$

The quantity K_M is the magnetic force constant analogous to the electric force constant K that occurs in Coulomb's law. The value of K_M is

$$K_M = 10^{-7} \text{ N/A}^2 = 10^{-7} \text{ T-m/A} \quad (13.11)$$

As we will see, the value of K_M is a consequence of the way the ampere is defined in the metric system of units.

Notice that Eq. 13.10 gives only the contribution of the field strength ΔB due to the current element $I \Delta L$. Although we can express the field in terms of $I \Delta L$, we must realize that a steady current can flow only if we have a complete circuit. Thus, the actual field B at any point such as P is always the vector sum of all the small field contributions ΔB, with the sum carried out over the entire circuit. Usually, this summing process requires the methods of calculus, but in the next subsection we discuss a case for which the summation is uncomplicated.

The Field Due to a Current Loop

Figure 13.23 shows a length of wire that has been bent into a circular loop with a radius a. We wish to

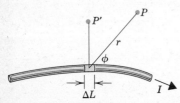

Figure 13.22 The magnetic field strength ΔB at P due to the current element $I \Delta L$ is given by Eq. 13.10. At P' the angle ϕ is 90° and the sin ϕ factor in Eq. 13.10 is unity.

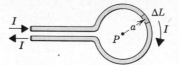

Figure 13.23 The strength of the magnetic field at P is $2\pi K_M I/a$.

determine the field strength at P, the center of the loop, when a steady current I flows around the loop. First, notice that the wires that lead the current into and out of the loop are parallel and closely spaced. Because the current flows in opposite directions in these two wires, the magnetic fields are opposed, and because each wire carries the same current, the field strengths are equal; consequently, the fields cancel. Therefore, the field at the center of the loop is due entirely to the current in the loop; there is no contribution from the external circuit.

Every part of the loop, such as that labeled ΔL in Fig. 13.23, is at a distance a from the center P. Moreover, the current flowing through each segment of the loop is perpendicular to the direction from the segment to the center. Thus, the contribution from each segment is (Eq. 13.10 with $\sin \phi = 1$ and $r = a$)

$$\Delta B = K_M \frac{I \Delta L}{a^2}$$

Now, the total field B is equal to the sum of all the partial fields ΔB:

$$B = \sum \Delta B = \sum K_M \frac{I \Delta L}{a^2}$$

Because K_M, I, and a^2 are all constant, we can rewrite the expression B as

$$B = K_M \frac{I}{a^2} \sum \Delta L$$

The sum of all the segments ΔL around the loop is the circumference of the loop, $2\pi a$. Therefore,

$$B = K_M \frac{I}{a^2} \times 2\pi a$$

$$= 2\pi K_M \frac{I}{a} \quad \text{(center of a circular loop)} \quad (13.12)$$

If the loop is actually a thin coil that consists of N individual loops of wire (Fig. 13.24), then the field strength at the center is N times larger than that given by Eq. 13.12.

Figure 13.24 The field strength at the center of a double loop of wire is twice that due to a single loop of wire.

● *Example 13.4*

What is the strength of the magnetic field at the center of a five-loop coil of wire that has a diameter of 40 cm and carries a current of 50 A?

Using Eq. 13.12 and supplying a factor $N = 5$,

$$B = 2\pi K_M \frac{NI}{a} = 2\pi \times (10^{-7} \text{ T-m/A}) \times \frac{5 \times (50 \text{ A})}{0.2 \text{ m}}$$
$$= 7.85 \times 10^{-4} \text{ T} = 7.85 \text{ G} \quad \blacksquare$$

The Field Due to a Solenoid

The magnetic field due to a circular loop is not uniform; this is easy to see in Fig. 13.8. We can, however, produce a uniform field by stretching a coil of wire into a long helical spiral, as shown in Fig. 13.25. Inside such a coil (which is called a *solenoid*) the field lines are straight and uniform. Within the solenoid (at positions not too close to the ends), the field strength is

$$B = 4\pi K_M nI \quad \text{(solenoid)} \tag{13.13}$$

where n is the number of turns of wire per meter of solenoid length. Notice that the field strength does not depend on the size of the solenoid (if the length is large compared with the diameter) nor does it depend on position within the solenoid. The magnetic field within a solenoid is *uniform*.

The Field Due to a Long, Straight Wire and the Definition of the Ampere

Another situation in which it is useful to know the magnetic field produced by a current is the case of a long straight wire. Figure 13.26 shows a point P that lies at a perpendicular distance r_0 from a straight wire that extends indefinitely to the right and to the left and carries a steady current I. What is the strength of the magnetic field at P? The wire is represented as a series of current elements. The field at P is equal to the sum of the fields ΔB_n due to all of the individual current elements $I\Delta L_n$, extending in both directions from ΔL_0. Each ΔB_n can be calculated using Eq. 13.10. Notice that each current element makes a different angle ϕ_n with the line to P; moreover, there is an infinite number of current elements. The evaluation of the sum therefore requires the methods of calculus and we will not pursue these techniques here. The result of such a calculation is that the strength of the field at a distance r from a long straight wire carrying a steady current I is

$$B = 2K_M \frac{I}{r} \quad \text{(long straight wire)} \tag{13.14}$$

Notice that the strength of the field due to a *current element* decreases as $1/r^2$ (Eq. 13.10), whereas that due to a *long straight wire* decreases only as $1/r$ (Eq. 13.14). Can you see why the field strength falls off more slowly with distance in the latter case?

● *Example 13.5*

At a distance of 15 cm from a long straight wire that carries a steady current the strength of the magnetic field is measured to be 5 G. What current flows in the wire?

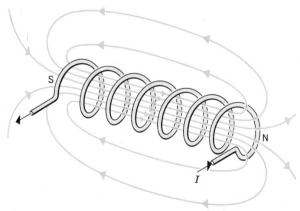

Figure 13.25 A *solenoid* consists of a long, helical spiral of wire. When the wire carries a steady current, the magnetic field within the solenoid is uniform.

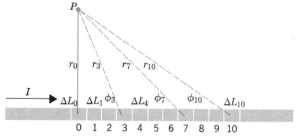

Figure 13.26 The magnetic field at P is the sum of the fields due to all of the tiny current elements. When this summation is carried out, the result is $B = 2K_M I/r_0$.

Solving Eq. 13.14 for I,

$$I = \frac{Br}{2K_M} = \frac{(5 \times 10^{-4}\text{ T}) \times (0.15\text{ m})}{2 \times (10^{-7}\text{ T-m/A})} = 375\text{ A} \blacksquare$$

Suppose that two long straight current-carrying wires lie parallel to one another. Each wire produces a magnetic field at every point occupied by the other wire and this field exerts a force on the moving charges in that wire. Depending on the directions of current flow, the two wires will experience either a mutual attraction or a mutual repulsion. The magnitude of the force can be obtained by substituting Eq. 13.14 for the field strength B into Eq. 13.5 for the magnetic force F_M. In doing this we notice that the symbol I in the two equations actually refers to different currents. In Eq. 13.14, I is the current in the wire producing the field; in Eq. 13.14, I is the current in the wire experiencing the force. We will label these currents I_1 and I_2, respectively. We also notice that in Eq. 13.5, $\sin\theta = 1$ because the magnetic field and the current are in mutually perpendicular directions. Then, we have

$$F_M = 2K_M L \frac{I_1 I_2}{r} \quad \text{(parallel wires)} \quad (13.15)$$

where r is the distance between the parallel wires.

If the wires are separated by a distance of 1 m and if each wire of the pair carries a current of 1 A, the mutual force between the wires (per unit length of wire) is

$$\frac{F_M}{L} = 2K_M \frac{I_1 I_2}{r} = 2 \times (10^{-7}\text{ N/A}^2) \times \frac{(1\text{ A}) \times (1\text{ A})}{1\text{ m}}$$
$$= 2 \times 10^{-7}\text{ N/m}$$

The *ampere* is now defined in terms of the force per unit length between a pair of parallel wires carrying the same current. In fact, *an ampere is that current which, when flowing in each of two long straight parallel wires separated by a distance of 1 m, causes each wire to exert on the other wire a force of 2×10^{-7} N for each meter of length of the wires.* Using this definition of the ampere, the constant K_M is determined; Eq. 13.15 then shows that K_M has the value 10^{-7} N/A².

You will sometimes see the equations we have been discussing written in a different way. For example, Eq. 13.15 is sometimes expressed as

$$F_M = \frac{\mu_0}{2\pi} \frac{LI_1 I_2}{r}$$

The constants K_M and μ_0 are related in the following way:

$$\mu_0 = 4\pi K_M = 12.57 \times 10^{-7}\text{ N/A}^2$$

The equations we have been discussing involve the calculation of magnetic fields and magnetic forces as they exist in free space or vacuum (or, very closely, in air). The quantity μ_0 that appears in these equations is called the *permeability of free space*.

Electromagnetism and Light

Although we originally specified the magnitude of the *coulomb* in terms of the electron charge (Section 12.1), we can now give a proper definition based on the ampere: *a coulomb is the amount of charge that passes a particular point in a circuit during a time of 1 s when a steady current of 1 A is flowing.* That is, 1 C = (1 A) × (1 s).

Now that we have defined the coulomb in a proper way, we can continue to work backward. We see that our definition of the coulomb determines the constant K in Coulomb's law: $K = 9.00 \times 10^9$ N-m²/C², which is the value indicated in Eq. 12.3. We are now in a position to make an interesting observation. Suppose that we take the ratio of the Coulomb force constant K to the magnetic force constant K_M. We find

$$\frac{K}{K_M} = \frac{9.00 \times 10^9\text{ N-m}^2/\text{C}^2}{10^{-7}\text{ N/A}^2} = (3.0 \times 10^8\text{ m/s})^2$$

The quantity on the right-hand side of this equation is the square of the speed of light! This equality can hardly be accidental; we must conclude that there is some connection between electricity and magnetism, on the one hand, and light, on the other. This connection was formulated in a precise way by the great Scottish mathematical physicist James Clerk Maxwell (1831–1879) during the 1860s. Maxwell succeeded in developing a unified theory of electromagnetism which includes the propagation of electromagnetic radiation, such as radio waves, heat radiation, and light. We will learn more about electromagnetic radiation in Chapter 15.

13.6 Magnetic Materials

The Effect of Magnetic Fields on Matter

Different types of matter respond in different ways when they are placed in a magnetic field. Suppose that we establish a reference field (in vacuum) by allowing a steady current I to flow in a solenoidal coil. We denote this reference field by B_0. Now, sup-

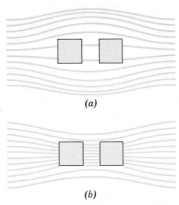

Figure 13.27 (a) Magnetic field lines are expelled from a diamagnetic material and the field strength at the position of the gap is decreased. (b) Magnetic field lines are concentrated by a paramagnetic material and the field strength at the position of the gap is increased. Both effects are exaggerated in these diagrams.

pose we insert into this field a split cylinder of some material and measure the field strength in the narrow gap between the two halves (Fig. 13.27). This type of experiment reveals that all matter can be divided into three magnetic classes:

Diamagnetic materials: These substances (e.g., antimony, bismuth, copper, graphite) cause the field lines to be expelled from the region of the material. Consequently, the field strength B in the gap is *less* than the reference field strength B_0; however, the effect is very small. Typically,

$B = B_0(1 - 10^{-6})$

Notice that B differs from B_0 by only about one part per million (1 ppm); nevertheless, this tiny difference has important physical consequences. The effect of a diamagnetic material is shown in Fig. 13.27a.

Paramagnetic materials: These substances (e.g., aluminum, glass, liquid oxygen, rubber, zinc) cause the field lines to concentrate in the sample. Thus, the field strength in the gap is *greater* than B_0. Again, the effect is not large. Typically,

$B = B_0(1 + 10^{-4})$

The effect of a paramagnetic material is shown in Fig. 13.27b.

Ferromagnetic materials: These substances (limited to iron, cobalt, nickel, and certain alloys and compounds of these metals) cause the field strength in the gap to be much greater than B_0. Typically,

$B = B_0(10^3)$

Diamagnetism

The diamagnetic, paramagnetic, and ferromagnetic effects that are found for different materials have separate and quite distinctive origins. Only the diamagnetic effect is common to all substances, but it is overwhelmed when a material exhibits paramagnetism of ferromagnetism.

All materials contain electrons that are bound to and circulate around atoms. A moving electron is equivalent to a current, so each orbiting atomic electron generates a magnetic field of the type shown in Fig. 13.10. In an ordinary sample of material, these individual electron-produced fields do not combine to produce a net magnetism because the orbits of the various atomic electrons are oriented at random. When an external magnetic field is applied, however, the electrons experience a magnetic force. This causes their motions either to be speeded up or slowed down, depending on the orientation of the orbit with respect to the field. The effect of the magnetic force is to *decrease* the electron-produced fields that are in the same direction as the external field and to *increase* those fields that are in the direction opposite to that of the external field. The net result is a *decrease* in the magnetic field within the sample compared with the original field. This amounts to the expulsion of field lines from the region occupied by the diamagnetic sample, as shown in Fig. 13.27a.

Paramagnetism

Various experiments that were carried out during the 1920s revealed that the electron possesses a previously unsuspected property in addition to mass and charge. The electron was discovered to have an intrinsic *angular momentum* in the same way that it has an intrinsic mass and an intrinsic charge; the electron simply *has* these properties. In classical terms, we can picture an electron as a tiny ball that is spinning on its axis; a spinning mass has angular momentum. But the electron possesses charge as well as mass. A spinning charge is equivalent to a current; therefore, an electron is a tiny magnet with a field shape similar to that of a current loop (Fig. 13.28).

Of course, the electron is not a tiny ball and it has no "axis" on which to spin. In the modern view of electrons and atoms (i.e., according to *quantum theory*), we say only that an electron possesses an *intrinsic* angular momentum and an *intrinsic* magnetism just as it possesses an *intrinsic* mass and an *intrinsic* charge. Thus, we do not inquire *how* the electron comes to have angular momentum and magnetism; instead, we view these properties in the same way that we view mass and charge. Neverthe-

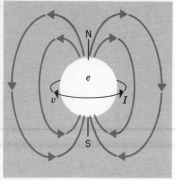

Figure 13.28 An electron can be pictured as a tiny spinning ball. Because the electron carries a charge, the spinning produces a magnetic field. (Compare Fig. 13.10.)

applied, the unpaired electron spins tend to align with the field direction, thereby increasing the field strength compared with that of the applied field.

Ferromagnetism

Diamagnetism and paramagnetism are quite small effects, the magnitudes of which range from 1 to 100 parts per million. Ferromagnetism, on the other hand, is a huge effect, in which the field enhancement can be a factor of 10^3 or greater. Consequently, some of the most important applications of magnetism involve ferromagnetic materials.

The ferromagnetism of a substance is the result of unpaired electron spins and the extraordinary way that the spins are aligned in crystals of the material. If a piece of iron is cleaned with acid and examined under a microscope, it is easy to see that the sample consists of tiny crystals of irregular shape that are packed tightly together. Each microcrystal consists of one or more regions (called *domains*) in which the electron spins are so strongly coupled that they are all aligned in the same direction. Each iron atom has four unpaired electrons, so the net magnetic effect of these electrons in a domain (which consists of billions of atoms) is large. In unmagnetized iron, however, the domains are oriented at random and the sample has no net magnetism.

Now consider what happens to the domains in a sample of iron when it is subjected to an external magnetic field. Figure 13.29 shows in a schematic way the changes that can occur in the domains of a single iron crystal. In Fig. 13.29a we see the domains in a crystal when there is zero external field. In each domain the arrow indicates the direction of the magnetic axis (S pole to N pole) of the field due to the aligned electrons. When an external field is applied, the domains can respond in two ways. First, as shown in Fig. 13.29b, the domain whose magnetic

less, it is often helpful to use the spinning ball model to provide a convenient picture of various physical effects; we will do so here and in later chapters.

In all diamagnetic materials, in the absence of an external magnetic field, the magnetic effects of the orbital electron motions all sum to zero. Moreover, the electrons all exist in coupled pairs with their directions of spin opposed; consequently, the spin magnetism is also zero. When a diamagnetic substance is subjected to a magnetic field, the orbital electron magnetism acts to decrease the magnetic field strength, as we discussed in the preceding subsection. The pairing of the spins, however, is unaffected by an external field and so the spin magnetism maintains its net zero value.

In paramagnetic materials (and also in ferromagnetic materials), some of the electrons are *unpaired* so that there is a net spin magnetism associated with each atom. If there is no external field applied to a paramagnetic substance, the spin directions are oriented at random and the net result is still zero. But when an external magnetic field is

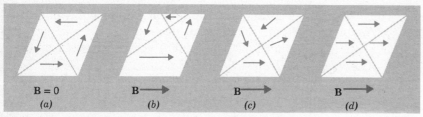

Figure 13.29 The effect of an external magnetic field on the domains in a crystal of iron. (a) Zero applied field and zero net magnetism. (b) The domain that is aligned with the field direction grows at the expense of the other domains, producing a net magnetism. (c) The domains are rotated to align more nearly with the direction of the external field, producing a net magnetism. (d) Saturation of the domain field.

Figure 13.30 The "magnet switch yard" at the National Bureau of Standards electron accelerator laboratory. The electron beam enters from the left and the five large "H" magnets are used to steer the beam to any of three experimental rooms located to the right of the area pictured.

National Bureau of Standards

axis is in the direction of the field can grow at the expense of the neighboring domains. That is, some of the atoms along the border of the lower domain are induced to change direction and become a part of this domain. The result is that the crystal acquires a net magnetism in the direction of the external field.

The second type of response to an applied field is shown in Fig. 13.29c. Here, the domains maintain their sizes, but their magnetic axes are rotated to lie more nearly along the direction of the external field. Again, the crystal acquires a net magnetism.

If a sufficiently strong field is applied, essentially all of the domains will become aligned (Fig. 13.29d) and the sample will be magnetically saturated. In the arrangement shown in Fig. 13.27b, the maximum field strength that can be achieved in the gap between two pieces of silicon steel (a common iron alloy used in motors and transformers) is about 1.6 T or 16 kG. This field can be obtained for a reference field (no material in the solenoid) of about 12 G; thus, the enhancement factor is greater than 10^3. Consequently, when large magnetic fields are required, iron or some alloy of iron is made into a configuration such as that shown in Fig. 13.9. The current flowing in the wire that is coiled around the iron causes the magnetization of the iron and produces a large field in the gap. Figure 13.30 shows several large "H" magnets in use at the National Bureau of Standards.

13.7 Electromagnetic Induction

Induced Currents Due to the Relative Motion of Wires and Fields

Thus far we have considered only the effects that are produced when moving charges and currents interact with *static* magnetic fields. We now turn to a discussion of *time-dependent* phenomena, including cases in which there is relative motion between current-carrying wires and magnetic fields or in which a magnetic field varies with time.

In the same way that a moving charge experiences a force in a magnetic field, a current-carrying wire (equivalent to a line of moving charges) that passes through a magnetic field, as in Fig. 13.31, will also be acted on by a magnetic force. If we disconnect the external source (for example, a battery) that supplies current to the wire, then, of course, the magnetic force will disappear. Now, suppose that we move the current-free wire through the magnetic field in the manner illustrated in Fig. 13.32. This motion will produce a magnetic force on the charges in the wire and they will begin to move along the wire. As long as the wire is in motion in the field, these charges will continue to flow; that is, a current has been *induced* in the wire. The phenomenon of

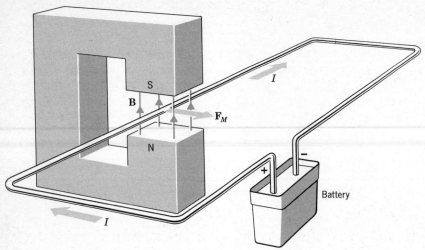

Figure 13.31 A static magnetic field exerts a force on a current-carrying wire.

electromagnetic induction was discovered in 1831 by the English physicist, Michael Faraday (1791–1867).

In order for a current to be induced in a wire it is not crucial that the field be stationary while the wire moves. Actually, it is only the *relative* motion of wire and field that is important. Therefore, the situation shown in Fig. 13.33, in which a magnetic field moves to the *right* across a stationary wire is entirely equivalent to the case in which the wire moves to the *left* through a stationary field, as in Fig. 13.32. In both cases the induced current flows in the same direction.

Another situation in which a current is induced by the relative notion between a magnetic field and a wire is shown in Fig. 13.34. Here a bar magnet is thrust into a wire loop that is connected to a current-measuring device (an *ammeter*). In order to find the direction of the magnetic force F_M on the charges in the wire, we need to know the direction of the velocity v of the charges relative to the field. Because the magnet is being thrust into the loop, the direction of v is *opposite* to that of the motion of the magnet, as shown in Fig. 13.34. Application of the right-hand rule for F_M (for example, at the top of the wire loop, as indicated in Fig. 13.34) shows that a current will flow in a clockwise sense when viewed with the N pole of the magnet approaching. If the magnet continues its motion, the induced current will drop to zero when the magnet is centered in the loop because at that position the motion of the wire (and the

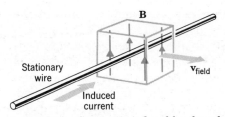

Figure 13.33 Current is induced by the relative motion of wire and field. This situation is entirely equivalent to that shown in Fig. 13.32. Only a section of the complete loop of wire is shown; compare Figs. 13.31 and 13.32.

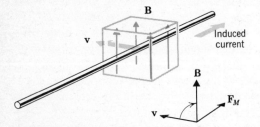

Figure 13.32 The motion of a wire through a magnetic field induces a current to flow in the wire. The section of wire shown is only a portion of the complete loop of wire that is necessary in order for a steady current to flow; compare Fig. 13.12.

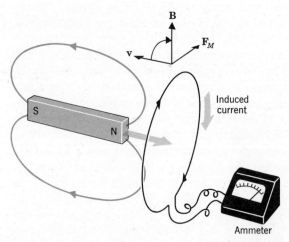

Figure 13.34 The relative motion between the field of the bar magnet and the wire loop induces a current in the wire which is detected by the ammeter.

charges it carries) is directly *along* the field line; when **v** is parallel to **B**, the magnetic force vanishes. As the S pole of the magnet passes through the loop, an induced current will flow again, but now the direction of flow will be opposite to that shown in Fig. 13.34. (Why?)

Currents Induced by Time-varying Fields

Consider a loop of wire that is connected to a battery through a switch. If the switch is open, no current flows and there is no magnetic field. Closing the switch causes the current to flow. But the current does not instantaneously attain its final steady value. Instead, the current is zero at the exact instant that the switch is closed and builds up to its final value during a certain short interval of time. Similarly, the magnetic field that is due to the flow of current starts at zero when the switch is closed and builds up to its final value just as does the current. Therefore, at any particular position in the vicinity of the wire, the magnetic field increases with time during the interval required for the current to attain its final steady value.

We can describe the situation in a pictorial way by saying that the circular magnetic field lines originate at the wire (beginning at the instant that the switch is closed) and spread out into space until the final steady field configuration is attained. The "movement" of the field lines in this outward expansion is similar in its effect to the physical movement of a magnet. Therefore, current will be induced in a wire that lies in the path of the "moving" field lines. Figure 13.35 shows such a situation; the field that expands from the right-hand loop when the switch is closed induces a current to flow in the left-hand loop. As soon as the current reaches its final steady value in the right-hand loop, the field ceases to expand and the induced current drops to zero. If the switch is now opened, the magnetic field will collapse and a current will be induced in the left-hand loop but in a direction opposite to that for the case of the expanding field.

Lenz's Law

Because of the magnetic force on the current-carrying wire in Fig. 13.31, the wire will begin to move. As soon as the wire has a velocity relative to the field, there will be a new magnetic force, $\mathbf{F}'_M$, on the charged particles due to their motion in the direction of the original $\mathbf{F}_M$. Application of the right-hand rules shows that $\mathbf{F}'_M$ is in the direction *opposite* to the direction of current flow. That is, there is an induced current that tends to oppose the original current flow. This is a general result—if any electromagnetic change A causes an effect B, then B will always induce a reaction C that tends to oppose A. This principle was discovered by Heinrich Lenz (1804–1865), a German physicist, and is known as Lenz's law. This law can always be used to predict the direction of current flow induced in a circuit due to external changes.

Although Lenz's law is reminiscent of Newton's third law, it is, in fact, a completely distinct statement. Actually, Lenz's law is simply a statement of energy conservation applied to induced currents. (If induced currents *aided* one another, in opposition to Lenz's law, then the currents would grow without the benefit of any energy input to the system. This is a clear violation of energy conservation.)

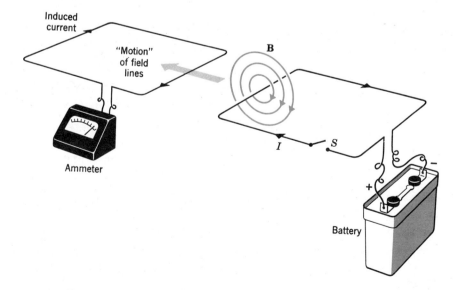

Figure 13.35 When the switch S is closed, a current begins to flow in the right-hand loop of wire. A magnetic field expands from this wire as the current builds up; the "moving" field lines induce a current in the left-hand loop.

Magnetic Flux and Faraday's Law

In each of the situations we have been discussing we can summarize the result by stating that the effect of induction is to produce an *EMF* and that the EMF causes the current to flow. We now wish to express the connection between induction and EMF in a quantitative way. One of the important quantities we must define is the *amount* of magnetic field in a particular region. We can think of the "amount" of the field in terms of the number of magnetic field lines that pass through a surface. We call this the *magnetic flux;* the greater the number of field lines that pass through a particular surface, the greater is the magnetic flux through that surface. Suppose that we consider a uniform magnetic field **B** (or, if the field is not uniform, we restrict our attention to a region sufficiently small that the field is essentially uniform). We select a plane surface whose area is A and which lies perpendicular to the field **B**, as in Fig. 13.36. The magnetic flux Φ that passes through this surface is defined to be the product of the field strength B and the surface area A:

$$\Phi = BA \qquad (13.16)$$

In the event that the surface is not oriented perpendicular to the field, the magnetic flux is calculated by considering only the perpendicular component of **B**; that is,

$$\Phi = B_\perp A \qquad (13.16a)$$

In the metric system, the unit of magnetic flux is the *weber* (Wb), named for the German physicist, Wilhelm Eduard Weber (1804–1891). Then, the magnetic field strength B is measured in Wb/m²; we have called this unit the tesla (T):

$$1 \text{ Wb/m}^2 = 1 \text{ T} = 10^4 \text{ G} \qquad (13.17)$$

Faraday found that he could summarize the results of his experiments concerning electromagnetic induction with the following statement:
The EMF generated in a circuit is equal to the rate of change of the magnetic flux through that circuit.

In equation form we can write

$$\mathcal{E} = \frac{\Delta \Phi}{\Delta t} \qquad (13.18)$$

where $\mathcal{E}$ is the EMF and where $\Delta \Phi$ is the change in the magnetic flux that takes place during the time interval Δt.

Faraday's law tells us that the effect of changing the magnetic flux through a circuit is the same as inserting a battery into the circuit. In Fig. 13.37a we have a circuit that encloses an area A in a magnetic field **B**. If the flux Φ through the circuit changes at the rate of 1 Wb/s, the induced EMF will be 1 V. While the flux is changing at this rate (and only while it is changing at this rate), current will flow in the circuit exactly as if a battery with a voltage of 1 V were inserted into the circuit (Fig. 13.37b).

The magnitude of the induced EMF can be increased if the circuit consists of a coil of wire instead of a single loop as shown in Fig. 13.37a. If there are N turns of wire in the coil, all with the same area, the induced EMF will be N times the value given by Eq. 13.18.

The equation that represents Faraday's law is often written with a negative sign:

$$\mathcal{E} = \frac{\Delta \Phi}{\Delta t} \qquad (13.18a)$$

The purpose of the negative sign is to remind us that the induced EMF *opposes* the flux change. That is, the negative sign in Faraday's law states Lenz's law!

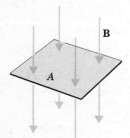

Figure 13.36 The magnetic flux through the surface is $\Phi = BA$.

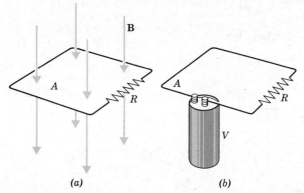

Figure 13.37 While the flux $\Phi = BA$ through the circuit is changing (a), the current flow will be the same as if a battery with a voltage $V = \Delta\Phi/\Delta t$ were inserted into the circuit (b).

Example 13.6

In Fig. 13.37a, suppose that the circuit has an area of 0.6 m² and consists of 20 loops of wire. The resistance in the circuit is $R = 12\ \Omega$. If the magnetic field strength is reduced at a constant rate from 8000 G to 3000 G in 2 s, how much current will flow during the field change?

Faraday's law can be written as

$$\mathcal{E} = N\frac{\Delta \Phi}{\Delta t} = NA\frac{\Delta B}{\Delta t}$$

because only the field strength changes, not the area of the circuit. Then, expressing ΔB in tesla, $\Delta B = 0.8\ \text{T} - 0.3\ \text{T} = 0.5\ \text{T}$, we have

$$\mathcal{E} = NA\frac{\Delta B}{\Delta t}$$
$$= 20 \times (0.6\ \text{m}^2) \times \frac{0.5\ \text{T}}{2\ \text{s}} = 3\ \text{V}$$

This means that the voltage across the resistor is also 3 V, so

$$I = \frac{V}{R} = \frac{3\ \text{V}}{12\ \Omega} = 0.25\ \text{A}$$

This current flows only while the field is changing.

Example 13.7

We next consider an example that corresponds to the situation in Fig. 13.32.

Suppose that a uniform magnetic field, $B = 0.4\ \text{T}$, exists over an area that is 0.2 m × 0.2 m, as shown in the diagram. A circuit, consisting of a rectangular loop of wire and a resistor $R = 8\ \Omega$, encloses the region of the field. This entire circuit is now moved to the right with a velocity $v = 20\ \text{m/s}$ so that one section of the loop cuts through the field. During the time that the wire is moving through the field, what current flows through R?

First, notice that the size of the *circuit* is irrelevant. The induction effect takes place only in the section of wire that cuts through the field, so it is

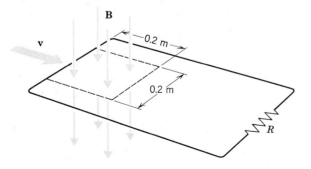

only the size of the *field region* that is important. The induced EMF is

$$\mathcal{E} = \frac{\Delta \Phi}{\Delta t} = B\frac{\Delta A}{\Delta t}$$

because B remains constant and the area A of the field inside the loop changes. The time Δt during which the enclosed area changes from 0.2 m × 0.2 m = 0.04 m² to zero is

$$\Delta t = \frac{s}{v} = \frac{0.2\ \text{m}}{20\ \text{m/s}} = 0.01\ \text{s}$$

Then,

$$\mathcal{E} = (0.4\ \text{T}) \times \frac{0.04\ \text{m}^2}{0.01\ \text{s}} = 1.6\ \text{V}$$

The voltage across the resistor is also 1.6 V, so that the current flowing through the resistor is

$$I = \frac{V}{R} = \frac{1.6\ \text{V}}{8\ \Omega} = 0.2\ \text{A}$$

(In which direction does this current flow?) ∎

In the next chapter we will discuss several applications of electromagnetic induction.

Summary of Important Ideas

Like magnetic poles *repel;* unlike magnetic poles *attract.*

A freely-suspended magnet aligns itself from S pole to N pole *along* the magnetic field lines.

The magnetic field lines in the space surrounding a magnet have the direction from the N pole to the S pole.

The S pole of the Earth is near its *north* geographic pole.

The poles of a magnet have no independent existence; N and S poles *always* occur together.

The direction of the magnetic field lines surrounding a current-carrying straight wire or loop of wire can be determined by using the *right-hand rule*.

Magnetic field lines are *continuous;* they have no beginning and no end. This is equivalent to the statement that *magnetic monopoles do not exist.*

The magnetic force $\mathbf{F}_M$ on a moving positively charged particle is *perpendicular* to the plane defined by $\mathbf{v}$ and $\mathbf{B}$. The direction of $\mathbf{F}_M$ is the same as the direction of advance of a right-hand screw when turned in the sense that rotates $\mathbf{v}$ toward $\mathbf{B}$.

Magnetic field strengths are measured in *tesla* (T) or in *webers per square meter* (Wb/m²) or in *gauss* (G): $1\ \text{T} = 1\ \text{Wb/m}^2 = 10^4\ \text{G}$.

Magnetic fields have no existence independent of electric charges.

A charged particle that moves in a uniform static magnetic field executes a *circular orbit* (or moves in a *helical path*). The magnetic force is always perpendicular to the motion of the particle; therefore, the field *does no work* on the particle.

The magnetic field within a long solenoid is uniform.

The *ampere* is defined in terms of the force between two long straight parallel wires that carry the same current.

When a piece of matter is inserted into an existing magnetic field B_0, the field strength within the substance can be slightly less than B_0 (*diamagnetic* (materials), slightly greater than B_0 (*paramagnetic* materials), or very much greater than B_0 (*ferromagnetic* materials). Diamagnetism is due to the magnetic effect of orbital atomic electrons. Paramagnetism and ferromagnetism are due to the magnetic effect of unpaired electron spins.

When the *magnetic flux* ($\Phi = BA$) through a circuit changes with time, a current is induced to flow in that circuit. This is called *electromagnetic induction*. *Faraday's law* states that the EMF induced is $\mathcal{E} = -\Delta\Phi/\Delta t$.

Lenz's law states that any electromagnetic change that causes an effect results in the effect inducing a reaction opposed to the original change.

◆ Questions

13.1 Sketch the magnetic field lines for the two pairs of bar magnets shown in the diagram.

(a) [S N] [N S]

(b) [S N] [S N]

Question 13.1

13.2 A current-carrying wire lies in a north-south direction. A compass is placed immediately above the wire and the N pole points eastward. In what direction are the electrons in the wire moving?

13.3 Explain why a bar magnet that is free to move will align itself with the direction of a magnetic field but will not undergo any net displacement.

13.4 An electron moves in an eastward direction near the equator. In what direction does the Earth's magnetic field exert a force on the electron?

13.5 A wire lies in a north-south direction and a current flows north in the wire. A positively-charged particle moves in the vicinity of the wire. In what direction will $\mathbf{F}_M$ act if (a) the particle is over the wire and moves north, (b) the particle is east of the wire and moves toward the wire, and (c) the particle is west of the wire and moves away from the wire?

13.6 An electron is projected into a current loop exactly along the axis. Describe the motion of the electron. What difference will there be if the electron's velocity vector is at a slight angle with respect to the axis of the loop?

13.7 It has been proposed that the Earth's magnetic field is due to a ring of electron current that flows in the molten metallic interior of the Earth. In what direction would the electrons have to flow in order to give the correct polarity for the Earth's field?

13.8 Two identical cardboard tubes are wound with wire in exactly the same way. The tubes are placed end-to-end and equal currents are passed through the wires. The currents circulate about the tubes in the same way. Will there be attraction or repulsion between the tubes?

13.9 When current is caused to flow through a loose helical coil of wire (such as the solenoid in Fig. 13.25), do the coils tend to contract or expand?

13.10 Two long wires lie parallel and carry equal currents in opposite directions. Sketch the lines of **B** in a plane that is perpendicular to the wires. Will the wires be mutually repelled or attracted?

13.11 How could you make a compass without using any ferromagnetic material (such as lodestone or other iron-containing mineral or alloy)?

13.12 An unmagnetized iron rod is held with its long dimension parallel to the Earth's field. One end is tapped with a hammer. It is found that the rod now possesses a weak magnetism. Explain why.

13.13 The atoms in any sample are continually being jiggled about because of thermal agitation. If a paramagnetic substance is in a magnetic field, does thermal motion tend to enhance or destroy the alignment of the electron spins with the field direction? How do you suppose that the magnitude of the paramagnetic effect changes as the temperature is lowered?

13.14 Discuss the effect illustrated in Fig. 13.35 in terms of magnetic flux and Faraday's law instead of "moving" field lines.

13.15 One end of a bar magnet is thrust into a wire loop. The induced current in the wire flows in the clockwise direction as viewed by looking along the direction of motion in the magnet. Which pole of the magnet was thrust into the loop? In what direction will the current flow if the magnet is *withdrawn* from the loop?

13.16 A steady current I flows in the wire loop (1) in the direction shown. If loop (1) is moved toward loop (2), in what direction will the induced current flow in loop (2)?

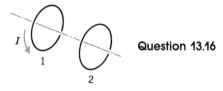

Question 13.16

13.17 A *short length* of wire is originally at rest in a static magnetic field. A mechanical force is suddenly applied to the wire, causing it to move through the field at a constant velocity. Explain why current will flow in the wire for a short time but will then stop even though the wire continues to move.

13.18 A bar magnet is dropped through a horizontal loop of wire with the N pole entering first. Describe the induction of current in the wire. Use Lenz's law to determine whether the magnet will experience an acceleration greater than or less than g while passing through the loop.

★Problems

Section 13.3

13.1 What is the maximum force that a 1-T magnetic field can exert on an electron whose energy is 10 keV? What is the minimum force and under what conditions would it be attained?

13.2 An electron moves with a velocity $v = 2.5 \times 10^7$ m/s in a magnetic field of 50 G. The velocity vector **v** makes an angle of 45° with respect to **B**. What is the magnetic force on the electron?

13.3* A 1-MeV proton is moving horizontally in an eastward direction near the Earth's equator where the magnetic field strength is 0.3 G. By approximately how much will the proton be deflected from its original line of motion after traveling 10 m? In what direction is the deflection?

13.4 A 1-g mass is moving eastward at the equator (where $B = 0.3$ G) with a velocity $v = 0.1$ c. What charge must the mass have if the upward magnetic force cancels the downward gravitational force?

13.5 A power line running east-west is carrying a current of 10^3 A. The Earth's magnetic field at the location of the power lines is 0.60 G in a north-south plane and makes an angle of 70° with the horizontal (see diagram). What is the magnitude and direction of the force on a 1-m length of the power line?

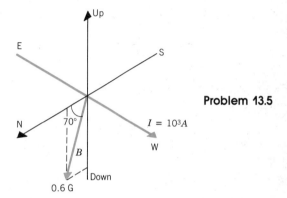

Problem 13.5

13.6 A long horizontal wire passes directly through a magnetic field whose direction is vertical. The region of the field has a diameter of 0.3 m and $B = 500$ G. If a current of 12 A is passed through the wire, what will be the magnetic force on the wire?

13.7 A current of 5 A flows in a wire that has 0.5 m of its length in and perpendicular to a magnetic field. The wire experiences a force of 0.25 N. What is the strength of the magnetic field?

13.8 A 1.5-m segment of a wire lies at an angle of 45° with respect to a magnetic field whose strength is 0.6 T. When a current of 30 A flows in the wire, what is the magnetic force on the 1.5 m segment?

13.9* A 10-turn stiff loop of wire in the form of a square with sides of length 15 cm carries a current of 10 A. It is placed in a uniform magnetic field of 300 G directed parallel to one side of the square loop. What is the net force on the loop? What torque is exerted on the loop? Describe the initial motion that the loop would follow if released to move freely.

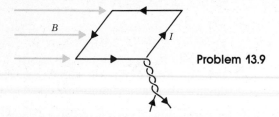

Problem 13.9

13.10 The weight of the rectangular loop of 10 turns shown in the diagram is first balanced by adding weights to the balance pan when $B = 0$. When the uniform field **B** is turned on and a current of 0.5 A flows in the coil, an additional mass of 8 g must be added to regain balance. What is the magnitude of **B**?

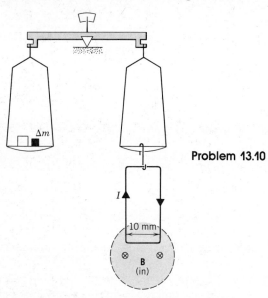

Problem 13.10

Section 13.4

13.11 There are sharp boundaries between a field-free region of space and a region containing a uniform magnetic field with the dimensions shown in the diagram. A charged particle enters the field region (from the field-free region) and moves perpendicular to the field lines. Describe the subsequent motion

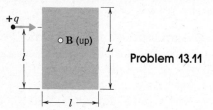

Problem 13.11

of the particle for the cases in which the orbit radius, R, has the values: (a) $R < \frac{1}{2}l$, (b) $\frac{1}{2}l < R < l$, and (c) $R > L$.

13.12* Refer to Problem 13.11 with $\ell = 1$ m, $L = 2$ m, and $B = 1000$ G. The particle is a 1-MeV proton. Through what angle has it been deflected as it leaves the field region?

13.13 An electron is accelerated, starting from rest, by falling through a potential difference of 1000 V. The electron then enters a magnetic field and is found to execute a circular orbit with a radius of 0.2 m. What is the strength of the magnetic field?

13.14* A proton moves in a helical path in a uniform magnetic field of 1000 G, as shown in Fig. 13.21. It requires 0.3 μsec for the proton to drift a distance of 0.3 m along the field direction. The radius of the helix is 0.1 m. What is the *speed* of the proton?

13.15 Two particles have the same momentum but one particle carries twice the charge of the other. What will be the ratio of their orbit radii in the same uniform magnetic field?

13.16 A singly charged carbon ion ($^{12}C^+$) is found to have the same orbit as a 2-MeV proton in a certain magnetic field. What is the energy of the carbon ion?

13.17 A cosmic ray proton with an energy of 10^{18} eV behaves as if its mass were approximately 10^9 times its mass when at rest (because of the relativistic increase of mass with velocity). The velocity of such a proton is essentially the velocity of light. Calculate the radius of the orbit that a 10^{18}-eV proton would execute in galactic space where the average magnetic field is about 3×10^{-10} T. Compare the result with the size of the local Galaxy.

13.18 A particle of mass m and charge q moves in a circular orbit in a magnetic field of strength B. Show that the time required to complete an orbit does not depend on the velocity of the particle.

13.19 An electron moves in a circular path in a uniform magnetic field $B = 10{,}000$ G. What is the period of its orbit? Refer to Problem 13.18.

13.20 A device for determining the atomic mass of elements is the mass spectrograph shown

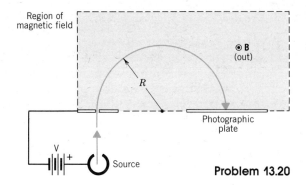

Problem 13.20

schematically in the diagram. Positive ions that are produced in the source and accelerated through the potential difference V enter a uniform magnetic field **B**. The ions strike a photographic plate allowing the radius R to be determined. Show that

$$M = \frac{eB^2}{2V} R^2$$

for singly-charged ions of mass M.

13.21 It is desired that 40-MeV α particles in a cyclotron should execute an orbit with a radius of 1 m. What magnetic field strength is required?

13.22* A certain cyclotron can produce 10-MeV protons. If the field strength is 10^4 G, how much time is required for each revolution? If the accelerating voltage across the "dee" gap is 40 kV, how many revolutions are necessary to accelerate protons to the final energy? How long does the acceleration process take?

13.23 In a certain region of space, an electric field **E** and a magnetic field **B** are mutually perpendicular. A charged particle enters the field region with a velocity **v** that is perpendicular to both **E** and **B**. Describe the motion of the particle if $v = E/B$.

13.24 An electron is moving along the $+x$-axis with a velocity $v = 4 \times 10^5$ m/s. When the electron is at the origin, what are the magnetic field strengths at (a) the point (0, 2 cm), (b) the point (2 cm, 2 cm), and (c) the point (2 cm, 0)?

Section 13.5

13.25* A certain copper wire has a resistance of 5.3×10^{-3} Ω/m. An 80-m length of this wire is wound into a thin coil that has a radius of 0.2 m. If the ends of the wire are connected to the terminals of a 12-V battery, what will be the strength of the magnetic field at the center of the coil?

13.26* A current I flows in a single circular loop of wire that has a loop radius a. Show that the magnetic field on the axis of the loop a distance d from the center of the loop is $B = 2\pi K_M I a^2/(a^2 + d^2)^{3/2}$.

13.27* Charges are placed around the rim of a plastic disc (diameter = 0.2 m) so that the charge density is 3×10^{-4} C/m. The disc is now rotated about its axis at a rate of 12 revolutions per second. What is the magnetic field strength at the center of the disc?

13.28 A 60-m length of wire is wound into a thin circular coil that has a diameter of 15 cm. What is the strength of the magnetic field at the center of the coil when it carries a current of 6 A?

13.29 A solenoid consists of 650 turns of wire on a form that has a length of 0.45 m. What is the magnetic field within the solenoid when a current of 3.8 A flows in the wire?

13.30 A solenoid has a diameter of 8 cm and a length of 40 cm. How many turns must there be in the winding if the field strength in the solenoid is to be 300 G when the current is 6 A?

13.31 A 100-m length of wire is completely and uniformly wound around a hollow cardboard cylinder whose diameter is 4 cm. The distance between the first and last turn is 20 cm. What is the magnetic field strength inside the cardboard cylinder when a current of 8 A is flowing in the wire?

13.32 What is the strength of the magnetic field at a distance of 6 cm from a long straight wire that carries a current of 40 A?

13.33 Two long straight wires are separated by a distance of 10 cm. The wires carry currents of 4 A in opposite directions. What is the magnetic field strength midway between the wires?

13.34 Two straight wires are parallel and are separated by 15 cm over a distance of 5.6 m. When one wire carries a current of 125 A, it is found that this wire exerts a total force of 0.075 N on the other wire. What current flows in the second wire?

13.35 The *toroidal coil* shown in the diagram may be considered a long solenoid curved in a radius R so that its two ends meet. If the total number of turns is N and cross-sectional radius r is small compared to the toroid radius R, show that B is given by

$$B = \frac{2K_M NI}{R}$$

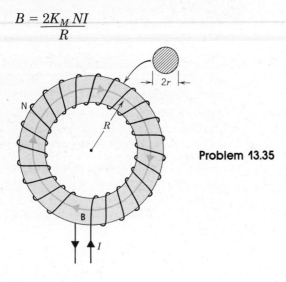

Problem 13.35

Section 13.6

13.36 The *permeability number* κ (or the *relative permeability*) of a substance is equal to the ratio of the magnetic field strength B that exists in a sample of the material to the field strength B_0 that would exist in the same location if the sample were not present (vacuum condition). Thus, for a typical paramagnetic material, $\kappa \cong 1.0001$, whereas for ferromagnetic materials, κ is a few thousand at relatively modest field strengths. (Actually, κ for ferromagnetic materials depends on the field strength B_0.) Suppose that a wire is wound around a cylindrical piece of annealed iron so that there are 230 turns per meter. When a current of 0.5 A flows through the wire, the magnetic field strength within the iron is 0.75 T. What is the value of B_0? What is the permeability number of the iron at this value of B_0? (This example actually gives the *maximum* value for κ for annealed iron.)

13.37 The core of the toroidal coil of Problem 13.35 is made of iron with relative permeability $\kappa = 300$. The total number of turns is $N = 500$ and $R = 8$ cm. What is the magnetic field strength per ampere of current, B/I?

Section 13.7

13.38 A circular coil of wire consists of 12 loops, each with a radius of 40 cm. A magnetic field that is directed perpendicular to and passes through the coil is changing at a rate of 0.02 T/s. What EMF is induced in the coil?

13.39 A total of 250 turns of wire are wound around the edge of a square board whose dimensions are 40 cm $\times$ 40 cm. The plane of the board makes an angle of 45° with respect to the field lines of a magnetic field whose initial strength is 4500 G. If the field strength is changed uniformly to 1800 G in 1.6 s, what EMF is produced in the wire coil during the change?

13.40 A coil consists of 40 turns of wire wound around a square frame that is 20 cm on a side. How rapidly must this coil be withdrawn from a perpendicular magnetic field with a strength of 0.6 T in order to produce an EMF of 3.2 V?

13.41* In the circuit shown in the diagram, the conducting rod PQ is caused to slide over the wires with a velocity **v**. A magnetic field **B** that is perpendicular to the plane of the circuit exists over the entire area of the circuit. Consider the work per unit charge done in moving the rod the distance Δs during the time Δt against the magnetic force; thus, obtain the EMF induced in the wire. Next, apply Faraday's law to the situation and show that the same result is obtained.

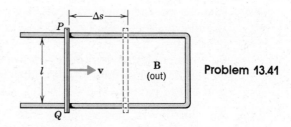

Problem 13.41

13.42 Suppose that the U-shaped conductor shown in the diagram for Problem 13.33 has a width of 25 cm and is located entirely within a perpendicular field of 2500 G. If the straight rod moves across the U-shaped conductor with a velocity of 4.5 m/s, what EMF is induced in the circuit?

13.43 A certain solenoid consists of 400 turns of wire and has a length of 0.8 m. In the middle of the solenoid another wire, insulated from the solenoid winding, makes 6 turns around the solenoid. The diameter of this coil is 8 cm. What average EMF is induced in the secondary coil when the current in the solenoid is increased from 2.2 A to 6.4 A during a 4-s interval?

13.44 A 20-turn circular coil of wire has a diameter of 0.5 m. This coil is originally aligned parallel to a certain magnetic field. The coil is then rotated until it lies perpendicular to the field. During the rotation, which requires 0.2 s, the average EMF generated in the coil is measured to be 3.93 V. What is the strength of the magnetic field? (This "flip-coil" method is sometimes used to measure magnetic field strengths.)

14 Applied Electromagnetism

In the two preceding chapters we discussed the basic ideas of electricity and magnetism. In this chapter we examine some of the ways these ideas are put to use in practical situations involving electric circuits and electric devices. We distinguish between two different types of current flow in the circuits we will study: *direct current* (DC) and *alternating current* (AC). We begin by analyzing several different kinds of DC circuits and DC devices. Then, we see how to generate AC and how various circuit elements behave in AC circuits.

14.1 Electric Circuits Containing Batteries and Resistors

Ohm's-Law Circuits

In Section 12.7 we learned how to calculate the total effective resistance in series connections of resistors and in parallel connections of resistors. We begin our survey of practical electric systems by treating resistor arrangements that are more complex than those in the simple series and parallel circuits we discussed earlier.

Suppose that we have a combined series-parallel connection of resistors, as shown in Fig. 14.1a. We wish to find the total resistance R_t in the circuit and the value of the current I. First, we look at the two parallel combinations, A and B. We can use the standard expression (Eq. 12.53) to calculate the total

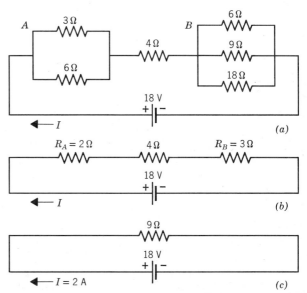

Figure 14.1 Steps in reducing a series-parallel combination of resistors to a single equivalent resistor. In dealing with such circuits, always begin by reducing the parts of the circuit that consist of parallel connections.

resistance of each of these parts of the circuit:

For A: $\dfrac{1}{R} = \dfrac{1}{3} + \dfrac{1}{6} = \dfrac{2+1}{6} = \dfrac{3}{6} = \dfrac{1}{2}$

so that $R_A = 2\ \Omega$.

For B: $\dfrac{1}{R_B} = \dfrac{1}{6} + \dfrac{1}{9} + \dfrac{1}{18} = \dfrac{3+2+1}{18} = \dfrac{6}{18} = \dfrac{1}{3}$

so that $R_B = 3\ \Omega$. Figure 14.1b shows each of the two parallel combinations replaced by a single resistor equal to the effective resistance of that part. The circuit is now a simple series circuit. Therefore, the total resistance is

$R_t = 2\ \Omega + 4\ \Omega + 3\ \Omega = 9\ \Omega$

Figure 14.1c shows the series combination replaced by a single resistor equal to the total effective resistance in the circuit, that is, 9 Ω. Then, the current I is given by

$I = \dfrac{V}{R_t} = \dfrac{18\text{ V}}{9\ \Omega} = 2\text{ A}$

Circuits that contain resistors connected in the general way illustrated in Fig. 14.1a can always be analyzed by using the technique employed in this example. Begin with the parallel combinations and reduce these to the equivalent resistances. Proceed until all parallel combinations are eliminated and a simple series circuit remains. Then, sum the individual resistance values.

Figure 14.2a shows a simple circuit that consists of two batteries and two resistors. How much current flows in this circuit? First, note that we can change the location of either battery or either resistor without affecting the current in the circuit. (Can you see why?) Thus, we can represent the circuit as in Fig. 14.2b. Here, we have the two resistors in series; the total resistance is clearly 4 Ω and this is indicated in Fig. 14.2c. What about the two batteries? In Fig. 14.2b we see that the batteries *oppose* one another; that is, the terminals of the two batteries are connected to the circuit in opposite directions. The 16-V battery attempts to force a current flow in the clockwise direction around the circuit, whereas the 24-V battery attempts to force a current flow in the counterclockwise direction. It is easy to see that the battery with the larger voltage will dominate. The net result is that the two opposing

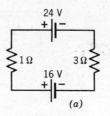

(a)

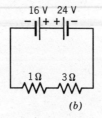

(b)

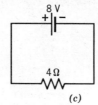

(c)

Figure 14.2 Steps in reducing a two-battery, two-resistor circuit to a simple circuit. Notice that the two batteries *oppose* one another, so that the net effect is the same as a battery with a voltage equal to the *difference* in voltage of the two batteries.

batteries have the same effect as a single battery with voltage equal to 24 V − 16 V = 8 V; this is indicated in Fig. 14.2c. Notice that the orientation of this equivalent battery is the same as that of the dominant 24-V battery. We can now conclude that the current in the circuit is $I = 8\text{ V}/4\ \Omega = 2\text{ A}$.

● Example 14.1

How much current flows in the circuit shown?

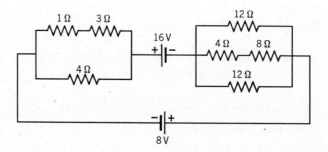

First, we notice that the resistor combination on the left is actually two 4-Ω resistors in parallel; the effective resistance of this pair is 2 Ω. Similarly, the combination on the right is actually three 12-Ω resistors in parallel; the effective resistance of this group of resistors is 4 Ω. Next, notice that the two batteries *aid* one another, so that they are equivalent to a single 24-V battery. Therefore, we can represent the circuit as in the following diagram:

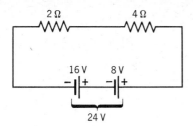

We see that the original circuit is equivalent to a 6-Ω resistor connected to a 24-V battery. Consequently, the current in the circuit is $I = 24\text{ V}/6\text{ Ω} = 4$ A. ∎

Kirchhoff's Circuit Rules

If we attempt to apply the straightforward reasoning we have used in the preceding discussion to analyze the circuit shown in Fig. 14.3, we find that we can make no progress. The problem is that the resistor R is not connected to the other resistors in any simple way; R is neither in series nor in parallel with the other resistors. In order to analyze circuits of this type, as well as more complex circuits, we must use *Kirchhoff's circuit rules*.

Kirchhoff's rules are actually quite simple, and they represent a systematic way of analyzing the behavior of circuits. The first rule merely states that when a current flows through a circuit, we can neither gain nor lose charge during the process; that is,

> I. At any point in a circuit, the sum of the currents entering that point must equal the sum of the currents leaving that point.

Figure 14.4 shows two examples of this rule. In each of the cases illustrated, the total current flowing into point a is equal to the total current flowing away from a.

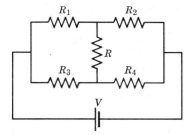

Figure 14.3 The total effective resistance in this circuit cannot be obtained by successively reducing the circuit to a simple series circuit.

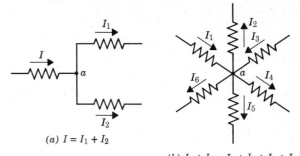

(a) $I = I_1 + I_2$

(b) $I_1 + I_3 = I_2 + I_4 + I_5 + I_6$

Figure 14.4 Kirchhoff's rule I states that the total current flowing into point a must equal the total current flowing away from that point. (Point a is called a *branch point*.)

Kirchhoff's second rule deals with the way the electric potential changes around a circuit. The rule states that the net change in potential (or voltage) in going completely around a circuit is zero. That is, if we begin at a particular point in a circuit and move around the loop of the circuit, measuring each potential increase and each potential drop, when we return to the starting point, the net sum of all the potential changes will be zero; thus,

> II. The sum of all the potential increases around a circuit must equal the sum of all the potential drops.

Notice that Kirchhoff's first rule is a consequence of the conservation of charge. If you think about the way the second rule is formulated, you can see that it follows from the conservation of energy.

We can see how Kirchhoff's second rule works by looking at the potential changes around the circuit shown in Fig. 14.5. Let us start at point a and move around the circuit in the same direction as the indicated current flow. (We could proceed equally well in the opposite direction.) We have four separate elements in the circuit and so we will have four potential changes:

1. $a \rightarrow b$. In this section of the circuit the current flows from a toward b. Now, current always

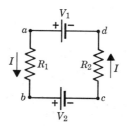

Figure 14.5 Kirchhoff's rule II states that the sum of all the potential increases around the circuit is equal to the sum of all the potential drops.

flows from a point of high potential to one of low potential. (Why?) Therefore, in moving from a to b the potential must *decrease*. The amount of the potential drop in this section is equal to IR_1, the voltage across the resistor R_1.

2. $b \to c$. In moving from b to c we pass through the battery V_2 from the positive terminal to the negative terminal. Now, the positive terminal of a battery has a potential that is *higher* than that of the negative terminal. Therefore, in moving from b to c the potential *decreases* and the amount of the potential drop is V_2.

3. $c \to d$. As in the section $a \to b$, the potential decreases in moving from c to d. The amount of the potential drop is IR_2.

4. $d \to a$. In moving from d back to the starting point a, we pass through the battery from the negative terminal to the positive terminal. This means that the potential *increases* by an amount V_1.

Summarizing our findings:

$a \to b$: Potential drop, IR_1

$b \to c$: Potential drop, V_2

$c \to d$: Potential drop, IR_2

$d \to a$: Potential increase, V_1

According to Kirchhoff's second rule, we can now write

$$V_1 = V_2 + IR_1 + IR_2$$

Notice that the circuit in Fig. 14.5 is exactly the same as that in Fig. 14.2a. If we supply the values of the resistors, the voltages, and the current from the previous example, we find

$$24 \text{ V} = 16 \text{ V} + (2 \text{ A}) \times (1 \text{ }\Omega) + (2 \text{ A}) \times (3 \text{ }\Omega)$$
$$= 16 \text{ V} + 2 \text{ V} + 6 \text{ V} = 24 \text{ V}$$

and Kirchhoff's second rule is verified for this case.

● **Example 14.2**

Find the current in each of the loops of the circuit illustrated.

In order to use Kirchhoff's circuit rules, we must first identify and assign directions to the currents in the various loops of the circuit; this has been done in the diagram. It does not matter whether we guess

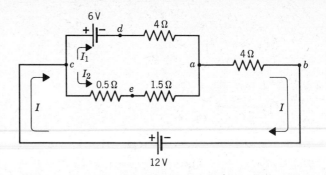

the correct directions for the currents. If any direction is chosen incorrectly, the answer we obtain from the analysis will be *negative*, indicating that the actual current is in the direction *opposite* to that chosen. (Notice that this flexibility is the same as we discovered in dealing with forces and torques in Section 5.3; see Example 5.6.)

We have three unknown currents (I, I_1, and I_2), so we need three independent equations involving these quantities in order to obtain a solution. From Kirchhoff's first rule applied at point c (or at point a), we obtain one equation:

$$I = I_1 + I_2 \tag{1}$$

The other two equations are obtained by applying the second rule. We choose two loops, $abeca$ and $aecda$. In the equation for each loop we write a potential increase as a positive quantity and a potential drop as a negative quantity. Then, for the first loop,

$$(a \to b) + (b \to c) + (c \to e) + (e \to a) = 0$$
$$-4I + 12 \text{ V} - 0.5I_2 - 1.5I_2 = 0 \tag{2}$$

and for the second loop,

$$(a \to e) + (e \to c) + (c \to d) + (d \to a) = 0$$
$$1.5I_2 + 0.5I_2 - 6 \text{ V} - 4I_2 = 0 \tag{3}$$

Now, substitute $I = I_1 + I_2$ into (2) and rearrange (3) to obtain

$$-4I_1 - 6I_2 + 12 = 0 \tag{2'}$$

$$-4I_1 + 2I_2 - 6 = 0 \tag{3'}$$

Then, subtracting (3') from (2'), we find

$$-8I_2 + 18 = 0$$

Solving for I_2,

$$I_2 = \tfrac{18}{8} = 2.25 \text{ A} \tag{4}$$

Substituting this value for I_2 into (3) and solving for I_1, we have

$I_1 = \frac{1}{4}(2I_2 - 6) = \frac{1}{4}(4.50 - 6) = -0.375 \text{ A}$ (5)

Finally,

$I = I_1 + I_2 = -0.375 + 2.25 = 1.875 \text{ A}$ (6)

Notice that I_1 is negative, indicating that our original choice for the direction of I_1 was incorrect. ∎

If we need to solve for n unknown quantities in a circuit, we need n independent equations. We obtain these from Kirchhoff's rules. The first rule will give us a number of equations equal to one fewer than the number of branch points in the circuit. (In Example 14.2 there are two branch points a and c. We obtain only a single equation, however, because the equation for a is the same as the equation for c.) The second rule yields the remaining equations. In choosing the circuit loops to apply the second rule, each loop must contain a difference of at least one circuit element from every other loop. (In Example 14.2 we chose two loops, $abcea$ and $aecda$. There is one other possibility, namely, $abcda$. But this loop contains no new circuit element not already contained in the other two loops. Write down the Kirchhoff equation for this third loop and verify that it is equal to the sum of equations (2) and (3) in Example 14.2—that is, it is not an independent equation.)

Battery EMF

In our preceding discussions, we have implicitly assumed that the voltage between the terminals of a battery in a circuit is constant and does not depend on the current that flows through the battery. Within a battery the chemical reactions do provide a constant EMF and this EMF is responsible for the flow of current when the battery is connected to a circuit. But a battery is not an ideal device and in addition to a source of EMF, every battery contains a certain amount of internal resistance. We can represent a battery as a pure (resistanceless) EMF connected in series with a resistor r, as shown in Fig. 14.6. The points a and b represent the terminals of the battery and the voltage between these points is V_{ab}. If there is no connection to the battery, no current flows, there is no voltage drop across r, and the terminal voltage is equal to the battery EMF; that is $V_{ab} = \mathcal{E}$. However, when the resistor R is connected to the battery, the current I is equal to the EMF divided by the total resistance in the circuit, that is, $r + R$. Thus,

$$I = \frac{\mathcal{E}}{r + R} \quad (14.1)$$

Now, the potential drop across R is IR and this is just the terminal voltage V_{ab}; therefore,

$$V_{ab} = IR = \mathcal{E} \times \frac{R}{r + R} \quad (14.2)$$

A 12-V battery (i.e., a battery with an EMF of 12 V) might have an internal resistance of only 0.02 Ω. If a 4-Ω resistor is connected to such a battery, the current will be

$$I = \frac{\mathcal{E}}{r + R} = \frac{12 \text{ V}}{0.02 \text{ Ω} + 4 \text{ Ω}} = \frac{12 \text{ V}}{4.02 \text{ Ω}} = 2.985 \text{ A}$$

or, essentially 3 A, the value that would be calculated by neglecting the internal resistance. The terminal voltage of the battery while delivering this current is

$$V_{ab} = IR = (2.985 \text{ A}) \times (4 \text{ Ω}) = 11.94 \text{ V}$$

Thus, the terminal voltage is nearly equal to the EMF of the battery.

As a battery ages, its internal resistance increases. Suppose that the 12-V battery eventually has an internal resistance of 9.5 Ω. Then, when it is connected to a 4-Ω resistor, the terminal voltage will be

$$V_{ab} = \mathcal{E} \times \frac{R}{r + R} = (12 \text{ V}) \times \frac{4 \text{ Ω}}{4.5 \text{ Ω}} = 10.67 \text{ V}$$

which is a substantial decrease from the original value.

● **Example 14.3**

What value of resistance R must be chosen so that a 12-V battery with an internal resistance $r = 0.2$ Ω will deliver maximum power when connected to R?

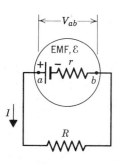

Figure 14.6 A battery can be represented as a pure EMF in series with a resistor r (the *internal resistance* of the battery).

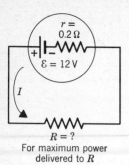

For maximum power delivered to R

When the resistor R is connected to the battery, the current is

$$I = \frac{\mathcal{E}}{r + R}$$

Therefore, the power delivered to R is (see Eq. 12.60)

$$P = I^2 R = \frac{\mathcal{E}^2 R}{(r + R)^2}$$

We now substitute the values for $\mathcal{E}$ and r, and we calculate the power P for various values of R between 0 and 0.5 Ω. The results are indicated in the graph on the following page. We see that the power delivered to R is a maximum (180 W) when $R = 0.2$ Ω. That is, the power is maximum when R (the *internal resistance*). This same kind of result is found also for AC circuits where it is called *impedance matching*.

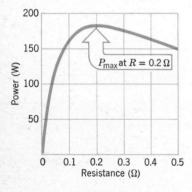

Notice that the graph indicates zero power when $R = 0$. This does not mean that there is no current flow and therefore no power developed by the battery. It means only that there is no power delivered by the battery to an external resistance (because there is none!). Indeed, when $R = 0$—that is, when a resistanceless wire is connected between the battery terminals—a very large current will flow: $I = \mathcal{E}/r = 12 \text{ V}/0.2 \text{ Ω} = 60$ A. The power developed will be $P = I^2 r = (60 \text{ A})^2 \times (0.2 \text{ Ω}) = 720$ W. Needless to say, a battery cannot sustain this kind of power drain for very long! All of the power is dissipated within the battery and results in a substantial heating effect. If a battery is *short circuited* in this way, it will quickly "burn out." ∎

Null-current Circuits

The terminal voltage of a battery depends on the amount of current that flows through the battery. Only if there is zero current will the terminal voltage (to which we have direct access) equal the battery EMF (to which we do not have direct access). Therefore, if we wish to know the EMF of a battery, we must make the measurement under a condition of zero current through the battery. A simple method of accomplishing this is to use a *potentiometer circuit* (Fig. 14.7). The potentiometer itself is simply a long resistor wire with a sliding contact that can be positioned at any point between the ends of the wire.

In Fig. 14.7 the sliding contact divides the potentiometer into two sections with resistances R_1 and R_2. As the sliding contact is moved, the total resistance of the potentiometer, $R_1 + R_2$, remains constant. The battery located at the top of the diagram (usually called the *working battery*) has a terminal voltage V and causes a current I to flow through the potentiometer wire. We are not concerned about the EMF or internal resistance of this battery because its only function is to provide a constant current I through R_1 and R_2.

This battery with the unknown EMF $\mathcal{E}_x$ and the standard cell with EMF $\mathcal{E}_s$ can be connected to the potentiometer circuit through the sensitive current-detecting meter G by means of the switch S. (The

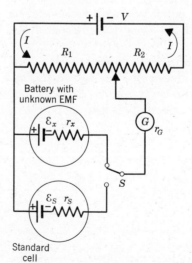

Figure 14.7 A potentiometer circuit for determining the EMF of a battery.

meter G is a *galvanometer*, named in honor of the Italian scientist, Luigi Galvani, 1737–1798). Notice that both $\mathcal{E}_x$ and $\mathcal{E}_s$ are connected in the circuit to *oppose* the voltage of the working battery. When the switch is in the position shown in Fig. 14.7, the galvanometer will indicate zero current if

$$\mathcal{E}_x = IR_1$$

(Verify this by applying Kirchhoff's second rule to the lower portion of the circuit. Notice that the galvanometer has an internal resistance r_G.) Once this condition has been established by moving the sliding contact until the proper value for R_1 is determined, the standard cell is switched into the circuit by means of S. The sliding contact is moved until the galvanometer again reads zero. This new value of R_1 is R_{1s}, and we have

$$\mathcal{E}_s = IR_{1s}$$

Notice that when there is no current flowing in the lower portion of the circuit, the current flowing in the upper loop is $I = V/(R_1 + R_2)$, and $R_1 + R_2$ remains constant as the sliding contact is moved.

The value of the unknown EMF is now obtained by dividing the equation for $\mathcal{E}_x$ by that for $\mathcal{E}_s$ and solving for $\mathcal{E}_x$. We find

$$\mathcal{E}_x = \frac{R_1}{R_{1s}} \mathcal{E}_s \qquad (14.3)$$

Another circuit that employs a potentiometer and the null-current technique is the *Wheatstone bridge* (Fig. 14.8), used for making precision resistance measurements. These are four resistance elements in this circuit—R_1 and R_2, the two sections of the potentiometer; R_x, the unknown resistance; and R_s, the standard resistance. The sliding contact is moved until there is zero current through the galvanometer. Then, we know that the potential at points a and b is exactly the same, so we can write two equations for the potential drops in the two parts of the circuit separated by the galvanometer connection *ab*:

$$I_1 R_1 = I_x R_x \quad \text{and} \quad I_2 R_2 = I_s R_s$$

(These equations represent Kirchhoff's second rule for the two loops of the circuit.) The fact that there is zero current flowing between *a* and *b* means that $I_1 = I_2$ and $I_x = I_s$ (Kirchhoff's first rule). Using these equalities and taking the ratio of the equations above, we find

$$\frac{R_x}{R_s} = \frac{R_1}{R_2} \qquad (14.4)$$

If the potentiometer has uniform resistance per unit length, the ratio R_1/R_2 is equal to the length ratio ℓ_1/ℓ_2 (see Fig. 14.8). Then, the value of the unknown resistance is equal to the standard resistance R_s multiplied by a ratio of measured lengths:

$$R_x = \frac{\ell_1}{\ell_2} R_s \qquad (14.5)$$

14.2 Capacitance and Inductance in DC Circuits

Simple RC Circuits

The circuits we have been discussing have consisted only of resistors and batteries. We now allow for two additional types of circuit elements, namely, *capacitors* (which we have already introduced in Section 12.5) and *inductors* (which we will define in the next subsection).

Suppose that we connect a resistor R and a capacitor C in series with a battery whose terminal voltage is V, as in Fig. 14.9. (We assume here and in most of our subsequent discussions that the internal resistance of the battery is negligibly small compared to any other resistance in the circuit.) How will the charge Q build up the capacitor when the switch S_1 is closed (with S_2 open)? The initial rate of

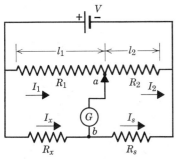

Figure 14.8 A Wheatstone bridge for measuring the resistance of a resistor.

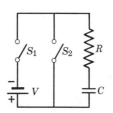

Figure 14.9 A simple *RC* circuit. The capacitor is charged by closing switch S_1.

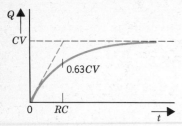

Figure 14.10 The accumulation of charge on the capacitor in the circuit of Fig. 14.9. The charge Q approaches the limiting value CV. The initial slope of the curve (dashed line) is equal to the current $I_0 = V/R$. The time $t = RC$ is the *time constant* of the circuit.

Table 14.1 Charging a Capacitor in an *RC* Circuit

Time (in Units of *RC*)	Charge (in units of *CV*)
0	0
0.5	0.393
0.693	0.500
1	0.632
2	0.865
3	0.950
4	0.982
5	0.993
6	0.998
7	0.999

charge accumulation (i.e., the initial *current*) is determined only by the battery voltage V and the resistance R. As charge is deposited on the capacitor, however, it becomes more difficult to add to the charge because of the repulsive effect of the previously deposited charge. The charge Q approaches a maximum value equal to CV, but, strictly speaking, this value is reached only after an infinite time. Figure 14.10 shows how the accumulated charge increases with time. The slope of the curve at any instant is the rate at which the charge is changing, that is, the *current*. At the time $t = 0$, the slope is equal to the initial current, $I_0 = V/R$, and as time increases, the current continually decreases.

The curve of Q versus t in Fig. 14.10 is of a type that is encountered frequently in a variety of physical situations. We will only note that this curve is described by an *exponential function*, but we will not inquire into the mathematical details. The time required for the charge Q to reach $0.63CV$ is called the *time constant* or the *characteristic time* of the circuit and is equal to the product RC. Notice that RC is the time at which the capacitor would be fully charged if the initial charging rate (dashed line in Fig. 14.10) were maintained. (Can you see why the time required to accumulate a definite fraction of the maximum charge must be proportional to both R and C?) After a time $5RC$, the Q has increased to within 0.7 percent of the maximum value CV (see Table 14.1). For times equal to or longer than about $5RC$, we can usually consider the capacitor to be fully charged.

If the switch S_1 in Fig. 14.9 is opened after the capacitor has been charged to some value of charge Q_0, this charge will remain on the capacitor. If we now close the switch S_2, what will happen to the charge? When the circuit is completed by closing S_2, the voltage that exists between the plates of the capacitor is placed directly across the resistor. Consequently, a current will flow. But as the current flows, the charge on the capacitor is depleted, the voltage across the resistor is decreased, and the current diminishes. The net result is that the charge on the capacitor decreases with time following a curve that is the inverse of the charging curve in Fig. 14.10. The curve that represents the discharging of a capacitor is shown in Fig. 14.11. The slope of the dashed line again gives the initial rate of discharge, that is, the initial current, $I_0 = V/R = Q_0/RC$. If this rate of discharge were maintained, the charge on the capacitor would become zero at $t = RC$.

● *Example 14.4*

A 4-μF capacitor is charged through a 100 Ω resistor by a 12-V battery. (a) What is the time constant of the circuit? (b) What is the maximum charge that can accumulate on the capacitor? (c) How long will be required for the charge to accumulate to 90 percent of the maximum value?

(a) The time constant is

$$RC = (100 \text{ }\Omega) \times (4 \times 10^{-6} \text{ F}) = 4 \times 10^{-4} \text{ s} = 0.4 \text{ ms}$$

(b) The maximum charge is

$$Q_{max} = CV = (4 \times 10^{-6} \text{ F}) \times (12 \text{ V}) = 4.8 \times 10^{-5} \text{ C}$$

(c) According to Table 14.1, the time at which the charge will reach 90 percent of its maximum value lies between $2RC$ and $3RC$. By plotting the

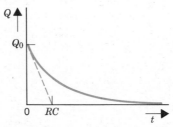

Figure 14.11 The discharge of a capacitor C through a resistor R.

information in the table, we can determine that the 90 percent point occurs at $2.3RC$, so that

$t(\text{for } Q = 0.9CV) = 2.3RC = 2.3 \times 0.4 \text{ ms} = 0.92 \text{ ms}$ ∎

Simple RL Circuits

We know that a magnetic field is produced with a current flows through a wire. If the wire is in the form of a coil or a solenoid, each field line links several turns. This linkage is particularly important when there is a change in the current flowing in a coil. What happens when the switch S is closed and completes the circuit in Fig. 14.12? (The coil of wire in Fig. 14.12 is labeled L, the symbol customarily used to indicate *inductance* or an *inductor*; we will define inductance shortly.)

When current begins to flow in the coil, a magnetic field begins to build up. The buildup of the field around one of the turns means that the adjacent turns are subjected to a changing magnetic field and, in accordance with Faraday's law of induction (Section 13.7), an EMF is developed in these turns. Lenz's law tells us that this EMF is in the direction to oppose the flow of current from the battery. For this reason the induced EMF is called the *back EMF*. Because it is opposed by the current due to the back EMF, the current through the coil rises slowly to its maximum value, V/R. (We assume that the coil has no resistance, so the total resistance in the circuit is that of the resistor R.) The variation of the current I with time is shown in Fig. 14.13. This curve is described by an exponential function and has exactly the same shape as the Q versus t curve for an RC circuit (Fig. 14.10).

The property of a coil of wire that determines how effective the coil will be in producing a back EMF is called the *inductance* L (and the coil is called an *inductor*). If a back EMF of 1 V is generated in a coil when the current through the coil is changing at a rate of $\Delta I/\Delta t = 1$ A/s, the coil is said to have an inductance $L = 1$ henry (H). That is,

$$\text{EMF} = -L \frac{\Delta I}{\Delta t} \tag{14.6}$$

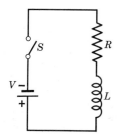

Figure 14.12 A simple RL circuit.

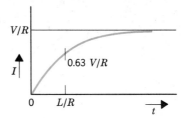

Figure 14.13 The current in the circuit of Fig. 14.12 as a function of time after the switch S is closed. The time constant is L/R.

where the negative sign represents Lenz's law. The characteristic time of a series RL circuit is L/R (see Fig. 14.13).

The inductance of a single coil of wire is often referred to as the *self-inductance* in order to distinguish it from the *mutual inductance* that exists between two adjacent circuits in which the currents are changing. The unit of inductance is named in honor of the great American physicist, Joseph Henry (1797–1878). Henry actually discovered electromagnetic induction before Michael Faraday did so, but Faraday published his results first. However, Henry is credited with the discovery of *self*-induction.

The exponential function that describes the charging and discharging of a capacitor and the buildup of current through an inductor is also found in the mathematical treatments of other physical phenomena. For example, the curve in Fig. 14.11 is exactly the same as the curve that describes the decay of a radioactive substance and the curve that represents the decrease in atmospheric pressure with altitude.

• Example 14.5

A solenoid has a length ℓ of 20 cm, a cross-sectional area A of 2.5 cm^2, and consists of a total of 2000 turns of wire. What is the inductance of this solenoid?

First, we obtain a general expression for the inductance of a solenoid. According to Eq. 13.13, the magnetic field within a solenoid that consists of n turns per meter and that carries a current I is

$$B = 4\pi K_M n I \tag{1}$$

The magnetic flux Φ through the solenoid is the product of B and the total area of the coils. This total area is the area A of each turn multiplied by $n\ell$, the number of turns in the coil:

$$\begin{aligned}\Phi &= BA_{\text{total}} = B(n\ell A) \\ &= 4\pi K_M n^2 I \ell A\end{aligned} \tag{2}$$

Now, when this flux changes, as EMF is generated (Eq. 13.18a):

$$\text{EMF} = -\frac{\Delta \Phi}{\Delta t} = -4\pi K_M n^2 \ell A \frac{\Delta I}{\Delta t} \quad (3)$$

The second equality can be written because only the current I changes. According to Eq. 14.6, the coefficient of $-\Delta I/\Delta t$ is the inductance L:

$$L = 4\pi K_M n^2 \ell A \quad \text{(solenoid)} \quad (4)$$

The value of n, the number of turns per meter, is $n = (2000 \text{ turns})/(0.2 \text{ m}) = 10^4 \text{ m}^{-1}$. Then, substituting for the other quantities in (4), we find

$$L = 4\pi \times (10^{-7} \text{ H/m}) \times (10^4 \text{ m}^{-1})^2 \times (0.2 \text{ m})$$
$$\times (2.5 \times 10^{-4} \text{ m}^2)$$
$$= 6.3 \times 10^{-3} \text{ H} = 6.3 \text{ mH}$$

(Notice that we expressed K_M as 10^{-7} H/m; verify that this is correct. ■

Effects of Electric Current in the Human Body

All of the normal functions of the body involve electric currents. Muscular activity, including respiration and the beating of the heart, is controlled by electric currents, and the information acquired by the various sensory elements in the body is transmitted to the brain by electric currents. Although electric currents are essential in the functioning of the body, currents from external sources flowing through the body's vital organs can cause damage or even death.

The magnitude of the current that flows through the body from an external source is determined by Ohm's law and therefore depends on the applied voltage and the resistance of the body. For DC and low frequency (household) voltages, the resistance of the skin at the contact location is the chief factor that limits the current. (At high frequencies, the internal body resistance is most important.) Consequently, in most situations the amount of current that will flow in the body depends critically on the condition of the skin at the contact locations. Dry skin will have a high resistance, but damp or wet skin will have a low resistance because the ions in the moisture allow the easy passage of the current into the body. Between extremities of the body (e.g., from foot to hand or from one hand to the other), the resistance can be 10^5 Ω or higher if the skin is dry, but it can be 1 percent of this value if the skin is wet. The total resistance of the body between normal, perspiring hands is typically 1500 Ω.

In the two situations mentioned above, the peak currents expected from contacts with 120-V household AC circuits are

Dry skin: $I_0 = \dfrac{120 \text{ V}}{10^5 \text{ Ω}} = 1.2$ mA

Wet skin: $I_0 = \dfrac{120 \text{ V}}{1500 \text{ Ω}} = 80$ mA

A current of 1 mA passing through the body will barely be noticeable, but a current of 80 mA will usually be fatal unless it is immediately stopped and the victim is promptly treated.

The parts of the body that are most sensitive to electric currents are the brain, the chest muscles and nerve centers that control respiration, and the heart. Let us examine the behavior of the heart and its response to externally applied electric currents.

The pumping action of the heart is due to the regular contraction of the muscles that comprise the heart walls. The rhythmical contraction of these muscles is caused by the propagation of an action potential through the cardiac fibers. Most of

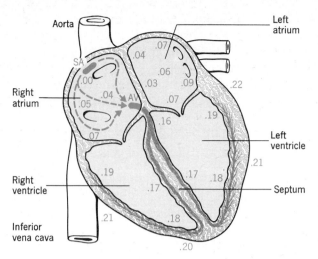

Figure 1 The numbers indicate the time intervals in seconds required for the propagation of the action potential that begins at the SA node to other parts of the heart. [Adapted from A. C. Guyton, *Textbook of Medical Physiology*, 5th ed.; Saunders, Philadelphia, 1976; Fig. 14.6.]

these fibers have the capability of *autoexcitation;* that is, the action potential is self-triggering in these fibers, a process that results in automatic heart action. The region of the heart that displays autoexcitation to the greatest extent is the *sinoatrial* (SA) node (see Fig. 1). The action potential that starts here propagates within about 40 ms to the atrioventricular (AV) node, and within about 90 ms the signal has extended throughout the atria. Figure 1 shows the time intervals required for the propagation of the original SA impulse to the rest of the heart.

In response to the action potential, the atria begin to contract. Notice that there is a delay of approximately 0.1 s between the time that the AV node receives the action potential and time that this node passes along the impulse to the upper end of the septum. This delay allows the atria to empty their contents into the ventricles before the ventricles begin to contract. The action potential propagates down the septum and then upward along the ventricular walls. The signal travels so rapidly that the response of the muscles is a coordinated contraction that completes the pumping cycle. A recovery time of about 0.8 s is required before the next SA action potential can again trigger a ventricular contraction. The result is a self-excited pumping action that produces approximately 72 heart beats per minute.

The spread of the changing potential of the heart membranes due to the propagation of the action potential results in a changing electric field around the heart. This field can be detected by electrodes placed on the skin at various locations over the body. By recording the time variation of the electric signals from these electrodes the performance of the heart can be monitored and the effects of disease or disorder can be detected. This technique is called *electrocardiography* (EKG).

If an electric current from an external source is passed through the heart, the cardiac fibers will be induced to "fire" at random and action potentials will begin to propagate in all directions throughout the heart. The result will be an uncoordinated contraction of the ventricles and no pumping action will occur. This effect is called *ventricular fibrillation*. Once the heart fibrillates, it will continue even after the offending current ceases to flow. The heart can be induced into ventricular fibrillation by currents as low as 50 to 100 μA. Such currents *directly through the heart* correspond to currents about 1000 times greater between skin contact locations. That is, when currents greater than about 100 mA flow from one extremity to another, ventricular fibrillation will probably result.

When ventricular fibrillation begins, it almost never ceases spontaneously. Within 1–2 minutes the ventricular muscles become weakened, because they have not been resupplied with coronary blood, to the extent that they cannot be forced back into normal contraction and death results. If measures are taken before this point is reached, regular heart action can often be restored. A common technique for defibrillation is to pass a large current (about 10 A) through the heart for a few milliseconds. (This can be accomplished by discharging a capacitor through two electrodes that are placed on the body above and below the heart; AC can also be used.) This current causes a uniform polarization (actually, a *de*polarization, as we have used the term in the essay on page 271) of the heart membranes and places all of the fibers in the same part (the recovery part) of the action potential cycle. Thus, all the fibers return to the normal resting condition at the same time and the next autoexcited signal from the SA node may begin regular coordinate contractions. Sometimes 10 to 20 of these defibrillating shocks are necessary to start the heart functioning again.

If heart disease impairs the functioning of the SA node to the extent that it no longer provides reliable timing for the pacing of the heart, an artificial pacemaker can be used to give the heart regular stimulation. Such a pacemaker is an electronic device that produces short pulses at regular intervals (72 min^{-1}); these pulses are applied through electrodes to the heart and initiate and control the heartbeat. In a hospital situation, the pacemaker signals may be delivered to the heart through catheters. For ambulatory patients, a miniaturized battery-operated pacemaker can be implanted under the skin. These devices function reliably, but the battery must be replaced every few years. Great care must always be exercised when dealing with electrodes that contact the heart because even very small current through the heart can induce fibrillation.

Currents in the body that are insufficient to cause ventricular fibrillation may induce respiratory arrest by inhibiting the action of the nerve centers that control breathing. This effect can persist even after interruption of the current. Respiratory paralysis can result from body currents in the range from 25 to 100 mA. Even currents as low as 10 mA can so contract the chest muscles that breathing is stopped. The best procedure to follow in treating a victim of such a shock is to apply artificial respiration.

Some of the effects of body currents are listed in the table below.

Current [60 Hz, 1 s]	Effect
0–0.5 mA	None
0.5–2 mA	Threshold of feeling
2–10 mA	Pain; muscular contractions
10–20 mA	Increased muscular effects, some injury; about 16 mA is "let-go" current, above which a person cannot release electrodes
20–100 mA	Respiratory paralysis
100 mA–3 A	Ventricular fibrillation; fatal unless resuscitation occurs immediately
above 3 A	Cardiac arrest; heart can be restarted if shock is very brief; severe burns

Shock Hazards in Ordinary Situations

The potential for electric shock exists in many everyday situations. Electric devices that are improperly designed or faulty or handled carelessly are capable of delivering lethal shocks. Let us examine what can happen with a household electric heater. Figure 2a shows an ordinary resistance heater in proper working order. One of the AC power leads is connected to ground, as indicated. Now, suppose that some fault develops such that the high-voltage lead becomes connected to the metal case of the heater. This could happen if the insulation of the lead is frayed to the point that the wire actually touches the case; or a bridge of dust and dirt could build up, providing

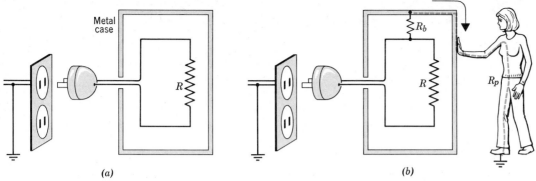

Figure 2 (a) A resistance heater in proper working order. (b) A resistance R_b connects the high-voltage lead with the case and allows a current (dotted path) to flow through a person who touches the case.

a path for the current to the case. Figure 2b shows a resistance R_b due to a bridge of material connected the high-voltage lead to the case. If a person who is grounded touches the heater case, this will complete an alternate path for the current, as indicated by the dashes in the diagram. If the person is standing in a bathtub or on a damp floor or gripping a grounded pipe, the body resistance R_p will be low and a subtantial current will flow through the body.

Many electric devices still use the *two-wire* system illustrated in Fig. 2. Any device that employs this system is a potential shock hazard and should be treated accordingly. To correct this situation, a *three-wire* system that incorporates a grounding wire is coming into use. All new household wiring must be of this type, and many appliances are being manufactured to take advantage of the increased safety of the three-wire system. Figure 3 shows how the additional grounding wire protects the user. The third wire in the heater is connected to the case, and when the heater plug is inserted into the AC receptacle, this wire is grounded. Then, if any fault develops, the current is routed to ground (dashed path) and not through the user because the resistance between the case and ground is essentially zero.

Special care is required in hospital situations to guard against possible leakage currents. If a patient has catheters inserted into his or her body, particularly in the region of the heart, even small leakage currents that find their way to the internal electrode system could result in a fatal shock. In the early days of heart catheterization, many of the deaths of heart patients were attributed to heart failure, but they were probably the result of ventricular fibrillation produced by small leakage currents accidentally introduced directly into the heart by catheters. The safety measures now in effect have reduced the incidence of such accidents to very low levels.

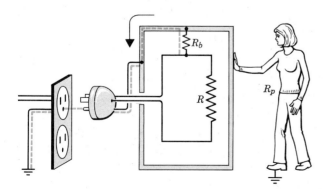

Figure 3 A *three-wire* system incorporating a ground wire protects the user from shock.

References

C. F. Dalziel, "Electrical Shock Hazard," IEEE Spectrum 9, 41 (February 1972).

A. C. Guyton, *Textbook of Medical Physiology*, 5th ed., Saunders, Philadelphia, 1976, Chapter 14.

D. E. Tilley and W. Thumm, *Physics for College Students with Applications to the Life Sciences*, Cummings, Menlo Park, Cal., 1974, Chapter 10.

■ Exercises

1. A defibrillating device delivers a shock to the heart area by discharging a capacitor that is charged to 5000 V. The body resistance between the electrodes is 500 Ω. What is the current as the capacitor begins to discharge? After 6 ms, the voltage on the capacitor has dropped to 250 V. What is the capacitance of the device? How much energy is delivered to the body during the discharge?

2. If Fig. 2b suppose that the body resistance of the person is $R_p = 10^3$ Ω and that the fault resistance is $R_b = 10^4$ Ω. What current will flow through the person? Will she receive a bad shock? Suppose that the connection from the high-voltage lead to the case is *direct* because of frayed insulation (i.e., $R_b = 0$). What current will now flow through the body? How bad will this shock be?

14.3 Electric Devices

The Torque on a Current Loop—Electric Motors

The way in which a current-carrying loop of wire behaves in a magnetic field is essential to the operation of all types of electric motors and electric meters. In Fig. 14.14 we have a current I that flows around a wire loop in the direction *abdc*. Look at the section *ab*. By using the right-hand rule for determining the direction of the magnetic force, we see that the force $\mathbf{F}_M$ on the section *ab* is *downward*. Similarly, we find that the magnetic force on the section *cd* is upward. These forces act in opposite directions, and if the magnetic field is uniform, the forces are equal in magnitude. But they do not act along the same line, so the loop experiences a *torque*. Notice that the forces acting on the section *bc* and *da* are both *inward*; these forces do act along the same line and therefore cancel.

The torque that is exerted on the loop causes it to rotate. Regardless of the orientation of the loop in the magnetic field, the force vectors $\mathbf{F}_M$ remain the same. However, as the rotation of the loop carries the section *ab* downward and the section *cd* upward, the torque *decreases* because the forces act more nearly along the same line. When *ab* is in the most downward position, the torque becomes zero because the force vectors are directly opposed.

If the magnetic forces on the sections of the current-carrying loop are to produce continued rotation, we must devise some method for reversing the direction of current flow just as the loop passes through the point of zero torque. This is easily accomplished by sending the current to the loop through a *split ring*, as indicated in Fig. 14.15. The connection from the battery to the ring is made through a pair of brushes that slide across the rotating ring. (In some

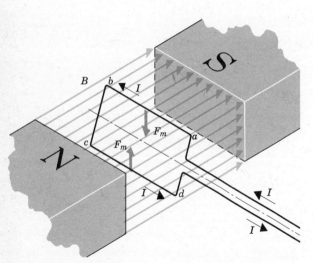

Figure 14.14 A magnetic field exerts a torque on a current-carrying loop of wire.

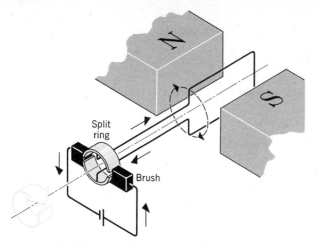

Figure 14.15 Schematic diagram of a DC electric motor.

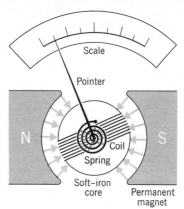

Figure 14.16 The important elements of the meter movement of a galvanometer.

cases the "brushes" are actually wire brushes, but often one finds these contactors made of graphite.) By using the split ring to reverse the current in the loop every half cycle, we can produce continuous rotation. This is the principle of the *direct-current* (DC) *motor*. The electric motor was invented by Joseph Henry and the electric generator was invented by Faraday; these two devices constitute the basis for our entire electric industry.

In the next section we discuss *alternating current* (AC), current that reverses direction at regular intervals. Notice that if we attach the leads from the loop in Fig. 14.15 directly to a source of alternating current (without using the split ring), we have automatically provided for the current reversal necessary for motor operation. Thus, we have an *AC motor*. The rate at which the loop rotates in a simple AC motor is the same as the rate at which the current changes direction. That is, the loop (or *armature*) rotates in synchronism with the alternating current and the motor is called a *synchronous* motor.

The Galvanometer

The magnetic force that is exerted on a current-carrying loop of wire by a magnetic field is directly proportional to the current and to the magnetic field strength. Therefore, we can use the deflection of a current-carrying loop in a constant magnetic field as a measure of the current in the loop. Figure 14.16 shows a typical meter movement of a current-measuring device called a *galvanometer*. Basically, the meter is a loop of wire in a magnetic field, just as we have in Fig. 14.14. But there are several important refinements in the meter illustrated in Fig. 14.16. First, the magnetic torque is enhanced by using a coil consisting of many turns of wire wound around a soft-iron core. Also, the pole pieces of the permanent magnet are shaped so that the field lines are radial instead of parallel. Finally, a spring is provided to exert a restoring torque on the coil that is proportional to the coil rotation. All of these features combine to produce an angular deflection of the pointer that is proportional to the current through the meter coil. (If the deflection is not exactly linear with current, the scale is calibrated against standards so that it will yield accurate readings.)

A carefully constructed galvanometer is a very sensitive device; that is, an exceedingly small current (19^{-6} A = 1 μA, or even less) will cause a full-scale deflection of the pointer. Even so, the galvanometer movement is the basic component of *ammeters* that are capable of measuring large currents and of *voltmeters* that measure voltages. We will now see how this is done.

Ammeters and Voltmeters

A typical galvanometer of medium sensitivity will have an internal resistance of a few ohms and will deflect full scale when a current of about 10^{-3} A = 1 mA flows through the coil. If a current in excess of 1 mA is allowed to flow through such a meter, the movement can be damaged or destroyed. Therefore, in order to measure currents that are larger than the basic full-scale current, the excess current must be diverted around the meter movement. We can do this by placing a small resistance R_s (called a *shunt* resistance) in parallel with the galvanometer resistance R_g, as indicated in Fig. 14.17. Then, if a large current I is allowed to flow between the terminals a and b, most of the current will pass through R_s and the meter movement is protected. By adding a shunt resistor whose resistance is small compared to the coil resistance, a galvanometer is converted into an *ammeter*. The scale is appropriately labeled to read the current I directly.

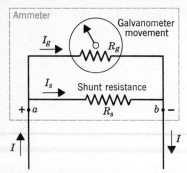

Figure 14.17 An ammeter consists of a galvanometer movement and a shunt resistance R_s which is small compared to the coil resistance R_g.

● *Example 14.6*

Suppose we have a galvanometer that deflects full scale when a current of 10 mA flows through the coil. The internal resistance of the galvanometer is 5 Ω. What shunt resistance R_s is required to convert the galvanometer into an ammeter that reads full scale for a current of 2 A?

The ammeter circuit is the same as that shown in Fig. 14.17. First, we note that a current of 10 mA through the 5-Ω coil corresponds to a voltage across the meter movement of (0.010 A) × (5 Ω) = 0.050 V or 50 mV. (We would say that this galvanometer has a "50-mV movement.") Next, we note that

$$I_s = I - I_g = 2 \text{ A} - 0.010 \text{ A} = 1.99 \text{ A}$$

The value of the shunt resistance R_s must be such that a current of 1.99 A will flow through the resistor when the voltage across R_s is 0.050 V. Therefore,

$$R_s = \frac{0.050 \text{ V}}{1.99 \text{ A}} = 0.025 \text{ Ω}$$

Notice that the total resistance of the ammeter is very small. (The shunt resistance is in parallel with the coil resistance so that the total resistance is slightly less than 0.025 Ω.) Therefore, when this small resistance is inserted into a circuit to measure the current, the circuit is disturbed hardly at all. ■

Some ammeters are constructed so that any one of several different shunt resistors can be selected by a switch. Then, each position of the switch corresponds to a different current value for a full-scale deflection. Such meters are called *multirange* meters (Fig. 14.18).

A galvanometer can also be modified in such a way that the *voltage* across a particular circuit element can be determined. The type of connection is shown in Fig. 14.19. Again, we wish to protect the galvanometer coil from excessive current. But now we accomplish this by placing a *large* resistance R_m in *series* with R_g instead of a *small* resistance in *parallel* as in an ammeter. The meter resistance R_m must be large compared to that of the R_g and to that of the resistor R across which the meter is connected, so that the drain of current from the circuit is small: A large value of R_m not only protects the meter movement, but, because it is connected in parallel with the circuit resistor R, there is very little disturbance of the current flow in the circuit.

The galvanometer is still a current-measuring device, so when it is converted to a voltmeter by adding a series resistance, we must make the connection between the meter current I_g and the voltage V_{ab} across the circuit resistor: $V_{ab} = I_g R_m$. We know the value of R_m, and I_g is measured, so V_{ab} is determined. The scale of the meter is labeled to read directly in volts. Multirange voltmeters can be constructed by arranging for different resistances R_m to be switched into the circuit. In fact, many commercial meters (of the type called *multimeters*) are designed so that they can be used as multirange ammeters, multirange voltmeters, and multirange ohmmeters.

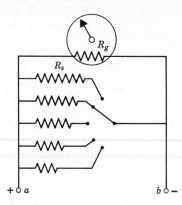

Figure 14.18 Using several different shunt resistances makes a multirange ammeter.

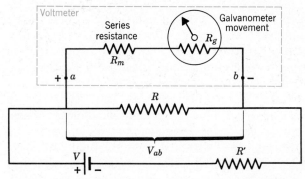

Figure 14.19 A voltmeter consists of a galvanometer movement and a series resistance R_m which is large compared to R_g and to the circuit resistor R.

● *Example 14.7*

What series resistance R_m is necessary to convert the galvanometer in Example 14.6 into a voltmeter with a full-scale deflection for a voltage of 20 V?

The current flow through the meter movement (and, hence, also through R_m) is 0.010 A for a full-scale deflection. Therefore, the total resistance of the meter, $R_m + R_g$, must be

$$R_m + R_g = \frac{V}{I} = \frac{20\text{ V}}{0.010\text{ A}} = 2000\text{ }\Omega$$

We know that $R_g = 5\text{ }\Omega$, so

$R_m = 2000\text{ }\Omega - R_g = 1995\text{ }\Omega$ ■

14.4 Alternating Current

AC Generators

All of the circuits we have been discussing involve currents that flow only in a single direction (DC). An important class of circuits operates on current that reverses direction at regular intervals—these are *alternating-current* (or *AC*) *circuits*. Before we look at some of the circuits themselves we will discuss the way that alternating current is generated.

Figure 14.20 shows a simple AC generator. The wire loop and the slip rings (through which contact to the external resistor is made) are connected to a rigid shaft that is rotated by some mechanical device. The armature is a real AC generator always consists of a coil of wire wound on an iron core in order to increase both the effective loop area NA (see Example 13.6) and the magnetic flux. As the armature rotates, the induced EMF causes the current to flow first in one direction and then in the other direction. A current reversal takes place during each half-cycle of rotation. The brushes that contact the

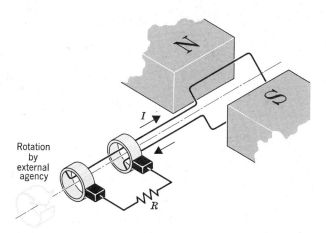

Figure 14.20 The essential components of an AC generator.

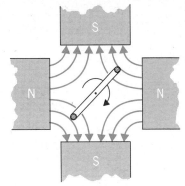

Figure 14.21 A four-pole AC generator. For the same rate of rotation of the drive shaft, the AC frequency in this case is twice that of the two-pole generator in Fig. 14.20.

slip rings allow the alternating current to be applied directly to an external circuit or to be transmitted to a power distribution network.

In the simple generator shown in Fig. 14.20, the current reverses direction during each 180° of rotation. Thus, the AC frequency is the same as that of the rotating drive shaft. Most AC generators contain more than a single pair of pole pieces. In Fig. 14.21 we show a generator field that is produced by two pairs of poles. As the armature rotates in this field, a current reversal occurs for each 90° of rotation. (Can you see why?) Large AC generators contain many poles so that the desired AC frequency can be maintained for drive-shaft rotation rates that are low. The standard AC frequency in the United States is 60 cycles per second. We usually express frequencies in units of *hertz* (Hz), so that the usual AC frequency is given as 60 Hz.

Let us now obtain a mathematical description of the way the EMF and the current produced by an AC generator vary with time. Refer to Fig. 14.22, which shows the plane of the armature coil making an angle θ with respect to the field lines. How does the EMF change as the drive shaft is rotated and θ is increased at a uniform rate? According to Faraday's law of induction (Eq. 13.18a), $\mathcal{E} = -\Delta\Phi/\Delta t$, and the magnetic flux is given by (Eq. 13.16a), $\Phi = B_\perp A$. In the case of the generator, the area A of the armature

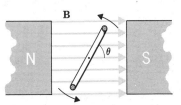

Figure 14.22 The rotation of an armature coil in the magnetic field of an AC generator. The component of **B** perpendicular to the plane of the coil varies sinusoidally with time.

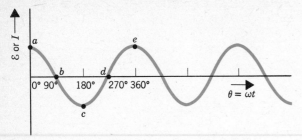

Figure 14.23 The EMF and the current induced in the armature coil of an AC generator vary sinusoidally with time.

coil remains constant, but the component of the magnetic field perpendicular to the plane of the coil depends on the time. In fact, $B_\perp$ varies *sinusoidally* with time, so the EMF and the current have this same variation. Thus, the curve of $\mathcal{E}$ or I versus θ (or time) has the shape of a sine (or cosine) curve. In Fig. 14.23 we see this variation of $\mathcal{E}$ and I. When $\theta = 0°$ (point a), the coil cuts across the field lines at the maximum rate (i.e., $\Delta\Phi/\Delta t$ is maximum even though $\Phi = 0$ at this point) and the EMF and the current are maximum. When $\theta = 90°$ (point b), the coil is moving along the field lines so that the rate of change of the flux is zero, as are $\mathcal{E}$ and I. Between $\theta = 90°$ (point b) and $\theta = 270°$ (point d), the direction of the induced current is opposite to that in the first part of the cycle. Accordingly, the values of $\mathcal{E}$ and I in this region are plotted as negative quantities in Fig. 14.23. By the time point e is reached, the armature has returned to its original position and the process begins to repeat itself. The sequence *abcde* represents one complete cycle of the system. Altogether, we can express the time variation of $\mathcal{E}$ and I as

$$\mathcal{E} = \mathcal{E}_0 \cos \omega t = \mathcal{E}_0 \cos \frac{2\pi t}{\tau}$$

$$I = I_0 \cos \omega t = I_0 \cos \frac{2\pi t}{\tau}$$
(14.7)

where the angle θ increases uniformly at the rate ω rad/s and has a piece $\tau = 2\pi/\omega$ (see Eq. 2.32 or Eq. 11.21); thus, $\theta = \omega t = 2\pi t/\tau$. The quantities $\mathcal{E}_0$ and I_0 represent the maximum values attained by the EMF and the current, respectively (points a and e in Fig. 14.23). Ordinary household current has a frequency of 60 Hz ($\tau = 1/60$ s) and a peak voltage[1] of about 170 V. Special circuits with a higher voltage are usually provided for heavy-duty appliances such as electric clothes dryers and air conditioners.

The Heating Effect of Alternating Current

The amount of electric power that is dissipated in a resistor R is I^2R and does not depend on the direction in which the current is flowing; the same power is expended by a current $+I$ as by a current $-I$. If the current is AC, then to calculate the power dissipated as heat in a resistor, we must take account of the fact that the magnitude of the current (or, more properly, the magnitude of I^2) changes with time. That is, we want the average value of I^2:

$$(I^2)_{\text{ave}} = I_0^2 \left(\cos^2 \frac{2\pi t}{\tau} \right)_{\text{ave}}$$

where the average is to be taken over one complete cycle.

Now, we can prove mathematically (or we can see in Fig. 14.24b) that the average value of the square of a cosine function is $\tfrac{1}{2}$. Therefore,

$$(I^2)_{\text{ave}} = \tfrac{1}{2} I_0^2 \qquad (14.8)$$

IF we take the square root of $(I^2)_{\text{ave}}$ we have the *root-mean-square* current (compare the discussion of rms velocity in Section 10.2). The rms current is equal to the DC current that would have the same heating effect when flowing in the same resistor R.

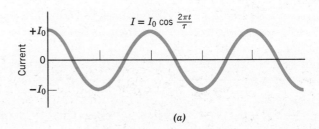

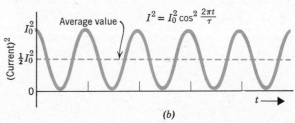

Figure 14.24 (a) The alternating current through a resistor varies sinusoidally with time. (b) The square of the current is always positive and has an average value equal to $\tfrac{1}{2}I_0^2$.

---
[1] The *effective* voltage of a household circuit is about 120 V; see the following paragraph.

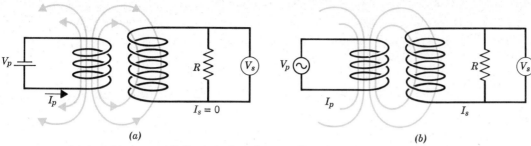

Figure 14.25 (a) A steady current I_p through the primary coil produces a constant magnetic flux through the secondary coil; consequently, the secondary current I_s is zero. (b) An alternating current through the primary coil produces a changing flux through the secondary coil and induces an alternating current I_s to flow. (Notice the special symbol used to indicate a source of alternating EMF.)

Thus, the effective current is

$$I_{\text{rms}}(\text{or } I_{\text{eff}}) = \frac{1}{\sqrt{2}} I_0 = 0.707\, I_0 \tag{14.9}$$

Then, the power dissipated is

$$P = I_{\text{eff}}^2 R = \tfrac{1}{2} I_0^2 R \tag{14.10}$$

The voltage across the resistor also has a sinusoidal variation (Eq. 14.7), so

$$V_{\text{eff}} = \frac{1}{\sqrt{2}} V_0 \tag{14.11}$$

Ordinary household AC has $V_0 = 170$ V, so the effective voltage[2] is $V_{\text{eff}} = 120$ V. This is why we refer to a "120 V line." Now, using Ohm's law we can write

$$P = \frac{V_{\text{eff}}^2}{R} = \frac{1}{2}\frac{V_0^2}{R} = \frac{1}{2} I_0 V_0 \tag{14.12}$$

● *Example 14.8*

What is the peak alternating current that causes a power expenditure of 60 W in a light bulb that has a resistance of 100 Ω?

Using Eq. 14.10,

$$I_0 = \sqrt{\frac{2P}{R}} = \sqrt{\frac{2 \times (60\text{ W})}{100\ \Omega}} = 1.10 \text{ A} \quad \blacksquare$$

Transformers

When a current flows in a coil of wire, the field lines will link any other coil that is placed near the current-carrying coil (Fig. 14.25a). If the first coil (the *primary*) carries a steady direct current I_p, no EMF can be induced in the secondary coil. However, suppose that an alternating current flows in the primary coil (Fig. 14.25b). Now, we have a changing magnetic field surrounding the primary circuit and a changing magnetic flux through the secondary coil. Under these conditions there is an EMF induced in the secondary, and a current I_s flows through the resistor. (We say that there is a *mutual inductance* between the two coils.) The voltage and the current in the primary circuit vary sinusoidally with time, and so do the voltage and the current in the secondary circuit. A pair of coils coupled by a time-varying magnetic field is called a *transformer*. Because the coupling in Fig. 14.25b takes place through air, this particular device is called an *air-core* transformer.

It is easy to see that many of the field lines produced by the current in the primary coil of Fig. 14.25b no not link with the secondary coil. This unlinked flux cannot contribute to the current induced in the secondary. Therefore, an air-core transformer of this design is a very inefficient device. Almost all ordinary transformers incorporate an iron core in order to improve the efficiency. A simple way of doing this is shown in Fig. 14.26.

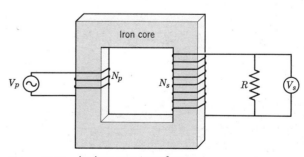

Figure 14.26 An iron-core transformer.

---

[2]This voltage is sometimes decreased by as much as 5 percent to reduce the power in peak-load situations (*brownouts*). Any further decrease would tend to damage rotating electric machinery. Some power companies maintain a slightly different voltage (e.g., 117 V).

In an iron-core transformer, essentially all of the field lines are confined within the iron. Then, the flux through each turn of the primary coil is the same as the flux through each turn of the secondary coil, and, of course, the flux changes in the same way in each coil. The EMF in a coil is equal to the rate of change of the *total* flux through the coil, $N \, \Delta\Phi/\Delta t$, where N is the number of turns of wire in the coil and Φ is the flux through a single turn. The voltage V_p applied to the primary coil (N_p turns) is, therefore, $V_p = N_p \, \Delta\Phi/\Delta t$. Similarly, the voltage V_s generated across the secondary coil is $V_s = N_s \, \Delta\Phi/\Delta t$. Because $\Delta\Phi/\Delta t$ is the same in each section of a transformer, we can equate the ratio of the voltage across one coil to the number of turns in the coil to the same ratio for the other coil:

$$\frac{V_p}{N_p} = \frac{V_s}{N_s} \qquad (14.13)$$

Therefore, if we wish the secondary voltage V_s to be larger than the primary voltage V_p, we must have a greater number of turns in the secondary coil than in the primary coil. A transformer constructed in this way is called a *step-up* transformer. In a *step-down* transformer there are more primary turns than secondary turns.

If a transformer is perfectly efficient (that is, if there are no losses due to resistance heating and no leakage of magnetic flux), the rate at which energy is delivered to the primary must equal the rate at which energy is extracted from the secondary. In other words, the input (or primary) power P_p must equal the output (or secondary) power P_s:

$$P_p = V_p I_p = V_s I_s = P_s \qquad (14.14)$$

In a well-designed transformer the power losses may be no more than a percent or two.

●*Example 14.9*

A step-down transformer has 2500 turns in the primary coil and 500 turns in the secondary coil. When the primary is connected to a source of 120-V AC, what power is delivered to an 8-Ω resistor connected to the secondary?

The voltage across the resistor is obtained by solving Eq. 14.13 for V_s:

$$V_s = \frac{N_s}{N_p} \times V_p = \frac{500}{2500} \times 120 \text{ V} = 24 \text{ V}$$

This is the *effective* voltage across the resistor, so the power delivered is

$$P = \frac{V_s^2}{R} = \frac{(24 \text{ V})^2}{8 \, \Omega} = 72 \text{ W} \quad \blacksquare$$

Most of the electric power that is used in the world is carried many kilometers from the power plant to the customer via transmission wires. Obviously, it is desirable to transmit this electric power as efficiently as possible. The long wires that are used have a certain resistance, so I^2R heating losses are inevitable. Not much can be done about decreasing the resistance of the transmission wires (but see the comments in Section 20.4 about superconductors), so any gain in efficiency must be the result of lowering the current I. This can be accomplished through the use of transformers. At the power station, a transformer is used to step up the generator output voltage to a very high value, at least 20,000 V and sometimes 600,000 V. The corresponding current is lowered by the same factor that the voltage is raised. Thus, only small currents flow in the high-voltage transmission wires. (These wires are suspended by means of elaborate insulators from special towers that can be seen throughout the countryside.) At the user's end of the transmission line, the voltage is stepped down (usually, in several stages) to the 120 V or 240 V that is needed by the customer.

●*Example 14.10*

A generating plant delivers 1000 MW of power into a distribution network at an effective voltage of 500,000 V. The output power is distributed equally among ten transmission lines. (a) If there is a 2 percent loss of power in each line, what is the effective resistance of each line? (b) How much current can be delivered to consumers from each line when the voltage is stepped down to 120 V?

(a) The effective voltage current in one of the outgoing transmission lines (each of which carries 100 MW of power) is

$$I = \frac{P}{V} = \frac{10^8 \text{ W}}{5 \times 10^5 \text{ V}} = 200 \text{ A}$$

If 2 percent of the power is lost to resistance heating, then

$$R = \frac{0.02 \, P}{I^2} = \frac{2 \times 10^6 \text{ W}}{(200 \text{ A})^2} = 50 \, \Omega$$

(b) Each line delivers 0.98×10^8 W at 120 V.

Therefore,

$$I = \frac{P}{V} = \frac{0.98 \times 10^8 \text{ W}}{120 \text{ V}} = 8.2 \times 10^5 \text{ A} \quad \blacksquare$$

Inductive and Capacitance Reactance

In Section 14.2 we studied the effects of capacitors and inductors in DC circuits. If a voltage V is placed across a series connection of a resistor R and an inductor L (which has zero resistance), we know that the current will rise to a value $I = V/R$ after an interval of a few times the characteristic time L/R (see Fig. 14.13). Now, suppose that we replace the battery in Fig. 14.12 with a source of alternating EMF (Fig. 14.27). If we measure the current in this circuit, we find that I_{eff} is much less than the DC current V/R. For some reason, the effective resistance of the RL combination is much greater in an AC circuit than in a DC circuit. Moreover, if we repeat the measurement using an alternating EMF with a higher frequency, we find that the effective resistance is even greater.

It is easy to see why this is so. Look at Fig. 14.13 and think about what happens when the voltage applied to the RL combination varies with time. We begin with a certain polarity of the applied voltage, and the current begins to increase in the direction determined by this polarity. If the period of the alternating voltage is much less than the characteristic time L/R, then the current will not increase very much before the polarity is reversed and the current attempts to build up in the opposite direction. The higher frequency of the AC source, the smaller will be the increase in the current before the next reversal.

The proportionality factor connecting the current and the voltage across a pure resistor is the resistance R, and this resistance does not depend on the frequency of the applied voltage. We can also write an Ohm's-law type equation connecting the current and the voltage across an inductor:

$$V = IX_L \quad (14.15)$$

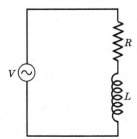

Figure 14.27 An alternating current RL series circuit. The current in the circuit is less than V/R.

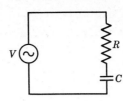

Figure 14.28 An alternating current RC series circuit. At very high AC frequencies the current in the circuit is essentially unaffected by the presence of the capacitor.

The quantity X_L is called the *inductive reactance*. We have already argued that X_L must increase with frequency; in fact,

$$X_L = 2\pi\nu L = \omega L \quad (14.16)$$

where ν is the frequency (in Hz) of the applied EMF; ω is the angular frequency, $\omega = 2\pi\nu$. Because X_L stands in Eq. 14.15 just as would a resistance value, inductive reactance clearly is measured in ohms. Notice that X_L becomes zero for the DC case ($\nu = 0$), because a pure inductor has no resistance.

Next, let us make the same type of analysis for an RC circuit (Fig. 14.28). Figure 14.10 shows the way that charge builds up on a capacitor to which is applied a DC voltage. Remember, current is the rate of flow of charge; therefore, the current in the RC circuit at any instant is equal to the slope of the Q versus t curve. In Fig. 14.10 we see that the current is large for times very soon after the voltage is applied; for times long compared with the characteristic time RC, the current is small and approaches zero. From this observation we can conclude how the current behaves in an RC circuit under AC conditions. If the frequency is *high*, so that the period is much shorter than RC, we expect the current to be high; that is, we expect the *capacitive reactance* X_C to be *small*. Conversely, if the frequency is *low*, we expect X_C to be *large*. Because there is no current flow through a capacitor in a DC circuit (for times long after the circuit is completed), the capacitive reactance must be infinite for zero frequency. Summarizing in the way we did for the case of an inductor, we can write

$$V = IX_C \quad (14.17)$$

$$X_C = \frac{1}{2\pi\nu C} = \frac{1}{\omega C} \quad (14.18)$$

Phase Relationships

When a sinusoidal (AC) voltage is impressed on a circuit consisting of resistors, capacitors, and inductors, the current that flows in the circuit is also sinusoidal. However, the current zeros, maxima, and minima occur at different times from the corresponding features in the applied voltage. A simple way of expressing this fact is to write for the voltage.

$$V(t) = V_0 \cos \omega t \quad (14.19a)$$

and for the current

$$I(t) = I_0 \cos(\omega t - \delta) \quad (14.19b)$$

For the simple circuit consisting of a resistor R, an inductor L, and a capacitor C, all connected in series, the quantity δ, called the *phase angle*, is given by the trigonometric expression

$$\tan \delta = \frac{X_L - X_C}{R} \quad (14.20)$$

with

$$-\frac{\pi}{2} \leq \delta \leq +\frac{\pi}{2}$$

Figure 14.29 shows the time variation of the voltage and current in an inductive (i.e., $X_L > X_C$) RLC series circuit. The current is said to *lag* the voltage by the phase angle δ, because the current reaches its zero, maximum, and minimum value at times δ/ω *later* than the voltage. In a capacitive (i.e., $X_C > X_L$) RLC series circuit the current would *lead* the voltage. Note also that for a circuit consisting of only a pure inductor $\delta = \pi/2$, and the current lags the voltage by a whole quarter cycle. For a pure capacitor $\delta = -\pi/2$, and the current leads the voltage by a whole quarter cycle. For a pure resistor, $\delta = 0$ and the current and voltage are *in phase*.

The instantaneous power delivered to the series RLC circuit is[3]

$$P(t) = V(t)I(t) = V_0 I_0 \cos \omega t \cos(\omega t - \delta)$$
$$= V_0 I_0 (\cos^2 \omega t \cos \delta + \cos \omega t \sin \omega t \sin \delta) \quad (14.21)$$

The first term in the expanded equation is always positive and represents power dissipated in the resistor. The second term is oscillatory and represents the surging of energy back and forth between the electromagnetic fields associated with the capacitive and inductive parts of the circuit. The average value of the oscillating $\cos \omega t \sin \omega t$ term vanishes. The average value of the $\cos^2 \omega t$ term is just $\frac{1}{2}$, hence the average power is

$$P = \frac{1}{2} I_0 V_0 \cos \delta, \quad (14.22)$$

compare with Eq. 14.12. Note that if the circuit consists of just a resistor R, then $\delta = 0$ and Eq. 14.22 is exactly equal to Eq. 14.12. The quantity $\cos \delta$ is called the *power factor*.

Impedance and Resonant Circuits

Suppose we have a circuit that consists of an inductor and a capacitor connected together in series. If we apply an alternating voltage to this circuit, current will flow, and we can define an effective overall resistance Z according to Ohm's law:

$$I_0 = \frac{V_0}{Z} \quad (14.23)$$

The quantity Z is called the *impedance* of the circuit and is related to the resistance and the reactances in the following way:

$$Z = \sqrt{R^2 + (X_L - X_C)^2} \quad (14.24)$$

The form of this expression for the impedance of the circuit is a result of the fact that the voltage and

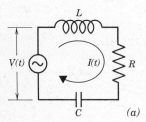

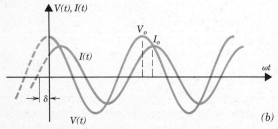

Figure 14.29 *(a)* A series RLC circuit with impressed AC voltage $V_0 \cos \omega t$. *(b)* The current $I(t)$ shown lagging the voltage by the phase angle δ.

---

[3] We have used the trigonometric identity $\cos(A - B) = \cos A \cos B + \sin A \sin B$ for the term $\cos(\omega t - \delta)$.

current do not have the same simple relationship for inductors and capacitors that they do for resistors. If the applied AC voltage is given by Eq. 14.19a and the resulting current by Eq. 14.19b, then in view of Eqs. 14.23 and 14.24

$$\left.\begin{array}{l} V(t) = V_0 \cos \omega t \\ I(t) = \dfrac{V_0}{\sqrt{R^2 + (X_L - X_C)^2}} \cos(\omega t - \delta) \\ \tan \delta = (X_L - X_C)/R \end{array}\right\} \quad (14.25)$$

● *Example 14.11*

A resistor ($R = 40\ \Omega$), an inductor ($L = 2$ mH), and a capacitor ($C = 5\ \mu$F) are connected in series to a source of alternating voltage. What is the impedance of this circuit at frequencies of 60 Hz and 20 kHz?

(a) For $\nu = 60$ Hz, we have

$R = 40\ \Omega$

$X_L = 2\pi\nu L = 2\pi \times (60\ \text{Hz}) \times (2 \times 10^{-3}\ \text{H}) = 0.75\ \Omega$

$X_C = \dfrac{1}{2\pi\nu C} = \dfrac{1}{2\pi \times (60\ \text{Hz}) \times (5 \times 10^{-6}\ \text{F})}$
$= 530\ \Omega$

$Z = \sqrt{R^2 + (X_L - X_C)^2}$
$= \sqrt{(40)^2 + (0.75 - 530)^2}$
$= 531\ \Omega$

(b) For $\nu = 20$ kHz $= 2 \times 10^4$ Hz, we have

$R = 40\ \Omega$

$X_L = 2\pi\nu L = 2\pi \times (2 \times 10^4\ \text{Hz}) \times (2 \times 10^{-3}\ \text{H})$
$= 251\ \Omega$

$X_C = \dfrac{1}{2\pi \times (2 \times 10^4\ \text{Hz}) \times (5 \times 10^{-6}\ \text{F})} = 1.6\ \Omega$

$Z = \sqrt{R^2 + (X_L - X_C)^2}$
$= \sqrt{(40)^2 + (250 - 1.6)^2}$
$= 253\ \Omega$

Notice that at the lower frequency, the capacitive reactance is much greater than the inductive reactance, whereas the reverse is true at the higher frequency.

● *Example 14.12*

Given an *RLC* series circuit, which for a particular frequency ω has values $R = 25\ \Omega$, $X_L = 500\ \Omega$, and $X_C = 450\ \Omega$. IF $V_0 = 100$ V find the magnitude of the peak current, the phase angle δ, and the power dissipated in the resistor.

We have

$Z = \sqrt{R^2 + (X_L - X_C)^2}$
$= \sqrt{(25)^2 + (500 - 450)^2}$
$= 55.9\ \Omega$

Thus, the peak current is

$I_0 = V_0/Z = 100\ \text{V}/55.9\ \Omega = 1.79\ \text{A}$.

Using Eq. 14.20 or 14.25

$\tan \delta = (X_L - X_C)/R = (500 - 450)\Omega/25\ \Omega = 2$

or

$\delta = +63.4°$.

The average power dissipated is

$P = \dfrac{I_0 V_0}{2} \cos \delta$

$= \dfrac{(1.79\ \text{A})(100\ \text{V})}{2} \cos 63.4°$

$= 40\ \text{W}$ ■

In Example 14.11 we have seen how the impedance of a circuit can change with the frequency of the applied voltage. If we wish to know the current that flows in an *RLC* circuit, we must calculate the impedance Z and use Eq. 14.25 to find I. For a particular value of the peak applied voltage V_0, the peak current will be $I_0 = V_0/Z$ and this current will depend on the frequency. Notice that the inductive and capacitive reactances occur in the expression for Z as the combination $X_L - X_C$. Therefore, at some frequency the two reactances will be equal, the impedance Z will be a minimum, and the peak current I_0 will be a maximum. This is a *resonance* condition (see Section 11.2), and the resonant frequency ν_R is found by equating X_L and X_C:

$$2\pi\nu_R L = \dfrac{1}{2\pi\nu_R C}$$

Solving for ν_R, we find

$$\nu_R = \dfrac{1}{2\pi\sqrt{LC}} \quad (14.26)$$

We should note that at resonance $X_L = X_C$ also gives $\delta = 0$ for the phase angle. This implies that the *RLC* circuit in series resonance behaves as if it consisted only of the pure resistor R.

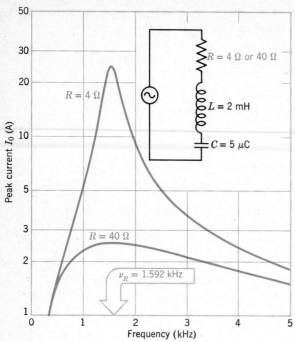

Figure 14.30 Resonance curves for an *RLC* circuit. The current scale is logarithmic.

● *Example 14.13*

What is the resonant frequency for the circuit in Example 14.11?

The values of L and C are $L = 2$ mH and $C = 5\ \mu$F, respectively. Therefore,

$$\nu_R = \frac{1}{2\pi \times \sqrt{(2 \times 10^{-3}\ \text{H}) \times (5 \times 10^{-6}\ \text{F})}}$$
$$= 1592\ \text{Hz} = 1.592\ \text{kHz} \quad \blacksquare$$

The way in which the current varies with frequency for the circuit in Example 14.11 is shown in Fig. 14.30. Notice that the curve for $R = 40\ \Omega$ has a broad peak with a maximum value at ν_R. If we reduce the resistance in the circuit from 40 Ω to 4 Ω, we find the more sharply peaked curve shown in the diagram. (This is the same as reducing the mechanical friction in the case illustrated in Fig. 11.14.) If the resistance were removed entirely from the circuit (this is not actually possible in practice), the impedance would vanish at ν_R and the current would become infinite.

Sharply peaked resonant circuits are used in electronic devices such as radio and television receivers. In these circuits they function to pass the signal with the particular frequency of the station desired and to exclude all other signals with different frequencies.

14.5 Electronics

Converting AC to DC— Rectifier Circuits

The most efficient way to produce and to deliver electric power is in the form of AC. But many types of electric and electronic circuits require DC. Often, such devices are powered by batteries; however, if a circuit requires a relatively high input power, a battery is not a practical source. The problem of supplying DC to a device is solved by producing and transmitting the electric power as AC and then converting to DC, when necessary, within the device itself. A circuit that performs this function is called a *rectifier*.

The basic component of any rectifier circuit is a device through which current will flow only in one direction. These devices are called *diodes* and they can be of vacuum-tube construction or they can be made from solid-state materials.

Most modern electronic circuits employ *solid-state* (or *semiconductor*) diodes to rectify the input AC. We discuss these devices in Chapter 20 where we see how the atoms and electrons behave during the conduction of electric current through semiconductor materials. For now, we summarize the important characteristics of a solid-state diode by means of a graph that relates the voltage applied to the diode to the current through the diode. A circuit element that obeys Ohm's law will have an I versus V diagram that is a straight line (dashed line in Fig. 14.31). A diode belongs to a class of circuit elements that are *nonohmic* in their behavior. An I versus V curve for a semiconductor diode is shown in Fig. 14.31. In the voltage region on the right-hand side of the figure, where we say that the voltage is applied to the diode in the "forward" direction, the current flows readily. For the opposite polarity of the voltage

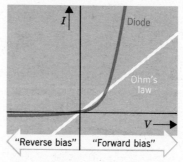

Figure 14.31 Current-voltage characteristics of a solid-state diode. The behavior of an Ohm's-law (linear) circuit element is shown for comparison.

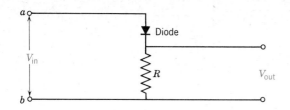

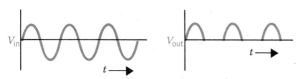

Figure 14.32 Schematic diagram of a semiconductor diode half-wave rectifier circuit. Current flows through the diode and a voltage is developed across the resistor R only when point a is at a positive potential with respect to point b.

(i.e., when the voltage is applied in the "reverse" direction), there is very little current flow.

The simplest way to use a semiconductor diode in a rectifier circuit is illustrated in Fig. 14.32. The symbol used for the diode includes an arrowhead that points in the direction of current flow through the diode (the "forward" direction). When point a has a positive potential with respect to point b, current will readily flow through the diode and through the resistor, thereby producing an output voltage across the resistor R. When the polarity is reversed, however, essentially no current will flow through the diode and no output voltage will be developed across the resistor. Thus, when an alternating voltage is applied across a diode, current flows only during one half of each cycle.

One common form of a semiconductor rectifier circuit is the *full-wave bridge rectifier* shown in Fig. 14.33. During the first half of the AC cycle, the point a is positive with respect to c. Therefore, current flows along the path $a \rightarrow b \rightarrow d \rightarrow c$. During the second half of the AC cycle, the point c is positive with respect to a. Therefore, current flows along the path $c \rightarrow b \rightarrow d \rightarrow a$. That is, the current through the load resistor R is in the same direction during both halves of the cycle. The voltage V_{bd} across the resistor is shown in the lower portion of the diagram. (Compare the voltage graph of the half-wave rectifier in Fig. 14.32.)

The graphs of the output voltages from the rectifier circuits in Figs. 14.32 and 14.33 are not true DC patterns. Instead, the voltage that results is really characterized as a *pulsating* DC voltage. In order to smooth the output, it is customary to use some sort of *filter network*, such as that shown in

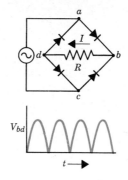

Figure 14.33 A full-wave bridge rectifier. The voltage across the load resistor R always has the same polarity.

Fig. 14.34. The pulsating voltage across the load resistor R is inhibited in its excursions by the reactances of the inductors and capacitors in the LC filter. The inductors work against the rapid change in voltage and the capacitors, which accumulate charge during the voltage peaks, release it during the voltage minima. More elaborate filtering and regulation in high quality electronic circuits produce steady DC voltages with almost no residual "ripple."

In addition to being the basic components in rectifier circuits, diodes are used in many other types of circuits where the unidirectional flow of current is desired.

Amplification Circuits

Diodes perform the specific function of permitting current to flow through them in only one direction. But within the diode itself there is no means to control the amount of current. To do this we need a *third* section of the device. Such electronic elements—the ones most commonly used today—are *transistors*. We briefly discuss transistor amplifier circuits in Section 20.3.

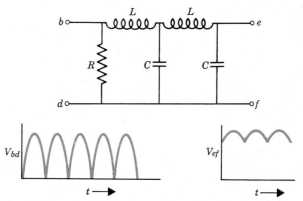

Figure 14.34 An LC filter network connected to the load resistor of a rectifier circuit smooths the pulsating voltage V_{bd} and produces a DC voltage V_{ef} with only a small amount of ripple.

Advantages of Semiconductor Circuit Elements

Semiconductor devices (diodes and transistors) have several important advantages compared to vacuum tubes. Semiconductor circuit components can be made very much smaller than vacuum tubes because the latter require glass or metal envelopes and relatively bulky bases. Moreover, a vacuum tube requires a heater element, and the fact that tubes operate at high temperatures is an important limitation on the density with which vacuum tubes can be packaged and on their useful lifetimes. A semiconductor diode or transistor that is used in a lower-power circuit operates near room temperature and because it has no fragile internal parts, the device will have an extremely long lifetime. Finally, semiconductor diodes and transistors are generally much less expensive than their vacuum-tube counterparts.

Summary of Important Ideas

Circuits that contain *series-parallel connections* of resistors can be analyzed by successively reducing all parallel combinations to equivalent single resistances. The final step is adding all of the resulting series-connected resistances.

Complex circuits can be analyzed by using *Kirchhoff's circuit rules:* I. At any point in a circuit, the sum of the currents entering that point must equal the sum of the currents leaving that point. (Conservation of charge.) II. The sum of all the potential increases around a circuit must equal the sum of all the potential drops. (Conservation of energy.)

A real battery has *internal resistance* which influences the *terminal voltage* of the battery when connected in a circuit. A battery will deliver *maximum power* to a circuit when the load resistance is equal to the internal resistance.

A *potentiometer circuit* can be arranged in which zero current flows through a battery so that the battery EMF can be compared directly with that of a standard cell.

A *Wheatstone bridge* is a null-current circuit that is used for the precision comparison of a resistance with a standard resistance.

When a battery is connected to a circuit consisting of a series connection of a *capacitor* and a *resistor,* the *charge* on the capacitor builds up slowly. The charge reaches 0.63 of its maximum value, CV, after a time $t = RC$ (called the *characteristic time* or the *time constant* of the combination).

When a battery is connected to a circuit consisting of a series connection of an *inductor* and a *resistor,* the *current* in the circuit builds up slowly. The current reaches 0.63 of its maximum value, V/R, after the characteristic time L/R.

The design of *electric motors* depends on the fact that a magnetic field exerts a *torque* on a current-carrying loop or coil of wire. This same fact is used in the design of current-detecting devices (*galvanometers*).

An *ammeter* is a galvanometer that is shunted with a low resistance to divert most of the current away from the delicate galvanometer movement. When inserted into a circuit, the low total resistance of an ammeter does not disturb the current in the circuit.

A *voltmeter* is a galvanometer that is in series with a high resistance to prevent any large current from flowing through the meter movement. When connected in parallel with a resistor in a circuit, the high total resistance of a voltmeter does not appreciably disturb the current in the circuit.

Alternating current is produced by mechanically rotating a coil of wire in a magnetic field, thereby converting mechanical energy into electrical energy. The EMF of an AC generator varies sinusoidally with time.

In terms of the power that can be delivered to a circuit element (e.g., the heating of a resistor), the *effective current* I_{eff} in an AC circuit is equal to $1/\sqrt{2}$ times the peak current I_0 in the circuit. The power delivered is $P = \frac{1}{2} I_0^2 R$.

A *transformer* is a device used to step up or step down the voltage in an AC circuit by means of electromagnetic induction. The power in the primary and secondary parts of a transformer circuit are equal.

In an AC circuit an *inductor* has a *reactance* (a resistance to current flow) given by $X_L = 2\pi\nu L$, where ν is the frequency of the alternating current. In an AC circuit a *capacitor* has a *reactance* given by $X_C = 1/2\pi\nu C$.

The total effective resistance (or *impedance*) in a series *RLC* circuit is $Z = \sqrt{R^2 + (X_L - X_C)^2}$. If the applied voltage is $V(t) = V_0 \cos \omega t$ then, the current is $I(t) = I_0 \cos(\omega t - \delta)$ where, the phase angle δ is given by the expression $\tan \delta = (X_L - X_C)/R$ and where, $I_0 = V_0/Z$.

Maximum current will flow in the circuit when $X_L = X_C$, that is, when $\nu = \frac{1}{2} \pi \sqrt{LC}$; this is the *resonance* condition.

The average power dissipated in the *RLC* circuit is

$$P = \frac{I_0 V_0}{2} \cos \delta.$$

Diodes (either vacuum-tube or semiconductor type) permit current to flow through them in only one direction. Diodes are the essential components in all *rectifier circuits* that convert AC to DC.

◆ **Questions**

14.1 In a Wheatstone bride circuit, current is continually being drawn from the battery and, consequently, the terminal voltage decreases slowly with time. Does this affect the accuracy of the measurements? Explain.

14.2 After the switch S in Fig. 14.12 has been closed for some period of time, the current is essentially at its maximum value, V/R. Suppose that we now take a wire and connect it across the battery, thereby effectively shorting it out of the circuit. Make a sketch of I versus t for the current that now flows through R.

14.3 In Fig. 14.14, as the loop turns, the force $\mathbf{F}_M$ remains constant but the torque exerted on the loop does not. Explain why. In Fig. 14.16 the torque on the coil remains constant as the coil turns through its entire range (defined by the extent of the scale). Explain why.

14.4 In the type of multirange meter shown in Fig. 14.18, the switch is of the *make-before-break* type. That is, when the arm is moved from one position to another, contact is made with the new resistor before contact is broken with the old resistor. Why is it important to use this type of switch in meter circuits?

14.5 The current through and the voltage across a circuit element are to be measured. Explain carefully why the ammeter is connected in *series* with the circuit element, whereas the voltmeter is connected in *parallel*.

14.6 A resistor R is connected to a battery. Draw a diagram showing how a voltmeter and an ammeter should be incorporated into the circuit to determine the value of R.

14.7 What mechanical modification could be made to the arrangement of the AC generator in Fig. 14.20 to convert it into a DC generator?

14.8 If you connect an ammeter to an AC motor, you will notice that there is an initial surge of current when the motor is turned on and that after a few moments the current becomes steady at a value lower than the initial peak. Explain why this is so.

14.9 What is one advantage of having many poles in an AC generator?

14.10 How would you design an ammeter to measure alternating current? (Refer to Fig. 14.16.) What feature of this meter movement makes it unsuitable for measuring AC? How could you modify the movement to eliminate this feature? Hint: An AC ammeter does not use a permanent magnet.)

14.11 Electric motors are designed to be operated with either a DC power source or with an AC power source. A resistance heater, on the other hand, will operate with either type of source (if it does not have a blower motor!) Explain the difference in these two situations.

14.12 When the demand for electricity begins to exceed the ability of the power company's generators to product it, the company will usually lower the voltage delivered to the power network. Why is this done?

14.13 The first commercial electric power companies distributed DC. Why has the use of DC been almost entirely abandoned in favor of AC for distributing electric power?

14.14 If an AC power line is subject to momentary surges of voltage (voltage spikes), a capacitor is sometimes connected across the line. Why do you suppose this is done?

14.15 A coil, designed for use in the laboratory, consists of a few dozen turns of heavy-gauge copper wire. A student decides to connect this coil across a 120-V line, but he selects a pair of DC terminals instead of AC terminals. What do you suppose happens?

14.16 A capacitor C is charged by a battery and then the battery is removed. If an inductor L is now connected across the capacitor, explain what happens.

14.17 A machine shop with many motors represents a very inductive load on the power lines. The power company claims it should receive a higher rate for the actual power consumed by the machine shop citing the increased losses in the power transmission lines as justification. Explain. Is there anything you would suggest to the shop management to improve its power utilization?

14.18* Refer to Fig. 14.27. Draw a sine curve (V versus ωt) that represents the voltage across the inductor L. Now, argue how the current I through the inductor will vary with time. (The current is proportional to the *rate of change* of the applied voltage. That is

$$V_L = -\lim_{\Delta t \to 0} \left(L \frac{\Delta I}{\Delta t} \right)$$

from Eq. 14.6.) Add to the graph a curve that represents I versus ωt. The peak values of V and I do not occur at the same times. We say that "current lags voltage" for an inductor. What is the angle of this "lag"? Make a similar analysis for an RC circuit and show that "current leads voltage" in this case. Here we have $Q = CV$ and hence, $\Delta Q = C\Delta V$ and

$$\frac{\Delta Q}{\Delta t} = C \frac{\Delta V}{\Delta t},$$

also,

$$I = \lim_{\Delta t \to 0} \frac{\Delta Q}{\Delta t}$$

14.19* When an AC voltage is applied to a series RLC circuit, the average input power is dissipated in the resistor. According to Eq. 14.22, $P = \frac{1}{2} I_0 V_0 \cos \delta$ yet it should also be given by Eq. 14.10, $P = \frac{1}{2} I_0^2 R$. Show that these two equations are consistent. (Hint: The expression for the phase angle δ can be depicted graphically with the right triangle figure).

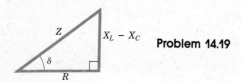

Problem 14.19

★ **Problems**

Section 14.1

14.1 What is the effective resistance between points a and b for the resistor network illustrated here?

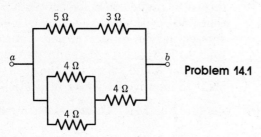

Problem 14.1

14.2 What current flows through the 1-Ω resistor in this circuit?

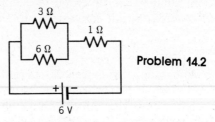

Problem 14.2

14.3 The terminal voltage of the battery in the circuit illustrated is V and remains constant. When the variable resistor R is set at R_1, the ammeter reads 4 A. When R is increased by 14 Ω (i.e., $R_2 = R_1 + 14$ Ω), the current measured by the ammeter drops to 0.5 A. Find V and R_1.

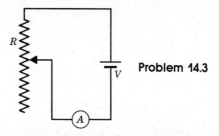

Problem 14.3

14.4* Six light bulbs are connected in parallel across a 120-V DC line, and each bulb uses 40 W of power. What is the resistance of each bulb? What current flows through each bulb? The bulbs are now connected in series and the entire combination is connected across the same power line. What current flows through each bulb? What is the power used by each bulb? Why is this value different from 40 W?

14.5 Find the total current I that flows through the battery in the circuit illustrated here. What is the current in the 4-Ω resistor?

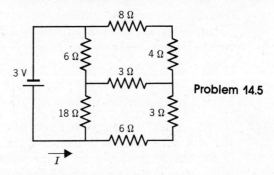

Problem 14.5

14.6 By how much does the current measured by the ammeter vary as the resistor R is changed from 0 to 12 Ω?

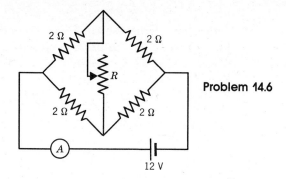

Problem 14.6

14.7 Find the current through each of the resistors in the circuit illustrated there.

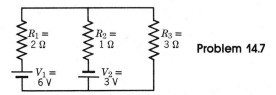

Problem 14.7

14.8 Find the current in each of the resistors in the circuit illustrated here.

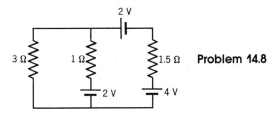

Problem 14.8

14.9 For the circuit illustrated here, find the three currents indicated and find the potential difference between a and b.

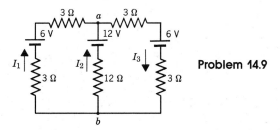

Problem 14.9

14.10 Find each of the currents in the circuit illustrated here. (Assume the currents have the direction shown.)

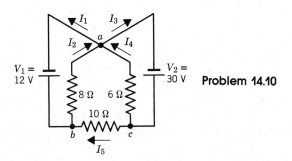

Problem 14.10

14.11 In the circuit illustrated, find the current in each of the resistors and the current through the 12-V battery.

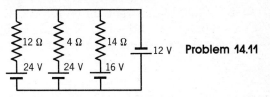

Problem 14.11

14.12* In the circuit illustrated, find the current in each of the resistors and find the eventual charge on the capacitor. (Hint: What current flows through the 6-Ω resistor?)

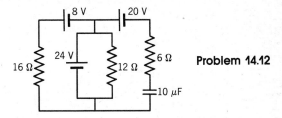

Problem 14.12

14.13 An automobile battery has an EMF of 12 V, but when it delivers 18 A to the starter circuit, the terminal voltage is 11.2 V. What is the internal resistance of the battery?

14.14 A battery has an EMF of 12 V and an internal resistance of 0.4 Ω. Two resistors ($R_1 = 1\ \Omega$, $R_2 = 2\ \Omega$) are connected in parallel and the combination is connected to the battery. What current flows through each resistor? What is the voltage between the terminals of the battery?

14.15 When a 1-Ω resistor is connected to a certain battery, a current of 4.6 A flows through the resistor. When a 2-Ω resistor is substituted, the current is 2.6 A. What is the EMF of the battery and the internal resistance of the battery?

14.16 The potentiometer in a Wheatstone bridge circuit (Fig. 14.8) consists of a wire with a length of 0.85 m. The standard resistor has a resistance of 1.086 Ω. When the galvanometer indicates that a null condition has been reached, the sliding contact is 17.6 cm from the end that is common with the standard resistor. What is the value of the unknown resistance?

Section 14.2

14.17 A 100-μF capacitor is charged to a potential of 40 V and then is discharged through a

20-kΩ resistor. How much charge remains on the capacitor 6 s after the connection to the resistor is made?

14.18 When the current through a coil of wire is changing at the rate of 3 A/s, and EMF of 0.48 V is produced across the coil. What is the inductance of the coil?

14.19 In Fig. 14.12, $R = 600\ \Omega$, $L = 150$ mH, and $V = 12$ V. What is the current in the circuit 0.5 ms after the switch S is closed?

14.20* The switch in Fig. 14.12 is closed at $t = 0$. At $t = 3$ s, the current has risen to $\frac{3}{4}$ of its final value. Use Table 14.1 and estimate the time constant of the circuit. If the resistance in the circuit is 4 Ω, find the value of the inductance.

14.21 A 4000-turn solenoidal coil has a length of 50 cm and a volume of 200 cm³. What is the inductance of the solenoid? If the current through the solenoid changes at the rate of 240 A/s, what EMF is developed across the coil?

14.22 A certain DC motor has a coil resistance of 4 Ω. When this motor is operating from a 120-V line, it draws a current of only 1 A. Describe what is happening.

Section 14.3

14.23 A certain galvanometer has an internal resistance of 12 Ω and will deflect full scale when a voltage of 80 mV is applied across its terminals. How can this galvanometer be made to function as an ammeter with a full-scale deflection for a current of 10 A?

14.24 A galvanometer has an internal resistance of 50 Ω and a full-scale deflection current of 10 mA. What resistance must be placed in series with this galvanometer to convert it into a voltmeter with a full-scale reading of 30 V?

14.25 Refer to the galvanometer in the preceding problem. What shunt resistances must be placed in parallel with this galvanometer to produce a multirange ammeter with full-scale readings of 300 mA, 3 A, and 10 A?

14.26 The diagram shows a galvanometer used to measure resistance in an arrangement called an ohmmeter. When the terminals ab are shorted R_0 is adjusted to give full-scale deflection. This setting corresponds to $R_X = 0$. With terminals ab open, zero galvanometer current corresponds to $R_X = \infty$. If $V = 3.0$ V, the galvanometer sensitivity is 1 mA full scale, and $R_g = 50\ \Omega$. What should the value of R_0 be to zero the ohmmeter? What is the value of the resistance being measured R_X if when it is connected across the terminals ab, the galvanometer reads half scale?

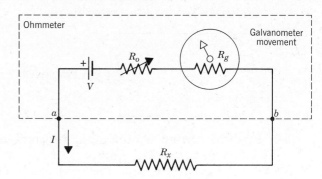

Problem 14.26

Section 14.4

14.27 The armature of an AC generator (Fig. 14.20) rotates at a rate of 12 Hz and the current through R is measured to be 4 A. What will be the current if the rotation rate is increased to 30 Hz?

14.28 A color television set requires a peak current of 4 A from an ordinary household circuit (120 V AC). If the set is operated an average of 4 h each day, what is the monthly cost of operation if the power company charges 3.5 cents per kWh?

14.29 An electric heater is rated at 1400 W and operates from a 120 V AC line. What is the resistance of the heater?

14.30 The peak output voltage of a certain AC electric generating station is 100,000 V. Power is delivered to a users' network through a transmission line that has a resistance of 4 Ω and carries a peak current of 1000 A. What is the average power output of the generating station? What percentage of the output is lost due to Joule heating in the transmission line?

14.31 In order to power a household device it is necessary to step down the usual voltage

(120 V) to 12 V. A transformer has 600 turns on the primary. How many turns are needed on the secondary to provide the desired voltage?

14.32 The primary of a transformer is connected to an ordinary 120-V line. What is the effective current in the primary when 30 W of power is drawn from the secondary?

14.33 A transformer that is to be connected to a 120-V power line has 800 turns on the primary. The secondary has several sets of windings so that voltages of 240 V, 24 V, and 6.3 V can be obtained. Describe the secondary windings. (Can you think of two different arrangements for the secondary windings?)

14.34 A transformer has $N_p = 2500$ and $N_s = 50$. The peak voltage of the EMF placed across the primary is 170 V. The secondary is connected to a lamp bulb whose resistance is 0.8 Ω. What peak voltage is developed across the lamp filament? What peak current flows through the filament? What power is delivered to the lamp?

14.35 An alternating EMF ($V_0 = 170$ V, $\nu = 60$ Hz) is applied to a (resistanceless) solenoid. It is found that a peak current of 200 mA flows through the coil. What is the self-inductance of the solenoid?

14.36 What is the reactance of a 40 μF capacitor at a frequency of 60 Hz? At 60 MHz?

14.37 A coil has an inductance $L = 0.6$ H. When this coil is connected to a source of 400 HZ AC with $V_0 = 180$ V, the peak current is 80 mA. What is the DC resistance of the coil?

14.38 A certain resistor dissipates 10 W when connected to an ordinary household circuit ($V_0 = 170$ V), $\nu = 60$ Hz). What is the value of the resistance and what peak current flows through the resistor? Next, a 2-μF capacitor is connected in series with the resistor. What power is now dissipated by the resistor?

14.39* A resistor ($R = 400$ Ω), a capacitor ($C = 0.8$ μF), and an inductor ($L = 140$ mH) are connected in series to an AC power supply ($V_{eff} = 40$ V, $\nu = 400$ Hz). What is the impedance of the RLC combination? What peak current flows in the circuit? At what AC frequency will the current be a maximum?

14.40 A coil is wound with a length of wire that has a total resistance of 0.8 Ω. The inductance of the coil is 0.2 mH. What is the impedance of the coil at $\nu = 1200$ Hz?

14.41* A 3-Ω resistor is connected in series with an inductor L and the combination is connected across a 120-V, 60-Hz line. A peak current of 30 A flows through the circuit. What size capacitor should be placed in series with the resistor and inductor in order to increase the peak current to 34 A?

14.42* A 150-W light bulb is connected to a 120-V, 60-Hz power line. If a 500-μF capacitor is inserted in series with the light bulb, will there be any significant decrease in the light output of the bulb? What do you suppose would happen if the capacitor were connected in *parallel* with the light bulb?

14.43* A solenoid has a volume of 80 cm^3 and has 30 turns of wire per centimeter. What size capacitor must be connected in series with this solenoid so that the combination will resonate at 2200 Hz?

14.44 A 1000 Ω resistor and a 2 μF capacitor are connected in series to a household AC power source ($V_0 = 170$ V, $\nu = 60$ Hz). If the voltage may be written as $V(t) = V_0 \cos \omega t$ and the current as $I(t) = I_0 \cos(\omega t - \delta)$, determine I_0 and δ. Is the current leading or lagging the voltage? How much power is drawn from the line?

14.45 An RLC series circuit consisting of $R = 50$ Ω, $L = 100$ mH, and $C = 10$ μF is connected across the household line ($V_0 = 170$ V, $\nu = 60$ Hz). Find the current in the circuit. What is the power factor? How much power is delivered to the circuit? Find the magnitude of the peak voltage across each of the circuit elements.

14.46* A machine shop with a very inductive load has a power factor of $\cos \delta = 0.50$ when drawing 150 kW from a $V_{eff} = 220$ V line. What is the effective current? What is the resistive component of the load? If the capacitive component of the load is zero, what is the inductive reactance of the load? (Hint: Refer to Question 14.19.)

14.47 In an experiment using the circuit shown in Fig. 14.9 it is determined that the time constant of the resistor-capacitor combination is 1.0 ms. What is the phase angle δ if this series RC combination is connected to an AC voltage source with a frequency of 150 Hz?

15

Waves, Sound, and Radiation

In Chapter 11 we discussed the oscillatory motions of single particles. We now turn our attention to cases in which the motion of any given particle in a collection of particles influences and is influenced by the motion of its neighbors. An important case of this type of *cooperative phenomenon* is that of *wave motion*.

Examples of wave motion are to be found virtually everywhere. Water waves are familiar to all of us, as are sound waves that travel through air. Waves set up in the strings of pianos and violins produce musical tones that we hear. Electromagnetic waves in the form of radio waves and light are in the space all around us. Generally, we classify waves in two groups—mechanical waves and electromagnetic waves. Mechanical waves (including water waves, sound waves, and waves on strings) are characterized by the oscillatory motions of *material particles*. Electromagnetic waves (including radio and television waves, heat radiation, visible light, and X rays) are characterized by oscillations of the *electromagnetic field*. Thus, the presence of matter is required for the propagation of mechanical waves; electromagnetic waves, on the other hand, can propagate through empty space (as well as through matter).

15.1 Traveling Waves on Strings and Springs

The Propagation of Wave Pulses

One of the easiest ways to produce a simple wave motion is by vibrating the end of a long string. Figure 15.1 shows the result of a single up-and-down motion applied to a string. In Fig. 15.1*b*, the end of the string begins its upward motion; this results in an upward force being applied to the section of string immediately adjacent to the end and this section also begins to move upward. In Fig. 15.1*c* the end of the string has reached its highest position and is instantaneously at rest; nevertheless, the section to the right of the end is still experiencing the result of displacing the end and this section is moving upward. In Fig. 15.1*d* the end of the string has returned halfway to its starting point, and in Fig. 15.1*e* it has come to rest in its original position. The downward motion of the end (Figs. 15.1*d* and 15.1*e*) is transmitted to the adjacent sections of the string which then also move downward.

This type of motion of the segments of string continues after the end is again at rest. Each string

352 Waves, Sound, and Radiation

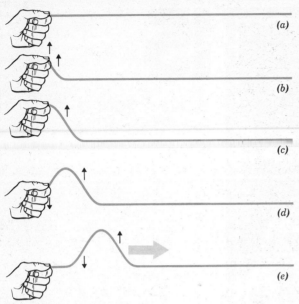

Figure 15.1 A wave pulse that propagates along a string is produced by a single up-and-down movement of one end of the string. After the end has been returned to its original position, the wave pulse still propagates to the right.

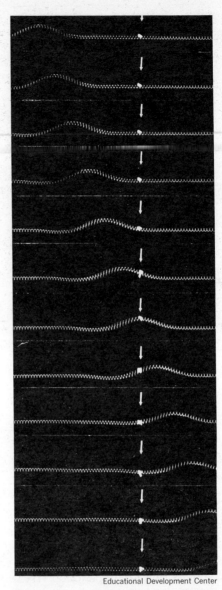

Figure 15.2 A wave pulse travels along a coiled spring from left to right. The various pictures are individual frames from a film of the motion recorded by a movie camera. A ribbon is tied to one of the coils (indicated by the arrow) in order to illustrate the motion of a particular small portion of the spring.

segment is pulled upward and then downward by the action of the segment to its left. Similarly, each segment exerts first an upward force and then a downward force on the segment to its right. In this way the *wave pulse* formed by the single up-and-down excursion of the end propagates along the string to the right.

Figure 15.2 shows a wave pulse propagating along a coiled spring. A ribbon, tied to one of the coils, provides a marker for observing the motion of a particular portion of the spring. It is evident that no portion of the spring moves very far in the horizontal direction and yet the pulse travels along the spring by virtue of the fact that the particles at the front of the pulse are forced to move upward and those at the rear are forced to move downward to their original positions (see Fig. 15.3).

By noting the position of the wave pulse at various instants of time, we can determine the propagation velocity of the pulse in the same way that we would determine the speed of a moving particle (see Fig. 15.4). If the pulse travels a distance Δx in a time interval Δt the propagation velocity is

$$v = \frac{\Delta x}{\Delta t} \qquad (15.1)$$

On what factors will the velocity of propagation of a wave pulse depend? First of all, the velocity should *not* depend on the length of the string along which the pulse travels nor should it depend on the amplitude or the shape of the pulse. (Recall the discussion in Section 11.2 concerning the fact that the period of SHM does not depend on the amplitude.) But the velocity is likely to depend on the density ρ

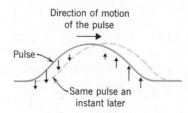

Figure 15.3 The relation between the motion of the portions of the spring and the forward motion of a pulse.

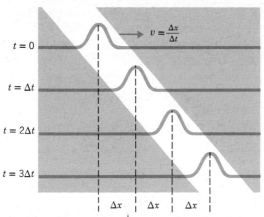

Figure 15.4 The wave pulse travels toward the right with a propagation velocity, $v = \Delta x/\Delta t$.

and the cross-sectional area A of the string and on the tension T in the string. In fact, as we will see in Example 15.1, the velocity of a wave pulse on a string is given by

$$v = \sqrt{\frac{T}{\rho A}} \qquad (15.2)$$

● *Example 15.1*

Argue on the basis of the physical dimensions of the quantities involved that Eq. 15.2 for the velocity of a wave pulse on a string is correct.

If we indicate the dimensions of a quantity by square brackets around the symbol for the quantity, we can write

$[v] = \text{m/s} = \text{m-s}^{-1}$

$[T] = \text{N} = \text{kg-m/s}^2 = \text{kg-m-s}^{-2}$

$[\rho] = \text{kg/m}^3 = \text{kg-m}^{-3}$

$[A] = \text{m}^2$

We must arrange T, ρ, and A in some combination that has the same dimensions as v, that is, m/s. If this combination involves T to some power a, ρ to some power b, and A to some power c, we can write

$\text{m-s}^{-1} = [T]^a \times [\rho]^b \times [A]^c$
$= (\text{kg-m-s}^{-2})^a \times (\text{kg-m}^{-3})^b \times (\text{m}^2)^c$

Collecting terms on the right-hand side, we have

$\text{m-s}^{-1} = \text{kg}^{a+b} \times \text{m}^{a-3b+2c} \times \text{s}^{-2a}$

Now, for this relationship to be valid, the dimensions on each side must be the same; for example, the power to which m is raised on the left-hand side (i.e., 1) must equal the power to which m is raised on the right-hand side (i.e., $a - 3b + 2c$). Writing these equalities for all three quantities gives us three equations connecting a, b, and c:

$a + b = 0$

$a - 3b + 2c = 1$

$2a = 1$

The last equation yields $a = \frac{1}{2}$; then the first equation gives $b = -\frac{1}{2}$. Inserting these values into the second equation gives $c = -\frac{1}{2}$. Thus, we have altogether,

$v = T^a \rho^b A^c = T^{1/2} \rho^{-1/2} A^{-1/2}$

which we can write as

$$v = \sqrt{\frac{T}{\rho A}}$$

We have succeeded in obtaining (not really *deriving*) Eq. 15.2 by using a powerful technique called *dimensional analysis*. This method of analysis permits us to obtain the connection among various physical quantities merely by requiring that the expression have the correct dimensions. Notice that dimensional analysis cannot produce any *numerical* factors that might enter into the expression. For example, if we had used dimensional analysis to obtain an expression for the period of a simple pendulum, we would have missed the factor 2π. In the case of the motion of wave pulses on a string, however, a rigorous analysis shows that the numerical factor is unity, so that the result of dimensional analysis in this case is (fortuitously) correct in magnitude.

● *Example 15.2*

How fast will pulses propagate along a string that has a mass per unit length of 0.1 kg/m when it is under a tension of 40 N?

First notice that if the string has a mass m and a length ℓ, we can write

$m = \rho V = \rho(A\ell)$

where A is the cross-sectional area and $A\ell$ is the volume V. Then,

$\rho A = m/\ell$

and Eq. 15.2 for the velocity can be expressed as

$$v = \sqrt{\frac{T}{\rho A}} = \sqrt{\frac{T}{m/\ell}}$$

Substituting the values for T and m/ℓ,

$$v = \sqrt{\frac{40 \text{ N}}{0.1 \text{ kg/m}}} = \sqrt{400} = 20 \text{ m/s} \blacksquare$$

Describing Traveling Waves

In Fig. 15.1 we showed how to produce a single wave pulse that propagates along a string. Now, let us consider what happens when we continue to move the left end on the string up and down. In particular, we wish to study the case in which the end is driven by the action of a simple harmonic oscillator. As we learned in Section 11.2, a simple harmonic oscillator moves in such a way that the displacement is described by a sinusoidal function. Figure 15.5 shows how the motion of the string develops when the end is driven sinusoidally. A series of identical pulses (first, with a displacement in a positive sense, then, with a displacement in a negative sense) travels along the string. This series of propagating pulses constitutes a *traveling wave*.

The velocity of the traveling wave is the same as that of a single pulse, $v = \Delta x / \Delta t$. (That is, each pulse travels with the same velocity, so the shape of the train of pulses does not change along the string.) The distance between any two successive maxima of the string is called the *wavelength* λ of the disturbance that propagates along the string. The motion of the string follows the motion of the driving force supplied by the simple harmonic oscillator. Thus, the period τ of the wave motion is the same as that of the oscillator. During each time interval τ the wave moves forward by an amount equal to the wavelength λ of the wave (see Fig. 15.5). Therefore, when $\Delta t = \tau$, we have $\Delta x = \lambda$, so that Eq. 15.1 can be expressed as

$$v = \frac{\lambda}{\tau} = \lambda \nu \qquad (15.3)$$

where we have also used the connection between the period τ and the frequency ν, that is, $\nu = 1/\tau$, as expressed by Eq. 11.17.

From Fig. 15.5 we can see that the displacement of the driven end of the string is described by

$$y(t) = y_0 \sin \frac{2\pi}{\tau} t \qquad (15.4)$$

At any position along the string, the displacement as a function of *time* will be described by a similar sinusoidal expression. For example, if we pick a particular spot, $x = x_0$, along the string and begin measuring its motion at time $t = 0$ when the displacement is at the maximum value y_0, then we find (Fig. 15.6a)

$$y(t) = y_0 \cos \frac{2\pi}{\tau} t \qquad \text{(at } x = x_0\text{)} \qquad (15.5)$$

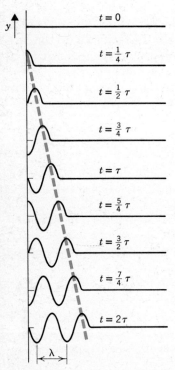

Figure 15.5 The generation of a traveling wave in a length of string. The left end of the string is driven up and down in a sinusoidal manner and the motion is propagated along the string, forming a sinusoidal traveling wave. The dashed line shows the steady motion forward of the initial pulse on the string.

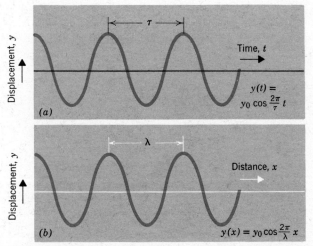

Figure 15.6 (a) The displacement of a given point of the string as a function of time. (b) The displacement of the string as a function of distance for a given instant. Both curves are sinusoidal curves.

Similarly, we can take a "snapshot" of the waveform at some instant $t = t_0$. We then find that the displacement varies with distance according to

$$y(x) = y_0 \cos \frac{2\pi}{\lambda} x \quad \text{(at } t = t_0\text{)} \tag{15.6}$$

This function is shown in Fig. 15.6b.

Suppose that a traveling wave is generated in the laboratory by moving the end of a string up and down with SHM. The wave progresses from left to right along the x-axis with the velocity v. Now, suppose that an observer moves toward the right with this same velocity and views the wave. To this observer, the wave appears stationary. If the coordinates in the reference frame moving with the observer are indicated by primes, the observer describes the waveform by

$$y' = y_0 \cos \frac{2\pi}{\lambda} x'$$

If we start the observer moving in such a way that the y and y' axes coincide at $t = 0$, *we have*

$$x' = x - vt$$

(Notice that the vertical motion of the string is the same whether viewed by an observer stationary in the laboratory or moving with the waveform; that is, $y' = y$ because the motion is along the x-axis.) Then, substituting for x', the displacement can be expressed in the laboratory coordinates, x and y, as

$$y(x,t) = y_0 \cos \frac{2\pi}{\lambda}(x - vt)$$

Using Eq. 15.3 for v, this equation becomes

$$y(x,t) = y_0 \cos 2\pi \left(\frac{x}{\lambda} - \frac{t}{\tau}\right) \tag{15.7}$$

This expression describes the propagation of one-dimensional waves. If the position x is fixed, the displacement executes a simple harmonic vibration; if the time t is fixed, the spatial variation of the displacement is a sinusoid.

Transverse and Longitudinal Waves

In the wave motions we have been discussing, the particles in the strings or springs move at right angles to the direction of propagation of the wave. Such

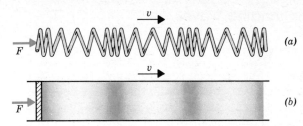

Figure 15.7 (a) Longitudinal compressional waves in a coiled spring are initiated by the application of a periodic driving force at one end. (b) Longitudinal waves (sound waves) in a column of gas are initiated by the application of a periodic force to a piston located in one end.

waves are therefore called *transverse* waves. Wave motion is also possible in which the particles move back and forth along the direction of wave motion. Such waves are called *longitudinal* waves. Examples of this type of wave motion are the *compressional waves* that can be propagated along a spring (Fig. 15.7a) and the *sound waves* that can be propagated along an air column in a pipe (Fig. 15.7b). Generally, whether a wave motion is transverse or longitudinal depends on the elastic properties of the medium. If the medium tends to restore a transverse displacement of a portion of the medium, then transverse wave propagation is possible. Similarly, if the medium tends to restore a longitudinal displacement, then longitudinal wave propagation is possible. A string is a floppy structure and cannot be compressed; therefore, only transverse waves can be propagated along a string. On the other hand, there are no forces within a gas that tend to restore a transverse displacement; consequently, only longitudinal waves propagate through gases (and liquids). Both transverse and longitudinal waves can be propagated through solids. (Can you see why?)

When a longitudinal wave is propagated along a spring (Fig. 15.7a), portions of the spring are alternately compressed and extended, and the variation of "coil density" along the spring is sinusoidal. Similarly, in a sound wave the density of the gas molecules (or the pressure) varies sinusoidally along the column, and we have a regular alternation of *compressions* and *rarefactions*. Because some physical property (such as the density of particles or the pressure) in longitudinal waves varies with distance and time in exactly the same way as in transverse waves, these two different types of wave motion are described by identical mathematical expressions. For example, in Eq. 15.7 we need only call $y(x, t_0)$ the pressure in a column of gas in order to represent the propagation of a sound wave with a frequency $\nu = v/\lambda$.

15.2 Standing Waves

Superposition

We have thus far been considering wave motion on *unterminated* strings; that is, the strings have been assumed to extend indefinitely far in the direction of the wave propagation. An important class of problems involves wave motion on strings that are attached to rigid supports. In these cases we must deal with the combination of different wave forms propagating along the same string due to the reflection of the waves from the terminated end (or ends) of the string.

If two wave pulses are traveling along the same string, how do they interact with one another? The answer is extremely simple: each pulse propagates along the string in a way that is completely independent of the other pulse! That is, each pulse displaces the string in a certain way and the combined effect of the two pulses is a simple sum of the individual displacements. Thus, waves on strings obey a *principle of superposition* similar to that we discussed in Section 12.4 for the electric field.

Figure 15.8 is a photographic record of two pulses on a coiled spring that approach and pass one another. After passing, they continue in opposite directions without having suffered any change in shape or size (or velocity). When the two pulses pass one another, the displacement of the spring is the *sum* of the two individual displacements. If these individual displacements are of the same shape and size but are of opposite signs, then a *cancellation* will result at the moment of passing (see the 5th frame of Fig. 15.8 and Fig. 15.9c). If the pulses are of the same sign, then they will *add* at the moment of passing.

The combination of wave pulses on a string is one example of the phenomenon of *interference* (see also Section 15.4). When the pulses *add* together, we say that the interference is *constructive;* when the pulses *cancel* one another, we say that the interference is *destructive.*

Reflection of Wave Pulses

Consider a string or a spring that is terminated at one end; that is, the end of the string is attached to a rigid wall and cannot be moved. Suppose that we send a wave pulse along such a terminated string, as in Fig. 15.10a. When the pulse reaches the terminated end, the wave motion will not simply stop (because energy must be conserved); instead, the pulse will be *reflected* and will proceed back along the

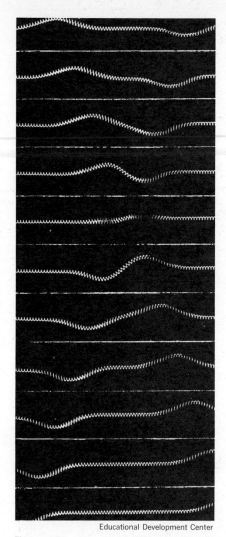

Educational Development Center

Figure 15.8 The superposition of two almost identical wave pulses traveling along a coiled spring. At the moment of passing (fifth frame), they almost cancel each other. Notice, however, that a blurring of the photograph indicates that, although the spring has almost no net displacement, there is a substantial vertical *velocity* at two positions.

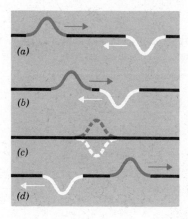

Figure 15.9 A schematic representation of the superposition of two identical pulses as shown photographically in Fig. 15.8. In (c) the two pulses exactly cancel.

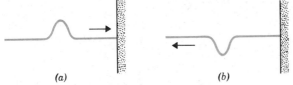

Figure 15.10 (a) A wave pulse travels along a string toward the terminated end. (b) After reaching the end, the pulse is reflected and travels back along the string in the opposite direction with the opposite sign of the displacement. The motion of the wave pulse is very similar to that of a ball rebounding from a wall.

string in the opposite direction (Fig. 15.10b). Notice that the reflected wave pulse has the same size and shape as the initial pulse, but the *sign* of the displacement is reversed. This reversal in the sign of the displacement is due to the fact that the pulse exerts an *upward* force on the wall (just as it would on the next segment of string), and the wall exerts a *downward* reaction force on the string. Because the wall does not move, this reaction force manifests itself in a downward displacement of the reflected pulse. Figure 15.11 is a photographic record of a pulse traveling along a coiled spring; at the terminated end the pulse is reflected with the attendant change in sign of the displacement. In this case, the reflected pulse is slightly smaller than the initial pulse because there are frictional losses that occur when the pulse is reflected at the termination.

Sinusoidal Waves on Terminated Strings

What happens when a traveling sinusoidal wave is incident on the terminated end of a string? Reflection takes place, and the terminated end remains motionless throughout. (Any position along a vibrating string that remains motionless is called a *node*; a terminated end always corresponds to a node.) The pattern of wave motion on the string is then a combination of the incident and reflected waves. The most interesting case is that in which *both* ends of the string are terminated and are necessarily nodal points. If the string is then vibrated in such a way that the wavelength of the wave motion is exactly equal to the length of the string, the successive reflections from the terminated ends will preserve the shape of the wave pattern. Figure 15.12 shows the time development of such a wave. Between the times $t = 0$ and $t = \frac{1}{4}\tau$, the effect of reflection from the terminated ends is to reduce the *size* of the pattern, but the *shape* is unaltered. Between $t = \frac{1}{4}\tau$ and $t = \frac{1}{2}\tau$, the size increases in the opposite sense until the wave form is the mirror image of that at $t = 0$. This sequence of changes repeats itself indefinitely (in the absence of frictional losses). Such wave forms are called *standing waves*.

The requirement for the existence of a standing wave is that the terminated ends of the string correspond to nodes of the wave motion. (The positions of maximum motion are called *antinodes*.) If the length L of the string is equal to the wavelength λ of the wave motion, this condition is satisfied (Fig. 15.12). However, there are additional wavelengths that also meet the nodal requirement. The standing wave with the longest wavelength (lowest frequency) for a particular string has $\lambda = 2L$ (Fig. 15.13a). The other diagrams in Fig. 15.13 show the succession of wave patterns that correspond to standing waves on the same string. The connection between the wave-

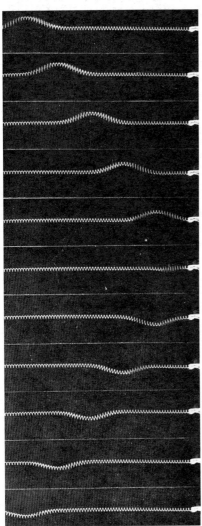

Educational Development Center

Figure 15.11 Photographic record of a wave pulse reflected from the terminated end of a coiled spring. The reflected pulse is slightly smaller than the initial pulse because of losses that occur at the termination.

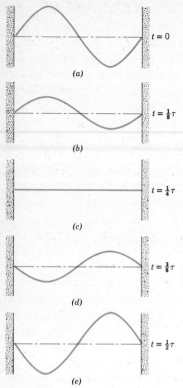

Figure 15.12 Time development of a sinusoidal displacement on a string between a pair of fixed supports.

length λ and the string length L is

$$n\frac{\lambda}{2} = L \qquad n = 1, 2, 3, \ldots \quad (15.8)$$

The lowest frequency with which a standing wave can be set up on a particular string is called the *fundamental frequency*, ν_1 (Fig. 15.13a). If the wavelength of the wave is reduced to one half that of the fundamental, the frequency must double because the propagation speed is the same for all waves (see Eq. 15.3). Therefore, the next lowest frequency for a standing wave has frequency $\nu_2 = 2\nu_1$ (Fig. 15.13b). In fact, any standing wave has a frequency that is an integer multiple of the fundamental frequency. Thus, the standing-wave frequencies are ν_1, $2\nu_1$, $3\nu_1$, $4\nu_1$, and so forth. The waves with frequencies $\nu_2 = 2\nu_1$ and higher are called *harmonics* of the fundamental. In music, the waves with these higher frequencies are referred to as *overtones*. The frequency ν_2 corresponds to the second harmonic or first overtone; the frequency ν_3 corresponds to the third harmonic or second overtone; and so forth. (The numbering of overtones proceeds in order whether all of the harmonics are present or not. If, for some reason, the second harmonic of the fundamental produced by a device were not allowed, then the first overtone would correspond to the *third* harmonic—or the next one that is allowed.)

Standing waves can be set up for any type of wave motion if proper reflection points or surfaces are provided. We can produce standing waves for sound waves, radio waves, light waves, and all other forms of electromagnetic waves.

• **Example 15.3**

A certain string has a linear mass density of 0.1 kg/m. When a length of this string is stretched between two supports that are 4 m apart, the string tension is 10 N. What is the lowest frequency standing wave that can be set up on this string?

According to Eq. 15.3, $\nu = v/\lambda$, where $\lambda = 2L$ for the lowest frequency (Fig. 15.13a). Thus,

$$\nu_1 = \frac{v}{\lambda} = \frac{v}{2L}$$

Using the result in Example 15.2 for v, we can write

$$\nu_1 = \frac{1}{2L}\sqrt{\frac{T}{m/\ell}} = \frac{1}{2 \times (4\text{ m})}\sqrt{\frac{10\text{ N}}{0.1\text{ kg/m}}} = 1.25\text{ Hz}$$

■

15.3 Sound

The Speed of Sound Waves

Compressional waves that are due to a vibrating source and that are capable of producing a sensation in the auditory system are called *sound waves*. Ordinary sound waves are usually produced in air, but

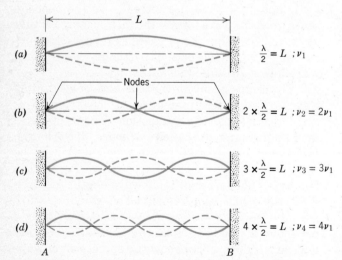

Figure 15.13 Standing waves on a stretched string. All such waves must have nodes at the termination points.

sounds propagating through water can be heard by an underwater listener. The human ear is responsive to these waves if they have frequencies in the range from about 15 Hz to about 20,000 Hz. Generally, an individual's ability to hear high-frequency sounds deteriorates with age. A young person may be able to hear 20,000-Hz sounds, but when this same person reaches middle age, he or she may be unable to hear sounds with frequencies above 12,000 or 14,000 Hz. Similarly, persons who are exposed to very loud sounds for long periods of time (e.g., persons who work near high-powered aircraft or jackhammers, and rock music performers or fans) will often experience impaired hearing, with the most pronounced effect in the higher frequencies. Also, many species of animals and insects can detect sounds with much higher frequencies than those that can be heard by humans. (Bats and dolphins, for example can hear sounds with frequencies above 100,000 Hz.) Nevertheless, for our purposes we consider the range of sound frequencies to be from 15 Hz to 20,000 Hz.

The speed with which sound propagates through air is easy to measure. One simple method is to station yourself a known distance from a large flat wall, such as the side of a building. Then, make a sharp sound and, using a stop watch, measure the time required between the instant the sound is made and the instant you hear the echo. This method, suitably refined, will yield a result of about 330 m/s (about 1100 ft/s) for the speed of sound at a temperature of 0°C. At this temperature the wavelengths in air corresponding to the extremes of "normal" sound frequencies are:

Longest: $\lambda = \dfrac{v}{\nu} = \dfrac{330 \text{ m/s}}{15 \text{ Hz}} = 22 \text{ m}$

Shortest: $\lambda = \dfrac{v}{\nu} = \dfrac{330 \text{ m/s}}{20{,}000 \text{ Hz}} = 0.0165 \text{ m} = 1.65 \text{ cm}$

The velocity with which a sound wave will propagate depends on the elastic properties of the medium, whether that medium is air, a steel bar, or a string. In a medium that cannot support any shear stress, such as a liquid or a gas, the appropriate elastic property on which the speed of sound depends is the *bulk modulus* (Section 11.1). Using the same type of dimensional analysis that we employed in Example 15.1, it is easy to show that the speed of sound in a medium with a bulk modulus B is (see Problem 15.16)

$$v = \sqrt{\dfrac{B}{\rho}} \qquad (15.9)$$

(In this case we are again lucky to find that the numerical factor which results from the rigorous analysis is unity.)

In Example 11.3 we found that the bulk modulus for a gas is equal to the pressure P. Therefore, we might expect that the speed of sound in a gas is equal to $\sqrt{P/\rho}$. But in Example 11.3, we used Boyle's law, $PV = $ constant, to determine that $B = P$. Boyle's law, remember, is valid only under *isothermal* (constant temperature) conditions. In a sound wave, however, the wave motion is so rapid and the heat conductivity is so low that there is insufficient time for the heat produced in the compressed regions to be conducted to the rarefied regions. That is, in a sound wave in a gas, the gas undergoes compressions and expansions *adiabatically*, not isothermally. According to Eq. 10.2, the relationship between pressure and volume in an adiabatic process in a gas is $PV^\gamma = $ constant, where γ is the ratio of specific heats, c_p/c_v, and is equal to 1.40 for air (see Table 9.6). (Air is a mixture of nitrogen and oxygen; $\gamma = 1.40$ for both N_2 and O_2.) Using the adiabatic relation instead of Boyle's law changes the expression for the bulk modulus to $B = \gamma P$. Then, the equation for the speed of sound in a gas becomes

$$v = \sqrt{\dfrac{\gamma P}{\rho}} \quad \text{(for gases)} \qquad (15.10)$$

The pressure P in this expression is, of course, the *average* pressure in the gas.

If we use the ideal gas law, $PV = nRT$ (Eq. 10.8), then we can rewrite Eq. 15.10 as (see Problem 15.17)

$$v = \sqrt{\dfrac{\gamma RT}{M}} \quad \text{(for gases)} \qquad (15.11)$$

where M is the molecular mass of the gas expressed in units of kg/mol.

● *Example 15.4*

What is the speed of sound in air at 0°C?

Air consists of approximately 80 percent nitrogen and 20 percent oxygen. The masses of the nitrogen and oxygen molecules are 28 u and 32 u, respectively. Therefore, the mean molecular mass for air is

mean molecular mass $= (0.8 \times 28 \text{ u})$
$+ (0.2 \times 32 \text{ u}) = 28.8 \text{ u}$

Therefore, $M = 28.8$ g/mol $= 0.0288$ kg/mol. Then, we have

$$v = \sqrt{\frac{1.40 \times (8.314 \text{ J/mol-K}) \times (273 \text{ K})}{0.0288 \text{ kg/mol}}}$$

$= 332$ m/s   at 0°C

At any other temperature T, the speed of sound in air can be obtained by multiplying this result by $\sqrt{T/273 \text{ K}}$. For example, at an altitude of 100,000 ft (30.5 km), the temperature is about $-50°$C or 223 K. Then,

$$v = (332 \text{ m/s}) \times \sqrt{\frac{223 \text{ K}}{273 \text{ K}}} = 300 \text{ m/s}$$

For a solid substance in the shape of a rod or bar, the appropriate elastic constant that determines the speed of sound is Young's modulus. Therefore, sound waves in long, thin pieces of material propagate with a speed given by

$$v = \sqrt{\frac{Y}{\rho}} \qquad (15.12)$$

(In bulk samples of solid materials, shear stresses can occur, and the speed of sound will depend on the elastic response to shear as well as on the bulk modulus. We will not discuss this type of sound propagation.)

The speed of sound in various media is given in Table 15.1.

● *Example 15.5*

What is the frequency of a 2-cm sound wave in sea water?

From Table 15.1, $v = 1.53 \times 10^3$ m/s in sea water, so

$$\nu = \frac{v}{\lambda} = \frac{1.53 \times 10^3 \text{ m/s}}{0.02 \text{ m}}$$

$= 7.6 \times 10^4$ Hz $= 76$ kHz

which is an *ultrasonic* wave (that is, above the human audible range). A 2-cm sound wave in *air* would be audible. (What is the frequency?) ∎

For rapidly moving aircraft and rockets, it is often more important to know the speed of the object compared to the speed of sound in the medium than it is to know the speed in m/s. (Frictional effects depend critically on how close the speed of an object is to the speed of sound.) Consequently, aircraft and rocket speeds are frequently given in *Mach num-*

Table 15.1 The Speed of Sound in Various Media

Medium	Temperature (°C)	Speed (m/s)
Air	0	332
Carbon dioxide	0	259
Hydrogen	0	1284
Water (distilled)	25	1498
Water (sea)	25	1531
Mercury	25	1450
Aluminum (rolled)[a]	20	5000
Copper (rolled)[a]	20	3750
Lead (rolled)[a]	20	1210
Steel (mild)[a]	20	5200

[a] For a thin rod or bar of the material.

bers. The Mach number is the ratio of the speed of the object to the speed of sound at the location of the object. Thus, Mach 0.5 means half the speed of sound and Mach 2.0 means twice the speed of sound. According to the result of Example 15.4, an aircraft flying at a speed of Mach 1.0 at an altitude of 100,000 ft would be moving with a speed of 300 m/s (670 mi/h). However, if the aircraft were flying at Mach 1.0 at sea level where the temperature is 0°C, its speed would be 332 m/s (742 mi/h).

The Time Variation of Sound Waves

It is easy to observe simple standing-wave patterns on a long string or rope, but the vibrations of a violin string or a piano string are much too rapid and too complex to be analyzed visually. And, of course, we cannot see at all the vibration patterns of air molecules responding to a sound source. We can, however, convert any type of vibration or sound wave into a visible pattern by displaying the wave displacement as a function of *time* on an *oscilloscope*. The sound to be studied is picked up by a microphone which produces an electrical signal that is used as the input to the oscilloscope. The basic component of an oscilloscope is a *cathode-ray tube* (Fig. 15.14), which is similar to a television picture tube. A beam of electrons is produced within the tube and when these electrons strike the fluorescent screen on the tube face, they produce a visible spot of light. As the electron beam is scanned across the screen by an internal electronic circuit, the input signal from the microphone causes up-and-down deflections of the beam and a curve is traced on the screen. (The oscilloscope is an electrical analog of the mechanical vibration-tracing apparatus we studied in Fig. 11.2.) If the sound picked up by the microphone consists of a pure frequency, the signal displayed on the oscilloscope face will be a sinusoidal pattern corresponding

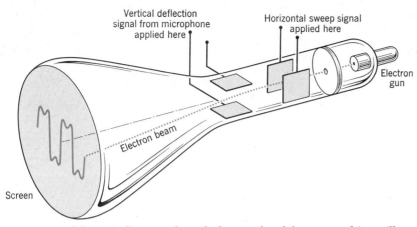

Figure 15.14 Schematic diagram of a cathode-ray tube of the type used in oscilloscopes. The electron beam is swept horizontally across the screen by a varying voltage from an internal source applied to the horizontal deflection plates. The signal to be studied is applied to the vertical deflection plates.

to the particular frequency of the sound (Fig. 15.15a).

Suppose, now, that instead of a sound with pure frequency, we have a sound that consists of two different frequencies—for example, a certain fundamental frequency and its third harmonic. In this case, the input signal consists of the frequencies ν_1 and $3\nu_1$. The third harmonic is shown in Fig. 15.15b.

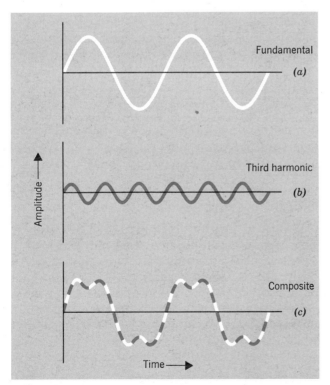

Figure 15.15 The combination (or superposition) of two waves, the fundamental (a) and the third harmonic (b).

The display that will appear on the oscilloscope is the superposition of these two signals. This composite signal is shown in Fig. 15.15c (see also the signal in Fig. 15.14).

Beats

When two waves with quite different frequencies are combined, a pattern such as that shown in Fig. 15.15c results. If the frequencies of the combining waves are very close, an interesting effect occurs. Suppose that one tone generator produces a steady sound with a frequency $\nu_1 = 1000$ Hz and a second generator produces a steady sound with the same amplitude but with a frequency $\nu_2 = 996$ Hz. An observer who listens to the tones separately will probably not be able to distinguish between the two frequencies. (A frequency difference of 4 parts in 1000 is near the resolution limit of the human ear, at least for those persons not specially trained.) But when the tones are produced together, the observer will hear an "average frequency" of 998 Hz. In addition, he will hear superimposed on this tone a modulation of the intensity at a rate equal to the difference of the two frequencies. That is, the observer will hear a *beat* with a frequency of $1000 - 996 = 4$ Hz.

Figure 15.16 shows how beats occur. The first two diagrams (Figs. 15.16a and b) represent the sinusoidal variation with time of two waves with pure frequencies ν_1 and ν_2. Figure 15.16c shows these two wave forms on the same scale. The superposition of the two waves is shown in Fig. 15.16d. Here, we see a wave with a high frequency upon which is superimposed a low-frequency modulation of the amplitude (dashed curve). When a piano tuner adjusts one of

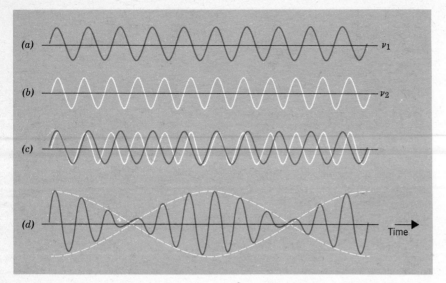

Figure 15.16 Two waves with slightly different frequencies combine to produce beats.

the strings of a piano, he listens for the beat between the sound from the piano string and a standard tone. He then varies the tension in the string (thereby changing the frequency) until the beat disappears; the string and the standard are then vibrating with the same frequency.

The time variation (at a particular point) of the displacement of each wave is given by Eq. 15.5:

$$\left.\begin{array}{l} y_1 = y_0 \cos 2\pi\nu_1 t \\ y_2 = y_0 \cos 2\pi\nu_2 t \end{array}\right\} \quad (15.13)$$

The combined wave, corresponding to Fig. 15.16d, has the form

$$y_1 + y_2 = y_0(\cos 2\pi\nu_1 t + \cos 2\pi\nu_2 t) \quad (15.14)$$

We can now use the trigonometric identity,

$$\cos A + \cos B = 2 \cos \tfrac{1}{2}(A + B) \cos \tfrac{1}{2}(A - B) \quad (15.15)$$

to express Eq. 15.14 as

$$y_1 + y_2 = 2y_0 \cos 2\pi \frac{(\nu_1 + \nu_2)}{2} t \cos 2\pi \frac{(\nu_1 - \nu_2)}{2} t \quad (15.16)$$

In this equation, the first cosine term involves the average frequency, $\bar{\nu} = \tfrac{1}{2}(\nu_1 + \nu_2)$, and the second cosine term involves the difference frequency $\Delta\nu = \nu_1 - \nu_2$. Therefore, we can write

$$y_1 + y_2 = \left(2y_0 \cos 2\pi \frac{\Delta\nu}{2} t\right) \cos 2\pi\bar{\nu}t \quad (15.17)$$

We can interpret this expression in the following way. The main variation with time is contained in the term $\cos 2\pi\bar{\nu}t$, which oscillates with the average frequency $\bar{\nu}$. The term in parentheses can be considered to be the amplitude of the wave, but an amplitude that varies slowly with time. (The variation is *slow* because $\Delta\nu$ is small compared to $\bar{\nu}$.) The sound intensity heard by an observer will be maximum whenever the amplitude term is maximum ($+2y_0$) or minimum ($-2y_0$). (Can you see why this is so?) The variation between $+2y_0$ and $-2y_0$ occurs *twice* during each complete cycle. Consequently, the beat frequency ν_b is *twice* the frequency of the amplitude term ($\tfrac{1}{2}\Delta\nu$):

$$\nu_b = 2 \times (\tfrac{1}{2}\Delta\nu) = \Delta\nu = \nu_1 - \nu_2 \quad (15.18)$$

Thus, an observer hears a beat with a frequency exactly equal to the difference in frequency of the two tones. The human ear can easily detect a beat that has a frequency up to 8 or 10 Hz.

Musical Sounds

The sound of any musical instrument consists of a fundamental and several different harmonics. Such sounds are more complex than the signal shown in Fig. 15.15c; for example, the wave pattern of a violin playing the note A is shown in Fig. 15.17. Such a sound consists of a rich mixture of harmonics; violin and piano notes may contain 15 or 20 harmonics. It is the harmonic or overtone structure of a musical sound that is responsible for the quality or *timbre* of the sound.

One way to indicate the harmonic structure of a sound is in a display of the *frequency spectrum*. Fig-

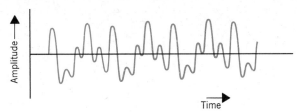

Figure 15.17 The wave pattern of a violin A note. Can you identify the fundamental period?

ure 15.18a shows the spectrum of the simple composite wave in Fig. 15.15c. This wave consists of a fundamental with frequency ν_1 and a third harmonic with frequency $3\nu_1$; the relative amplitudes of these two waves is 7:2. This information—the frequencies involved and their relative amplitudes—is shown graphically in Fig. 15.18a. The much more complicated frequency spectrum of the violin tone of Fig. 15.17 is shown in Fig. 15.18b.

The sound that we hear and identify as a violin A note can be synthesized by combining the pure notes generated by individual vibrating strings or electrical tone generators. If the mixture of the fundamental and harmonics is the same as that produced by the violin, the resulting tone will sound exactly the same as that from a violin. Indeed, electric organs synthesize tones in just this way. Such organs can generate tones that duplicate a variety of different instruments by combining pure notes in different ways. It is difficult, however, to produce exactly the correct mixture of pure notes, and an artificial violin tone from an electric organ sounds slightly different than a real violin tone. The ear is quite sensitive to the presence of extra harmonics or to the absence of those that should be present but are missing.

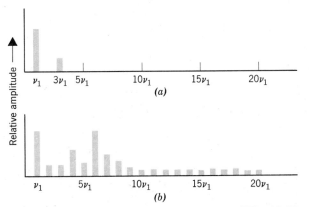

Figure 15.18 (a) Frequency spectrum for wave of Fig. 15.15c. Only two frequencies, ν_1 and $3\nu_1$, are present in this wave. (b) Frequency spectrum for a violin playing the A note (Fig. 15.17). At least 20 harmonics are present in this complex sound. Notice that the sixth harmonic is particularly evident.

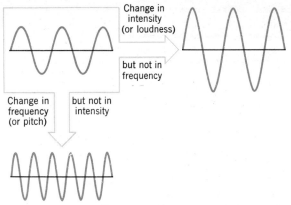

Figure 15.19 The frequency and the intensity of a sound wave do not depend on one another.

Loudness

There are two qualities of every sound that we easily perceive: the *pitch* (or frequency) and the *loudness* (which is closely associated with intensity). The *intensity* of a sound wave is measured in terms of the amount of energy per second (i.e., the *power*) the wave delivers to a unit area of a surface. (Accordingly, sound intensity is measured in watts per square meter, W/m^2.) The frequency and the intensity are independent aspects of a wave. That is, we can have a high-frequency or a low-frequency sound wave with the same intensity; or, we can have a loud or a soft sound with the same frequency (see Fig. 15.19).

The human ear is sensitive over an incredibly large range of sound intensity. The loudest sound that the ear can tolerate has an intensity about 10^{12} times that of the softest sound the ear can perceive! Because this range is so large, it proves convenient to compare sound intensities on a logarithmic scale. The unit of sound intensity level is called the *bel* (B), in honor of Alexander Graham Bell (1847–1922), the inventor of the telephone. If a sound has an intensity level that is one bel (1 B) greater than another sound, the ratio of the two intensities is 10. A difference of 2 B means an intensity ratio of 10^2; a difference of 3 B means an intensity ratio of 10^3; and so forth.

A bel is a rather large unit, so we usually refer to sound intensity levels in terms of the *decibel* (dB): 10 dB = 1 B. A difference of 20 dB (= 2 B) means a sound intensity ratio of 10^2; and so forth. We can express this in equation form by writing

$$\Delta\beta = 10 \log \frac{I_2}{I_1} \qquad (15.19)$$

Table 15.2 Sound Intensity Ratios Expressed in dB

I_2/I_1	$\Delta\beta$ (dB) = 10 log (I_2/I_1)
1	0
2	3.01
3	4.77
4	6.02
5	6.99
6	7.78
7	8.45
8	9.03
9	9.54
10	10.00

where $\Delta\beta$ is the difference in the *sound intensity level* (measured in dB) for the sounds with intensities I_1 and I_2. A factor of 2 in sound intensity (i.e., $I_2/I_1 = 2$) corresponds to a difference $\Delta\beta$ of approximately 3 dB. Some additional intermediate values are given in Table 15.2. Because the logarithm of the product of two numbers is equal to the sum of the logarithms of the numbers, $\log AB = \log A + \log B$, we can use Table 15.2 to help us find $\Delta\beta$ for any intensity ratio. For example, if $I_2/I_1 = 24$, we can write

$$\Delta\beta = 10 \log 24 = 10 \log 6 + 10 \log 4$$
$$= 7.78 + 6.02 = 13.8 \text{ dB}$$

If your pocket calculator possesses the log function, you could calculate the result directly. (A review of logarithms will be found in the Appendix.)

● *Example 15.6*

A chorus consists of 36 individuals all of whom can sing with the same intensity. A soloist sings a certain passage and then is joined by the remainder of the chorus for a repeat of the passage. What is the difference in the sound intensity level in the two cases?

The intensity of the soloist is I_1 and that of the chorus is $I_2 = 36I_1$. Therefore,

$$\Delta\beta = 10 \log 36 = 15.56 \text{ dB} \quad \blacksquare$$

The threshold of hearing corresponds to a sound intensity of about 10^{-12} W/m²; the corresponding intensity level is designated 0 dB. The loudest tolerable (but painful) sound has an intensity level of 120–130 dB (1–10 W/m²) and depends somewhat on frequency. Even this extreme sound intensity corresponds to a very small movement of the air molecules (about 10^{-5} m; see Problem 15.27) and to an

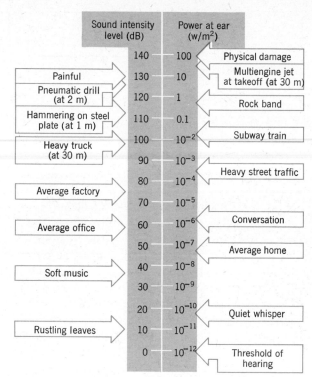

Figure 15.20 The decibel scale of sound intensity level and some common sounds. Each difference of 10 dB corresponds to a factor of 10 in sound intensity.

increase in air pressure of only about 0.03 percent (see Problem 15.28). Figure 15.20 illustrates some of the sound intensities to which we are exposed, from the gentle rustle of leaves to the ear-splitting roar of a powerful jet aircraft.

The Vibrations of Air in Pipes

In Fig. 15.13 we showed the way that transverse standing waves with various frequencies (all interger multiples of the fundamental frequency) could be set up on a vibrating string. The same type of effect occurs for longitudinal sound waves in air columns. Suppose that we have a long cylindrical pipe in which there is a movable piston. Suppose also that we have a tone generator capable of producing sound with a pure frequency. (A simple device of this type is a *tuning fork,* which is a U-shaped piece of metal that emits a pure tone when struck a sharp blow.) We place the tone generator above the open end of the pipe, as in Fig. 15.21a, and listen to the intensity of the sound. By moving the piston within the pipe, we can locate a position for which the sound intensity is particularly high. In this case the sound waves from the generator set up a standing wave within the pipe; the sound waves are therefore

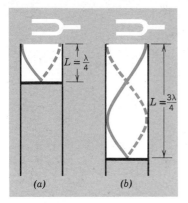

Figure 15.21 The intensity of the sound from a tone generator (here, a tuning fork) will be enhanced when a standing wave is set up in the air column. This will occur when (a) $L = \lambda/4$, (b) $L = 3\lambda/4$, and for additional lengths in this series. The curves represent the amplitudes of the longitudinal motion of the air molecules; there is maximum motion at the open end of the pipe and a note at the closed end.

reinforced by the reflected waves and we hear an increase in the sound intensity. The *shortest* air column for which this reinforcement occurs has a length L equal to $\frac{1}{4}$ of the wavelength of the sound wave, $L = \lambda/4$ (Fig. 15.21a). If the piston is moved downward, it is found that another intensity peak occurs when $L = 3\lambda/4$ (Fig. 15.21b). In fact, standing waves will be set up for the series of lengths, $\lambda/4$, $3\lambda/4$, $5\lambda/4$, and so forth.

The effect we have just examined is another example of the phenomenon of *resonance* (Section 11.2). The characteristic frequency of the tone generator is equal to that of the air column in the pipe when $L = \lambda/4$ (or the other lengths in the series). Consequently, there is an efficient transfer of energy from the generator to the air column and the sound intensity is enhanced. In fact, an air column in a pipe that is closed at one end will resonate at a series of wavelengths λ given by (see Fig. 15.22)

$$\lambda = \frac{4L}{n} \qquad n = 1, 3, 5, \ldots$$

(pipe closed at one end) (15.20)

The frequencies corresponding to these wavelengths are

$$\nu = \frac{v}{\lambda} = \frac{nv}{4L} \qquad n = 1, 3, 5, \ldots \qquad (15.21)$$

If we designate the fundamental frequency ($n = 1$) by ν_1, the sequence of standing-wave frequencies is ν_1, $3\nu_1$, $5\nu_1$, and so forth. Thus, the air column in a

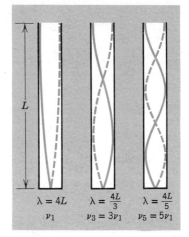

Figure 15.22 The first three standing-wave patterns in the air column in a pipe that is closed at one end.

pipe that is closed at one end will support standing waves only with the *odd* harmonics of the fundamental. This particular sequence of resonant frequencies is dictated by the fact that there must be a node at one end of the pipe (the closed end) and an antinode at the other end (the open end).

A pipe that is open at both ends can support only those standing-wave patterns that have antinodes at both ends. The three lowest-frequency waves that satisfy this condition are shown in Fig. 15.23. The wavelengths are given by

$$\lambda = \frac{2L}{n} \qquad n = 1, 2, 3, \ldots$$

(pipe open at both ends) (15.22)

and the corresponding frequencies are

$$\nu = \frac{v}{\lambda} = \frac{nv}{2L} \qquad n = 1, 2, 3, \ldots \qquad (15.23)$$

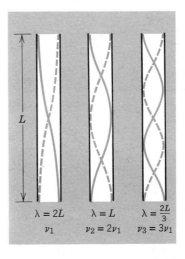

Figure 15.23 The first three standing-wave patterns in the air column in a pipe that is open at both ends.

This sequence of frequencies is ν_1, $2\nu_1$, $3\nu_1$, and so forth. Thus, the air column in a pipe that is open at both ends can support standing waves with *all* harmonics of the fundamental. Notice that ν_1 for a pipe closed at one end is equal to $\frac{1}{2}$ the value of ν_1 for a pipe open at both ends: ν_1 (closed) $= \frac{1}{2}\nu_1$ (open).

Many types of musical instruments depend on the resonance characteristics of air columns to produce their characteristic sounds. Organ pipes, for example, are excited by blowing air through an opening in a specially shaped section at the open end of the pipe. The vibrations associated with this rushing air have many frequencies. However, because the pipe resonates at only certain frequencies, the wave motions that take place with these frequencies are enhanced and we hear these sounds. Usually, the sound from an organ pipe consists mainly of the fundamental, but if the velocity of the air blowing through the opening is increased, higher harmonics can be excited and the quality of the sound is distinctly different.

● *Example 15.7*

What is the frequency of the first overtone that can be heard from an organ pipe that is closed at one end and has a length of 2.5 m if the temperature of the air is 20°C?

For a pipe closed at one end, the first overtone corresponds to the third harmonic. We use Eq. 15.21 with $n = 3$:

$$\nu = 3\nu_1 = \frac{3v}{4L}$$

According to Example 15.4, the speed of sound in air at 20°C is

$$v = (332 \text{ m/s}) \times \sqrt{\frac{293 \text{ K}}{273 \text{ K}}} = 344 \text{ m/s}$$

Therefore, the frequency is

$$\nu = \frac{3 \times (344 \text{ m/s})}{4 \times (2.5 \text{ m})} = 103 \text{ Hz} \quad \blacksquare$$

The Doppler Effect

We are all familiar with the fact that the frequency of the sound from a siren on a moving vehicle changes dramatically when the vehicle passes. When the sound source is approaching, the apparent frequency is high, and, when the source passes and is moving away, the apparent frequency is low. The dependence of the frequency of a wave disturbance

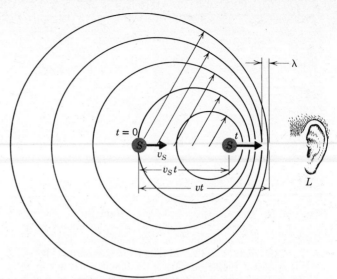

Figure 15.24 The source of sound waves S moves toward the listener L with a velocity v_s. The waves become "bunched up" and the listener perceives a *higher* frequency of sound than he would if the source were at rest. If the listener were at the *left*, he would hear a *lower* frequency. (The arrows directed upward and to the right indicate the radii to which the waves emitted at various points have propagated by the time t.)

on the relative motion of the source and the observer is termed the *Doppler effect*, after the Austrian physicist, Christian Johann Doppler (1803–1853), who extensively studied this phenomenon.

Figure 15.24 shows the spherical sound waves (of constant frequency and amplitude) that are emitted by a source S, which moves at constant velocity v_S toward a listener L. (We are interested here, not in the three-dimensional character of the waves, but only in those portions of the waves that move directly from the source to the listener.)

In the reference frame of the source, the frequency of the waves is ν_S. Therefore, during a time t, the source will emit $\nu_S t$ waves. Consider the interval from time zero until time t. The wave emitted at $t = 0$ will travel a distance vt during this interval, where v is the velocity of the waves in the particular medium. (The velocity of waves in a medium depends only on the mechanical properties of that medium and does not depend in any way on the velocity of the source.) When the last of the $\nu_S t$ waves has been emitted, the source will have traveled a distance $v_S t$. Hence, the $\nu_S t$ waves occupy the distance $vt - v_S t$ and so their wavelength λ_L as determined by the listener, is

$$\lambda_L = \frac{\text{distance}}{\text{no. of waves}} = \frac{vt - v_S t}{\nu_S t} = \frac{v - v_S}{\nu_S}$$

The frequency ν_L, again as determined by the listener, is

$$\nu_L = \frac{v}{\lambda_L} = v \times \frac{\nu_S}{v - v_S} = \nu_S\left(\frac{v}{v - v_S}\right)$$

so that, dividing through by v,

$$\nu_L = \frac{\nu_S}{1 - v_S/v} \quad (S \text{ moving toward } L) \quad (15.24)$$

This expression shows that the frequency of the sound heard by L is *increased* over that which he would hear if the source were at rest. If the source moves *away from* the listener, the sign of v_S is changed and the negative sign in Eq 15.24 becomes a positive sign. Thus, the frequency is *lowered*. This effect with water waves is illustrated in Fig. 15.25.

The expression we have just derived for the apparent frequency ν_L is valid for the case in which the source is moving toward (or away from) the listener. We can use a similar procedure to obtain the equivalent expression in the event that the listener is moving toward (or away from) the source. The result is (see Problem 15.37)

$$\nu_L = \nu_S\left(1 + \frac{v_L}{v}\right) \quad (L \text{ moving toward } S) \quad (15.25)$$

As in the previous discussion, we need only to change the sign of v_L to make the equation applicable for the case in which the listener moves away from the source.

The Doppler effect applies to all types of wave motions—water waves and sound waves, as well as radio waves, light, and other electromagnetic radiations. (However, the Doppler effect for electromagnetic waves is more complex than for mechanical waves.) *Doppler radar*, for example, is now used extensively by law enforcement officers to identify motor vehicles that are exceeding the speed limit.

● **Example 15.8**

An airplane is flying at Mach 0.5 (half the speed of sound) and carries a sound source that emits a 1000-Hz signal. What frequency sound does a listener hear if he is in the path of the airplane?

$$\nu_L = \frac{\nu_S}{1 - v_S/v}$$

But a speed of Mach 0.5 means that $v_S/v = 0.5$; therefore,

$$\nu_L = \frac{\nu_S}{1 - 0.5} = 2\nu_S$$

so that the listener hears a 2000-Hz sound. What frequency does the listener hear after the airplane has passed?

$$\nu_L = \frac{\nu_S}{1 + v_S/v} = \frac{\nu_S}{1 + 0.5} = \frac{2}{3}\nu_S$$

so that the listener hears a 667-Hz sound. ■

Shock Waves and Sonic Booms

Figures 15.24 and 15.25 show what happens when the speed of the source through the medium is *less* than the wave speed in that medium. Next, let us suppose that we increase the speed of the source until it is just *equal* to the wave speed. In this case (Fig. 15.26), the source keeps pace with the outgoing spherical waves in the direction of motion. The waves can never move ahead of the source and they continue to pile up. As more and more waves are formed, the "pileup" region extends farther and farther from the source in the direction perpendicular to its motion. The result is a wave front that is a sheet or a plane, with the source located at its center.

Finally, let us see what happens when the speed of the source *exceeds* the wave speed in the medium.

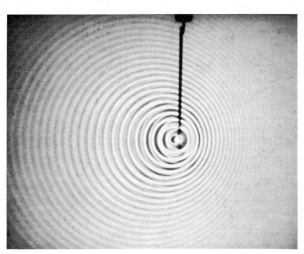

Educational Development Center

Figure 15.25 Photograph of water waves produced by a moving source. The dark line in the photograph is a vibrating rod that is moved through the water from left to right. An observer on the right would measure a shorter wavelength (higher frequency) than if the source were at rest.

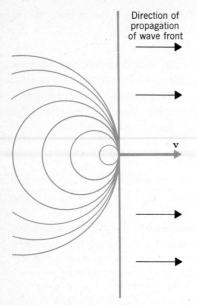

Figure 15.26 When the source moves with a speed exactly equal to the wave speed, the waves pile up and form a plane that extends perpendicular to the direction of motion of the source.

In this case (Fig. 15.27), the source runs ahead of the outgoing waves, and the pileup of the waves produces a wave front (actually, a cone-shaped wave front) that makes a certain angle with the direction of motion. The faster the source moves, the greater is the degree to which the wave front is bent back.

It is easy to see the relationship between the angle of conical wave front and the speed of the object by referring to Fig. 15.28. During the time t, the wave emitted when the source was at O has propagated to Q, a distance vt, where v is the speed of sound. During this same time interval, the source itself has moved from O to P, a distance $v_S t$, where v_S is the velocity of the source. Therefore the cone angle θ is obtained from

$$\sin\theta = \frac{vt}{v_S t} = \frac{v}{v_S} \tag{15.26}$$

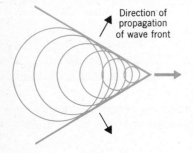

Figure 15.27 When the speed of the sound source exceeds the wave speed, a conical wave front is formed which trails behind the source. This front is a *shock wave*.

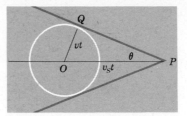

Figure 15.28 Geometry for Eq. 15.26.

Notice that $1/\sin\theta = v_S/v$ is the *Mach number* of the moving object.

When an object, such as a bullet or an aircraft, moves through the air with a speed greater than the speed of sound, the motion is said to be *supersonic*. (At speeds below the speed of sound, the motion is *subsonic*.) Along the cone-shaped wave front that is produced by a supersonic object, the air is highly compressed. This moving sheet of high-pressure air is called a *shock wave*. Actually, there are *two* shock waves produced by an supersonic object—one from the forward part of the object and one from the rear. Figure 15.29 shows the shock waves produced by a supersonic aircraft. The graph at the bottom of the diagram indicates how the air pressure varies between the shock waves. The air pressure rises sharply along the leading shock wave, which is due to the forward part of the aircraft. The pressure then

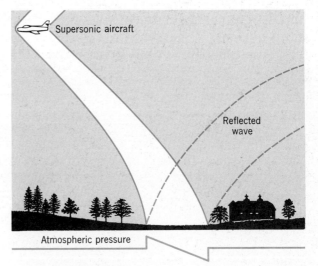

Figure 15.29 A supersonic aircraft produces shock waves that are heard as a *sonic boom*. The graph at the bottom of the diagram shows that the air pressure rises sharply along the shock wave formed by the forward part of the aircraft. The pressure then falls below normal atmospheric pressure and again rises sharply along the shock wave formed by the rear of the aircraft. (The wavefronts are curved because the speed of sound depends on the air temperature and the temperature varies with altitude.)

falls, becoming less than normal atmospheric pressure. The pressure again rises sharply along the trailing shock wave, which is due to the rear of the aircraft. This second rise brings the pressure back to normal.

If you are unfortunate enough to find yourself along the ground track of a supersonic aircraft, you will hear and feel the effect of the rapid pressure changes due to the shock waves. We refer to the sound produced by a supersonic aircraft as a *sonic boom*. Actually, you will hear only a single *crack* or *boom* when the shock waves pass because the time between the two sudden pressure increases is typically only about $\frac{1}{50}$ of a second.

The Ear and Hearing

The human ear, together with the human eye, are two of the most marvelous and efficient devices ever constructed for sensing information carried by waves. In the case of the ear, impinging sound waves are coupled by the various parts of the ear to the sensory nerves that are associated with hearing. Figure 1 shows the outer portion of the human ear. Sound waves enter the system through the *ear canal,* which is a kind of closed-end acoustic pipe with a length of approximately 2.7 cm. The air column in the ear canal will resonate for a sound wavelength equal to 4 times its length, corresponding to the situation in Fig. 15.21a. The fundamental resonant frequency is $\nu = v/4L = 3$ kHz. This resonance effect accounts in part for the fact that the human ear is most sensitive for the range of frequencies near 3 kHz (see Fig. 4).

At the end on the ear canal is the *tympanic membrane* (or *ear drum*). The pressure variations of the incident sound wave are transmitted by the ear drum to the system of small bones (*ossicles*) in the middle ear (Fig. 2). These bones are named according to their shapes: the *malleus* (hammer), the *incus* (anvil), and the *stapes* (stirrup). The ossicles serve to match the response of the outer ear to that of the inner ear of *cochlea*.

The middle ear is an air-filled cavity. Any pressure difference between the outer ear and the middle ear will tend to displace the tympanic membrane. This is how sound information is transmitted to the bones in the middle ear. But any significant change in the ambient air pressure (e.g., due to a change in altitude) will distort the membrane and can cause pain or even rupture the ear drum. To relieve such pressure differences there is a small tube (the *Eustachian tube*) that connects the middle ear cavity with the *pharynx* (a part of the upper throat). The Eustachian

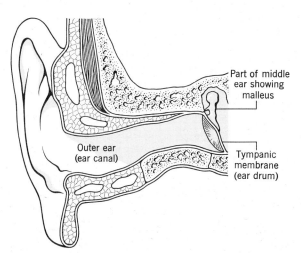

Figure 1 Schematic diagram of the outer parts of the human ear. (After Ackerman).

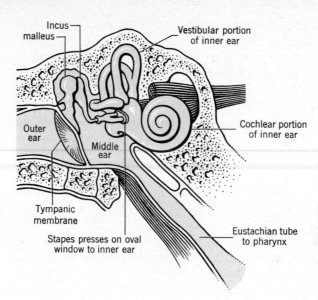

Figure 2 Schematic diagram of the middle and inner parts of the human ear. The vestibular apparatus of the inner ear is associated with maintaining positional equilibrium and is not involved in hearing. (After Ackerman.)

tube is sufficiently large to allow the slow passage of air to adjust for gross pressure changes but is sufficiently small that the rapid pressure changes due to sound waves can still be transmitted across the tympanic membrane. (The type of pressure changes encountered in air travel can produce particularly severe pain if the Eustachian tube is blocked and constricted in any way—as might be the case if an individual is suffering from a bad cold.)

The part of the inner ear that is involved in hearing is the coiled, fluid-filled tube called the *cochlea* (Fig. 2). The vibratory motion of the stapes is transmitted to the cochlear fluid through a small membrane-covered opening called the *oval window*. The passage of sound energy through the fluid in the ducts in the cochlea sets into motion the *basilar membrane* which, in turn, excites the cells of the hairlike fibers that are connected to the end organ of the auditory nerve (the *organ of Corti*). The electrical signals that are generated in this process are transmitted to the brain and a sensation of *hearing* results.

The passage of a sound signal from the outside air to the cochlear fluid is accompanied by a considerable pressure amplification. First, in the important middle range of frequencies from 500 Hz to 5000 Hz (which encompasses almost all speech tones), the resonance effect in the ear canal produces a pressure gain of about a factor of 2. Next, the leverage system of the ossicles produces a mechanical advantage of about 2, which results in a pressure gain of the same factor. Finally, the largest part of the amplification is due to the fact that the area of the oval window (3.2 mm^2, on the average) is about 20 times smaller than the area of the tympanic membrane (66 mm^2, on the average). The net effect is an overall pressure amplification in the system of about $2 \times 2 \times 20 = 80$, or perhaps slightly more (about 20 dB) in favorable circumstances.

When a sound signal is transmitted through the oval window to the cochlear fluid, a disturbance is propagated along the basilar membrane. This membrane is thin near the oval window and thickens along its entire length (Fig. 3). The thin regions respond best to high-frequency signals, whereas the thick regions are most responsive to signals with low frequencies. Consequently, the particular part of the basilar membrane that is stimulated by a sound wave depends on the frequency of that wave. The correlation between the wave frequency and position on the basilar membrane is indicated in Fig. 3. Thus, electrical signals that the auditory nerve sends to the brain indicate sound frequency in terms of the location of the stimulated fibers.

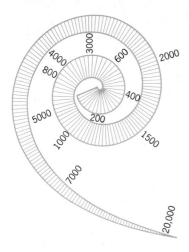

Figure 3 The basilar membrane thickens from about 0.04 mm at the oval window to about 0.5 mm at the apex. The numbers indicate the frequencies of the incident sound wave and are located at the positions of maximum response.

The response of the human ear to a sound wave depends both on the frequency and the intensity level of the sound. For example, the faintest 1000 Hz sound that a "normal" era can detect corresponds to an intensity level of 0 dB (or 10^{-12} W/m^2). On the other hand, this same ear will perceive a 100-Hz sound only if the intensity level is about 37 dB (5×10^{-9} W/m^2). A plot of the intensity levels for sounds of various frequencies that are barely audible results in the *threshold of hearing* curve shown in Fig. 4 and labeled 0 loudness level. This figure also shows the corresponding curves for other loudness levels. For example, if an individual listens to a 1000-Hz tone at an intensity level of 60 dB, he will experience the same sensation of loudness for a 100-Hz tone if that tone has an intensity level of about 71 dB. These two sounds are said to have the same *loudness level* and the value is chosen to be that of the 1000-Hz tone. Notice that the ear's response is much less frequency dependent as the threshold of feeling is approached.

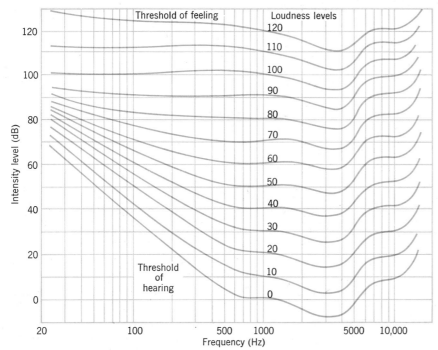

Figure 4 Loudness contours for pure tones. An intensity level of 0 dB corresponds to a power of 10^{-12} W/m^2.

References

E. Ackerman, *Biophysical Science,* Prentice-Hall, Inc., Englewood Cliffs, N.J., 1962, Chapter 1.

C. von Békésy, "The Ear," *Scientific American,* August 1957.

F. R. Hallett, P. A. Speight, and R. H. Stinson, *Introductory Biophysics,* Halsted (Wiley), New York, 1977, Chapter 2.

T. S. Littler, *The Physics of the Ear,* Macmillan, N.Y., 1965.

O. Stuhlman, Jr., *An Introduction to Biophysics,* Wiley, New York, 1943, Chapter VII.

A. L. Stanford, *Foundations of Biophysics,* Academic Press, New York, 1975, Chapter 6.

■Exercises

1. At the threshold of hearing, the ear can detect a 1000-Hz sound with an intensity level of 0 dB (Fig. 4). To what pressure variation on the ear drum does this correspond? Express the result in N/m^2 and also as a fraction of normal atmospheric pressure. (Refer to Problem 15.32.)

2. Use the formula given in Problem 15.31 to calculate the amplitude of excursion of the tympanic membrane for an incident 1000-Hz sound at the threshold of hearing (Fig. 4). Compare this result with the size of a typical air molecule (10^{-10} m).

3. Many high-fidelity systems have *loudness* controls which are used to alter the high- and low-frequency output levels depending on the overall level at which the music is played. Explain why this is desirable.

4. What is the power of a 60-Hz tone that has the same loudness as a 3000-Hz tone with an intensity level of 25 dB?

5. The scale of the graph in Fig. 15.20 is actually in *loudness units* instead of intensity level units. Use the information in Fig. 15.20 and Fig. 4 to estimate the pressure of a 5-kHz sound wave that will do physical damage to the ear drum. Compare this result with atmospheric pressure. (Refer to Problem 15.32.)

15.4 Diffraction and Interference

Plane Waves

The waves we have been discussing move along straight lines guided by strings or springs or pipes; such waves are called *one-dimensional waves*. There are also important types of wave motions that take place in two or three dimensions. Suppose we place a long, straight board in still water, with its longest dimension horizontal, so that part of the board is submerged. Now we oscillate the board back and forth with a regular periodic motion that is transverse to the board's longest dimension (Fig. 15.30). Each time the board moves, it piles up the water in front of it and pushes the water forward. As a result, a series of crests are propagated across the water with troughs between them. This type of movement constitutes a wave motion and the distance between successive crests (or troughs) is the wavelength of the disturbance. Waves that propagate in this manner are called two-dimensional *plane waves*.

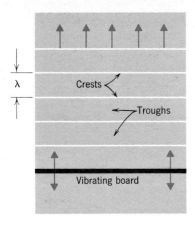

Figure 15.30 A vibrating board in an expanse of still water produces two-dimensional plane waves.

Huygen's Principle

Suppose that a plane wave of wavelength λ is incident on a panel into which is cut a slot of width d, where d is small compared to λ (see Fig. 15.31). Will the wave that emerges from the slot be confined to the narrow region of width d? The answer to this question is "no," because waves exhibit the phenomenon of *diffraction*. A convenient way to view this effect is to use a clever construction invented by the Dutch mathematical physicist, Christiaan Huygens (1629–1695), who first formulated the wave theory of light. According to *Huygen's principle*, the manner in which a wave front of arbitrary shape will advance can be determined by considering every point on a given wave front at any instant to be the source of a circular wave (or spherical wave in the case of three dimensions). Therefore, by drawing a series of circular waves emanating from a given wave front and then constructing the envelope of these waves, the shape and position of the entire wave at a later time can be found. Huygen's method is illustrated in Fig. 15.32. This type of construction is clearly correct for the cases of plane or circular waves (try the constructions), and it can be shown to be valid in general.

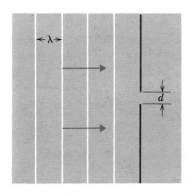

Figure 15.31 A plane wave is incident on a panel into which is cut a slot. Will the emerging waves be confined to the narrow region of width d?

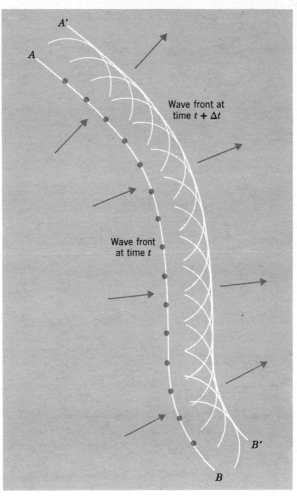

Figure 15.32 Huygens' construction for determining the advance of a wave front of arbitrary shape. Every point on the wave front AB at a time t is considered to be the source of a circular wave. By drawing these circular waves and constructing the envelope, the wave front $A'B'$ at time $t + \Delta t$ is determined.

We are now in a position to predict the form of the wave motion on the right-hand side of the panel in Fig. 15.31. Every point of the slot is to be considered as a source of circular waves propagating to the right.[1] If the slot width is small compared to the wavelength of the incident plane wave ($d \ll \lambda$), then the slot is essentially a point source. Therefore, to the right of the panel we would expect only circular waves emanating from the slot. Figure 15.33 shows the construction of the wave pattern according to Huygens' principle and a photograph of water waves that exhibit the predicted diffraction effect.

---

[1] A more detailed analysis is required to show that there are no backward-going waves even though the statement of Huygens' principle would allow such waves.

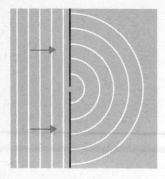 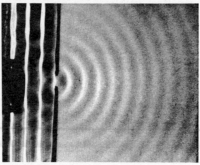

Figure 15.33 Pattern of waves produced by a plane wave incident on a slot. Even though the width of the slot is actually not much smaller than the wavelength of the incident wave, nevertheless, a circular wave pattern results. (The dark rectangle at the left of the photograph is a part of the mechanism that produces the plane wave.)

If the width of the slot is large compared to the wavelength of the incident disturbance ($d \gg \lambda$), then the Huygens' construction shows that the wave pattern to the right of the panel is essentially a plane wave, with curved ends (Fig. 15.34). Thus, the wave is largely unaffected by the presence of the panel but the circular wave effect persists near the edges of the slot.

Two-source Interference

By combining the principle of superposition with Huygens' principle, we are able to explain a variety of important and interesting *interference* effects that occur in wave motions of all sorts. Any wave motion in which the amplitudes of two or more waves combine will exhibit *interference*. Wave pulses of opposite signs that are traveling along a string will *cancel* when they pass (Figs. 15.8 and 15.9); we call this effect *destructive interference*. If the pulses are of the same sign, they will *add* when passing; this is *constructive interference*.

In two or three dimensions, a pattern of constructive and destructive interference will be developed throughout a region that supports the propagation of waves from two or more sources. For example, consider two sources of circular water waves of the same wavelength and separated by a certain distance. At certain definite positions on the surface of the water, the amplitudes of the waves from the two sources will have the same sign (the waves will be *in phase*) and constructive interference will result; that is, at these positions the disturbance of the water will be enhanced. At other positions the waves will have opposite signs (the waves will be *out of phase*) and destructive interference will result; that is, at these positions the water will remain calm. Figure 15.35 is a photograph of the interference pattern produced by two sets of circular water waves.

If we allow a plane wave to strike a panel in which there are two slots (each with $d \ll \lambda$), Huygens' principle tells us that these slots will act as separate sources of circular waves. Therefore, the

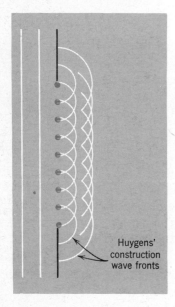 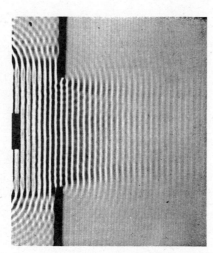

Huygens' construction wave fronts

Figure 15.34 If the width of the slot is large compared to the wavelength of the incident plane wave, the wave passes through the slot almost unaffected. There is a noticeable diffraction effect only near the edges of the slot.

Figure 15.35 The pattern of constructive and destructive interference produced by two sets of circular water waves.

situation should be exactly the same as in Fig. 15.35, and a similar interference pattern should result. Figure 15.36 shows the geometry of such a case and the photograph in Fig. 15.37 demonstrates the validity of Huygens' principle in predicting interference effects.

The Interference Condition

Two separate sets of circular waves emanate from a pair of source slits, S_1 and S_2, as in Fig. 15.38. How can we determine whether constructive or destructive interference will occur at a point P that is at a distance L_1 from S_1 and at a distance L_2 from S_2? The nature of the interference depends on whether

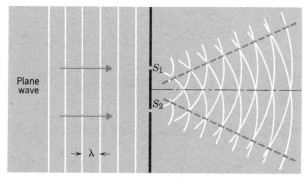

Figure 15.36 A plane wave is incident on a panel containing two slots (S_1 and S_2, each with $d \ll \lambda$). The circular waves emanating from these slots produce an interference pattern similar to that shown in Fig. 15.35. The dot-dash line is the central maximum and the dashed lines show the positions of secondary constructive interference.

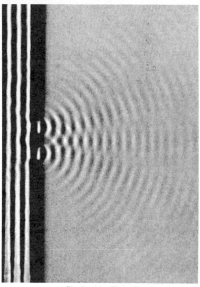

Figure 15.37 Photograph of the interference pattern in water waves produced in a geometry similar to that shown in Fig. 15.36.

the waves arrive at P *in phase* or *out of phase*. The wave that emanates from S_2 must travel a greater distance than the wave from S_1; this path difference is $L_2 - L_1$. Now, if there are exactly an integer number of wavelengths in the distance $L_2 - L_1$, the two waves will arrive at P with their maxima together (i.e., *in phase*), and constructive interference will result. (Refer to Fig. 15.36 and verify that at any point along the dashed lines the difference in the distances from S_1 and S_2 is always *one* wavelength; for the central maximum, the distances are equal.) Therefore, the condition for constructive interference is

$$\frac{L_2 - L_1}{\lambda} = N, \qquad N = 0, 1, 2, \ldots$$

(constructive) (15.27)

Similarly, if the distance $L_2 - L_1$ contains an *odd* number of *half* wavelengths (i.e., an integer number

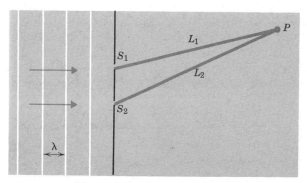

Figure 15.38 The condition for constructive interference at P is that $(L_2 - L_1)/\lambda$ is an integer number.

of wavelengths plus one-half wavelength), the waves will arrive at P with their maxima displaced from one another by $\frac{1}{2}\lambda$. Hence, the waves will be *out of phase* and destructive interference will result. This condition is

$$\frac{L_2 - L_1}{\lambda} = N + \tfrac{1}{2}, \qquad N = 0, 1, 2, \ldots$$

(destructive) (15.28)

If $L_2 - L_1$ is between $N\lambda$ and $(N + \tfrac{1}{2})\lambda$, the oscillations of the medium will have amplitudes smaller than the maximum amplitude.

Young's Double-slit Experiment

The first clear demonstration that light is a wave phenomenon (and therefore exhibits interference effects as mechanical waves are known to do) was made by the English physicist, Thomas Young (1773–1829), early in the 19th century. Young prepared a plane wave by allowing sunlight to pass through a small hole. This light was then directed toward a panel into which had been cut two narrow slots.[2] The light incident on the two slots was therefore in phase—that is, the slots were illuminated with *coherent* light. The experimental arrangement was the same as that shown in Figs. 15.37 and 15.38 except that the slots were much narrower. On a screen placed some distance away, Young found alternating bright and dark lines corresponding to the positions of constructive and destructive interference.[3]

We can account for the variation of the light intensity on the viewing screen in the following way. Refer to Fig. 15.39. Light waves with a definite wavelength λ are incident on the pair of narrow slits, A and B, which are separated by a distance D. The interference pattern is to be viewed on a screen that is a distance L away. (The diagram has been distorted for clarity; we must actually have L much greater than D.) At a position P on the screen, there will be a bright line, resulting from constructive interference, if the path difference l between the waves arriving from A and B is an integer number of wavelengths of the incident light. That is, for a bright line,

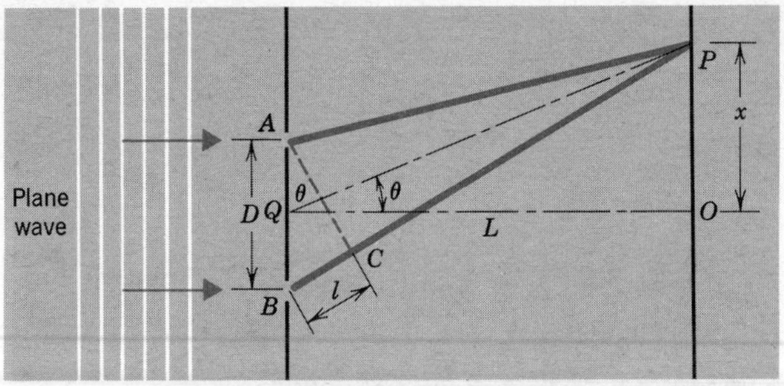

$$l = N\lambda, \qquad N = 0, 1, 2 \ldots$$

Now, if $L \gg D$, the angle $\angle ACB$ will be essentially a right angel and $\overline{AC}$ will be approximately perpendicular to $\overline{PQ}$; therefore, the angle $\angle BAC$ will be approximately equal to θ, the angle that specifies the position P on the screen. Consequently, the path difference l is given by

$$l \cong D \sin \theta$$

Therefore, bright lines on the screen will be observed at positions for which θ follows the relation

Bright lines (constructive interference):

$$N\lambda \cong D \sin \theta, \qquad N = 0, 1, 2 \ldots \quad (15.29)$$

There is a central maximum for $N = 0$ (that is, there is no path difference, $l = 0$) and there are secondary maxima on either side that correspond to larger values of N.

Dark lines in the interference pattern occur when l is equal to an odd number of half wavelengths. That is,

Dark lines: (destructive interference):

$$(N + \tfrac{1}{2})\lambda \cong D \sin \theta, \qquad N = 0, 1, 2 \ldots \quad (15.30)$$

Photographs of the interference patterns obtained for slits with three different separations are shown in Fig. 15.40.

Figure 15.39 The geometry of Young's double-slit experiment. The viewing screen is located at a distance that is large compared to the slit separation, i.e., $L \gg D$.

---

[2] For example, two pinholes in a sheet of black paper or two scribe marks on a smoked glass plate.

[3] Interference can be observed if sunlight (which consists of light with many frequencies) is used, but the effect is much more striking if monochromatic light (that is, light with a *single* frequency) is used. See Question 15.16 and Problem 15.49.

15.4 Diffraction and Interference

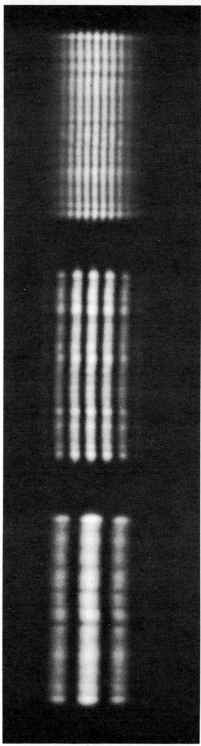

Figure 15.40 Double-slit interference patterns for slits with three different separations. Which case corresponds to the wide spacing and which to the narrow spacing?

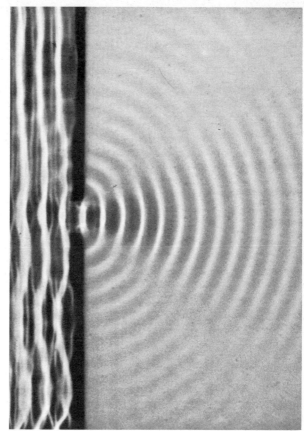

Figure 15.41 Interference pattern produced by water waves incident on a single slit. There is a broad central maximum with weaker secondary maxima on either side.

midpoint between the slits (Q) to P is approximately equal to L. Therefore, $\sin \theta \cong x/L$, and Eq. 15.29 becomes

$$x \cong \frac{N\lambda L}{D} \qquad (15.31)$$

Because N increases by unity in going from one bright line to the next, the spacing between adjacent bright lines (or dark lines) is

$$\Delta x \cong \frac{\lambda L}{D} \qquad (15.32)$$

● *Example 15.9*

The spacing between adjacent bright lines on the screen can be calculated as follows by referring to Fig. 15.39. Because $L \gg D$, the distance from the

In a double-slit experiment, $D = 0.1$ mm and $L = 1$ m. If yellow light is used, what will be the spacing between adjacent bright lines?

The wavelength of yellow light is approximately 6×10^{-7} m (see Section 15.5). Therefore, the

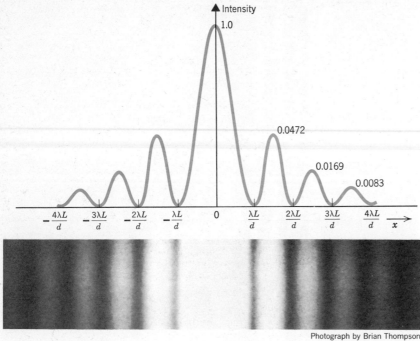

Figure 15.42 The diffraction pattern produced by a single slit. The photograph has been overexposed in order to reveal the secondary maxima; therefore, the central maximum is "washed out" and does not appear in its full intensity. The intensity graph at the top is only schematic; the secondary maxima have the peak intensities indicated (relative to an intensity of 1.0 for the central maximum).

spacing is

$$\Delta x \cong \frac{\lambda L}{D}$$

$$= \frac{(6 \times 10^{-7} \text{ m}) \times (1 \text{ m})}{10^{-4} \text{ m}}$$

$$= 0.006 \text{ m} = 6 \text{ mm}$$

Thus, the spacing between lines is about $\frac{1}{4}$ of an inch. ∎

Single-slit Diffraction

If we allow a plane wave to illuminate a *single* slit of narrow width, we also find a regular pattern of constructive and destructive interference. Figure 15.41 shows a photograph of water waves incident on a slit which has a width only slightly larger than the wavelength of the incident wave. It is apparent that there is a broad central maximum in the pattern of transmitted waves. Closer examination will reveal that there are also weak secondary maxima with interference minima between them.

Light waves exhibit a similar effect. Figure 15.42 shows a photograph of the diffraction pattern on a screen located behind a single slit illuminated with light. This photograph clearly shows that the central maximum is considerably broader than the secondary maxima that lie on either side. We can locate the positions of the interference minima by using the following argument. In Fig. 15.43 we see a plane wave incident on a slit that has a width d. Each point along the slit opening can be considered to be a source of outgoing waves. We identify a few of these points with dots in Fig. 15.43 and number them from 1 (just inside the top of the opening) to 14

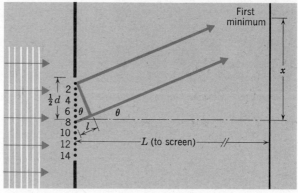

Figure 15.43 Geometry for the discussion of the single-slit interference pattern.

(just inside the bottom of the opening). The middle of the opening lies between points 7 and 8. The screen on which the interference pattern is to be observed is at a distance L that is very large compared to d. Therefore, the outgoing rays that emanate from any point in the opening and strike a particular point on the screen are essentially parallel.

In Fig. 15.43 we have two parallel rays, one leaving point 1 in the opening and the other leaving point 8. If we choose the angle θ to correspond to the first interference minimum above the central maximum on the screen, then the difference in path length l between the pair of rays must equal one-half wavelength of the incident radiation:

$$l = \tfrac{1}{2}\lambda$$

If the rays from points 1 and 8 arrive at the screen out of phase, then so do all other pairs of rays from similar points in the opening: 2 and 9, 3 and 10, 4 and 11, ... and 7 and 14. Thus, all of the rays cancel in pairs and the angle θ corresponds to the direction of an interference minimum.

Now, in Fig. 15.43 we also see that

$$l \cong \tfrac{1}{2} d \sin\theta$$

(The points in the opening are spaced very closely, so the error in this expression of one-half of a point-to-point interval is not important.) The distance x along the screen from the central maximum to the first minimum is

$$x = L \tan\theta$$

Because $L \gg d$, θ is a small angle, so that $\sin\theta \cong \tan\theta$. Then, combining the three equations we find

$$x \cong \frac{\lambda L}{d} \qquad (15.33)$$

It is easy to see that successive interference minima will be located at multiples of $\lambda L/d$ from the central maximum. This results in the central maximum having a "width" about twice that of each secondary maxima (see Fig. 15.42). Notice also that the secondary maxima do not have symmetric shapes, so the positions of maximum intensity are not located midway between the intensity minima (except for the central maximum).

Multiple-slit Diffraction

We have been considering the effects of the interference and diffraction of waves (including light) that have a single wavelength λ. Many types of sources (including light sources) produce waves with a mixture of wavelengths. We can use interference effects to measure the various wavelengths from a particular source; that is, we can analyze the wavelength *spectrum* by measuring the positions of interference maxima and minima. The difficulty with using single or double slits for this purpose is that the various intensity maxima for a given frequency of light are relatively broad so that the maxima for one frequency blend into those for another, not-too-different frequency. That is, the ability of a single or double slit to distinguish between (or to *resolve*) two nearby frequencies is poor. This defect can be overcome by using a diffraction panel into which are cut many slits; as the number of slits is increased, the resolution improves. Figure 15.44 shows a compari-

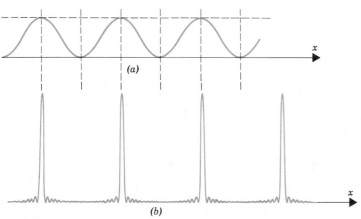

Figure 15.44 Intensity patterns for light of a single frequency passed through *(a)* a double slit and *(b)* a diffraction grating consisting of 20 slits.

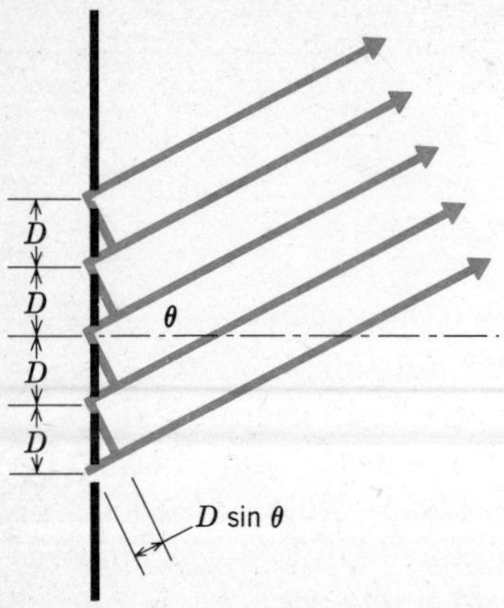

Figure 15.45 A multiple slit or diffraction grating. In order for constructive interference to result, the difference in path length $D \sin \theta$ between the waves from any pair of slits must be an integer number of wavelengths. This is the same condition as that for a double slit (Eq. 15.29).

$$D \sin \theta = N\lambda \qquad N = 0, 1, 2, \ldots \qquad (15.34)$$

where D is the distance between successive slits in the grating and is usually called the *grating spacing*. The quantity N is called the *order* of the maximum. The interference maxima appear on the screen at different locations depending on the value of N.

son of the intensity maxima for a double slit and for a set of 20 slits. There is a dramatic increase in the sharpness of the lines in the latter case.

If the slit system consists of a very large number of slits, it is usually referred to as a *diffraction grating*. Such gratings can be prepared, for example, by scratching fine lines through the emulsion of a photographic film. The film is then developed and mounted between pieces of glass; the unscratched portion of the film is dark and the scratches are clear, forming the series of slits.

The condition for intensity maxima in the light that passes through a diffraction grating is the same as that for a double slit. (This can be seen in Fig. 15.45.) Thus,

● **Example 15.10**

Orange light has a wavelength of 6.1×10^{-7} m and green light has a wavelength of 5.4×10^{-7} m (see Table 15.3). If orange and green light are passed through a diffraction grating that has 4000 lines/cm, what will be the angular spacing between the orange and green maxima (a) in first order and (b) in third order?

The quantity D that we need in Eq. 15.34 is

$$D = \frac{1}{4000 \text{ cm}^{-1}} = \frac{1}{4 \times 10^5 \text{ m}^{-1}} = 2.5 \times 10^{-6} \text{ m}$$

(a) In first order,

$$\sin \theta_1 = \frac{\lambda}{D}$$

so that

$$\theta_1 \text{ (orange)} = \sin^{-1}\left(\frac{6.1 \times 10^{-7} \text{ m}}{2.5 \times 10^{-6} \text{ m}}\right) = 14.12°$$

$$\theta_1 \text{ (green)} = \sin^{-1}\left(\frac{5.4 \times 10^{-7} \text{ m}}{2.5 \times 10^{-6} \text{ m}}\right) = 12.47°$$

Table 15.3 Some Typical Radiations in and near the Visible Spectrum

Name	Wavelength (m)	(Å)[a]	(nm)[b]	Frequency (Hz)
Near infrared	1.0×10^{-6}	10,000	1,000	3.0×10^{14}
Longest visible red	7.6×10^{-7}	7,600	760	3.9×10^{14}
Orange	6.1×10^{-7}	6,100	610	4.9×10^{14}
Yellow	5.9×10^{-7}	5,900	590	5.1×10^{14}
Green	5.4×10^{-7}	5,400	540	5.6×10^{14}
Blue	4.6×10^{-7}	4,600	460	6.5×10^{14}
Shortest visible blue	4.0×10^{-7}	4,000	400	7.5×10^{14}
Near ultraviolet	3.0×10^{-7}	3,000	300	1.0×10^{15}
X ray (long wavelength)	3.0×10^{-9}	30	3	1.0×10^{17}
X ray (short wavelength)	1.0×10^{-11}	0.1	0.01	3.0×10^{18}

[a] The *meter* is the standard unit of length in the SI system. However, a few non-SI units are so frequently used that the SI Commission considers them acceptable. The *angstrom* (Å) is so commonly used in optical physics that it is an "acceptable" unit.
[b] The *nanometer* (1 nm = 10^{-9} m = 10^{-1} Å) is a unit gaining wide use.

Thus,

$$\Delta\theta_1 = \theta_1 \text{ (orange)} - \theta_1 \text{ (green)}$$
$$= 14.12° - 12.47° = 1.65°$$

(b) In third order,

$$\sin\theta_3 = \frac{3\lambda}{D}$$

so that

$$\theta_1 \text{ (orange)} = \sin^{-1}\left(\frac{3 \times 6.1-10^{-7} \text{ m}}{2.5 \times 10^{-6} \text{ m}}\right) = 47.05°$$

$$\theta_3 \text{ (green)} = \sin^{-1}\left(\frac{3 \times 5.4 \times 10^{-7} \text{ m}}{2.5 \times 10^{-6} \text{ m}}\right) = 40.39°$$

Thus,

$$\Delta\theta_3 = \theta_3 \text{ (orange)} - \theta_3 \text{ (green)}$$
$$= 47.05° - 40.39° = 6.66°$$

Notice that the angular spacing (the *dispersion*) increases with the order. For high resolution measurements, an experimenter will usually choose to use the highest order that is practical. ∎

Our discussion of the interference effects produced by single, double, and multiple slits has been phrased primarily in terms of *light* waves. We must remember, however, that these effects are in no way unique to light waves—diffraction and interference effects are common features of *all* types of wave motion. The important characteristic of a wave that determines the extent to which it will exhibit diffraction and interference is its *wavelength*. The effects are most pronounced when the openings through which the wave must pass have sizes and spacings that are *comparable* with the wavelength of the particular wave. Thus, an optical diffraction grating is completely unsuited for studying interference effects of sound waves. But a wall into which are cut a number of slots with separations of 10 cm or 20 cm would act as a diffraction grating for middle-frequency sound waves.

15.5 Electromagnetic Radiation

Energy Transfer by Accelerated Charges

From the discussions in Chapter 13, we know that a steady current flowing in a wire will produce a static (that is, unchanging) magnetic field in the vicinity of the wire. If a charged particle moves in such a field, the force exerted on the particle by the field is always at right angles to the direction of motion of the particle. Therefore, although the field can accelerate the particle (by causing it to move in a circular orbit), the *speed* of the particle is not changed. A static magnetic field can do no *work* on the particle. That is, there can be no *energy* transferred to the particle from the moving charges in the wire via the intermediary of the field.

Now let us consider the case in which the current in the wire is allowed to vary so that the fields are no longer static. In Section 13.7 we learned that a changing magnetic flux induces an electric field (an EMF) to which charged particles will respond and from which they can gain energy. Therefore, energy can be transferred from a wire carrying a changing current (changing flux) to a charged particle in the vicinity of the wire, although no such transfer can occur for the case of a wire carrying a steady current (static flux). The essential difference in the two situations is that the charges that move in the wire in the case of the changing current undergo accelerations, whereas there is no acceleration of the moving charges in the case of the steady current. *Only accelerating charges can produce energy transfers through the electromagnetic fields that they generate.*

Electromagnetic Waves

In Section 14.4 we discussed ways of producing alternating current. Sinusoidal current variations clearly constitute a case of accelerated motion of the charges in the wire and can therefore lead to the transfer of energy to charged particles outside the wire.

The generation of an electromagnetic field by the changing current in a wire antenna is similar to the production of a traveling wave in a string by applying an oscillation to one end, as in Fig. 15.5.

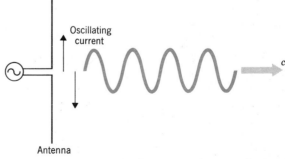

Figure 15.46 The sinusoidal variation of current flowing in a wire produces sinusoidal variations in **E** and **B** which are propagated through space with the velocity c. Here, the source of the electromagnetic wave is a simple antenna.

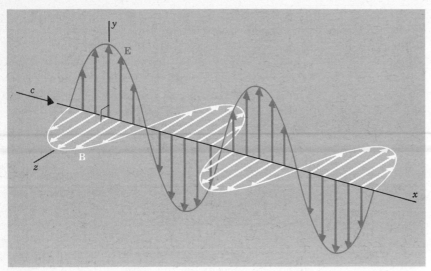

Figure 15.47 At a large distance from the antenna (located in the negative x-direction), the variation of **E** and **B** with distance at a given instant is sinusoidal. Furthermore, the electromagnetic wave is a plane wave and has $\mathbf{E} \perp \mathbf{B}$ at every point and for all values of the time.

Just as a sinusoidal displacement in the string is propagated along the string, the sinusoidal variation of the field vectors, **E** and **B**, is propagated through space. This is shown schematically in Fig. 15.46.

The similarity between the mechanical and electromagnetic cases goes further. For the mechanical wave motion shown in Fig. 15.6, the displacement as a function of time at any position is sinusoidal, and the displacement as a function of distance at any instant of time is also sinusoidal. The same is true for the electromagnetic case. In an electromagnetic wave there is a sinusoidal variation in space and in time of the electromagnetic field quantities, **E** and **B**, as shown in Fig. 15.47.

In a region of space that is far from the antenna (located in the negative x-direction in Fig. 15.47), and at a given instant of time, there will be essentially no variation of the field vectors over the y-z plane. That is, the condition of the electromagnetic field is uniform over a plane that is perpendicular to the direction of propagation of the wave away from the antenna.[4] The propagating electromagnetic wave is therefore a *plane wave*. Electromagnetic waves have the following important properties:

1. The electromagnetic field vectors, **E** and **B**, in a *plane wave* are everywhere mutually perpendicular:

---
[4]But this uniformity exists only at large distances from the wire that carries the changing current; near the wire the electromagnetic field is quite complicated.

$$\mathbf{E} \perp \mathbf{B} \quad \text{(plane wave)} \quad (15.35)$$

2. The direction of propagation of an electromagnetic wave is given by a right-hand rule: the direction of propagation is the same as the direction of advance of a right-hand screw when turned in the sense that carries **E** into **B** (Fig. 15.48). That is, the direction of propagation of the wave is the same as the direction of $\mathbf{E} \times \mathbf{B}$. Even though the directions of the vectors **E** and **B** change with time and with position, the wave always propagates in the same direction. Application of the right-hand rule to the various portions of the wave in Fig. 15.47 shows this to be the case.

3. Electromagnetic waves are *transverse waves*. In Fig. 15.47 notice that the vectors **E** and **B** are always *perpendicular* to the direction of propagation of the wave; at large distances from the source, the field vectors never have any component in the direction of propagation.

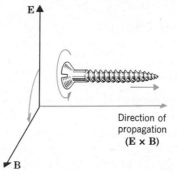

Figure 15.48 The direction of propagation of an electromagnetic wave is the same as the direction of advance of a right-hand screw when it is turned in the sense that carries **E** into **B**. This is the direction of $\mathbf{E} \times \mathbf{B}$.

4. The velocity of propagation in empty space of all types of electromagnetic waves (light, radio waves, X rays, etc.) is $c = 3 \times 10^8$ m/s. (We will discuss the velocity of light in more detail in Section 16.1 and again in Section 17.1.)

5. Electromagnetic waves carry *energy* and *momentum*. The electric and magnetic field amplitudes for an electromagnetic plane wave have a fixed relationship to each other, that is $|\mathbf{B}|$ is proportional to $|\mathbf{E}|$. The energy and momentum carried by the wave is proportional to $\mathbf{E} \times \mathbf{B}$ and hence, proportional to E^2 in magnitude. Thus, the intensity of an electromagnetic wave is proportional to the square of the electric field strength.

Electromagnetic Energy and Momentum

When electric charges undergo accelerations, a time-varying electromagnetic field is produced and electromagnetic waves are propagated outward from the source. It is easy to see that this *electromagnetic radiation* carries both *energy* and *momentum*. Consider a charged particle ($+q$) that is in the path of an electromagnetic wave (Fig. 15.49). If the particle is initially at rest, the magnetic field **B** will have no influence on the particle, but the electric field **E** will exert a force on the particle and will give it an acceleration in the direction of **E**. As soon as the particle has acquired a velocity in this direction, the magnetic field will exert a force on the particle. Application of the right-hand rule for the magnetic force ($\mathbf{F} \propto \mathbf{v} \times \mathbf{B}$) shows that this force is in the direction of propagation of the wave. Furthermore, the force will be in the same direction during the next halfcycle of the wave when the directions of **E** and **B** are both reversed. (Try the right-hand rule again.) Now, it is true that the force exerted on the particle by the electric field is perpendicular to the direction of propagation of the wave, but the direction of this force reverses every half-cycle and thus cancels when averaged over a complete cycle. The magnetic force, on the other hand, is always in the same direction. Therefore, the average force on the particle is the magnetic force. If a charged particle, initially at rest, can be given an acceleration by an electromagnetic wave, then both energy and momentum have been transferred to the particle from the wave. We must conclude that electromagnetic waves carry both energy and momentum.

The Spectrum of Electromagnetic Radiation

In 1862, James Clerk Maxwell, predicted on the basis of his theory of electromagnetism that electromagnetic *waves* should exist. His calculations showed that these electromagnetic waves should propagate with a velocity that was the same as that previously found for the propagation of light in air (or empty space). This fact immediately suggested that light is just a particular form of an electromagnetic wave. Electromagnetic waves (apart from *light*) were not observed until 1887 when Heinrich Hertz produced waves with wavelengths of 10–100 m by forcing a charged sphere to spark to a grounded sphere. Hertz's experiments indicated that these waves were identical in every respect to light waves, except that the wavelength was much longer.

Following the pioneering experiment of Hertz, electromagnetic waves were generated with an ever-increasing range of frequencies. Indeed, it seemed that waves with *any* frequency could be produced if some method of driving electric charges with the appropriate oscillation frequency could be found. This is, in fact, the situation today: there appears to be no physical limitation on the frequency of electromagnetic waves—all we require is a suitable source. Electronic methods have been used to generate electromagnetic waves with frequencies up to about 10^{12} Hz ($\lambda = 0.3$ mm). In this range of frequencies we classify the radiation as either *radiofrequency* (RF) waves or as *microwaves* (see Fig. 15.50). In the former category are the standard broadcast, FM, TV, air and marine, and amateur broadcast bands. Radar and point-to-point relay signaling use microwaves.

In order to generate radiation with frequencies above the microwave range, direct electronic methods are no longer useful, and we employ *atomic* radiations. *Infrared* or *heat radiation* lies in the frequency range between microwaves and the narrow band of frequencies that constitute *visible* radiation. At still higher frequencies are *ultraviolet* radiation and *X rays*. The limit on the frequency that can be generated by atomic systems lies near 10^{20} Hz; radi-

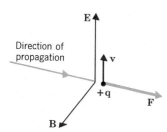

Figure 15.49 An electromagnetic wave always exerts an average force on a charged particle that is in the direction of propagation of the wave. (Does the direction of the force depend on the sign of the charge?)

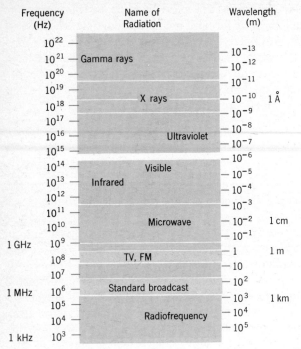

Figure 15.50 A part of the spectrum of electromagnetic radiation. The classification terms used for electromagnetic radiations are not well-defined; there is actually considerable overlapping of adjacent categories.

ation with higher frequencies (*gamma rays*) are produced within *nuclei* and by interactions involving very high-speed particles.

The fact that names have been assigned to these various bands of radiation must not obscure the essential feature of electromagnetic waves, namely, that all of these radiations are *identical* in character and differ only in their *frequency*. Thus, the radio waves that are broadcast from a 100-m radio tower are exactly the same as the penetrating gamma rays that originate in a nucleus whose diameter is only 10^{-14} m; the only difference between the two radiations is a factor of 10^{15} or so in frequency. The way in which an electromagnetic wave interacts with matter *does* depend on its frequency, even though its fundamental characteristics are the same. Thus, the eye is sensitive only to visible light, whereas the skin can sense heat radiation. Radio waves will be stopped by a thin sheet of metal, but X rays and gamma rays will pass through. The great variety of ways that electromagnetic waves interact with matter makes these radiations extraordinarily useful.

Summary of Important Ideas

A wave propagates along a string or through a medium with a speed that depends on the *elastic properties* of the medium.

The *speed* v of a wave in any medium is related to the *wavelength* λ and the *frequency* ν according to $v = \lambda \nu$.

A medium that is set into motion by the vibrations of a simple harmonic oscillator will support *traveling waves* that vary *sinusoidally in time and space*.

Waves in a medium are *transverse* or *longitudinal* depending on whether the elastic properties of the medium will tend to restore transverse or longitudinal displacements. Waves on strings are transverse; waves in liquids and gases are longitudinal; waves in solids can usually be both transverse and longitudinal; electromagnetic waves are transverse.

Waves of all types obey the principle of *superposition*.

Standing waves are produced by the successive reflections of the waves from the ends of the medium when the length L of the medium and the wavelength λ of the wave are related by $n(\lambda/2) = L$, where n is an integer. The standing wave corresponding to $n = 1$ is called the *fundamental*. The $n = 2$ wave is the *second harmonic* (or *first overtone*), and so forth.

The *speed of sound* in a gas is equal to $v = \sqrt{\gamma P/\rho}$, where $\gamma = c_p/c_v$ is the ratio of specific heats for the gas, P is the pressure, and ρ is the density. For air at 0°C, $v = 332$ m/s (about 1100 ft/s).

Beats are the slow variations in amplitude of a wave that consists of the superposition of two waves with nearly equal frequencies. The *beat frequency* is equal to the difference in frequency between the combining waves.

The *loudness* or *intensity* of sound is measured on a logarithmic *decibel* (dB) scale. The difference in the sound intensity level $\Delta\beta$ between two sounds with intensities I_1 and I_2 is $\Delta\beta = 10 \log(I_2/I_1)$ dB. The response of the human ear spans a range of about 120 dB between the threshold of hearing (10^{-12} W/m^2) and the threshold of pain (1 W/m^2).

The standing waves that can be set up in the air column in a pipe closed at one end correspond to the fundamental and the *odd harmonics*. (There must be a molecular displacement *node* at the closed end and an *antinode* at the open end.) The standing waves that can be set up in the air column in a pipe open at both ends correspond to the fundamental and *all* of the harmonics. (There must be displacement antinodes at both ends of the pipe.)

The observed frequency of a sound wave depends on the relative motion of both the source and the observer with respect to the medium. All types of wave motion exhibit this *Doppler effect*. (The Doppler effect for electromagnetic radiation is more complex than that for sound.)

If the source of a sound wave moves through a medium with a speed in excess of the speed of sound in that medium, a *shock wave* (or *sonic boom*) is produced.

The propagation of a wave in space can be determined by considering every point on a given wave front to be the source of out-going spherical waves (*Huygens' principle*).

Diffraction causes waves to "bend around" obstacles and to spread out upon passing through narrow apertures.

When waves from two (or more) sources arrive at a particular point, *constructive interference* results if the waves are *in phase* and *destructive interference* results if the waves are *out of phase*.

Electromagnetic waves are produced only by electric charges that are undergoing *accelerations*.

In a *plane electromagnetic wave* the field vectors, **E** and **B**, are mutually perpendicular.

An electromagnetic wave possesses and can transfer both energy and momentum. The intensity of the wave is proportional to the square of the electric field strength of the wave.

◆ Questions

15.1 A wave pulse is sent down a long string. Is there any way that a second pulse, started a certain time after the first pulse, can be made to overtake the first pulse (before reflection from the other end occurs)?

15.2 A transverse pulse is moving along an elastic cord. You wish to increase the speed of the pulse, so you increase the tension by pulling on the end at which the pulse was started. Explain what happens.

15.3 Transverse waves are possible in solids but not in gases. What is the physical difference in the two situations that accounts for this fact?

15.4 All sound waves are compressional waves, but not all compressional waves are sound waves. Why?

15.5 Explain how the string in Fig. 15.9c, which has no displacement at any position, can have a displacement at a later time. Describe the physical situation that allows a displacement to be generated from a condition of no displacement.

15.6 Figure 15.10 shows the reflection of a pulse from the terminated end of a string. Suppose, instead, that the end of the string is attached to a small ring (with negligible mass) and that this ring can slide frictionlessly up and down on a vertical pole. Describe the reflection of a wave pulse from this *free end* of the string.

15.7 Explain why small rooms are not generally suited for listening to high-fidelity music.

15.8 What do you think would be your bodily reaction to an intense sound wave of wavelength $\lambda = 50$? (What is the frequency of the wave?)

15.9 Sound waves do not propagate for great distances through air; eventually, the waves "die out." What happens to the *energy* in the sound wave?

15.10 Compare the average velocity of molecules in air (Table 10.1) with the velocity of sound in air. Why should these two numbers be so close? Why should the sound velocity be *less* than the molecular velocity? (Consider how sound waves are propagated through air.)

15.11 Some swimming pools are equipped with underwater speakers so that swimmers can continue to listen to their favorite music even while submerged. Is it necessary to adjust the frequencies of the sounds emitted by these speakers in order to take account of the difference in the speed of sound in water compared to that in air? What will be different about the sound waves in water?

15.12 According to Eq. 15.11, the speed of sound in a gas depends on the temperature of the gas. Explain physically why this is reasonable.

15.13 A violin string is set into vibration by the slow, steady pull of the bow across the string. A piano string, on the other hand, is set into vibration by a sharp, hard blow. Do you expect the frequency spectrum for a piano A note to be more or less complex than that for a violin (shown in Fig. 15.18b)?

15.14 A pianist and a violinist play the same note on their instruments and notice a beat between the sounds with a frequency of 2.5 Hz. How can they determine whose instrument is producing the higher-frequency sound?

15.15 Is there a Doppler effect when the sound source moves perpendicular to the line connecting the source and the observer?

15.16 Suppose that a double slit is illuminated with *white light* (that is, light containing *all* frequencies) instead of *monochromatic light* (light of a *single* frequency). Describe the interference pattern that would be produced.

15.17 A sound wave in air and an electromagnetic wave in air each have a wavelength of 10 cm. Classify the two waves (audible, inaudible; light, radio wave, etc.). Why are two waves of the same wavelength so different in their properties?

15.18 The Sun radiates huge amounts of electromagnetic energy. We know that such radiation carries momentum. Does the momentum of the Sun change with time? Explain.

★Problems

Section 15.1

15.1 An elastic cord has a mass per unit length of 0.050 kg/m. (a) With what speed will waves propagate along this cord when the tension is 80 N? (b) If the tension is increased to 120 N, what will be the new wave speed?

15.2 A solid rubber "rope" has a density ρ and a cross-sectional area A. When the rope is under a tension T, waves propagate along the rope with a velocity v. The rope is now split in half; ρ remains the same for each section but A is halved. What tension must now be applied to each section in order to have the same wave velocity as before? Explain why your result is physically reasonable.

15.3 An elastic cord has a density $\rho = 0.8$ g/cm^3 and a cross-sectional area of 0.5 cm^2. When the cord is subjected to a tension of 100 N and one end is vibrated up and down with a frequency of 20 Hz, what is the wavelength of the waves that propagate along the cord?

15.4 A siren operates by compressed air being blown through a series of holes that are drilled in the periphery of a disc that is rotated at high speed. Suppose that 400 holes are equally spaced at a distance of 8 cm from the center of the disc and that the disc is rotated at a rate of 240 rev/min. What is the frequency of the siren tone?

15.5* The wave equation for the displacement $y(x,t)$ on a string is given by

$$y(x,t) = 1.5 \cos (x/s - 4t) \text{ cm}$$

where x is in centimeters and t in seconds. What is the amplitude of the wave? What is the wavelength? What is the wave velocity? In what direction along the x-axis is the wave traveling?

Section 15.2

15.6 Waves travel with a velocity of 20 m/s on a certain taut string, which is attached to two supports that are separated by 2 m. What are the frequencies of the first four standing waves (starting with the longest wavelength) that can be set up in the string? Which of these modes of vibration will produce an audible sound?

15.7 A string has a length of 80 cm and a mass of 2 g. When this string is stretched between two fixed supports, what must be the tension so that the fundamental vibration has a frequency of 50 Hz?

15.8 A string stretched between two supports is vibrating in its second harmonic mode when the tension is 12 N. What tension is required to excite the third harmonic mode?

15.9 A violin string has a length of 36 cm and a mass of 0.42 g. If the string tension is 90 N, what are the lowest three harmonics?

15.10* A string on a cello has a length of 70 cm and the tension is adjusted to sound concert A ($\nu_A = 440$ Hz) as the fundamental. By what amount must the length be shortened by "fingering," with the original tension, to change the fundamental to C' ($\nu_{C'} = 523.3$ Hz)?

15.11* A metal rod of length L is clamped at one end and then is set into transverse vibration. Sketch the motion of the rod for the two modes of vibration with the lowest frequencies. (Hint: Is there a node or an antinode at the free end of the rod?)

Section 15.3

15.12 A flash of lightning is observed and 8 s later a clap of thunder is heard. About how far away was the lightning?

15.13* A stick of dynamite is exploded on the surface of the sea ($T = 25°$C). The sound is propagated through the water as well as the air ($T = 25°$C). At a distance of 3 km, which signal will be heard first? What will be the time interval between the arrival of the two signals?

15.14 An organ note ($\lambda = 11$ m) is sustained for 1 s. How many full vibrations of the wave have been emitted? (Assume $T = 0°C$.)

15.15 At a height of 90 km above the Earth's surface, the atmosphere is composed primarily of oxygen and the temperature is approximately $-90°C$. What is the speed of sound under these conditions?

15.16 In the stratosphere (the region from about 10 km to about 45 km above the Earth's surface) the minimum speed of sound is approximately 296 m/s. What is the minimum temperature in the stratosphere? (Assume that the composition of the air in the stratosphere is the same as that near the surface of the Earth.)

15.17 What is the speed of sound in helium at $20°C$? (Use Table 9.6.)

15.18 Use the information in Table 11.1 and determine the speed of sound in water.

15.19 Compare the rate of change of the speed of sound with temperature (m/s-deg) for hydrogen gas and air near $0°C$. What property of the gases accounts for the difference?

15.20 Use dimensional analysis to derive Eq. 15.9.

15.21 Use the ideal gas law to convert Eq. 15.10 into Eq. 15.11.

15.22 What is the wavelength of a 10-kHz sound wave in a bar of iron? What would be the wavelength of the same wave in air?

15.23 What is the wavelength of a 4 kHz sound wave propagating through a rolled copper bar? (Use Table 15.1.)

15.24 What is Young's modulus for rolled lead? (Use Table 15.1.)

15.25 An aircraft model in a wind tunnel is studied by forcing carbon dioxide at $100°C$ past the model at a velocity of 400 m/s. What is the Mach number for this situation?

15.26* Sketch a sinusoidal wave which has a wavelength λ and another of the same amplitude which has a wavelength $\lambda/4$. Add the two waves graphically to produce the wave that results from the superposition of one on the other and show that a "modulated" wave is obtained.

15.27 A piano tuner listens to the beat between the A key and his 440 Hz standard. He hears a beat frequency of 4 Hz. What are the possible frequencies for the piano note? He now slowly *tightens* the string and hears a beat frequency of 3 Hz, then 2 Hz. What was the original piano frequency? Should the piano tuner now tighten the string more or loosen it?

15.28 Zero decibels corresponds to a wave intensity of 10^{-12} W/m². What is the sound intensity level in dB of a sound whose intensity is 6×10^{-8} W/m²?

15.29 What is the intensity in W/m² of a sound wave whose intensity level is 67 dB?

15.30* An explosive charge is detonated at a height of several kilometers in the atmosphere. At a distance of 400 m from the explosion, the sound intensity is 10 W/m². Assuming that the atmosphere is perfectly homogeneous, what will be the sound intensity level (in dB) at a distance of 4 km from the source? (In passing through air, sound is absorbed at a rate of approximately 7 dB/km.)

15.31* The intensity I (in W/m²) of a sound wave is given by the expression, $I = 2\pi^2 \rho v \nu^2 A^2$, where A is the amplitude. Calculate the maximum excursion of air molecules in a 1-kHz sound wave in air that is just painful (120 dB).

15.32 The *overpressure* produced by a sound wave (i.e., the maximum pressure excursion above normal atmospheric pressure) is given by $\Delta P = 2\pi \rho v \nu A$. Use the expression for I given in the preceding problem and find the overpressure for a sound wave in air with an intensity of 120 dB. (You do not need the result of Problem 15.31 in order to calculate the overpressure; solve the equation in Problem 15.31 for the amplitude A and substitute into the expression for ΔP.)

15.33 Figures 15.22 and 15.23 represent the amplitudes of motion of the air molecules in the pipes. Construct a similar set of diagrams that represent the amplitudes of the *pressure* variations.

15.34 A pipe that is open at both ends has a length of 11 m. What is the series of frequencies that will cause the air column in this pipe to resonate? Now, suppose that one end of the pipe is closed. What is the new series of resonant frequencies?

15.35 A pipe that is open at both ends will resonate with a certain series of frequencies. Which of these frequencies are the same as those that would result if one end of the pipe were closed?

15.36 The temperature of the air is a factor in determining the pitch of an organ pipe. Consider a pipe that is closed at one end and is designed to produce a fundamental frequency of 528 Hz when the air temperature is 20°C. By how much will the pitch change if the temperature rises (during the summer) to 30°C?

15.37 The first overtone produced by a pipe of length L_1 that is open at both ends has the same frequency as the first overtone produced by a pipe of length L_2 that is closed at one end. What is the ratio L_1/L_2 of the lengths of the pipes?

15.38* The speed of sound in air is measured by placing a tuning fork whose natural frequency is 400 Hz over a pipe whose lower end is closed by a movable piston (as in Fig. 15.21). Resonance is obtained when the length of open pipe is 63.75 cm. The piston is slowly lowered, and the next resonance is found when the length is 106.25 cm. What is the speed of sound deduced from these observations?

15.39 As you stand by a railroad track, a train approaches you at a speed of 60 mi/h and its whistle emits a 2-kHz note. What frequency do you hear? When the train passes, what frequency do you hear?

15.40 As a train passes you, you hear the frequency of its whistle drop from 1000 Hz to 800 Hz. What is the speed of the train in mi/h?

15.41 Follow the method used to obtain Eq. 15.24 and derive Eq. 15.25.

15.42* A supersonic aircraft is flying parallel to the ground. When the aircraft is directly overhead, an observer sees a rocket fired from the aircraft. Ten seconds later the observer hears the sonic boom, followed 2.8 s later by the sound of the rocket engine. What is the Mach number of the aircraft?

Section 15.4

15.43 Show by means of constructions that Huygens' principle yields the correct results for the propagation of *plane* waves and *circular* waves (in *two* dimensions).

15.44 A measurement will show that the waves in Fig. 15.35 have a wavelength that is $\frac{1}{3}$ of the distance between the two sources. Use a compass and construct two sets of circular waves that duplicate the conditions of Fig. 15.37. Show that the regions of constructive and destructive interference are the same as those in the figure.

15.45 Two equal strength microwave sources with $\lambda = 40$ cm are driven in phase (i.e. both sources reach maximum positive amplitudes at the same time). What are the possible distances separating them if there is no radiation emitted along the line joining them? When this condition is met, what is the intensity along the perpendicular bisector to the line connecting the sources relative to the intensity from one source alone? (These conditions exist for a so called *broadside* antenna arrangement.)

15.46 Two sources of waves with $\lambda = 2$ m are separated by a distance of 6 m. Along what lines (measured relative to the line perpendicular to the line connecting the two sources, refer to Fig. 15.36) will the interference be constructive if the sources are in phase? (Consider the values N = 1, 2, 3; use a graphical construction.)

15.47 Two narrow slits are so close together that a direct measurement of their separation is difficult to make. By illuminating these slits with light ($\lambda = 5 \times 10^{-7}$ m) it is found that on a screen 4 m away adjacent bright lines in the interference pattern are separated by 2 cm. What is the separation of the slits?

15.48* Blue light ($\lambda_B = 4 \times 10^{-7}$ m) and yellow light ($\lambda_Y = 6 \times 10^{-7}$ m) simultaneously illuminate a pair of slits whose separation is 0.02 mm and bright diffraction lines (blue and yellow) are formed on a screen that is

2 m away. If the central lines of the two colors are numbered "zero," what will be the numbers of the lines at the first point beyond the central lines where a blue line falls at exactly the same place as a yellow line? How far is this point from the central lines?

15.49 Light waves have the same *frequency* whether they travel through air (essentially the same as empty space as far as light waves are concerned) or through some medium such as glass or water. But in glass or water, the velocity of light is only about 0.7 c and so the *wavelength* of the light is different. Suppose that the double-slit apparatus of Example 15.9 were immersed in water. What would be the spacing between adjacent bright lines on the screen?

15.50 Microwaves ($\lambda = 0.5$ cm) are incident on a pair of slits that are separated by 0.25 m. Describe the intensity pattern that would be found by moving a microwave detector along a screen that is 15 m from the slits.

15.51 Yellow light ($\lambda = 6 \times 10^{-7}$ m) illuminates a single slit whose width is 0.1 mm. What is the distance between the two dark lines on either side of the central maximum if the diffraction pattern is viewed on a screen that is 1.5 m from the slit?

15.52 Light with a wavelength of 6000 Å is incident on a pair of narrow slits that are separated by a distance of 0.15 mm. The light is projected onto a screen that is 2 m away. What is the distance between the two dark lines on either side of the central maximum? The region between the two slits is now removed, making a single slit with a width equal to the previous separation. What is the distance now between the two dark lines on either side of the central maximum?

15.53 The red light from a helium-neon laser has a wavelength of 6328 Å. What is the angle of the second-order maximum for this light when a diffraction grating with 4500 lines/cm is used?

15.54 The second-order maximum of the light from a certain source is at an angle of 36° when passed through a grating with 5000 lines/cm. What is the wavelength of the light? What is the color of the light? (Use Table 15.3.)

15.55 A microwave beam with $\lambda = 1$ cm is incident normally on a sheet of metal into which are cut slots at intervals of 6 cm. Where is the first-order diffraction maximum for these waves?

15.56 When using a diffraction grating with 5000 lines/cm, the second-order maximum for red light ($\lambda = 7.2 \times 10^{-7}$ m) coincides with the third-order maximum for light with what wavelength?

Section 15.5

15.57 What is the wavelength of the 25-MHz radiation that WWV used to broadcast time signals? What is the frequency of radiation in the 10-m shortwave broadcast band?

15.58 A proton oscillates back and forth across the diameter of a nucleus (about 10^{-14} m) with a velocity $v \cong 0.05$ c. What is the approximate frequency of the emitted radiation? (The frequency of the radiation is equal to the frequency of the oscillation.) How would you classify this radiation?

15.59 An electromagnetic plane wave has its electric field oscillating in the *yz*-plane and its magnetic field in the *xy*-plane. When the electric field has its maximum positive value, the magnetic field also has its maximum positive value. Make a sketch similar to Fig. 15.47 for this situation. In what direction is the plane wave traveling?

15.60 An electromagnetic plane wave is traveling in the positive *z*-direction and the electric field is in the *xy*-plane. Make a sketch similar to Fig. 15.47 for this situation.

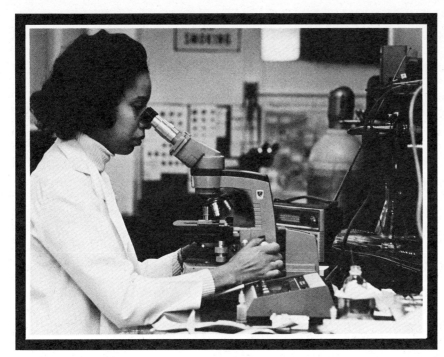

16

Light and Optics

In the preceding chapter we discussed the wave character of electromagnetic radiations, including light. We also learned that waves tend to bend around corners—that is, waves exhibit *diffraction*. This effect occurs for all types of waves—mechanical waves, sound waves, electromagnetic waves. But our experience tells us that light does not bend around corners, that light travels in straight lines. The reason that we do not observe diffractive effects for light under ordinary circumstances is that diffraction is appreciable only when the wavelength of the wave is comparable to the size of the diffracting object or to the size of the region that we examine. The wavelength of visible light, however, spans only the range from about 4×10^{-7} m to 7.5×10^{-7} m. Therefore, to observe diffractive effects for light we must use exceedingly narrow slits or we must look very carefully at the region adjacent to an illuminated sharp edge. We do not meet these requirements in our ordinary viewing.

Under almost all everyday conditions, then, we can analyze the behavior of light as if it *does* travel in straight lines. That is, we can use the idea of narrow light beams or *light rays* to describe most common situations. But we must keep in mind that light is actually a wave phenomenon and that a light ray is merely a line that is drawn to represent the direction of advance of the wavefront of the light wave. Indeed, to understand a phenomenon as basic as *refraction*, we must return to the wave description, because the ray picture provides no insight regarding the physical reason for the effect.

In this chapter we use ray diagrams whenever convenient to describe the way light interacts with mirrors and lenses. For some of the discussions we return to the fundamental description of light in terms of waves.

16.1 The Speed of Light

The question of whether light travels with a finite or infinite speed was debated even by the ancient Greeks. Although the value of the speed of light was of obvious importance in physical theory, the first reasonably accurate measurement was not made until the middle of the nineteenth century. Around 1600 Galileo had attempted to determine the speed of light between two hilltops separated by about 3 km. He and a companion, each equipped with a lantern and a cover, took up positions on the hilltops one night. Galileo reasoned that he would determine the speed of light by estimating (he had no accurate clock) the time required for light to pass from one hill to the other and back again. He would uncover his lantern and his companion, seeing the light from Galileo's lantern, would uncover his own lantern. Then, Galileo would estimate the time between the uncovering of his lantern and the first sight of the light from his companion's lantern. It did not take long for Galileo to realize that he was observing a delay caused, not by the speed of light, but by the combined reaction times of himself and his companion! The speed of light is far too great to measure in this crude way. (The time required for the round-trip of a light signal between the two hilltops used by Galileo is about 2×10^{-5} s, whereas the human response time for a muscular action following a visual signal is typically 0.2 s.)

Some estimates of the speed of light were obtained before the end of the seventeenth century, but it was not until 1849 that a reliable measurement was made. Armand Fizeau (1819–1896), a French physicist, was the first to make a direct measurement of the speed of light by using a train of pulses that had been generated by mechanically "chopping" a light beam into short segments. Fizeau directed a beam of light through the gaps in a rapidly rotating toothed disc. If the disc were turned at the proper rate, the light that passed through one gap would travel to a mirror placed on a hilltop some distance away where it would be reflected and would return to pass through the next gap as it rotated into place. Thus, the speed of light was obtained from the angular frequency of the disc and the total path length. Fizeau's result for the speed of light was about 5 percent higher than that presently accepted.

The technique of using a mechanically "chopped" beam for determining the speed of light was carried to its height by Albert A. Michelson (1852–1931), one of the most renowned American scientists of his era. His measurements of the speed of light began in 1878 and continued on and off for the remainder of his life. In Michelson's experiments, a rapidly rotating eight-sided mirror was used to "chop" the beam, and during the period 1923–1927 he located the rotating mirror at the observatory on Mount Wilson and the other mirror at a station on Mount San Antonio, 35 km away (see Fig. 16.1). These measurements resulted in a value for the speed of light of 2.99798×10^8 m/s, only about 20 parts per million different from the presently accepted value.

Today, the most precise method for determining the speed of light involves the use of laser beams. The best current results give the speed of light as 2.99792458×10^8 m/s, with an uncertainty of only one digit in the last decimal, a precision of three parts per *billion*. In our discussions here we will denote the speed of light (in vacuum) by the symbol c and we will use the approximate value,

$$c \approx 3.00 \times 10^8 \text{ m/s} \tag{16.1}$$

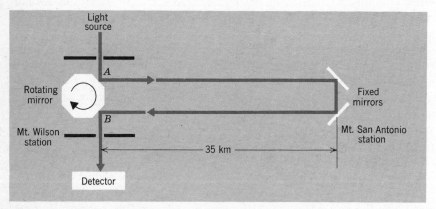

Figure 16.1 Michelson's method for measuring the speed of light. A light ray from the source is flashed toward the Mt. San Antonio station from the side A when the rotating mirror is in the position shown. If the rotation speed is such that the mirror makes exactly ⅛ of a revolution during the time that the light travels to the mirrors on Mt. San antonio and back, the pulse will be reflected from the side B and will enter the detector. The rotation speed is adjusted until the maximum light signal is detected. The speed of light is therefore determined from the rotation frequency of the mirror and the distance between the two stations.

16.2 Reflection and Refraction

Reflection of Light from a Flat Mirror

It was known even in ancient times that when a light ray is incident on a flat polished surface, the ray is *reflected* in such a way that the angle of reflection ϕ_r is equal to the angle of incidence ϕ_i (Fig. 16.2). Notice that these angles (as well as those involved in refractive effects) are measured from a line that is *perpendicular* to the reflecting surface. To an observer at position B in Fig. 16.2, the light appears to originate *behind* the mirror and to come to his eye along the path $A'B$ instead of along the actual path through A.

Figure 16.3 shows an object located in front of a flat mirror. Several of the rays that originate at the tip of the object are traced through their paths, which involve reflection at the mirror surface. Each ray is reflected with $\phi_r = \phi_i$ (check this in the figure). If we trace each reflected ray backward, we find that they all diverge from a common point behind the mirror. The same is true for the light rays from every other point on the object. Therefore, any observer located to the left of the mirror will see the light from the object just as if it had originated behind the mirror. That is, he will see behind the mirror the reflected *image* of the object. Moreover, the observer will observe this image to be located exactly as far *behind* the mirror surface as the object is located in *front* of the surface. (Can you see why this is so?)

Refraction

When light is incident on a *transparent* material, the light divides, at the surface, into two parts (Fig. 16.4). One part of the light is reflected, and this ray

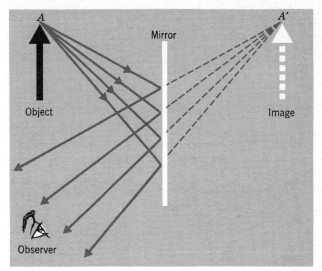

Figure 16.3 The reflected image of the object is located behind the mirror.

obeys the reflection rule just discussed. The remainder of the light is transmitted through the material. Notice in Fig. 16.4 that the transmitted ray does not have the same direction as the incident ray. The bending of a light ray as it passes from one medium to another is called *refraction*.

A light ray that passes from a medium such as air into a medium such as glass will be bent *toward* the line that is perpendicular to the surface (Fig. 16.4a); that is, $\theta < \phi$. If the light is incident on the glass-air surface from the glass side, the transmitted ray will be bent *away from* the perpendicular line (Fig. 16.4b); that is, $\theta > \phi$.

We can understand the phenomenon of refraction by referring to the wave characteristics of light. In Fig. 16.5 we have a light wave, represented by the series of wavefronts, AA', BB', and CC', that is incident on an air-glass surface from the air side. Remember, the direction of a light *ray* is always perpendicular to the direction of the associated *wavefront*. Therefore, in Fig. 16.5 the arrow aa' represents a ray of the incident light and corresponds to the direction of propagation of the wave. The right-hand end C of the wavefront CC' is incident on the surface when the end C' is still some distance from the surface. Within glass, light travels with a speed that is *less* than the speed in air (or vacuum).[1] Therefore, while the lefthand end of the wavefront

---

[1] This was first demonstrated in 1853 for the case of air and water by the French physicist Jean Bernard León Foucault (1819–1868). Foucault used a rapidly rotating mirror for his measurements of the speed of light. This technique was later refined by Michelson.

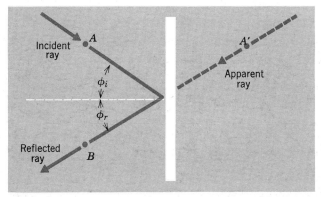

Figure 16.2 A light ray undergoes reflection at a flat surface with $\phi_r = \phi_i$.

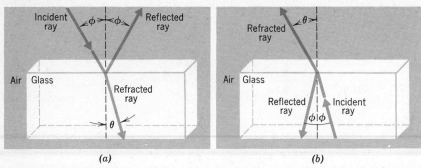

Figure 16.4 (a) When light passes from air into glass, the transmitted ray is bent *toward* the perpendicular line, so that $\theta < \phi$. (b) When light passes from glass into air, the transmitted ray is bent *away from* the perpendicular line, so that $\theta > \phi$.

travels from C' to the surface at D', the right-hand end of the same wavefront moves the shorter distance from C to D. As a result, the wavefronts DD', EE', FF', and GG', all of which lie entirely within the glass, propagate in a direction, represented by the arrow gg', that is different from the incident direction aa'. Notice also that the wavelength of the wave in the glass is *less* than the wavelength in air. (Can you see why this must be the case if the speed of propagation is reduced in passing from air into glass? The *frequency* of the wave remains the same. Again, why?)

The ratio of the speed of light in vacuum to the speed in a medium is called the *index of refraction* of the medium. This index is usually represented by the symbol n:

index of refraction:

$$n = \frac{\text{speed of light in vacuum}}{\text{speed of light in medium}} = \frac{c}{v} \quad (16.2)$$

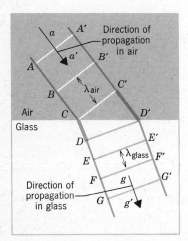

Figure 16.5 Because light travels with a lower speed in glass than in air, a light wave incident on an air-glass surface will have its direction of propagation shifted. The wave will be *refracted*.

As we will see later in this chapter, the index of refraction for a particular medium depends slightly on the wavelength of the light. When a single number is given for n, this number usually refers to the value for yellow light (the middle of the visible spectrum). Values of n for some substances are given in Table 16.1. Notice that the value for air is so close to 1 that we can usually assume $n = 1$ without significant error. Notice also that $n \cong \frac{4}{3}$ for water and $n \cong \frac{3}{2}$ for plastics and ordinary glasses.

Snell's Law

We can easily derive the expression that relates the angle of incidence to the angle of refraction for a light ray that passes from one optical medium to another. Refer to Fig. 16.6, which is the same as Fig. 16.5 except that the important angles are now indicated. The light originates in the medium with index of refraction $n_1 = c/v_1$ and passes into the medium with $n_2 = c/v_2$. During the time t that the right-hand end of the wavefront moves from C to D, the left-hand end moves from C' to D'. Thus,

Table 16.1 Indexes of Refraction for Various Substances for Yellow Light

Substance	n (for 20°C)
Air	1.00027
Methyl alcohol	1.3294
Water	1.3330
Lucite	1.4913
Fused silica	1.4584
Sodium chloride	1.5443
Carbon disulfide	1.6280
Crown glass (typical)	1.520
Flint glass (light)	1.575
Flint glass (heavy)	1.650
Flint glass (heaviest)	1.890
Diamond	2.4175

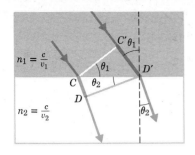

Figure 16.6 Angles involved in the refraction of a beam of light at the interface between two optical media. CC' and DD' are wavefronts.

$$C'D' = v_1 t = \frac{ct}{n_1}$$

$$CD = v_2 t = \frac{ct}{n_2}$$

Now, notice that we can write

$$CD' = \frac{C'D'}{\sin \theta_1} = \frac{CD}{\sin \theta_2}$$

Substituting the expressions for $C'D'$ and CD into this equation, we have

$$\frac{ct/n_1}{\sin \theta_1} = \frac{ct/n_2}{\sin \theta_2}$$

Canceling the factor ct and taking the reciprocal, we find

$$n_1 \sin \theta_1 = n_2 \sin \theta_2 \qquad (16.3)$$

This equation is known as *Snell's law*, in honor of the Dutch mathematician, Willebrord Snell (1591–1626), who discovered the result experimentally in about 1621.

●*Example 16.1*

A light ray is incident at an angle of 25° on the surface between air and water. What angle does the refracted ray make with the perpendicular to the surface when the ray is incident from (a) the air side and (b) the water side?

(a) In this case, $n_1 = 1.000$, $n_2 = 1.333$, and $\theta_1 = 25°$. Then, using Eq. 16.3,

$$1.000 \times \sin 25° = 1.333 \times \sin \theta_2$$

or,

$$\theta_2 = \sin^{-1}\left(\frac{\sin 25°}{1.333}\right) = \sin^{-1}(0.3170) = 18.5°$$

(b) In this case, $n_1 = 1.333$, $n_2 = 1.000$, and $\theta_1 = 25°$; thus,

$$1.333 \times \sin 25° = 1.000 \times \sin \theta_2$$

or,

$$\theta_2 = \sin^{-1}(1.333 \times \sin 25°) = \sin^{-1}(0.5634) = 34.3° \blacksquare$$

You have probably noticed that when you observe an object immersed in water, the apparent depth of the object is not the actual depth. A swimmer who is standing waist deep in water appears, to an observer out of the water, to have unnatural, stubby legs. This effect is due to the refraction of light as it passes through the air-water interface. In Fig. 16.7 we see why the apparent depth is less than the actual depth. Each ray that originates at the object is refracted away from the perpendicular upon emerging from the surface. An observer (in air) sees these rays diverging from the apparent position of the object. The ratio of the actual depth to the apparent depth is equal to the index of refraction of the medium (see Problem 16.9).

Refraction in Flat Plates

Suppose that a light beam originates in a medium with index of refraction n_a, passes through a flat plate of material with index n_b, and finally into a medium with index n_c (see Fig. 16.8). If the angle of incidence is θ_a, what will be the final angle of refraction θ_c? Because the plate of material with index n_b has parallel surfaces, the angle θ_b with which the ray enters this material is the same as the angle

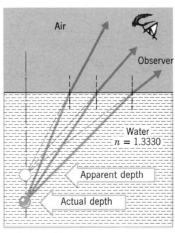

Figure 16.7 Because of refraction at the surface, an object immersed in water appears to an observer in air to be at a depth that is less than the actual depth.

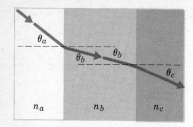

Figure 16.8 The angle θ_c does not depend on the intermediate medium if it consists of a flat plate of material with parallel surfaces.

with which the ray is incident on the third medium. Therefore, we can write for the two refractions.

$$n_a \sin \theta_a = n_b \sin \theta_b \quad \text{and} \quad n_b \sin \theta_b = n_c \sin \theta_c$$

so that

$$n_a \sin \theta_a = n_c \sin \theta_c \qquad (16.4)$$

Thus, we see that the angle θ_c of the final refracted ray does not depend at all on the intermediate medium. Indeed, we could have had any number of flat plates made from different materials between the initial and final media without affecting the angle θ_c.

The result we have just obtained means that when a light ray passes through a flat plate immersed in a medium, the exit angle is the same as the angle of incidence (Fig. 16.9). In such a case, the ray is displaced parallel to its original path but there is no net deviation in the direction of the ray. This is the situation, for example, for a light ray passing through a glass plate with air on both sides.

Internal Reflection

When light is incident from air on a glass surface, there will be a ray transmitted into the glass for any angle of incidence. In the reverse situation, however, when the light is incident on the surface from the glass side, there will be no transmitted ray if the angle of incidence is greater than a certain angle called the *critical angle* θ_c. To see why this is true, we need only to use Snell's law to calculate θ_2 when $n_2 < n_1$.

Figure 16.10 shows light incident on an air-glass interface from the glass side for several different angles of incidence. The glass has $n_1 = 1.520$, and Snell's law gives $\theta_2 = 69.4°$ when $\theta_1 = 38°$. Increasing θ_1 to 40° results in an increase of θ_2 to 77.7°. When $\theta_1 = 41.14°$, we find $\theta_2 = 90°$. Any further increase in θ_1 makes $\sin \theta_2 > 1$, an impossible situation. There is no refracted ray when θ_1 exceeds the *critical angle* $\theta_c = 41.14°$. For all such angles of incidence, the light is entirely reflected at the interface (Fig. 16.10d). We call this effect *internal reflection* (or *total* internal reflection).

For light incident on a medium with index n_2 from a more optically dense medium with index $n_1 > n_2$, the critical angle is equal to the angle θ_1, for which $\theta_2 = 90°$. In general, Snell's law gives

$$\theta_1 = \sin^{-1}\left(\frac{n_2}{n_1} \sin \theta_2\right)$$

when $\theta_2 = 90°$, then $\sin \theta_2 = 1$, so that

$$\theta_c = \sin^{-1}\left(\frac{n_2}{n_1}\right) \qquad (16.5)$$

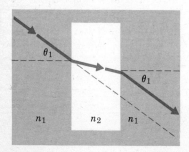

Figure 16.9 A light ray that passes through a flat plate immersed in a medium will be displaced, but it will not be deviated in direction.

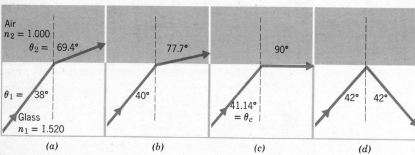

Figure 16.10 Light incident on an air-glass boundary for several angles near the critical angle $\theta_c = 41.14°$. For any incident angle greater than θ_c, the light is entirely reflected at the surface (d).

● *Example 16.2*

What is the critical angle for a water-lucite boundary?

The critical angle exists only for the case in which the light is incident from the medium with the greater optical density. Therefore, using Table 16.1, $n_1 = n(\text{lucite}) = 1.4913$ and $n_2 = n(\text{water}) = 1.3330$. Then,

$$\theta_c = \sin^{-1}\left(\frac{1.3330}{1.4913}\right) = \sin^{-1} 0.8939 = 63.4° \quad \blacksquare$$

The critical angle for the interface between air and a material with index of refraction $n = \sqrt{2} = 1.414$ is 45°. Essentially, all types of glass have indexes of refraction that are greater than $\sqrt{2}$ and therefore have critical angles that are less than 45°. Consequently, a wedge-shaped piece of glass (called a *prism*) that is cut with angles of 45° can be used as a mirror, as shown in Fig. 16.11. In this illustration we see a light ray that is reflected four times in passing through three 45° prisms. Such combinations of prisms are used in binoculars to increase the light path without increasing the overall length of the instrument. (This is the reason that most binoculars have a bulge on the side of each barrel.)

When light is projected into the end of a small-diameter glass or plastic rod, the angles with which the various rays are incident on the walls of the rod are greater than the critical angle. Consequently, the rays are internally reflected and propagate within the rod even though it is twisted and curved (Fig. 16.12). Such rods are called *light pipes* and have a number of interesting practical applications. For example, a physician can use a light pipe to see inside a patient's stomach. Communications signals

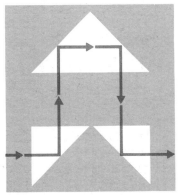

Figure 16.11 A light ray undergoes four internal reflections in passing through a combination of 45° prisms. Notice that the direction of the emerging ray is the same as that of the incident ray.

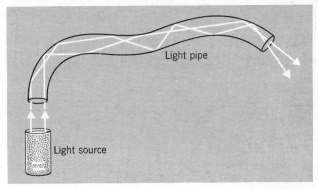

Figure 16.12 Light rays can be piped along a small-diameter rod of glass or plastic by repeated internal reflections.

can be piped over long distances by using newly developed materials with very small diameters and very low loss rates. Bundles of these rods (*optical fibers*) are beginning to replace wires in certain types of communications systems (Fig. 16.13).

16.3 The Formation of Images by Lenses and Mirrors

Focusing by Refraction

One of the most common and most useful of optical devices is the *lens*. A lens is usually in the form of a circular piece of relatively thin glass whose surfaces are sections of spheres (one surface may be plane). If the thickness of the lens along the axis is greater than the thickness at the edge, the lens is *convex* (Fig. 16.14a, b, c). If the thickness along the axis is less than the thickness at the edge, the lens is *concave* (Fig. 16.14d, e, f). In Fig. 16.14, lens (b) is *planoconvex* and lens (e) is *planoconcave*. As we will see, convex lenses cause the light that passes through them to *converge*, whereas concave lenses cause the light to *diverge*.

Figure 16.15 shows a plane wave of light incident on a convex lens. The wavefronts are perpendicular to the axis of the lens. The wave first strikes the lens at its thickest point and within the lens the speed of the wave is less than its speed in air. Consequently, the outer parts of the wave move ahead of the more slowly moving central part and the wavefront becomes distorted. Upon emerging from the lens, the wavefront once again travels uniformly with its original speed. But the outer parts of the wave are now moving *inward* and converge at the point F, the *focal point* of the lens.

Figure 16.15 is a *wave* diagram that shows the converging effect of a convex lens. Figure 16.16 is the equivalent *ray* diagram. Notice that the incident plane wave is represented by a number of rays that

Figure 16.13 An optical fiber of the type used in communications systems. Notice the glow at the handheld end where the light emerges from the fiber.

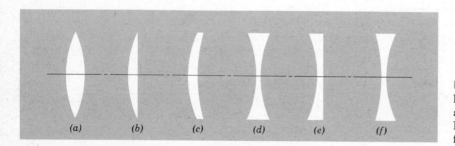

Figure 16.14 Different types of thin lenses. (a), (b), and (c) are *convex* lens, and (d), (e), and (f) are concave lenses. Lenses (b) and (e) have one plane surface.

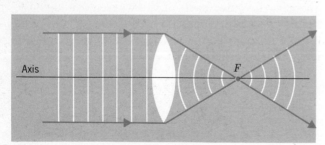

Figure 16.15 A plane light wave is incident on a convex lens and converges toward the focal point F.

are *parallel* to the axis. The *focal length f* of the lens is the distance from the lens to the focal point F.

In Fig. 16.16, parallel rays are incident from the left and are brought to a focus at the right of the lens. Similarly, if rays are incident from the right, they will be brought to a focus at the left of the lens. Thus, a lens has *two* focal points, which are located at equal distances from the lens.

If a ray follows a certain path in passing through a lens (or a lens system), another ray can pass through the lens in the reverse direction follow-

16.3 The Formation of Images by Lenses and Mirrors 399

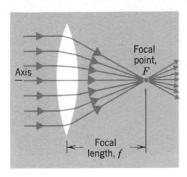

Figure 16.16 A ray diagram showing the focusing effect of a convex lens.

ing the same path. For example, in Fig. 16.16 we have a number of rays that are originally parallel, and, on passing through the lens, converge at the focal point F. If we place a light source at F, the rays that radiate outward and pass through the lens will emerge as parallel rays at the left of the lens. Light rays follow *reversible* paths.

As we consider light rays passing through lenses (and later, reflecting from curved mirrors), we require the rays to lie close to the lens axis (these are the *paraxial* rays). We do this to avoid the distortion introduced by the rays far from the axis. The figures have been enlarged to show off-axis rays but these rays have been drawn to conform to the behavior of the paraxial rays in order to make the ray diagrams easier to understand.

Images Formed by Convex Lenses

We now have all the information we need about the way light rays are influenced by convex lenses to describe the formation of *images* by these lenses. Suppose that we have an object located some distance to the left of a convex lens, as shown in Fig. 16.17. We can easily trace through the lens three rays from some point on the object, such as the tip. First, the ray OA is incident on the lens parallel to the axis; therefore, we know that this ray passes through the focal point F at the right of the lens. Second, the ray OB passes through the central part of the lens. In this region the two lens surfaces are essentially parallel, and as we discovered earlier (Fig. 16.9), a ray that passes through a medium with parallel surfaces is undeviated. Moreover, we are considering only lenses that are *thin*, so the amount of displacement experienced by the central ray is negligibly small. Thus, the ray OBI is a straight line. Third, the ray OC that passes through the focal point F' emerges from the lens parallel to the axis. All three of these rays converge to a point, thereby determining the position of the image of the tip of the object. Notice that the image position is determined by any two of the three rays. In making graphical constructions, however, it is always advisable to use the third ray as a check.

The image formed in Fig. 16.17 by the converging of the rays is called a *real* image because it can actually be projected onto a screen or onto the film in a camera. Also, the image extends below the axis, whereas the object is above the axis. An image of the type shown in Fig. 16.17 is called a *real, inverted* image.

In Fig. 16.17 the object is located farther from the lens than the focal distance. Figure 16.18 shows what happens when the object is placed closer to the lens than the focal point. When we trace the three rays in this case, we find that they do not converge to a point at the right of the lens. Instead, the rays *diverge* from a point to the *left* of the lens. This image cannot be projected onto a screen because the light rays do not pass through the image. However, the image can be seen by an observer; to him the rays *appear* to originate in an image located behind the lens, but, in fact, they do not. This type of image is a *virtual* image. The image is on the same side of the axis as the object, so we call this a *virtual, erect* image. (Compare the image in Fig. 16.3.)

In Fig. 16.18 notice that the image is *larger*

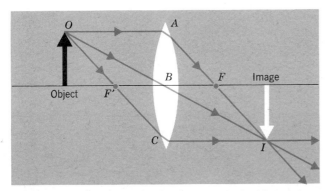

Figure 16.17 The image of an object can be located by tracing these three rays through the lens.

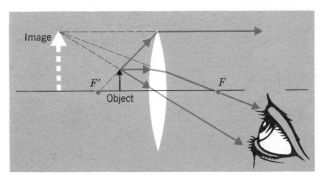

Figure 16.18 A virtual, erect image results when the object is located inside the focal point of the lens.

than the object. Thus, a convex lens serves as a simple *magnifying glass* if the object is placed closer to the lens than the focal point.

Images Formed by Concave Lenses

A *convex* lens causes a parallel bundle of incident light rays to *converge* toward the focal point (Fig. 16.16). A *concave* lens, on the other hand, causes a parallel bundle of incident rays to *diverge,* as shown in Fig. 16.19. Rays that are incident on this lens are bent away from the axis, and to an observer at the right of the lens, the rays appear to have originated at F. We call F the "focal point" of the lens even though the rays do not actually pass through this point. (Notice that the symbol F represents the "focal point" for rays incident from the left and F' represents the corresponding point for rays incident from the right for both converging and diverging lenses. Because the focal point F of a diverging lens is located on the same side of the lens as the incident light, we say that such a lens has a *negative* focal length, whereas a converging lens has a *positive* focal length.)

Figure 16.20 shows how a virtual image is formed by a concave lens. Notice carefully how each of the three rays is constructed. One ray is the central ray that passes through the lens undeviated, and the other two rays follow the rule that parallel rays appear to diverge from the focal points. (The middle ray is the *reverse* of the ray that follows the rule, but reversed rays are possible rays.) A concave lens alone *never* forms a real image, regardless of where the object is located.

The Lens Equation

We have seen how to locate by graphical means the position of the image formed by a lens when the lo-

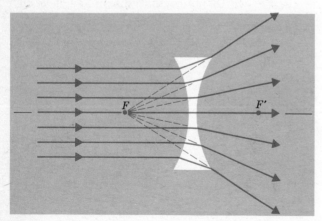

Figure 16.19 A concave lens causes a parallel bundle of incident rays to diverge, apparently from the point F.

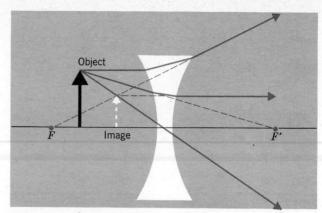

Figure 16.20 The formation of a virtual image by a concave lens.

cation of the object and the focal length of the lens are known. We can also easily derive an expression that relates these same quantities by referring to Fig. 16.21. The symbols have the following meaning:

f: focal length of the lens; positive for a converging lens and negative for a diverging lens.

x_0: distance from the lens to the object; always positive (for a single lens).

x_i: distance from the lens to the image; positive if the image is on the side of the lens opposite the object, negative if the image is on the same side of the lens as the object.

h_0: height of the object; positive when the image is above the lens axis.

h_i: height of the image; positive when the image is above the lens axis.

In the ray diagram in Fig. 16.21 we see two pairs of shaded triangles. Each pair consists of *similar* triangles; that is, the corresponding angles in each triangle of a pair are equal, so we can write

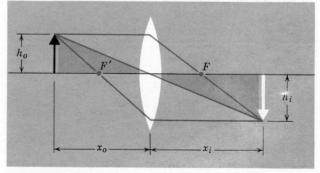

Figure 16.21 Ray diagram for deriving the lens equation.

equalities connecting ratios of the sides. For the triangles on the object side of the lens,

$$\frac{-h_i}{h_0} = \frac{f}{x_0 - f}$$

where we attach a negative sign to h_i because $h_i < 0$ (the image lies *below* the lens axis and the ratio $f/(x_0 - f)$ is positive). Also, for the triangles on the image side of the lens,

$$\frac{-h_i}{h_0} = \frac{x_i - f}{f}$$

Equating these two expressions for $-h_i/h_0$, we have

$$\frac{f}{x_0 - f} = \frac{x_i - f}{f}$$

from which

$$x_0 x_i - f x_i - f x_0 = 0$$

Finally, dividing each term by $x_0 x_i f$ and rearranging, we find

$$\frac{1}{x_0} + \frac{1}{x_i} = \frac{1}{f} \qquad (16.6)$$

This is the *lens equation*.

● *Example 16.3*

An object is placed 12 cm to the left of a lens that has a focal length of 8 cm. Where is the image located?

Inserting $x_0 = 12$ cm and $f = 8$ cm into Eq. 16.6, we find

$$\frac{1}{x_i} = \frac{1}{f} - \frac{1}{x_0} = \frac{1}{8} - \frac{1}{12} = \frac{3 - 2}{24} = \frac{1}{24}$$

so that

$$x_i = 24 \text{ cm}$$

● *Example 16.4*

Suppose that the object is the preceding example is moved closer to the lens so that $x_0 = 6$ cm. What is the new image distance?

In the same way as before,

$$\frac{1}{x_i} = \frac{1}{f} - \frac{1}{x_0} = \frac{1}{8} - \frac{1}{6} = \frac{3 - 4}{24} = -\frac{1}{24}$$

so that

$$x_i = -24 \text{ cm}$$

The negative sign means that the image is located to the *left* of the lens. This is the case illustrated in Fig. 16.18; the image is a *virtual* image.

● *Example 16.5*

The lens equation is useful also for locating the image formed by a diverging lens if we are careful to express the focal length as a negative quantity. Consider the case illustrated in Fig. 16.20: let $f = -16$ cm and $x_0 = 12$ cm. Where is the image located?

We have

$$\frac{1}{x_i} = \frac{1}{f} - \frac{1}{x_0} = -\frac{1}{16} - \frac{1}{12} = -\frac{3 + 4}{48} = -\frac{7}{48}$$

so that

$$x_i = -\frac{48}{7} = -6.86 \text{ cm}$$

The negative sign means that the image is located to the *left* of the lens (as indicated in Fig. 16.20) and is therefore a *virtual* image.

Notice that the lens equation will always give $x_i < 0$ when $f < 0$ and $x_0 > 0$. Thus, a diverging lens always produces a virtual image on the same side of the lens as the object.

● *Example 16.6*

The location of the image position for a combination of two or more lenses can be found by proceeding in stages. First, locate the image position for the first lens. Then, use this result as the object position for the next lens; and so forth. (Even a virtual image formed by the first lens can serve as an object for the second lens. For example, the virtual image produced by a magnifying glass is the object for the eye lens which then focuses a real image on the retina.)

An object is located 6 cm to the left of a converging lens with a focal length $f_1 = 4$ cm. A second lens, a diverging lens with $f_2 = -8$ cm, is located 16 cm to the right of the first lens. Where is the image located? (See the diagram on page 402.)

For the image due to the first lens alone, we have

$$\frac{1}{x_i} = \frac{1}{f_1} - \frac{1}{x_0} = \frac{1}{4} - \frac{1}{6} = \frac{3 - 2}{12} = \frac{1}{12}$$

$$x_i = 12 \text{ cm}$$

The image produced by the first lens is real and lies 12 cm to its right and 16 cm − 12 cm = 4 cm to the left of the second lens. Therefore, the object distance for the second lens is $x_0' = 4$ cm. (The image produced by the first lens is labeled *intermediate image*

Lens diagram for Example 16.6.

in the diagram. Notice that this real image is used as a normal object for the second lens. Notice also that the line drawn from the tip of the intermediate image through the center of the diverging lens is a *construction line* only, used to locate the image, and does not represent an actual ray.) Then,

$$\frac{1}{x_i'} = \frac{1}{f_2} - \frac{1}{x_0'} = -\frac{1}{8} - \frac{1}{4} = -\frac{3}{8}$$

$$x_i' = -\frac{8}{3} = -2.67 \text{ cm}$$

Thus, the image is *virtual* and is located 2.67 cm to the *left* of the second lens. ∎

Magnification

The ratio of the height of the image formed by a lens to the height of the object is called the *magnification* M of the lens:

$$M = \frac{h_i}{h_0} \qquad (16.7)$$

For the real image formed by a single converging lens, h_i is negative (Fig. 16.17) and so the magnification is also a negative number.

The size of the image formed by a lens depends on the location of the object relative to the focal point of the lens. In fact, we can see immediately from the shaded similar triangles in Fig. 16.22 that

$$\frac{h_0}{x_0} = \frac{-h_i}{x_i}$$

where we have attached a negative sign to h_i for the same reason as before. Using this relationship, the magnification can be expressed as

$$M = \frac{h_i}{h_0} = -\frac{x_i}{x_0} \qquad (16.8)$$

This quantity is sometimes called the *lateral magnification* (in order to distinguish it from the *angular magnification* which we will define shortly).

● *Example 16.7*

What is the magnification for the cases in (a) Example 16.3 and (b) Example 16.5?

(a) We have $x_0 = 12$ cm and $x_i = 24$ cm, so

$$M = -\frac{x_i}{x_0} = -\frac{24 \text{ cm}}{12 \text{ cm}} = -2$$

That is, the image is twice as large as the object and is inverted (a *real* image).

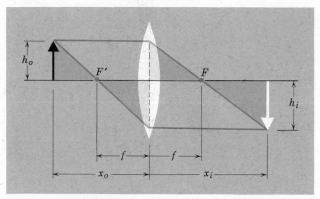

Figure 16.22 Ray diagram for determining the magnification of the lens.

(b) We have $x_0 = 12$ cm and $x_i = -6.86$ cm, so

$$M = -\frac{x_i}{x_0} = \frac{6.86 \text{ cm}}{12 \text{ cm}} = 0.57$$

That is, the image is a little more than half the size of the object and is erect (a *virtual* image), as shown in Fig. 16.20.

• **Example 16.8**

The magnification of a two-lens system is equal to the product of the individual magnification values (why?). What is the magnification of the lens combination in Example 16.6?

We have

$$M = M_1 M_2 = \left(-\frac{x_i}{x_0}\right) \times \left(-\frac{x_i'}{x_0'}\right)$$

$$= \frac{12}{6} \times \frac{-2.67}{4} = -1.33$$

Thus, the virtual image is inverted and is $\frac{1}{3}$ larger than the object. ∎

Equation 16.8 for the magnification M is useful when we are dealing with *real* images. However, when we consider virtual images that are viewed by the eye, another definition of magnification (i.e., *angular* magnification) proves to be more useful. Figure 16.23 shows an object that is viewed by an eye at two different distances. At the greater distance, the object subtends an angle θ_0 at the eye, and the image has a size h_i on the retina. When the object is moved closer to the eye, the viewing angle increases and the image size increases in the same proportion (if the angle is relatively small). That is, the apparent size of the object depends on the viewing angle:

$$\frac{h_i'}{h_i} = \frac{\theta_0'}{\theta_0}$$

The size of an image on the retinal surface cannot be increased indefinitely by bringing the object closer because the human eye cannot focus properly on an object that is located less than about 25 cm away. (This distance is called the *near point* of the eye.) By using a lens, however, a virtual image is formed which can be viewed clearly by the eye even though the object is closer than the near point. The increase in apparent size that is provided by the lens is conveniently expressed in terms of the *angular magnification* Γ of the lens.

Figure 16.24a shows an object located at the near point of an eye; the viewing angle is θ_0. The retinal image has a maximum size for an unaided eye because the object cannot be brought closer without deteriorating the focus. Figure 16.24b shows how the object can be brought closer to the eye and still viewed clearly by using a converging lens. The object is placed inside the focal length of the

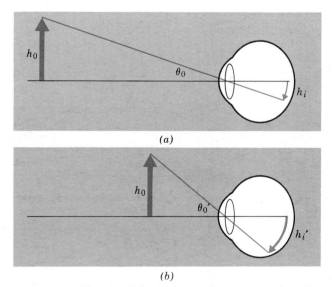

Figure 16.23 The size of the image on the retinal surface of the eye depends on the viewing angle.

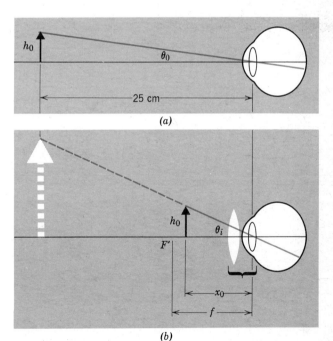

Figure 16.24 (a) The eye is focused on the tip of an object that is located at the near point, 25 cm away. (b) The eye is focused on the tip of the image, also located at a distance of 25 cm. In (b) the object is much closer, so the viewing angle and the size of the retinal image are much greater.

lens in order to provide a virtual image that the eye can see. The object distance x_0 is then adjusted until the image is located at the near point of the eye; this distance gives the maximum viewing angle θ_i and the maximum size of the retinal image. The angular magnification of the lens is equal to the ratio of the viewing angles:

$$\Gamma = \frac{\theta_i}{\theta_0}$$

We assume that the lens is sufficiently close to the eye that we can measure all distances from a single point (the lens of the eye) without significant error. Referring to Fig. 16.24a, we see that the tangent of the viewing angle θ_0 is equal to $h_0/25$ cm. We restrict our attention to small angles so that $\tan \theta_0$ is approximately equal to θ_0; similarly for θ_i. Then we can write

$$\tan \theta_0 \cong \theta_0 = \frac{h_0}{25}$$

$$\tan \theta_i \cong \theta_i = \frac{h_0}{x_0}$$

The angular magnification becomes

$$\Gamma = \frac{\theta_i}{\theta_0} = \frac{h_0/x_0}{h_0/25} = \frac{25}{x_0} \qquad (16.9)$$

According to the lens equation with $x_i = -25$ cm,

$$\frac{1}{x_0} = \frac{1}{f} - \frac{1}{x_i} = \frac{1}{f} + \frac{1}{25} = \frac{25 + f}{25f}$$

Substituting this expression into Eq. 16.9 for Γ gives

$$\Gamma = \frac{25}{x_0} = 25 \times \frac{25 + f}{25f} = \frac{25 + f}{f}$$

or,

$$\Gamma = \frac{25}{f} + 1 \qquad (16.10)$$

where the focal length f is measured in cm.

Notice that if the object is located at the focal point F', so that $x_0 = f$, then the image distance becomes infinitely large. (An object or image at infinity can be viewed clearly by the relaxed eye.) Equation 16.9 shows that the angular magnification now becomes

$$\Gamma = \frac{25}{f} \qquad \text{(for } x_0 = f\text{)} \qquad (16.11)$$

That is, the magnification of the object can be shifted between $25/f$ and $(25/f) + 1$ by moving the object slightly in the vicinity of the focal point of the lens. This effect is most easily observed for a lens with a low magnification ($\Gamma = 2$ or 3).

Focusing by Reflection

Focusing and image formation can be achieved by *mirrors* as well as by lenses. For example, Fig. 16.25 shows a bundle of parallel rays incident on the surface of a curved mirror. Upon reflection at the surface, the rays all pass through a common point, the focal point F. Actually, if the surface of the mirror is spherical, the focusing effect is good only for rays whose angles of incidence on the surface are small. We will restrict our attention to this case in our discussions here. (The focusing is exact only for a mirror with a *paraboloidal* surface, formed by rotating a parabola around its axis of symmetry, and then only for rays that are parallel to the axis.)

We can locate the position of the focal point F for a spherical mirror by referring to Fig. 16.26. The

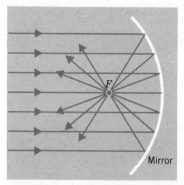

Figure 16.25 A bundle of parallel rays incident on a concave spherical mirror are brought to a focus at F (if the angles of incidence of the rays on the surface are all small).

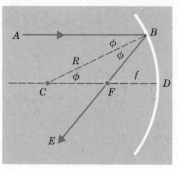

Figure 16.26 A ray AB that is incident on a concave spherical mirror in a direction parallel to the principal axis CD is reflected through the focal point F.

point C is the center of curvature of the spherical surface, so the line CB is a radius of the sphere and is perpendicular to the surface at B. The ray AB, parallel to the principal axis CD, is incident on the surface at B with the angle ϕ. The reflected ray BE also makes an angle ϕ at the surface. Moreover, because the incident ray is parallel to the line CD, the angle $\angle BCD$ is also equal to ϕ. The triangle $\triangle CBF$ is therefore an isosceles triangle in which $CF = FB$. Now, we restrict the angle ϕ to be small (it is exaggerated in the diagram); then, point B is close to point D, and without appreciable error, $FB = FD$. Thus, $CF = FD$, and the point F lies midway between C and D. That is, the focal length $f = FD$ is one half of the radius of curvature, $R = CB = CD$:

$$f = \frac{1}{2}R \quad \text{(for a spherical mirror)} \quad (16.12)$$

This expression, derived for a concave mirror, is also valid for a convex mirror (with a negative sign).

Image Formation by Spherical Mirrors

The location of the position of an image formed by a spherical mirror proceeds in a way that is analogous to the method we have used to locate images formed by lenses. In Fig. 16.27 we have an object at a distance x_0 from the surface of a spherical mirror. Two rays are drawn to locate the image. The ray AB is parallel to the principal axis CD; this ray is reflected at B and passes through F. The ray AH passes through the center C and is therefore perpendicular to the surface at H; this ray is reflected back along its original path. The intersection of the two rays at G defines the position of the image of the tip of the object. The image is real and inverted. (What is the

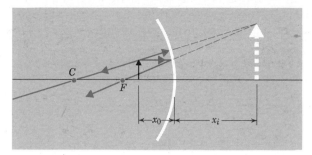

Figure 16.28 An enlarged, erect virtual image results when the object is located inside the focal point of a concave mirror.

third ray that could be used to locate the image?) A concave mirror always forms a real image if the object is located outside the focal point (just as does a convex lens).

Two additional cases of image formation by spherical mirrors are shown in Figs. 16.28 and 16.29. In Fig. 16.28 we have an object located closer to the mirror surface than the focal point. Just as in the case of a convex lens, this situation results in an enlarged, erect, virtual image. Mirrors of this type are used in various applications as *magnifying mirrors*.

Figure 16.29 shows a *convex* mirror, which produces an erect virtual image of reduced size for any location of the object (again, analogous to the case of a diverging lens). Mirrors of this type are sometimes used as automobile rear-view mirrors because the reduced magnification provides a wider field of view than does a flat mirror.

By considering the similar triangles in a ray diagram for a spherical mirror we can derive a *mirror equation* just as we obtained the lens equation from the diagram in Fig. 16.21. In fact, we find that the mirror equation is exactly the same as the lens equation:

$$\frac{1}{x_0} + \frac{1}{x_i} = \frac{1}{f} \quad \text{(for a spherical mirror)} \quad (16.13)$$

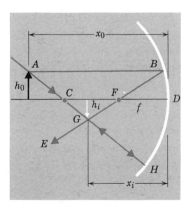

Figure 16.27 Locating the position of the image formed by a concave spherical mirror by means of a ray diagram. The image at G is real and inverted.

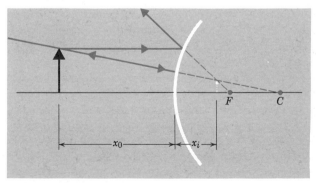

Figure 16.29 A convex mirror produces a virtual image for any object location.

406 Light and Optics

The conventions regarding the signs of the various quantities in this equation are as follows.

f: The focal length is *positive* for a *concave* mirror and *negative* for a *convex* mirror.

x_0: The object distance is always *positive*.

x_i: The image distance is *positive* when the image is on the *same* side of the mirror as the object, *negative* when on the side *opposite* the object.

The magnification M of a mirror is defined in the same way as for a lens:

$$M = \frac{h_i}{h_0} = -\frac{x_i}{x_0} \tag{16.14}$$

● *Example 16.9*

In Fig. 16.28, $f = 6$ cm and $x_0 = 4$ cm. Where is the image and what is the magnification?
Using Eq. 15.13,

$$\frac{1}{x_i} = \frac{1}{f} - \frac{1}{x_0} = \frac{1}{6} - \frac{1}{4} = \frac{2-3}{12} = -\frac{1}{12}$$

$x_i = 12$ cm

The image is located to the right of the mirror and is a *virtual* image. Also,

$$M = -\frac{x_i}{x_0} = -\frac{(-12 \text{ cm})}{4 \text{ cm}} = 3$$

The image is three times larger than the object. A dentist might use such a mirror to examine the inner surfaces of a patient's teeth. ■

Aberrations in Lenses and Mirrors

The lenses and mirrors we have been discussing are not perfect optical devices. No lens or mirror system can render a perfect image for an object with finite size. The difficulties encountered in producing images with real optical systems take several different forms, involving various aberrations and distortions. Each of these defects can be attacked and minimized by altering the curvature of the lens (or mirror) surface or by using appropriate combinations of lenses, but none can be completely eliminated for all situations. We will discuss here only one of the several different types of aberrations, namely, *spherical aberration*. (In the following section we mention *chromatic aberration*, which results from a variation of the refractive index with the wavelength of light.)

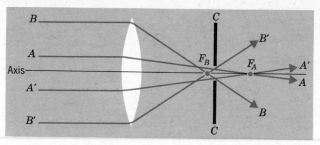

Figure 16.30 Light rays that pass through a lens with spherical surfaces will be focused at different positions, depending on their distance from the lens axis. The plane CC locates the *circle of least confusion*, the best focal position for the lens.

If we confine our attention to light rays that are incident on lenses or mirrors with very small angles of incidence and that lie near the optic axis, the focusing effect can be excellent and sharp images can be produced. However, off-axis rays and rays with large angles of incidence will not be focused at the same position as the central, low-angle rays. Figure 16.30 shows two pairs of parallel rays, AA' and BB', incident on a spherical lens. The rays AA' lie close to the axis of the lens and are focused at F_A. The rays BB' are incident on the lens near the outer edge and are focused at the point F_B, which is closer to the lens than the focal point for the central rays. There is no single point at which the entire bundle of parallel incident rays is brought to a focus. This effect is due entirely to the spherical character of the lens surface; the focal points for rays that are at different distances from the lens axis can be found by tracing the rays through the lens using Snell's law.

Suppose that the light passing through the lens is viewed by placing a screen perpendicular to the lens axis at the right of the lens. If the focus were perfect, there would be one position for the screen at which a point of light would be observed. However, for any location along the axis, the light on the screen will actually form a small circle. At some position CC, the spot of light will have a minimum size; this is called the *circle of least confusion*. The plane CC defines the plane of best focus for the lens.

Spherical aberration also occurs for mirrors. Figure 16.31 shows that parallel rays incident far from the axis of a spherical mirror do not focus at the same position as the central rays. The envelope of the reflected rays is called a *caustic*. This light pattern is easily observed when light is reflected from the inner surface of a cup or drinking glass.

Spherical aberration of lenses and mirrors can be minimized by changing the shape of the surface. For example, if the radius of curvature of the lens surface on the side opposite the incident light is made larger than that of the other surface, the

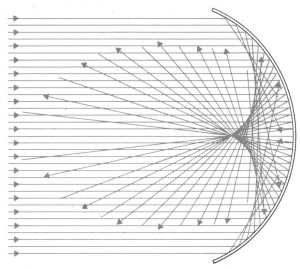

Figure 16.31 The effect of spherical aberration on rays incident parallel to the axis of a spherical mirror.

spherical aberration will decrease. If the surface of a mirror is paraboloidal instead of spherical, there will be no spherical aberration whatsoever for incident rays that are parallel to the optic axis. (But the aberration will still exist for rays with other angles of incidence.)

16.4 Color and Spectra

Dispersion of Light by a Prism

When light passes from air into glass (or the reverse), the amount of refraction that takes place depends on the index of refraction for the glass. As we pointed out in Section 16.2, the value of n for a substance varies slightly with the wavelength of the light. For all transparent materials, n is smaller for long-wavelength light than for short-wavelength light (see Table 16.2). Suppose that a light beam consisting of two colors, red (long wavelength) and blue (short wavelength), is incident on a glass prism, as

Table 16.2 Index of Refraction of Crown Glass for Various Wavelengths of Light

λ (Å)	Color	n
3610	Violet	1.539
4340	Blue	1.528
4860	Green	1.523
5890	Yellow	1.517
6560	Orange	1.514
7680	Red	1.511

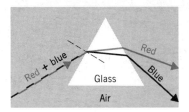

Figure 16.32 Because the index of refraction of the glass is slightly different for red and blue light, the colors are separated (dispersed) in passing through the prism.

in Fig. 16.32. If we trace the rays through the prism, the smaller index of refraction for red light causes the red ray to undergo less refraction than the blue ray. Consequently, when the rays emerge from the prism, they no longer travel in the same direction. The prism has transformed a wavelength difference for the light into a spatial separation. We say that the light has been *dispersed* by the prism.

Sunlight or the radiation from an incandescent lamp consists of light with all wavelengths. Such a mixture of colors is perceived by the eye to be *without color*, that is, *white light*. If a beam of white light is incident on a prism, the light will be dispersed into its component wavelengths and the entire rainbow of the visible spectrum of colors will be revealed (Fig. 16.33). We assign color names to the wavelength regions of the visible spectrum according to the schedule shown schematically in Fig. 16.34.

Chromatic Effects in Lenses

Light that consists of more than a single wavelength will be dispersed by a simple lens in the same way that it is dispersed by a prism. If a parallel beam of

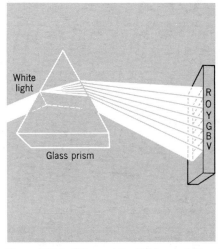

Figure 16.33 When white light, consisting of all colors, passes through a prism, the light is dispersed into a spectrum of colors. Red light (R) is refracted least and appears at the top of the spectrum in this diagram followed by orange (O), yellow (Y), green (G), blue (B), and violet (V).

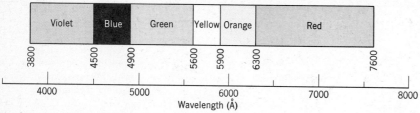

Figure 16.34 Wavelengths of colors in the visible spectrum. (The dividing points between colors are not always given in exactly this way.)

light consisting of a mixture of red and blue light is incident on a lens, the dispersive effect will result in different focal points for the two colors. As we see in Fig. 16.35, the red focal point F_R lies farther from the lens than the blue focal point F_B. Thus, if an object is illuminated by white light, a series of slightly displaced, differently colored images will be formed. This effect is called *chromatic aberration*. Images subject to chromatic aberration appear to be surrounded by colored fringes.

We can bring the red and blue components of a light beam to the same focal point by using two lenses, one converging and one diverging, made from glasses with different refractive indexes. Figure 16.36 shows the result of a properly constructed pair of lenses. The dispersion that is introduced by the converging lens is exactly compensated by the diverging lens. In this diagram the lenses are shown separated, but generally this type of lens pair is cemented together and becomes a *color-corrected doublet* or *achromatic lens*. Spherical aberration can also be reduction in achromatic lens.

A lens that is made from a single piece of glass will have a slightly different focal length for each color of light. A typical camera lens has a focal length of 50 mm (measured for yellow light). If this lens has no color correction, the focal length for orange light will be about 50.2 mm and the focal length for blue light will be about 49.8 mm. This type of variation in focal length with color is shown in Fig. 16.37. An achromatic lens, on the other hand, will have the same focal length for two different colors, but *only* for these two colors. In Fig. 16.37 we see that the focal-length curve for a particular achromatic lens has the same value (50.0 mm) only for blue and orange light. However, the variation with wavelength of the focal length for the achromatic lens is much less severe than is that for an uncorrected lens. Even though the focal length is exactly the same only for two colors, there is a vast improvement in the color performance of the lens.

A lens that consists of three elements can have the same focal length for three colors. This type of lens is called an *apochromatic lens* (or *apochromat*) and is the most common type of color-corrected lens found in high quality cameras. The focal-length curve for an apochromat is also shown in Fig. 16.37. The focal length is 50.0 mm for red, green, and in-

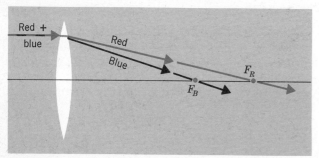

Figure 16.35 The result of dispersion in a lens results in different focal points for different colors (chromatic aberration).

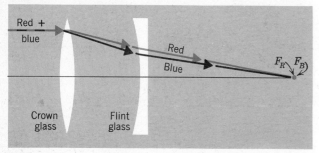

Figure 16.36 By using glasses with different refractive indexes and by constructing the lenses with the proper curvatures, the red and blue focal points can be made to coincide exactly. A lens system that is corrected for two colors is called an *achromatic lens*.

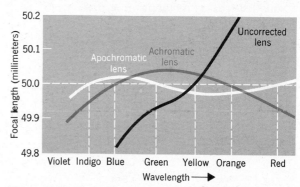

Figure 16.37 Focal length versus wavelength for an uncorrected lens and two different color-corrected lenses. The flatter the curve, the better the color performance of the lens will be.

digo light. Notice that this curve has less variation than the curve for the achromatic lens; an apochromatic lens has a correspondingly better color performance than an achromatic lens. (An apochromat is also corrected for spherical aberration at two wavelengths.) A higher degree of color correction could be achieved by using even more elements in the lens, but the improvement is usually not worth the additional expense in constructing a more complex lens.

Spectra

In 1802 the English scientist William Wollaston (1766–1828) discovered that when sunlight passes through a slit and is dispersed by a prism, there occur superimposed on the continuous band of colors a number of sharp dark lines that correspond to individual images of the slit. These lines are now called *Fraunhofer lines* in honor of the German physicist Joseph von Fraunhofer (1787–1826) who made an extensive study of the solar spectrum and cataloged the wavelengths of 576 dark lines. Fraunhofer showed that these lines originate in some process on the Sun, not on the Earth. But no one had any clear idea as to the origin of the dark solar lines.

Fraunhofer lines are dark regions in an otherwise bright continuous spectrum of colors and therefore represent the *absence* of light at particular wavelengths. That is, some effect acts to remove light with certain discrete wavelengths from the white light that is generated in thermal processes within the Sun. We can easily duplicate this type of effect in the laboratory. Suppose that we have a light source that produces white light. If we pass this light through a prism, we will observe a continuous spectrum of colors. Now, suppose that we have a sealed glass cell containing some sodium metal. We heat the metal until it vaporizes and fills the cell with sodium vapor. Next, we interpose the cell between the light source and the prism. When we examine the dispersed light, we now find that there are two dark lines in the yellow part of the spectrum. That is, the sodium atoms have absorbed a part of the light passing through the cell, but the absorption has been limited to two particular wavelengths of yellow light. *Absorption lines* are *dark* lines.

The light that is absorbed by the sodium atoms is quickly reemitted. But this emitted light is radiated in all directions, so little is reintroduced into the original beam. (Thus, the dark lines are not filled in; they remain dark.) If we move the prism to one side of the cell (out of the direct beam from the source), we will find that the light from the cell consists of two yellow lines. These are *emission lines* from the sodium atoms and they are *bright* lines.

In Chapter 19 we continue the discussion of line spectra, a subject that remained mysterious until the development of quantum theory in the 1920s. For now we need only to point out that each chemical element emits (and absorbs) light at its own characteristic wavelengths. Figure 16.38 shows portions of the bright-line spectra of several elements. These spectra were recorded photographically in a *spectrograph* that dispersed the light by means of a diffraction grating (see Section 15.5). The line spectrum of an element is a kind of unique "fingerprint" that permits the unambiguous identification of the element on the basis of the wavelengths of its emitted (or absorbed) light. Spectrographic measurements constitute one of the most powerful techniques at our disposal for performing chemical analyses.

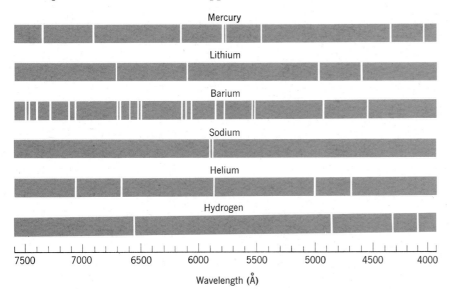

Figure 16.38 Bright-line spectra for several elements.

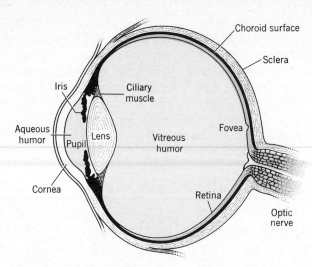

Figure 1 Schematic diagram of the human eye.

The Eye and Vision

The human eye is a marvelous optical instrument. It can produce sharp images for a wide range of object distances; it can function over an extremely wide range of light intensity; it can detect small differences in color; and it requires very little maintenance. Modern technology has been unable to develop an optical instrument with comparable performance.

The eye of an adult human is an almost spherical structure with a diameter of approximately 23 mm (Fig. 1). The outer protective coating of tough elastic tissue is called the *sclera*. At the front of the eye, the sclera leads into the *cornea* which is a transparent structure that admits light into the eye. After entering the cornea, incident light passes through a chamber filled with a clear liquid called the *aqueous humor*. The light next passes through and is limited by the pupillary aperture in the *iris* before entering the *lens* of the eye. Behind the lens is a jellylike substance called the *vitreous humor*. Finally, at the rear of the eye is a dense collection of light-sensing elements (*rods* and *cones*) that constitute the *retina*.

The muscles of the eye perform two mechanical functions that relate to the optical properties of the eye. The purpose of the optical elements of the eye is, of course, to produce on the retinal surface an image of the object viewed. A normal relaxed eye will focus parallel light rays on the retina. In fact, the rays from an object as close as 6 m are essentially parallel, so a relaxed eye will produce a sharp retinal image for an object at any distance from about 6 m to infinity. If an object is closer to the eye than 6 m, the focal length of the lens must be changed in order to produce a sharp image. This is accomplished by the *ciliary muscles* (see Fig. 1) which distort the lens into a shape that has a greater curvature than the relaxed lens. This action is called *accommodation*. The curvature of the lens cannot be increased indefinitely, and the eye can no longer accommodate an object that is closer than a certain point called the *near point* of the eye. For adults, the distance of the near point is about 25 cm, but the actual distance varies from person to person and depends particularly on age. A 10-year-old child may have a near point of only 7 cm; at 20 years, the near point may have increased to 10 or 12 cm; and by age 40 or 45, the distance may be about 25 cm. If the distance to the near point increases beyond about 25 cm, it no longer is convenient to read or to do close work; eyeglasses are then required to assist the eye in its focusing. The deterioration of the eye's accommodation is due to the gradual decrease in the elasticity of the lens.

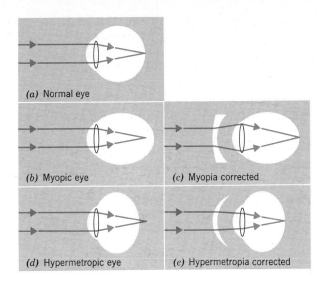

Figure 2 Two common eye defects and their correction by means of auxiliary lenses.

The second muscular function in the eye is the control of the pupil size. The human eye is sensitive over an intensity range of nearly 10^{10}, the upper extreme corresponding to an intensity at which the eye experiences discomfort (such as when gazing momentarily directly at the Sun). When the level of illumination is low, the pupil is dilated (opens) to a diameter of 7 or 8 mm; when the incident light is intense, the pupil is constricted to a diameter of 2 or 3 mm. The pupil therefore acts in the same way as the adjustable aperture in a camera. However, the amount of light entering the eye can be regulated by the pupil only over a range of about 10:1, which is small compared to the sensitivity range of the eye. The most effective way that the eye responds to changes in illumination is by adaptive processes in the retina. However, these processes are relatively slow, so the rapid response of the pupil probably is an important part of the initial adaptation to light intensity changes.

Parallel light rays will be focused by a relaxed eye lens only if the focal length of the lens (actually, the lens-cornea combination) is equal to the diameter of the eyeball* (Fig. 2a). If the diameter is greater than the focal length, the rays will converge too rapidly and will focus in front of the retina, producing a blurred image on the retina (Fig. 2b). This type of eye defect is called *myopia* (shortsightedness) and is corrected by a lens that diverges the rays before they enter the eye (Fig. 2c).

The defect called *hypermetropia* (longsightedness) is the opposite of myopia, namely, the diameter of the eyeball is too small to permit parallel rays to come to a focus before the retina is reached (Fig. 2d). Hypermetropia is corrected by using a converging lens (Fig. 2e).

A third type of eye defect occurs when there is an asymmetry in the curvature of the lens-cornea system. This defect, called *astigmatism,* makes it impossible for the eye to focus simultaneously on both horizontal and vertical lines. Correction for astigmatism is afforded by a lens that has different curvatures along different planes, thereby compensating for the eye's lack of symmetry by introducing the opposite asymmetry.

As a person ages, the ability of the eye to accommodate lessens and a converging corrective lens is required to restore the ability to focus on nearby objects. This effect is called *presbyopia* (old sight) and the correction is the same as that for hypermetropia. But a person suffering with presbyopia requires an assisting lens

---

*See footnote on page 412.

only for reading because the distant vision is unaffected. Current practice is to use "half glasses" to correct presbyopia. With these glasses, the person looks down and through the glasses for reading, whereas he looks over the glasses for distant viewing.

It is customary to discuss the human eye and correcting eyeglasses in terms of the *power* of the lenses. If the focal length f of a lens is measured in meters, the *power* of the lens is defined to be $1/f$, measured in units called *diopters*. Thus, a lens that has a focal length of 4 cm has a power of $1/(0.04 \text{ m}) = 25$ diopters.

The power of the eye's lens system is maximum when the eye is fully accommodated. In this situation the object distance (for an adult normal eye) is 25 cm and the image distance is the diameter of the eyeball,* 23 mm. Then,

$$\frac{1}{f} = \frac{1}{x_0} + \frac{1}{x_i} = \frac{1}{0.25 \text{ m}} + \frac{1}{0.023 \text{ m}} = 47.5 \text{ diopters} \tag{1}$$

Most of the power of the human eye is contributed by the cornea and relatively little by the lens proper (see Exercises 1 and 3).

Suppose that a person suffering from presbyopia is able to focus on an object when it is no closer than 60 cm. What power eyeglasses are needed to correct this condition? First, we notice that the power of the lens at maximum accommodation is

$$\frac{1}{f_1} = \frac{1}{0.60 \text{ m}} + \frac{1}{0.023 \text{ m}} = 45.1 \text{ diopters}$$

Now, the result of Problem 16.24 is that the focal length f of a close combination of two lenses with focal lengths f_1 and f_2, respectively, is

$$\frac{1}{f} = \frac{1}{f_1} + \frac{1}{f_2}$$

That is, the *power* of the combination is equal to the sum of the powers of the individual lenses. (This result is the reason for introducing the idea of lens power.)

In the case of the presbyopic eye, we need a correcting lens with a power $1/f_2$ which, when added to the power of the eye, $1/f_1 = 45.1$ diopters, will produce a normal result, that is, $1/f = 47.5$ diopters. Thus,

$$\frac{1}{f_2} = \frac{1}{f} - \frac{1}{f_1} = 47.5 \text{ diopters} - 45.1 \text{ diopters} = 2.4 \text{ diopters}$$

(Eyeglass powers are usually specified in increments of $\frac{1}{4}$ diopter, so the prescription in this case would be for a lens with a power of $2\frac{1}{2}$ diopters.)

References

G. H. Begbie, *Seeing and the Eye,* Natural History Press, Garden City, N.Y., 1969, Chapter 3.

A. C. Guyton, *Textbook of Medical Physiology,* Saunders, Philadelphia, 5th ed., 1976, Chapter 58.

F. R. Hallett, P. A. Speight, and R. H. Stinson, *Introductory Biophysics,* Halsted (Wiley), New York, 1977, Chapter 3.

---

*The eye is a complicated structure and we should not really take the elementary view that the focal length is equal to the diameter of the eyeball. However, this simplification does not introduce appreciable error into the discussion in this essay or the next.

K. N. Ogle, *Optics: An Introduction for Ophthalmologists,* 2nd ed., Charles C. Thomas, Springfield, Ill., 1968.

A. L. Stanford, *Foundations of Biophysics,* Academic Press, New York, 1975, Chapter 7.

■ *Exercises*

1. Although the lens of the eye has rather strongly curved surfaces (see Fig. 1), the focusing effect of the lens is actually not very great and most of the focusing action in the eye is due to the cornea. Explain this effect. (The index of refraction of the lens material is 1.424, whereas that for the aqueous humor and vitreous humor is 1.336.)

2. One of the primary effects of the constriction of the pupil in high illumination conditions is to improve the sharpness of the retinal image. Discuss this effect in terms of the *depth of field* of the eye's lens system. (The depth of field of a lens system is the range in distance over which objects will be in sharp focus.) Make a sketch and explain how the depth of field depends on the pupil size.

3. An adult normal eye can focus on an object at any distance from 25 cm to infinity. What is the range of power of the eye? Consider the cornea and the lens to be a close combination of lenses. What is the power of the cornea? What are the maximum and minimum powers of the lens proper?

4. A certain adult has a normal cornea and lens, but the diameter of the eyeball is 24 mm. From what condition does this individual suffer? What is the power of the correcting lens that will permit normal vision?

16.5 Polarized Light

Orientation of the Electric Field Vector

Any electromagnetic wave (including light) consists of propagating oscillations in an electromagnetic field. In Fig. 15.47 we showed the oscillating nature of the electric field vector **E** and the magnetic field vector **B**, which together characterize the electromagnetic field. In this diagram we see that the vector **E** oscillates along the *y*-direction because the antenna that produces the wave (Fig. 15.46) is oriented along this direction. In any light source, however, there are many oscillating "antennas" (i.e., atoms), and so there are many individual waves with the electric field vectors oscillating in all possible directions. (We will direct our attention to the electric field vector for the remainder of this chapter because **E** is of primary importance in all of the phenomena we will be discussing.)

Figure 16.39a shows the electric field vectors for several individual components of a light beam that is propagating along the *z*-axis, toward the viewer. When the field vectors point in all directions, as they do in this example, we say that the light is *unpolarized*. This is the case for light from an ordinary source, such as an incandescent lamp or the Sun.

Any vector can be represented in terms of components that lie along mutually perpendicular axes. Therefore, we can simplify the picture in Fig. 16.39a by resolving all of the individual **E** vectors into *x*- and *y*-components. The result is shown in Fig. 16.39b. This diagram still represents unpolarized light because there is no preferred direction for the field vectors. In Fig. 16.39c, however, we have eliminated the oscillations along the *x*-direction and have remaining only those along the *y*-direction. This diagram represents light that is *polarized* along the *y*-direction. (This is exactly the case in Fig. 15.47.)

Optical Polarizers and Analyzers

In order to understand some of the properties of polarized light, it is helpful to look first at the behavior of transverse waves on a string. Suppose that we attempt to pass such waves through a slot in a barrier, as in Fig. 16.40. If the slot is perpendicular to the direction of displacement of the wave (Fig. 16.40a), the wave is blocked and is not transmitted

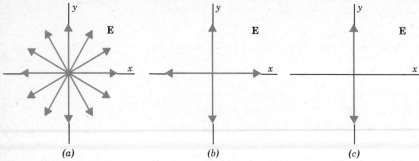

Figure 16.39 The electric field vectors as they appear to a viewer looking along the z-axis toward the approaching wave. (a) Unpolarized light, consisting of individual waves with **E** vectors oriented in all directions. (b) A simpler representation of unpolarized light; all of the **E** vectors have been resolved into components that lie along the x- and y-axes. (c) Polarized light with the **E** vectors all oriented along the y-direction.

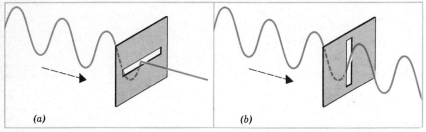

Figure 16.40 (a) A transverse wave whose displacement is vertical cannot pass through a horizontal slot in a barrier. (b) If the slot is in line with the direction of displacement, the wave is freely transmitted.

through the barrier. On the other hand, if the slot is in line with the direction of displacement of the wave (Fig. 16.40b), the wave is freely transmitted through the barrier.

If the part of the string that lies to the left of the barrier in Fig. 16.40b carries a complicated wave consisting of the superposition of vibrations with many different orientations, the slot will allow only the vertical component of each wave to pass. That is, all of the horizontal components will be filtered out and the remaining wave will be completely polarized in the vertical direction.

We might consider attempting to polarize electromagnetic waves by passing them through a wire grid. But, then, polarization will result only for wavelengths that are longer than the grid spacing. Thus, a wire grid will act as a polarizer for microwaves or television signals, but the spacing cannot be made sufficiently small to polarize light.

An effective optical polarizer can be constructed from a "grid" that consists of aligned molecules instead of wires. The first material of this type to be produced was *H-sheet*[2] in which the polarizing agents are long molecules containing many iodine atoms. These molecules can be made to lie close together in almost perfect parallel alignment. The result is a grid with exceedingly narrow spacing that will pass light with only a particular orientation.

When unpolarized light is passed through H-sheet, the transmitted light is polarized in the direction of the polarization axis of the material, as indicated in Fig. 16.41. (The polarization axis of H-sheet is *perpendicular* to the direction of the long molecules. Can you see why? Refer to Question 16.13.) If this now-polarized light is incident on another piece

---

[2]The first sheet-type polarizer, called J-sheet, was invented in 1928 by E. H. Land, who later developed the popular "picture-in-a-minute" system of photography. The polarizing agents in J-sheet were microcrystals, which tended to produce a haziness in the transmitted light because of their size. This difficulty was overcome in the mid-1930s by the development of H-sheet with its aligned molecular polarizers. In 1937, Land formed the Polaroid Corporation to manufacture and distribute his polarizing materials.

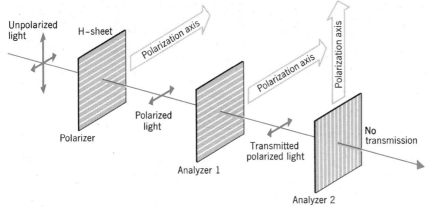

Figure 16.41 The fraction of an unpolarized light beam that is transmitted through a piece of H-sheet is polarized in the direction of the polarization axis of the material. This now-polarized light will be freely transmitted through another piece of H-sheet if it has the same orientation, but there will be no transmission through a sheet that has a perpendicular orientation.

of H-sheet that has the same orientation, the beam will be transmitted with very little loss in intensity. However, if the polarized beam is next incident on a sheet with its polarization axis rotated by 90°, the beam will be absorbed and there will be no transmitted light (Fig. 16.41). The first sheet of material through which the beam passes is called the *polarizer;* the subsequent sheets are called *analyzers*.

If you place two pieces of H-sheet together and slowly rotate one with respect to the other, you will see the transmitted light change from bright to dark and back to bright with every 180° of rotation. However, no polarizer or analyzer can be made absolutely perfect. The beam transmitted through a polarizer is never 100 percent polarized, nor is a beam reduced to exactly zero intensity by a perpendicular polarizer-analyzer combination. A perfect polarizer would transmit 50 percent of an unpolarized beam; H-sheet can be made to transmit about 38 percent of the incident light. By adding absorptive substances, the transmission can be reduced to any value. (Additional absorption is desirable, for example, in the manufacture of Polaroid sunglasses.)

The Law of Malus

The manner in which the intensity of the transmitted light for a polarizer-analyzer pair varies as the angle between their polarization axes is rotated was discovered as early as 1809 and is called the *law of Malus*. Unpolarized light enters the polarizer in Fig. 16.42 from the left and emerges polarized in the direction of the polarization axis of the polarizer. At this point the light intensity is reduced to $\frac{1}{2}$ of the incident unpolarized light I_0. The analyzer-polarization axis is shown in Fig. 16.42 set at an angle θ with that for the polarizer. The electric field strength vector $\mathbf{E}$ incident on the analyzer is replaced by its vector components, one parallel to the analyzer polarization axis, $\mathbf{E}'$, and the other in the perpendicular direction, $\mathbf{E}''$. Only the component $\mathbf{E}'$ is transmitted, thus

$$E' = E \cos \theta \tag{16.15}$$

The final transmitted intensity passed by the analyzer is thus,

$$I = \frac{I_0 \cos^2 \theta}{2} \tag{16.16}$$

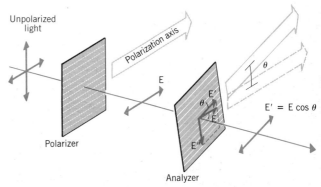

Figure 16.42 A polarizer-analyzer pair set with an angle θ between the polarization axes. The electric field vector $\mathbf{E}$ incident on the analyzer is decomposed into parallel and perpendicular components $\mathbf{E}'$ and $\mathbf{E}''$. The component $\mathbf{E}'$ is transmitted.

Example 16.10

What angle θ should be set between the polarization axes of a polarizer-analyzer pair of H-sheets if the final transmitted light has an intensity $\frac{1}{4}$ of the initially incident unpolarized light? Assume the transmission through an H-sheet to be 0.80 of ideal.

The transmitted intensity is reduced to

$$\frac{0.80}{2} = I_0$$

on passage through the polarizer. It is further reduced by the factor $0.80 \cos^2 \theta$ in passing through the analyzer, thus, we have

$$\tfrac{1}{4} = \tfrac{1}{2}(0.80)^2 \cos^2 \theta$$

Solving for θ gives $\theta = 27.9°$ ∎

Polarization by Reflection and Scattering

There are several natural processes that produce polarized light. For example, when light is reflected by a transparent material such as water or glass (but *not* by a metal surface), the reflected light is partially polarized, with the electric vector parallel to the surface. The polarization of the reflected beam is 100 percent at one particular angle of incidence, called *Brewster's angle* θ_B, which occurs when $\tan \theta_B = n$. (The index of refraction of the medium of incidence is assumed to be unity.) For ordinary glass with $n = 1.520$, a reflected light beam will be completely polarized when the incident angle is equal to $\tan^{-1} 1.520 = 56.7°$. For water ($n = 1.333$), Brewster's angle is 53.1°. The reflected light is polarized with the electric vector parallel to the surface. Therefore, if this light is viewed through a piece of H-sheet with the polarization axis vertical, the intensity will be much reduced. This is the way Polaroid sunglasses reduce (or sometimes completely eliminate) the glare of light reflected from water or glass.

When sunlight is scattered by air molecules in the atmosphere, the scattered light is partially polarized. The polarizing effect is particularly strong when sunlight is scattered through 90° (Fig. 16.43). On a clear day, when there is little dust or water vapor in the air, if the Sun is near the horizon, the light from overhead can be polarized to the extent of 70 percent.

The human eye is not adapted to the detection of polarized light. Most persons see absolutely no difference between polarized and unpolarized light. (A few persons can perceive polarization effects—such

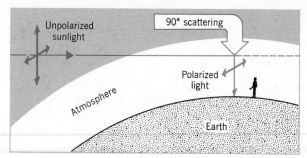

Figure 16.43 When unpolarized sunlight is scattered by the air molecules in the atmosphere, the scattered light is partially polarized. When the scattering angle is 90°, the percentage of polarization is maximum (about 70 percent).

as those associated with partially polarized sunlight—even with the unaided eye, but the effect is always small.) Certain insects, however, have eyes that are much more sensitive to polarized light than is the human eye. Some types of ants and beetles, and particularly the honeybee, can sense the direction of polarization of polarized light. Indeed, the honeybee uses the polarization of scattered sunlight for navigation purposes. It has been demonstrated that bees navigate by reference to the Sun. But if the Sun is obscured by clouds, they can continue to navigate properly if any piece of blue sky is visible. It has been convincingly shown that bees detect the polarization of the sunlight scattered from the patch of blue sky and from this they infer the position of the Sun!

16.6 Optical Instruments

Microscopes

The simplest type of optical instrument is the *magnifying glass*, which we discussed in Section 16.3. We commonly find these magnifiers with $\Gamma = 2$ or 3. If the magnifying power is greater, then the various aberrations and distortions to which an inexpensive lens is subject will seriously degrade the image. With careful design the magnifying power can be increased to about 10. For applications that require higher magnification, it is necessary to add more lens elements; the instrument then becomes a *microscope*.

Figure 16.44 shows the lens arrangement in an elementary, two-lens microscope. The lens that is nearer the object to be magnified is called the *objective lens;* the lens that is nearer the eye is called the *eyepiece* (or *ocular*). Notice that the object is located just *outside* the focal point of the objective lens so that a *real* image is formed between the two lens. This *intermediate image* is located just *inside* the

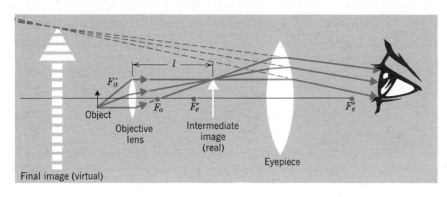

Figure 16.44 Schematic diagram of a simple, two-lens microscope.

focal point of the eyepiece so that a final *virtual* image is formed. The eyepiece is actually a simple magnifier used to view the real image formed by the objective lens.

The magnification produced by the two-lens microscope illustrated in Fig. 16.44 is obtained in the following way. First, the objective lens produces an enlarged image of the object; the magnification M_0 of the objective lens is defined in Eq. 16.8. Next, the eyepiece uses this image as an object and forms an enlarged virtual image for the eye to view; the magnifying power Γ_e of the eyepiece is defined in Eq. 16.10. The overall magnification of the microscope is, then,

$$\text{magnification} = M_0 \times \Gamma_e \qquad (16.17)$$

Thus, we see that the eye views an image enlarged by a factor equal to the product of the lateral magnification M_0 of the objective and the angular magnification Γ_e of the eyepiece.

If the lens separation is adjusted until the virtual image is at the near point of the eye, the magnification will be maximum. However, if the intermediate image is allowed to fall exactly at the focal point of the eyepiece, the image will be at infinity and the rays viewed by the eye will be parallel. This is the condition of minimum eye strain and is the usual way in which a microscope is used. Then, the angular magnification of the eyepiece is $\Gamma_e = 25/f_e$, where f_e is the focal length of the eyepiece. Also, the object is located almost at the focal point of the objective lens, so we can express the magnification of this lens as $M_0 = -l/f_0$, where l is the image distance (see Fig. 16.44) and where f_0 is the focal length of the objective. Therefore, we have

$$\text{magnification} \cong -\frac{l}{f_0} \times \frac{25}{f_e} \qquad (16.18)$$

Because all of the quantities in this expression are positive, the negative sign indicates that the image is inverted.

● *Example 16.11*

A lens with a focal length of 0.8 cm is used as the objective in a microscope whose eyepiece has a focal length of 2.0 cm. What is the magnification of this microscope?

In order to use Eq. 16.18 we need a value for l. In most microscopes the lenses are arranged so that $l = 15\text{–}20$ cm. Let us use $l = 18$ cm. Then,

$$\text{magnification} \cong -\frac{18 \text{ cm}}{0.8 \text{ cm}} \times \frac{25 \text{ cm}}{2.0 \text{ cm}} \cong -280$$

This is usually written as 280×, dropping the negative sign. ■

Refracting Telescopes

A simple refracting telescope consists of an objective lens and an eyepiece, just as does a microscope. In a microscope the object is placed near the focal point of a short focal length objective lens. In a telescope, the focal length of the objective lens is large and the object distance is so great that we consider the incident rays to be parallel, as shown in Fig. 16.45. The objective lens forms a real intermediate image at the focal point F_0. This image lies just inside the focal point F_e' of the eyepiece so that an enlarged virtual image is formed for the eye to view.

The angular magnification of the telescope illustrated in Fig. 16.45 is the ratio of the image angle θ_i to the object angle θ_0:

$$\text{angular magnification} = \frac{\theta_i}{\theta_0}$$

Because we are dealing with small angles, the object angle θ_0 is approximately equal to the height h of the intermediate image divided by the focal length

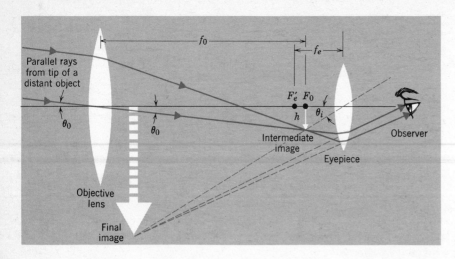

Figure 16.45 Schematic diagram of a simple refracting telescope.

f_0 of the objective: $\theta_0 \cong h/f_0$. If the focus of the instrument is adjusted for minimum eye strain, the intermediate image will be located at the focal point F_e' of the eyepiece and the rays viewed by the eye will be parallel. (The length L of the telescope then becomes $L = f_0 + f_e$.) Under this condition, $\theta_i \cong -h/f_e$. (θ_i is negative because the angle lies *below* the optic axis of the instrument; compare Fig. 16.45 with Fig. 16.24.) Thus,

$$\text{angular magnification} = \frac{\theta_i}{\theta_0} \cong \frac{-h/f_e}{h/f_0} = -\frac{f_0}{f_e}$$

(16.19)

Again, the overall negative sign indicates an inverted image. In order for the magnification to be large, we need a small value for f_e; this is the same requirement as for a microscope (see Eq. 16.18). But for a telescope to have a large magnification, the focal length f_0 of the objective needs to be large; this is opposite to the requirement for a microscope.

The simple telescope shown in Fig. 16.45 forms an *inverted* image. This is of no particular consequence for astronomical observing, but it makes this design unsuited for terrestrial viewing. In order to form a final erect image, another lens element of a system of prisms is required. The latter alternative is used in the construction of binoculars. (The prism system also decreases the overall length of the instrument to reasonable dimensions for holding by hand—see Fig. 16.11.)

Reflecting Telescopes

If a refracting telescope is to be used in observing weak light sources, as is frequently the case in astronomical research, the lens diameter must be made large in order to increase the light-gathering capability of the instrument. Large glass lenses are difficult to manufacture without flaws, they are awkward to handle, and they are subject to cracking due to temperature changes. Glass lenses are also subject, in some degree, to chromatic aberration (Section 16.4). For these reasons, large-diameter refracting telescopes are not practical instruments, and very few are still in use for astronomical research.[3]

Isaac Newton understood the problems associated with refracting telescopes and in 1667 he devised a new type of telescope in which the main focusing element is a curved mirror instead of a lens. A diagram of a Newtonian *reflecting telescope* is shown in Fig. 16.46. As we pointed out in Section 16.3, the exact focusing of parallel light rays by a curved mirror requires that the mirror surface be paraboloidal. However, if the radius of curvature is large and only a small part of the surface is actually used (this is the usual case in astronomical telescopes), a spherical mirror surface is adequate for all except the most demanding purposes.

Figure 16.46 shows parallel light rays from a distant source incident on the mirror of a reflecting telescope. Before the reflected rays are brought to a focus, they are intercepted by a small mirror located within the telescope tube and are deflected through a hole in the tube to a viewing system or photographic plate.

The largest American reflecting telescope is the 200-in. (5-m) diameter instrument located on Mount Palomar in southern California. An even larger telescope (6-m diameter) has recently been put into service by Soviet astronomers in the Caucasus Mountains. It is unlikely that any larger telescopes

---

[3] The largest refractor still in service is the 40-in. diameter instrument at the Yerkes Observatory of the University of Chicago.

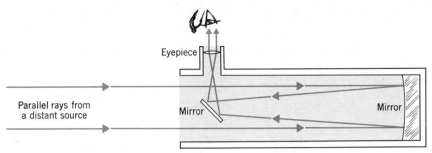

Figure 16.46 Schematic diagram of a Newtonian reflecting telescope. Different lenses are used at the eyepiece position, depending on whether the image is to be recorded photographically or viewed with the eye.

will ever be constructed. More information of higher quality can be obtained by using smaller instruments mounted on artificial satellites or on the Moon. These telescopes can operate free from the disturbing influences of the Earth's atmosphere, effects that limit the usefulness of Earth-based instruments.

Resolution

When two point sources of light are viewed with some sort of optical instrument, such as a telescope, two separate images will be seen if the sources are sufficiently far apart. However, when the sources are moved closer together, we know from experience that the images will eventually merge together into a single spot even though the sources do not actually coincide. How close will the sources be when the light pattern can just be resolved into two distinct images?

The physical limit to the resolution of two light sources is set by *diffraction* effects. As we pointed out in Section 15.4, light that passes through a single slit will produce a diffraction pattern consisting of a broad central maximum and secondary maxima of lesser intensity (Fig. 15.42). If we replace the slit with a circular aperture, the same general type of diffraction pattern is found (although the positions of the minima and the intensity ratios for the various maxima are slightly different). In particular, if we use a lens in the circular aperture to focus incident parallel rays from two distant sources, fuzzy images will be formed (Fig. 16.47). As the sources are moved closer together, the broad central maxima overlap and finally merge. The two sources will be just distinguishable when the central maximum for one source falls at the position of the first minimum for the other source. Then, the angular separation α of the sources is given by

$$\alpha = \frac{1.22\,\lambda}{d} \tag{16.20}$$

where λ is the wavelength of the light and d is the diameter of the lens. One way to improve the resolv-

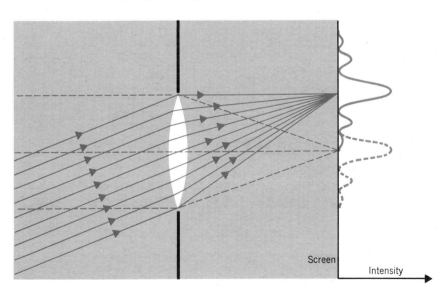

Figure 16.47 Because of diffraction, a lens does not focus the light from a distant source as a point. Instead, a fuzzy spot is formed. If two sources are close together, the diffracted images will merge.

ing power of an instrument is to use light with a short wavelength. Thus, the image of an object produced with blue light will be sharper than the image produced by red light.

● **Example 16.12**

What is the diffraction-limited resolution of the 200-in. Mount Palomar telescope for yellow light?

According to Eq. 16.20,

$$\alpha = \frac{1.22\,\lambda}{d} = \frac{1.22 \times (6 \times 10^{-7}\ \text{m})}{5\ \text{m}}$$

$$= 1.5 \times 10^{-7}\ \text{rad} = 0.03\ \text{arc sec}$$

Although this is the *diffraction*-limited resolution of the Palomar instrument, other factors (such as atmospheric turbulence) conspire to reduce the actual resolution to about 1 arc sec or 5×10^{-6} rad. With this resolution, the Palomar telescope can distinguish two points at a distance l that have a linear separation $z = \alpha l$. For example, on the surface of the Moon ($l = 3.8 \times 10^8$ m), we have

$$z = (5 \times 10^{-6}\ \text{rad}) \times (3.8 \times 10^8\ \text{m}) = 1900\ \text{m}$$

Thus, the linear resolution at the distance of the Moon is about 2 km. The main advantage of a large telescope such as the Palomar instrument is not that the diffraction-limited resolution is small (other factors negate this), but that the light-gathering power is large. ∎

Resolving Power of a Microscope

The diffraction-limited resolving power of a circular aperture (or a telescope) is given by Eq. 16.20, which expresses the angular separation α of two distant point sources of light that can just be distinguished. In referring to the performance of a microscope, however, it is customary to express the resolving power in terms of the distance separating two point objects that can just be seen as two objects instead of one. Now, an object that is viewed with a microscope must be illuminated with light from some source. Usually, a condensing lens is used to converge the light on the object. In this case, the resolving power becomes (we only state the result because the derivation is complicated)

$$\text{R.P.} = \frac{1.22\,\lambda}{2n \sin \beta} \qquad (16.21)$$

where n is the index of refraction of the medium between the object and the objective lens and where β is the angle between the lines drawn from the object to the center and to the edge of the objective lens (Fig. 16.48).

An object that is viewed with a microscope must be placed in a definite position relative to the objective lens; this position is determined by the focal length of the lens (see Fig. 16.44). For this reason, the angle β in Eq. 16.21 is fixed for a particular lens. It proves convenient to specify for a lens the value of $n \sin \beta$—this quantity is called the *numerical aperture* (N.A.) of the lens and is a measure of the effective angular opening presented by the lens for gathering light from the object:

$$\text{N.A.} = n \sin \beta \qquad (16.22)$$

The greater the value of N.A., the greater will be the amount of light gathered and the better will be the resolving power of the lens.

We can now express the resolving power of a microscope as

$$\text{R.P.} = \frac{0.61\,\lambda}{\text{N.A.}} \qquad (16.23)$$

This formula strictly applies only to a single lens, not to a microscope as a whole. However, the performance of a microscope is determined almost exclusively by the resolving power of the objective lens. Most microscope objectives are stamped with the magnification value and the N.A. value.

The resolving power of a lens will be best (that is, *smallest*) when the numerical aperture is as large as possible. The maximum practical N.A. is obtained when β is about 70°. Then, if the space between the object and the lens is filled with oil (called *immersion oil*, $n = 1.55$), the numerical aperture becomes

$$\text{N.A. (max)} = n \sin \beta = 1.55 \times \sin 70° = 1.46$$

If light with $\lambda = 6 \times 10^{-7}$ m is used, the resolving power is

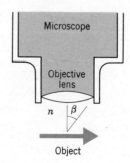

Figure 16.48 Geometry for determining the angle β in the expressions for R.P. and N.A.

$$\text{R.P.} = \frac{0.61\,\lambda}{\text{N.A.}}$$

$$= \frac{0.61 \times (6 \times 10^{-7}\,\text{m})}{1.46}$$

$$= 2.5 \times 10^{-7}\,\text{m} = 2500\,\text{Å}$$

This result is for the optimum case. Common values for objectives are (mag. = 10×, N.A. = 0.25) and (mag. = 40×, N.A. = 0.65); the corresponding resolving powers are 14,640 Å and 5630 Å.

A microscope adjusted for maximum magnification presents to the viewer an image that lies at the near point of the eye. At this distance, most persons can resolve two points separated by 2×10^{-4} m (0.2 mm). The viewer's eye therefore places a limitation on the maximum useful magnification of a microscope; this magnification is

$$\frac{\text{maximum useful}}{\text{magnification}} = \frac{2 \times 10^{-4}\,\text{m}}{2.5 \times 10^{-7}\,\text{m}} = 800\times$$

(This corresponds to using a 100× objective and an 8× ocular). If a magnification greater than about 800× is used, the viewer will only see a magnified diffraction pattern surrounding the image.

Visual Acuity

The light-sensing part of the eye consists of a collection of sensors called *rods* and *cones*, which are packed closely together on the retinal surface. The human eye contains about 6×10^6 cones and about 120×10^6 rods. The rods and cones are connected to *bipolar cells* which, in turn, are connected to about a million neurons called *ganglion cells*. The axons of the ganglion cells, which are spread out over the retinal surface, collect together and pass out of the eye as the *optic nerve* (see Fig. 1 in the essay on page 410).

The rods and cones are not distributed uniformly over the retinal surface. At the point where the optic nerve passes through the retina, rods and cones are completely absent; consequently, light that is incident on this area (the *blind spot*) causes no visual sensation. When we fixate on an object by looking directly at it, the light rays from the object converge at the position of the *fovea*, a small pitlike depression in the retina that contains only cones (see Fig. 1 in the previous essay). (Notice in the figure that the fovea does not lie on a line that passes along the axis of the lens. The line from the center of the lens to the fovea—the *visual axis*—is at an angle of 3–5° with respect to the lens axis.) The fovea has an area of about $\frac{1}{20}$ mm^2 and contains about 10,000 cones; the average density of cones over the rest of the retina is about 5000 per mm^2. The rods begin to appear just outside the fovea and the density rapidly increases to about 150,000 per mm^2; the density gradually falls to about 30,000 per mm^2 at the edges of the retinal surface.

Rods and cones perform different functions. Rods are much more sensitive in low-light conditions than are cones. On the other hand, rods do not readily distinguish color; essentially all of our information concerning color is derived from cones.

Our ability to see can be described in terms of how well we can distinguish two closely spaced points (this we call the *acuity* of the eye). We have seen, in Eq. 16.20, that the diffraction-limited resolving power of an aperture with a diameter d is

$$\alpha = \frac{1.22\,\lambda}{d} \tag{1}$$

An eye with a pupil diameter of 6 mm should therefore be able to resolve two point light sources ($\lambda = 6000$ Å) that have an angular separation of

$$\alpha = \frac{1.22 \times (6 \times 10^{-7}\,\text{m})}{6 \times 10^{-3}\,\text{m}} = 1.2 \times 10^{-4}\,\text{rad} \quad \text{or} \quad 25\text{ arc sec}$$

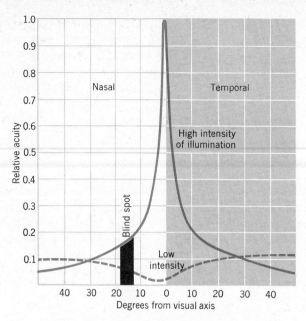

Figure 1 Relative visual acuity for images formed at various angles with respect to the visual axis (the line from the center of the eye lens to the fovea). "Nasal" means the side toward the nose and "temporal" means the side toward the temple. Notice that the acuity in the fovea falls to a very low value when the illumination is poor because the cones have sensitivity and there are no rods in the fovea. (After Stuhlman.)

Tests reveal that the resolution of the human eye is almost this good. An eye that is able to resolve two point sources with an angular separation of 3×10^{-4} rad or 1 arc min is said to have a *visual acuity* of 1. Most persons with "normal" eyesight have a visual acuity that is near 1; some persons have an acuity of 2, corresponding to a resolution of 30 arc sec. (Robert Hooke first pointed out, in 1671, that the human eye should be able to resolve the components of a double star if they are separated by an angle of 1 arc min.)

The resolving power of the eye depends on the distribution of light-sensing elements on the retina as well as on diffraction effects. Clearly, if two point sources are to be resolved, the images on the retina cannot evoke equal responses from adjacent cones. Instead, there must be an unexcited (or, at least, less excited) cone between the two cones that produce full response. The average diameter of a foveal cone is about 1.5 μm (1.5×10^{-6} m). Therefore, the separation of two excited cones with one unexcited cone between them is 3 μm. The angular separation of these cones, measured from the eye lens, is $\alpha = 3\ \mu\text{m}/23\ \text{mm} = 1.3 \times 10^{-4}$ rad, a value very near the diffraction limit. The spacing of the light receptors in the fovea is matched closely to the physical properties of the pupil aperture.

Maximum acuity is possible only for light that strikes the fovea, where the density of receptors is exceptionally large. Thus, the highest acuity is obtained when viewing objects on the visual axis and within a very narrow cone. If an image is formed on the retina outside the fovea, the acuity will be considerably decreased because of the low density of receptors. Acuity is also influenced by the level of illumination. If the light is intense, the low-sensitivity cones in the fovea are effective and the acuity is high. However, if the light level is low, the cones fail and because there are no rods in the fovea, the acuity in this region falls to a very low value (about $\frac{1}{20}$). Outside of the fovea, the rods provide an acuity that is greater than that in the fovea for low light levels but is substantially less than the high value realized with good illumination. Figure 1 shows the relative visual acuity as a function of the angle relative to the center of the fovea and for two levels of illumination. Note the occurrence of the blind spot, corresponding to light striking the region where the optic nerve passes through the retina.

Visual acuity is routinely measured by using an eye chart (a *Snellen* chart) that consists of rows of letters with different sizes. The row that is labeled 20 FT (or 6 M) consists of letters that have heights of 5 arc min and stroke widths of 1 arc min.

These letters should be discernible to persons with visual acuity 1 at a distance of 20 ft (6 m). If no smaller letters can be read, the person's eyesight is said to be 20/20 (or 6/6). If the smallest letters that can be read at 20 ft are those that a person with normal vision would read at 40 ft, the eyesight rating is 20/40. Or, if the 15-ft letters can be read at 20 ft, the rating is 20/15. Notice that the eyesight ratio is actually equal to the visual acuity number.

Another way to gauge visual acuity is in terms of how precisely two lines can be aligned (as in reading a vernier scale). In resolving two point sources of light, only two receptors are involved. But in aligning two lines, many receptors in a row are producing responses, so the information delivered to the brain is considerably increased. Consequently, it is possible to judge the alignment with much higher precision than would be expected on the basis of the resolution of points. Two lines can be aligned to within about 2 arc sec, corresponding to a visual acuity of 30.

References

G. H. Begbie, *Seeing and the Eye,* Natural History Press, Garden City, N.Y., 1969, Chapters 4 and 5.

F. R. Hallett, P. A. Speight, and R. H. Stinson, *Introductory Biophysics,* Halsted (Wiley), New York, 1977, Chapter 3.

A. L. Stanford, *Foundations of Biophysics,* Academic Press, New York, 1975, Chapter 7.

■ Exercises

1. From the information given in the essay about the size of the fovea and the density of cones, calculate the average distance between cone centers in the fovea.

2. The fovea has an area of about $\frac{1}{20}$ mm². Maximum acuity is therefore confined to a forward cone with what included angle? To what linear dimension does this correspond at a distance of 100 m?

3. What is the expected visual acuity if an image is formed on a part of the retina where the receptor spacing is 6 μm?

Summary of Important Ideas

The *speed of light* in vacuum is $c = 3.00 \times 10^8$ m/s. In any medium the speed v is less than c, and the ratio c/v is the *index of refraction n* of the medium.

When light is reflected from a surface, the *angle of reflection* is equal to the *angle of incidence*.

When light crosses a boundary between two optical media, the light is *refracted* according to Snell's law: $n_1 \sin \theta_1 = n_2 \sin \theta_2$.

If a light beam is incident on a boundary between two media from the side of the medium with the higher optical density, the beam will undergo *total internal reflection* if the angle of incidence exceeds the critical angle $\theta_c = \sin^{-1}(n_2/n_1)$, where $n_1 > n_2$.

A *convex* lens causes incident light to *converge* to a *focus*. A *concave* lens causes incident light to *diverge*.

A *convex* lens will form a *real* (*virtual*) image of an object located *outside* (*inside*) the focal point of the lens. A *concave* lens will form a *virtual* image of an object located at any position.

The object position x_0, the image position x_i, and the focal length f for a lens are related by the *lens equation*, $f^{-1} = x_0^{-1} + x_i^{-1}$.

The *magnification M* of a lens is $M = -x_i/x_0$. The *angular magnification* Γ of a lens that forms a virtual image at infinity is $\Gamma = 25/f$, where the focal length f is measured in cm.

Light rays can be focused by *paraboloidal mirrors* (or, less well, by spherical mirrors). The focal length of a spherical mirror is one half the radius of curvature. The *mirror equation* is the same as the lens equation.

Both lenses and spherical mirrors are subject to *spherical aberration* which results in imperfect focusing.

The *index of refraction* of a medium depends to a

small extent on the *wavelength* of the incident light. Because of this effect, a prism will *disperse* white light into a rainbow spectrum of colors and a simple lens exhibits *chromatic aberration*.

The light emitted by the atoms of each chemical element consists of *characteristic discrete wavelengths*. In *emission* the light is in the form of a spectrum of *bright* lines; in *absorption* the light displays *dark* lines superimposed on a continuous background.

Light from an ordinary source can be *polarized* by passing the beam through a substance that transmits only radiation with the electric field vector in a particular direction. Light can also be polarized by reflection from a transparent material at *Brewster's angle* and by scattering at 90° from air molecules. The law of Malus governs the transmitted intensity of polarized light through an analyzer; the intensity varies with $\cos^2 \theta$, where θ is the angle between the polarizer-transmission axis and the polarization of the incident wave.

A simple *microscope* (*telescope*) is a two-lens instrument with an *eyepiece* and with an *objective lens* that has a short (long) focal length.

The *diffraction-limited resolution* of a lens with a diameter d for light with wavelength λ is given in terms of the minimum angle α at which a pair of pointlike sources can just be distinguished: $\alpha = 1.22 \lambda/d$.

◆ **Questions**

16.1 Water waves exhibit refraction. The speed of propagation of water waves is greater in deep water than in shallow water. Argue that when water waves approach a shore, refraction causes the waves to come in parallel to the shoreline.

16.2 A sphere of plastic whose refractive index is 1.333 is placed in a container filled with water. Is it possible to determine the location of the sphere by visual inspection? Explain.

16.3 In Fig. 16.11 the light ray enters (and leaves) each prism perpendicular to the surface. Why do you suppose the prisms are arranged this way?

16.4 Parallel light rays are incident on a convex lucite lens that is immersed in a certain liquid. It is found that the rays are *diverged* by the lens. What can you say about the index of refraction of the liquid?

16.5 A *projector* is a device used to form an enlarged image on a screen of a transparent object such as a slide. Make a sketch of a projector system, indicating the light source, slide, lens, and screen. Indicate the focal point of the lens. (What type of lens must be used?)

16.6 Photographs can actually be taken with a camera that has no lens. Instead of the *lens* in such a camera there is a tiny *hole*. Explain how a *pinhole camera* works. (*Nautiluses*, a type of shelled *Mollusca*, have optic systems that use the pinhole effect instead of a lens.)

16.7 It is possible to construct a diverging lens that will form a (virtual) image that has the same size as the object? Explain.

16.8 Argue that the image formed by a convex spherical mirror is always virtual, always erect, always reduced, and always located inside the focal point, regardless of the position of the object.

16.9 A spherical surface with a radius of 2 m is to be used as the reflector in a searchlight. Where should the light source be located?

16.10 Does the fact that light can be polarized prove that light waves are *transverse*? Explain.

16.11 A fisherman wishes to look into a body of water to locate fish but he is hampered by reflected glare. To eliminate the glare, he uses Polaroid sunglasses. At what angle above the water surface should he look for the glare to be minimum?

16.12 A light beam consists of two components of equal intensity, one of which is polarized and one of which is unpolarized. This beam is passed through a piece of H-sheet and the intensity of the transmitted beam is measured. Make a sketch of the measured intensity as the H-sheet is rotated through 360°.

16.13 Suppose that an unpolarized radio wave is incident on a grid of parallel wires. The distance between the wires in the grid is small compared to the wavelength of the wave. Which component of the wave—parallel or perpendicular to the wires—will be transmitted through the grid? (Which component will be absorbed by inducing a current in the wires? Why will the other component *not* induce a current?)

16.14 Astronomical photographs are often made

using a blue filter. Why do you suppose this is done?

★Problems

Section 16.1 and 16.2

16.1 Two flat mirrors are placed together, edge to edge, with an angle of 90° between them. A light beam is incident on one of the mirrors with an angle ϕ. Trace the beam through two reflections and find the final direction of the beam.

16.2 What is the speed of light in (a) water and (b) heavy flint glass?

16.3 What is the speed of light in methyl alcohol? In diamond? (Use Table 16.1.)

16.4 Light pulses are started simultaneously down two 1-km tubes. One of the tubes is evacuated and the other contains air at standard conditions. What time interval separates the arrival of the two pulses at the opposite ends of the tubes?

16.5 Above a calm sea, the air density (and also the refractive index) sometimes decreases with altitude in a smooth way. Use a diagram and show that when this condition prevails it is possible to see a ship that is actually over the horizon. (This phenomenon is called *looming*.)

16.6 A light ray is incident from air on a piece of heavy flint glass ($n = 1.650$) at an angle of 37°. What angle does the refracted ray make with the normal to the surface?

16.7 A pan that is filled with water has a bottom surface that acts as a mirror. (The water surface and the mirror surface are parallel.) A light beam is incident on the water surface at an angle of 40°. Trace all of the refracted and reflected rays and show them in a diagram. At what angle does the primary ray emerge from the water?

16.8 A light ray enters a piece of lucite from air. Within the lucite the ray makes an angle of 32° with the normal to the surface. At what angle does the incident ray strike the surface? (Use Table 16.1.)

16.9* Refer to Fig. 16.7. Show that the ratio of the actual depth of the object in the water to its apparent depth (as seen by an observer in air) is equal to the index of refraction of the medium. Assume that the angles involved are sufficiently small that $\sin\theta \cong \tan\theta$.

16.10* A point object (in air) is located a distance d from a plate of material whose surfaces are parallel and which has a thickness l and an index of refraction n. An observer (in air) is on the opposite side. How far behind the plate does the object appear to the observer? Sketch the situation, showing several light rays.

16.11 A light ray is incident at an angle of 40° on a flat plate of glass with a thickness of 1.2 cm and an index of refraction $n = 1.60$ (see Fig. 16.9). By how much is the emerging ray displaced from the direction of the incident ray?

16.12 For what range of incident angles will a light beam originating in crown glass be totally reflected from a surface of methyl alcohol? (Use Table 16.1.)

16.13 Suppose that you lie on the bottom of a swimming pool and look upward. You will see a bright circle at the water surface. Why? What is the angular diameter of the circle (as seen by your eye)?

16.14 A beam of light enters the flat end of a lucite light pipe, as shown in the diagram. What is the maximum angle of incidence θ that will result in total internal reflection within the light pipe?

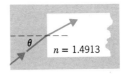

Problem 16.14

Section 16.3

16.15 A convex lens made from lucite is immersed in carbon disulfide. Make a sketch showing the behavior parallel light rays that are incident on the lens. (Refer to Table 16.1.)

16.16 Make a series of sketches that show the position and size of the image formed by a converging lens with focal length f when the object is located at (a) $x_0 > 2f$, (b) $x_0 = 2f$, (c) $f < x_0 < 2f$, (d) $x_0 = f$, and (e) $x_0 < f$. Characterize each image (real or virtual, erect or inverted, enlarged or reduced).

16.17 What type of lens or lens system must be contained in the box in the diagram if (a) ray A emerges as ray C and ray B emerges as ray D, (b) ray A emerges as ray D and ray B emerges as ray C? Sketch the diagram for each case.

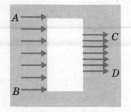

Problem 16.17

16.18* Parallel rays of light are incident on a glass sphere whose radius is 20 cm and whose refractive index is 1.50. Trace one off-axis ray through the sphere and find the location of the focal point.

16.19 A diverging lens has a focal length $f = -20$ cm. What is the position of the image of an object located at $x_o = 15$ cm?

16.20 A lens with a focal length of 40 cm (sign of f unknown) forms an image at $x_i = -20$ cm. Locate the object for the two possible types of lens.

16.21 The lens of a copy machine has a focal length f. Where should a document be located if the size of the copy is to be the same as the original?

16.22* In a two-lens combination, the first lens has a focal length $f_1 = 10$ cm and the second lens has a focal length $f_2 = 8$ cm. The lens are separated by a distance of 4 cm. Find the position of the image of an object placed 24 cm in front of the first lens.

16.23* Two lenses, each with $f = 20$ cm, are located on the same axis with a separation of 40 cm. Find the position and the size of the image of a 4-cm object located at $x_o = 60$ cm with respect to the nearer lens.

16.24* When two lenses are placed close together, they act as a single lens. Most lenses in quality optical instruments, such as cameras and microscopes, are combinations of two or more individual lenses. Consider two lenses, with focal lengths f_1 and f_2, that are separated by a distance d. Let parallel rays of light from a distant object be incident from the left on this combination. The image that would be formed by the first lens alone would be at a distance f_1 to the right of this lens. Let $d < f$, so that the image lies to the right of the second lens. This image is the *object* for the second lens. The object distance for the second lens is $f_1 - d$, which must be used as a *negative* quantity because the object lies to the right of the lens. Now, derive an expression for the focal length f for the combination. (Carefully sketch a ray diagram for the situation.) Show that when the separation d becomes very small ($d \to 0$), we have

$$\frac{1}{f} = \frac{1}{f_1} + \frac{1}{f_2}$$

Use this expression to find the focal length of a combination consisting of two lenses with focal lengths of (a) 4 cm and 5 cm, (b) 4 cm and -5 cm.

16.25 Optometrists express the *power* of a lens in terms of *diopters,* a unit that is equal to the reciprocal of the focal length of the lens in meters. What is the power of the lens that must be used in contact with a 60-diopter lens to produce an overall power of 80 diopters? (Use the result of the preceding problem.)

16.26 A camera lens has a focal length of 50 mm. What is the size of the image on the film of a 5-m high monument that is at a distance of 20 m?

16.27 A camera lens has a focal length of 25 mm and forms an image on the focal plane that is twice the size of the object. What are the object and image distances?

16.28 A camera with a 50-mm lens is to be used to photograph an object that is 1.6 m high. What should be the camera-to-object distance if the image is to just fill the frame (35 mm)?

16.29* In a certain slide projector the object, a transparent slide 35 mm wide, is located 10 cm behind the lens. The lens has a variable focal length in order to allow the slide image to fill the viewing screen for a range of projector-screen distances. If the screen has a width of 1.5 m, what range of focal lengths are required if the screen is to be placed from 3 to 5 m from the projector?

16.30 A certain magnifying glass has a maximum

value of Γ equal to 6. (a) What is the focal length of the lens? (b) At what distance from the lens must the object be placed to achieve this magnification?

16.31 A telephoto camera lens has a focal length of 600 mm. If this lens is used to photograph the Moon, what will be the diameter of the image on the film?

16.32 If your eye can focus clearly on an object only if it is no closer than 40 cm, what will be the maximum angular magnification you can realize with a lens that has a focal length of 8 cm?

16.33 A lens has a focal length of 10 cm. What is the maximum angular magnification of the lens? At what distance from the lens must the object be located in order to achieve this magnification? How close to the focal point is the object?

16.34 Use a ray diagram to find the position and the size of the image of a 2-cm object placed 20 cm in front of a convex spherical mirror whose radius of curvature is 15 cm. Check your results by using the mirror equation.

16.35 An image formed by a spherical mirror is to be one-half the size of the object. Where must the object be located for (a) a convex mirror and (b) a concave mirror? Is the image, in each case, real or virtual?

16.36 An object is placed 5 cm in front of a concave mirror with a radius of curvature of 20 cm. Locate and characterize the image.

16.37 A 4-cm object is placed 60 cm from a concave mirror whose focal length is 20 cm. Find the position and the size of the image.

16.38 A polished metal sphere has a diameter of 1 m. What is the location of the Sun's image formed by this surface?

16.39 Characterize the image formed by a concave spherical mirror of an object located at the center of curvature of the mirror. What is the analogous case for a convex lens?

Section 16.4

16.40 Plot the data in Table 16.2 and draw a smooth curve through the points. (Choose an appropriate scale!) Use the curve to estimate the refractive index of crown glass for wavelengths of 4000 Å and 6985 Å.

16.41 A certain mixture of elements is heated and the emitted light is spectrographically analyzed. Bright lines with the following wavelengths (among others) are found: 4554 Å, 4603 Å, 4934 Å, and 6708 Å. What elements are in the sample? (Use Fig. 16.38.)

16.42* A narrow pencil of light enters a 2-cm thick flat plate of crown glass, see Fig. 16.9, with angle $\theta_1 = 60°$. Determine the *difference* in the deviation of the light from the original direction, for the violet and red components of the incident light. Use Table 16.2.

16.43* What is the *difference* in the critical angles for violet and red light, when light is incident on the air-glass boundary for crown glass; see Fig. 16.10. Use Table 16.2.

Section 16.5

16.44 At what angle must a light beam be incident on a surface of carbon disulfide for the reflected light to be completely polarized? (Use Table 16.1.)

16.45* Show that when a light ray is incident on a substance at Brewster's angle, the angle between the reflected and transmitted rays is 90°. Can you guess why this condition leads to polarization of the reflected ray?

16.46* Two pieces of H-sheet are set with polarization axes at right angles, so that no light is transmitted through the pair. A third piece of H-sheet is placed between the original pair allowing some of the incident unpolarized light to now pass through the combination of the three sheets. At what angle should the polarization axis of the interposed sheet be set to allow the largest intensity of transmitted light through the entire combination?

Section 16.6

16.47 A simple microscope has an eyepiece with a magnification Γ = 10. What must be the focal length of the objective if the overall magnification is to be 150×?

16.48* A microscope consists of an eyepiece with a focal length of 1.5 cm and an objective lens with a focal length of 3 cm. The distance between the two lenses is 20 cm. What is the angular magnification of the eyepiece when the image is located at a distance of 25 cm?

Where is the image formed by the objective lens with respect to the eyepiece?

What is the image distance for the objective lens? Where is the object located?

What is the overall magnification of the microscope?

16.49 Suppose that you have a lens with a focal length of 2 cm and wish to build a telescope with an angular magnification of -80 ($80\times$). What additional lens do you need? Which lens should be the objective lens? What should be the separation of the two lenses?

16.50 Show in a sketch how you would add a third lens to the telescope illustrated in Fig. 16.45 in order to produce an erect virtual image.

16.51 What is the diffraction-limited resolution of the 40-in. Yerkes Observatory telescope? (Assume yellow light.) At what distance could the disc of a star with the same diameter as the Sun just be discerned? Are there any stars that are this close or closer to the Sun?

16.52 American television sets produce pictures consisting of 525 lines. A 23-in. (diagonal measure) television screen has a vertical height of approximately 16 inches. From how far away must you view this screen if the individual lines are not to be discernible? (Assume that the diameter of the eye's pupil is 5 mm.)

16.53 A radiotelescope is tuned to operate with 10-cm radiation. What must be the diameter of the radiotelescope if the angular resolution is to be as good as that of the human eye for yellow light? (Assume that the diameter of the eye's pupil is 5 mm.)

16.54 A microscope has an objective lens with a focal length of 2.5 mm and a diameter of 6 mm. What is the resolving power of this microscope if used with an object in air and with blue-green light (4900 Å)?

17 Relativity

Isaac Newton was aware of the fact that the equations of dynamics are equally valid in all inertial reference frames. That is, if $F = ma$ is valid in one reference frame, then it is also valid in any reference frame that moves at constant velocity with respect to the first frame (see Section 3.6). On the other hand, the theory of electromagnetic radiation that was being perfected in the last half of the nineteenth century seemed to require a *particular* reference frame for the proper description of electromagnetic waves. It was fashionable to imagine space to be filled with a mysterious medium, called the *ether,* which constituted the necessary reference frame and provided a "substance" through which the waves could propagate.

This artificial theory, based on a never-well-defined ether, was more trouble than it was worth. Despite the efforts of many able physicists and mathematicians, the ether theory was never completely successful in the interpretation of the various experimental results regarding the propagation of light. In 1905, a crucial new idea was contributed by Albert Einstein (1879–1955). In a single bold stroke, Einstein swept away the ether theory with all its complications and replaced it with only two postulates, which he used as a foundation to construct a beautiful theory that is a model of logical precision. In Einstein's theory it was postulated that *all* physical laws, not just those of mechanics, are valid in all inertial reference frames. This new way of looking at the physical world provided the key link between mechanics and electromagnetism; the two great theories of classical physics had been placed on a common footing.

When we first encounter them, the ideas of relativity seem somewhat strange and forced. But relativistic effects are important only when velocities approaching the velocity of light are encountered, and our intuition is based on our everyday experience in which we never meet situations involving such high velocities. Perhaps if we were reared in a much faster-moving world, relativistic concepts would be natural and easy to accept. Nevertheless, various predictions of the theory have been completely verified by a multitude of experiments. As far as we know today, relativity theory is the only correct description of phenomena that take place at high speeds. Moreover, when the speeds involved are low, the Newtonian equations are recovered. Thus, the theory is complete and all encompassing.

17.1 Light Signals—The Basis of Relativity

Einstein's Postulates

Einstein's great triumph was the realization that it was possible to remove all of the apparent discrepancies between the dynamics of mechanical and electromagnetic systems by basing a new theory on only two postulates:

> 1. All physical laws are the same in all inertial reference frames.
> 2. The velocity of light (in vacuum) is the same for any observer in an inertial reference frame regardless of the relative motion between the light source and the observer.

Figure 17.1 Albert Einstein as a young man (1905).

The theory that is based on these postulates and that applies to all nonaccelerating systems is called the *special theory of relativity*. (The more complicated situation of accelerating systems is the subject of the *general theory of relativity*, which is described briefly in Section 17.5.) It is indeed remarkable that such a far-reaching theory—one that forced a complete reexamination of the traditional views concerning the fundamental concepts of space and time and that has had such a profound effect on the interpretation of atomic, nuclear, and astrophysical effects—can be built on only two postulates as simple as those given by Einstein. If it is the goal of physical theory to formulate the laws of Nature with brevity and economy of assumptions, then relativity theory is surely the showpiece of science.

Einstein's first postulate is not too difficult to accept. As we have mentioned, Newton was aware that the laws of mechanics are the same for all inertial observers. For example, if you throw a ball inside an airliner moving with constant velocity, you will observe the same parabolic path that you would see if the throw were from the Earth's surface. To Einstein, there was no distinction between mechanical and electromagnetic systems. (All matter consists ultimately of charged particles and all charge is carried by material particles; moreover, electromagnetic radiation is produced only by the acceleration of charge.) Einstein therefore extended Newton's idea to the laws of electromagnetism.

According to the first postulate, there is no single preferred reference frame for the description of physical events—all nonaccelerating reference frames are equally valid. Thus, there is no reference frame at "absolute rest" and there exists no "absolute motion." We can never state that one object is "at rest" and another object is "in motion;" we can only say that there is a *relative motion* between the objects. (We will sometimes refer to an Earth-based observer as "stationary" and to an observer in a moving system as "moving," but we understand that it is only the relative motion that is physically meaningful.)

Einstein's second postulate—that the velocity of light is the same for all inertial observers—runs counter to our everyday experience. Accordingly, we devote the following paragraphs to an examination of this new idea.

The Velocity of Light is Constant

Suppose that an airplane moves with a velocity $v = 400$ km/h with respect to a stationary Earth-based coordinate system (Fig. 17.2). If a rocket mounted on the airplane is fired in the forward direction with a velocity $u = 600$ km/h *with respect to the airplane*, what will be the velocity of the rocket as measured by a ground-based observer? According to our Newtonian ideas, the answer is simply $V = v + u = 1000$ km/h.

Will there be any difference in the way in which we make our analysis if we increase the velocity of the airplane to one-half the velocity of light and fire a light pulse instead of a rocket? According to Einstein's second postulate, we cannot simply add the two velocities to obtain $0.5c + c = 1.5c$ for the velocity of the light pulse as measured by the observer on

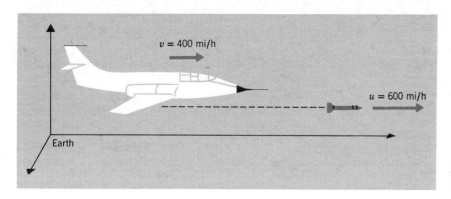

Figure 17.2 To an Earth-based observer, the velocity of the rocket is $V = v + u = 1000$ km/h.

the ground. Einstein's statement is that the light pulse travels with the velocity $c = 3 \times 10^8$ m/s with respect to the airplane *and* with respect to the ground! The velocity of the source (the airplane) does not influence the velocity of the light pulse as measured *by any observer* in an inertial reference frame.

This prediction can be put to experimental tests. A particularly elegant demonstration was recently made in the following way. A large accelerator was used to produce a beam of extremely high speed neutral π mesons (π° mesons or neutral *pions*). These π° particles have the property that they decay spontaneously into energetic photons (or *gamma rays*). The velocity of the particles in the beam was measured to be $0.99975c$, and when the decays took place, the velocity of the photons was also measured. The result was that the photons traveled with a velocity that was equal to c within 1 part in 10^{14} (see Fig. 17.3). This is a direct and striking confirmation of Einstein's postulate.

Further evidence comes from astronomical observations. Many of the stars in the sky (probably more than half in our Galaxy) are actually *binary stars*, two stars orbiting around one another. In some cases only one star of a pair shines brightly whereas the other is dim or even dark. In Fig. 17.4 we show a bright-dark pair of stars orbiting around their common center of mass. We imagine that we are looking toward the binary pair along a line, from the right-hand side of the page, that passes through the plane of rotation. At one point in its orbit (Fig. 17.4a), the bright star will be moving directly *away from* us, and at another point (Fig. 17.4b), it will be moving directly *toward* us. If the velocity of the emitted light were to depend on the motion of the source, then in Fig. 17.4a the velocity of the light traveling toward us would be $V = c - v$, whereas in Fig. 17.4b, the velocity would be $V = c + v$. At other positions in the orbit the velocity would have intermediate values. In this situation, the "fast" light (Fig. 17.4b) could overtake the "slow" light (Fig. 17.4a) emitted at an earlier time. In fact, we would see a curious pattern of light sources suddenly appearing and disappearing, sometimes with several distinct sources in view simultaneously! No observa-

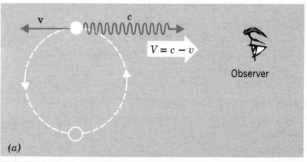

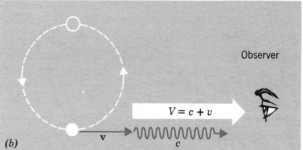

Figure 17.4 Suppose that (a) the bright star of a binary pair, when moving away from the observer, emits light that travels with a velocity $V = c - v$ and (b) when moving toward the observer, the star emits light that travels with a velocity $V = c + v$. No evidence has ever been found that substantiates this variation of the velocity of light.

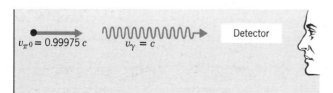

Figure 17.3 Neutral pions (π°) traveling with a velocity $v_\pi = 0.99975c$ decay into gamma rays that travel with a velocity $v_\gamma = c$ with respect to the laboratory observer.

tion of a binary star system has ever revealed such fantastic behavior. All observations are consistent with the postulate that the velocity of light does not depend on the relative motion of the source and the observer.

The Difference between Light Waves and Waves in Material Media

Suppose that we have two coordinate systems, K and K', that have a relative velocity v. Suppose also that there is a body of water that is stationary with respect to system K. At the instant that K' passes K and the two origins coincide (Fig. 17.5a), a stone is dropped into the water and waves begin to spread out in circles from the origin O. Figure 17.5b shows the situation a short time later. Of course, the K observer sees the waves spreading out from his origin. But what does the K' observer see? Because the water is a *material* medium and defines a specific reference frame for any observer, moving or stationary, the observer in K' will also see waves spreading out from O (not from his own origin O'). Everyone has observed this or an equivalent phenomenon and the reason for this effect is clear.

Let us repeat this experiment using light waves instead of water waves. A light source is located at O in the system K and as K' moves past K, a light pulse is emitted the instant O and O' coincide (Fig. 17.6a). Because the source is at rest in his system, it is clear that the K observer will see a spherical light wave radiating outward from O. The K' observer sees a light source approaching him with a velocity v and at the instant the source is at O', a light pulse is emitted. Because the velocity of light is independent of the velocity of the source, the situation, as observed in K', is exactly the same as if the source were located in a stationary position at O'. That is, the K' observer also sees a spherical light wave radiating from his origin O', not from O (Fig. 17.6b). Therefore, each observer sees exactly the same thing—a spherical light wave spreading out uniformly in all directions from his origin with the velocity c.

At first thought this seems to be a surprising (if not fantastic) result. What is the crucial difference between water waves and light waves that causes such a remarkable disparity in the results of the two similar experiments? The point is actually quite simple. In the case of the water-wave experiment there is a tangible, material medium (the water) that is stationary in the K system. The water wave propagates by virtue of the motions of the water molecules. These material particles "belong" to K and both observers perceive this fact. In the light-wave experiment, however, there is no material medium involved—the light wave propagates through empty space. Because there is no matter to be identified as "belonging" to one system or the other, each observer views the light wave relative to his own coordinate system and therefore sees light waves radiating outward with spherical wave fronts.

17.2 Time and Length in Special Relativity

Time Dilation

In what way does the fact that two observers are in relative motion affect their measurements of time intervals? In order to examine this question we must have an appropriate clock. Let us prepare a "standard clock" in the following way. At a distance L from the origin along the y-axis we place a mirror M, as in Fig. 17.7a. At the origin we place a light flasher and a light detector. The standard unit of time will be the interval required for light to travel from the flasher to the mirror and back to the detector. If the K observer operates his flasher at $t = 0$, he finds that the light pulse makes the round trip and returns to the origin after an interval of time,

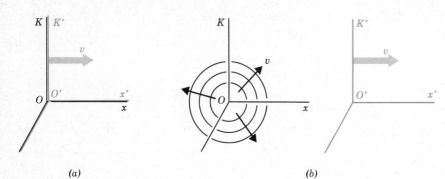

Figure 17.5 Water waves are seen by both observers (in K and in K') as propagating outward from the origin O.

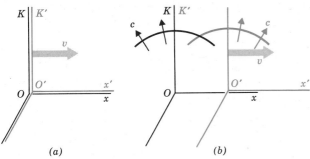

Figure 17.6 A light pulse is emitted from the common origin of K and K' as they pass. Each observer then sees a spherical light wave radiating outward from his or her origin with the velocity c.

$$\Delta t = \frac{2L}{c} \tag{17.1}$$

Now, an identical clock is installed in the K' system and the K observer views its operation. The K' system moves past the K system at a velocity v along the x-axis. At the instant when O and O' coincide ($t' = 0$), the flasher of the K' clock is operated. Because the K' system is moving relative to K, the K observer notices that the light flash must travel from O' to M' on a slanted path that is longer than the path traveled by the light flash in the K clock. When the light flash reaches M', this time interval corresponds to one half of a standard interval of the K' clock, that is, $\frac{1}{2}\Delta t'$ (Fig. 17.7b). The complete standard interval $\Delta t'$ ends when the reflected light flash again reaches O' having traveled the path OPO' (Fig. 17.7c). During this time interval the origin O' has moved a distance $v\Delta t'$ from O'. We can see that the K observer believes that the K' clock runs slowly because the light signal moves along a path that is *longer* than the path for the light pulse in the K clock.

In order to compare the intervals Δt and $\Delta t'$, we use the Pythagorean theorem for the triangle $OM'O'$ (Fig. 17.7b) or OPP' (Fig. 17.7c). Thus,

$$(\tfrac{1}{2}c\Delta t')^2 = (\tfrac{1}{2}v\Delta t')^2 + L^2$$

or, using Eq. 17.1 for L,

$$(\tfrac{1}{2}c\Delta t')^2 = (\tfrac{1}{2}v\Delta t') + (\tfrac{1}{2}c\Delta t)^2$$

canceling the factor $(\tfrac{1}{2})^2$ and transposing the term $(v\Delta t')^2$, we find

$$(c^2 - v^2)\Delta t'^2 = c^2\Delta t^2$$

so that

$$\Delta t'^2 = \frac{c^2\Delta t^2}{c^2 - v^2} = \frac{\Delta t^2}{1 - \dfrac{v^2}{c^2}}$$

or finally,

$$\Delta t' = \frac{\Delta t}{\sqrt{1 - \dfrac{v^2}{c^2}}} = \frac{\Delta t}{\sqrt{1 - \beta^2}} \tag{17.2}$$

where we have used the customary notation, $\beta = v/c$.

The standard time interval $\Delta t'$ of the K' clock, *as viewed by the K observer*, is *longer* than the interval of the K clock. (The K observer must draw this conclusion because he views the light flash in the K' clock traveling a greater distance than the light flash travels in his own clock.) Therefore, the observer in K finds that the K' clock runs *slower* than his clock. Of course, if the K' observer views the K clock, he concludes that the K clock runs slower than his. Therefore, we can state that *any observer will find that a moving clock runs slower than an identical clock that is stationary in his reference frame*.

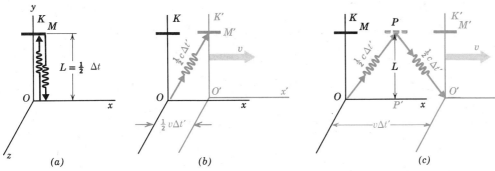

Figure 17.7 According to the observer in K, it requires a shorter time for the round trip OMO of the light signal in K than for the round trip OPO' in K'. The K observer concludes that the clock in K' runs slower than the clock in K.

This behavior of moving clocks is called the *time dilation effect*. *(To dilate* means to enlarge or expand; the time interval is expanded when the clock runs more slowly.) Although we imagined the experiments performed with light-pulse clocks, any other kind of clock would have led to the same conclusion (but the analysis would have been more difficult).

The speed v of the moving clock is always less than c. But when v is a large fraction of c, the denominator of the time dilation expression becomes very small. This means that $\Delta t'$ becomes large compared to Δt. For example, when v is 87 percent of the speed of light, $\Delta t'$ is twice as large as Δt. That is, the duration of a "tick" of the moving clock is twice as long as that of a "tick" of the observer's own clock. Thus, according to the stationary observer, the moving clock runs at a rate that is half the rate of the stationary clock.

Time Dilation in Pion Decay

We can illustrate the time dilation effect by considering the motion of the short-lived pions. When viewed at rest, pions have an average lifetime of $\tau_\pi = 2.6 \times 10^{-8}$ s before decaying into other elementary particles. Pions are produced in copious quantities by the interaction of high-energy protons with matter and therefore are relatively easy to study.

If pions move with a velocity of $0.75c$, the average distance they would travel before decay is $l_\pi = v\tau_\pi = 0.75 \times (3 \times 10^8 \text{ m/s}) \times (2.6 \times 10^{-8} \text{ s}) = 5.85$ m. At the Columbia University cyclotron a beam of pions with $v = 0.75c$ was produced and it was found that the average distance these particles traveled before decay was not 5.85 m but 8.5 ± 0.6 m. We can account for this difference in terms of the time dilation effect. Because the pions are moving in the laboratory system (corresponding to our K system), the laboratory observer sees any clock in the system moving with the pions (the K' system) running slowly. But the decay rate of the pions is a type of clock and so the laboratory observer will find the average lifetime of the pions to be longer than τ_π. In fact,

$$\tau_{\text{lab}} = \frac{\tau_\pi}{\sqrt{1-\beta^2}} = \frac{2.6 \times 10^{-8} \text{ s}}{\sqrt{1-(0.75)^2}} = 3.9 \times 10^{-8} \text{ s}$$

Therefore, the average distance in the laboratory that the pions will travel before decay is

$$l_{\text{lab}} = v\tau_{\text{lab}} = 0.75 \times (3 \times 10^8 \text{ m/s})$$
$$\times (3.9 \times 10^{-8} \text{ s}) = 8.8 \text{ m}$$

which agrees with the measured value of 8.5 ± 0.6 m.

Length Contraction

Let us now view the decay of the pions from the standpoint of an observer moving with the pions. In the pion rest-frame, the pions will live an average of 2.6×10^{-8} s and will travel an average distance of 5.85 m. But we know that the average distance *as measured in the laboratory* is 8.8 m. The distance through which the pions moved before decay is the *same physical* distance whether measured in the pion frame or in the laboratory frame. That is, this distance corresponds to the distance between two marks on the laboratory floor and these marks remain in the same position regardless of any motion of the observer.

We must conclude that not only are *time* measurements different in two reference systems in relative motion but so are *distance* measurements. The observer in the pion frame sees the laboratory moving past him with a velocity of $0.75c$. The length of 8.8 m, as measured by an observer in the laboratory frame, appears to the observer in the pion frame to be only 5.85 m. That is, *there is a contraction of length due to relative motion*. The length l' of an object in motion relative to an observer is contracted compared to the length l of an identical object at rest with respect to the observer:

$$l' = l\sqrt{1-\beta^2} \qquad (17.3)$$

A measurement made by the observer in the pion frame of the 8.8 m laboratory length gives $l' = (8.8 \text{ m}) \times \sqrt{1-\beta^2} = 5.85$ m.

It should be noted that the contraction of length takes place only *along* the direction of relative motion; the dimensions of an object *transverse* to the direction of motion are unaffected. (This is why the mirror in the light clock of Fig. 17.7 was placed on the y-axis and not on the x-axis; the observers in both the K and K' systems measure the same distance L for the distance from the origin to the mirror.)

● *Example 17.1*

An observer moves past a meter stick with a velocity that is one-half the velocity of light. What length does he measure for the meter stick?

$$l' = l\sqrt{1-\beta^2} = (1 \text{ m}) \times \sqrt{1-(0.5)^2}$$
$$= (1 \text{ m}) \times \sqrt{0.75} = 0.866 \text{ m} \quad \blacksquare$$

In order to obtain numerical results for problems that involve the relativistic factor $\sqrt{1-\beta^2}$; the following approximate expressions can be used when the velocity v is very small compared to c:

If $\beta \ll 1$ ($v \ll c$): $\sqrt{1-\beta^2} \cong 1 - \frac{1}{2}\beta^2$ (17.4a)

$$\frac{1}{\sqrt{1-\beta^2}} \cong 1 + \frac{1}{2}\beta^2 \quad (17.4b)$$

When $v \cong c$ and $\beta \cong 1$, we can write

$$\sqrt{1-\beta^2} = \sqrt{(1+\beta)(1-\beta)} \cong \sqrt{2(1-\beta)} \quad (17.4c)$$

● *Example 17.2*

Suppose that the velocity of the observer relative to the meter stick in the previous example is reduced to 30 m/s (about 67 mi/h). What length does he now measure for the meter stick?

$$l' = (1 \text{ m}) \times \sqrt{1 - \left(\frac{30 \text{ m/s}}{3 \times 10^8 \text{ m/s}}\right)^2}$$

$= (1 \text{ m}) \times \sqrt{1 - 10^{-14}}$

$\cong (1 \text{ m}) \times (1 - 0.5 \times 10^{-14})$

$= 0.999999999999995$ m

It is easy to see that the relativistic contraction of length is of little practical consequence in everyday matters! ■

Are Relativistic Effects REAL?

Is it really true that moving clocks run more slowly than stationary clocks and that moving meter sticks contract? If observers A and B are in relative motion, B will find that A's meter stick is shorter than his own and that A's clock runs more slowly than his own. But if we join A and observe his meter stick and clock, we see nothing unusual. The reason is that the length contraction effect and the time dilation effect are the result of using length and time standards in one's own reference frame to measure lengths and times in a moving frame. The observer who is at rest relative to a meter stick or clock finds no contraction or dilation effect at all. So, are these effects *real*?

It seems only reasonable to define *reality* in terms of *measurements* that we can make. Then, relativistic effects are certainly real. Every experiment that has ever been carried out involving the measurement of length and time (and, as we will see, also *mass*) in rapidly moving systems has verified the predictions of relativity theory. Relativistic effects are not optical illusions nor are they some kind of scientific hocus pocus. Relativistic effects are *real*.

The Twin Paradox

One of the results of relativity theory that has been much discussed (and misunderstood) in recent years is the so-called "twin-paradox." Suppose that there are twins, Al and Bob, and that Bob is an astronaut. Bob embarks on a space journey to a star that is 10 light years distant; Al remains on Earth. If Bob's spaceship travels at a velocity of $0.99c$ relative to the Earth, according to Al the trip will require a time[1]

$$\Delta t = \frac{10 \text{ L.Y.}}{0.99c} \cong 10 \text{ years}$$

An equal time will be required for the return journey, so Bob will arrive back on Earth when Al is 20 years older than when Bob departed.

In Bob's spaceship, however, the Earth and the star appear to be moving with a velocity of $0.99c$ relative to Bob. Therefore, the Earth-star distance is contracted to

$$l' = (10 \text{ L.Y.}) \times \sqrt{1 - (0.99)^2} = 1.4 \text{ L.Y.}$$

According to Bob's clock, the trip will require only 1.4 years and he will return to Earth after having aged by 2.8 years. When he again greets his brother, Bob discovers that his twin is $20 - 2.8 = 17.2$ years *older* than he is! But we know that all motion is relative. Therefore, if the trip is viewed from Bob's reference frame, he sees Al (and the Earth) go on a round trip journey. Hence, Al's clock should run more slowly than Bob's and when Al returns (along with the Earth), Bob should find that his twin is *younger* than he is. Thus, the paradox.

The "paradox" rests on invoking the symmetry of the situation. It should not matter which twin takes the trip and which remains at home. But it *does* matter, because *Al (the stay-at-home) is always in an inertial reference frame whereas Bob (the traveler) has undergone accelerations.* In leaving the Earth, Bob was accelerated to $0.99c$; he was accelerated when he turned around at the star; and he was accelerated again when he returned to Earth and landed. Therefore, the situation is *not* symmetric between Al and Bob. Is is Bob or Al who experiences a force during the events we are discussing? Because

---

[1] Because 1 L.Y. is the distance traveled by light in 1 year, the quantity 1 L.Y./c is just *1 year*.

inertial reference frames are not involved throughout, the analysis must be carried out quite carefully. A proper calculation (which can be made within the context of special relativity if appropriate care is exercised) does in fact show that Bob ages less rapidly than his twin.

Since 1971, a number of experiments have been performed comparing high-precision atomic clocks, two in flight aboard airplanes flying in opposite directions and one stationary on the ground. In all cases, the differences in clock readings (on the order of 10^{-7} s) agreed with the predictions of relativity.

Time Travel

Because of the time dilation effect, we can imagine the exciting possibility of traveling to distant stars. If the trip is made at a velocity sufficiently close to the velocity of light, the traveler can easily cross vast distances of space within a time short compared to his lifetime. But he would return to a different Earth—one that has progressed (?) by hundreds or even thousands of years during his absence.

Table 17.1 lists the times of travel for a trip to a star 1000 L.Y. away for various speeds of the spaceship. Two columns of times are given—one according to the space travelers' clock and one according to the Earth observer's clock. Notice that doubling the speed from one-half the speed of light to just under the speed of light causes the travel time to decrease by one-half, according to the Earth observer. But, because of the time dilation effect, this same speed change causes a much greater reduction in travel time according to the space travelers.

It must be emphasized that the space travelers sense nothing unusual about the passage of time during their high-speed voyage. To them, events seem to happen at the normal rate. According to their clocks, they age in a perfectly natural way. It is only to an observer moving with a high speed relative to the spaceship (for example, an Earth observer) that the activity in the spaceship seems slowed down. Thus, the space traveler ages much more slowly than his Earthbound friends.

If the spaceship makes the trip to a star 1000 L.Y. away at a speed of 99.999 percent of the speed of light, the one-way passage requires 4.5 years (see Table 17.1). If the return trip is made at the same speed, the space travelers will arrive back at Earth 9 years after their departure, according to their own clocks. But they will find an Earth that has aged by 2000 years! Relativity theory shows us the way to skip over hundreds or thousands of years, to become time travelers voyaging into the future. Such excursions, however, are one-way trips. There is no way to reverse the procedure and to travel backward in time. Time moves only forward, and relativistic time travel permits us to go only into the future, not into the past.

Do not plan your trip into the future just yet, however. Even though relativity theory shows us how time travel is possible *in principle,* we now have no idea how we might accomplish a high-speed space voyage. In order for the time dilation effect to be useful for long-distance space trips, the speed of the spaceship must be very close to the speed of light. Table 17.1 shows that a visit to a star 1000 L.Y. away becomes feasible only if a speed of at least 99.99 percent of the speed of light is reached. At the present time we have absolutely no idea how to provide the incredibly huge amount of energy that would be necessary to propel a spaceship to such a high speed. The best that we have been able to do so far is to accelerate *atomic particles* to these speeds. We do not know how to project a grain of sand, much less a spaceship, to a speed of 99.99 percent of the speed of light!

The Velocity Addition Rule

How can the statement that the velocity of light is independent of the motion of the source be consistent with the fact that all ordinary mechanical velocities simply *add* algebraically, as in Fig. 17.2? Einstein showed that the simple addition formula for mechanical velocities is not correct and must be modified. If two velocities v_1 and v_2 are to be added, the sum is

$$V = \frac{v_1 + v_2}{1 + \dfrac{v_1 v_2}{c^2}} \tag{17.5}$$

Table 17.1 Travel Times to a Star 1000 Light Years Away

	Travel Time (years)	
$\beta = \dfrac{v}{c}$	According to Space Traveler's Clock	According to Earth Observer's Clock
0.10	9950	10 000
0.50	1732	2000
0.90	484	1111
0.99	142	1010
0.999	45	1001
0.9999	14	1000.1
0.99999	4.5	1000.01
0.999999	1.4	1000.001

In the event that v_1 and v_2 are small compared to the velocity of light (the case for all ordinary mechanical situations), the term $v_1 v_2/c^2$ is much less than unity and can be neglected. Then, the sum velocity is $V = v_1 + v_2$, identical to the result that would be obtained by applying Newtonian reasoning.

If one of the velocities is the velocity of light, $v_1 = c$, then

$$V = \frac{c + v_2}{1 + \frac{cv_2}{c^2}} = \frac{c + v_2}{1 + \frac{v_2}{c}} = \frac{c + v_2}{\left(\frac{c + v_2}{c}\right)} = c$$

This result insures that the velocity of light is the same for all observers because no matter what velocity v_2 is added to c, the addition rule always yields c.

● *Example 17.3*

A spaceship moving away from the Earth at a velocity $v_1 = 0.75c$ with respect to the Earth, launches a rocket (in the direction *away* from the Earth) that attains a velocity $v_2 = 0.75c$ with respect to the spaceship. What is the velocity of the rocket with respect to the Earth?

$$V = \frac{v_1 + v_2}{1 + \frac{v_1 v_2}{c^2}} = \frac{0.75c + 0.75c}{1 + \frac{(0.75c)(0.75c)}{c^2}}$$

$$= \frac{1.5c}{1 + 0.5625} = 0.96c$$

Therefore, in spite of the fact that the simple sum of the two velocities exceeds c, the actual velocity relative to the Earth is slightly less than c. ∎

17.3 Variation of Mass with Velocity

A Collision Experiment

The first postulate of relativity theory is that all physical laws must be the same in all inertial reference frames. One of these laws is the conservation of linear momentum and we shall now make use of the invariability of this law to assess the effect of motion on mass.

Consider two observers who are stationed in two reference frames, K and K', that are in relative motion with the velocity v, as in Fig. 17.8. In each reference frame there is a stationary mass m_0. (That the two masses are in fact identical can be established beforehand by a balance comparison when the masses are at rest relative to one another.) The positions of the masses are such that when the reference frames pass one another, a grazing collision of the masses takes place. That is, each mass receives a small velocity at right angles (that is, *transverse*) to the direction of the relative velocity of K and K'. (In such a collision neither mass will receive any appreciable longitudinal velocity relative to its own reference frame.) Therefore, after collision, the situation is that shown in Fig. 17.9. The mass in K has a transverse velocity u and a transverse momentum p as measured by the observer in K; similarly for the mass in K' (but with primed quantities). Each observer uses a meter stick and a clock to measure the transverse velocity of his mass in his reference frame. Each observer obtains a numerical result for the velocity of his mass, which result he then communicates to his colleague in the other reference frame. They are both happy to note that the results are identical and congratulate themselves on having verified the conservation of linear momentum in the collision. In order to check the results, they decide to repeat the experiment twice again—once, K will observe K' making his measurements, and then K' will observe K making his measurements.

On the first rerun, K confirms that the meter stick used by K' is properly calibrated (transverse dimensions are unaffected by relative motion; contraction of length takes place only *along* the direction of relative motion), but that his clock runs

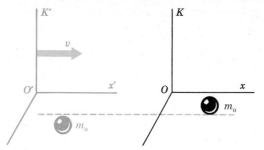

Figure 17.8 A mass m_0 is stationary in each reference frame. The relative velocity is v.

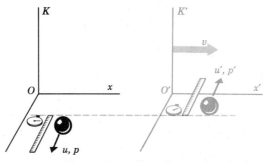

Figure 17.9 The grazing collision between the masses produces for each mass a velocity and a momentum transverse to the direction of relative motion.

slowly. Therefore, when K' reported that his mass traveled 1 meter in T seconds, K concludes that by *his* clock it required a time *greater* than T seconds for the 1-meter trip. Thus, K calculates that the velocity of the mass in K' is *smaller* than the value u' reported by K'—in fact, smaller by the time-dilation factor $\sqrt{1-\beta^2}$. If the velocity is smaller and if conservation of linear momentum is still to hold, then the mass used by K' must be (so argues K) *larger* than that used by K—in fact, larger by the amount of $1/\sqrt{1-\beta^2}$.

Of course, during the second rerun of the experiment, K' draws exactly the same conclusions about the measurements of K. Both observers therefore agree that the mass of an object in motion is greater than the mass of an identical object that is at rest. The increase of mass with velocity (just as length contraction and time dilation) is symmetrical between the two reference frames in relative motion.

The mass of an object as measured in a reference frame at rest with respect to the object is denoted by m_0 and is called the *rest mass* or *proper mass*. Then, the mass m as measured by an observer moving with a velocity v relative to the object is

$$m = \frac{m_0}{\sqrt{1-\beta^2}} \qquad (17.6)$$

We must conclude from this equation that no material particle can attain or exceed the velocity of light because if $v = c$, the term $\sqrt{1-\beta^2}$ vanishes and m becomes infinite. An infinite mass is a meaningless concept and we are therefore forced to accept the conclusion that material particles *always* move with velocities that are less than the velocity of light. And, because of the velocity addition rule (Eq. 17.5), the velocity of a material particle is less than c in *any* reference frame.

● *Example 17.4*

What is the velocity of an elementary particle whose mass is 10 times its rest mass?

The mass m is 10 m_0, so

$$10 m_0 = \frac{m_0}{\sqrt{1-\beta^2}}$$

$$\sqrt{1-\beta^2} = \frac{1}{10}$$

Squaring, we have

$$1 - \beta^2 = 0.01$$
$$\beta^2 = 1 - 0.01 = 0.99$$
$$\beta = \sqrt{0.99} = 0.995$$

Therefore,

$v = 0.995\, c$ ∎

Particles with Varying Mass

The difference between the mass m and the rest mass m_0 is quite small unless the relative velocity v is greater than a few percent of the velocity of light. Therefore, the relativistic increase of mass with velocity is undetectable for all everyday velocities; it is only when we deal with elementary particles that have been given high velocities in accelerators that we encounter appreciable mass increases. Figure 17.10 shows Eq. 17.6 in graphical form; indicated on the curve are points corresponding to one everyday object (an automobile traveling at 50 mi/h) and three high-velocity elementary particles. Notice that a 1-GeV proton has *twice* the mass of a proton at rest, whereas the mass increase of a 50-mi/h automobile is insignificantly small. Table 17.2 gives the result of Eq. 17.6 in more detail.

The increase in mass of an object in motion with a velocity *small* compared to the velocity of light is

$$\Delta m = m - m_0 = m_0 \left(\frac{1}{\sqrt{1-\beta^2}} - 1 \right)$$

Because $v \ll c$ so that $\beta \ll 1$, we can use Eq. 17.4b to write

$$\Delta m \cong m_0 [(1 + \tfrac{1}{2}\beta^2) - 1] = m_0 \times \tfrac{1}{2}\beta^2$$

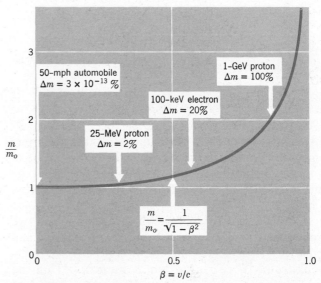

Figure 17.10 The relativistic increase of mass with velocity. The energies given for the elementary particles are the kinetic energies.

Table 17.2 Increase of Mass with Velocity

$\beta = \dfrac{v}{c}$	$\dfrac{m}{m_0} = \dfrac{1}{\sqrt{1-\beta^2}}$
0	1
0.10	1.005
0.50	1.16
0.90	2.29
0.99	7.09
0.999	22.4
0.9999	70.7
0.99999	224
0.999999	707

Therefore, the fractional increase in mass is

$$\frac{\Delta m}{m_0} \cong \tfrac{1}{2}\beta^2 \qquad (v \ll c) \tag{17.7}$$

● *Example 17.5*

What is the fractional increase of mass for a 600-mi/h jetliner?

$$v = 600 \text{ mi/h} \cong 2.7 \times 10^2 \text{ m/s}$$

$$\beta = \frac{v}{c} = \frac{2.7 \times 10^2 \text{ m/s}}{3 \times 10^8 \text{ m/s}} \cong 10^{-6}$$

Therefore,

$$\frac{\Delta m}{m_0} \cong \tfrac{1}{2}\beta^2 \cong 0.5 \times 10^{-12}$$

so that the mass is increased by only a trivial amount. ∎

Experimental Tests of Relativistic Mass Increase

There are two ways to demonstrate in striking fashion that the relativistic formula for mass increase is correct. First, consider a particle, such as an electron, that falls through a certain potential difference and acquires a high velocity. If we project this electron into a magnetic field, we find that the radius of the orbit is *greater* than that calculated from the simple formula based on Newtonian dynamics (Eq. 13.9) if m_0 is substituted for the mass. Indeed, a measurement of the orbit radius can be used to determine the mass. The results of measurements made by just this technique are shown in Fig. 17.11; the theoretical prediction and the experimental results are clearly in excellent agreement.

Next, consider an elastic collision between a moving particle and an identical particle at rest. From the discussion in Example 7.9, we know that

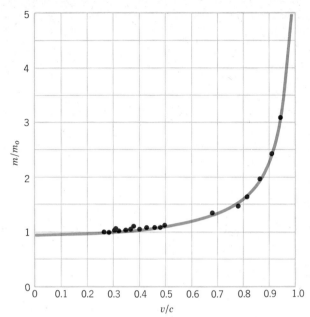

Figure 17.11 Variation of electron mass with velocity. The data represent measurements made during the period 1901–1909 and the curve is the prediction of relativity theory. This early experiment provided a striking confirmation of this prediction of the theory.

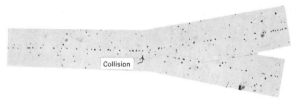

"The Study of Elementary Particles by the Photographic Method" by Powell

Figure 17.12 Photograph of the tracks left in a nuclear emulsion by a high-velocity electron colliding with an electron in an atom of the emulsion material. The angle between the outgoing particles is much less than 90°, in agreement with the relativistic prediction.

Newtonian dynamics predicts that the velocity vectors of the particles after the collision should be at *right angles* (see Fig. 7.11). However, when a fast electron or proton collides with a similar particle at rest, it is found that the angle between the velocity vectors after collision is *smaller* than 90° indicating that the incident particle had a mass *greater* than the struck particle (Fig. 17.12).

17.4 Mass and Energy

Einstein's Mass-Energy Relation

If $v \ll c$, Eq. 17.6 can be written approximately as

$$m \cong m_0(1 + \tfrac{1}{2}\beta^2) \qquad (v \ll c)$$

Multiplying both sides of this equation by c^2 and noting that $c^2\beta^2 = v^2$, we find

$$mc^2 \cong m_0 c^2 + \tfrac{1}{2} m_0 v^2 \qquad (v \ll c) \tag{17.8}$$

The term $\tfrac{1}{2} m_0 v^2$ is just the Newtonian result for the *kinetic energy*. The term $m_0 c^2$ is clearly some *intrinsic* aspect of the object because it depends only on the *rest* mass. We call this quantity the *rest energy* of the object. The sum of the *rest* energy and the *moving* energy (that is, the kinetic energy) is the *total energy* of the object:

$$\underset{\text{(total energy)}}{mc^2} = \underset{\text{(rest energy)}}{m_0 c^2} + \underset{\text{(kinetic energy)}}{KE} \tag{17.9}$$

If v is not small compared to c, there are additional terms on the righthand side of Eq. 17.8 that result from the expansion of $1/\sqrt{1-\beta^2}$. Nonetheless, the difference between the total energy and the rest energy is still the kinetic energy and Eq. 17.9 is correct. This equation is just the expression of the Einstein mass-energy relation:

$$\mathscr{E} = mc^2 \tag{17.10}$$

where $\mathscr{E}$ is the *total* energy (rest energy + kinetic energy) of the object.

● *Example 17.6*

What is the mass of an electron that has a kinetic energy of 2 MeV?

First, we calculate the rest energy of an electron:

$$m_0 c^2 = (9.11 \times 10^{-31} \text{ kg}) \times (3 \times 10^8 \text{ m/s})^2$$

$$= (8.2 \times 10^{-14} \text{ J}) \times \left(\frac{1 \text{ MeV}}{1.6 \times 10^{-13} \text{ J}}\right)$$

$$= 0.511 \text{ MeV}$$

The kinetic energy expressed in units of $m_0 c^2$ is

$$KE = 2 \text{ MeV} = (2 \text{ MeV}) \times \left(\frac{m_0 c^2}{0.511 \text{ MeV}}\right)$$

$$\cong 4 \, m_0 c^2$$

Therefore,

$$mc^2 = m_0 c^2 + KE$$
$$\cong m_0 c^2 + 4 \, m_0 c^2 = 5 \, m_0 c^2$$

Hence, the mass of a 2-MeV electron is approximately 5 times the mass of an electron at rest. ∎

The relativistic increase in mass is appreciable for an electron even when the kinetic energy is rather low because the electron rest energy is only 0.511 MeV or 511 keV. Therefore, the relativistic equations must be used whenever the electron KE is greater than a few tens of keV. Protons, on the other hand, have a rest energy of 938 MeV and so even a 10-MeV proton is "nonrelativistic" because its mass increase is only about 1 percent of its rest mass.

● *Example 17.7*

What is the mass-energy conversion factor that relates the atomic mass unit and MeV?

From Eq. 1.2 we have

$$1 \text{ u} = 1.6605 \times 10^{-27} \text{ kg}$$

Therefore,

$$(1 \text{ u}) \times c^2 = (1.6605 \times 10^{-27} \text{ kg}) \times (3 \times 10^8 \text{ m/s})^2$$

$$= (1.4945 \times 10^{-10} \text{ J}) \times \left(\frac{1 \text{ MeV}}{1.6022 \times 10^{-13} \text{ J}}\right)$$

$$= 931.5 \text{ MeV}$$

Or, using more precise values for the constants, we can express c^2 as $c^2 = 931.481$ MeV/u. ∎

Einstein's equation, $\mathscr{E} = mc^2$, states that an object with a mass m has an equivalent energy $\mathscr{E}$, and, conversely, that an entity with a total energy has an equivalent mass m. (The equation does *not* state that mass and energy are the same thing, only that mass has equivalent energy and that energy has equivalent mass.)

Whenever an object or a system experiences a change in energy, there must accompany this change an equivalent change in mass (and vice versa); that is, $\Delta \mathscr{E} = \Delta m \times c^2$. Einstein's relativity therefore merges two of our conservation laws (mass is conserved; energy is conserved) into a single, all-encompassing law:

The total amount of mass-energy in an isolated system remains constant.

According to Eq. 17.10, there is a truly enormous amount of energy associated with even small amounts of mass. Suppose, for example, that 1 kg of matter could be converted entirely into energy. This process would yield 9×10^{16} J or about 3×10^{10} kWh! This is approximately one-half of the energy consumption of the United States per day.

The Limitation on Extracting Useful Mass-Energy

Ordinarily, the entire mass-energy of a given substance is not available to be transformed into useful energy.[2] We cannot destroy neutrons and protons—we can only release energy by rearranging them into forms that have different total mass. This is another conservation law: the total number of neutrons and protons remains constant.

One method of rearranging neutrons and protons within nuclei is the *fission* process in which a heavy nucleus (suitably jostled by a neutron) breaks into two fragments. The difference between the initial mass and the final mass is released as kinetic energy. (However, the total number of neutrons and protons remains the same.)

● *Example 17.8*

How much energy is released in the fission of 1 kg of ^{235}U?

The amount of mass-energy that is converted to kinetic energy in the fission process is approximately 200 MeV per nucleus. (This is only about 0.1 percent of the total mass-energy of a uranium nucleus; the other 99.9 percent remains in the masses of the neutrons and protons and is therefore not available for conversion into kinetic energy.) In 1 kg of ^{235}U there are approximately 2.5×10^{24} atoms. Therefore, the total energy release is

$\mathscr{E} = (200 \text{ MeV}) \times (2.5 \times 10^{24})$

$\phantom{\mathscr{E}} = 5.0 \times 10^{26}$ MeV

$\phantom{\mathscr{E}} = (5.0 \times 10^{26} \text{ MeV}) \times (1.6 \times 10^{-13} \text{ J/MeV})$

$\phantom{\mathscr{E}} = 8.0 \times 10^{13}$ J

We can convert this into another widely used unit by noting that the explosion of 1 ton of TNT releases approximately 4.1×10^9 J. Thus, the fission of 1 kg of ^{235}U releases an amount of energy

$$\mathscr{E} = \frac{8.0 \times 10^{13} \text{ J}}{4.1 \times 10^9 \text{ J/ton TNT}} \cong 20 \text{ kilotons TNT}$$

This is approximately the yield of the original atomic bomb of 1945. ■

---

[2]We mean to exclude the matter-antimatter annihilation that powers many science-fiction spacecraft.

The fission process in any heavy nucleus produces approximately 200 MeV of energy. The *total* mass-energy of a ^{235}U nucleus is approximately 2.2×10^5 MeV. Comparing these two energies, we see that fission is approximately 0.1 percent efficient in the conversion of total mass-energy into useful energy. On the other hand, chemical burning of fuels such as coal extracts only about 10^{-9} of the total mass-energy as useful energy because there is relatively little energy stored in molecules in the form of electrical energy in the bonds. Therefore, fission is approximately 10^6 times more efficient than fuel burning in the generation of energy.

We return to the discussion of nuclear fission and nuclear energy in Chapter 21.

17.5 General Relativity

The Relativity of Accelerated Motion

Thus far we have considered only motions that take place with constant velocity. If we wish to treat the case of accelerated motion, then the special theory of relativity is no longer adequate and we must turn to the *general theory of relativity*. As we shall see, the general theory is more than a relativistic description of abstract accelerated motion, it is a theory of *gravitation*.

Although the special theory is supported by experimental tests of a wide range of predictions, the general theory enjoys much less in the way of experimental verification. In fact, there are only a few predictions that the theory in its current form can make and for none of these has a really definitive experimental test been carried out at the present time. Nevertheless, the issues that are raised by the general theory are so profound that a brief survey of the theory is warranted.

The first postulate of special relativity is that all physical laws are the same in all inertial reference frames. The general theory makes a much more sweeping statement:

All physical laws can be formulated in such a way that they are valid for any observer, no matter how complicated his motion.

If we allow an observer to undergo a complicated accelerated motion, it is clear that the mathematical expression of physical laws in his reference frame will also be complicated. Indeed, the general theory involves the use of an awesome arsenal of mathematical techniques. In spite of this fact, we can still gain some appreciation of the general theory without sophisticated mathematics.

The Principle of Equivalence

The first important aspect of the general theory has to do with the equivalence of gravitational fields and accelerated motion. If we are in a laboratory on the Earth, as in Fig. 17.13a, a mass that is released will accelerate downward due to the gravitational attraction of the Earth. Now, let us move this laboratory into space, away from the gravitational influence of the Earth or any other body, and attach it to a rocket that is accelerating, as in Fig. 17.13b. If the magnitude of the rocket's acceleration a is equal to the acceleration due to gravity g and if the rocket pushes on the floor of the laboratory, the floor will be accelerated toward a mass that is released. Insofar as observations of the motion of the mass relative to the floor are concerned, the accelerated motion in the two cases will be exactly the same. If the laboratory has no windows, the observer can never distinguish between an acceleration due to gravity and an acceleration due to a push by a rocket.

Einstein incorporated this reasoning into his general theory by postulating the *principle of equivalence*:

In a closed laboratory, no experiment can be performed that will distinguish between the effects of a gravitational field and the effects due to an acceleration with respect to some inertial reference frame.

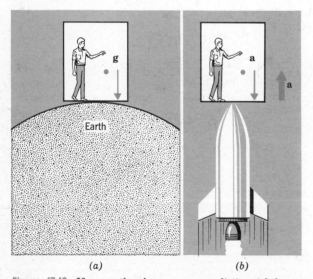

Figure 17.13 If $a = g$, the observer cannot distinguish between the acceleration produced by gravity and that produced by the push of the rocket. (The observer's box laboratory has no windows.)

Notice that in the "experiment" of Fig. 17.13 we deal with *gravitational* mass in the first case and with *inertial* mass in the second case. If there were any difference between these two types of mass, then sufficiently precise measurements would permit the observer to decide whether he is in the Earth's gravitational field or accelerating in space. Therefore, the equivalence principle requires $m_G = m_I$; experiments verify this equality to within 1 part in 10^{11}. In spite of the high precision of this result, an even sharper confirmation of the assertion of the equivalence principle would be desirable in order to provide a greater support for this fundamental postulate of the general theory.

When Einstein first formulated his general theory of relativity (more than 50 years ago), he proposed two experimental tests of the theory. These tests deal with anomalies in the motions of the inner planets, particularly Mercury, and with the behavior of electromagnetic waves, particularly light, in the vicinity of massive objects such as the Sun. Next we briefly describe these two tests and the modern refinements.

The Precession of the Perihelion of Mercury's Orbit

In Section 6.2 it was mentioned that the perihelion of Mercury's orbit has been observed to move in space (i.e., to *precess*) at a rate that is larger than that predicted on the basis of Newtonian dynamics. After subtraction of the calculable perturbations due to the other planets, there remains a net precession of 43.11 ± 0.45 seconds of arc per century. If the special theory of relativity is used to calculate the effects of time dilation and mass increase with velocity, one finds a precession that is only one-half the observed value. Einstein was able to obtain, on the basis of his general theory, the value 43.03 seconds of arc per century. The excellent agreement between the calculated and the observed values is one of the outstanding successes of the general theory.

The Bending of Light Rays by the Sun

The general theory predicts that a light ray passing close to a massive object will follow a path that is slightly curved. We can understand this result qualitatively if we recall that electromagnetic radiation, including light, has *energy* and that energy has equivalent *mass*. Therefore, a gravitational field will affect a light ray and cause it to be bent in much the same way that a fast particle would bend when passing by a massive object. Because light travels at such an enormous velocity there is only a brief time during which the "attraction" is effective, and hence the deflection is small even for a passage near such a massive object as the Sun. (As in the case of Mercury's precession, the special theory, through the relation $\mathscr{E} = mc^2$, predicts for the deflection only one-half the value that results from the general theory.)

This prediction can be tested by observing the shift in the apparent position of a star when its light passes close to the Sun, as shown in Fig. 17.14. Because of the brightness of the Sun, measurements of the effect are carried out by comparing the apparent positions of stars during a solar eclipse with the positions six months later when the Sun is not in that part of the sky and ordinary night photographs of the stars can be taken. Such measurements are clearly difficult to make. (Not the least of the problems is the fact that there is a sudden decrease in temperature at the onset of the eclipse with resulting thermal contractions in the photographic apparatus!) However, the apparent shifts in position of hundreds of stars have been made by this technique and the average result is approximately 2 seconds of arc for the deflection; the general theory predicts 1.75 seconds of arc. Unfortunately, the uncertainty in the measurements is about 10 percent and there are some conflicting results, so that we cannot view this test as definitive.

Recent experiments of this type have used radio signals instead of light. When Mars moved behind the Sun in 1976, signals from the Viking spacecraft passed close to the Sun and were detected on Earth. A time delay in these signals (equivalent to the bending of light rays) was found to agree with the prediction of the theory to better than 1 percent.

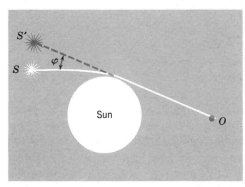

Figure 17.14 The light from a star S is bent on passing close to the Sun, thus causing a shift in the apparent position of the star to S'.

The Gravitational Red Shift

When a mass is released in a gravitational field, it accelerates downward and gains kinetic energy. Similarly, a light pulse—because it has an equivalent mass—will gain energy by falling in a gravitational field. The increase in kinetic energy of a massive particle results from the increase in its velocity. Because light always travels with the velocity c, an increase in energy can occur only by virtue of an increase in the *frequency* of the light wave. The converse is also true—if a light pulse is projected *opposite* to the gravitational field vector, it will *lose* energy and the frequency will decrease. Visible light emerging from the Sun will have its frequency *lowered* or, equivalently, the wavelength will be *increased* or shifted to the *red* part of the spectrum. The amount of the shift is small, but by comparing the wavelengths of certain solar radiations with the wavelengths of the radiations from the same types of atoms in an Earth laboratory, the predicted shift has been verified to a precision of about 10 percent.

We can derive the expression for the fractional change in frequency due to the gravitational red shift from Newtonian principles combined with the relativistic mass-energy relation. As we will see in the next chapter, the energy of a light proton (or a light wave) is proportional to its frequency. Therefore, the fractional frequency change is equal to the fractional energy change:

$$\frac{\Delta \nu}{\nu} = \frac{\Delta \mathscr{E}}{\mathscr{E}}$$

Now, $\mathscr{E} = mc^2$ and the *change* in energy due to a change in position by an amount h in a uniform gravitational field g is $\Delta \mathscr{E} = mgh$. Therefore,

$$\frac{\Delta \mathscr{E}}{\mathscr{E}} = \frac{mgh}{mc^2} = \frac{gh}{c^2}$$

so that

$$\frac{\Delta \nu}{\nu} = \frac{gh}{c^2} \qquad (17.11)$$

A verification of this expression has been carried out in the laboratory by allowing electromagnetic radiations from nuclei to fall through a distance of 74 ft = 22.6 m. The predicted fractional frequency change is

$$\frac{\Delta \nu}{\nu} = \frac{(9.8 \text{ m/s}^2) \times (22.6 \text{ m})}{(3 \times 10^8 \text{ m/s})^2} = 2.5 \times 10^{-15}$$

Figure 17.15 Albert Einstein in his later years.

This incredibly small frequency change has been measured to a precision of 1 percent (!) using a particular kind of nuclear frequency selector (based on the *Mössbauer effect*). In order to do as well in the calculation of the national debt (about 10^{12} dollars), your figures would have to be accurate to $1/400$ of a cent!

The periodic nature of an electromagnetic or light wave constitutes a kind of *clock*. Consider two observers, one in the strong gravitational field near the Sun and the other in a weak field in a space laboratory. Both observers have identical atomic light sources and standard clocks.[3] Each observer counts the number of oscillations in his source for a predetermined interval of time according to *his* standard clock. When the results are compared, it is found that the observer in the strong field has measured a greater number of oscillations (that is, a greater light frequency) than the observer in the weak field. They conclude that the clock in the strong field must have run more slowly in order for more oscillations to have occurred in the standard time interval. Notice that, unlike the case of time dilation is special relativity, *both* observers agree that the clock in the strong field is slow—there is no symmetry in the results.

Because the effects of gravity and acceleration are indistinguishable, we must conclude that an accelerating clock runs more slowly than a clock in an inertial reference frame. (Compare the discussion of the "twin paradox.")

---

[3]Things are not quite as straightforward as implied here. The discussion of clocks (even *light* clocks) in the context of general relativity is extremely complicated.

Black Holes and Related Stellar Objects

A bizarre possibility in the interaction of light with gravitation has been discovered in astronomy. It involves the case shown in Fig. 17.14 when the gravitational field of a star is able in effect to bend a light ray into a circle.

Stars such as our sun maintain a relatively stable size by having the internal contracting gravitational forces balanced by the outward directed stream of radiation produced by the nuclear reactions proceeding in the deep stellar interior. Eventually such stars will consume the nuclear fuel of light elements in the ongoing reactions and a severe gravitational collapse will take place. The final fate of the star under such a collapse depends largely on its mass. One possible stellar fate of massive stars with mass greater than 3 solar masses is to collapse to radii on the order of a few kilometers! The gravitational field at the surface of such a star can be sufficiently large to prevent the escape of any light (or anything else) from its surface. The condition for the ratio of stellar mass to radius, M/R, for this to happen is

$$2G\,M/R > c^2 \tag{17.12}$$

where G is the universal constant of gravitation and c is the speed of light. The result is the creation of a *black hole*. The star disappears! Its presence may only be inferred indirectly. It may, for example, form a binary system with an ordinary star. The continual leaking of matter from the ordinary star to its massive black hole companion results in the production of radiation (X rays) which can be detected. It is estimated that there are many millions of black holes in our universe.

What is the final fate of stars less massive than those that become black holes? Model calculations suggest that stars with a mass less than about 1.2 solar mass, contract into very dense and only faintly luminous objects called *white dwarfs*. A white dwarf with one solar mass would have a radius less then one percent of our Sun. Under these conditions the electrons of the constituent atoms are stripped loose from their nuclei and form a degenerate sea of electrons. It is the hydrostatic pressure of this electron sea that halts the gravitational collapse.

In another scenario, stars on the way to developing into white dwarfs may accrete sufficient matter from a nearby binary partner or a nearby supernova explosion to continue the gravitational contraction beyond that leading to a stable white dwarf. Such stars with masses below about three solar masses contract into *neutron stars* with very small radii. One with a solar mass would have only a radius of about 10 kilometers. Under these conditions of enormous densities, the atomic electrons are literally squeezed into the nuclei converting protons into neutrons. Thus, a pure degenerate neutron star interior is formed. Light leaving such stars may only leave within a relatively narrow cone perpendicular to the stellar surface. Light leaving at larger angles to the surface normal are simply bent back into the surface by the strong gravitational field. The light source for a neutron star is closely associated with the position of the star's magnetic field which is seldom aligned with the rotation axis. As a consequence, the emitted light from the rotating magnetic field source is directed toward the Earth once each time the star turns on its axis, much the same effect as for light coming from a light house with a rotating beacon.

The conservation of angular momentum would require neutron stars to rotate very rapidly. For example, a star such as the Sun with a rotational period of approximately 27 days would, on collapse to a neutron star, rotate with a period of about 0.05 seconds! Such rapidly pulsating light sources are called *pulsars*. The pulsar in the Crab nebula has a period of only 0.033 seconds.

Summary of Important Ideas

The *special theory of relativity* rests on two postulates:

1. The laws of physics are the same in all inertial reference frames.
2. The velocity of light in vacuum is the same for all observers.

An observer who is in motion relative to a clock will find that clock to run more *slowly* than an identical clock at rest in his frame of reference (*time dilation*).

A measurement by an observer of the length of an object that is in motion relative to the observer will give a *smaller* result than a measurement carried out by an observer at rest with respect to the object (*length contraction*). (The contraction is only in the dimension along the direction of relative motion; the transverse dimensions are unaffected.)

No *material particle* can have a velocity with respect to any reference frame that is equal to or greater than the velocity of light; no *signal* can be transmitted at a velocity greater than c.

A particle moving with respect to an observer has a mass *greater* than that of an identical object at rest with respect to the observer.

The *total energy* of an object is the sum of its *rest energy* $m_0 c^2$ and its *kinetic energy;* this total energy is $\mathcal{E} = mc^2$.

The *principle of equivalence* states that the effects of gravity and accelerated motion are indistinguishable.

The predictions of the *general theory of relativity* have been verified for the precession of the perihelion of Mercury's orbit, and the bending of light rays that pass near the Sun.

The *gravitational red shift* is an immediate consequence of the principle of equivalence and the fact that light has "mass." That clocks run slowly in a gravitational field is due to the same effect.

◆ Questions

17.1 The "writing speed" of a cathode-ray oscilloscope is the speed with which an electron beam can trace a line on the screen. A certain manufacturer claims that the writing speed of his oscilloscopes is 6×10^8 m/s. Explain why his claim can be true.

17.2 Using the results of relativity theory, comment on this anonymous limerick:

> There was a young lady named Bright,
> Who could travel much faster than light
> She departed one day,
> In a relative way,
> And returned on the previous night.

17.3 Suppose that the velocity of light suddenly became 30 mi/h. Describe a few of the effects this would have on everyday events.

17.4 Discuss the implications for relativity theory if the velocity of light were infinite.

17.5 An observer views the light from two identical light sources, one of which is stationary in his laboratory and one of which is moving toward him with a high velocity v. The observer measures the velocity of the light from each source and obtains c in each case. But there *is* a physical difference in the light from the two sources (one source is moving relative to the observer and one is not)—what is it? Is there any way to distinguish the light from one source compared with that from the other source?

17.6 A bicycle rider pedals past you at a velocity of 2.5×10^8 m/s. Make a sketch of the way the bicycle would appear to you. Would the rider think you were your usual self?

17.7 In Section 10.2 we considered the connection between the *average velocity* of molecules in a quantity of matter and the *temperature* of the material. In this chapter we learned that there is an upper limit (i.e., c) to the velocity of any material object. Should we conclude, then, that there is an upper limit for allowable temperatures.

★Problems

Section 17.2

17.1 The star nearest the Earth, *Alpha Centauri*, lies at a distance of 4.3 L.Y. If a space traveler were to make a trip from Earth to *Alpha Centauri* at a uniform velocity of 0.95 c, how long would it take according to an Earth clock? How long would it take according to the space traveler's clock?

17.2 An astronaut orbits the Earth with a speed of 8000 m/s. After a mission of 10 days, by how much does his biological clock differ from those of his Earthbound companions? (Consider the motion to be linear.)

17.3 *Muons* are elementary particles that are formed high in the atmosphere by the decay of pions which result from the interaction of cosmic rays with the gases in the atmosphere. The velocities of these muons are near the velocity of light ($v \cong 0.998\ c$). From laboratory experiments we know that the average lifetime of a muon is 2.2×10^{-6} s (in its own rest frame). Show that muons formed at an altitude of 8000 m can reach the surface of the Earth in spite of their short lifetime but that they would not do so without relativistic effects.

17.4* Could an astronaut, with a remaining life expectancy of 40 years, make a trip (at least in principle) to a galaxy that is 10^{10} light years distant? (This galaxy would be near the limit of the Universe that is visible from Earth.) If so, at what velocity would his spaceship have to travel?

17.5* In a paper published in 1905, Einstein stated: "We conclude that a balance clock at the equator must go more slowly, by a very small amount, than a precisely similar clock situated at one of the poles under otherwise identical conditions." Neglect the fact that the equator clock is actually rotating with the Earth and consider the velocity to be uniform (that is, consider the problem within the context of *special* relativity). Show that after a century the clocks will differ by approximately 0.0038 s.

17.6* A distant star is receding from us at a speed of 0.8 c. The light intensity from this star is observed to vary with a period of 5 days. What is the time between intensity maxima in the reference frame of the star?

17.7 An observer measures the length of a moving meter stick and finds a value of 0.5 m. How fast did the meter stick move past the observer?

17.8 A highway billboard is in the form of a square, 5 m on a side, and stands parallel to the highway. If a traveler passes the billboard at a speed of 2×10^8 m/s, what will be the dimensions of the billboard as viewed by the traveler?

17.9 A spaceship is moving away from the Earth with a velocity of 1.5×10^8 m/s and launches a small rocket toward the Earth with a velocity of 2.5×10^8 m/s with respect to the mother ship. What is the velocity of the rocket as determined by an observer on the Earth? (Notice that the *signs* of the two velocities are different.)

17.10 An observer sees one spaceship moving away from him at a velocity 0.9 c and another spaceship moving in the *opposite* direction at the same velocity. What does he conclude is the relative velocity of the two spaceships? (Consider this carefully; the Einstein velocity addition rule is *not* involved in the answer. Why?) Does this result agree with that obtained by the occupants of the spaceships? Why?

17.11* A rocket ship traveling with a velocity of 1.5×10^8 m/s with respect to the Earth fires a small rocket in the direction of its motion with a velocity of 10^5 m/s relative to the rocket ship. With what precisions would a measurement on the Earth have to be carried out in order to distinguish between the Einstein velocity addition rule and that based on Newtonian mechanics?

17.12 A rocket is moving with a velocity of $0.6\,c$ directly away from the Earth. A missile that has a length of 5 m (as measured by the rocket crew) is launched from the rocket toward the Earth with a velocity of $0.385\,c$ with respect to the Earth. What is the length of the missile as determined by an Earth observer?

17.13* If the missile in the preceding problem is launched 10 s after the rocket launch (according to a clock in the rocket), how long after the rocket launch (according to an Earth clock) will the missile strike the Earth?

Section 17.3

17.14 A certain ball has a mass of 5 kg. If this ball moves past an observer with a velocity of 1.8×10^8 m/s, what mass does the observer measure for the ball?

17.15 A rocket has a mass of 5×10^6 kg. What must be its velocity with respect to an observer so that the observer will measure a relativistic mass increase of 1 g?

17.16 What is the fractional increase in mass of an electron moving with a velocity of 10^8 m/s?

17.17 An observer measures the mass m and the length l of a rod that is stationary. He therefore concludes that the linear mass density of the rod is $\rho = m/l$. Now, suppose that he repeats the measurements while the rod is moving past him with a velocity v. (The motion is along the length of the rod.) What will the observer find for the linear mass density? What would he find if the direction of motion were perpendicular to the length of the rod?

17.18 A meter stick has a rest mass of 2 kg. When this meter stick moves past an observer (with its long dimension in the direction of motion), he measures the mass to be 6 kg. What does the observer measure for the length of the meter stick?

17.19 At what velocity is the mass of an object equal to three times its rest mass?

17.20 A meter stick is in motion in the direction along its length with a velocity sufficient to increase its mass to twice the rest mass. What is the apparent length of the meter stick?

17.21* Electrons emerge from the Stanford Linear Accelerator (SLAC) with velocities only 1.5 cm/s slower than the velocity of light. How long does a 1-km flight path in the laboratory appear to these electrons? What is the mass of these electrons? Compare this mass with that of an iron nucleus.

Section 17.4

17.22 A 100-W light bulb operates for 1 year. What is the mass equivalent of the radiated energy?

17.23 What is the velocity of a particle that has a kinetic energy equal to its rest energy?

17.24 How much energy is required to double the velocity of an electron which moves initially with a velocity of 10^8 m/s?

17.25 To melt 1 kg of ice at 0°C requires an input of 80 Cal of heat. How much greater is the mass of the water compared to the 1-kg mass of the ice?

17.26 What increase in mass results when an electron is accelerated from rest by falling through a potential difference of 25,000 V?

17.27* The explosion of a star at the edge of the observable Universe (10^{26} m) produces neutrons with energies of 10^6 J. A free neutron will undergo radioactive β decay, forming a proton and an electron, with a lifetime of about 1000 s. Could any of the neutrons emitted in the explosion survive to reach the Earth? Explain.

17.28* Compete the following table:

	KE (MeV)	v/c
Electrons	0.1	—
	—	0.99
Protons	—	0.1
	1000	—

(Use the value 937 MeV for the rest energy of the proton.)

17.29 Through what potential difference must an electron fall (starting from rest) in order to achieve a velocity of $0.5\,c$?

17.30* Calculate the orbit radius of a 2-MeV electron in a magnetic field of 1500 G.

17.31* What is the lifetime of a pion that has a kinetic energy of 200 MeV?

17.32 A light pulse is directed parallel to the Earth's surface. After traveling 10 km, how far will it have fallen? (Make a simple calculation by assuming the light has equivalent mass and consequently experiences an acceleration equal to g. This will actually give *half* the value that is obtained in the general theory.)

17.33* What must be the magnitude of a uniform gravitational field in which the frequency of a light pulse changes by 1 part in 10^6 by falling through a distance of 100 m? Is a field of this size available in the solar system? Suppose that we carry out the experiment in the Earth's field over a distance of 1 km. What will be the fractional frequency shift?

17.34 To how small a radius would a star of five solar masses have to contract to become a black hole?

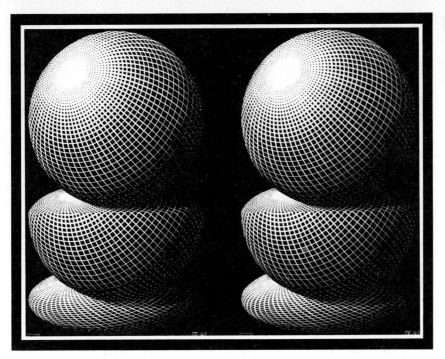

18
The Quantum

At the end of the nineteenth century many scientists viewed physics as a closed subject. What else that was significant and fundamental could be learned about the physical aspects of Nature? The laws of mechanics and the theory of universal gravitation had been established for more than 200 years. Maxwell's theory of electromagnetism was complete. It was understood that matter consists of atoms. Thermodynamics had recently been placed on a firm foundation with the development of the statistical approach to systems with large numbers of particles. The great conservation principles—of energy, linear momentum, angular momentum, mass, and electric charge—were well established and understood. What else of *real* importance could be discovered?

In spite of the general complacency among nineteenth-century physicists concerning the status of the subject, nevertheless, there *were* problems lurking about. It was soon to become clear that these were not all trivial problems and that the very heart of nineteenth-century physics was coming under violent attack. First, it was Einstein's relativity theory that forced a new way of thinking about such fundamental concepts as space and time. Even before this revolution could be digested, a new series of equally far-reaching questions was being asked about the nature of radiation and matter, how did they differ and in what ways were they the same—what was the inner structure of atoms—what was the origin of the newly discovered *radioactivity?* The answers to these questions began to emerge in the early years of the twentieth century and culminated with the development of modern *quantum theory*.

18.1 Radiation and Quanta

The Spectrum of Radiation from Heated Objects

When a material is heated it becomes incandescent and emits electromagnetic radiation with a continuous range of wavelengths. If the temperature is sufficiently high, enough radiation will fall in the visible region of the spectrum that we can perceive the object to glow. Figure 18.1 shows the shapes of the radiation spectra from objects at three different temperatures. Notice that the wavelength λ_m at the maximum of the distribution decreases with increasing temperature. Notice also how the intensity of the radiation increases with temperature. This

450 The Quantum

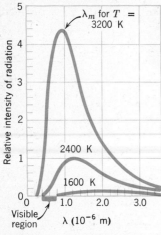

Figure 18.1 Radiation spectra for objects at various temperatures. Notice that the region of visible light covers only a small part of the wavelength range (from about 0.4×10^{-6} m to about 0.7×10^{-6} m).

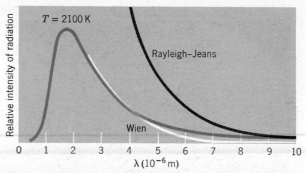

Figure 18.2 Comparison of an experimental thermal radiation spectrum (brown curve) with the results of the theories of Wien and Rayleigh (the latter as modified by James Jeans).

increase follows the Stefan-Boltzmann radiation law in which the radiated power is proportional to T^4 (Eq. 9.14).

If the object has $T = 1600$ K, we can see only the tiny fraction of the emitted light that falls in the long-wavelength part of the visible region—the object appears red in color. For $T = 3200$ K, however, a much larger fraction of the radiation matches the sensitive range of the eye and we see the object glowing with an intense yellow light. For very high temperatures, most of the emitted radiation is in the ultraviolet or X-ray regions of the spectrum.

Measurements of the radiation spectra from objects at various temperatures have shown that the relationship connecting λ_m and the absolute temperature T is

$$\lambda_m = \frac{2.9 \times 10^{-3}}{T} \text{ m-K} \qquad (18.1)$$

This expression is called *Wien's law*, in honor of its discoverer, Wilhelm Wien (1864–1928).

Many attempts were made to explain the *shape* of the thermal radiation spectrum in terms of classical electromagnetic theory. Two of the attempts were those made by Wien and by Lord Rayleigh (1842–1919). The predictions of these theories are compared with an experimental spectrum in Fig. 18.2. It is clear that neither is a satisfactory representation of the experimental data: Wien's result fails at long wavelengths whereas Rayleigh's result agrees with experiment only in the long wavelength region. Thus, the classical theories were found inadequate to explain the thermal radiation spectrum, and so remained the situation until a bold and imaginative step forward was made by the German physicist Max Planck (1858–1947).[1]

Planck's Quantum Hypothesis

In 1900 Planck undertook to accomplish what Wien and Rayleigh and others had failed to do—to explain in detail the shape of the radiation spectrum for *all* wavelengths, not just in the long- or short-wavelength regions. Planck found a formula that did in fact reproduce the complete shape of the spectrum, but there was no theoretical justification for the formula whatsoever. In order to provide a basis for his formula, Planck was forced to make an unorthodox assumption. He concluded that the exchange of energy between radiation and matter does not take place in a continuous way. Instead, Planck hypothesized, energy exchange takes place by means of individual units of energy called *quanta*. The amount of energy associated with a quantum of frequency ν is a certain constant times ν; that is,

$$\mathscr{E} = h\nu \qquad (18.2)$$

where h is the proportionality constant, now known as *Planck's constant*. The value of the constant h is extremely small:

$$h = 6.626 \times 10^{-34} \text{ J-s} \qquad (18.3)$$

Therefore, each quantum has associated with it only a very small amount of energy. It is not surprising, then, that when large amounts of radiation are involved, the discrete nature of the energy is not evi-

---

[1] Max Planck was the winner of the Nobel Prize in 1918 for his quantum explanation of the thermal radiation spectrum.

dent because a change of only a few quanta is completely negligible.

In formulating his theory of thermal radiation, Planck did not draw on any *direct* evidence for radiation quanta; he only inferred the existence of quanta because this concept appeared necessary to explain the shape of the radiation spectrum. Consequently, Planck's method of reproducing the spectrum was regarded by most physicists as an interesting trick of no fundamental importance. In 1900 it was still generally believed that all physical processes were *continuous,* and even Planck did not go so far as to suggest that all electromagnetic radiation was quantized; his hypothesis applied only to exchanges of energy in certain situations. Thus, the one single great idea that was necessary to understand the nature of radiation—the quantum hypothesis—was incompletely developed and was generally ignored.

Planck's idea remained in limbo for several years before Einstein found the quantum hypothesis necessary for the explanation of the photoelectric effect (described in the next section). Einstein extended Planck's idea and postulated that *all* electromagnetic radiation occurs in the form of quanta (or *photons*). Finally, Planck's great contribution was realized and he was awarded the 1918 Nobel Prize in physics for his introduction of the quantum concept and his explanation of the thermal radiation spectrum.

The energy associated with the quanta of electromagnetic radiation of various types is given in Table 18.2.

● *Example 18.1*

In order to see how small is the amount of energy associated with a single quantum of light, consider the following case. Suppose that a 1-gram mass is allowed to fall through a height $H = 1$ cm. (You would hardly feel the impact if the mass were dropped on your hand.) If all of the energy acquired in the fall could somehow be converted to yellow light ($\lambda = 6 \times 10^{-7}$ m), how many photons would be emitted?

The energy acquired in the fall is just the potential energy:

$PE = mgH$
$= (10^{-3} \text{ kg}) \times (9.8 \text{ m/s}^2) \times (10^{-2} \text{ m})$
$= 9.8 \times 10^{-5}$ J

The energy of each yellow photon is

$$\mathscr{E} = h\nu = \frac{hc}{\lambda}$$

Table 18.1 The Fundamental Atomic Constants

Velocity of light	$c = 2.9979 \times 10^8$ m/s
Mass of electron	$m_e = 9.1096 \times 10^{-31}$ kg
Charge of electron	$e = 1.6022 \times 10^{-19}$ C
Planck's constant	$h = 6.6262 \times 10^{-34}$ J-s

$$= \frac{(6.6 \times 10^{-34} \text{ J-s}) \times (3 \times 10^8 \text{ m/s})}{6 \times 10^{-7} \text{ m}}$$

$= 3.3 \times 10^{-19}$ J/photon

Therefore, the number of photons is

$$\text{No. of photons} = \frac{9.8 \times 10^{-5} \text{ J}}{3.3 \times 10^{-19} \text{ J/photon}}$$

$$= 3 \times 10^{14} \text{ photons}$$

Thus, only a tiny amount of energy is associated with an individual photon. ■

The Fundamental Atomic Constants

With Planck's constant h we have completed the introduction of the fundamental atomic constants: c, m_e, e, and h (Table 18.1). All the other important numbers that occur in atomic theory are either multiples or combinations of these four constants. Because of the largeness of c and the smallness of m_e, e, and h, the physical phenomena that depend on these quantities do not reveal themselves in everyday occurrences. Therefore, the true nature of these phenomena was not appreciated until the modern era when sufficiently delicate instruments were developed for measuring effects taking place at the atomic level.

18.2 The Photoelectric Effect

Ejection of Electrons by Light

As early as 1887 it had been observed by Heinrich Hertz, the discoverer of electromagnetic waves, that the intensity of the spark between two high-voltage electrodes was increased when the electrodes were exposed to ultraviolet (UV) light. Almost immediately, this effect was pursued by others and it was found, for example, that a clean zinc plate acquires a positive electric charge when irradiated by UV light (Fig. 18.3). These two effects can both be explained in terms of the ejection of electrons from the materials by the incident light. In Hertz's experiment, the ejected electrons assisted in the conduction of the spark between the electrodes and therefore, the act

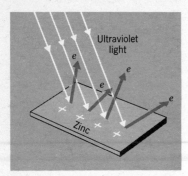

Figure 18.3 The photoelectric effect. Ultraviolet light incident on a metal plate, such as zinc, ejects electrons and the plate acquires a positive charge.

of irradiating the electrodes with UV light increased the intensity of the spark. In the zinc-plate experiment the removal of electrons by the action of UV light leaves behind a positively charged plate. This phenomenon of electron ejection by light is called the *photoelectric effect*. Thus, Hertz, who believed the wave theory of light was a "certainty," accidentally uncovered the first piece of evidence for the particlelike aspect of light.

The qualitative explanation of the photoelectric effect in terms of the removal of electrons by the action of light was in no way revolutionary. Light was known to consist of oscillating electric and magnetic fields that carry electromagnetic energy. It was entirely consistent with classical theory that these waves could transfer energy to electrons in the metal and when an electron had acquired sufficient energy from the wave it could escape from the metal. But further experiments showed that several aspects of the photoelectric effect did not agree with the predictions of classical electromagnetic theory.

The Results of Photoelectric Experiments

The photoelectric effect can be studied by using apparatus similar to that shown schematically in Fig. 18.4. Two plates, *A* and *B*, are contained within an evacuated tube and are connected to a source of variable voltage and a sensitive ammeter. Plate *A* is the surface from which photoelectric emission takes place (the *photoemissive* surface) and is coated with the metal to be studied, for example, lithium, sodium, potassium, or cesium.

When Plate *A* is irradiated with UV light of a definite frequency, energetic photoelectrons are emitted. These electrons are collected by plate *B* and the ammeter registers the photoelectric current. Now, suppose that we apply a negative potential to plate *B* and increase this potential until we find the

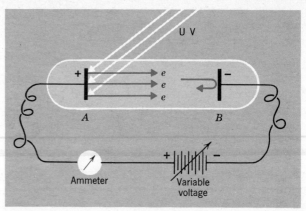

Figure 18.4 Apparatus for the study of the photoelectric effect. The motion of the energetic electrons emitted from plate *A* is retarded by the negative potential of plate *B* relative to *A*.

value V_s for which all of the photoelectrons are stopped before they reach the plate. Thus, the potential V_s is the potential (the *stopping potential*) that just causes the photoelectric current to become zero. If an electron with a kinetic energy KE is just stopped by the retarding potential V_s, then $KE = eV_s$; that is, eV_s is the *maximum* kinetic energy of the photoelectrons ejected from plate *A*. (We deal with the maximum KE because these electrons come from the surface of the photoemissive material and do not suffer energy losses in collisions with atoms of the material.)

Measurements of this type have led to the important result that the KE of the photoelectrons increases in direct proportion to the frequency of the incident radiation. Furthermore, there was found to be a *threshold frequency* ν_0 below which no photoelectrons could be produced *regardless of the intensity of the incident radiation* (Fig. 18.5). It was also observed that the number of photoelectrons ejected is directly proportional to the *intensity* of the UV radiation for any frequency of radiation greater than ν_0.

The experimental facts regarding the photoelectric effect that must be explained are these:

1. The intensity of the incident UV radiation determines the *number* of photoelectrons ejected, but the intensity does *not* determine the stopping potential required to stop the electrons (that is, the *energy* of the electrons does not depend on the *intensity* of the UV radiation).

2. The *maximum energy* of the electrons depends only upon the *frequency* of the UV radiation.

3. For each material there exists a definite threshold frequency ν_0, and UV radiation with a lower frequency will *not* eject photoelectrons no matter

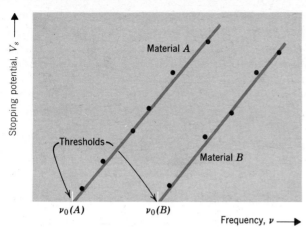

Figure 18.5 The maximum stopping potential that can be overcome by photoelectrons is directly proportional to the UV frequency for all frequencies above the threshold at $\nu = \nu_0$. The dots show the results of various measurements made with discrete UV frequencies for two different materials. The two lines have the same slopes but different thresholds.

what the intensity of the radiation. However, if $\nu > \nu_0$, then even the most feeble UV radiation will cause photoelectrons to be ejected (albeit in small numbers) with *no time delay* between the turning on of the UV source and the appearance of the first photoelectron.

Einstein's Explanation of the Photoelectric Effect

According to the classical theory, UV radiation delivers electromagnetic energy to a photoemissive surface and, when sufficient energy has been concentrated in one electron, this electron is ejected. An increase in the UV intensity causes more electrons to be ejected. Therefore, the classical theory is capable of explaining the first part of fact No. 1, above. But this theory is completely inadequate to account for the remaining facts. For example, if the UV intensity is low, the classical theory would predict that it would require a certain time to concentrate sufficient energy in any one electron to cause its ejection; this prediction is contrary to the experimental result (fact No. 3).

In 1905 Einstein offered a theory that provides an explanation of the entire set of facts concerning the photoelectric effect. Einstein's photoelectric theory was beautifully simple, with the same brevity and elegance that characterized his relativity theory (proposed in the same year).[2] Einstein extended Planck's neglected hypothesis of quantized energy exchanges and proposed that *all* electromagnetic radiation was in the form of discrete bundles of electromagnetic energy called *quanta* or *photons*. He further proposed that when a photon interacts with matter it behaves as a *particle* and delivers its energy, not to the material as a whole or even to an atom as a whole, but to an *individual* electron. The occurrence of a threshold is due to the fact that a certain amount of energy must be supplied to the electron in order to free it from the material (even if it receives no kinetic energy). Furthermore, different materials have different values for the threshold energy.

According to Einstein, then, the kinetic energy of a photoelectron must be the difference between the energy of the incident UV photon and the minimum energy necessary to free the electron from the material (called the *work function* of the material). That is,

(electron *KE*)
    = (photon energy) − (work function)   (18.4)

When $h\nu$ is used for the photon energy (Eq. 18.2), this equation becomes

$$KE = h\nu - \phi \qquad (18.5)$$

where ϕ is the work function for the particular material.[3] Each quantity in this equation is measured in energy units, usually joules (J) or electronvolts (eV).

As a mechanical analog to this explanation of the photoelectric effect, consider a ball of mass m that is at rest in a trough, as in Fig. 18.6. If a sufficient amount of energy $\mathscr{E}$ is supplied to the ball, it will roll up the side of the trough of height H and "escape" with a final velocity v. The energy equation for the process is

$$\tfrac{1}{2}mv^2 = \mathscr{E} - mgH$$

---

[2] Also the same year of his theory of Brownian motion—truly a remarkable performance.

[3] This identification of ϕ is correct for an isolated photoemissive material. However, in a circuit such as that in Fig. 18.4 it is necessary to take account of the *contact potential* that exists when the cathode B and the anode A are connected (by a wire). In this situation, ϕ is actually the work function of the *collector* (the cathode, B) and not the *photoemissive surface* (the anode, A). We can avoid the effect of the contact potential by making both the anode and the cathode from the same material.

Figure 18.6 A mechanical analog of the photoelectric effect. Sufficient energy must be supplied to the ball for it to "escape" from the trough and proceed away with a velocity v.

In this expression, mgH is the "work function" (that is, the potential energy barrier that must be overcome) and $\mathscr{E}$ is equivalent to the photon energy.

Einstein's photoelectric equation, Eq. 18.5, is in complete agreement with the experimental results illustrated in Fig. 18.5. If we use $KE = eV_s$ and express the work function as the energy of the incident photons at threshold, $\phi = h\nu_0$, Eq. 18.5 becomes

$$eV_s = h\nu - h\nu_0$$

or,

$$V_s = \frac{h}{e}(\nu - \nu_0) \qquad (18.6)$$

which is exactly the equation of the lines in Fig. 18.5. Notice that the slope of the two lines is equal to the ratio h/e of Planck's constant to the electron charge.

Careful measurements of the photoelectric effect for several elements were made by the American physicist Robert A. Millikan (1868–1953), beginning in 1916. These experiments showed that the slopes of the lines of V_s versus ν were indeed consistent with Einstein's theory, which predicted the value h/e for the slope of all such lines. When the results of Millikan's earlier experiments on the determination of the electronic charge (see Section 12.4) were combined with his h/e measurements, this provided the best available value for Planck's constant. (More recently, different methods have been used for precise determinations of h.)

For his detailed explanation of the photoelectric effect (*not* for his relativity theory!), Einstein was awarded the Nobel Prize in 1921. Millikan received the 1923 Prize for his experimental work on the determinations of e and h.

● *Example 18.2*

What will be the maximum kinetic energy of the photoelectrons ejected from magnesium (for which $\phi = 3.7$ eV, see Table 18.3) when irradiated by UV light of frequency 1.5×10^{15} Hz? (Remember, 1 Hz = 1 s^{-1}.)

The energy of a photon with frequency 1.5×10^{15} Hz is

$$h\nu = (6.6 \times 10^{-34} \text{ J-s}) \times (1.5 \times 10^{15} \text{ s}^{-1})$$

$$= (9.9 \times 10^{-19} \text{ J}) \times \left(\frac{1}{1.6 \times 10^{-19} \text{ J/eV}}\right)$$

$$= 6.2 \text{ eV}$$

Therefore, the maximum kinetic energy is

$$KE = h\nu - \phi$$
$$= 6.2 \text{ eV} - 3.7 \text{ eV}$$
$$= 2.5 \text{ eV}$$

● *Example 18.3*

When silver is irradiated with ultraviolet light of wavelength 1000 Å, a potential of 7.7 volts is re-

Table 18.2 Some Properties of Different Types of Electromagnetic Radiation

Type of Radiation	Typical Wavelength	Frequency (Hz)	Photon Energy
Gamma rays	10^{-13} m	3×10^{21}	12 MeV
X rays	10^{-11} m	3×10^{19}	120 keV
UV radiation	10^{-7} m = 1000 Å	3×10^{15}	12 eV
Visible light (yellow)	6×10^{-7} m = 6000 Å	5×10^{14}	2 eV
Infrared radiation (heat rays)	10^{-5} m	3×10^{13}	0.12 eV
Microwaves	10^{-2} m	3×10^{10}	1.2×10^{-4} eV
Radio waves	300 m	10^6	4×10^{-9} eV[a]

[a] This wavelength is so long and the energy is so small that the quantum character of the radiation is unimportant. That is, the energy of a single photon of this radiation is so small that it can have no measurable effect in any experiment we might consider here.

Table 18.3 Photoelectric Properties of Some Elements

Element	Work Function, ϕ (eV)	Threshold Frequency (Hz)	Threshold Wavelength (Å)	
Cesium	1.9	4.6×10^{14}	6500	Visible light
Potassium	2.2	5.3	5600	
Sodium	2.3	5.6	5400	
Calcium	2.7	6.5	4600	
Magnesium	3.7	8.9×10^{14}	3400	Ultraviolet
Silver	4.7	11.4	2600	
Nickel	5.0	12.1	2500	

quired to stop completely the photoelectrons. What is the work function of silver?

First, the energy of a 1000-Å photon is

$$h\nu = \frac{hc}{\lambda} = \frac{(6.6 \times 10^{-34} \text{ J-s}) \times (3 \times 10^8 \text{ m/s})}{(1.0 \times 10^{-7} \text{ m}) \times (1.6 \times 10^{-19} \text{ J/eV})}$$
$$= 12.4 \text{ eV}$$

Then,

$$\phi = h\nu - eV_s$$
$$= 12.4 \text{ eV} - 7.7 \text{ eV}$$
$$= 4.7 \text{ eV}$$

(Note that we have used the fact that the *energy* of an electron that is completely stopped by a *potential* of 7.7 volts is just 7.7 electron volts.) ∎

18.3 Waves or Particles? Two Crucial Experiments

The Wave-Particle Dilemma

The wave character of light was established in the early years of the 19th century when a series of interference and diffraction experiments disproved the competing particle theory of light. But Einstein's photoelectric theory revived the notion that light behaves as a particle—at least, in its interactions with atomic electrons. Does this mean that we are forced to abandon the wave theory and return to the old particle theory? Or is there some peculiar feature of light that presents a two-sided appearance, sometimes wavelike and sometimes particlelike? If so, then how do we know when to expect one feature and not the other?

These were the questions about light that were raised by the proposal of the quantum nature of electromagnetic radiation. An equally important question concerned the properties of particles: If light plays a dual role of particle and wave, can an electron—an object that we have always considered to be a *particle*—also behave as a *wave*?

These questions were finally resolved in the 1920s by a series of experiments performed in America and in Britain that showed unequivocally that light *and* electrons could each exhibit properties associated with both waves and particles. This wave-particle duality was then incorporated as an essential feature into the emerging *wave mechanics* or *quantum theory*. The first of these crucial experiments demonstrates the wave character of electrons and the second concerns the particle behavior of radiation.

The de Broglie Wavelength

Einstein's explanation of the photoelectric effect had clearly demonstrated that electromagnetic radiation has properties that closely resemble those of material particles. In 1924 a young Frenchman, Louis Victor de Broglie (1892–   ), proposed in his doctoral thesis that, in view of the particlelike behavior of waves, there should also be a wavelike behavior of particles.

In order to describe a wave, there must be a definable wavelength for the propagation. Consider a photon that has an energy $\mathcal{E}$. According to the Einstein mass-energy relation (Eq. 17.10), a mass m has a mass-energy $\mathcal{E} = mc^2$. Or, conversely, an entity that possesses a total energy $\mathcal{E}$ has associated with it an equivalent mass $m = \mathcal{E}/c^2$. And because electromagnetic radiation travels with a velocity c, the momentum of the radiation (that is, equivalent mass × velocity) must be

$$p = m \times c = \frac{\mathcal{E}}{c^2} \times c = \frac{\mathcal{E}}{c}$$

Then, using $\mathcal{E} = h\nu$ and $\nu/c = 1/\lambda$, we can express the momentum as

$$p = \frac{\mathcal{E}}{c} = \frac{h\nu}{c} = \frac{h}{\lambda}, \quad \text{or} \quad \lambda = \frac{h}{p}$$

De Broglie argued that the wavelength of a material *particle* should follow exactly the same prescription:

de Broglie wavelength: $\quad \lambda = \dfrac{h}{p} \quad$ (18.7)

Within three years after de Broglie's ingenious proposal (which earned him the 1929 Nobel Prize), the wave properties of electrons had been demonstrated in experiments by Davisson and Germer in America and by Thomson in Britain which showed the diffraction of electrons. These experiments are described briefly in connection with Fig. 18.9 on page 462.

● *Example 18.4*

What is the wavelength of a 10-eV electron?

For an electron of this low energy we can use the nonrelativistic expression without appreciable error. Since $KE = \frac{1}{2}mv^2$ and $p = mv$, we have

$$p = \sqrt{2 m_e KE}$$

so that

$$\lambda = \frac{h}{p} = \frac{h}{\sqrt{2 m_e KE}}$$

$$= \frac{6.62 \times 10^{-34} \text{ J-s}}{\sqrt{2 \times (9.11 \times 10^{-31} \text{ kg}) \times (10 \text{ eV}) \times (1.60 \times 10^{-19} \text{ J/eV})}}$$

$$= 3.88 \times 10^{-10} \text{ m} = 3.99 \text{ Å}$$

This wavelength is comparable with atomic sizes.

A *photon* with this same wavelength would have an energy

$$\mathcal{E} = h\nu = \frac{hc}{\lambda}$$

$$= \frac{(6.62 \times 10^{-34} \text{ J-s}) \times (3 \times 10^8 \text{ m/s})}{(3.88 \times 10^{-10} \text{ m}) \times (1.60 \times 10^{-19} \text{ J/eV})}$$

$$= 3.2 \text{ keV}$$

Such an energetic photon is an *X ray*. ∎

The Electron Microscope

In order to examine the structures of the smallest living things, microscopes with high magnification and good resolving power are necessary. We have seen, in Section 16.6, that optical microscopes using midrange visible light are limited to magnifications of about $800\times$ and have resolving powers no better than about 2500 Å. These values are inadequate for viewing small features of cells.

The resolving power of a microscope is given by Eq. 16.23:

$$\text{R.P.} = \frac{0.61 \lambda}{\text{N.A.}} \tag{1}$$

We can improve the resolving power by increasing the numerical aperture N.A. or by decreasing the wavelength λ. Unfortunately, no microscope design can make N.A. much larger than 1, so we must concentrate on decreasing λ. Microscopes using short-wavelength ultraviolet light have been constructed, but a much greater increase in R.P. can be achieved by making use of the wave properties of material particles, especially electrons.

If a beam of electrons, instead of light, were used in a microscope, we could take advantage of the fact that the electrons carry charge to guide and focus them with electric and magnetic fields. And, because the wavelength of a rapidly moving electron is very small, the diffraction effects that limit the usefulness of an optical microscope would be almost entirely absent. Thus, an *electron microscope* should be far superior to an optical microscope.

The de Broglie wavelength of an electron that has been accelerated from rest through a potential of V volts is

$$\lambda = \frac{h}{p} = \frac{h}{mv} = \frac{h}{\sqrt{2emV}} = \frac{12.3}{\sqrt{V}} \text{ Å} \qquad (2)$$

(This expression ignores relativistic effects and is therefore valid only for accelerating potentials of about 50 kV or less. For the relativistic expression, see Exercise 1.)

For $V = 50$ kV, a typical value for an electron microscope, we have $\lambda = 0.055$ Å. This is 10^5 times smaller than the wavelength of visible light and 50 to 100 times smaller than the interatomic spacings in solid matter. It would therefore appear that a 50-kV electron microscope would be capable of resolving all possible detail of the atomic and molecular structure of matter. Unfortunately, the electric and magnetic lenses that must be used to focus the electrons in an electron microscope are subject to aberrations, just as optical lenses are. Spherical aberration is a particularly severe problem in electron microscopes, and the only effective way to reduce this defect is by limiting the electron trajectories to lie near what would be called the central ray in an optical microscope. That is, the angle β in Eq. 16.21 is made small with the result that the numerical aperture (Eq. 16.20) is also small. Most electron microscopes have N.A. values in the range 0.01 to 0.001. The value for a special high-resolution instrument might be N.A. = 0.02; then, the resolving power (Eq. 16.23) becomes (for $V = 50$ kV)

$$\text{R.P.} = \frac{0.61\,\lambda}{\text{N.A.}} = \frac{0.61 \times (0.055 \times 10^{-10} \text{ m})}{0.02} = 1.7 \text{ Å} \qquad (3)$$

Even this resolving power is difficult to achieve in practice. Most commercial 50-kV microscopes have R.P. = 5 to 10 Å. Some specialized instruments have been able to achieve 2 Å in favorable circumstances.

Electric and magnetic lenses are used in electron microscopes to focus the electrons in the same way that glass lenses are used in optical microscopes to focus light. All electron microscopes use electric lenses in the *electron gun*, the device that provides the focused beam of rapidly moving electrons that probes the specimen. Most instruments employ magnetic lenses for focusing the beam after it leaves the electron gun. (Some electron microscopes use electric lenses throughout, but generally, such high field strengths are necessary to focus fast electrons with these lenses that arc-over is often a problem.)

It is easy to understand how electric and magnetic lenses work. Figure 1a is a schematic view of an electric lens. Two cylindrical pipes are separated by a narrow gap. The pipe at the right is maintained at a higher positive potential than the one at the left. Electrons from the gun are accelerated into the left-hand pipe and then drift through the pipe at constant velocity (there is no field within the pipe except near the ends). When the electrons reach point A, they experience an electric field $\mathbf{E}_A$ that is directed upward and to the left. Because the electrons carry a negative charge, they are acted upon by a force that is downward and to the right, as indicated in the diagram, until they are midway through the gap. At point B the electrons experience a field $\mathbf{E}_B$ that is directed downward and to the left; the electrons now react to a force that is upward and to the right. But the electrons at B are more energetic than they were at A because they have been accelerated through most of the potential difference that exists between the pipes. Consequently, the electrons are deflected less by the right-hand part of the gap field and converge to a focus at F. Figure 1b is the optical analog of the electric lens in Fig. 1a. Notice that the effect of the left-hand part of the gap field is the same as that of a strong converging lens;

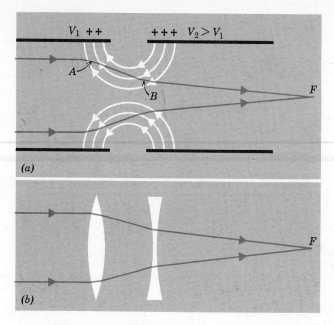

Figure 1 (a) Schematic of an electric lens, showing the focusing action on a beam of electrons. (b) Optical analog of the electric lens in (a).

the effect of the right-hand part of the gap field is the same as that of a weak diverging lens.

Figure 2 shows a magnetic lens which consists of a tightly packed coil of wire encased in an iron shield. Notice the narrow gap in the shield which results in the concentration of the external magnetic field in the vicinity of the center of the coil. At point P the electrons moving to the right experience a magnetic field $\mathbf{B}_P$ that is directed downward and to the right. These electrons are therefore acted upon by a force $\mathbf{F}_P = -e\mathbf{v} \times \mathbf{B}_P$ that is directed *out* of the plane of the page. This force causes the electron trajectory to rotate in a clockwise sense (looking along the lens axis in the direction of the electron beam). The electrons now have a velocity component out of the plane of the page and therefore are acted upon by a force that has a component directed toward the axis of the lens. Thus, the electrons are focused. (Verify that an electron moving along *any* trajectory will experience this same clockwise rotation and same focusing effect.) It is easy to trace the electrons through the right-hand part of the field and to show that they eventually come back to the axis of the lens (see Exercise 3). The rotation of the electron trajectory within the field of the lens is shown in Fig. 3. Because of the rotation of the beam trajectory in passing through a magnetic lens, the image produced will likewise be rotated with respect to the object. In a microscope this effect is not important.

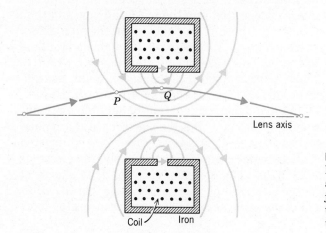

Figure 2 Schematic of a magnetic lens, showing the focusing action on a beam of electrons. The electron trajectory is actually rotated in passing through the lens, as shown in Figure 3.

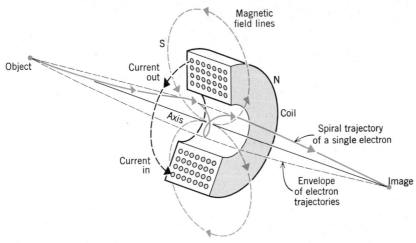

Figure 3 Spiral trajectory of an electron passing through a magnetic lens. (Adapted from G. A. Meek, *Practical Electron Microscopy for Biologists,* Wiley, New York, 1970.)

Figure 4 shows the lens system in the RCA EMU-3 electron microscope. (This general type of instrument has been manufactured by RCA since 1941.) Electrons are emitted from a heated oxide-coated filament and are accelerated toward the anode by a potential of up to 50 kV. The magnetic condensing lens concentrates the

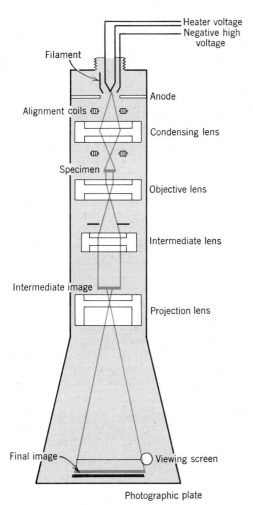

Figure 4 Schematic of the lens system in an electron microscope (RCA Model EMU-3). All lenses after the electron gun are magnetic. (Some electron microscopes have no lens corresponding to the intermediate lens shown here.)

Figure 5 Electron micrograph of a nearly perfect crystal of tobacco necrosis virus protein. The magnification is 110,000×. Notice the hole in the lattice above and to the left of the center, where one molecule on the top face of the crystal is missing. (From L.W. Labaw and R. W. G. Wyckoff, *J. Ultrastructure Research*, 2 (8), 1958.) (See Plate XVa in R. W. G. Wyckoff, *The World of the Electron Microscope*, Yale University Press, New Haven, Conn., 1958.)

beam on the specimen. More electrons are scattered out of the beam by the thicker parts of the specimen than are scattered by the thinner parts. The transmitted beam therefore has spatial differences in density that correspond to the features of the specimen. The final image is produced by two stages of magnification. The objective and intermediate lenses produce a real intermediate image, and the projection lens forms the final image, which can be viewed by means of a fluorescent screen or photographed on a special electron-sensitive plate.

The magnification achieved by an electron microscope can be as high as 10^6. But, as with optical microscopes, resolving power is a much more significant measure of the performance of the instrument. (Huge magnifications can be produced by photographically enlarging the image on the negative.) Usually, biological specimens are viewed with magnifications in the range from 10^3 to 10^5 and with resolving powers from 5 Å to 50 Å or so. Figure 5 shows an electron micrograph of a crystal of tobacco virus in which the magnification is ×110,000.

Even more spectacular effects can be achieved by using modern *scanning electron microscopes* (SEMs). In an SEM, the beam is scanned over the specimen in the same way that an electron beam is scanned over the face of a television picture tube. The micrographs taken in this way have a remarkable three-dimensional appearance.

References

G. A. Meek, *Practical Electron Microscopy for Biologists,* Wiley, New York, 1970.

A. L. Stanford, *Foundations of Biophysics,* Academic Press, New York, 1975, Chapter 11.

R.W.G. Wyckoff, *The World of the Electron Microscope,* Yale University Press, New Haven, Conn., 1958

■ *Exercises*

1. Show that the wavelength of an electron falling through a potential V volts is given by the relativistic expression,

$$\lambda(\text{Å}) = \frac{12.3}{\sqrt{V + 10^{-6} V^2}}$$

What error is introduced in λ by using Eq. 2 for $V = 10^5$ V?

2. An electron passes through a narrow gap that separates a region at a potential V_1 from a region at a potential V_2. The electron undergoes *refraction* in the process. The electron has a velocity v_1 in region 1 and makes an angle θ_1 with respect to the normal to the plane of the gap. When in region 2, the electron has a velocity v_2 and makes an angle θ_2 with the normal. Show that the angles θ_1 and θ_2 are related by a "Snell's law" equation of the form

$$V_1^{1/2} \sin \theta_1 = V_2^{1/2} \sin \theta_2$$

3. Refer to Fig. 2 and describe the trajectory of the electrons and the force they experience at point Q. Justify the direction of the trajectory of the electrons to the right of Q as shown in the figure.

4. Refer to Fig. 5 and estimate the size of the tobacco virus molecule. What are the dimensions of the crystal shown? What is the size of the finest detail that can be seen in the micrograph? (That is, what is the resolving power?)

Diffraction of X Rays and Electrons

When electrons are allowed to fall through a potential of several thousand volts and strike a metal target, extremely short-wavelength radiation is produced. Many attempts were made in the early 1900s to measure the wavelengths of these X rays by conventional diffraction experiments using finely ruled gratings (see Section 15.4). These experiments were only marginally successful until Max von Laue (1879–1960) realized that the regular planes of atoms in crystals, with spacings of only a few Ångstroms, could be used as diffraction gratings. In 1912 von Laue succeeded in obtaining interference patterns of X rays diffracted from calcite crystals. The spacing between the planes of atom in the crystal could be calculated approximately from a knowledge of the properties of the material. The measured diffraction patterns then showed that typical wavelengths for X rays were of the order of 1 Å, very much shorter than UV wavelengths.

The diffraction of X rays is illustrated schematically in Fig. 18.7 where the dots represent the ordered array of atoms in a simple crystalline lattice. The rows of atoms are spaced a distance d apart and therefore form a kind of grating through which the radiation can pass. In the direction specified by the angle θ with respect to the direction of the incident beam, the two scattered rays differ in their path lengths by an amount $d \sin \theta$. Therefore, on a photographic plate placed some distance away, the interference between the two rays will be *constructive* if $d \sin \theta$ is an integer number of wavelengths of the

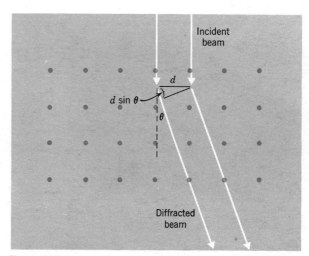

Figure 18.7 An incident beam of X rays (or electrons) is diffracted by the grating formed by the regular array of atoms in a crystal.

"Fundamentals of College Physics" by W. Wallace McCormick
(a)

Wollan, Shull, & Marney, Physics Rev. 73, 527 (1948)
(b)

Figure 18.8 Laue patterns for the diffraction of (a) X rays and (b) neutrons by single cubic crystals of rock salt (Na CL). The light spots are the exposed regions of the photographic plates caused by constructive interference. Note the similarity of the interference patterns. The central bright spot in each pattern is due to the undeflected portion of the beam.

radiation; the interference will be *destructive* if $d \sin \theta$ is an odd number of half wavelengths. Consequently, there will be exposed and unexposed regions that alternate on the photographic plate. Because the crystal is a *three*-dimensional array of atoms, the directions in space in which the interference is constructive are quite limited. As a result, the exposed regions are *spots* rather than lines. Figure 18.8 shows such a spot pattern (called a *Laue pattern*). Measurements of the distances between the spots can be used to determine the wavelength of the incident radiation if the atomic spacing in the crystal is known. (Actually, the method is now used in reverse: X rays of known wavelength are employed to study the details of crystal structure by diffraction.)

Suppose that we use a *foil* of material for the diffraction experiment instead of a single crystal. Foils usually consist of large numbers of tiny crystals (*micro*crystals), which are randomly oriented. Therefore, there will be a random distribution of the angles of incidence of the X-ray beam upon the individual crystals. The interference pattern observed for the transmitted beam of X rays will then correspond to a larger number of spot patterns that are located at random angles around the direction of the incident beam. That is, each spot in the single-crystal Laue pattern will produce a *circular* pattern of constructive interference. Such a circular diffraction pattern is shown in Fig. 18.9a.

X ray diffraction patterns obtained with DNA constituted important evidence leading to the discovery of the double-helix structure for DNA.

According to de Broglie's hypothesis, electrons should exhibit wave properties and therefore should be capable of producing diffraction patterns in the same way that X rays produce these patterns. In 1927 George P. Thomson (1892–1975) observed elec-

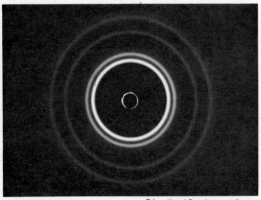

Educational Development Center
(a)

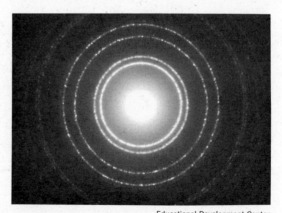

Educational Development Center
(b)

Figure 18.9 X-ray and electron diffraction in aluminum, showing the similarity in the diffraction patterns. The X-ray photograph (a) was made with X rays of wavelength 0.71 Å. The electron diffraction photograph (b) was obtained with 600-eV electrons ($\lambda = 0.50$ Å). The electron pattern has been enlarged × 1.6 in order to facilitate the comparison.

tron diffraction by passing electrons through foils. At almost the same time, C. J. Davisson (1881–1958) and L. H. Germer (1896–1971) directed electrons onto the surfaces of single crystals and found interference peaks in the distribution of the reflected electrons. Davisson and Thomson shared the 1937 Nobel Prize for these experiments that established the wave character of electrons.

The correspondence between X-ray diffraction and electron diffraction is strikingly illustrated in Fig. 18.9. Here are photographs of diffraction patterns obtained by passing X rays and electrons of approximately the *same wavelength* through a thin aluminum foil. Also, Fig. 18.8 shows the Laue patterns produced by X rays and by neutrons diffracted by rock salt. The close similarity between the patterns in each pair of photographs is dramatic proof of the identical wave properties of electromagnetic radiation and matter.

Compton Scattering

The second of the crucial experiments dealing with the question of wave-particle duality involves the *scattering* of photons. If Einstein's interpretation of the photoelectric effect is correct and a photon interacts with a single electron to eject it, then it should be possible to observe a *scattering* process in which a photon and an electron interact in much the same way as do two colliding balls. In 1924, the American physicist Arthur H. Compton (1892–1962) demonstrated just such a process.

In describing the Compton scattering experiment, we wish to simplify the situation and treat the electrons as *free* particles. But we know from the results of photoelectric experiments that electrons are *bound* in matter with energies of several eV. In his experiment, Compton used photons (X rays) of energy 17.5 keV. Because this energy is so much greater than the binding energy of the electrons, we make no appreciable error by treating the electrons a free particles.

We proceed by considering a photon of energy $h\nu$ incident on an electron at rest (Fig. 18.10), and we treat the problem in a strictly classical way. (But relativistic effects must be taken into account when we deal with fast-moving electrons.) We have available three equations that represent the following statements:

1. *Conservation of energy:* The energy before collision (the photon energy $h\nu$ plus the electron rest energy) must equal the energy after collision (the energy $h\nu'$ of the scattered photon plus the total energy of the recoiling electron.)

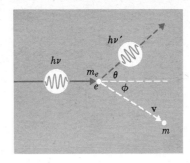

Figure 18.10 Schematic diagram of the Compton scattering of a photon of energy $h\nu$ by an electron.

2. *Conservation of linear momentum in the direction of the incident photon:* The momentum of the incident photon, $p = \mathcal{E}/c = h\nu/c$, must equal the sum of the components of the electron momentum and the momentum of the scattered photon in the direction of the incident photon.

3. *Conservation of linear momentum in the direction transverse to the direction of the incident photon:* Because there is no momentum in the transverse direction prior to collision, there can be no net transverse momentum after collision. Therefore, the transverse momentum components of the scattered photon and the recoiling electron must be equal and opposite.

By writing down these equations and performing the necessary (but tedious) algebra, we find that the frequency ν' of the scattered photon is related to the frequency ν of the incident photon and to the angle of scattering θ according to:

$$\frac{1}{\nu'} = \frac{1}{\nu} + \frac{h}{m_e c^2}(1 - \cos\theta) \qquad (18.8)$$

or, converting to wavelengths by using $\lambda = c/\nu$, we have

$$\lambda' = \lambda + \frac{h}{m_e c}(1 - \cos\theta) \qquad (18.9)$$

The quantity $h/m_e c$ is known as the *Compton wavelength of the electron* and has the value

$$\frac{h}{m_e c} = 2.426 \times 10^{-12} \text{ m} \qquad (18.10)$$

This is the wavelength of a photon whose energy is $m_e c^2$, the rest-mass energy of an electron.

Compton measured λ' as a function of θ for several different initial wavelengths λ and found agreement with Eq. 18.9, thus demonstrating that photons behave as *particles* not only in the photoelectric effect but in scattering processes as well. Compton

Example 18.5

A 100-keV X ray is Compton scattered through an angle of 90°. What is the energy of the X ray after scattering?

Dividing Eq. 18.8 by h, we have

$$\frac{1}{h\nu'} = \frac{1}{h\nu} + \frac{1}{m_e c^2}(1 - \cos\theta)$$

Using the values $m_e c^2 = 511$ keV and $\cos 90° = 0$, we find

$$\frac{1}{h\nu'} = \frac{1}{100 \text{ keV}} + \frac{1}{511 \text{ keV}}$$

or,

$$h\nu' = \frac{(100 \text{ keV}) \times (511 \text{ keV})}{(100 \text{ keV}) + (511 \text{ keV})} = \frac{51100}{611} \text{ keV}$$

$$= 84 \text{ keV}$$

The electron, of course, carries off the remainder of the incident energy:

$$KE = 100 \text{ keV} - 84 \text{ keV} = 16 \text{ keV} \quad \blacksquare$$

18.4 The Basis of Quantum Theory

Photons

The various experiments we have just described conclusively demonstrated the dual nature of radiation and matter—an electron can propagate as a wave and light can interact as a particle. We have already learned how to describe the classical aspects of electrons and electromagnetic radiation, but how do we describe "light particles" and "electron waves"?

We have often referred to the *frequency* of an electromagnetic wave. However, we have been a bit cavalier with the terminology because, in fact, *the frequency of a real electromagnetic wave does not exist*. That is, such radiation never possesses a single, precisely defined frequency. In order to define *the* frequency of a wave, the wave must be absolutely uniform in its properties throughout space. Thus, a wave of *absolutely pure* frequency (Fig. 18.11) must have infinite extent. But all electromagnetic oscillators, be they radio antennas or atoms, oscillate only for finite periods of time. Therefore, the radiation is necessarily of finite extent and cannot have a single, precise frequency. *Real* radiation is always equivalent to a combination (or *superposition*) of oscillations with various frequencies.

Figure 18.11 A wave that has a pure or precise frequency must be the same everywhere; that is, it must have infinite extent.

If these frequencies lie in a range around a central frequency, the effect is to produce *constructive* interference in one region of space and *destructive* interference everywhere else. The appearance of such a superposition of waves is shown schematically in Fig. 18.12. This localized bunch of oscillations is called a *wave packet* and, for the case of electromagnetic radiation, the packet (i.e., a *photon*) propagates as a unit with the velocity of light.

The range of frequencies that exists in typical light photons is quite small. For example, consider yellow light of nominal frequency 5×10^{14} Hz. In the emission of such light from an atom, the range of frequencies is only about 2 parts in 10^6; that is, $\Delta\nu/\nu \cong 2 \times 10^{-6}$. The corresponding range of wavelengths is only 0.01 Å; that is, the average wavelength is 6000 Å with a spread in wavelength from 5999.99 Å to 6000.01 Å. No spectral line is absolutely sharp; there is always a small natural width.

Figure 18.12 shows only 7 oscillations within the wave packet; for a typical light photon, such as that just described, the wave packet would consist of approximately 6×10^5 oscillations. This photon, containing such a large number of oscillations, retains many of its wavelike characteristics; but it still is a discrete unit and therefore interacts individually with electrons, for example, in Compton scattering or in the photoelectric effect.

Electron Waves

Let us consider a very simple type of experiment involving electrons. Suppose that we direct a beam

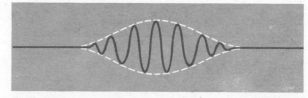

Figure 18.12 A wave packet (or photon) is localized in space because the superposition of waves of various frequencies produces constructive interference in one region and destructive interference everywhere else. Real photons contain 10^5–10^6 oscillations rather than the few sketched here.

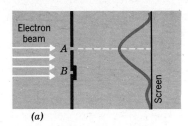
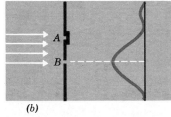
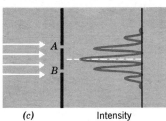

Figure 18.13 (a, b) Intensity patterns of electron impacts on a screen for a single open slit. (c) Intensity pattern for a double slit. This pattern is not the same as the sum of the two previous patterns. This pattern is drawn for the case in which the distance between the two slits is three times the width of each slit. Notice that increasing intensity is plotted to the left.

of electrons toward a panel in which two slits have been cut. In Fig. 18.13a we show the situation for the case in which the lower slit (labeled B) is blocked off so that all of the electrons that penetrate the panel must go through slit A. At a certain distance behind the panel we place a screen for the purpose of observing the positions of impact of the various individual electrons. Such a screen can be made from a material that *scintillates;* that is, a flash of light is produced for every electron impact. Or, we could use some sort of electron detector, such as a Geiger counter, and by moving the detector up and down along the screen, record the number of electron impacts as a function of position along the screen. In any event, the essential quantity that we can measure by using one of several available techniques is the distribution of electron impacts along the screen. The curve in Fig. 18.13a represents such an *intensity* distribution; it shows that the peak of the distribution is directly in line with the slit. If the slit is *narrow*, the distribution will be *broad* (compare Fig. 15.42).

If we repeat the experiment with slit B open and slit A closed, we will, of course, find exactly the same type of distribution centered now around the position in line with slit B.

What will happen if we open *both* slits? If electrons behaved as little balls, we could think of the experiment in terms of projecting a stream of marbles at the pair of slits: some marbles would go through slit A and, because of scattering at the slit edges, would produce a pattern on the screen like that shown in Fig. 18.13a and some would go through slit B and produce the pattern of Fig. 18.13b. The net result would be the *sum* of the two intensity patterns. But this expectation based on classical reasoning is not borne out by experiment for the case of electron beams. Actually, a much more complex pattern is observed (Fig. 18.13c). The maximum intensity is found *midway* between the two slits and there are several subsidiary maxima, falling off in intensity uniformly on each side of the central maximum. Notice also that, at some points on the screen, the intensity has actually been *decreased* by the opening of an additional slit. We can only draw the conclusion that electrons behave as waves in this experiment and produce exactly the same type of interference effects that are produced by light waves.

Self-Interference and Probability

Suppose that we repeat the double-slit experiment with electrons (we could equally well use photons), but now we make the incident beam exceedingly weak—so weak, in fact, that at any one instant there can be only a single electron in the vicinity of the apparatus. How will the pattern on the screen be affected? When any individual electron passes through the apparatus, it will not smear itself out in conformity with the curve of Fig. 18.13c because each individual electron interacts with the screen at a definite *point*. We can determine the position of the impact by observing the light flash that is produced, but *before* the impact has occurred, *we have absolutely no way in which to predict the point at which it will occur.*

After the first 10 electrons have passed through the apparatus, the results might be similar to those illustrated in Fig. 18.14a where each box represents a light flash observed at the corresponding position on the screen. With this small number of events, the pattern is only a scatter of boxes. After 40 events (Fig. 18.14b) definite structure has begun to emerge; there are pronounced valleys between the groups of boxes. When a large number of events has occurred (in this case, thousands), it becomes possible to represent the results with a curve that encloses the distribution of boxes (Fig. 18.14c). For any particular electron, then, we can refer only to the *probability* that it will strike the screen at a given point. The height of the curve at any position along the screen is proportional to the probability that an electron will interact with the screen at that position. The

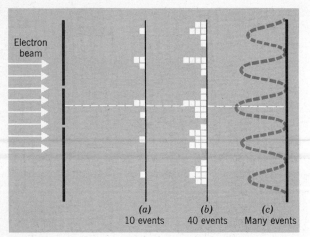

Figure 18.14 Distribution of light flashes on a screen after (a) 10, (b) 40, and (c) many electrons have passed through the apparatus. Each box represents one light flash observed at the corresponding position on the screen. The point of occurrence of any particular event cannot be predicted; only the probability (the dashed curve) can be stated.

probability is highest for the midpoint and is least for the valleys between the various maxima. The dashed curve in Fig. 18.14c is called the *probability curve* or the *intensity curve* that describes the distribution of events. This curve is exactly the same as that predicted by the wave theory of diffraction for radiation with the same wavelength.

We are therefore led to the view that *individual* electrons (or photons) exhibit *interference,* but only when large numbers of particles are involved does it become possible to identify the diffraction pattern with that predicted by the wave theory. When a single electron or photon is in the apparatus at a given instant, it ha only *itself* with which to interfere. This *self-interference* results from that portion of the wave that goes through slit *A* interfering with the portion that goes through slit *B*.

From these arguments we can draw two important conclusions that are crucial for the development of quantum theory:

1 Individual electrons or photons have the *wave*like property that they can interfere with themselves.

2 Individual electrons or photons have the *particle*like property that they interact with matter at discrete points but the prediction of where such interactions will take place can be made only in the probabilistic sense.

The probability (or intensity) curve that describes the interactions of electrons or photons with matter is identical with the prediction of wave theory and corresponds to the pattern of individual interactions that would occur for a very large number of events.

Is the Wave-Particle Duality *Real*?

How can we reconcile the fact that electrons and photons both appear sometimes as particles and sometimes as waves? Does each exist as part wave and part particle? Or are they both capable of transforming back and forth between these two different descriptions? When is a wave a *wave* and when is it a *particle*? We have become accustomed to think of waves and particles in *classical* (i.e., *Newtonian*) terms. A wave is a propagating disturbance in a medium—a wave is not a thing that can be localized in space. A particle, on the other hand, is a material object that can definitely be localized in space. But photons and electrons are *not* classical entities—they are *quantum* things. We must not force classical descriptions on nonclassical entities. A photon is not a wave and it is not a particle, nor is it a mixture of the two—a photon is a *photon*.

When we perform experiments with photons and electrons, we must use apparatus of some sort. If we choose to perform an experiment in which a beam of photons is directed through a narrow slit, then, because of the nature of the apparatus, we have guaranteed that we will observe a diffraction pattern that is characteristic of waves. If we use apparatus designed to study particles (for example, a photoelectric tube), we must obtain a particle result. It is the way in which we make our measurements that determines whether we obtain a wave or particle result.

Max Born (1882–1970), the originator of the probabilistic interpretation of quantum events, has said, "Every process can be interpreted either in terms of particles or in terms of waves, but . . . it is beyond our power to produce proof that it is actually particles or waves with which we are dealing, for we cannot simultaneously determine all the other properties which are distinctive of a particle or of a wave, as the case may be. We can, therefore, say that the wave and particle descriptions are only to be regarded as complementary ways of viewing one and the same objective process. . . ."

18.5 The Quantization of Momentum and Energy

The Free Particle

A quantum *free* particle can move through space just as a classical free particle can without any restriction on the value of the energy that it can have.

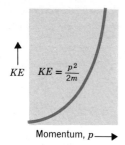

Figure 18.15 For a free particle, the curve that relates the kinetic energy to the momentum is a parabola, and every point on the curve refers to an allowed energy and the corresponding allowed momentum.

Thus, the particle wave can have any wavelength $\lambda = h/p$, and it can have any kinetic energy, which, in a nonrelativistic situation, is expressed as

$$KE = \frac{1}{2}mv^2 = \frac{1}{2}\frac{(mv)^2}{m} = \frac{p^2}{2m} \qquad (18.11)$$

The relationship between the kinetic energy and the momentum, therefore, is *parabolic*, as shown in Fig. 18.15 every point on the curve represents an allowed energy and the corresponding allowed momentum.

There is no difference between the results of classical mechanics and quantum theory for the allowed energies of a free particle—all energies are allowed. However, if we *confine* the particle in some way by subjecting it to forces that restrict its motion, the two theories no longer yield the same results.

Particle Bound in a One-dimensional "Box"

In many situations we must deal, not with *free* particles, but with particles that are constrained in some way to remain in a certain region of space. A simple example of this situation is that in which a particle is required to move along a straight line (e.g., the *x*-axis) between the points $x = 0$ and $x = L$. Classically, we can think of this particle as bouncing between a pair of unyielding walls, always maintaining straight-line motion in the $+x$- or $-x$-direction. From the standpoint of classical theory, there is again no restriction on the energy that the particle can have. Energy and momentum are still related by Eq. 18.11 and any combination of *KE* and *p* that satisfies this condition is allowed.

Now consider a quantum particle (for example, an electron) that is required to move in the same way. We must think of *this* particle in terms of its wave character and, in particular, we must examine the conditions imposed on the electron wave by the presence of the walls. It proves convenient to describe the electron terms of a *wave function*, denoted by the symbol ψ. The value of the wave function depends on position and in the one-dimensional case is written as $\psi(x)$. The square of the wave function, $|\psi(x)|^2$, is proportional to the probability that a measurement will reveal the electron to be located at the particular position specified by *x*. In fact, the wave function itself, $\psi(x)$, has no physical significance and cannot be directly measured; only $|\psi(x)|^2$ can be determined by experiment and is therefore physically meaningful.[4]

If the particle is confined to a "box," then there is zero probability of finding the particle outside the box, and hence, the wave function must be zero in this region. Because the wave function cannot change abruptly from some nonzero value just inside the box to zero just outside the box, the wave function must also be zero exactly at the boundary walls. With these facts in mind, the solution to the problem is now quite simple because it is exactly the same as the problem of standing waves on a string (Section 15.2). The wave function for the particle (just as the displacement or amplitude of the string) must be zero at $x = 0$ and $x = L$. That is, standing de Broglie waves must be fitted into the box, and this can be accomplished only when an integer number of wavelengths is equal to $2L$. Thus,

$$n\lambda_n = 2L, \qquad n = 1, 2, 3, \ldots \qquad (18.12)$$

The first 4 allowed ψ-waves are illustrated in Fig. 18.16.

The probability of finding the particle at a particular point within the box is proportional to $|\psi(x)|^2$. This quantity is shown in Fig. 18.17 for the case $n = 4$. Notice that there are 4 regions in which there

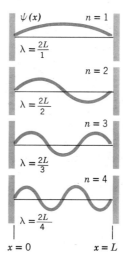

Figure 18.16 For a quantum particle bound in a one-dimensional "box," standing de Broglie waves (ψ-waves) must be fitted into the "box." The allowed wavelengths are therefore integer fractions of $2L$.

---

[4]In classical physics the wave amplitude *does* have physical meaning and can be measured. In both classical and quantum physics the square of the wave amplitude is proportional to the *intensity* of the wave.

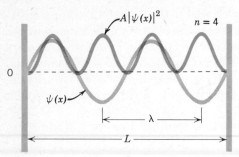

Figure 18.17 The probability amplitude $\psi(x)$ and the intensity $A|\psi(x)|^2$ for the case $n = 4$. The particle is most likely to be found at discrete positions within the "box."

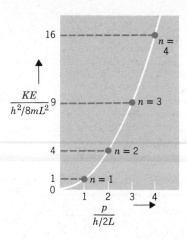

Figure 18.18 Only discrete energies (indicated by the horizontal lines and the dots) are allowed for a quantum particle in a box.

is a high probability of finding the particle and that there is zero probability not only at the walls but also at certain points *within* the box. Clearly, this result is contrary to that of classical mechanics.

Allowed Energies in the "Box"

We can now calculate the energies corresponding to the allowed wavelengths by using the de Broglie relation and Eq. 18.12. The allowed momenta are

$$p_n = \frac{h}{\lambda_n} = n\frac{h}{2L}, \qquad n = 1, 2, 3, \ldots \quad (18.13)$$

and the corresponding energies are

$$KE_n = \frac{p_n^2}{2m} = n^2\left(\frac{h^2}{8mL^2}\right), \qquad n = 1, 2, 3, \ldots \quad (18.14)$$

Therefore, we have the important result that only certain discrete energies and momenta are allowed; energy and momentum are *quantized*. Instead of the classical result in which every point of the *KE* versus *p* parabola corresponds to a possible energy-momentum combination, the quantum result shows that only certain of these points are in fact allowed (see Fig. 18.18).

Another important point to realize is the fact that *zero* kinetic energy for the particle is *not* allowed—the particle cannot be at rest in the box. The state of rest requires zero momentum and, hence an infinitely long wavelength. Such a de Broglie wave cannot fit into any finite box and therefore is not allowed. In general, no quantum system (with the exception of the ideal free particle) can possess zero kinetic energy. Even at the absolute zero of temperature, when according to classical theory all motion must cease, a quantum system still possesses a *zero-point energy*.

A particle in a box (either one dimensional or three dimensional) can be considered to be in a *potential energy well*; that is, the interior of the box corresponds to some finite potential (e.g., *zero*) whereas the walls and the exterior are at *infinite* potential. Thus, the particle can never escape the box because, in order to do so, it must have infinite energy. *Real* potential wells of course, are never infinite and a real particle can always escape from such a well if sufficient energy is supplied. For example, the electrons in a piece of metal find themselves in a certain type of potential well, but if sufficient energy is transferred to such an electron by an ultraviolet photon, the electron can escape from the metal—this is the photoelectric effect.

In the next chapter we will study another important case of a finite potential—the hydrogen atom.

18.6 The Uncertainty Principle

Where Do The Electrons "Really" Go?

When a series of individual electrons, spaced in time, are incident on a double slit, it is necessary to treat each electron as if its "probability wave," described by $\psi(x)$, goes through *both* slits. The interference of the two portions of each wave at the screen determines the distribution of light flashes that is observed. But an electron is a *particle* and is indivisible; no one has ever observed a *part* of the mass or a *part* of the charge of an electron. Our intuition tells us that an electron cannot go through two separated slits—it must go through one or the other. Can we perform an experiment to determine where the electron "really" goes? Let us alter the double-slit experiment by placing a thin detector behind one of the slits so that the electrons will still be able to pass through but will give a signal when this occurs. Then, we allow electrons to enter the apparatus one at a time and we record only those flashes of light on the screen that are accompanied by a signal that indicates the electron went through the slit with the detector. What pattern of flashes do we find? We find exactly the *single*-slit pattern again (Fig. 18.13a)! The act of determining which slit the electron went through has destroyed the double-slit interference effect.

Perhaps the trouble with our experiment was the detector. Perhaps it was not sufficiently "thin" and actually disrupted too severely the electron trajectories passing through it. We can repeat the experiment using a source of light behind one of the slits. Whenever we detect a photon that is scattered by an electron passing through the slit, we arrange to record the position of impact of the electron on the screen. But this technique is no more successful than the first; the results of this experiment are the same as before. In fact, whatever method we devise to indicate that the electrons have gone through a particular slit, the result is always that the interference effect is destroyed.

So our intuition is wrong. Our insistence in thinking of electrons in terms of classical particles leads to inconsistencies. Because of the difficulties that arose in applying classical reasoning to individual events in the atomic domain, the German theorist Werner Heisenberg (1901–1976) concluded that there must be a general principle of Nature that places a limitation on the capabilities of all experiments. This principle, formulated in 1927, is known as the *uncertainty principle*.

According to Heinsenberg's principle, it is impossible to build a detector to determine through which slit the electron passed without destroying the interference pattern. That is, there can be no device that can reveal the presence of an electron with a sufficiently delicate touch that the interference pattern will be unaffected—the act of "looking at" an electron with even a *single* photon is sufficient to change the wave function of the electron and disrupt the interference pattern.

Statement of the Uncertainty Principle

The concept of a *particle* is something that is localized in space. According to classical theory a particle has, at a given instant, both a well-defined position and a well-defined velocity. Let us attempt to apply this reasoning to an elementary particle, such as an electron.

What is the restriction, if any, on determining the location of an electron with an imaginary super-high-power microscope? Because we use light (or higher energy photons, such as X rays) to determine position with a microscope, we must remember the wave nature of such radiation. In fact, we cannot locate the electron in this way to better than about one wavelength of the radiation; that is, the uncertainty in the position Δx is approximately equal to one wavelength: $\Delta x \cong \lambda$. But in the act of observing the position of the electron, at least one photon must have been scattered by the electron. Thus, the uncertainty in the momentum of the electron will be equal to the momentum imparted to it by the photon, which (using the de Broglie relation) will be $\Delta p_x \cong h/\lambda$. The product of Δx and Δp_x is, therefore, $\Delta x \, \Delta p_x \cong \lambda \times (h/\lambda)$. Thus,

$$\Delta x \, \Delta p_x \cong h \qquad (18.15)$$

There are many other methods for determining this expression for Heisenberg's uncertainty principle and there are many experiments (including *thought* experiments) that can be devised to test its validity. *No* experiment has ever been found to be in disagreement with the uncertainty principle.

It is important to realize that the uncertainty principle refers to the *predictability* of events. *After* an electron goes through the slit in Fig. 18.14a and strikes the screen, we know where it did so from the position of the light flash, but before the event takes place, we can only give the *probability* that a light flash will be observed at a particular point. Quantum theory cannot predict the result of any single event, but the *average* of a large number of events (e.g., the position of the peak of the probability or intensity curve) can be predicted with precision. This is the essential meaning of the uncertainty principle.

The uncertainty principle is not to be thought of as some mysterious device conceived by Nature to prevent Man from probing too deeply into her methods of making atoms behave properly. Rather, the uncertainty principle is just one manifestation of the wave-particle duality of radiation and matter. Waves cannot be localized in space and so any measurement of the position of a wavelike object must be subject to uncertainty. The Heisenberg principle gives a quantitative description of this uncertainty.

● *Example 18.6*

The position of a free electron is determined by some optical means to within an uncertainty of 10,000 Å or 10^{-6} m. What is the uncertainty in its velocity? After 10 s how well will we know its position?

Nonrelativistically, we can express the uncertainty relation as

$$\Delta p_x = m_e \times \Delta v_x \cong \frac{h}{\Delta x}$$

so that

$$\Delta v_x \cong \frac{h}{m_e \Delta x} = \frac{6.6 \times 10^{-34} \text{ J-s}}{(9.1 \times 10^{-31} \text{ kg}) \times (10^{-6} \text{ m})}$$

$$\cong 700 \text{ m/s}$$

Therefore, after 10 s the electron could be anywhere within a distance of 7×10^3 m or 7 km! The act of locating the electron at one instant to within a distance as small as 10^{-6} m stringently limits our knowledge of where the electron is at future times. ■

Another Form of the Uncertainty Principle

Suppose that we try to measure the frequency of a photon by counting the number of cycles that are observed in a time interval Δt. Because the photon is not of infinite extent, it does not have pure frequency; instead, it is a mixture of frequencies (see Fig. 18.12). Consequently, we will never be able to get an exact count on the number of oscillations that occur each second. We can reasonably assume that our count will be uncertain by about one cycle. That is, the uncertainty in the frequency measurement is

$$\Delta \nu \cong \frac{1 \text{ cycle}}{\Delta t} = \frac{1}{\Delta t} \text{ Hz}$$

The corresponding uncertainty in the photon energy will be h times $\Delta \nu$:

$$\Delta \mathcal{E} = h \, \Delta \nu \cong \frac{h}{\Delta t}$$

so that

$$\Delta \mathcal{E} \, \Delta t \cong h \qquad (18.16)$$

This statement of the uncertainty principle has exactly the same content as our previous statement, $\Delta x \, \Delta p_x \cong h$. In 1928, Niels Bohr summarized the conclusions that had been reached concerning indeterminism in quantum theory by stating that *if an experiment allows us to observe one aspect of a physical phenomenon, it simultaneously prevents us from observing a complementary aspect of the phenomenon*. This statement is known as Bohr's *principle of complementarity*. The complementary features to which the principle applies may be the position and momentum of a particle, the wave and particle character of matter or radiation, or the energy and time interval for an event. Quantum indeterminacy as stated in the Heisenberg uncertainty principle is included in the more general principle of complementarity.

● *Example 18.7*

In emitting a photon, an atom radiates for approximately 10^{-9} s. What is the uncertainty in the energy of the photon?

$$\Delta \mathcal{E} \cong \frac{h}{\Delta t}$$

$$\cong \frac{6.6 \times 10^{-34} \text{ J-s}}{10^{-9} \text{ s}} \times \frac{1}{1.6 \times 10^{-19} \text{ J/eV}}$$

$$\cong 4 \times 10^{-6} \text{ eV}$$

If the photon has a nominal wavelength of 6000 Å, the energy is 2 eV (see Table 18.2), so that the *relative* uncertainty in the energy is

$$\frac{\Delta \mathcal{E}}{\mathcal{E}} \cong \frac{4 \times 10^{-6} \text{ eV}}{2 \text{ eV}} = 2 \times 10^{-6} \quad ■$$

Does the Uncertainty Principle Destroy Quantum Theory?

What good is a theory that cannot answer such a simple question as "which slit did the electron go through?" There is really no basis to fault quantum theory for its inability to answer such a question because a theory can address itself only to questions that can be settled by experiments. There is no place in a physical theory for a quantity that cannot be defined by a measurement. Because no *measurement* can decide which slit a particular electron went through in a double-slit interference experiment, we cannot expect any *theory* to provide the answer. A theory can be no better than the measurements to which it applies. The usefulness of quantum theory is in no way vitiated by the uncertainty principle—indeed, the uncertainty principle is the cornerstone on which quantum theory is constructed.

Summary of Important Ideas

The details of *thermal radiation* and the *photoelectric effect* can be explained only if electromagnetic radiation occurs in discrete packets or *photons*. The energy of a photon is $\mathcal{E} = h\nu$.

The fundamental atomic constants are: the velocity of light c, the mass of the electron m_e, the charge of the electron e, and Planck's constant h.

Depending on the type of measurement that is made, electrons and photons can exhibit properties of either *waves* or *particles*.

Any piece of matter has associated with it a *de Broglie wavelength* $\lambda = h/p$. Radiation has associated with it an equivalent mass $m = \mathscr{E}/c^2$ and momentum $p = \mathscr{E}/c$.

In a double-slit experiment, *no* measurement can be made to determine through which slit the photon or electron went without destroying the double-slit interference pattern.

Because of the wave nature of radiation and particles, we can never predict the exact behavior of any particular photon or particle; we can only predict the *average* behavior of large numbers of photons or particles. Individual events can be dicussed only in terms of *probabilities*.

The wave function $\psi(x)$ that describes a particle or a photon is a *probability amplitude;* only the *square* of the wave function, which is proportional to the *intensity,* can be measured and therefore has physical meaning.

Except for the (ideal) free particle, all quantum systems are constrained to have only certain discrete energies and momenta.

The *Heisenberg uncertainty principle* expresses the fact that we cannot simultaneously measure with arbitrarily high precision *complementary* aspects of a particle or a photon (such as *momentum* and *position* or the *energy* of an event and the *time* interval during which it took place.) The position-momentum form of the uncertainty principle is $\Delta x \, \Delta p_x \cong h$.

◆ Questions

18.1 An object at room temperature emits electromagnetic energy. Why can't you see the furniture in a closed room at night?

18.2 Why is a red light used as a "safe light" by photographers when they are working with undeveloped film in a darkroom?

18.3 Describe an experiment to distinguish between an X ray whose wavelength is $\lambda_X = 10^{-10}$ m and an electron whose de Broglie wavelength is $\lambda_e = 10^{-10}$ m. What experiments would *not* be suitable?

18.4 Discuss some of the changes in everyday events that would result if Planck's constant were suddenly increased to 1 J-s.

18.5 Refer to Fig. 18.17 where it is shown that the probability of finding a particle at a given position in a "box" is a maximum at certain positions and is *zero* at other positions. How is it possible for the particle to "move" from one position of maximum probability to another since in order to do so it must pass through a position that it is not allowed to occupy? Explain the situation carefully in terms of measurements that can be made.

18.6 According to John A. Wheeler, the following two items are complementary in the sense of Bohr's principle of complementarity: (a) the use of a word to convey *information,* (b) the analysis of the *meaning* of the word. Discuss complementarity in this case.

18.7 Are *angular position* and *angular momentum* complementary quantities in the sense of Bohr's complementarity principle? If so, state an "uncertainty principle" for these quantities.

18.8 One of the reasons cited to justify the construction of expensive high-energy accelerators is that particles (and photons) of high energy are needed to probe the detailed structure of nuclei and nucleons. Why is this so?

18.9 Choose a quantity that was once thought to be continuous and discuss how it was shown to be discrete. Choose a quantity that was once thought to be discrete and discuss how it was shown to be "fuzzy."

18.10 Discuss the proposition that we shall eventually be able to overcome the limitations now set by the uncertainty principle and shall then be able to discuss microscopic phenomena with the same kind of deterministic approach that is appropriate for macroscopic mechanical phenomena (i.e., the approach of Newtonian dynamics).

★ Problems

Section 18.1

18.1 The Earth absorbs solar radiation, is thereby heated, and re-radiates energy. At what wavelength is the radiation most intense if the surface temperature is 27°C?

18.2 The surface temperature of the Sun is approximately 5800 K. At what wavelength is the emitted light most intense? To what color does this correspond?

18.3 The radiation from a certain thermal source is maximum for a wavelength of 1.8 μm. What is the temperature of the source?

18.4 What is the frequency of radiation whose wavelength is 1 Å? What energy is carried by such a photon?

18.5 Determine the energy of photons of visible light with wavelengths given in Table 16.2. Give your answers in both joules and electron-volts.

18.6 Show that the wavelength λ (in nm) of a photon with an energy $\mathcal{E}$ (in eV) is given by $\lambda(\text{nm}) = 1240/\mathcal{E}(\text{eV})$. (This is often a useful relationship.)

18.7 What is the wavelength and frequency of a photon having an energy of 10^{-12} J? To what portion of the electromagnetic spectrum does this photon belong? Consult Fig. 15.50.

18.8 Under certain conditions the retina of the human eye can detect as few as five photons of blue-green light ($\lambda = 5 \times 10^{-7}$ m). What is the corresponding amount of energy received by the retina in joules and in eV? If five such photons strike the eye and are absorbed each second, what is the rate of energy transfer in watts?

Section 18.2

18.9* A beam of UV radiation ($\lambda = 1000$ Å) delivers 10^{-6} W to a certain photoemissive surface. How many photons strike the surface each second? If 1 percent of the photons eject photoelectrons, what will be the resulting electron current?

18.10 The threshold wavelength for the photoelectric effect on a certain material is 3000 Å. What is the work function of the material?

18.11 A potential of 2.7 V is required to stop completely the photoelectrons from a certain material when the electrons are ejected by 2100 Å radiation. What is the work function of the material?

18.12 The work function for platinum is 5.32 eV. What is the longest wavelength photon that can eject a photoelectron from platinum?

18.13 The work function of barium is 2.48 eV. What is the maximum kinetic energy of a photoelectron ejected from barium by photons of wavelength 2000 Å?

(Wavelengths of radiation in or near the visible part of the spectrum are sometimes expressed in units of *nanometers*: 1 nm = 10^{-9} m = 10 Å.)

18.14 The longest wavelength radiation that will eject photoelectrons from tungsten has $\lambda = 270$ nm. What is the work function of tungsten in eV? What potential is necessary to stop completely the photoelectrons if the incident radiation has $\lambda = 200$ nm?

18.15 What potential is necessary to prevent the flow of photoelectric current from a calcium surface when the incident radiation has a wavelength of 300 nm and 400 nm?

18.16* A mercury arc lamp is used as the source of UV radiation to study the photoelectric effect on lithium. By using various filters, discrete wavelengths can be isolated from the spectrum. The following wavelengths were used and the corresponding voltages were found necessary to stop completely the photoelectrons:

λ (Å)	V_s (volts)
2536	2.4
3132	1.5
3663	0.9
4358	0.35
5770	(no photoelectrons)

Plot the data on an appropriate graph. (Be certain to plot V_s versus ν—not versus λ.) Find the work function of lithium. Calculate the experimental value of h/e and compare with the value computed from the constants listed in Table 18.1

Section 18.3

18.17 What is the mass equivalent of a photon of wavelength 6×10^{-7} m (yellow light)? How many such photons would be required to make up the rest-mass energy of one electron?

18.18 A 20-keV X ray is Compton scattered by a free electron. What is the wavelength of the incident X ray? If the X ray is scattered through an angle of 90°, what is the final wavelength? What is the final energy of the X ray? What is the kinetic energy of the recoil electron?

18.19 A photon whose wavelength is 0.8 Å is Compton scattered through 90°. What is the energy of the scattered photon and how much energy is imparted to the electron?

18.20 What is your wavelength if you are running at a speed of 10 m/s? What is the significance of such a wavelength? Could you ever be "diffracted"?

18.21 Through what potential difference must an electron fall (starting at rest) in order that its wavelength be 1.6 Å?

18.22* Complete the following table:

	Energy (eV)	λ (m)
Electron	1	—
	—	1×10^{-10}
	1000	—
Photon	1	—
	—	1×10^{-7}
	1000	—

18.23 A proton, an electron, and a photon all have de Broglie wavelengths of 1 Å. If they all leave a given point at $t = 0$, what are the arrival times at a point 10 m away?

18.24 What is the velocity of a helium atom that has a wavelength of 1 Å?

18.25 When a neutron is in thermal equilibrium with objects at room temperature it has an energy of $\frac{3}{2}kT$; such neutrons are called *thermal* neutrons. What is the wavelength of a thermal neutron?

Section 18.4

18.26 What is the energy of a photon whose wavelength is the size of an atom (10^{-10} m), the size of a nucleus (5×10^{-15} m)?

18.27 In producing a photon of green light ($\lambda = 5000$ Å), an atom radiates for 10^{-9} s. How many oscillations are there in the proton? What is the "length" of the photon?

18.28 A beam of electrons ($v = 10^6$ m/s) passes through a 0.01 mm slit. What is the width of the central diffraction maximum on a screen 1 m away?

18.29 An electron diffraction experiment is performed with a crystal whose atomic planes are spaced 1 Å apart (see Fig. 18.7). The first diffraction minimum is found at an angle of 30° with respect to the direction of the incident beam. What is the electron wavelength and energy?

Section 18.5

18.30 What is the minimum kinetic energy (in eV) of an electron confined to a one-dimensional box with a size equal to the diameter of a typical atom ($L = 1$ Å)?

18.31 What is the minimum kinetic energy (in MeV) of a proton confined to a one-dimensional box with a size equal to the diameter of a typical nucleus ($L = 10^{-14}$ m)?

18.32 What is the minimum kinetic energy in eV of an electron confined to a one-dimensional box with a size of 1 cm? What would be the value of 1 cm? What would be the value of the quantum number n in Eq. 18.14 if the confined electron had a kinetic energy of 1 eV?

18.33 The allowed quantized energies of a harmonic oscillator consisting of a pendulum bob of mass m swinging on the end of a string of length l is

$$E = (n + \tfrac{1}{2})h\nu_0 \qquad n = 0, 1, 2, 3, \ldots$$

with

$$\nu_0 = \frac{1}{2\pi}\sqrt{\frac{g}{l}}$$

and h is Planck's constant. A small mass bob is suspended from a string of length $l = 50$ cm and is set into oscillation by a displacement of 10 from the vertical. What is the energy of the bob? What is the frequency ν_0? What is the value of the quantum number n? If n were increased by one, what difference in initial angular displacement would this correspond to? On the basis of these answers, would you judge quantum effects to be important for laboratory situations involving ordinary-sized objects?

Section 18.6

18.34 A baseball ($m = 0.15$ kg) is thrown through a 0.2-m wide slot in a fence. What transverse velocity could the baseball acquire by passing through the slot?

18.35 Estimate the answer to Problem 18.28 using the uncertainty principle.

18.36 An electron is localized in the x-direction to within 1 mm. How precisely can its x-velocity be known?

18.37 A proton in a nucleus is localized to within a distance approximately equal to the nuclear radius. What is the approximate uncertainty in the velocity of a proton in an iron nucleus ($R \cong 6 \times 10^{-15}$ m)? What is the corresponding uncertainty in energy? Express the result in MeV. (A nonrelativistic calculation is adequate for the accuracy desired.)

18.38 It is known that an electron is within an atom that has a diameter of 3 Å. What is the uncertainty in energy (in eV) of the electron? Comment on the result.

18.39 A certain atomic energy level has a mean lifetime of 10^{-8} s. (That is, after the level is excited by some means, it requires, on the average, a time of 10^{-8} s before it spontaneously radiates a photon.) What is the uncertainty in the energy of the emitted radiation?

19
Atoms and Atomic Radiations

In the preceding chapter we used the results of several crucial experiments concerning electrons and photons to build the basic framework of quantum theory. The actual development of quantum theory followed a much more tortuous path and utilized the results of many more experiments. In this chapter we sketch a parallel line of development based on the interpretation of experiments with more complicated systems, namely atoms. This phase of the campaign was directed toward understanding the optical spectra of atoms. Niels Bohr's early work on this problem set the stage for the tremendous outpouring of theoretical and experimental results that, in the short space of four years from 1924 to 1928, firmly established quantum theory as a proper description of atomic processes.

We can give in this chapter only a brief account of some of the more important lines of reasoning that so rapidly produced the new theory of how Nature behaves in the atomic domain. This quantum theory is still with us, improved by the inclusion of relativistic effects and aided by sophisticated methods of computation; it has proved to be the most precise description of natural phenomena that science has ever known. At the end of this chapter we shall discuss some of the important and interesting ways in which quantum theory has been applied to modern atomic problems.

19.1 Atomic Models

Thomson's Model

By 1902 a sufficient number of experiments had been performed to provide convincing evidence that the electron is one of the universal and fundamental constituents of all matter. Sir J. J. Thomson (1856–1940) showed, on the basis of classical electromagnetic theory, that the size[1] of the electron must be about 10^{-15} m. Furthermore, kinetic theory of the 19th century had shown that atoms have sizes of a few Ångstroms (that is, a few times 10^{-10} m). Thomson therefore reasoned that the positive electrical charge of an atom, which was required to balance the negative charge of the electrons and thereby render the atom electrically neutral, should be contained in the vast regions of space in the atom unoccupied by the tiny electrons. In 1906 Thomson proposed a model in which an atom was considered to contain electrons, equal in number to the chemical atomic number of the element, in which the total charge of the electrons was neutralized by a positively charged medium that contained most of the mass of the atom (Fig. 19.1). Because in this model the electrons were embedded in a positive sea in

---

[1]Recent attempts to measure the "size" of an electron indicate an upper limit of 10^{-18} m.

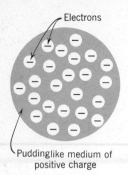

Figure 19.1 Thomson's "plum-pudding" model of the atom. The electron "raisins" are embedded in the positively charged "plum pudding." The diameter of the system is a few Ångstroms.

much the same way as raisins in a plum pudding, the irreverent appellation of "plum-pudding model" was applied to Thomson's scheme.

Although Thomson's model contained attractive features from the standpoint of the involvement of electrons in atomic structure, it survived only until 1911 when Ernest (later Lord) Rutherford succeeded in demonstrating that the positively charged portion of an atom is not distributed throughout the atom but is concentrated in a massive core of exceedingly small size—the *nucleus* of the atom.

Rutherford's Model

One of the most influential physicists of the early twentieth century was Ernest Rutherford (1871–1937).[2] Working first at McGill University in Canada and then moving to Manchester in 1907, Rutherford exhaustively studied the newly discovered radiations from radioactive substances. He was particularly interested in the positively charged radiations called *alpha rays* or α *particles*. By 1908 he had shown conclusively that α rays are the nuclei of helium atoms carrying a charge of $+2e$.

Almost immediately on arriving in Manchester, Rutherford began a systematic investigation of the scattering of α particles in matter. He had learned that a *single* α particle, when it strikes a zinc sulfide screen, produces a visible flash of light. Therefore, the apparatus to a study α-particle scattering was arranged as shown schematically in Fig. 19.2. Alpha particles from a radioactive source were confined to a narrow cone by a lead collimator. After scattering by the gold foil, the α particles struck a zinc sulfide screen and were detected by observing the light flashes with a small microscope. The detector could be rotated in order to measure the relative number of α particles scattered at various angles θ.

---

[2]Lord Rutherford was the key figure in unraveling the mysteries of radioactivity and in establishing the nuclear model of the atom. For his work on radioactivity, Rutherford was awarded the 1908 Nobel Prize (in chemistry).

According to Thomson's atomic model, α particles should be able to pass freely through the gold atoms; only occasionally should an α particle be slightly deflected by the Coulomb field of the electrons (Fig. 19.3). It was therefore expected that a beam of α particles would be somewhat "smeared out" in passing through the thin foil and that average scattering angles of a few degrees would result. Indeed, small-angle scattering was observed, but, most unexpectedly, it was found that about 1 α particle in 20,000 of those incident was turned completely around by a sheet of gold only 4×10^{-7} m thick and emerged from the side facing the source. Rutherford commented: "It was quite the most incredible event that has ever happened to me in my life. It was almost as incredible as if you had fired a 15-inch shell at a piece of tissue paper and it came back and hit you."

It took several years (until 1911) for Rutherford to convince himself that he understood completely the meaning of the unexpected α-particle scattering that was observed at large angles of deflection. He concluded that the only way in which it is possible to account for the experimental results is to assume that the positive charge of the atom is concentrated in a small volume at the center of the atom instead of distributed throughout the atom as in Thomson's model. Thus, Rutherford proposed the *nuclear* model of the atom.

In a collision between an α particle and an atom, the α particle should suffer very little deflection from the atomic electrons (just as in the scattering from a Thomson atom), but according to Rutherford, when the trajectory brings the α particle close to the nucleus, the intense electrical repulsion can cause a considerable change in the direction of motion of the α particle. Typical encounters between α particles and atoms are shown in Fig. 19.4.

By assuming that the repulsive Coulomb force, which acts between an α particle and an atomic nucleus maintains its $1/r^2$ character even down to extremely small intra-atomic distances (about 10^{-14} m), Rutherford was able to derive an expression for the distribution of scattered particles in α particle-nucleus collisions. He showed that his nuclear model of the atom predicts the probability for scattering at an angle θ to be inversely proportional to the fourth power of the sine of $\theta/2$, that is, to $1/\sin^4(\theta/2)$. This function is shown in Fig. 19.5 in which each unit on the vertical scale is a factor of 10. From this figure it can be seen that the probability for scattering at angles greater than 90° (i.e., into the *backward* direction) is exceedingly small compared to small-angle scattering. In fact, the ratio of

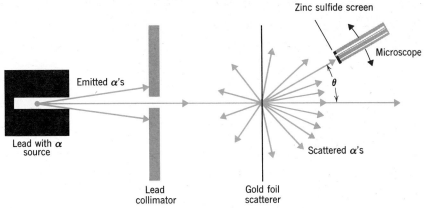

Figure 19.2 Schematic representation of the apparatus used by Rutherford to study the scattering of α particles. The entire apparatus was contained within an evacuated chamber in order to prevent the absorption of the α particles in air. Working under Rutherford's direction, Geiger and Marsden performed the experiments.

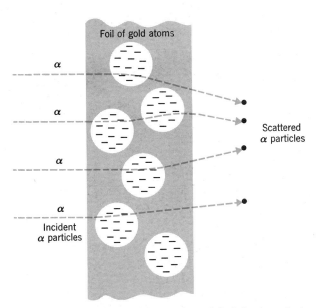

Figure 19.3 According to Thomson's model of the atom incident α particles should suffer small-angle deflections but none should be scattered through large angles by the gold atoms in a foil. The experiments of Rutherford and his co-workers showed this view of the scattering of α particles by atoms to be incorrect.

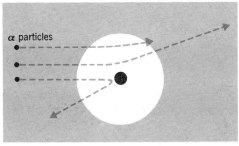

Figure 19.4 According to Rutherford's nuclear model of the atom, an α particle should usually pass through the atom with only little deflection, but occasionally the direction of motion of the α particle will bring it sufficiently close to the nucleus to cause a large-angle scattering.

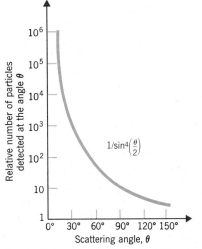

Figure 19.5 The relative probability (or number of light flashes expected per unit time in the detector) for Rutherford scattering at various angles is proportional to $1/\sin^4(\theta/2)$.

scattering at $\theta = 120°$ compared to that at $\theta = 5°$ is less than 10^{-5}.

The careful scattering measurements that were made by Geiger and Marsden in Rutherford's laboratory verified the nuclear model in every respect—it was conclusively demonstrated that atoms consist of nuclear cores of extremely small dimensions (about 10^{-14} m) surrounded by atomic electrons.

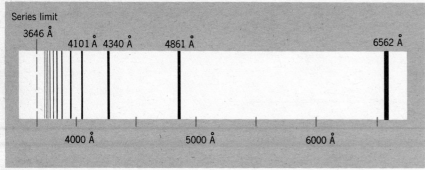

Figure 19.6 The Balmer series in the spectrum of hydrogen. The series limit (3646 Å) corresponds to substituting $n = \infty$ in the Balmer formula (Eq. 19.1).

19.2 The Hydrogen Atom

Spectroscopic Results

In the middle of the eighteenth century it was discovered that the light from flames does not consist entirely of a continuous spectrum (such as the thermal radiation spectrum, Fig. 18.1) but that there are certain discrete parts of the spectrum that are more intense than the background continuum and stand out clearly as *lines*. Line spectra from a variety of sources (including the Sun) were investigated extensively during the latter half of the 19th century and *atomic spectroscopy* became a highly developed field of study.

It has been discovered in 1885 by Johann Balmer (1825–1898), a Swiss music teacher with an interest in numbers, that the wavelengths of the lines in the optical spectrum of hydrogen (Fig. 19.6) could be represented by a simple mathematical expression. Although Balmer originally presented his formula in terms of wavelength λ, it was soon appreciated that the expression was more revealing if stated in terms of $1/\lambda$. In this form, Balmer's formula for the hydrogen spectrum becomes

$$\frac{1}{\lambda} = R\left(\frac{1}{2^2} - \frac{1}{n^2}\right), \quad n = 3, 4, 5, \ldots \quad (19.1)$$

The substitution of the series of numbers 3, 4, 5, . . . for n in this expression permits the calculation of the wavelengths of the various spectral lines. The quantity R is called the *Rydberg constant*, after the Swedish spectroscopist who made extensive investigations of atomic spectra. The value of R for the hydrogen spectrum is now known to be

$$R = 10{,}967{,}758 \text{ m}^{-1} \text{ (hydrogen)} \quad (19.2a)$$

although, of course, in Balmer's time the value was not known so precisely. The inverse of R is also useful in many calculations:

$$\frac{1}{R} = 911.76 \text{ Å (hydrogen)} \quad (19.2b)$$

How supremely successful the Balmer formula is in representing the hydrogen spectrum is demonstrated in Table 19.1. In this table the wavelengths calculated by Balmer are compared with the wavelengths obtained from measurements made by the Swedish physicist, Anders Jonas Ångström (1814–1874). These four lines lie in the visible part of the spectrum. On the basis of his formula, Balmer also predicted several other lines that should occur in the ultraviolet region. Indeed, in the spectra of some stars, as many as 50 lines in the Balmer series have been observed.[3]

Although Balmer's formula was exceptionally accurate in reproducing the observations, no one understood why this should be so. It was almost 30 years before Neils Bohr provided the first glimmer of understanding.

Table 19.1 Comparison of Hydrogen Spectral Lines with Calculations from the Balmer Formula

Line Designation	n	λ (Computed by Balmer)	λ (Observed by Ångstrom)
H_α	3	6562.08 Å	6562.10 Å
H_β	4	4860.80 Å	4860.74 Å
H_γ	5	4340.0 Å	4340.1 Å
H_δ	6	4101.3 Å	4101.2 Å

---

[3]The tenuous atmospheres of stars provide more favorable sources of hydrogen radiation than do laboratory sources.

Bohr's Interpretation of Energy States

In 1913 Bohr took a bold and surprising step in an attempt to interpret the spectroscopic results for the hydrogen atom. He has accepted Rutherford's model of the atom with its nuclear core and outer electrons. According to classical theory, a system consisting of a massive, positively charged core and light, negatively charged electrons can be stable only if the electrons are in motion. Thus, an atom should be similar to a miniature solar system with a nuclear "Sun" and "planetary" electrons. The analogy would be expected to be quite good (after all, the electric and gravitational forces both depend on $1/r^2$) were it not for the fact that classical theory also predicts that accelerating electric charges radiate energy in the form of electromagnetic waves. Therefore, the orbiting "planetary" electrons would be expected to lose their motional and electric energy by radiation and rapidly fall toward the nucleus. A calculation based on classical electromagnetic theory shows that the electron in a hydrogen atom will radiate all of its energy in a small fraction of a second. But, of course, the atom does not do this. What is wrong with the classical model?

It was Bohr's audacious proposal that classical electromagnetic theory simply does not apply to an electron circulating in an orbit around a nucleus. At the same time he reasoned that the two terms in Balmer's formula refer to the total energies of two *allowed* orbits (or energy states) of the electron in the hydrogen atom.

In order to follow Bohr's idea, we first convert Balmer's formula for the wavelength into an energy equation. Multiplying $1/\lambda$ by hc and using $c/\lambda = \nu$, we have

$$\frac{1}{\lambda} \times hc = h \times \frac{c}{\lambda} = h\nu \qquad (19.3)$$

where $h\nu = \mathscr{E}$ is the energy of the photon with wavelength λ. Therefore, multiplying Eq. 19.1 by hc, we can write

$$\frac{hc}{\lambda} = h\nu = hcR\left(\frac{1}{2^2} - \frac{1}{n^2}\right) \qquad (19.4)$$

This is an energy equation, and Bohr interpreted the right-hand side as the difference between two energy terms that are characteristic of the atom. Bohr assumed that a hydrogen atom can exist only in certain configurations, each with its own particular discrete energy. Then, as the atom changes from a configuration (or *state*) of a higher energy $\mathscr{E}_n$ to one of lower energy $\mathscr{E}_{n'}$, a photon will be emitted with energy equal to the energy difference $\Delta\mathscr{E}_{nn'}$ between the states:

$$h\nu = \mathscr{E}_n - \mathscr{E}_{n'} = \Delta\mathscr{E}_{nn'} \qquad (19.5)$$

Bohr realized that this expression is the same as the Balmer energy equation (Eq. 19.4) if we identify

$$\mathscr{E}_n = -\frac{hcR}{n^2} \quad \text{and} \quad \mathscr{E}_{n'} = \mathscr{E}_2 = -\frac{hcR}{2^2} \qquad (19.6)$$

(The energies $\mathscr{E}_n$ and $\mathscr{E}_{n'}$ are *negative* because the system is *bound* in these states; see the discussion in Section 7.6.) In each of these energy states the electron moves in an orbit (in Bohr's view, a *classical* circular orbit) with a particular radius and velocity appropriate for that energy.

Bohr's new hypothesis was that an electron can exist in one of the discrete energy states in an atom, executing a radiationless (but otherwise classical) orbit; radiation occurs only when the electron makes a transition between two allowed orbits. These transitions occur spontaneously and continue sequentially until the atom is in its state of lowest allowed energy.

Having challenged the applicability of classical electromagnetic theory in the atomic domain, how was Bohr to support his claim?

Bohr's Angular Momentum Hypothesis

In order to justify his interpretation of the hydrogen spectrum in terms of radiations accompanying the transitions between allowed energy states, Bohr sought to calculate the energies of these states. He was able to obtain a set of discrete allowed states only by making the drastic assumption that *angular momentum is quantized*. By specifying that the angular momentum must be an integer multiple of $h/2\pi$, Bohr was finally able to derive the Balmer formula. (The combination $h/2\pi$ is used so frequently in atomic theory that it is given a special symbol: $h/2\pi = \hbar$ pronounced "h-bar".) Bohr's condition is, therefore (see Eq. 4.6),

$$L = m_e v r = n\hbar, \qquad n = 1, 2, 3, \ldots \qquad (19.7)$$

n is called the *principal quantum number* for the particular state.

Numerically,

$$\hbar = 1.0546 \times 10^{-34} \text{ J-s}$$
$$= 6.583 \times 10^{-16} \text{ eV-s} \qquad (19.8)$$

Next, we use Bohr's angular momentum condition in the calculation of the energies of states in the hydrogen atom. The magnitude of the electric force between the orbiting electron and the nucleus in the hydrogen atom is

$$F_E = K\frac{qQ}{r^2} = K\frac{e^2}{r^2} \qquad (19.9)$$

Because the nucleus is much more massive than the electron, we can consider the nucleus to remain in an essentially fixed position as the electron executes its orbit. The force on the electron is equal to its mass multiplied by its acceleration, and the acceleration is just the *centripetal* acceleration, $a_c = v^2/r$; thus,

$$F_E = m_e a_c = \frac{m_e v^2}{r} \qquad (19.10)$$

Equating the right-hand sides of these two expressions for F_E and solving for v, we find

$$v = \sqrt{\frac{Ke^2}{m_e r}} \qquad (19.11)$$

Bohr's hypothesis of quantized angular momentum (Eq. 19.7), when combined with Eq. 19.11 for v, leads to

$$L = m_e v r = m_e r \times \sqrt{\frac{Ke^2}{m_e r}} = e\sqrt{Km_e r} = n\hbar \qquad (19.12)$$

Squaring the last part of this equation and solving for r, we have

$$r_n = \frac{n^2 \hbar^2}{Km_e e^2} \qquad (19.13)$$

where we have attached a subscript n to the radius in order to indicate that this is the value for a particular value of n (=1, 2, 3, . . .).

● **Example 19.1**

What is the radius of the first Bohr orbit for hydrogen?

$$r_{n=1} = a_1 = \frac{\hbar^2}{Km_e e^2}$$

$$= \frac{(1.05 \times 10^{-34} \text{ J-s})^2}{(9 \times 10^9 \text{ N-m}^2/\text{C}^2) \times (9.1 \times 10^{-31} \text{ kg}) \times (1.6 \times 10^{-19} \text{ C})^2}$$

$$= 0.53 \times 10^{-10} \text{ m} = 0.53 \text{ Å} \qquad \blacksquare$$

Next, for the total energy of the nth orbit, we write, using Eq. 19.11 for v and Eq. 12.4 for the electric potential energy,

$$\mathcal{E}_n = KE + PE = \frac{1}{2}m_e v^2 - K\frac{e^2}{r_n}$$

$$= \frac{1}{2}m_e\left(\frac{Ke^2}{m_e r_n}\right) - K\frac{e^2}{r_n}$$

$$= -\frac{1}{2}K\left(\frac{e^2}{r_n}\right)$$

Substituting for r_n, we have, finally,

$$\mathcal{E}_n = -\frac{1}{2}\left(\frac{K^2 m_e e^4}{\hbar^2}\right)\left(\frac{1}{n^2}\right), \qquad n = 1, 2, 3, \ldots \qquad (19.14)$$

Thus, by using the quantization condition on the angular momentum, Bohr succeeded in obtaining a set of discrete energy states. If we use modern values to calculate the combination of constants in this expression, we obtain

$$\mathcal{E}_n = -(13.6 \text{ eV}) \times \frac{1}{n^2}, \qquad n = 1, 2, 3, \ldots \qquad (19.14a)$$

The energies associated with the first few hydrogen atom states are

$n = 1$: $\mathcal{E}_1 = -13.6$ eV
$n = 2$: $\mathcal{E}_2 = -3.39$ eV
$n = 3$: $\mathcal{E}_3 = -1.51$ eV
$n = 4$: $\mathcal{E}_4 = -0.85$ eV

The lowest possible energy state for an atom (or any other system) is called its *normal* or *ground* state. For the hydrogen atom, the state with $n = 1$ is the ground state.

The difference in energy between a state with principal quantum number n and one with n' is

$$\Delta \mathcal{E}_{nn'} = \mathcal{E}_n - \mathcal{E}_{n'}$$
$$= \frac{1}{2}\left(\frac{K^2 m_e e^4}{\hbar^2}\right)\left(\frac{1}{n'^2} - \frac{1}{n^2}\right) \quad (19.15)$$

which we can now express as

$$\Delta \mathcal{E}_{nn'} = (13.6 \text{ eV}) \times \left(\frac{1}{n'^2} - \frac{1}{n^2}\right) \quad (19.15a)$$

If an electron makes a transition between these states (n to n', where $n > n'$), the wavelength of the radiation will be given by

$$\frac{1}{\lambda_{nn'}} = \frac{\nu_{nn'}}{c} = \frac{h\nu_{nn'}}{hc} = \frac{\Delta \mathcal{E}_{nn'}}{hc}$$
$$= \frac{1}{2}\left(\frac{K^2 m_e e^4}{\hbar^2}\right) \times \frac{1}{hc}\left(\frac{1}{n'^2} - \frac{1}{n^2}\right)$$
$$= \frac{K^2 m_e e^4}{4\pi \hbar^3 c}\left(\frac{1}{n'^2} - \frac{1}{n^2}\right) \quad (19.16)$$

Substituting numerical values for the fundamental constants in this expression, Bohr found that the value of the quantity multiplying the terms in parentheses is very close to the experimental value of the Rydberg constant determined from the hydrogen spectrum (Eq. 19.2). By applying the angular momentum quantization condition to a classical model of the atom, Bohr was able to duplicate the results of measurements of the hydrogen atom spectrum.

Again, using modern values for the constants (or using the result in Eq. 19.2b directly), we can express Eq. 19.16 as

$$\frac{1}{\lambda_{nn'}} = \left(\frac{1}{911.76 \text{ Å}}\right) \times \left(\frac{1}{n'^2} - \frac{1}{n^2}\right) \quad (19.16a)$$

Bohr's success in explaining the hydrogen spectrum on the basis of his half-classical, half-quantum model was not exactly hailed as a triumph. In fact, he was severely criticized for tampering with centuries of classical theories; even Bohr was at a loss to explain the fundamental significance of his curious mixture of classical dynamics and quantum hypotheses. It was more than 10 years before the development of the new quantum theory provided the proper explanation of Bohr's remarkable results.

Hydrogen Energy States

By substituting $n' = 2$ into Eq. 16.16, Bohr had been able to compute the wavelengths of the lines in the Balmer spectrum. By using $n' = 3$, he was able to account for a series of infrared lines discovered by Paschen on 1909. Bohr argued that other values of n' should give rise to additional spectral series. In fact, Lyman found the series of ultraviolet lines that was predicted for $n' = 1$, and in 1922 Bracket and Pfund discovered the infrared series for $n' = 4$ and $n' = 5$.

The first 5 orbits of the Bohr hydrogen atom are shown in Fig. 19.7 along with lines that represent the transitions of the spectral series for $n' = 1, 2,$ and 3. The energies associated with electrons residing in these orbits (or energy states) are shown in the energy level diagram in Fig. 19.8. Because the electron is *bound* to the nucleus, according to our standard convention all energies are *negative*. The amount of energy required to free an electron from an atom in a particular state (that is, to raise the energy to *zero*) is called the *binding energy* for that state.

● **Example 19.2**

Perform the calculation referred to in obtaining Eq. 19.14a and find the value of the binding energy of the hydrogen atom in its ground state. Using Eq. 19.14 with $n = 1$, we have

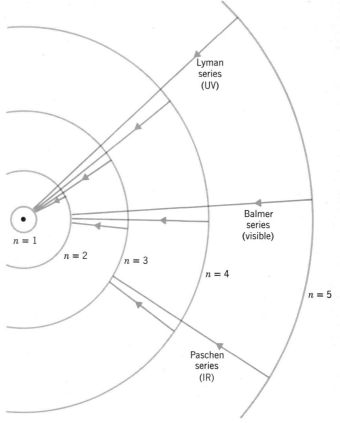

Figure 19.7 Orbits in the Bohr model of the hydrogen atom. Portions of three of the spectral series are shown.

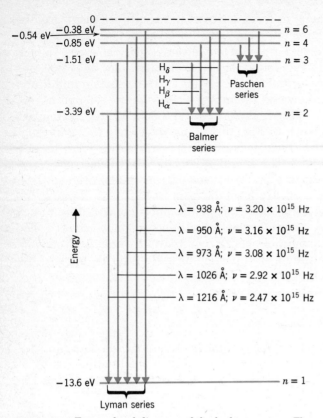

Figure 19.8 Energy level diagram of the hydrogen atom. The vertical lines represent transitions occurring in three of the spectral series.

$$\mathcal{E}_1 = -\frac{K^2 m_e e^4}{2\hbar^2} = -\frac{2\pi^2 K^2 m_e e^4}{h^2}$$

$$= -\frac{2\pi^2 \times (9 \times 10^9 \text{ N-m}^2/\text{C}^2)^2 \times (9.1 \times 10^{-31} \text{ kg}) \times (1.60 \times 10^{-19} \text{ C})^4}{(6.6 \times 10^{-34} \text{ J-s})^2}$$

$$= -2.18 \times 10^{-18} \text{ J} \times \left(\frac{1 \text{ eV}}{1.60 \times 10^{-19} \text{ J}}\right)$$

$$= -13.6 \text{ eV}$$

This (negative) energy is the *total* energy of the state, and so, the *binding energy* (that energy that must be supplied to raise the total energy is *zero* and thereby release the electron) is 13.6 eV. ∎

Two relationships that are useful in calculations of frequencies and wavelengths of photons are

$$\nu = (2.42 + 10^{14}) \times \mathcal{E} \quad \text{(in Hz or s}^{-1}\text{)} \quad (19.17a)$$

$$\lambda = \frac{12{,}400}{\mathcal{E}} \quad \text{(in Å)} \quad (19.17b)$$

when the photon energy $\mathcal{E}$ is expressed in eV.

De Broglie Waves in the Hydrogen Atom

Bohr's explanation of the hydrogen spectrum in terms of electrons orbiting with quantized angular momenta was tremendously successful in accounting for the observed spectral lines. But the quantization rule remained as an *ad hoc* hypothesis, not based on any deeper theory, until de Broglie made his famous proposal of the wave character of matter. As soon as it was appreciated that an electron with a momentum p had associated with it a wave of wavelength $\lambda = h/p$, it became clear why only certain orbits are available to the electron in a hydrogen atom.

Consider a hydrogen atom in a state labeled n. The de Broglie wavelength of the electron in this state is

$$\lambda_n = \frac{h}{p_n} = \frac{h}{m_e v_n} \quad (19.18)$$

According to Eq. 19.7, we can write the angular momentum of the electron as

$$L_n = m_e v_n r_n = n\hbar = \frac{nh}{2\pi}$$

Solving for $m_e v_n$,

$$m_e v_n = \frac{nh}{2\pi r_n}$$

and substituting this expression into Eq. 19.18, we find

$$\lambda_n = \frac{h}{(nh/2\pi r_n)} = \frac{2\pi r_n}{n}$$

or,

$$2\pi r_n = n\lambda_n \quad (19.19)$$

That is, the circumference of the nth orbit, $2\pi r_n$, is just n wavelengths of the electron wave.

Why is it so important that an allowed orbit contain exactly an integer number of wavelengths? The answer lies simply in the interference property of waves. If a wave requires a distance equal to the orbit circumference to complete exactly an integer number of cycles, this means that the wave will join smoothly onto itself and constructive interference or self-reinforcement will result (see Fig. 19.9a). On

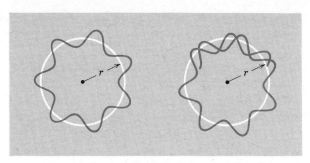

Figure 19.9 (a) Constructive interference results when an integer number of wavelengths are just fitted onto the circumference of an orbit so that this state of the system is maintained by self-reinforcement. (b) Destructive interference results when the condition $n\lambda_n = 2\pi r_n$ is not satisfied and the state is rapidly damped to zero amplitude.

the other hand, if we try to fit into a certain orbit a wave with a wavelength that is not an integer fraction of the circumference, *destructive* interference will result and the wave will rapidly damp to zero amplitude (see Fig. 19.9b).

De Broglie's proposal of the wave character of matter therefore explains in a simple and straightforward way the puzzling angular momentum quantization rule of Bohr. (This, in fact, was the successful application of de Broglie's wavelength-momentum condition.) But in spite of the increased understanding brought about by incorporating de Broglie's idea into the theory, Bohr's atomic model was still basically classical in character—electrons were still thought to move in orbits around the nuclear "Suns"—and no one understood why the electrons were prevented from radiating while in an allowed orbit. The merging of Bohr's and de Broglie's concepts provided much of the impetus that led within a short period of time to the development of the modern theory of atomic structure in quantum terms.

19.3 Angular Momentum and Spin

The Angular Momentum Quantum Number

Although Bohr's model of the hydrogen atom was capable of accounting for most of the lines in the hydrogen spectrum, Bohr was unable to extend the same ideas and apply them to the spectra of more complicated atoms. The theory was clearly in need of modification. The next step was taken by the German theorist, Arnold Sommerfeld (1869–1951).

In Bohr's theory, only *circular* electron orbits are considered. Each orbit is described by its particular value of the principal quantum number n and each orbit has a unique value of angular momentum, $L = n\hbar$ (Eq. 19.7). In 1915 Sommerfeld expanded Bohr's model by allowing *elliptical* orbits. Sommerfeld showed that for each value of n there are a number of states with the same energy that have different elliptical shapes and different values of the angular momentum. Sommerfeld introduced a new quantum number l to specify the angular momentum:

$$L = l\hbar, \qquad l = 0, 1, 2 \ldots \qquad (19.20)$$

It is a result of the details of the model that l can have only positive values from zero to $n - 1$. Thus, in the Bohr-Sommerfeld model (as in the complete quantum theory analysis), there is *zero* angular momentum associated with the state whose principal quantum number is 1 (that is, $l_{\max} = n - 1 = 0$ for $n = 1$), whereas for $n = 2$, l can have the values 0 or 1, and for $n = 3$, $l = 0, 1,$ or 2, etc.:

$$l = 0, 1, 2, \ldots, n - 1 \qquad (19.21)$$

In describing electron orbital angular momentum states it is customary to use a letter to denote the value of l. The convention is as follows:

$l = 0$	S state
$l = 1$	P state
$l = 2$	D state
$l = 3$	F state
$l = 4$	G state

This curious code is a holdover from the early days of spectroscopy when S, P, D, and F meant, respectively, that the spectral lines that are associated with transitions originating from these angular momentum states were classified as *sharp, principal, diffuse,* or *fundamental*. For higher values of l, the letter designations continue alphabetically.

In spite of its basic shortcomings, the Bohr-Sommerfeld theory was applied with considerable success to *one-electron systems* (such as the hydrogen atom, once-ionized helium, twice-ionized lithium, etc). But with more complicated electronic systems, insurmountable difficulties were encountered. Even with their always great ingenuity, neither Bohr nor Sommerfeld could juggle the orbits of complicated atoms to give the proper results. Only the application of the full power of quantum theory was eventually able to solve this enormous problem.

Magnetic Substates

An orbiting atomic electron acts as a tiny current loop and produces a magnetic field (see Fig. 13.10). If the atomic electron and its magnetic field are placed in an external magnetic field (Fig. 19.10), there will be an interaction between the two fields. By observing the spectral lines emitted by atoms when in a magnetic field, an important conclusion has been reached. The N-S axis of the electron orbit (which is the same as the direction of the angular momentum vector **L**) cannot point in any arbitrary direction with respect to the field direction; instead, *the component of **L** in the direction of the field is quantized.* That is, not only is the magnitude of **L** restricted to discrete multiples of $\hbar$, but so is the component of **L** that lies along the field direction (see Fig. 19.11). The quantum number m_l is used to specify the magnitude of the angular momentum component in the direction of the field. If this direction is called the z-axis, the quantization rule can be expressed as

$$L_z = m_l \hbar, \quad m_l = l, l-1, \ldots, 0, \ldots, -l+1, -l$$

(19.22)

where L_z is the component of **L** along the z-axis.

Because the component of **L** can be either *along* the direction of **B** or *opposite* to it, m_l can have both positive and negative values; $m_l = 0$ is always possible (**L** perpendicular to **B**) and the maximum and minimum values of m_l are $+l$ and $-l$, respectively.

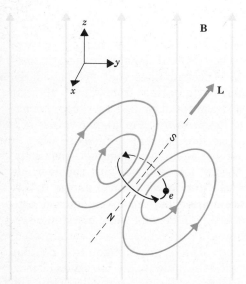

Figure 19.10 The magnetic field produced by an orbiting atomic electron interacts with an external field **B**. Only certain discrete directions of **L** relative to **B** are allowed.

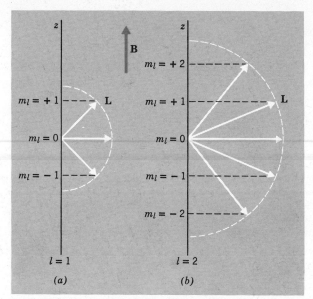

Figure 19.11 The angular momentum vector **L** can assume only those orientations in a magnetic field for which the component along **B** is an integer multiple of $\hbar$: $L_z = m_l \hbar$.

Thus, there are always $2l + 1$ allowed values of m_l (see Table 19.2). Each state characterized by a value of m_l is called a *magnetic substate*.

According to modern theory Eq. 19.20 should be modified to read

$$L^2 = \ell(\ell+1)\hbar^2$$

Eq. 19.22 is correct as it stands.

Spin

In spite of the obvious great success of the Bohr-Sommerfeld picture of atomic structure, there were still many unexplained facts even for relatively simple systems. For example, certain spectral lines that were expected to be single were found, on close examination, actually to be *doublets* spaced apart by only a very small energy.

In 1925, Goudsmit and Uhlenbeck argued that this effect indicates that an electron possesses angular momentum quite apart from the angular momentum that arises from orbital motion. In classical

Table 19.2 Magnetic Substances

State	l	m_l	$2l + 1$
S	0	0	1
P	1	+1, 0, −1	3
D	2	+2, +1, 0, −1, −2	5
F	3	+3, +2, +1, 0, −1, −2, −3	7

terms, we can picture an electron as a spinning, charged ball (Fig. 13.28)—the mechanical spin produces an angular momentum. This classical model has no meaning within the framework of quantum theory (where we speak only of the *intrinsic angular momentum* in exactly the same way that we speak of the intrinsic *mass* or the intrinsic *charge* of an electron). Nevertheless, the classical model is a convenient picture and is often used; indeed, it is customary to refer to instrinsic angular momentum simply as *spin*. The spin of an electron is

$$S = s\hbar \qquad s = \tfrac{1}{2} \qquad (19.23)$$

Unlike the orbital angular momentum quantum number l that can take on values $0, 1, 2, \ldots n - 1$, the *spin quantum number* s has a *single* value: $s = \tfrac{1}{2}$.

Measured values of angular momenta can differ only by integer multiples of the fundamental unit of angular momentum, $\hbar$. This is the reason that both L and L_z are proportional to $\hbar$. Similarly, S_z, the projection of the spin angular momentum **S** in a special direction (such as that of a magnetic field), can take on only values that differ by integer multiples of $\hbar$. But the magnitude of S_z cannot exceed $S = \tfrac{1}{2}\hbar$. Therefore, there are only *two* allowed values for S_z: $+\tfrac{1}{2}\hbar$ and $-\tfrac{1}{2}\hbar$. That is,

$$S_z = m_s\hbar, \qquad m_s = +\tfrac{1}{2}, -\tfrac{1}{2} \qquad (19.24)$$

Notice that $S_z = 0$ is not allowed.

By this stage in the development of the theory of atomic structure, the original Bohr theory had been modified almost beyond recognition. No longer were atomic electrons thought to be orbiting "planets" described by a single quantum number n. Instead, it was recognized (but not completely understood) that the wave property of electrons is important in atoms; furthermore it was known that *four* quantum numbers are required to specify the state of an atomic electron (see Table 19.3).

Again, modern theory modifies Eq. 19.23 to read

$$S^2 = s(s + 1)\hbar^2 = \frac{3}{4}\hbar^2$$

Eq. 19.24 is correct as it stands.

Total Angular Momentum

The spin and the orbital angular momentum of an electron combine to produce the *total* angular momentum: that is, $\mathbf{J} = \mathbf{L} + \mathbf{S}$. Because L and S are quantized, the total angular momentum J is also quantized:

$$J = j\hbar \qquad (19.25)$$

where j is the *total angular momentum quantum number* of the electron and can have only two possible values for a given l:

$$\left.\begin{array}{l} j = l + s = l + \tfrac{1}{2} \\ \text{or } j = l - s = l - \tfrac{1}{2} \end{array}\right\} \qquad (19.26)$$

except that for $l = 0$ only $j = \tfrac{1}{2}$ is possible.

Modern analysis requires modifying Eq. 19.25 to read

$$J^2 = j(j + 1)\hbar^2$$

For every nonzero *orbital* angular momentum, two *total* angular momenta are possible. For $l = 1$, we have $j = \tfrac{1}{2}, \tfrac{3}{2}$, for $l = 2$, we have $j = \tfrac{3}{2}, \tfrac{5}{2}$, and so forth. The value of j for a state is given as a subscript following the letter designation for the orbital angular momentum. Thus, a state with $l = 1, j = \tfrac{3}{2}$ is denoted by $P_{3/2}$, and $P_{1/2}$ means $l = 1, j = \tfrac{1}{2}$. Some of the states of the hydrogen atom are shown in Table 19.4; the numbers preceding the letter designations refer to the values of n.

The importance of electron spin in the matter of atomic spectra lies in the fact that the magnetic field associated with the spinning charge produces a dif-

Table 19.3 Quantum Numbers of Atomic States

Property	Quantum Number	Allowed Values
Energy	$n\ \left(E \propto -\dfrac{1}{n^2}\right)^a$	$1, 2, 3, \ldots$
Angular momentum	$l(L^2 = l(l + 1)\hbar^2)$	$0, 1, 2, \ldots, n - 1$
Projection of angular momentum	$m_l(L_z = m_l\hbar)$	$-l$ to $+l$ ($2l + 1$ values)
Spin projection	$m_s(S_z = m_s\hbar)$	$+\tfrac{1}{2}, -\tfrac{1}{2}$

[a]For one-electron systems.

Table 19.4 Some of the States of the Hydrogen Atom

n	l	Designation
1	0	$1\,S_{1/2}$
2	0	$2\,S_{1/2}$
	1	$2\,P_{1/2},\ 2\,P_{3/2}$
3	0	$3\,S_{1/2}$
	1	$3\,P_{1/2},\ 3\,P_{3/2}$
	2	$3\,D_{3/2},\ 3\,D_{5/2}$
4	0	$4\,S_{1/2}$
	1	$4\,P_{1/2},\ 4\,P_{3/2}$
	2	$4\,D_{3/2},\ 4\,D_{5/2}$
	3	$4\,F_{5/2},\ 4\,F_{7/2}$

ference in the energy of the state of the electron depending on the orientation of **S** with respect to **L**. The reason is that, from the standpoint of the electron's reference frame, the positively charged nucleus rotates around the electron and constitutes a current loop that produces an magnetic field at the position of the electron. The energy of interaction of the electron's spinning charge with this field depends on the orientation of the spin vector with respect to the field. Figure 19.12 shows a P ($l = 1$) state that is split into two substates, $P_{3/2}$ and $P_{1/2}$, because of this interaction.[4] Therefore, instead of a single spectral line, corresponding to a P → S transition, there are actually two closely spaced lines, corresponding to the $P_{3/2} \to S_{1/2}$ and $P_{1/2} \to S_{1/2}$ transitions. It is just such a situation that produces the famous doublet of yellow lines in the spectrum of sodium at 5890 Å and 5896 Å.

Fermions and Bosons

Thus far we have mentioned only electrons in the discussion of spin. In fact, all elementary particles, as well as aggregates of such particles (e.g., nuclei) have measurable intrinsic angular momentum properties. Some elementary particles, such as protons and neutrons, have the same spin as electrons ($s = \frac{1}{2}$). Pions have no intrinsic angular momentum ($s = 0$) and photons carry a single unit ($s = 1$). One group of elementary particles (the omega hyperons) has $s = \frac{3}{2}$. The spin of a nucleus is determined by the way the individual spins of the constituent protons and neutrons add together. If a nucleus consists of an even number of nucleons, the spin can be 0, 1, 2, . . . ; if the number is odd, the spin can be $\frac{1}{2}, \frac{3}{2}, \frac{5}{2}$, ... The spins of some elementary particles are listed on Table 19.5.

As we will see, in certain types of quantum situations, particles with half-integer spins ($s = \frac{1}{2}, \frac{3}{2}, \ldots$) have a fundamentally different behavior compared to those with integer spins ($s = 0, 1, 2, \ldots$). For some discussions, therefore, it is convenient to group together the particles in each of these classes. We refer to particles with half-integer spins as *Fermi particles* (or *fermions*), after Enrico Fermi (1901–1954), the great Italian-American physicist who was instrumental in establishing the theory of the emission of electrons (β particles) from radioactive nuclei. Particles with integer spins are called *Bose particles* (or *bosons*), after Indian physicist S. N. Bose (1894–1974) who first investigated the statistical properties of such particles (actually, photons).

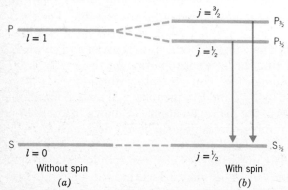

Figure 19.12 The existence of electron spin causes all states with nonzero orbital angular momenta to be split into two states with $j = l + \frac{1}{2}$ and $j = l - \frac{1}{2}$. The transitions from these two states to the $S_{1/2}$ state produce a close doublet in the spectrum. This is the situation in the sodium atom that produces the pair of yellow lines at 5890 Å and 5896 Å. The splitting of the $P_{1/2}$ and $P_{3/2}$ states compared to the splitting of the P and S states is greatly exaggerated in the figure.

Table 19.5 Spins of Some Elementary Particles

Classification	Particle	Spin
Fermions	Electron	$\frac{1}{2}$
	Neutrino	$\frac{1}{2}$
	Muon	$\frac{1}{2}$
	Proton	$\frac{1}{2}$
	Neutron	$\frac{1}{2}$
	Omega hyperons	$\frac{3}{2}$
Bosons	Pion (π meson)	0
	Kaon (K meson)	0
	Photon	1

---

[4]This interaction is called the *spin-orbit* interaction because it depends on whether the spin vector is parallel or antiparallel to the orbital angular momentum vector.

19.4 Quantum Theory of the Hydrogen Atom

The Development of Quantum Theory

By the mid-1920s it was generally recognized that the Bohr-Summerfeld ideas of atomic structure, including as they did both classical and quantum concepts, left much to be desired in terms of a complete and satisfying physical explanation of the properties of atoms. In 1925 to 1926 there emerged a new view of atomic processes that was based, not on a description of electron orbits and "jumping" electrons, but on the *wave* properties of electrons. The classical idea of orbits was abandoned and in its place came the *wave mechanics* or *quantum theory* of elementary processes. In 1925 Werner Heisenberg and Erwin Schrödinger (1887–1961) produced equivalent mathematical descriptions of electron behavior, and Goudsmit and Uhlenbeck introduced the concept of electron spin. In the following year Max Born (1882–1970) contributed the probability interpretation of wave functions. Immediately on the introduction of these fundamental new ideas, there began a furious outpouring of results using the new theory. By 1928 Pauli had formulated the basic principle that explains why atomic electrons are arranged in shells, Heisenberg had developed the uncertainty principle, and P.A.M. Dirac (1902–   ) has introduced relativity into quantum theory. With these advances, all of the necessary fundamental ideas concerning atomic structure were available and the answer to any specific question involved mainly a computational problem. (Of course, some of these "computational problems" were—and still are—extremely formidable.) The development of quantum theory was a gigantic step forward in our understanding of Nature; it is even more remarkable when it is realized that all of the crucial developments took place in such a short period of time. Quantum theory is certainly an equal, if not a greater tribute to the powers of the human intellect than was Newton's formulation of the law of universal gravitation and his explanation of planetary motion.

Hydrogen Atom Wave Functions

If we surrender the classical idea of electron orbits, how are we to understand the various energy states of the hydrogen atom? Schrödinger approached this problem by setting down an equation that is similar to the equation that describes the propagation of mechanical waves and in which he included a term to represent the effect of the electrostatic potential energy of the electron in the field of the nuclear proton. The solution of this equation showed that there are certain discrete energies allowed for the system. These energies, which emerge in a natural way from the Schrödinger equation, correspond precisely to the energies in the Bohr theory but there are no "orbits" for the electron. Instead, each energy state has associated with it a *wave function*. The square of this wave function is proportional to the probability for finding the electron at any particular position.

Schrödinger's solution to the hydrogen-atom problem is illustrated in Fig. 19.13, which shows the relative probabilities for finding the electron at a distance r for the nucleus for three different states. Notice that the position of maximum probability for the 1S state is $r = a_1$ (the radius of the first Bohr orbit). For the 2P state the maximum occurs at $r = 4a_1$ and for the 3D state it occurs at $r = 9a_1$. These positions are exactly the same as those for the corresponding Bohr orbits (for which $r = n^2 a_1$).

The probability curves in Fig. 19.13 show that the electron does not exist at any well-defined distance from the nucleus. There are no "orbits;" instead, one can specify only the probability of finding the electron at a particular distance from the nucleus.

The corrections cited for Eqs. 19.20, 19.23, and 19.25 stem from the Schrödinger theory modified to include spin.

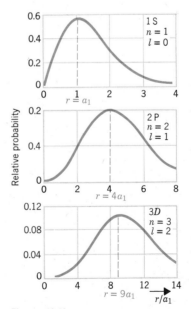

Figure 19.13 Relative probability of finding the electron in a hydrogen atom at a distance r from the nucleus. Notice that the distances are given in units of a_1, the radius of the $n = 1$ orbit in the Bohr model.

19.5 The Exclusion Principle and Atomic Shell Structure

Systematics of the Properties of the Elements

The Schrödinger-Heisenberg quantum mechanics of 1925 proved to be enormously successful in explaining the spectra of hydrogen and other one-electron systems. The introduction of spin permitted an understanding of some of the results for more complicated atoms. But at this stage it could not be claimed that the structure of atoms containing many electrons was understood in detail.

It was known, from the systematic study of atomic radiations, that there is a regular progression in the number of atomic electrons as one passes from element to element and that this electronic charge is balanced by an equal positive charge on the nucleus. Thus, hydrogen has one orbital electron and a charge of $+e$ on the nucleus; helium has two orbital electrons and a charge of $+2e$ on the nucleus. The number of electrons in the neutral atom (or, equivalently, the number of nuclear charges) is called the *atomic number* of the element and is denoted by Z.

It was also known that many of the physical and chemical properties of the elements could be organized in a systematic way and presented in the form of the *periodic table of the elements* (Fig. 19.14). In this table the elements are arranged according to *groups* and *periods*, with the members of each group having similar properties. Thus, the elements Li, Na, K, Rb, Cs, and Fr are all similar to hydrogen in that they participate in chemical reactions as if they have only a single effective electron (called a

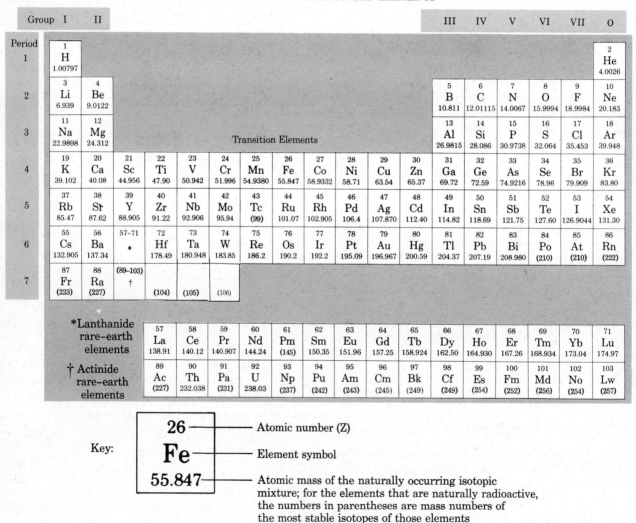

Figure 19.14 Periodic table of the elements.

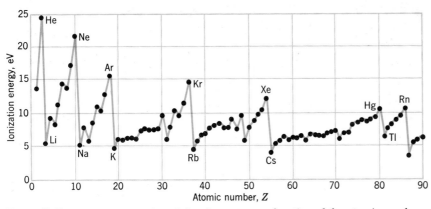

Figure 19.15 Ionization energies of the elements as a function of the atomic number.

valence electron). For example, the Group I elements combine readily in a one-to-one fashion with the Group VII elements for form such compounds as HF, HBr, NaCl, NaF, KCl, KBr. The Group 0 elements, the so-called *noble* or *inert gases,* do not readily combine with other elements;[5] these gases have no valence electrons.

The periodic table represents the cyclic behavior of many chemical and physical properties of the elements. Each of these cycles ends with a noble gas. Thus, the various periods terminate at the atomic numbers $Z = 2, 10, 18, 36, 54$, and 86. This periodicity is revealed in a striking way by the *ionization energies* of the elements. (This *ionization energy* is the minimum energy required to remove an electron from an atom and convert it into a singly charged ion.) Figure 19.15 shows that the ionization energy tends to be quite large for the noble gases and to be quite low for the element with the next higher atomic number (a Group I element with an easily removed valence electron). There is a more-or-less uniform increase in the ionization energy as we proceed across any given period from Group I to Group 0.

In spite of the obvious importance of the systematic behavior of elements with regard to both physical and chemical properties, there was no clue in either the old Bohr-Summerfeld model or in the early quantum theory as to the reason. All that could be said was that electrons seemed to exist in layers or *shells,* with each successive shell ending or closing with a noble gas so that there are no electrons (*valance* electrons) available to participate in chemical reactions. The elements at the beginning of each shell (the Group I elements) have one valence electron, the Group II elements have two valence electrons, and so on. The significance of this electronic shell structure and the meaning of the shell-closure numbers, 2, 10, 18, 36, 54, and 86 remained a mystery until the solution was given in a simple and elegant way by Wolfgang Pauli.

The Exclusion Principle

The key to the problem of atomic shell structure was discovered by Pauli in 1925. The closing of atomic shells implies that an arbitrarily large number of electrons cannot be placed in a given shell. Pauli realized that such a restrictive effect must have a truly fundamental cause, and his solution to the problem was the formulation of the following principle, known as the *exclusion principle:*

No two elements in an atom can have identical sets of quantum numbers.

That is, if one atomic electron is in a certain quantum state defined by a set of quantum numbers, n, l, m_l, and m_s, then other electrons in that atom are excluded from that particular quantum state.[6] It is most remarkable that the details of atomic structure can follow from a principle so simply stated.

Atomic Shell Structure

How many states are available to an electron in an atom? We limit our considerations now to the *ground states* of neutral atoms; that is, the Z electrons in the atom are arranged in the way that produces the *minimum* total energy for the system. For

---

[5] A limited number of inert gas compounds have been formed under special conditions.

[6] The exclusion principle is not limited to electrons (actually, all *fermions* obey the restrictions) nor to atomic phenomena. In the following chapter we will discuss phenomena of matter in bulk that require the exclusion principle for their explanation. *Bosons* do *not* obey the exclusion principle.

$n = 1$, only $l = 0$ is possible and, therefore, only $m_l = 0$ is possible. But there are two possible spin states, $m_s = +\frac{1}{2}$ and $m_s = -\frac{1}{2}$. Therefore, two electrons exhaust the $n = 1$ states and the first shell is filled for $Z = 2$ (helium), as shown in Fig. 19.16. The first shell, which contains only the two $n = 1$, S electrons is called the *K shell*.[7]

In order to form lithium ($Z = 3$), we must add the third electron in a state with $n = 2$ and with lithium we begin the L shell. For $n = 2$ we have available two 2S states ($l = 0$, $m_l = 0$, $m_s = \pm\frac{1}{2}$) and six 2P states (two states from $l = 1$, $m_l = \pm +1$, $m_s = \pm\frac{1}{2}$, two states from $l = 1$, $m_l = 0$, $m_s = \pm\frac{1}{2}$, and two states from $l = 1$, $m_l = -1$, $m_s = \pm\frac{1}{2}$). Therefore, the L shell has a total of 8 possible electron states and this shell consists of the 8 elements from $Z = 3$ (lithium) through $Z = 10$ (the noble gas, neon), as indicated in Fig. 19.16. All of the possible states for $n = 1$ and $n = 2$ are listed in Table 19.6.

Higher Shells

We would expect the third shell to consist of the entire complement of M states with $n = 3$: two 3S states, six 3P states, and ten 3D states, giving a total of 18 possible states. However, in the third shell an additional effect comes into play. Electrons in states with high angular momentum find themselves, on the average, much farther from the nucleus than the electrons in low angular momentum states. These distance electrons therefore do not react to a nucleus of charge $+Ze$; instead, they move under the influence of a lesser charge that results from the partial cancellation or shielding of the nuclear charge by the inner, low angular momentum electrons. The outer electrons experience a reduced force and are therefore only loosely bound to the atom. For the $n = 3$ (and higher) states this shielding effect produces an important change in the energetics of an

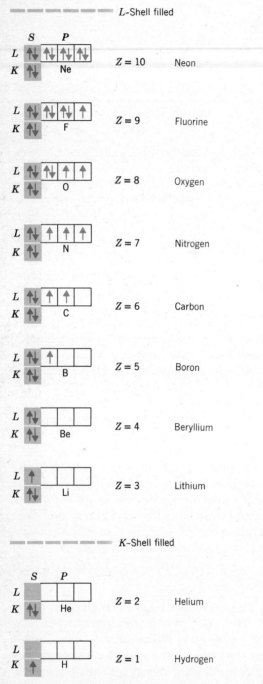

Figure 19.16 Filling of the first two shells of atomic electrons. In the first column are the S state (1S and 2S) and in the next three columns are the P states ($l = 1$, $m_1 = +1$, 0, −1). The boxes represent the magnetic substates; each substate contains two spin states, $m_s = \pm\frac{1}{2}$, which are represented by the arrows. The K shell is filled at helium and the next 8 electrons must be placed in the L shell, which is completed at $Z = 10$, neon.

Table 19.6 Electron States in the First Two Atomic Shells

Shell	n	l	m_l	m_s	No. of Electrons
K	1	0	0	$+\frac{1}{2}$	2 (1S)
		0	0	$-\frac{1}{2}$	
L	2	0	0	$+\frac{1}{2}$	2 (2S)
		0	0	$-\frac{1}{2}$	
		1	0	$+\frac{1}{2}$	6 (2P) — 8 total
		1	0	$-\frac{1}{2}$	
		1	+1	$+\frac{1}{2}$	
		1	+1	$-\frac{1}{2}$	
		1	−1	$+\frac{1}{2}$	
		1	−1	$-\frac{1}{2}$	

[7]It is customary to give letter designations to the sets of electrons with the same principal quantum number. Electrons with $n = 1$ are called K electrons; $n = 2$, L electrons; $n = 3$, M electrons; etc. The first two electronic shells contain only K electrons and L electrons, respectively. But as we shall see, the higher shells contain electrons with more than a single value of n.

atomic system. The energy of the 4S state is actually *lower* than that of the 3D state (i.e., the 4S state is more tightly bound). Consequently, the 4S state fills *before* the 3D state. Similarly, the 5S state fills before the 4D state. This distortion of the "normal" energy scheme is shown schematically in Fig. 19.17 where the third shell is seen to close at the 3P state and contains only 8 electrons. The fourth shell consists of the 4S, 3D, and 4P states and contains a total of 18 electrons. Similarly, the fifth shell contains 18 electrons.

The order of filling of the electron subshells is:

1S, 2S, 2P, 3S, 3P, 4S, 3D, 4P, 5S, 4D, 5P, 6S, 5D, 4F, 6P, 7S, 5F, 6D

The *transition elements* in the periodic table correspond to those atoms in which the shielding effect has displaced the subshells and a subshell with the "wrong" value of n is being filled. For example, with argon ($Z = 18$), the first 18 electron states (up to and including the 3P states) have been filled in the normal manner (see Fig. 19.17). But the 19th and 20th electrons, because of the shielding effect, go into the 4S states instead of the 3D state. With the addition of more electrons, we must "back up" and fill the 10 available 3D states before going on to the 4P states. These 10 elements, formed by "back filling" the 3D states, are the transition elements of period 4. The other transition elements of higher periods occur because of similar effects. When we reach $Z = 57$

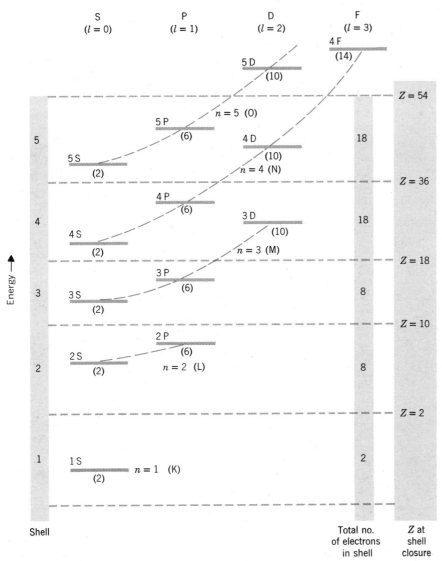

Figure 19.17 Distortion of the "normal" sequence of filling subshells because of the shielding of the nuclear charge by the inner, low angular momentum electrons. As a result of shielding, the high angular momentum electrons are raised in energy (i.e., are less tightly bound) and 3D electrons actually go into the fourth shell and the 4D electrons go into the fifth shell. (Energies are not to scale.)

(and also $Z = 89$), the "back filling" occurs in such a way that an entire series of elements (the *lanthanides*, beginning at $Z = 57$, and the *actinides*, beginning at $Z = 89$) fits into a single box in the first column of the transition elements. These series are therefore listed in separate rows at the bottom of the periodic table (Fig. 19.14).

The combination of the Pauli exclusion principle and the effect of shielding by the inner electrons accounts completely for the observed shell structure of atomic electrons.

19.6 Atomic Radiations

X Radiation

Most of the properties of atoms—chemical, electrical, magnetic, optical, etc.—depend on the configurations of the outermost electrons. Only in the event of a very energetic disturbance are the tightly bound inner electrons involved in the process. The reason is easy to see on the basis of energetics: it requires, for example, only 7.4 eV to remove the outermost electron from a lead atom ($Z = 82$), but an energy of 88 keV or 88,000 eV is necessary to remove one of the K electrons. The difference is smaller, of course, for atoms with lower atomic numbers but it is always much easier to remove an outer electron than an inner electron.

If sufficient energy is supplied to an atom by collision with a fast electron (as in an X-ray tube) or by irradiation with an energetic photon, then it is indeed possible to remove one of the inner K electrons (see Fig. 19.18a). The atom will not remain long in this condition with a vacancy in its K shell. It is energetically more favorable for an electron in a higher shell to make a transition and occupy the K-shell vacancy (Fig. 19.18b). It is most likely that an L electron will make this transition, emitting an energetic photon (called a K_α X ray) in the process. But then there is a vacancy in the L shell which is filled by an electron from one of the higher shells. Eventually, after this cascading of electrons and the emission of a series of X-ray photons, a free electron from the surroundings will be captured into the outer shell and return the atom to an electrically neutral condition.

Table 19.7 gives the K-shell ionization energy, the K_α X-ray energy, and the minimum ionization energy (i.e., for removal of an outer electron) for several elements. The K_α X ray, corresponding to a transition between the L and K shells, is always the most prominent feature of an atom's X-ray spectrum even though the transitions M → K (yielding a K_β X ray), N → K (yielding a K_λ X ray), and so on, are energetic.

Extensive studies of atomic X-ray spectra were made in 1913–1914 by H. G. J. Moseley (1887–1915), a brilliant student of Rutherford who met an untimely death in World War I. Moseley showed

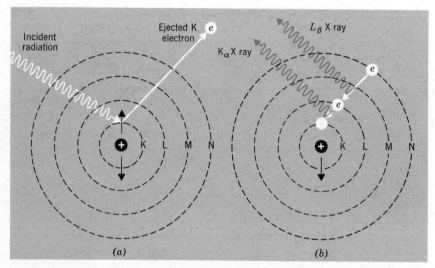

Figure 19.18 Schematic representation of the production of X rays. *(a)* A high-energy photon is incident on an atom; it penetrates to the innermost shell and ejects one of the two K electrons. *(b)* One of the L electrons then makes a transition, filling the vacancy in the K shell and emitting a K_α X ray in the process. Subsequently, an N electron makes a transition, filling the new vacancy in the L shell and emitting an L_β X ray. Finally, the vacancy in the outermost shell is filled by the capture of a free electron from the surroundings and the atom is again electrically neutral.

Table 19.7 Ionization and K_α X-ray Energies for Some Elements

Element	Z	K-shell Ionization Energy (keV)	K_α X-ray Energy (keV)	Minimum Ionization Energy (eV)
Al	13	1.56	1.49	6.0
Cu	29	8.99	8.06	7.7
Mo	42	20.0	17.5	7.4
Ag	47	25.5	22.1	7.6
W	74	69.6	59.3	8.1
Pb	82	88.1	75.0	7.4

from the systematics of his X-ray spectra that there is a direct connection between the X-ray energies and the atomic numbers of the emitting atoms. He found that the energy of the K_α X ray from an atom of atomic number Z is very closely given by

$$\mathscr{E}_{K_\alpha} = 10.26\,(Z-1)^2 \text{ eV} \tag{19.27}$$

We can understand Moseley's formula in the following way. A K_α transition involves an electron changing from an $n = 2$ state to an $n = 1$ state. Within an atom, this transition takes place inside the shell containing all of the atomic electrons except the one electron that occupies the K shell. (Remember, a K_α transition is one that proceeds to a vacancy in the K shell; the other K-shell electron remains in place during the transition.) Therefore, the effective charge on the nucleus is its total charge Ze less one unit due to the shielding effect of the K electron; that is, the effective charge is $(Z - 1)e$. Then, the energy of the transition can be calculated from the hydrogen-atom formula by substituting $(Z - 1)e$ for the effective charge of the nucleus. In obtaining Eq. 19.15, we used Ze with $Z = 1$ for the nuclear charge; the factor e^4 appears because we must square the nuclear charge Ze and the electron charge e. Consequently, the expression for the photon energy must be modified to

$$\mathscr{E}_{K_\alpha} = \tfrac{1}{2}\frac{K^2 m_e (Z-1)^2 e^4}{\hbar^2}\left(\frac{1}{1^2} - \frac{1}{2^2}\right)$$
$$= 10.2\,(Z-1)^2 \text{ eV}$$

The data for the elements up to an atomic number of about 50 are well represented by Moseley's formula with a numerical factor of 10.26 eV (instead of 10.2 eV as predicted by the simple Bohr-model expression).

Prior to Moseley's work, the elements had been placed in the periodic table according to increasing atomic *mass*, but this led to inconsistencies in certain cases. From the systematics of his X-ray spectra, Moseley established for the first time the atomic *numbers* of several elements. It then became clear that the inconsistencies occurred only in those cases for which the normal increase in atomic mass with Z was reversed[8] and that the problem could be resolved by ordering the elements in the periodic table according to atomic *number* instead of atomic *mass*. Moseley was also able to show that there were three elements then missing from the table of elements ($Z = 43$, 61, and 75). All three were subsequently discovered after Moseley had given the clue to their existence.

When a target material is bombarded by a beam of energetic electrons, radiations in addition to the discrete X rays are observed. These radiations arise because the electrons are slowed down in passing through the material and decelerating charges *radiate*. This type of radiation is called *bremsstrahlung* (or *braking radiation*). The bremsstrahlung spectrum is continuous (i.e., it contains no discrete lines) up to a maximum energy corresponding to the energy of the electron beam. Electrons that are accelerated through a potential difference of 15 kV and bombard a target of molybdenum will produce the spectrum indicated in Fig. 19.19. The spectrum is shown here on a *wavelength* scale and the minimum (or *cut-off*) wavelength of 0.83 Å for an electron energy of 15 keV is easily calculated using Eq. 19.17b, namely, $\lambda = 12{,}400/\mathscr{E}$ for $\mathscr{E}$ in eV and λ in Å. Figure 19.19 also shows the spectrum for an accelerating potential of 25 kV. The cutoff wavelength is now 0.50 Å. Notice that this spectrum contains the prom-

---

[8] Although the atomic mass is usually (except for hydrogen) approximately twice the atomic number, the order of increase of the two numbers is not always the same. For example, the atomic mass of cobalt ($Z = 27$) is greater than that of nickel ($Z = 28$); see Fig. 19.14 for other examples.

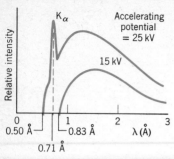

Figure 19.19 X-ray spectra produced by electron bombardment of molybdenum. The K_α X ray appears only for bombarding energies above 20 keV.

inent K_α line which did not appear in the 15-kV spectrum. The reason is that the energy required to produce a K-shell ionization in molybdenum is 20 keV (see Table 19.7), so the 25-keV electrons, but not the 15-keV electrons, can cause this ionization and the subsequent 17.5-keV K_α X ray.

Stimulated Emission of Radiation—Lasers

The probability that a transition between two states of an atom will take place within a certain specified time interval depends on the product of the wave functions describing those states and a quantity characteristic of the transition. Thus, from the viewpoint of quantum theory, there is no distinction between the transition from state A to B ($A \rightarrow B$) and the transition $B \rightarrow A$ because the same product of wave functions and transition quantity is involved in the description of each transition. The two equivalent processes, excitation and deexcitation, are shown schematically in Fig. 19.20 in each case the photon has an energy $h\nu = \mathscr{E}_B - \mathscr{E}_A$.

If we can arrange to have an atom in state B (the excited state) when a photon with an energy $h\nu = \mathscr{E}_B - \mathscr{E}_A$ is incident on that atom, then this photon will stimulate the deexcitation process to occur. (The photon cannot *excite* the atom because it is already excited, so it does the equivalent—is *deexcites* the atom.) This process is called *stimulated emission* and is illustrated schematically in Fig. 19.21.

The essential feature of the stimulated emission process that renders it both interesting and useful is that the incident photon and the stimulated photon proceed away from the atom *in phase* and traveling in the *same* direction. The two photons therefore reinforce one another. Such radiation is said to be *coherent*.

If we had a sample of atoms, some fraction of which were in the same excited state, then a single incident photon could begin triggering the deexcitation of these atoms by stimulated emission. Each stimulated photon could, in turn, cause another atom to emit a photon and the entire system would radiate its excitation energy almost at once with a single bundle of photons all in phase. What have we gained in such a process? Since all of the excited atoms would eventually have radiated away their

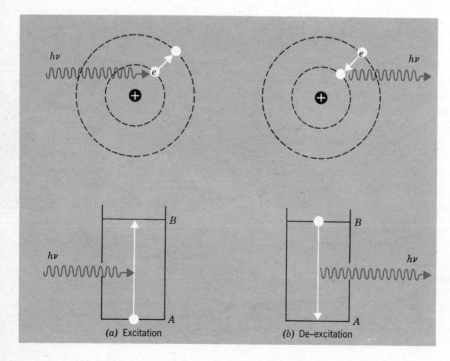

Figure 19.20 (a) An incident photon of energy $h\nu$ excites an atom by raising an electron to a higher energy state. (b) Deexcitation occurs when the electron returns to its ground-state configuration and a photon of energy $h\nu$ is emitted: According to the rules of quantum theory, the two processes are mathematically equivalent. The only difference between the two situations is that an energy $h\nu$ is absorbed in excitation and an energy $h\nu$ is emitted in deexcitation.

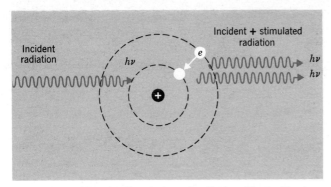

Figure 19.21 Stimulated emission of radiation. The incident photon of energy $h\nu$ finds the atom in an excited state and stimulates its decay. The two photons, each with energy $h\nu$, proceed away from the atom in phase.

excitation energy by spontaneous emission, we have done no more by stimulating the deexcitation than would have happened anyway. The difference is that in the spontaneous emission process, the photons are radiated in random directions and not in phase. Stimulated emission produces the photons essentially simultaneously and in phase.

How is it possible to take advantage of stimulated emission to produce an intense beam of coherent radiation? If the radiation is light, the device that accomplishes this is called a *laser,* an acronym for *l*ight *a*mplification by *s*timulated *e*mission of *r*adiation.[9]

In order to construct a practical laser, a number of problems must be solved. First, we must select an appropriate material and we must choose the pair of states between which the laser transition is to occur. Then, we must devise a method of exciting the atoms to the radiating state. Finally, we must arrange for the stimulated photons to be confined to a particular direction.

Why is it that an ordinary light source does not emit stimulated radiation in a coherent beam? The reason is that in all ordinary light sources there are, at any instant, far more ground-state atoms than there are excited atoms. Consequently, as a photon proceeds through the material, it is much more likely to encounter a ground-state atom and be absorbed by that atom than it is to encounter an excited atom and stimulate radiation by that atom. Because of this absorption effect, there is no opportunity for the multiplying effect of the stimulated emission to start.

In order to overcome absorption effects, we need a sample in which the number of excited atoms exceeds the number of ground-state atoms. Then, the stimulated emission process will be more likely and will dominate absorption. A sample in which this situation occurs is said to have a *population inversion.* That is, the normal population of states (many atoms in the ground state, few in the excited state) is inverted (many atoms in the excited state, few in the ground state). How can we do this? If the excited state has a lifetime typical of most excited states (a millionth of a second or so), then any atoms in this state will quickly radiate away their excess energy and return to the ground state. This will happen before stimulated emission is effective. To promote laser action, we must use an excited state that has a lifetime much longer than normal—a few tenths of a second or so. In this way, the atoms will remain in the excited condition sufficiently long that a beam of radiation can build up.

One way to accomplish a population inversion is to *pump* the atoms through a third excited state. Figure 19.22 shows three energy states of an atom: state A is the ground state and states B and C are excited states. The first step is the excitation of state C by photon absorption, electrical discharge, or collision. State C is a typical excited state and has a short lifetime. But state C is chosen so that when it radiates, it leaves the atom in state B, which has a long lifetime. State B is called a *metastable state.* The transition $B \to A$ is the laser transition. When a photon with energy $\mathscr{E}_B - \mathscr{E}_A$ encounters the atom in state B, it stimulates emission and the two photons proceed away from the atom in phase.

(Why do we not simply pump state B directly? Remember, the two transitions, $A \to B$ and $B \to A$, are equivalent. If state B has a long lifetime this means that the transition connecting A and B is weak. Therefore, the direct transition $A \to B$ is in-

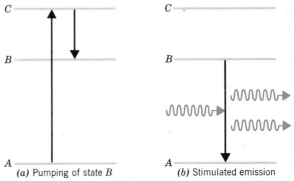

Figure 19.22 A three-state (or three-level) laser. The long-lived state that exhibits laser action (state B) is pumped by first exciting state C. The lifetime of state B is sufficiently long that a population inversion can be established.

---

[9]The first practical devices to be constructed utilizing this principle operated with microwaves and were called *masers* (*m*icrowave *a*mplification by *s*timulated *e*mission of *r*adiation).

hibited. However, once the atom is in state B, it can only radiate to state A; even if the transition is weak, it will eventually occur.)

If the high-energy state has a sharply defined energy (as does state C in Fig. 19.22), the pumping radiation must consist of photons with well-defined energy. A source of *white* light would not be suitable because such a source emits photons with a wide range of photon energies and so only a few of these can have the proper energy to be effective in pumping the atoms to the upper state. In 1960, Charles Townes and Arthur Shawlow of Columbia University called attention to an interesting property of ruby crystals that appeared to offer a solution to this problem. Ruby consists of aluminum oxide, a colorless substance, which contains a small amount of chromium as an impurity. The chromium impurity gives to ruby its characteristic red color. Figure 19.23 shows some of the energy states of the chromium atoms in ruby. The distinctive feature of this diagram is the fact that the energy states at $\mathcal{E}_2$ and $\mathcal{E}_3$ are actually *bands;* that is, the atom is not limited to a single well-defined energy but can exist with any energy within a range centered about $\mathcal{E}_2$ and $\mathcal{E}_3$. Because these bands are broad, the white light from the pumping source includes large numbers of photons whose energies fall within the range that permits the pumping of these energy bands. Each of the two bands radiates primarily to the state at $\mathcal{E}_1$. Hence, the laser transition is $\mathcal{E}_1 \to \mathcal{E}_0$ and the corresponding radiation is in the red part of the spectrum at 6934 Å.

The problem of directionality can be solved in the following way. A crystal of ruby is formed into a cylinder with the end surfaces accurately parallel (Fig. 19.24). One end is silvered to form a mirror while the other end is given only a partial coating of silver so that some of the radiation can escape from this end. The pumping is provided by a high-intensity discharge lamp that spirals around the cylindrical crystal. As soon as one photon is produced in the spontaneous transition $\mathcal{E}_1 \to \mathcal{E}_0$, this triggers the laser action. Those photons that move parallel to the cylinder axis are reflected at the ends and again transverse the crystal, stimulating the emission of additional photons. A fraction of this radiation escapes through the partially-reflecting surface and constitutes the laser beam. Most of the spontaneously emitted photons are not emitted parallel to the axis; these photons are reflected in the crystal and eventually escape through the sides. These spontaneous photons do not contribute to the beam, but a sufficient number of photons *are* reflected back and forth to sustain the laser action.

Energy is pumped at intervals[10] into the crystal by the light source and some fraction (usually very small) emerges as the laser beam; this radiation is coherent, has an almost pure frequency, and is highly directional. But in no sense is a laser a "source" of energy. In fact, only a very small fraction of the input energy appears in the beam. Some recently constructed lasers can produce bursts of radi-

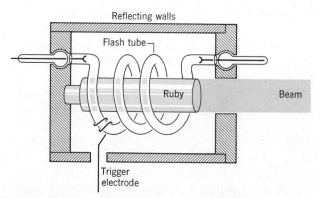

Figure 19.24 Schematic of a ruby laser system. The pumping radiation is furnished by a high-intensity source of white light. The stimulated photons are reflected back and forth between the parallel mirrors and build up the intensity of the radiation. The beam is formed by photons escaping through the partially reflecting surface.

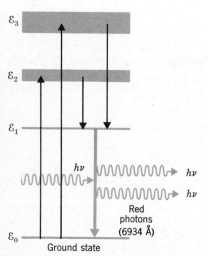

Figure 19.23 Some of the energy states of chromium atoms in a ruby crystal. The pumping radiation (upward arrows) excite the two energy bands, $\mathcal{E}_2$ and $\mathcal{E}_3$, which subsequently radiate to form the state $\mathcal{E}_1$, which exhibits laser action. The laser radiation (brown arrow) consists of red photons ($\lambda = 6934$ Å).

---

[10] Ruby and most other crystal lasers must be *pulsed* because continuous operation would result in overheating and the crystal would probably fracture. Some new artificially made crystals are less subject to heating effects than is ruby.

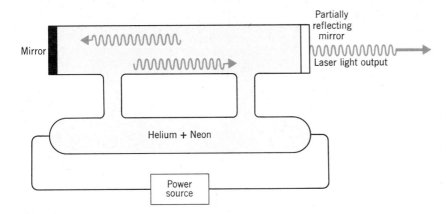

Figure 19.25 Construction of a helium-neon laser. The mirror and partially reflecting mirror are arranged the same way as in a ruby laser (Fig. 19.24). The pumping action is produced by the electrical current that flows through the tube when the power source is switched on.

ation in which tens of joules of energy is released within 10^{-12} s.

The most common and least expensive type of laser in use today is the helium-neon laser. Figure 19.25 shows in a schematic way the construction of this type laser. Basically, the laser consists of two glass tubes joined by a pair of short connecting tubes. The entire tube system is filled with a mixture of helium and neon gases at low pressure. One of the main tubes (the *laser tube*) has a mirror at one end and a partially reflecting mirror at the other end, just as in the case of a ruby laser. The other main tube (the *discharge tube*) has an electrode sealed in each end. These electrodes are connected to a power source.

The method used to pump a helium-neon laser is quite different from that used in a ruby laser. When the power source is switched on, an electric current flows through the gas from one electrode to the other. The gas atoms are excited by impacts with the moving electrons, and the excited atoms radiate spontaneously. The gas in the tube is seen to glow with a red color. This is, in fact, exactly the way a neon sign tube is made to glow. The region of excitation of the gas atoms is not confined to the discharge tube and spreads easily into the laser tube. The entire tube system glows with the same red color.

The transition that exhibits the laser action is a transition between two states in neon. But the neon is not excited directly. As shown in Fig. 19.26, the impacts of the electrons on the helium atoms cause the excitation of state B. Because the gas atoms are continually in motion, there are frequent collisions between atoms. If a helium atom in the excited state B collides with a neon atom in its ground state, the excitation energy of the helium atom can be transferred to the neon atom. Then, the neon atom is in the state C', which is a metastable state. This is the state that produces the laser photons when stimu-

lated into emission. The laser transition $C' \to B'$ produces red photons with a wavelength of 6328 Å.

Unlike a ruby laser, a helium-neon laser can be operated continuously. (Excitation by electrical current is more efficient than by the absorption of light from a flash lamp, so the heating problem is not severe.) This is a definite advantage in many situations.

Hundreds of materials—solids, liquids, and gases—have been found to exhibit laser action. These lasers have rapidly found an extraordinary number of applications in basic research, technology, and medicine. Some of the more spectacular uses have been in eye operations where a laser has been found ideal for depositing just the right amount of energy to "weld" a detached retina onto the choroid surface that lies beneath it. Micro-holes can be drilled in hard substances by laser beams, and the welding of materials that resist other methods can be accomplished with these devices. Modulated laser beams can carry an incredible number of communi-

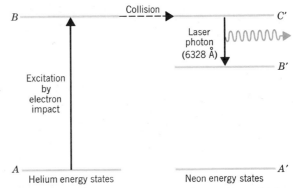

Figure 19.26 The atomic transition that produces the laser photons in a helium-neon laser involves a pair of states in neon. The radiating state C' is excited by a collision with a helium atom that is in the state B. In the collision process, the excitation energy is transferred from the helium atom to the neon atom.

cations channels and it is clear that the impact on the communications industry on the development of such devices will be enormous. Also, by reflecting a laser beam from a mirror placed on the Moon we are now obtaining information regarding the fluctuations of the Earth-Moon distance—information that will give important clues as to the geophysics of the Earth and of the Moon. Recently it has been found possible to "chop" a laser beam to give pulses of radiation with a duration of only 10^{-12} s. The use of such light pulses will provide new information on the interaction of radiation and matter.

Medical Uses of Lasers

The discovery and development of lasers has had a significant beneficial impact on medical research and practice. One of the important areas of biomedical research concerns the functions of the individual components of cells. These functions are studied by damaging a particular component and observing the subsequent behavior of the cell. If the cell is unable to perform one of its normal functions, it is reasonable to attribute this function to the damaged component. Before the advent of lasers, the damaging was done with a fine needle (or *microprobe*). Only large cell structures could be studied in this way, and the results were sometimes ambiguous. By using laser beams as probes, this line of research has been greatly refined. A laser beam can be focused to a diameter of 0.5 μm or smaller. A beam of this size can probe cell components too small for ordinary microprobes. For example, an important component of tissue cells are the *mitochondria,* which are roughly cylindrical in shape with diameters of about 0.5 μm and lengths up to 10 μm or so. A 0.5-μm laser beam can damage individual mitochondria, or even *parts* of these cell components!

Lasers are also used to perform delicate surgery and to seal off bleeding areas in difficult locations. For example, lasers have been used to control acute bleeding in hemorrhagic gastritis and stomach ulcers. In these procedures, a bundle of optical fibers (see Fig. 16.13) is inserted through the mouth into the stomach. Light is piped through part of the bundle to illuminate the region and the surgeon views the area through another part of the bundle. When the bleeding site is located, a burst of laser radiation directed toward it seals the area. The procedure is quick and the results are often permanent.

Surgery that has been performed by laser beams includes removal of vocal-cord lesions and cervical polyps. Certain types of cancers can also be controlled by laser surgery. These operations are carried out with no pain, bleeding, scars, or impairment of function. Laser beams can even obliterate birthmarks and erase tattoos.

Visible laser radiation passes through the eye to the retina just as ordinary light does. Therefore, laser beams are particularly suited for treating retinal disorders. The blood that supplies nutrients and oxygen to the inner layers of the retina is carried by the central retinal artery which enters the eye along with the optic nerve and then divides to supply the entire inner surface of the retina through a system of tiny blood vessels. In *diabetic retinopathy,* these blood vessels deteriorate and leak; moreover, new vessels grow on the surface of the retina and protrude and hemorrhage into the vitreous humor. Scar tissue forms in association with the new blood vessels and this may pull on the retina to the extent that it becomes detached from the choroid surface that lies between the retina and the sclera. By using a highly focused laser beam, the weakened blood vessels can be coagulated or welded, and the proliferating new ones can be destroyed. In patients who have received laser treatment for diabetic retinopathy, the incidence of vision loss has been reduced by 60 percent.

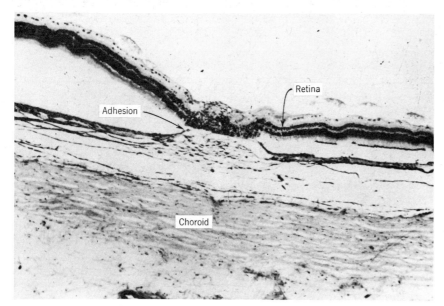

Courtesy of D. Smart and P. Hornby. International Research and Development Co., England

Figure 1 An adhesion between a detached retina (upper portion) and the choroid surface (lower portion) produced by the beam from a ruby laser. This technique was first used in the treatment of human patients in 1964, and since then thousands of successful treatments have been made.

Sometimes the retina does become detached from the choroid surface, most often due to the contracting of the fibrils of collagen in the vitreous humor which pulls the retina unevenly toward the interior of the eye. The outer layers of the retina, including the outer segments of the rods and cones, receive their nutrition, especially the oxygen, mainly through diffusion from the choroid blood vessels. If the retina becomes detached, the diffusion process can continue to supply a portion of the needs of the retina and it can resist degeneration for a number of days. If the damage is not promptly repaired, however, the retina will degenerate to the point that it is incapable of functioning even after surgical repair. For more than 10 years, lasers have been used routinely to repair detached retinas. A short burst of laser radiation (duration about 1 ms) is focused on the retina to damage some of the retinal material and the underlying choroid surface. The result of the "burn" is the formation of a weld that effectively bonds the two surfaces together (see Fig. 1). Because the necessary burst of radiation can be delivered in such a brief interval, the patient does not react to the light until after the exposure is completed; consequently, no head restraint is required during the operation. Furthermore, the patient experiences no discomfort.

Although laser beams can be used to produce healing effects in the eye, serious damage can result if too large a burst of radiation enters the eye. Many of the low-power lasers that are used in laboratories and classrooms have beam power levels of 1 mW, or sometimes more. It has been determined that an exposure of the eye to a 2-mW beam for a period of 1-2 s will result in a retinal burn. Because of the severity of the damage that can result from a retinal burn, every individual must be extremely cautious in the handling of lasers.

References

M. W. Berns, *Biological Microirradiation,* Prentice-Hall, Englewood Cliffs, N.J., 1974.

M. L. Wolbarsht, ed., *Laser Applications in Medicine and Biology,* Plenum, New York, 1971.

Exercise

1. The beam from a 10-mW He-Ne laser is focused to a spot that has a diameter of 0.4 mm for a certain surgical procedure. A 1.5-s burst of radiation is delivered by the laser. What is the total amount of energy (in J) in the burst? What is the power density (in W/m^2) in the spot? How many photons are delivered in the burst?

Summary of Important Ideas

Rutherford's analysis of α-particle scattering experiments showed conclusively that most of the mass of an atom is concentrated in a tiny, positively charged *nuclear* core.

In order to account for the lines of the hydrogen spectrum, Bohr found it necessary to postulate that each line corresponds to a transition between two allowed *discrete energy states* and that the angular momentum of the atom is limited to *discrete multiples of $\hbar$*. Bohr departed from classical electromagnetic theory by postulating that no radiation occurs except during the transition process.

According to the Bohr model, an integer number of de Broglie electron waves must exactly fit into every allowed electron orbit.

The complete specification of the quantum state of an electron in an atom requires *four* quantum numbers: $n, l, m_l,$ and m_s, which specify, respectively, the (gross) *energy*, the *angular momentum*, the *component of the angular momentum* in a particular direction, and the *orientation of the spin vector* relative to the angular momentum vector.

The *Pauli exclusion principle* states that no two electrons in an atom can have exactly the same set of four quantum numbers. Only *fermions* (particles with half-integer spins) obey the exclusion principle.

The occurrence of *electron shells* in atoms can be accounted for in terms of the *exclusion principle* and the effect of *shielding* by the inner electrons.

When a photon stimulates the emission from an atom of a photon with the same frequency, the two photons propagate away from the atom *in phase* and reinforce one another. The operation of *lasers* is based on this fact.

◆ Questions

19.1 According to the Bohr theory, the principal quantum number n determines what three physical properties of the atom?

19.2 List the total number of spectral lines that can result from the deexcitation of a hydrogen atom in the following states (a) $n = 3$, (b) $n = 4$, (c) $n = 5$. (Use the Bohr model.)

19.3 Among the various series of spectral lines of the hydrogen atom, the first to be discovered and studied was the Balmer series. Why do you suppose this was the case?

19.4 In footnote 3 on page 478 it was pointed out that stellar atmospheres provide better conditions for observing hydrogen line spectra than do laboratory sources. Explain why this is the case.

19.5 An *absorption spectrum* is one that results when "white" light (that is, light consisting of all frequencies) is passed through a substance. Absorption lines are *dark* lines on a background of "white" light. What lines are found in the absorption spectrum of hydrogen?

19.6 An omega hyperon (substituted for an electron in an atom) is in a state with $l = 2$. What are the possible values for the total angular momentum quantum number j?

19.7 A corollary to the exclusion principle is the principle of *indistinguishability* of elementary particles. This principle states, for example, that there is no way to distinguish any one electron from another electron. Contrast the situation in which one billiard ball collides with another billiard ball to that in which one electron collides with another electron. Can one measure the angle through which the *incident* object was scattered in both situations? (The billiard balls are *numbered*, but what about the electrons?)

19.8 What difference in atomic shell structure would there be if electrons were bosons instead of fermions?

19.9 What are the maximum values for the projections of the total angular momentum along the z-axis for L, M, and N electrons.

19.10 In what positions in the periodic table do you expect to find elements that have *low* photoelectric work functions? Compare you answer with the elements in Table 18.3.

19.11 What are the quantum numbers for the outermost or *valence* electron in a sodium atom in the ground state? Suppose that sufficient energy is supplied to a ground-state sodium atom to raise it to the first excited state. What are the quantum numbers for the electron now?

19.12 Repeat Question 19.11 for the case of potassium.

19.13 Examine Fig. 19.14 and find three cases in which the order of increase of atomic number does not follow the order of increase of atomic mass.

★ Problems

Section 19.1

19.1* An α particle of energy 5.3 MeV from a radioactive source of ^{210}Po approaches a gold nucleus "head on." How close to the nucleus can the α particle penetrate before being stopped and deflected backward? (That is, at what distance will the electrostatic potential energy equal the initial kinetic energy of the incident α particle?) It was from such a calculation that Rutherford was able to show that nuclei are much smaller than atoms.

19.2* In nuclei with low atomic numbers, α-particles of relatively low energy may approach close enough to initiate nuclear reactions in addition to scattering due to Coulomb forces. What minimum α-particle energy is required to approach the nitrogen nucleus to within 3.8×10^{-13} cm, thus enabling nuclear reactions to commence?

Section 19.2

19.3 What is the longest wavelength photon that can induce a transition in a hydrogen atom in its ground state? When that atom deexcites, in what series will the radiation be?

19.4 What is the longest wavelength photon that can ionize a hydrogen atom in its ground state? How would you classify this photon—visible, infrared, or ultraviolet?

19.5 What frequency must a photon have in order to raise a hydrogen atom from its ground state to the state with $n = 4$? Is this a "visible" photon?

19.6 What is the velocity of an electron in the second Bohr orbit in hydrogen (radius = 2.12×10^{-10} m)? What is the de Broglie wavelength of such an electron?

19.7 Show that the binding energy of an electron in any state of a hydrogen atom is equal to its kinetic energy in that state.

19.8* In stars the atoms often completely or almost completely stripped of their electrons. Consider a hydrogenlike sodium atom (i.e., a sodium nucleus with a single atomic electron). What is the binding energy of the electron? What is the energy and the wavelength of the photon that results from the $n = 2 \to n = 1$ transition? Classify this photon.

19.9 Insert the value $r = a_1 = 0.53$ Å into Eq. 19.11 and calculate the velocity of the electron in the ground state of the hydrogen atom. In deriving Eq. 19.11 (and the equations following), we used a nonrelativistic approach. Was this justified?

19.10 The *series limit* for the Balmer series is 3646 Å (see Fig. 19.6). What is the series limit for (a) the Lyman series and (b) the Paschen series?

19.11* What must be the temperature of a gas consisting of hydrogen *atoms* so that the average kinetic energy of the atoms is just sufficient to ionize any of the atoms that are in the $n = 2$ state?

19.12 According to quantum theory, there are no well-defined "orbits" for electrons in atoms. This conclusion is required by the uncertainty principle. Suppose that we confine an electron to a box with a size equal to that of the diameter of the hydrogen atom in the $n = 1$ Bohr orbit (i.e., $\Delta x = 1$ Å). This is not really very restrictive; if the electron actually were in an *orbit*, the uncertainty in its position would be much smaller than the orbit diameter! Nevertheless, use this value for Δx and compute the corresponding value for the kinetic energy (in eV) of the electron using the uncertainty principle. What conclusion can you draw concerning electron orbits in atoms?

19.13 How much energy is required to ionize a hydrogen atom in the $n = 3$ state?

19.14 A free electron ($KE = 0$) is captured by a

free proton into the $n = 2$ orbit. What is the wavelength and the energy of the photon emitted in this process?

19.15 The next series of hydrogen lines after the Paschen series is the *Brackett series*. What are the longest and shortest wavelengths in the Bracket series?

19.16 A beam of 12.5-eV electrons bombards a quantity of hydrogen gas and excites some of the atoms. What wavelengths of radiation will be observed?

19.17 What is the wavelength of the $n = 6$ to $n' = 3$ transition (one of the lines in the Paschen series)?

19.18 Add together the masses of a free proton and a free electron. By what fraction does this mass change if the two particles are combined to form a hydrogen atom in its ground state?

19.19* Extend the Bohr theory to the case of the He^+ ion (once-ionized helium). What is the radius of the first Bohr orbit? How much energy is required to remove the remaining electron? What are the wavelengths of the transitions $n = 2$ to $n = 1$ and $n = 3$ to $n = 2$?

19.20 An electron is attracted to a neutron only by the gravitational force (at least for distances greater than about 10^{-15} m). Construct a derivation parallel to that for that Bohr model of the hydrogen atom and obtain the radius of the smallest allowed orbit. Can such "atoms" play any important role in Nature? Explain. (Neglect magnetic effects.)

19.21 Sketch de Broglie wave pictures (similar to Fig. 19.9a) for the hydrogen atom in the states $n = 2, 3,$ and 4.

Section 19.3

19.22 Continue Table 19.4 for $n = 5$ and $n = 6$.

19.23 The two strong yellow lines in the spectrum of sodium correspond to the P-state transitions shown in Fig. 19.12. If the two wavelengths are 5889.96 Å and 5895.93 Å, calculate the energy splitting of the P-state doublet.

Section 19.5

19.24 Extend Table 19.6 to include the M electrons. How many are there? Which are not found in the third atomic shell?

19.25 Prove that the maximum number of electrons in an atom that can have the principal quantum number n is $2n^2$.

19.26 The valence electron of sodium can have the following energies: -5.1 eV, -3.0 eV, -1.9 eV, -1.6 eV, -1.4 eV, -1.1 eV, and so forth. Identify the state (n and l) corresponding to each energy. (Refer to Fig. 19.17.) What is the wavelength of the $4P \rightarrow 3S$ transition? All of the states with $l \geq 1$ are actually closely spaced doublets, the splitting being due to the spin-orbit effect. Identify the pair of states involved in the transition that produces the yellow doublet of lines. (Refer to Fig. 19.12.)

19.27 Construct a diagram similar to Fig. 19.16 for the hypothetical situation in which electrons have spin $\frac{3}{2}$. What elements will be in the K shell? With what element will the L shell close?

Section 19.6

19.28 The energy of the K_α X ray from an unknown sample is found to be 6.45 keV. What is the element?

19.29 What is the energy of the K_α X ray from silver ($Z = 47$)?

19.30 The K_α X ray in sodium is due to a transition from the L-shell at $\mathcal{E}_L = -63$ eV to the K-shell at $\mathcal{E}_K = -1072$ eV. Determine the energy of the sodium K_α X ray and compare your answer to that obtained by using Moseley's formula.

19.31 A tungsten target is bombarded with 65-keV electrons. Describe the radiation produced. The electron energy is increased to 75-keV. How does the spectrum change?

19.32 What is the cutoff wavelength of the bremsstrahlung radiation produced when a beam of 80-keV electrons strikes a target?

19.33* The angular divergence of the photon beam from a certain laser is 10^{-4} radians. Consider such a laser that emits 1 mW of radiation directed toward a target that is 1 km away. What is the diameter of the light spot on the target? What would be the power output of a source that emits isotropically (i.e., equally in all directions) and delivers the same power per unit area to the target?

19.34 Consider an atom with states of the following energies: -13.2 eV (ground state), -11.1 eV, -10.6 eV, -9.8 eV. Only the state at -11.1 eV exhibits laser action. The state at -10.6 eV radiates primarily to the state at -11.1 eV. The state at -9.8 eV radiates primarily to the ground state. What wavelength radiation would you use to pump the laser? What is the wavelength of the laser radiation?

19.35 What is the spatial length of a laser pulse that lasts 10^{-12} s? If red light is produced, how many oscillations occur within the pulse?

20
The Structure of Matter

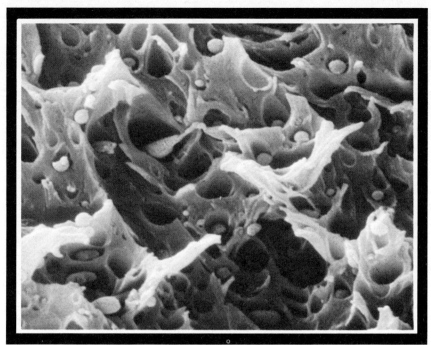

In order to understand the details of the microscopic world, the physicist first attacks the problems of the simplest atomic structures. In the preceding chapter we saw how the solutions to these problems have contributed to our understanding of the complicated electronic systems of atoms and the spectra that they produce—how the periodic table of elements has been explained—how such fundamental concepts as electron spin and the exclusion principle have emerged—and how our belief in the vast range of validity of quantum theory has been established.

But in our everyday experience we do not deal with *atoms*. The world around us is composed of aggregates of enormous numbers of atoms in the form of solids, liquids, and gases. Therefore our next step is to understand the way in which atoms interact with one another to form molecules and then bulk matter. In this transition from the microscopic to the macroscopic world we find that we do not leave quantum effects behind. Indeed, the quantum nature of matter profoundly influences a wide range of properties of bulk matter. The fact that copper is a good electric conductor whereas quartz and Teflon are extremely poor conductors is a quantum effect. The existence of the interesting *superconducting* materials, which are of increasing practical importance, is a beautiful example of quantum effects in bulk material. Even the individuality of human beings (indeed, *all* living things) is the result of differences in the structures of the giant molecules that carry genetic information and is therefore basically a quantum effect.

20.1 Bonds between Atoms

Ionic Binding

We are all familiar with the chemical compound *sodium chloride* (NaCl)—it is just ordinary *salt*. The basic unit of sodium chloride consists of two atoms, one of the metal sodium and one of the gas chlorine, and is typical of a large class of simple compounds. How are these two dissimilar atoms—a metal and a gas—bound together to form a stable substance such as NaCl? If we refer to the periodic table (Fig. 19.14), we see that sodium ($Z = 11$) is the first element in Period 3 and thus has one electron outside the closed L shell (i.e., Na is a Group I element). Chlorine ($Z = 17$) is the Period 3, Group VII element and therefore has 7 electrons outside the closed L shell or, equivalently, lacks one electron to fill the third shell.

The single M electron of sodium is relatively

easy to remove; it requires only 5.1 eV of energy to detach this electron and form a positively charged sodium ion, Na$^+$ (see Fig. 20.1a). An atom of chlorine, on the other hand, has an affinity for electrons and, if provided with a free electron, will absorb this electron into its outer shell, thus completely filling this shell. This process forms a negatively charged chlorine ion, Cl$^-$, and *releases* 3.7 eV of energy (see Fig. 20.1b).

The removal of an electron from sodium and the acquisition of an electron by chlorine are complementary situations. Sodium and chlorine can therefore exist together as a chemical compound, NaCl, in which the sodium electron is used to complete the outer electron shell of chlorine. But how can such a compound be stable since it requires 5.1 eV to remove the sodium electron and only 3.7 eV is gained by forming Cl$^-$? The answer lies in the fact that the electron transfer produces the ions Na$^+$ and Cl$^-$ and these ions are then attracted toward one another by the mutual electrostatic force. When the centers of the two ions are separated by a distance of about 11 Å, the electrostatic potential energy of the system has contributed the requisite 1.4 eV to effect the electron transfer. In fact, the electrostatic attraction pulls the ions even closer together and binds them more tightly. But the ions cannot approach more closely than a certain small distance (which turns out to be 2.4 Å) with all of the electrons in the lowest possible energy state; if the ions did approach more closely, two electrons with the same set of quantum numbers would be occupying the same region of space and this is prohibited by the exclusion principle. The only alternative would be for one or more electrons to be raised into a higher energy state, and under such conditions the system would no longer be bound. At the equilibrium separation of 2.4 Å, the binding energy of the ionic pair is 4.2 eV.

The binding together of atoms by virtue of the electrostatic attraction between ions formed by the *transfer* of an electron from one atom to the other is called *ionic binding*. In addition to NaCl, many other compounds are bound in this way, for example, NaBr, KCl, RbI, and LiF. The compound MgO is produced by the transfer of *two* electrons from magnesium to oxygen, forming Mg^{++} and O^{--}. The compound Na$_2$S is produced by the transfer of one electron from each of the two sodium atoms to the sulfur atom, forming 2 Na$^+$ and S^{--}.

Sodium chloride, together with most other ionic compounds, does not exist in Nature in the form of individual molecules. Instead, such substances occur as *ionic crystals*, highly organized systems of ions bound together in regular arrays by electrostatic forces. We discuss crystals again in the next section.

Covalent Binding

Chlorine gas occurs naturally as a two-atom molecule, Cl$_2$. How are the atoms bound together in this molecule? The mechanism must be quite different from that which binds sodium to chlorine because, in this case, the atoms are identical and each requires an additional electron to complete its outermost shell. The chlorine atoms solve the problem in the following way: Each atom *shares* one of its outer electrons with the other atom of the pair. By *sharing* an electron instead of *donating* it to the other atom, each atom manages to complete its outermost shell. This process is illustrated in Fig. 20.2 for the case of fluorine. The element fluorine is, like chlorine, a member of Group VII and therefore has an outer-shell structure that is the same as that of chlorine. The sharing of electrons to form molecular bonds is called *covalent binding*.

The hydrogen molecule, H$_2$, is also held together by covalent bonds. In Fig. 20.3 we show the orbit representation of the H$_2$ molecule. However, we should not lose sight of the fact that atomic electrons do not really move in well-defined orbits. Figure 20.4 shows a more realistic representation of the situation. In Fig. 20.4a we see the spherical electron

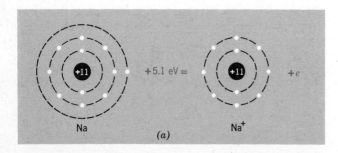

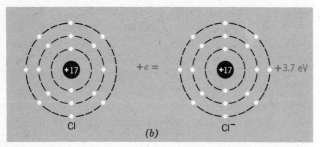

Figure 20.1 (a) It requires 5.1 eV of energy to detach the single M electron of sodium and form the ion, Na$^+$. (b) A chlorine atom will acquire an electron from its environment to fill the third shell and form the ion, Cl$^-$; in the process, 3.7 eV of energy is released. The two ions, Na$^+$ and Cl$^-$, attract one another and form the ionic salt Na$^+$Cl$^-$, (or, simply, NaCl).

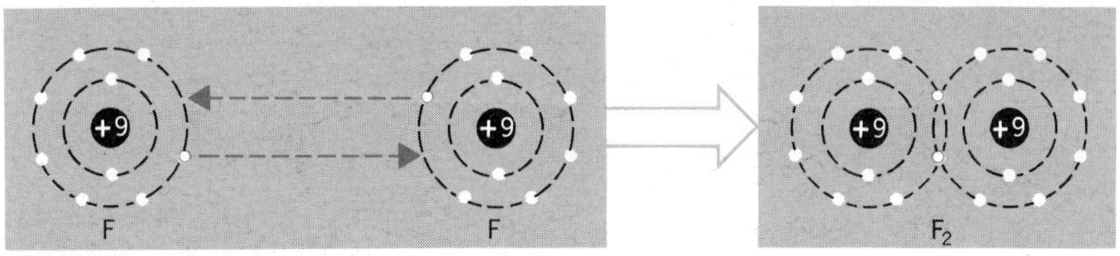

Figure 20.2 Two fluorine atoms share electrons in order to complete their outermost shells and form an F_2 molecule.

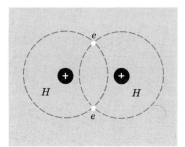

Figure 20.3 Schematic diagram of the way in which electron sharing by two hydrogen atoms completes the K shell of both atoms and forms a stable molecular structure.

cloud (the *probability* cloud) that surrounds the nuclear proton in each of the separated hydrogen atoms. In Fig. 20.4b we see the merging of the electron clouds when the H_2 molecule is formed. Each nuclear proton is attracted toward the concentration of negative charge between the atoms, thereby producing a strong molecular bond.

Among the large number of other molecules that are formed by covalent binding are water (H_2O) and ammonia (NH_3). Figure 20.5 illustrates schematically how these molecules employ covalent bonds. The box diagrams for oxygen and nitrogen, showing the electron spin states in the K and L shells are the same as those in Fig. 19.16. Oxygen has two unpaired electrons in the L shell while nitrogen has three such electrons. Each of these electrons can effect a covalent bond with a hydrogen atom, forming H_2O and NH_3. The spatial structures of these molecules are illustrated in Figs. 20.6 and 20.7. Of course, the molecules are not rigid as suggested in the diagrams, which are only schematic, but the *average* orientations of the electron clouds can be measured and have the directions shown. The peculiar values indicated for the bond angles are the result of the combination of the quantum requirements on the electron wave functions and the mutual electric repulsion of the hydrogen nuclei.

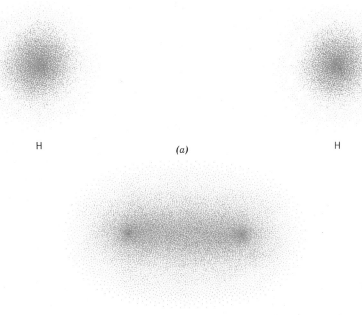

Figure 20.4 Electron clouds (*probability* clouds) for (a) a separated pair of hydrogen atoms and (b) a hydrogen molecule, H_2.

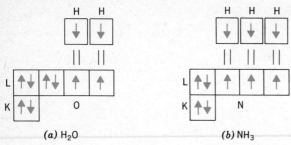

(a) H₂O (b) NH₃

Figure 20.5 Schematic representations of the covalent bonding in (a) water and (b) ammonia. The double lines indicate the bonds formed by the sharing of two electrons.

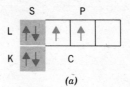 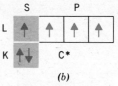

(a) (b)

Figure 20.8 (a) The electronic configuration of the ground state of the carbon atom. Very little energy is required to break the 2S electron pair and promote one of the electrons into the 2P sub-shell where it remains unpaired. In this excited atomic state (b) there are four unpaired electrons, all of which participate in covalent bonding.

Carbon Bonds

The electronic configuration of the carbon atom in its ground state, which is shown in Fig. 19.16 (and reproduced in Fig. 20.8a), indicates that there are two unpaired P electrons in the L shell. We would expect, therefore, that carbon atoms would participate in the formation of molecules by contributing two electrons toward covalent bonds. It is found, however, that carbon almost always appears in molecular structures with *four* equivalent covalent bonds. By the addition of only a small amount of energy (about 2 eV) to the carbon atom ground state, it is possible to break the 2S electron pair and promote one of the electrons into the 2P subshell (Fig. 20.8b). Thus, there are *four* unpaired electrons available for bonding if 2 eV can be supplied to the atom. Actually, the energy is supplied in the bonding process itself because the energy gained by making four covalent bonds, instead of two, more than compensates for the energy expended in breaking the 2S pair. Three of these bonds are made by P electrons and one by an S electron. This type of equivalent four-electron bonding of the carbon atom to other atoms is called SP^3 bonding or *hybridization*.

Carbon enters into an enormous number of chemical compounds via its particular type of covalent bond. In fact, the entire class of substances that we call *organic compounds* contain covalently bonded carbon atoms. One of the simplest compounds that contains carbon is *methane* or natural gas. In the methane molecule, each of the carbon bonds is used to bind a hydrogen atom to the molecule. The orbit representation of a methane molecule is shown in Fig. 20.9. Usually, we simplify this picture to

$$CH_4: \quad H-\underset{H}{\overset{H}{C}}-H$$

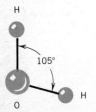

Figure 20.6 Representation of the water (H₂O) molecule. The straight lines indicate covalent bonds.

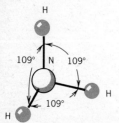

Figure 20.7 Representation of the ammonia (NH₃) molecule.

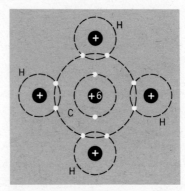

Figure 20.9 Orbit picture of the methane molecule, CH₄. The carbon atom participates in four covalent bonds.

where each of the short lines stands for a covalent bond, that is, a pair of electrons that are being shared.

In the molecule of carbon dioxide, CO_2, the carbon bonds are used in pairs instead of singly as in methane. As indicated schematically in Fig. 20.10, each of the oxygen atoms is joined to the carbon atom by *two* covalent bonds. Each of the curved bonding "sticks" in the diagram represents a pair of shared electrons. We represent these *double bonds* by a pair of short lines:

$$CO_2: \quad O=C=O$$

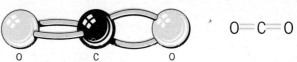

Figure 20.10 In the carbon dioxide molecule, CO_2, each oxygen atom is attached to the carbon atom by *two* covalent bonds.

Molecular Spectroscopy

The structures and functions of biologically significant molecules are important topics in the study of living things. Our knowledge of molecular structures derives primarily from spectral measurements—in the infrared as well as the visible parts of the spectrum. Atomic spectra are due entirely to electronic transitions. Molecular spectra, however, include features that are due to dynamical changes, for example, in the vibrations of the atoms relative to one another within a molecule and in the rotations of the molecule around its center of mass. Vibrations and rotations are more important than electronic transitions in determining the principal features of molecular spectra.

How many different ways can a molecule vibrate? Consider a molecule that consists of N atoms. Each atom can move in three dimensions; therefore, the molecule as a whole has $3N$ *degrees of freedom*. Three of these degrees of freedom correspond to the translation of the molecule through space and three more correspond to the rotations that the molecule can undergo about each axis in three dimensions. Thus, there remain $3N - 6$ degrees of freedom for vibrational motion. Let us examine the case of a simple diatomic molecule. When $N = 2$, the formula $3N - 6$ becomes zero, indicating that there is no vibrational degree of freedom. However, a *linear* molecule, such as a diatomic molecule, is a special case because, as we see in Fig. 1b, there are only *two* rotational degrees of freedom corresponding to rotations about the two axes that are perpendicular to the molecular axis. The atomic nuclei,

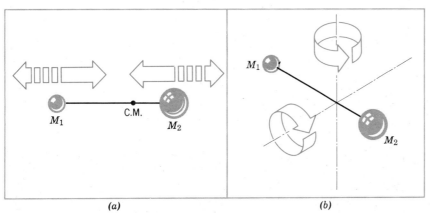

Figure 1 (a) One mode of vibration is available to a diatomic molecule. (b) The two possible rotational modes are indistinguishable and therefore have the same frequency.

M_1 and M_2, are essentially point masses, so there is no significance to rotation about the $M_1 - M_2$ axis. Consequently, there is one degree of freedom available for vibrational motion (Fig. 1a).

In our study of simple harmonic motion (Section 11.2), we found that the frequency of a harmonic oscillator is given by (see Table 11.3)

$$\nu = \frac{1}{2\pi}\sqrt{\frac{k}{M}} \tag{1}$$

We now have a slightly more complicated case in which two atoms vibrate in unison with respect to their common center of mass (Fig. 1a). For this situation the expression for the frequency becomes

$$\nu = \frac{1}{2\pi}\sqrt{k\left(\frac{1}{M_1}+\frac{1}{M_2}\right)} = \frac{1}{2\pi}\sqrt{\frac{k}{\mu}} \tag{2}$$

where

$$\frac{1}{\mu} = \frac{1}{M_1} + \frac{1}{M_2} \tag{3}$$

The quantity μ is called the *reduced mass* of the pair. A typical value of the force constant k for a pair of atoms in a molecule is near 10^3 N/m (see Exercises 1 and 2).

When discussing molecular spectra, it is customary to divide the frequency ν by c and to write $\nu/c = \bar{\nu}$. The quantity $\bar{\nu}$ is actually equal to $1/\lambda$ and is called the *wavenumber*, although it is often referred to simply as the "frequency." ($\bar{\nu}$ is the *number of waves* of the radiation per unit of length, usually, cm^{-1}.) Thus, we have

$$\bar{\nu} = \frac{1}{2\pi c}\sqrt{\frac{k}{\mu}} \tag{4}$$

We have one additional modification to the equations based on classical SHM, that is, the quantization of the energy. The vibrational energy of a quantum system is

$$\mathcal{E}_{vib} = h\nu = h\bar{\nu}c = \frac{h}{2\pi}\sqrt{\frac{k}{\mu}}(v+\tfrac{1}{2}) \tag{5}$$

from which

$$\bar{\nu} = \frac{1}{2\pi c}\sqrt{\frac{k}{\mu}}(v+\tfrac{1}{2}) \tag{6}$$

where v is the *vibrational quantum number* for the system. All integer values for v are allowed: $v = 0, 1, 2, 3, \ldots$. Another result of the quantum calculation is that in any vibrational transition, the quantum number v must change by exactly one unit. Thus, the *selection rule* for vibrators is $\Delta v = \pm 1$. Notice that the molecule has a residual energy even when it is in the lowest vibrational state. (This *zero-point energy* is analogous to that of a particle confined to a box—see Section 18.5.)

Equation 5 indicates that the energy levels of a quantum oscillator are equally spaced. Actually, a real molecule does not undergo true SHM. Instead, there are deviations that tend to make the level spacing decrease with increasing energy. However, this effect is not large if we confine our attention to small values of v.

Problem 20.2 concerns an elementary view of molecular rotation. Taking the result from this calculation and framing it in proper quantum notation, we can express the rotational energy of a diatomic molecule as

$$\mathcal{E}_{\rm rot} = \frac{\hbar^2}{2\mu d^2} J(J+1) \tag{7}$$

where d is the distance between the atomic nuclei. The quantity J is the *rotational quantum number*. (For other than simple diatomic molecules, the denominator of the expression for $\mathcal{E}_{\rm rot}$ becomes complicated.) The selection rule for J is $\Delta J = \pm 1$; transitions with $\Delta J = 0$ are allowed for certain types of molecules.

When a molecule undergoes a dynamical transition, it is usually the case that both the vibrational quantum number and the rotational quantum number will change. That is, the vibrations and the rotations of a molecule are coupled and we refer to a *vibration-rotation spectrum*. The energies of these transitions are only small fractions of an electronvolt and so they occur in the infrared part of the spectrum.

Vibration-rotation spectra are usually studied by a technique called *infrared absorption*, in which the sample is irradiated by a monochromatic beam of IR radiation and the transmitted intensity is measured as the frequency of the beam is changed. Sudden dips in the transmitted intensity correspond to absorption lines in the sample molecules.

Figure 2 shows the beautifully developed spectrum of methane gas in the frequency (wave number) range near 3000 cm^{-1} (corresponding to $\lambda = 3.3$ μm). The transitions in this spectrum are identified in the schematic energy level diagram in Fig. 3. The principal transition Q is that from the lowest member of the rotational band of states based on the $v = 0$ vibrational level to the lowest rotational state based on $v = 1$. All other transitions in the spectrum have $\Delta J = \pm 1$. The P branch corresponds to those transitions with $\Delta J = -1$; these transitions all have frequencies *lower* than that of Q. The R branch corresponds to those transitions with $\Delta J = +1$; these transitions all have frequencies *greater* than that of Q (see Fig. 2). Analysis of the spectral frequencies and frequency differences provides information about the distribution of mass within the molecule and about the strengths of the interatomic bonds.

The type of spectral analysis we have been discussing is applicable to relatively small molecules in the vapor phase. The simplest amino acid, glycine, consists of 10 atoms and therefore has 24 vibrational degrees of freedom; this is too large a num-

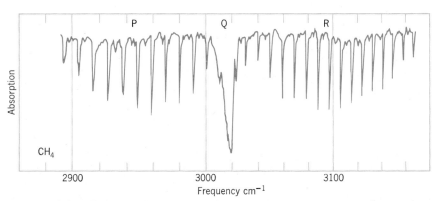

Figure 2   Infrared absorption spectrum of methane gas near $\bar{\nu} = 3000$ cm^{-1}. [From N. B. Colthup, L. H. Daly, and S. E. Wiberley, *Introduction to Infrared and Raman Spectroscopy*; Academic, New York, 1964, p. 25.]

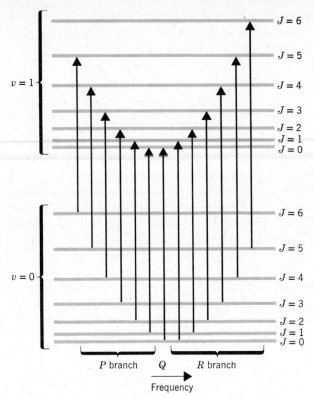

Figure 3 Portion of a molecular vibration-rotation energy level diagram showing the transitions that occur in the absorption spectrum of Fig. 2.

ber for a complete spectral analysis to succeed. Moreover, if the density of the molecules is high, as in the liquid or solid phases, the energy level structure of any particular molecule is perturbed by the electric fields of the neighbor molecules. Because of this effect, the rotational lines tend to be "smeared out" with the result that a group of lines such as that shown in Fig. 2 becomes a single broad absorption *band*. Therefore, IR absorption analysis of complex molecules takes a direction different from that used for small molecules in the vapor phase. Instead of attempting to locate the vibrations corresponding to each degree of freedom, the analysis centers on identifying the absorption bands that are due to the vibrations of particular types of interatomic *bonds*. The information needed to carry out this kind of analysis is obtained by studying the absorption spectra of simple molecules which contain only a limited number of bond types. For example, the absorption properties of the double carbon bond can be obtained by examining the absorption spectrum of ethylene, $H_2C{=}CH_2$. The table below lists the positions of the absorption bands for a few of the important types of bonds found in biological molecules. Notice that a definite frequency cannot be given for a particular bond because the value depends somewhat on the electric fields produced by the other parts of the molecule.

Because of the small mass of the hydrogen atom, the reduced mass of hydrogen plus any other type of atom in an organic molecule is always small. Thus, the vibration frequency for any hydrogen bond is always high (see Eq. 6). All fundamental vibrations that exhibit absorption bands with frequencies above 2500 cm^{-1} can be assigned with assurance to the stretching of a hydrogen bond between a hydrogen and another atom. The more massive the other atom, the higher will be the frequency; the table shows that the OH bond has the highest frequency, followed by the NH and then the CH bonds.

Bond	Absorption band (cm^{-1})
—O—H	3600–3700
$\geq$N—H	3300–3400
$\equiv$C—H	3260–3340 and 600–700
$=$C—H	3000–3100
$\geq$C—H	2850–3000 and 1350–1500
—C$\equiv$C—	2100–2140
$>$C$=$C$<$	1630–1690
$\geq$C—C$\leq$	900–1000

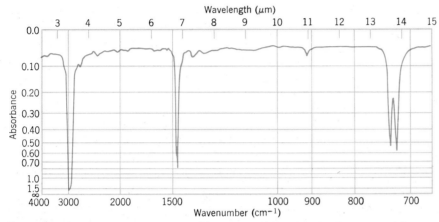

Figure 4 Infrared absorption spectrum for polyethylene. The *absorbance* is equal to $\log(I_0/I)$, where I_0 is the intensity of the incident radiation and I is the intensity of the transmitted radiation. [From R. P. Bauman, *Absorption Spectroscopy;* Wiley, New York, 1962, p. 337.]

Figure 4 shows the IR absorption spectrum for polyethylene from 4000 cm^{-1} to about 700 cm^{-1}. Polyethylene is a particularly simple organic compound whose molecule is a repeating chain of CH$_2$ groups:

The spectrum of this molecule is therefore relatively uncomplicated. Looking at the table of absorption frequencies, we can identify the very strong band near 3000 cm^{-1} and the slightly weaker band near 1460 cm^{-1} as due to two different modes of vibration of the CH bond. The band near 750 cm^{-1} is due to a CH$_2$ vibration (not listed in the table). The very weak absorption dip at 920 cm^{-1} is due to the C—C bond.

It is clear that the absorption spectra for complex biological molecules will be much more complicated than the polyethylene spectrum. Nevertheless, an enormous amount of progress has been made in analyzing the structures of large organic molecules by these methods. For example, the results of infrared absorption measurements represent some of the strongest experimental evidence for the existence of hydrogen bonds in protein and other large molecules. (A *hydrogen bond* is a bond between atoms due to the exposed positive charge of the hydrogen nucleus when the hydrogen electron is pulled toward another atom to provide a covalent bond. The positive charge of the hydrogen atom is then attracted to and forms a bond with the exposed negative charge of another polar molecule or submolecular group of atoms. Hydrogen bonds play a key role in holding together the two strands of DNA molecules.)

References

R. P. Bauman, *Absorption Spectroscopy,* Wiley, New York, 1962

C. E. Meloan, *Elementary Infrared Spectroscopy,* Macmillan, New York, 1963

R. E. Setlow and E. C. Pollard, *Molecular Biophysics,* Addison-Wesley, Reading, Mass., 1962, Chapter 7.

■*Exercises*

1 What are the force constants k for the C≡C, C=C, and C—C bonds? (Refer to the table.)

2 The fundamental vibration mode of the nitrogen molecule has a frequency of 2360 cm^{-1}. What is k for this molecule?

3 Carbon dioxide is a *linear* molecule, O=C=O. How many vibrational modes does CO_2 have? (First, how many *rotational* modes are there? Two of the vibrational modes have the same features but are turned 90° with respect to one another.) Sketch the various vibrational modes. The frequencies of the vibrations are 2349 cm^{-1}, 1388 cm^{-1}, and 667 cm^{-1}. Assign each frequency to a particular mode. Justify your assignment.

4 What is the amplitude of the fundamental vibration of the C—C bond? (Refer to Exercise 1 for the value of k. Write down the maximum potential energy in terms of the amplitude and equate to the energy of the $v = 0$ to $v = 1$ transition.)

5 In order to study complex molecules, we want to have the simplest spectra possible. This is why we use *absorption* spectra instead of *emission* spectra. Explain.

20.2 Crystals

Ionic Crystals

All the compounds formed from one Group I atom and one Group VII atom (such as NaCl, LiF, etc.), as well as many other two-atom compounds, arrange themselves into cubic structures when a bulk solid is formed. The tiny cubic crystals of common salt are well known. The atomic reason for this behavior was first proposed by William Barlow who, more than 80 years ago, visualized NaCl as consisting of a tightly packed cubic array of ball-like atoms (Fig. 20.11). Barlow's picture of the NaCl crystal structure was remarkably accurate. Modern methods of analyzing crystal structure, utilizing X-ray diffraction techniques, have shown that NaCl in solid form indeed consists of a cubic lattice of ions (Fig. 20.12). This type of X-ray crystalanalysis has also been used to

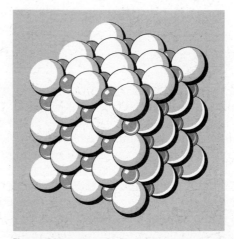

Figure 20.11 This arrangement of the sodium and chlorine atoms in solid NaCl was proposed by William Barlow in 1898. Barlow's scheme, which even shows the sodium ions to be smaller than the chlorine ions, has been proved correct by modern X-ray diffraction techniques.

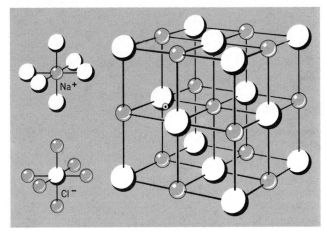

Figure 20.12 Solid sodium chloride is a cubic structure in which each sodium ion is immediately surrounded by 6 chlorine ions and each chlorine ion is immediately surrounded by 6 sodium ions. (The ions are shown smaller than their actual sizes relative to the cubic structure of the crystal in order to reveal more clearly the lattice structure.)

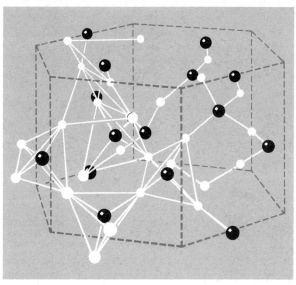

Figure 20.14 The hexagonal structure of the quartz crystal. The black spheres represent silicon atoms and the white spheres represent oxygen atoms. (Adapted from G. H. Wannier, Scientific American, December 1952.)

determine precisely the spacing between crystalline atoms; for NaCl the length of the side of a unit cube is 2.8 Å, slightly greater than the 2.4-Å separation of the ions in an isolated pair.

A *crystal* of sodium chloride is a much more stable configuration of the ions than a simple isolated pair. The extra stability, which amounts to 16.5 eV per ion pair, results from the fact that each Na^+ ion is surrounded by 6 Cl^- ions and *vice versa* (Fig. 20.12). The electrostatic energy of an ion in the field of its 6 neighbors provides the additional binding energy.

A crystal is any stable configuration of atoms in a regular, repeating array. In addition to the cubic pattern of the sodium chloride crystal, there are thirteen other basic crystal structures. For example, quartz (SiO_2) occurs as a *hexagonal* crystal (Figs. 20.13 and 20.14).

Covalent-bonded Crystals

Elemental carbon forms two different types of crystal structures by utilizing its covalent bonds in different ways. In the graphite form, the carbon atoms are arranged in planes of interconnecting hexagons with alternating single and double bonds (Fig. 20.15a). In diamond, the basic unit contains only 4 carbon atoms with all inter-atom connections made with single bonds (Fig. 20.15b). The atoms of graphite, illustrated in Fig. 20.16, lie in a series of stacked

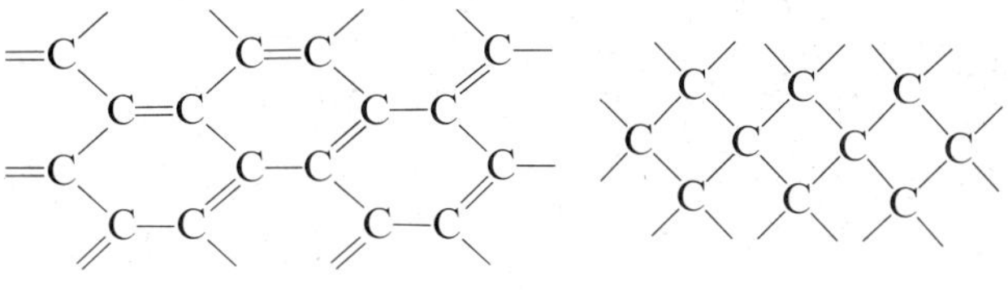

Figure 20.13 Quartz crystals.

Figure 20.15 The crystal structures of graphite and diamond differ in that double bonds are present in graphite and the basic crystal unit contains a ring of 6 atoms whereas the diamond crystal contains only single-bonded atoms arranged in groups of 4 atoms. The carbon atoms in graphite lie in planes, but in diamond they form a very stable three-dimensional structure (Figs. 20.16 and 20.17).

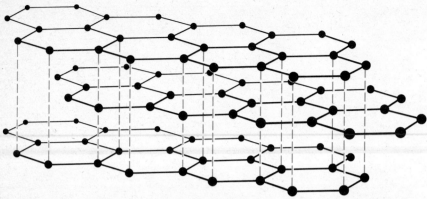

Figure 20.16 In the graphite form of carbon, the atoms are joined together to form planes. The binding between planes (dashed lines) is weak, so that the planes slip easily over one another.

planes. Because there are only weak forces between adjacent planes, it is quite easy to cleave graphite; thin strips can be removed from bulk graphite by using a sharp instrument such as a razor blade. The diamond crystal structure, on the other hand, is not planar. The four atoms to which any given atom is bound do not all lie in the same plane. This means that there is a strong connection between the atoms in adjacent planes, as shown in Fig. 20.17. Diamond is therefore a much more rigid structure than graphite (but diamond crystals can be cleaved along planes that define the sides of the unit cubes).

Metallic Crystals

Metals form still another type of crystal structure. It is a characteristic feature of metals that, in the bulk form, the electric fields in which the atoms find themselves are such that the outer electrons are no longer bound to particular atoms; these electrons are free to move throughout the material. (These are the *conduction electrons* to which we referred in Section 12.6 and which we will discuss further in the next section.) The free electrons constitute a kind of "sea" of negative electricity in which the positively charged metallic ions are bound, as illustrated schematically in Fig. 20.18. In ionic crystals and in crystals in which the atoms are bound by covalent bonds, each electron is associated with a particular atom or pair of atoms; there are no free electrons. Therefore, crystals such as NaCl or diamond are not good conductors of electricity. Metals, with free electrons in the inter-atomic spaces, are good conductors.

Metallic binding usually occurs only for atoms that have a small number of electrons in the outer shell. If there are too many electrons contributed by each atom, the exclusion principle forces some of these electrons into higher energy states and then the attractive forces are insufficient to cause metallic binding. Thus, the elements in Groups I and II of the periodic table, the transition elements, and some elements in Groups III and IV, form metallic crystals.

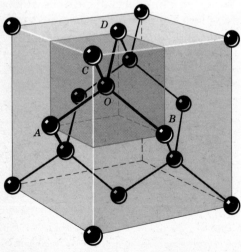

Figure 20.17 The three-dimensional structure of the diamond crystal. The 4 atoms (A, B, C, D) to which the atom O is connected do not all lie in the same plane; indeed, they are located at the diagonal corners of the unit cubic structure of the crystal. Therefore, there is a strong coupling between adjacent planes of atoms.

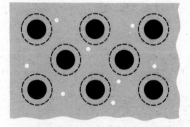

Figure 20.18 A typical metallic crystal. The outer electrons are detached from the metal atoms (the outer electron shell is indicated as empty in the diagram) and are free to move throughout the crystal.

20.3 Theory of Solids

Energy Bands in Solids

In the preceding chapter we discussed the behavior of electrons in *isolated* atoms. We shall now investigate how the behavior of these electrons is influenced and altered when the atoms are in a crystal of the bulk material.

Figure 20.19 shows the electrostatic potential in the vicinity of an isolated atom of lithium. The horizontal line labeled 0 indicates zero potential energy and the two lower lines indicate the 1S and 2S energy states. Two electrons are located in the 1S state and one in the 2S state. The 2S electron is bound by 5.4 eV; that is, the ionization of energy of an isolated lithium atom is 5.4 eV.

When lithium atoms are brought together to form a crystal, the net electrostatic field at any point in the crystal is the sum of all of the individual fields. As a result, the potential between the atoms never rises to zero potential. In fact, the potential at a particular point between atoms in a crystal is reduced substantially below the potential at the same distance from an isolated atom (Fig. 20.20). The reduction is so pronounced in the crystal that the 2S electron, which was *bound* in the isolated atom, no longer encounters a potential barrier that is sufficient to constrain it to the vicinity of any particular atom. The 2S electrons in a lithium crystal are *free* electrons or *conduction* electrons.

The conduction electrons "belong," not to individual atoms, but to the crystal as a whole. That is, the 2S electron wave functions are not localized but extend throughout the crystal. The 2S energy state of the isolated lithium atom becomes an energy "state" of the crystal. If there are N atoms in the crystal, there are N electrons in this "state." But we

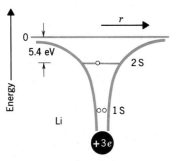

Figure 20.19 Schematic potential energy diagram for an isolated lithium atom. The outermost electron (i.e., the 2S electron) is bound by 5.4 eV; it requires about 75 eV to remove one of the 1S electrons and an additional 120 eV to remove the second 1S electron.

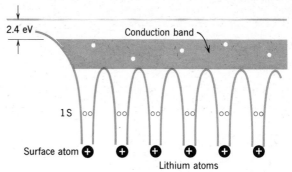

Figure 20.20 When lithium atoms are assembled in a crystal, the reduction of the potential between the atoms frees the 2S electrons which then partially fill the conduction band. (The energies are not to scale; a 1S electron is actually bound by 75 eV.) The electrons in the conduction band, shown schematically in the diagram, are not really localized; the electron wave functions extend throughout the crystal.

know that the exclusion principle does not permit more than two electrons to exist in any single energy state. Therefore, the atomic 2S state must expand into a series of closely spaced crystal states, each of which can accommodate two electrons. The spacing between these crystal states is so small[1] that the distribution of states is essentially continuous. As a result, the discrete atomic state becomes, in the crystal, an energy *band* (see Fig. 20.20). This band is the *conduction band* and the electrons that exist in this band are the *conduction electrons*.

The interesting physical properties of an element or compound in bulk form (for example, electrical resistance, thermal conductivity, magnetic properties) are determined in large measure by the details of the energy band structure. We shall investigate next how the crystal energy bands determine the electrical conductivity of the material.

Band Theory of Conductors and Insulators

In general, all the energy states of an atom appear as bands in a crystal, including all of those higher states that are empty when the atom is in its ground state. Thus, in a lithium crystal there are (unfilled) bands corresponding to the 2P, 3S, 3P, 3D, etc. states. The discrete atomic states can appear in the crystal as identifiable bands (Fig. 20.21b) or as overlapping bands in which it is no longer possible to specify from which subshell the electrons originated

---

[1]These states are spread over an energy of a few eV, and a total of $\frac{1}{2}N$ states are required to accommodate N electrons. For a 1-cm³ crystal, N is of the order of 10^{22}, so the spacing between the states is of the order of 10^{-22} eV.

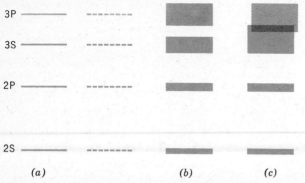

Figure 20.21 The discrete atomic energy states (a) correspond, in a crystal, to either individual bonds (b) or to overlapping bands (c). (The 1S state always lies at a much lower energy than the other states and so is not shown in this diagram.)

(Fig. 20.21c). Overlapping always occurs for the highest energy bands but the lower states usually remain as individual bands in the crystal. The details of the crystal lattice structure determine the excitation energy at which the overlapping begins.

The filling of the energy bands by electrons follows the same prescription as does the filling of atomic energy states. The 2S atomic state, for example, can accommodate two electrons and in a crystal consisting of N atoms, the 2S band can contain 2N electrons; the 3P band (if distinct from the 3S and 3D bands) can contain 6N electrons; and so forth. If a band is completely filled, no electron in this band can be given any additional energy unless it is given a sufficient amount to raise it to an unoccupied state in a higher band (Fig. 20.22a, b). Depending on the positions of the various bands, the amount of energy required to raise an electron from one band to another may be 5–10 eV. On the other hand, if the highest energy band that contains any electrons is only partially filled (e.g., the 2S band in lithium which contains N electrons and thus is half filled), there is an extremely large number of energy states within the band that are accessible to these electrons. Thus, an electron in a partially filled band can be given essentially *any* amount of additional energy as long as the total is less than the maximum energy allowed for the band (Fig. 20.22c).

If the highest occupied band is only partially filled, the electrons in this band can be made to drift in a particular direction by the application of an external electric field. The increase in energy brought about by this motion (because it is small) can be accommodated by the available energy states within the band. Materials that have partially filled bands can therefore conduct electricity and are called *conductors* (Fig. 20.23a).

If the highest occupied band is completely filled, increases in the energies of the electrons in that band are not allowed; then, an electric field will not result in the flow of electrons. Such materials resist the flow of electricity and are called *insulators* (Fig. 20.23b).

In most insulators it requires an appreciable increase in energy (5–10 eV) to raise an electron from the filled band across the forbidden energy region and into the empty conduction band. For example, in the diamond form of carbon there is a 5-eV energy gap between the filled $2P_{1/2}$ band and the empty $2P_{3/2}$ band[2] (Fig. 20.24). What electric field strength is necessary to raise an electron from the $2P_{1/2}$ band into the $2P_{3/2}$ band? In order to answer this question we must first realize that no *real* crystal is as perfect as the ideal crystal structures we have been discussing. There are always small amounts of impurities present in real crystals and there are always small imperfections in the lattice structure. These departures from the ideal crystal form prevent the electrons from moving unimpeded

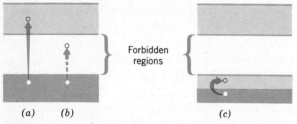

Figure 20.22 (a) An allowed transition of an electron from a filled band into an empty band; this type of transition usually requires 5–10 eV. (b) The transition of an electron from a filled band into the forbidden region is not allowed. (c) In a partially filled band an electron can make a transition into any unoccupied state that lies within the band; such transitions ordinarily involve only very small amounts of energy.

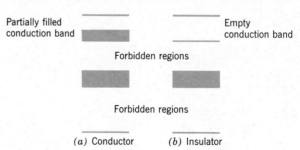

Figure 20.23 A conductor (a) is characterized by a partially filled band and an insulator (b) by a completely filled band above which is a forbidden region and, still higher, an empty band.

---

[2] In most materials that exhibit covalent bonding, the atomic fine-structure splitting (for example, the splitting of the 2P state into the $2P_{3/2}$ and $2P_{1/2}$ states) carries over into the bands of the solid.

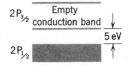

Figure 20.24 The energy bands of carbon in the form of diamond (not to scale). Because of the relatively large energy gap between the filled $2P_{1/2}$ band and the empty $2P_{3/2}$ band, diamond is a good insulator.

through the crystal. In fact, in even the purest crystals that have been made, an electron can travel only about 10^{-8} m before encountering one of these imperfections and being scattered with a consequent loss of kinetic energy. (The electron kinetic energy is converted into motional energy of the lattice, that is, into *heat*. This is the reason why all ordinary materials suffer a rise in temperature when they conduct an electric current.) Therefore, in order to gain 5 eV of kinetic energy in a distance of 10^{-8} m, an electron must be accelerated by a field of 5×10^8 V/m! This field is far greater than that which will cause an equivalent current to flow in metallic crystals of Li, Na, K, etc.[3] Diamond is therefore an extremely good insulator. Similar energy gaps in crystals such as quartz (SiO_2) and in materials such as Mylar and Teflon make these substances good insulators.

Semiconductors

Some materials that satisfy the requirement for classification as insulators, namely, a completely filled upper band, actually have a very small energy gap (an eV or less) separating the filled band from an empty conduction band. At low temperatures, essentially none of the electrons in these materials will have sufficient thermal energies to enable them to cross the energy gap into the conduction band. As the temperature is increased, however, more and more of the electrons, by virtue of thermal agitation, are found in the conduction band. Therefore, these materials—silicon and germanium are typical

members of this class—exhibit a weak electrical conductivity and are called *semiconductors*. This conductivity increases markedly with temperature; between 250 K and 450 K, the number of conduction electrons in silicon, for example, increases by a factor of 10^6. Of course, the degree to which a semiconductor material will conduct electricity depends on the magnitude of the energy gap; the smaller the gap, the greater the number of thermally excited electrons and, hence, the greater the electric conductivity. Some semiconductor energy gaps are compared with those of insulators in Table 20.1.

One of the methods of increasing the conductivity of a semiconductor is to illuminate the material with light. The absorption of the light raises some of the electrons into the conduction band and thereby increase the conductivity. Materials that exhibit this property are called *photoconductors*. Germanium, for example, shows a greatly increased conductivity when illuminated by photons of energy greater than 0.7 eV.

If it is desired to endow a semiconductor with increased conduction properties that are permanent, one cannot rely on the photoconduction process since this phenomenon is transient and disappears when the light source is removed. Nor is it usually convenient to operate the material at an elevated temperature so that thermal excitation will provide greater conductivity. An extremely effective way to change permanently the conduction properties of semiconductors is by the introduction of minute quantities of certain impurities into the crystal structure. These impurities are called *doping agents* or *dopants* and the doping of semiconductors such as silicon and germanium is now an essential part of the process for manufacturing transistors.

The semiconductor germanium is a Group IV element and has 4 outer or valence electrons. Germanium forms a diamondlike crystal structure in which each atom is connected to 4 other atoms by covalent bonds (Fig. 20.25a). Now, arsenic is a Group V element with 5 valence electrons and is adjacent to germanium in the periodic table. If an atom

---

[3]Even in an ideally pure metallic crystal there is some resistance to the flow of electrons because the lattice structure has vibrational motion (heat energy) that is characteristic of the temperature. Moving electrons can therefore lose energy in collisions with these vibrating atoms. At extremely low temperatures and for certain materials, this type of energy loss disappears and the material becomes *superconducting*. This phenomenon is discussed in the following section.

Table 20.1 Energy Gaps for Some Semiconductors and Insulators

Semiconductor	Energy Gap (eV)	Insulator	Energy Gap (eV)
Silicon	1.14	Diamond	5.33
Germanium	0.67	Zinc oxide	3.2
Tellurium	0.33	Silver chloride	3.2
Indium antimonide	0.23	Cadmium sulfide	2.42

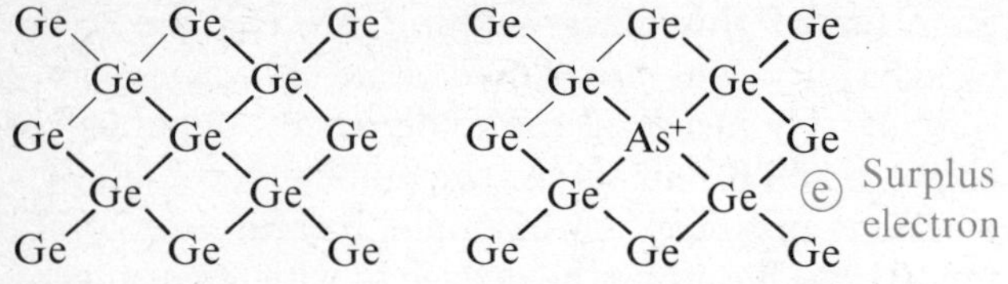

Ge   Ge   Ge   Ge   Ge   Ge
  Ge   Ge   Ge   Ge
Ge   Ge   Ge   Ge   As⁺   Ge
  Ge   Ge   Ge   Ge   ⓔ Surplus
Ge   Ge   Ge   Ge   Ge   Ge   electron
    (a)   (b)

Figure 20.25 (a) The diamondlike crystal structure of germanium. (b) Replacing one of the germanium atoms in the crystal by an atom of arsenic produces an arsenic ion (As⁺) and a surplus electron which is easily excited into the conduction band.

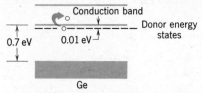

Figure 20.26 The energy states (donor states) contributed by the arsenic impurity atoms in a germanium crystal lie just below the conduction band. Electrons in these energy states can easily make transitions into the conduction band.

of arsenic replaces a germanium atom in a crystal, 4 of its valence electrons are used to duplicate the germanium bonds and there is one surplus electron. Thus, the arsenic atom in the crystal is actually an As⁺ ion and the extra electron remains loosely bound in the vicinity of the arsenic ion (Fig. 20.25b). The binding energy of the surplus electron to the As⁺ ion is extremely small, however—about 0.01 eV, an energy considerably smaller than the 0.7-eV gap energy in a pure germanium crystal. Therefore, the addition of only a small fraction of an eV of energy is sufficient to raise this electron into the conduction band. In fact, at room temperature almost all the surplus electrons contributed by the arsenic impurities are in the conduction band, whereas only a few of the germanium electrons have sufficient thermal energies to make the transition. Therefore, the presence of arsenic impurity atoms, even though their concentration may be only 1 part in 10^6 or 10^7, is the determining factor for conductivity. (In a 1-cm³ sample of arsenic-doped germanium there will be about 10^{16} *conduction* electrons contributed by the arsenic—this is quite a sufficient number to permit substantial conduction currents to flow.)

The energy states associated with the loosely bound electrons contributed by the arsenic ions appear in the germanium energy diagram as a set of additional states lying immediately below the lower boundary of the conduction band (see Fig. 20.26). Because the electrons that populate these states have been "donated" by the arsenic atoms, the states are called *donor states*. The semiconductor that results from doping by a donor-type impurity is called an *n*-type semiconductor.[4] Other Group V elements such as phosphorus or antimony can also be used to convert pure germanium into an *n*-type semiconductor.

If, instead of a Group V element, a Group III element, such as gallium, is introduced as an impurity into a germanium crystal, the effects are analogous. The difference in the two cases is that a gallium atom with three valence electrons, lacks one electron to effect the 4 double bonds that bind it to the 4 neighboring germanium atoms. In order to compensate for this deficiency, the gallium atom "steals" an electron from a germanium atom, thus producing a gallium ion (Ga⁻) and a germanium ion (Ge⁺). The germanium ion that lacks an electron is called an *electron hole*. The Ge⁺ ion can then "steal" an electron from another germanium atom; thus, in effect the hole moves from one position to another. This is illustrated schematically in Fig. 20.27.

Because the gallium impurity "accepts" electrons from the germanium atoms, these impurity atoms are called *acceptors*. The energy states associated with acceptor atoms lie very close to the upper energy of the filled band (Fig. 20.28). That is, it requires very little energy for the acceptor atom to "steal" an electron from a neighboring germanium atom and thereby create a hole.

Other Group III elements that serve as acceptors in germanium crystals are boron, aluminum, and indium. A semiconductor that is doped with acceptor atoms is called a *p*-type semiconductor.[5]

Semiconductor Devices

The great usefulness of semiconductor materials is the result of the interesting effects that occur when *n*-type and *p*-type materials are placed in contact. Figure 20.29a shows the two types of material with their charge carriers when separated. (Remember, the materials are electrically neutral as a whole; the *plus* and *minus* signs indicate only the charge *carriers*.) When the two types of material are placed in contact,[6] thereby forming a *p-n junction* (Fig.

---

[4]So called because the result is the contribution of mobile *negatively* charged carries of electricity.

[5]So called because the result is the contribution of mobile *positively* charged carriers of electricity (i.e., *holes*).

[6]One cannot simply press together the two materials because the irregularities of the crystal structures at the surfaces would prevent the desired effects from taking place. Instead, the dopants are introduced in the proper places while the crystal is being grown or are diffused from opposite sides into a grown crystal.

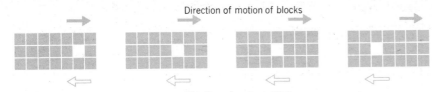

Figure 20.27 An array of blocks illustrates the movement of an electron hole in a crystal. As each block (or electron) moves to the right to fill the hole, the hole itself moves to the left.

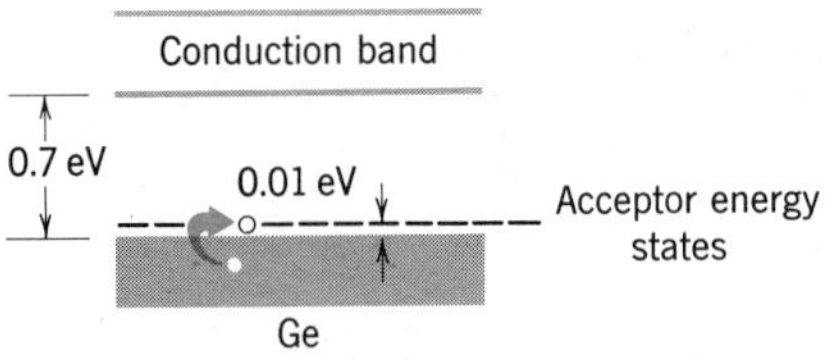

Figure 20.28 The energy states associated with acceptor atoms lie only slightly above the filled band. Thermally excited electrons can populate these states and leave holes in the filled band.

20.29b), some of the carriers of each sign diffuse across the boundary. This migration of electrons into the p-type material and holes into the n-type material builds up a positive charge in the n region and a negative charge in the p region. The amount of charge that can be accumulated in each region is limited by the fact that the free electrons tend to fill the holes, thereby eliminating both charge carriers.

This recombination process does not really diminish the number of electrons and holes because the thermal agitation in the material continually produces new electrons and holes. At equilibrium, the rate of thermal excitation is equal to the rate of recombination. The net result is that there exists a constant potential difference across the p-n junction.

Next, suppose that we connect a p-n junction to the terminals of a battery. How will the charge accumulation due to diffusion affect the flow of current through the crystal? Figure 20.30a shows what happens when the positive terminal of the battery is connected to the p-type material and the negative terminal is connected to the n-type material. The electrons are driven to the left (toward the positive terminal) and the holes are driven to the right (toward the negative terminal). The increased rate at which the charge carriers now move across the boundary means that the rate of recombination is

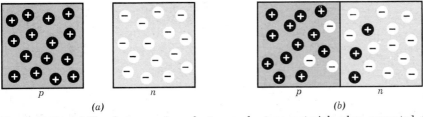

Figure 20.29 (a) The charge carriers of p-type and n-type materials when separated. (b) When the two materials are in contact, some of the charge carriers diffuse across the boundary into the other material.

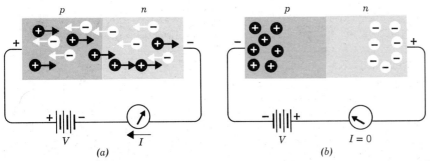

Figure 20.30 (a) With a battery connected to a p-n junction in this way, a current will flow through the crystal. (b) With the battery reversed, no current flows. A p-n junction will allow current to flow only in one direction.

much greater than in the no-voltage equilibrium condition. To offset these losses, new electrons and holes must be produced in both parts of the junction. The disappearance of electrons due to recombination in the *n* region is compensated by the inflow of electrons from the battery. Similarly, the holes that are lost in the *p* region are replaced by the electrons leaving the region on their way to the positive terminal of the battery.

If we now reverse the battery terminals and apply the opposite polarity to the junction, we find that the effect is quite different (Fig. 20.30b). In this case, the holes are drawn toward the negative terminal and the electrons are drawn toward the positive terminal. The region near the boundary is depleted of charge carriers and any new carriers that are created will not diffuse across the boundary because the applied voltage pulls the carriers in the opposite direction. Consequently, there is no current flow when the negative terminal of the battery is connected to the *p*-type material and the positive terminal is connected to the *n*-type material. We see therefore that a *p-n* junction is, in fact, a *diode* and conducts current only when a positive potential is placed on the *p* terminal. Semiconductor diodes are routinely used in rectifier circuits and in other applications that require unidirectional current flow (see Section 14.5).

A *transistor* is a device that consists of three semiconductor elements, often in the order *n-p-n* but for some applications in the order *p-n-p* (Fig. 20.31). These devices are used in electronic circuits to amplify and control electrical signals. The center (*p*) element of an *n-p-n* transistor is called the *base,* one of the *n* elements is the *emitter,* and the other is the *collector.*

Figure 20.32 shows a typical application of an *n-p-n* transistor. This type of circuit amplifies the input signal V_{in} and is called a *common-emitter amplifier* (because the emitter terminal is connected to both the input and output sections of the circuit). When a positive potential is applied to the base, the base-emitter combination acts as a *p-n* diode (Fig. 20.30a) and current flows through this part of the circuit from the base (*p*) to the emitter (*n*). This current depends strongly on the applied voltage; an increase of V_{in} from 0.15 V to 0.25 V, for example, will result typically in a 5- to 10-fold increase in the base current I_b.

In the base-collector part of the circuit, the posi-

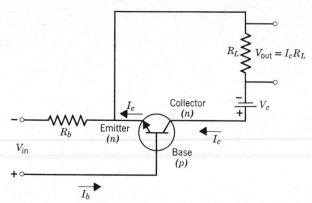

Figure 20.32 Schematic diagram of an *n-p-n* transistor amplifier circuit.

tive terminal of the battery V_c is connected to the collector. Because V_c is large compared to V_{in} (typically, $V_c = 9$ V whereas V_{in} is a volt or less), the base-collector section acts as a reverse-connected diode (Fig. 20.30b). Ordinarily, we would conclude that the collector current I_c would be small. But many electrons are being injected into the base by the emitter. (Remember, the direction of electron flow is opposite to that of the current.) These electrons readily diffuse across the narrow base region without being lost due to recombination with the holes in the *p*-type material of the base. That is, the electrons pass from the emitter to the collector and then into the external circuit; these electrons constitute the collector current I_c. In a typical transistor, the collector current will be several milliamperes when the base current is 0.1 mA. Thus, there is current amplification and, if the load resistor R_L is larger than R_b, there will also be voltage and power amplification.

The development of semiconductor devices, begun in 1949,[7] has completely revolutionized the electronics industry. The communications and computer industries now rely almost completely on semiconductor components. Recently, a tremendous advance has been made in the miniaturization of electronic devices through the introduction of *integrated circuits.* These tiny modules (called *IC chips*) are actually complete electronic subassemblies, consisting of large numbers of circuit elements (such as diodes, transistors, resistors, and capacitors) placed together on a single wafer of semiconductor material. Two such IC chips are shown in Fig. 20.33. The development of integrated circuits has produced not only savings in space and cost but has also increased the reliability of circuits and made the construction

(a)   (b)

Figure 20.31 (a) An *n-p-n* transistor. (b) A *p-n-p* transistor.

---

[7]John Bardeen, Walter Brattain, and William Shockley shared the 1956 Nobel Prize for the discovery and the development of the transistor.

Figure 20.33 Two modern integrated circuit chips, compared with a length of 22-gauge insulated wire, a size commonly used for wiring present-day computers. Each chip is 1.8 mm on an edge and contains 55 complete circuits utilizing 213 transistors. One of these units serves as a complete logic element in a high-speed computer. The etching of the semiconductor wafers to form the circuits is accomplished by using an electron microbeam. Even more sophisticated integrated circuits have been made, with densities of about 200,000 transistors per square inch.

of devices considerably easier. (The reduction of circuit size is clearly of importance in devices intended for use in spacecraft where small size and mass are highly desirable. But small size is of *crucial* importance in the construction of super high-speed computers where the limitation on speed is set by how fast it is possible to transfer a signal from one point to another; because the maximum signal transfer velocity is just c, small size means faster computation.)

20.4 Low Temperature Phenomena

Producing Low Temperatures

Most devices that provide low temperatures (including the household refrigerator) operate by driving a liquefiable gas through a cycle of compressions and expansions. If we do work on a gas by compressing it, the gas temperature rises. The additional thermal energy that is given to the gas can be removed by allowing it to increase the temperature of its surroundings. (In a common refrigerator, circulating air takes up the heat.) We now have a gas at high pressure but at (or near) room temperature. If we next expand the gas rapidly the temperature decreases and, if the decrease is sufficient, the gas will liquefy. High pressure steam, for example, if released from a boiler, will condense into water by virtue of the expansion process. A system whose function is to provide low temperatures to other objects does so by bringing the liquefied gas into thermal contact with the objects; the gas is then recycled through the system. On the other hand, if the purpose is to produce liquid gas for other uses, arrangements are made to remove the liquid gas from the system.

Compression-expansion systems can be used to liquefy all gases. Of course, if the gas becomes liquid only at a very low temperature (as do hydrogen and helium; see Table 20.2), then a particularly elaborate system is required. Helium (which becomes a liquid at 4.2 K) was first liquefied in 1908 by Kammerlingh Onnes of Leiden.

Once temperatures as low as those of liquid helium are available, it becomes possible to use other, less efficient methods to reduce the temperature still further. It is because helium remains in liquid form and does not solidify[8] (at atmospheric pressure) that extremely low temperatures (down to about 10^{-3} K) can be attained with relative ease.

Liquid Helium— the Superfluid

The liquid form of helium is a peculiar substance. In fact, liquid helium exists in *two* completely different forms. In the temperature range from 4.2 K down to

Table 20.2 Liquefaction (Boiling) Points of Some Gases at Atmospheric Pressure

Gas	Temperature, K
Chlorine	238.6
Xenon	166.1
Krypton	120.3
Oxygen	90.2
Argon	87.5
Nitrogen	77.4
Hydrogen	20.4
Helium	4.2

---

[8]Helium is unique in this regard.

a temperature of 2.18 K (called the λ *point*), helium behaves as an ordinary, classical fluid; in this region the liquid is referred to as *helium I*. Below the λ point, however, the properties of liquid helium (called *helium II* in this temperature region) take on an extraordinary character. For $T < 2.18$ K, helium consists of a mixture of fluids, one of which retains the classical properties exhibited by helium I whereas the other possesses new and remarkable properties that are responsible for the unusual behavior of the liquid at these low temperatures. The ordinary portion of the liquid is called the *normal component* and the extraordinary portion is termed the *superfluid component*.

One of the properties of superfluid helium is its ability to conduct heat with *absolutely no resistance*. That is, heat that is supplied at one point in the superfluid is conducted throughout the liquid *without any losses whatsoever*. Thus, we say that the heat conductivity of the superfluid is infinite—not just "very large" but *infinite!* We can verify this statement in a qualitative way (precise measurements can do so quantitatively) by observing the boiling action in liquid helium as the temperature is lowered through the λ point. If we supply a small quantity of heat at one point in a certain volume of liquid helium that is at a temperature above the λ point, the liquid boils at this point and vigorous bubbling is seen to take place. The reason for the localized boiling is that the liquid has a finite heat conductivity and is incapable of distributing the heat sufficiently rapidly so that the temperature of the entire liquid is raised uniformly. Instead, the temperature is increased only in the region near the point at which the heat is introduced and when the local temperature exceeds 4.2 K, the liquid boils and gas bubbles are formed. Now, let us lower the temperature of the system while continuing the introduction of heat at one point. When the λ point is reached, the boiling action suddenly ceases—no more bubbles are formed. The reason for the cessation of boiling is that below the λ point a portion of the liquid is in the superfluid form with infinite heat conductivity. As rapidly as the heat is introduced, it is distributed to all parts of the liquid and therefore the entire liquid is at exactly the same temperature. Under these conditions no gas bubbles can be formed.

Other experiments show that superfluid helium experiences absolutely no resistance to flow; that is, it has *zero* viscosity. Superfluid helium flows unimpeded through narrow channels or extremely fine capillary tubes that would block the passage of any ordinary fluid, including helium I.

Liquid helium II consists of both superfluid and normal components. The relative amount of the superfluid component as a function of the temperature can be determined in experiments that measure the resisting force on an object moved through the fluid. Only the normal component resists such motion; the object passes frictionlessly through the superfluid component. Such experiments have shown that the superfluid concentration increases with decreasing temperature from zero exactly at the λ point to essentially 100 percent for temperatures below about 1 K. Thus, for T less than about 1 K, helium is truly a superfluid.

Liquid helium possesses a number of spectacular properties, all of which are compatible with the two-fluid model in which the liquid consists of normal and superfluid components.

How can we explain the seemingly impossible properties of the superfluid component of helium? If there are to be viscous losses in the liquid or if there is to be a resistance to the conduction of heat, the motion of one part of the liquid must be different from that of the rest of the liquid; that is, the quantum state of some of the atoms must be different from that of others. An energy loss by one atom necessarily means an energy gain by another atom and it is precisely such energy transfers (and changes of quantum states) that are responsible for viscosity and for resistance to heat conduction.

Now, the helium atom has zero spin and so is a *boson;* bosons do not obey the exclusion principle. Therefore, a collection of helium atoms does not behave at all the way a collection of fermions would behave. The conduction electrons in a metal, for example, are forced, two-by-two, into discrete energy states and thus form the conduction *band* of states. As the temperature of a collection of helium atoms is lowered, the atoms have smaller and smaller energies. Eventually, at a sufficiently low temperature, all the atoms will be in the lowest possible energy state. *All* of the atoms can be in this state; the exclusion principle does not apply to helium atoms and force them into separate states. If all of the helium atoms are in the same quantum state, they must all have the same wave function. Thus, the atoms in superfluid helium all act together.[9] Consequently, one cannot supply heat to only one region of the superfluid—all atoms are influenced in the same way. Also, since there can be no energy transfers between atoms (because all of the atoms are in the *lowest* energy state), the viscosity is zero.

The explanation of the remarkable properties of superfluid helium is a striking example of the operation of quantum principles on a macroscopic scale.

---

[9]There is no "disorder" in superfluid helium; that is, the entropy is zero.

Although the basic ideas of the theory of superfluid helium have been firmly established, many details of the theory are still being investigated. It is not unlikely that liquid helium still holds some surprises for us.

Superconductors

When an electric field is applied to a conductor, the free electrons are set into motion and thereby produce a current. Resistance to the flow of current in a metallic crystal is caused in part by collisions of the electrons with impurities or with points of imperfection in the crystal lattice structure. However, even in an ideal crystal that is both pure and perfect, so that these sources of resistance are absent, the electrons cannot flow unimpeded because the thermal vibrations of the atoms provide sites from which the electrons can be scattered and with which they can exchange energy. As the temperature is decreased, the thermal vibrations are lessened and the motion of the electrons is less violently affected. Thus, the resistance to current flow decreases as the temperature decreases.

In 1911 Kammerlingh Onnes discovered that the element *lead* has the remarkable property that at a temperature of 7.2 K the electrical resistance suddenly becomes zero—not just "very small" but zero! At temperatures of 7.2 K or lower, lead is a *superconductor* (Fig. 20.34). In one experiment, a current of several hundred amperes was induced to flow in a highly refined sample of lead shaped into a ring, and this current was found to be still flowing, apparently undiminished, after a period of a year! The resistance of superconducting lead has been measured to be at least 10^{11} times smaller than the resistance of normal lead. There is every reason to believe that in pure samples of superconducting materials, the electrical resistance is indeed *zero*.

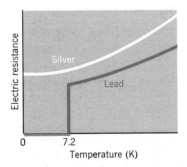

Figure 20.34 As the temperature is lowered, the electrical resistance of silver (a normal conductor) decreases, but the resistance does not decrease below a certain value, even as absolute zero is approached. On the other hand, lead (a superconductor) loses all electrical resistance below the critical temperature of 7.2 K.

Several elements and many alloys (over 1000 are known) have now been found to be superconductors at low temperatures.

The phenomenon of superconductivity is similar to that of superfluidity in that it is the result of macroscopic quantum effects. A great stride forward in the understanding of superconductivity was made in 1957 by John Bardeen, Leon Cooper, and Robert Schrieffer, whose theory is now referred to simply as the *BCS theory*.[10] The basic idea of the BCS theory is concerned with the fact that bosons do not obey the exclusion principle. In superconducting materials, the interaction of the conduction electrons with the vibrations of the atoms in the lattice overcomes the repulsive Coulomb force and results in a small net *attraction* between the electrons. Consequently, the electrons tend to group into *pairs* and a pair of electrons, with spins opposite, behaves as a *boson*. The electron pairs are not well localized—indeed, the spacing between pairs is less than the spacing between electrons in any given pair—but nevertheless the pairs are well defined in the quantum sense and they *do* act as bosons.

The net attraction between the electron pairs— the *pairing energy*—is very small and it does not require much agitation to break the pairs. Therefore, it is only at very low temperatures that the pairs can exist. Because they behave as bosons, the electron pairs all tend to collect in the lowest possible energy state as the temperature is reduced. When the critical temperature is reached (7.2 K for the case of lead), all of the pairs are in the lowest state and all have the same wave function that extends throughout the material. None of the pairs can change its energy state, and therefore the electrons all flow together and there is no dissipation of energy and no electrical resistance.

Certain aspects of superconductors are still not well understood and are currently under investigation, but the crucial point in the explanation of the phenomenon is contained in the BCS theory and we no longer consider superconductivity to be the great mystery that it once was.

Superconducting materials are beginning to be widely used in the construction of magnets for both research and technological applications. Electromagnets that produce strong magnetic fields are expensive to operate because of the substantial losses due to Joule heating in the windings. A conventional electromagnet that produces a field of 10^5

---

[10]The 1972 Nobel Prize in physics was awarded to Bardeen, Cooper, and Schrieffer for their theory of superconductivity. Bardeen, who shared the 1956 Prize, is the only scientist to have won two Prizes in the same field.

Figure 20.35 A superconducting magnet at the Argonne National Laboratory during final phase of construction. The magnet is 18 ft in diameter and contains 25 miles of specially fabricated superconducting ribbon. This magnet is now used in conjunction with a bubble chamber for elementary particle research.

gauss (about the largest field that can be achieved with this type of magnet) may require a megawatt of power to maintain the field. Furthermore, such a magnet requires a cooling system that uses thousands of gallons of water per minute to prevent the windings from melting because of the generation of heat by resistance effects. Magnets are now being used in which the windings are made from various superconducting materials operated at temperatures below the critical temperature. Once the current is established in the windings of such a magnet, it continues to flow without resistance losses. Therefore, the only expenditure of power in a superconducting magnet occurs when the current is first started; in order to maintain the current it is necessary only to insure that the temperature remains below the critical temperature for the material.[11]

Metallic alloys and compounds have been found to be more useful than pure elements in the construction of windings for superconducting magnets. A widely used material is a combination of niobium and tin, Nb_3Sn, which allows the production of fields up to 88,000 G (8.8 T).[12] By using V_3Ga, a combination of vanadium and gallium, it is expected that fields as large as 500,000 G (50 T) can be achieved. Figure 20.35 shows an extremely large superconducting magnet that is used in the elementary particle research at the Argonne National Laboratory. The windings of this magnet are made of Nb_3Ti (niobium and titanium) and a field of 20,000 G (2 T) can be obtained.

If the resistance losses in the transport of electric power could be eliminated or substantially reduced, enormous savings in cost would be realized. Therefore, the possibility of using superconducting materials for the construction of electric transmission lines is of great economic importance. Perhaps within the not-too-distance future we will begin to replace the huge steel towers that now carry our electric power with underground superconducting electric lines.

In order to facilitate the application of superconductors to practical problems, there is a continuing search for superconducting materials. Of particular interest is the search for materials that become superconductors at higher temperatures. Recently, it has been discovered that Nb_3Ge (niobium and germanium) becomes superconducting at 23 K. This temperature is above the boiling point of hydrogen (20 K), so that liquid hydrogen, instead of liquid helium, can be used to cool Nb_3Ge to the superconducting region. There is the exciting possibility that some special material can be developed which will

---

[11] Of course, no practical superconductor can be absolutely pure and so some energy losses do occur. But only very small amounts of input power are required to maintain the superconducting field.

[12] An upper limit to the field that can be produced by a superconductor exists because the electron pairing can be broken by high fields as well as by high temperatures.

be a superconductor at or near room temperature. Such a material could have an enormous impact on our electric industry.

Summary of Important Ideas

The *ionic binding* of two (or more) atoms results when an electron is transferred from one atom to another so that attractive electrostatic forces bind the atoms together. The binding is *strong* when the removal or the addition of only one or two electrons leaves a *closed* electron shell.

The *covalent binding* of atoms to form molecules results when two electrons are shared between the atoms. The binding is strong when the two electrons form a "spin up, spin down" pair.

The *exclusion principle* prevents more than two electrons ($m_s = \pm \frac{1}{2}$) from occupying a given energy state. In bulk material, these states are associated with the entire structure rather than with a single atom. Therefore, the atomic energy states are distributed over a certain energy range and become *energy bands*.

In a *conductor*, a portion of the highest energy band that contains electrons is available for electrons that acquire additional energy. (That is, transitions *within* the band are possible; such transitions require very little energy.)

In an *insulator*, the highest energy band that contains electrons is *completely filled*. Therefore, no transitions *within* the band are allowed and excitations are possible only when an electron is carried into the next empty band; such excitations require considerably more energy than excitations in conductors.

In a *semiconductor*, the highest energy band that contains electrons is completely filled, but the energy gap between this band and the next empty band is small thus allowing transitions to occur with thermal energies.

The conductivity of a *semiconductor* can be permanently enhanced by the addition of certain types of impurities (dopants) that produce additional energy states in the gap between the filled band and the conduction band.

In *n*-type semiconductors, the impurities provide for electrical conduction by electrons, and in *p*-type semiconductors, the conduction is by *holes*.

The strange properties of *liquid helium* II are the result of the fact that helium atoms are *bosons* and therefore do not obey the exclusion principle. Therefore, at low temperatures, the atoms all collect in the *same* energy state (the lowest possible) and have the same wave function. The atoms then all move together with no exchanges of energy and thus no energy losses occur.

The phenomenon of *superconductivity* is also the result of bosons (*electron pairs*) collecting in the lowest possible energy state and moving together without energy losses.

◆ Questions

20.1 Why are neutral atoms of lithium more chemically active than Li^+ ions?

20.2 Sodium and chlorine combine to form NaCl. Two chlorine atoms combine to form Cl_2. Why does sodium not form an Na_2 molecule?

20.3 Describe the way magnesium and chlorine combine to form a molecule. (Refer to Fig. 19.14 and decide how many electrons magnesium can contribute.)

20.4 Use diagrams similar to those in Fig. 20.5 and show schematically how the N_2 molecule is formed with three covalent bonds and how the O_2 molecule is formed with two covalent bonds.

20.5 Sketch the arrangement of the electrons in molecules of MgO and Na_2S.

20.6 What chemical compound do you expect to be produced by ionic binding between (a) calcium and fluorine, and between (b) potassium and sulfur?

20.7 Construct a schematic diagram to show the simplest way in which chlorine can combine with carbon. How does this molecule compare with methane?

20.8 Argue why it is not possible to form molecules consisting of two helium atoms (He_2). (Consider covalent bonding.)

20.9 Use the exclusion principle to show that three hydrogen atoms cannot be bound together by covalent bonding. (However, the H_3^+ ion can exist. Why?)

20.10 What would happen to the widths of the energy bands for a certain solid if sufficient pressure were applied to the solid to compress the atoms into a volume smaller than the original volume? (Use the uncertainty and exclusion principles.)

20.11 Do you expect the noble gases (in solid crystalline form) to be good electrical conductors? Explain.

20.12 Selenium is a semiconductor. What doping agent would you use to convert selenium into a *n*-type semiconductor? Into a *p*-type?

20.13 In natural helium, about one atom in 10^6 is the light isotope, ^{3}He, which has spin $\tfrac{1}{2}$. Do you expect pure liquid ^{3}He to be a superfluid? Explain.

20.14 Suppose that you have a ring of superconducting material and that the temperature is below the critical temperature. No current is flowing in the ring but the ring is in a certain magnetic field (produced by a conventional electromagnet). How would you start a superconducting current to flowing in the ring?

★ Problems

Section 20.1

20.1 The ionization energy of potassium is 4.3 eV and the electron affinity energy of chlorine is 3.7 eV. In KCl the separation between the K^+ and Cl^- ions is approximately 3 Å. What is the binding energy of KCl? (Begin by computing $PE_E = -Ke^2/R$; consider the ions to be point charges.)

20.2* A diatomic molecule can be pictured schematically as a "dumbell," two massive atoms connected by a weightless rod. Such molecules can execute rotational motion around an axis that is perpendicular to the connecting "rod." This rotational motion is quantized according to $L = l\hbar$. Show that the rotational kinetic energy is $KE = L^2/Md^2$, where M is the mass of each atom and where d is the separation distance. In the hydrogen molecule, $d = 0.742$ Å. Calculate the energy in eV of the photon that is emitted when the molecule deexcites from the $l = 2$ state to the $l = 1$ state. Sketch an energy level diagram showing the rotational levels of the hydrogen molecule. Compare these molecular levels with the hydrogen atomic energy levels.

Section 20.2

20.3* Use the fact that the density of KCl is 1.984 g/cm^3 to compute the spacing between the K^+ and Cl^- ions in the crystalline solid. Why is the separation much larger than that between the Na^+ and Cl^- ions in NaCl?

20.4* Prepare a graph that shows the energy of a Na^+-Cl^- ion pair as a function of the separation r of the ions. Extend the graph out to $r = 20$ Å or so. Consider the ions to be stationary at each distance, so the energy is equal to the electrostatic potential energy plus the work that was done in forming the ion pair. Show that the energy is zero for a separation of about 11 Å. What value does the energy approach for $r \to \infty$? What does this signify? What is the binding energy for $r = 2.4$ Å? Actually, you will obtain a binding energy for this separation that is slightly larger than the true value, 4.2 eV. The reason is that the overlapping of the electron structures of the two ions reduces the binding energy. The real energy-versus-separation curve goes through a minimum at $r = 2.4$ Å. Sketch in the part of the curve for $r < 2.4$ Å.

Section 20.3

20.5 Sulfur crystals are pale yellow and transparent. Sulfur is one of the better insulators. From this information alone, estimate the magnitude of the energy gap between the conduction band and the highest filled band in sulfur crystals.

20.6* The human nervous system (indeed, *any* nervous system) is composed of discrete units called *neurons*. There are approximately 10^{10} neurons in the human central nervous system. These neurons receive, process, and store or pass along the electrical signals that relate to all bodily functions. Electronic circuit models of neurons have been devised in which about 10 transistors together with other circuit elements such as resistors and capacitors perform the basic functions of a single neuron. Use the information in Fig. 20.32 and estimate the size of a computer constructed from integrated circuit modules that has the same capacity as the human nervous system. Would it be feasible to build such a computer? (Make some crude estimates of construction cost and time; assume that all of the components are in the form of IC chips such as those shown in Fig. 20.32 and that, because of mass-production economies, the cost per chip can be reduced—not unreasonably—to $0.50.

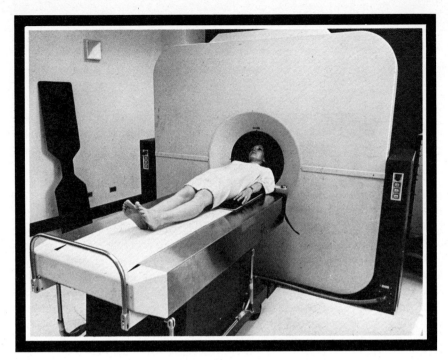

21
Nuclei and Particles

The development of nuclear physics has been intimately connected with the history of atomic structure theory and the emergence of the concepts of quantum theory. The first hint that there were unknown forces at work within the atom was the discovery of *radioactivity* in the 1890s. Rutherford used the α particles (helium nuclei) that are emitted in certain radioactive decay processes to probe the inner structure of atoms and, based on these findings, he developed the nuclear model of the atom. Bohr elaborated on this concept and formulated the first crude theory of atomic structure. The culmination of this work came in the 1920s with the development of modern quantum theory. During the 1930s, while the advances made in quantum theory were being consolidated, a heightened interest in the nucleus produced a series of discoveries that projected nuclear physics to the forefront of basic research activities. The discoveries of the neutron, artificial radioactivity, fission, and the mu meson, and the development of the first theories of β radioactivity and of nuclear structure—these were all products of the 1930s. The last 40 years has seen attention turn to studies of the details of nuclear structure and of the way in which nuclear and sub-nuclear particles interact at extremely high energies. Perhaps these investigations will lead eventually to a fundamental understanding of the way Nature behaves in the nuclear domain.

21.1 Nuclear Masses

Nuclear Binding Energies

At the very small distances that are characteristic of nuclear sizes, a strong nuclear force acts between pairs of nucleons to bind these particles into nuclei. The nuclear force is so strong that an input of several MeV of energy is required to remove a nucleon from a nucleus. By using the Einstein mass-energy relation, $\mathscr{E} = mc^2$, we can discuss the *binding energy* of a nucleus in terms of its *mass*.

Consider, first, an atom of deuterium. The deuterium nucleus (called a *deuteron*) consists of one proton and one neutron; this is the simplest nucleus that depends on the nuclear force for its existence. The mass[1] of a deuterium atom (m_D) is slightly *less*

---

[1] Isotopic masses are always listed in terms of the *atomic* mass instead of the *nuclear* mass, and we use the atomic masses here. Because the hydrogen atom and the deuterium atom each contain one electron, the difference in atomic masses is equal to the difference in nuclear masses.

than the combined masses of a hydrogen atom (m_H) and a free neutron (m_n):

$$m_H = 1.007\,825 \text{ u} \qquad m_D = 2.014\,102 \text{ u}$$
$$\underline{m_n = 1.008\,665 \text{ u}}$$
$$m_H + m_n = 2.016\,490 \text{ u}$$

Thus, for the mass difference, we have

$$(m_H + m_n) - m_D = 0.002\,388 \text{ u}$$

This result is, in fact, quite general: all nuclei have masses that are less than the masses of the constituent protons and neutrons in the free state. The magnitude of this mass difference for a particular nucleus is indicative of the degree of *binding* of the protons and neutrons in that nucleus. According to the Einstein mass-energy relation, a *mass* difference corresponds to an *energy* difference. For the deuteron, this energy difference (the *binding energy*) is

$$\mathcal{E}_b = [(m_H + m_n) - m_D] \times c^2$$

In order to calculate $\mathcal{E}_b$ in MeV we need to know the value of c^2 in units of MeV/u. Using $1 \text{ u} = 1.6605 \times 10^{-27}$ kg and multiplying each side by c^2, we find

$$(1 \text{ u}) \times c^2 = (1.6605 \times 10^{-27} \text{ kg})$$
$$\times (2.998 \times 10^8 \text{ m/s})^2$$
$$\times \left(\frac{1 \text{ MeV}}{1.6022 \times 10^{-13} \text{ J}}\right)$$
$$= 931.5 \text{ MeV}$$

Thus,

$$c^2 = 931.5 \text{ MeV/u} \qquad (21.1)$$

Physically, this equation means that the energy equivalent to 1 atomic mass unit is 931.5 MeV.

Returning to the calculation for the deuteron, we find

$$\mathcal{E}_b = (0.002\,388 \text{ u}) \times (931.5 \text{ MeV/u}) = 2.224 \text{ MeV}$$

This result means that it is necessary to supply 2.224 MeV of energy to a deuteron in order to separate it into a free proton and a free neutron (or, equivalently, to separate a deuterium atom into a hydrogen atom and a free neutron). If more than 2.224 MeV is supplied, the excess energy will appear in the form of kinetic energy of the proton and the neutron.

The nuclear binding energy can also be interpreted in the following way. If, for example, a slowly moving neutron (that is, a neutron with negligible kinetic energy) is captured by the proton in a hydrogen atom to form a deuteron, the initial mass-energy of the system, $(m_H + m_n)c^2$, is greater than the final mass-energy $m_D c^2$, and so the energy difference $\mathcal{E}_b$ is radiated in the form of a γ ray, as shown in Fig. 21.1.

The deuteron has an exceptionally low binding energy; for most nuclei, the binding energy per *particle* is approximately 8 MeV. Thus, to separate a nucleus of ^{20}Ne into 10 free protons and 10 free neutrons requires approximately $20 \times (8 \text{ MeV}) = 160$ MeV. The proton and the neutron are so similar in their properties (except for charge) that we frequently refer to these particles collectively as *nucleons*. Thus, Fig. 21.2 shows the *binding energy per nucleon*, $\mathcal{E}_b$. The decrease of $\mathcal{E}_b$ with the mass number for A greater than about 60 has, as we will see, great significance in the phenomenon of *fission*.

Nuclear Masses and Stability

Most of the nuclei that we find in Nature are *stable*. The nucleons in a stable nucleus are in the lowest possible energy state; such nuclei cannot spontaneously discharge a nucleon nor can they undergo radioactive decay. In order to separate a nucleon or a group of nucleons from a stable nucleus requires the *addition* of energy. For example, Table 21.1 shows that it requires at least 19.8 MeV to separate ^{4}He into any combination of nucleons but that only 7.55 MeV is required to separate the least bound particle (a proton) from ^{14}N. The ^{4}He nucleus (that is, the α particle) is therefore a particularly stable nucleus. Compare Fig. 21.2, which shows that the binding energy per nucleon of ^{4}He is substantially greater than that of any of its neighbors.) On the

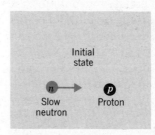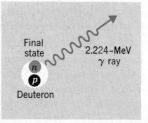

Figure 21.1 Schematic representation of the capture of a slow neutron (i.e., a neutron with negligible kinetic energy) by a proton to form a deuteron. The deuteron binding energy is radiated in the form of a γ ray.

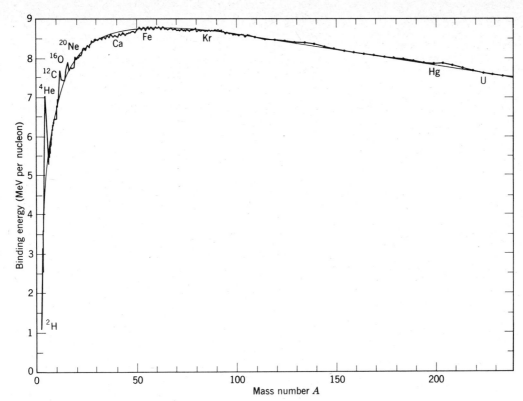

Figure 21.2 The binding energy per nucleon as a function of mass number. Some representative measured binding energies are shown as dots in the figure. The point for ^{4}He is far above those for neighboring nuclei because the ^{4}He nucleus (the α particle) is an exceptionally tightly bound group of nucleons.

Table 21.1 Separation and Binding Energies for ^{4}He and ^{14}N

Particle Group		Separation Energy (MeV)
^{4}He $\longrightarrow$	^{3}H + ^{1}H	19.814
	^{3}He + n	20.578
	^{2}H + ^{2}H	23.847
Total binding energy ($\longrightarrow$ 2^{1}H + 2n) = 28.296 MeV		
Binding energy per nucleon = $\dfrac{28.296}{4}$ = 7.074 MeV		
^{14}N $\longrightarrow$	^{13}C + ^{1}H	7.550
	^{12}C + ^{2}H	10.272
	^{13}N + n	10.553
	^{10}B + ^{4}He	11.613
	^{11}B + ^{3}He	20.736
	^{11}C + ^{3}H	22.736
Total binding energy ($\longrightarrow$ 7^{1}H + 7n) = 104.659 MeV		
Binding energy per nucleon = $\dfrac{104.659}{14}$ = 7.475 MeV		

other hand, the ^{5}He nucleus is *unstable* (i.e., *unbound*):

$m(^5\text{He}) = 5.012\ 297$ u

$m(^4\text{He}) = 4.002\ 603$ u
$m_n = 1.008\ 665$ u
$\overline{m(^4\text{He}) + m_n = 5.011\ 268\ \text{u}}$

Because the mass of ^{5}He is *greater* than the combined masses of ^{4}He and a neutron, ^{5}He is unstable and disintegrates spontaneously into ^{4}He + n, with the excess energy appearing as kinetic energy of the neutron and the ^{4}He nucleus. ^{5}He, therefore, does not occur in Nature.

Table 21.2 Atomic Masses of Some Light Elements ($Z \leq 4$)[a]

Nucleus	Z	Mass (u)	
n	0	1.008 665	
^{1}H	1	1.007 825	(stable)
^{2}H or ^{2}D (deuterium)	1	2.014 102	(stable)
^{3}H or ^{3}T (tritium)	1	3.016 050	
^{3}He	2	3.016 030	(stable)
^{4}He	2	4.002 603	(stable)
^{5}He	2	5.012 297	
^{6}He	2	6.018 893	
^{7}He	2	7.028 031	
^{5}Li	3	5.012 538	
^{6}Li	3	6.015 125	(stable)
^{7}Li	3	7.016 004	(stable)
^{8}Li	3	8.022 487	
^{6}Be	4	6.019 717	
^{7}Be	4	7.016 929	
^{8}Be	4	8.005 308	
^{9}Be	4	9.012 186	(stable)
^{10}Be	4	10.013 534	

[a] Compare the table inside the front cover which gives the masses of the naturally occurring isotopic mixtures.

Table 21.2 lists the masses of some of the light elements. From these data, it is easy to calculate that ^{5}He, ^{7}He, ^{5}Li, ^{6}Be, and ^{8}Be (and *only* these nuclei in the list) are unstable to the emission of nucleons or groups of nucleons:

$$^5\text{He} \longrightarrow {}^4\text{He} + n$$
$$^7\text{He} \longrightarrow {}^6\text{He} + n$$
$$^5\text{Li} \longrightarrow {}^4\text{He} + p$$
$$^6\text{Be} \longrightarrow {}^4\text{He} + 2p$$
$$^8\text{Be} \longrightarrow {}^4\text{He} + {}^4\text{He}$$

● *Example 21.1*

How much energy is required to break up a ^{12}C nucleus into three α particles? By definition, the atomic mass of ^{12}C is 12 u (exactly). Then,

$$3 \times m(^4\text{He}) = 3 \times (4.002\,603\text{ u}) = 12.007\,809\text{ u}$$

Therefore, the energy required is

$$\begin{aligned}\mathscr{E} &= [3 \times m(^4\text{He}) - m(^{12}\text{C})] \times c^2 \\ &= (12.007\,809\text{ u} - 12\text{ u}) \times c^2 \\ &= (0.007\,809\text{ u}) \times (931.5\text{ MeV/u}) \\ &= 7.274\text{ MeV}\end{aligned}$$

Conversely, we can conclude that when 3 α particles combine to form a ^{12}C nucleus, 7.274 MeV of energy is released. Just such a process is, in fact, important in the formation of carbon in stars. ■

The Limits of Stability

The decrease in binding energy with increasing mass number for heavy elements and the termination of nuclear stability near $A = 210$ are consequences of two effects: the *short-range* character of the nuclear force and the fact that protons and neutrons separately obey the *exclusion principle*. Because the nuclear force has a short range, there are attractive forces only between a given nucleon and its immediate neighbors. The Coulomb force, on the other hand, has a long range, so there is an electrostatic repulsive force between a given proton and *all* of the other protons in a nucleus. For sufficiently large atomic numbers, the Coulomb repulsion surpasses the attractive effects of the nuclear force. Thus, there must come a point at which it is no longer possible to form a stable nucleus by the addition of more protons. In fact, there are no *stable* nuclei with $Z > 83$.

What is the effect of the exclusion principle? If a pair of protons (spin *up*, spin *down*) occupies a particular energy state within a nucleus, an added proton (because of the exclusion principle) must go into a different energy state with *greater* energy. As more protons are added, they are forced into higher and higher energy states. Eventually, the energy of one of these states will exceed the separation energy and the nucleus will be unstable. Thus, the exclusion principle, as well as the Coulomb repulsion, acts to limit the number of protons in nuclei.

The exclusion principle governs the energy states that can be occupied by neutrons in the same way that it affects protons. As neutrons are added to a nucleus, they go two-by-two into higher and higher energy states, and eventually the nucleus reaches a point of instability.

The restrictions imposed by the short-range nuclear force and the exclusion principle mean that arbitrarily large stable nuclei cannot exist; in fact, *all* nuclei with $A > 210$ (and many with smaller mass numbers) are *unstable*. These instabilities take several forms; we shall discuss these in turn, beginning with β radioactivity in the following section.

Nuclear Shapes

Surprisingly, most nuclei are not spherical in shape. Instead nuclei are mostly either *oblate* (dishlike) or *prolate* (cigarlike) ellipsoids of revolution. It is not uncommon for major axes to be as much as 30 percent larger than minor axes. Even those nuclei that are spherical are readily deformable in excited modes.

In experimental situations involving unaligned nuclei the use of an effective spherical radius usually yields sufficiently satisfactory theoretical understanding of the process involved. Empirical evidence of this sort yields nuclear radii

$$R = R_0 A^{1/3}$$

where A is the nuclear mass number and R_0 is between 1.2×10^{-15} m and 1.4×10^{-15} m. The smaller value is more appropriate for the scattering of energetic electrons by nuclei while the larger value is indicated in the interaction of alpha particles with nuclei. Note that the above dependence of the radius on the mass number A leads to a constant nuclear density for all nuclei. Nuclei appear to be built up of neutrons and protons behaving as if they were close-packed spheres of about equal size. More sophisticated models also include a surface region of about 2×10^{-15} m in thickness in which the nuclear density smoothly decreases from its relatively constant interior value to zero.

21.2 Radioactivity

Types of Radioactive Decay

In 1896 the French physicist Henri Becquerel (1852–1908) discovered that certain naturally occurring minerals emit radiations of a type that had not previously been observed. Marie Curie, who later discovered the elements of radium and polonium, called this new phenomenon *radioactivity*, a name we still use. Within a few years after Becquerel's discovery, the radiations from radioactive substances were classified into three groups:

1 *Alpha rays*—massive, positively charged objects.

2 *Beta rays*—negatively charged objects of small mass.

3 *Gamma rays*—neutral rays with no detectable mass.

Detailed investigations of these radiations revealed that the beta rays are identical to electrons and that the alpha rays are nuclei of helium atoms. Gamma rays were found to have properties similar to light—the only difference being that the frequency of gamma rays is much higher than that of visible light.

Radioactivity is a *nuclear* phenomenon. Alpha and beta rays are emitted during the spontaneous disintegration of nuclei, and gamma rays result when the neutrons and protons within a nucleus spontaneously rearrange themselves (but without "disintegration").

The emission of an alpha ray (or α particle) by a nucleus necessarily changes both the atomic number and the mass number; that is, a new chemical element (a *daughter* element) is formed by the α decay of a *parent* nucleus. For example, when radium ($Z = 88$, $A = 226$) emits an α particle, radon ($A = 86, A = 222$) is formed, as indicated schematically in Fig. 21.3a. An α particle, therefore, has $Z = 2$ and $A = 4$—it is just the nucleus of a *helium* atom. Beta decay, on the other hand, does not involve the emission of a proton or neutron, so the mass number of the daughter nucleus is the same as that of the parent nucleus. But because the emitted particle carries a negative charge, the atomic number of the daughter nucleus is one unit *greater* than that of the parent. As shown in Fig. 21.3b, when ^{14}C ($Z = 6$, $A = 14$) decays by beta emission, ^{14}N ($Z = 7$, $A = 14$) is formed. Beta decay is therefore equivalent to the transformation of one of the nuclear neutrons into a proton. The β particles emitted by nuclei are exactly the same as atomic *electrons*. Gamma-ray (γ-ray) emission accompanies nuclear rearrangements that leave Z and A unaltered (Fig. 21.3c); γ rays are often emitted following α or β decay as the daughter nucleus readjusts its neutrons and protons into the lowest energy state.

The Half-Life

Because α and β emissions cause the parent nuclei to be transformed into different nuclear species, won't these disintegrations soon cause the complete depletion of the parent substance? Actually, this is not the case. Radioactive decay processes obey the following law: If we begin with an amount of a radioactive substance, then after a certain interval of time that is characteristic of the particular nucleus involved (called the *half-life* of the substance and denoted by $\tau_{1/2}$), *one-half* of the material will have disintegrated and one-half will remain. If we wait for another interval $\tau_{1/2}$, one-quarter of the original amount will remain. After each period of time $\tau_{1/2}$, there will remain one-half of the parent material that existed at the beginning of that time period. Radium-226, for example, has $\tau_{1/2} \cong 1600$ years; therefore, if we have 1 g of ^{226}Ra to start with, after 1600 years we shall have $\frac{1}{2}$ g, after 3200 years we

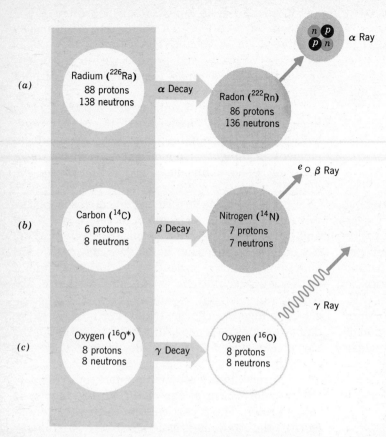

Figure 21.3 Examples of the three types of radioactive decay events, alpha and beta decay involve nuclear *disintegrations* (i.e., changes in species) while gamma decay results from intranuclear rearrangements. The excited ^{16}O nucleus that exists before γ emission takes place is indicated by ^{16}O*.

shall have $\frac{1}{4}$ g, after 4800 years we shall have $\frac{1}{8}$ g, and so forth (see Fig. 21.4). Thus, the amount of radium will approach zero only after an infinitely long time.

The particular shape of the radioactive decay curve results from the fact that the number of nuclei decaying per unit time interval is proportional to the number present. Thus, there is a continual decrease in the number of radioactive nuclei in the sample and a corresponding decrease in the number of decay events per second. The mathematical function that describes this behavior is an *exponential function* and is the same as that used to represent the discharge of a capacitor (Section 14.2). Indeed, Fig. 14.11 has exactly the same shape as Fig. 21.4.

Some typical half-lives for α and β decay are listed in Table 21.3.

β Radioactivity

Within a stable nucleus, neutrons and protons alike will remain unchanged indefinitely. But *outside* a nucleus, a neutron is a radioactive particle. Table 21.2 and Table 21.4 show that the mass of a *free* neu-

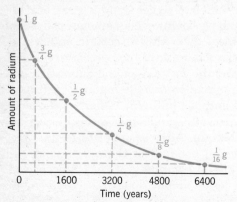

Figure 21.4 Radioactive decay curve of ^{226}Ra ($\tau_{1/2} \cong 1600$ y). In each interval of 1600 years, the amount of radium decreases by one-half.

Table 21.3 Some Radioactive Half-Lives

Nucleus	Type of Decay	Half-Life
Thorium (^{232}Th)	α	1.4×10^{10} y
Plutonium (^{238}Pu)	α	87.4 y
Uranium (^{229}U)	α	58 min
Carbon (^{14}C)	β	5730 y
Cobalt (^{60}Co)	β	5.24 y
Copper (^{66}Cu)	β	5 min
Krypton (^{93}Kr)	β	1.3 s
Californium (^{254}Cf)	spontaneous fission	54 d

Table 21.4 Properties of Nucleons

Property	Proton	Neutron
Mass, m	$m_p = 1.007\,276$ u[a]	$m_n = 1.008\,665$ u
Rest-mass energy, mc^2	938.259 MeV	939.553 MeV
Charge, q	$e = 1.602 \times 10^{-19}$ C	0
Spin, s	$\frac{1}{2}$	$\frac{1}{2}$
Half-life, $\tau_{1/2}$	∞ (stable)	10.6 min
Decay mode	none[b]	$n \to p + e + \nu$

[a]This is the mass of the *proton*; the mass of the hydrogen *atom* (see Table 21.2) is $m_H = 1.007\,825$ U.

[b]There is the possibility that the proton itself may be unstable. At present its half-life is known to be longer than 10^{30} years.

tron is greater than that of a hydrogen atom (that is, a proton and an electron). Therefore, it is energetically possible for a free neutron to separate into a proton and an electron, and, in fact, neutrons do undergo this type of decay process with a half-life of 10.8 min. The available decay energy in the disintegration of the neutron is

$$\begin{aligned}\mathcal{E}_{np} &= [m_n - (m_p + m_e)] \times c^2 \\ &= (m_n - m_H) \times c^2 \\ &= (1.008\,665\text{ u} - 1.007\,825\text{ u}) \times (931.5\text{ MeV/u}) \\ &= 0.782\text{ MeV}\end{aligned} \qquad (21.2)$$

The conversion of a neutron into a proton and an electron is the prototype of the class of nuclear disintegration processes known as β decay. Many nuclei (some of which occur naturally and others which must be produced artificially in the laboratory) are known to undergo β decay by the emission of electrons in exactly the same way that free neutrons decay.

All nuclear radioactive decay processes in which an electron is emitted can be considered to be the result of the β decay of a neutron *within* the parent nucleus. For example, consider the case of ^{8}Li (Fig. 21.5). The ^{8}Li nucleus consists of 3 protons and 5 neutrons and undergoes β decay with a half-life of 0.85 s. The ^{8}Li decay process transforms one of the 5 neutrons into a proton so that the new nucleus has 4 protons and 4 neutrons—this new nucleus is ^{8}Be.

In all β decay processes the electrons are not emitted alone. This conclusion was reached in the 1930s when studies of β decay events revealed that the β particles emerged from the decaying nuclei with a *distribution* of energies instead of a unique energy (Fig. 21.6). Because all nuclei of a particular isotope are identical, energy conservation in β decay can be valid only if some undetected additional particle carries away the remainder of the decay energy. Wolfgang Pauli made this proposal and Enrico Fermi named the unseen particle a *neutrino* (ν), meaning "little neutral one."

The neutrino that seemed necessary to understand the conservation of energy (and momentum) in β decay had to be endowed with unlikely properties—it must carry no electric charge and have no rest mass, it must have almost no interaction with ordinary matter, but it must carry energy and mo-

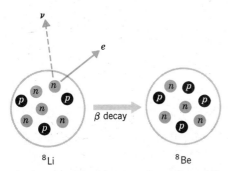

Figure 21.5 Schematic representation of the β decay of ^{8}Li. One of the ^{8}Li neutrons is transformed into a proton and the new nucleus is ^{8}Be.

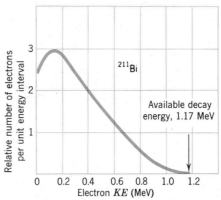

Figure 21.6 The electron energy spectrum for the β decay of ^{211}Bi. The electron energy has a maximum corresponding to the available decay energy, 1.17 MeV. For a particular decay event, the energy difference between 1.17 MeV and the electron energy is carried away by a neutrino.

mentum, and it must travel with the velocity of light![2] The neutrino was therefore a kind of "ghost" particle, but in 1953, Clyde Cowan and Fredrick Reines conducted a remarkable series of experiments in which the neutrino was actually detected. We will discuss these curious particles again in Section 21.9.

Nuclear β decay leads to an increase in the nuclear charge by one unit ($Z \rightarrow Z + 1$), but leaves unchanged the mass number A because the total number of nucleons remains the same. The β decay of a nucleus with atomic number Z and mass number A is energetically possible only if the atom with atomic number $Z + 1$ and mass number A (the *daughter* atom) has a mass smaller than that of the original (or *parent*) atom. That is, for β decay to occur, we must have

$$m(Z, A) > m(Z + 1, A)$$

If this relation is satisfied, a neutron in the nucleus (Z, A) can, in fact, transform into a proton (plus an electron and a neutrino) so that the nucleus $(Z + 1, A)$ results. The *decay energy* in such a case is the mass difference Δm multiplied by c^2 (compare Eq. 21.2):

$$\mathcal{E}_\beta = \Delta m \times c^2 = [m(Z, A) - m(Z = 1, A)] \times c^2 \quad (21.3)$$

Notice that because we use *atomic* masses here, it is unnecessary to include the electron masses in the expression for Δm. The final mass of the system consists of the mass of the *nucleus* $(Z + 1, A)$ plus the mass of the Z electrons that were the atomic electrons of the original atom (Z, A) plus the mass of the ejected electron. Altogether, these masses just total to the mass of the *atom* $(Z + 1, A)$.

Typical decay energies for β-radioactive nuclei are in the range 0.01–3 MeV (see Table 21.5). Except for the small amount of kinetic energy received by the recoiling daughter nucleus when the electron and neutrino are emitted, the decay energy $\mathcal{E}_\beta$ corresponds to the *maximum* kinetic energy that the electron can possess. For example, the maximum energy permitted electrons in the β decay of ^{211}Bi is 1.17 MeV, as indicated in Fig. 21.6. The *average* electron kinetic energy is approximately one-third the maximum energy.

Not all β-decay processes leave the residual nucleus in its ground state. In many instances, an excited state of the daughter nucleus is formed and one or more γ rays are emitted before the nucleus is left in its ground state.

Positron Decay

Figure 21.7 shows the relative masses of the various nuclei with $A = 65$. For this mass number, only ^{65}Cu is stable; ^{65}Co has a mass greater than that of ^{65}Ni and, therefore, undergoes β decay, forming ^{65}Ni; furthermore, ^{65}Ni has a mass greater than that of ^{65}Cu and undergoes β decay, forming stable ^{65}Cu. That is, two successive β decays transform a nucleus of ^{65}Co into ^{65}Cu. (The simultaneous emission of two electrons, which would transform ^{65}Co directly into ^{65}Cu, does not occur with a measurable rate.)

The $A = 65$ chart also shows that there are nuclei with $Z > 29$ that are more massive than ^{65}Cu. These nuclei cannot undergo normal β decay because each has a charge *greater* than that of the adjacent, less massive nucleus. A different type of β decay process is possible for these nuclei which allows them to rid themselves of one unit of the greater-than-normal positive charge that they carry. Nuclei such as ^{65}Ge, ^{65}Ga, and ^{65}Zn emit particles that are identical to electrons in every respect except that they have a *positive* charge. These particles are called *positrons* (β^+) and the decay process is called *positron decay*. We shall learn more about these interesting particles when we discuss the properties of elementary particles later in this chapter. For now it suffices to note that there are two complementary nuclear β-decay processes, which we label β^- decay and β^+ decay; these processes can be represented in the following way:

$$\beta^- \text{ decay: } n \longrightarrow p + e^- + \bar{\nu}_e \quad (21.4a)$$

$$\beta^+ \text{ decay: } p \longrightarrow n + e^+ + \nu_e \quad (21.4b)$$

We understand, of course, that these processes take place *within* nuclei; the electron (or positron) and the neutrino are *created* at the instant of disintegration and are immediately ejected from the nucleus (they do *not* preexist in the nucleus). (The different symbols for the neutrinos will be explained in Section 21.9.)

There is an additional process, called electron capture, which we have not discussed. In this type of transmutation, the nucleus captures one of its orbital electrons to convert a proton into a neutron and decrease the atomic number by one.

---

[2] At present all that is known is that the rest mass of the neutrino is less than about 30 eV/c^2. The neutrino may in fact have a rest mass, but if it does, it is very small indeed, perhaps 1 eV/c^2.

Table 21.5 Some Important β-Radioactive Nuclei

Nucleus	Half-Life	Type of Decay	Maximum Electron Kinetic Energy (MeV)	Remarks
^{3}H (tritium)	12.26 y	β^-	0.0186	Used in nuclear fusion devices (H-bombs)
^{14}C	5730 y	β^-	0.156	Used in archeological dating; also an important tracer in biochemical studies
^{22}Na	2.60 y	β^+	0.54	Useful source of positrons
^{24}Na	15.0 h	β^-	1.39	Used in medical diagnostics to follow the flow of sodium in the body
^{40}K	1.3×10^9 y	β^-	0.0118	Used in geological dating
^{60}Co	5.24 y	β^-	0.31	Accompanying γ rays used in medical therapeutics and for radiation processing of plastics, food, etc.
^{90}Sr	28.8 y	β^-	0.54	Important fission product (occurs in fallout from detonation of fission bombs)
^{131}I	8.05 days	β^-	0.61	Used in medical diagnostics and therapeutics, particularly in thyroid ailments

α Decay

Certain unstable nuclei, primarily those with mass numbers above 200, spontaneously emit helium nuclei (α particles). Table 21.6 lists some typical α-radioactive nuclei. The emission of an α particle by a nucleus decreases the original nuclear charge by 2 units and decreases the original mass number by 4 units. If a nucleus identified by (Z, A) has a mass greater than the sum of the masses of the nucleus $(Z - 2, A - 4)$ and a ^{4}He nucleus, the nucleus (Z, A) is unstable and can decay by the emission of an α particle. The available energy for the α-decay process is

$$\mathcal{E}_\alpha = [m(Z, A) - m(Z - 2, A - 4) - m(^4\text{He})] \times c^2 \tag{21.5}$$

● **Example 21.2**

What is the available energy for the α decay of ^{210}Po?

Alpha-particle emission from ^{210}Po leaves ^{206}Pb. Therefore, the pertinent masses are:

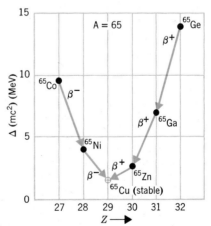

Figure 21.7 Nuclear β^- decay and positron (β^+) decay routes for nuclei with A = 65. There is a single stable nucleus with this mass number (^{65}Cu).

$m(^{210}\text{Po}) = 209.98287$ u $\quad m(^{206}\text{Pb}) = 205.97447$ u
$\qquad\qquad\qquad\qquad\qquad\quad m(^4\text{He}) = 4.00260$ u
$\qquad\qquad\qquad\quad\overline{m(^{206}\text{Pb}) + m(^4\text{He}) = 209.97707\text{ u}}$

Table 21.6 Some Typical α-Radioactive Nuclei

Nucleus	Half-Life	α-particle Kinetic Energy (MeV)[a]	Remarks
^{142}Ce	5×10^{15} y	1.5	Lightest naturally occurring α-radioactive nucleus
^{210}Po	138 days	5.30	Much used source of α particles
^{214}Bi	19.7 min	5.51	Also undergoes β decay
^{218}Po	3.05 min	6.00	Formed by two successive α decays starting with ^{226}Ra; also known as radium-A; used in original Rutherford scattering experiment
^{226}Ra	1620 y	4.78	α particles from this source first identified as helium nuclei (Rutherford)

[a]The *decay energy* is slightly higher (see Example 21.2 and Problem 21.13).

Hence,

$$\mathcal{E}_\alpha = (209.98287 \text{ u} - 209.97707 \text{ u}) \times c^2$$
$$= (0.00580 \text{ u}) \times (931.5 \text{ MeV/u})$$
$$= 5.40 \text{ MeV}$$

Actually, this decay energy is shared by the α particle and the ^{206}Pb nucleus (because the linear momenta of the two fragments must be equal and opposite). Consequently, the α particle emitted by ^{210}Po has a kinetic energy of 5.30 MeV and the recoil ^{206}Pb nucleus has a kinetic energy of 0.10 MeV. (See Problem 21.13.) ∎

Essentially all nuclei with A greater than about 100 are unstable with respect to breakup by the emission of α particles, but it is only for nuclei with A greater than about 200 that α decay is an important process. The heavier nuclei have α-decay half-lives sufficiently short that α-particle emission from a given sample occurs at a rate that permits observation, but the half-lives of the lighter nuclei tend to be so long that the α-decay process is unmeasurable, even though it is energetically allowed.

It is significant to note that these heavy nuclei emit only α particles, not protons, or neutrons, or deuterons, or other groups of small numbers of nucleons. The addition of about 7 MeV of energy is required to remove a nucleon from a nucleus with A near 200 and the addition of about 10 MeV is necessary to remove a deuteron from such a nucleus. Apart from fission into two fragments of roughly equal mass, α-particle decay is the only energetically possible process that results in the spontaneous emission of nucleons from heavy nuclei.

Radioactive Decay Chains

The heaviest elements found in Nature are uranium (U, $Z = 92$), protactinium (Pa, $Z = 91$), and thorium (Th, $Z = 90$). All of the isotopes of these elements are radioactive but each element has at least one isotope with a sufficiently long half-life that the element still exists in Nature. For example, ^{238}U has $\tau_{1/2} = 1.4 \times 10^9$ years. When these nuclei decay, they form new daughter elements that are also radioactive. Some of these nuclei are β radioactive and others emit α particles. A few can even decay by either α or β emission. A series of successive radioactive decays takes place that continues until a stable isotope of either lead (Pb, $Z = 82$) or bismuth (Bi, $Z = 83$) is formed. The stable isotopes of lead are ^{206}Pb, ^{207}Pb, and ^{208}Pb; the only stable isotope of bismuth is ^{209}Bi. These four nuclei are the termination points for all of the radioactive decay chains that originate with the long-lived heavy elements. One such decay chain begins with ^{238}U and ends with ^{206}Pb; this series of α and β decays is shown in Fig. 21.8.

The Valley of Stability

A convenient way to represent a number of nuclear properties is by the type of chart shown in Fig. 21.9. In this chart, every nucleus is assigned a position

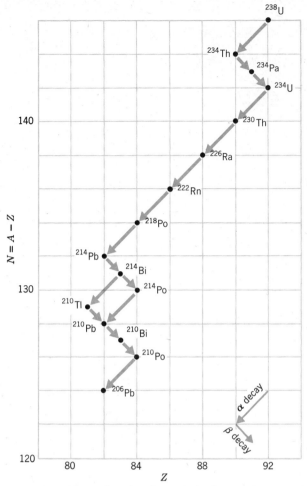

Figure 21.8 The radioactive decay chain that originates with ^{238}U and ends with ^{206}Pb. Notice that ^{214}Bi can undergo either α or β decay and that after one additional decay, both branches lead to ^{210}Pb. Each ^{214}Bi nucleus has a certain probability for decay by β-emission and a certain probability for decay by α-particle emission.

a sufficient time to be identified and therefore these nuclei are not shown in the chart.

A complete chart of the known nuclei would contain more than 1600 entries; some 330 nuclear isotopes have been found to occur naturally in the Earth (about 260 of which are stable) and almost 1300 have been produced artificially in the laboratory. A schematic chart of the nuclei known at present is shown in Fig. 21.10.

A number of interesting points appear on examination of the systematics of nuclear properties when the nuclei are arranged in this fashion:

1 The stable nuclei up to $Z \cong 20$ have approximately equal numbers of protons and neutrons ($Z \cong N$).

2 For Z greater than about 20, the stable nuclei tend to have an increasing preponderance of neutrons over protons; for example, uranium has $N/Z \cong 1.6$. The reason for this effect is easy to understand when we recall that the nuclear force has a *short* range whereas the Coulomb force has a *long* range. As more nucleons are added to form heavier nuclei, the average distance between nucleons becomes greater. Therefore, the long-range Coulomb repulsion becomes more effective relative to the short-range nuclear attraction and it becomes more and more difficult to add protons to a nucleus. For this reason it becomes energetically more favorable to add neutrons to heavy nuclei; consequently, N/Z increases with Z.

3 The stable nuclei are located along a narrow band of the N versus Z diagram, called the *valley of stability*. Along the lines of constant A, the nuclei on either side of the valley have slightly larger masses and undergo radioactive decay (by either β^- or β^+ emission) in order to reach a stable condition. (Of course, for the heavier nuclei, α decay is possible, and some nuclei undergo spontaneous fission in an effort to reach stability.)

4 Nuclei *above* the valley of stability are *neutron rich* and undergo β^- decay. An increase in the distance from the stable valley along a line of constant A causes: (a) the nuclei to become more massive (that is, less stable), (b) the β decay energy $\mathscr{E}_\beta$ to increase, and (c) the half-life to decrease. Sufficiently far from the valley, the neutron excess becomes very large and the instability increases to such a degree that the β^- decay process, which converts a nuetron into a proton, is replaced by the direct emission of a neutron.

according to the number of protons Z and the number of neutrons, $N = A - Z$, in the nucleus. For stable nuclei (the shaded boxes), the naturally occurring isotopic abundance is given, and for unstable nuclei, the mode of decay and the half-life is given. Nuclei that decay by particle emission (^{5}He, ^{5}Li, ^{6}Be, ^{8}Be, and ^{9}B, among those shown in the chart), have half-lives less than about 10^{-16} s and these half-lives are not shown. In such a chart, it is easy to trace the radioactive history of a particular nuclear species by following the arrows that indicate the decay modes. For example, ^{8}He β decays to ^{8}Li which, in turn, β decays to ^{8}Be which, finally, breaks up into two α particles.

The nuclei ^{4}H, ^{5}H, ^{2}He, ^{4}Li, etc., do not exist as configurations of nucleons that remain together for

540 Nuclei and Particles

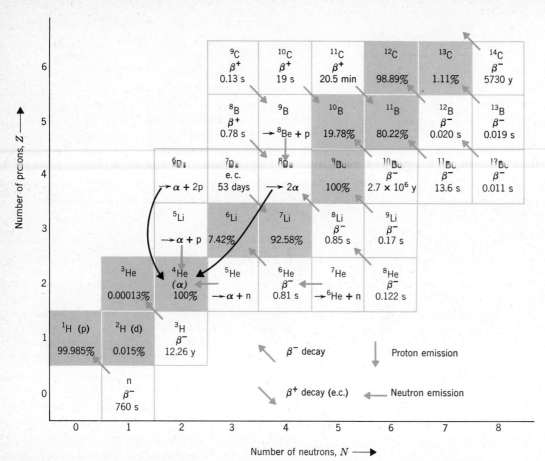

Figure 21.9 A portion of the chart of nuclei showing the light elements. For stable nuclei (shaded boxes), the naturally occurring isotopic abundance is given, for particle unstable nuclei, the mode of decay is given; and for β radioactive nuclei, the half-life is given.

5 Nuclei *below* the valley of stability are *proton rich* and undergo β^+ decay. Similar to the case above the valley, nuclei sufficiently far below the valley will be short-lived emitters of high-energy positrons or will decay by proton emission.

Nucleogenesis

Where did the elements corresponding to the valley of stability, shown in Fig. 21.10, come from? How did they arrive at the relative abundances now present in our universe? A relatively consistent explanation is now at hand.

In one model of cosmogony the universe began in a *big bang* fireball of enormous density and temperature. At first only protons, neutrons, and various elementary particles were present. The temperature was too high to allow more complicated nuclei to hold together. However, after about 100 seconds, when the temperature dropped to below 10^9 K, deuterium could be formed by a fusion process combining a neutron and a proton. Soon thereafter helium could be formed, but after a few minutes the temperature dropped to too low a value to permit any further nuclear fusion to occur. This portion of nucleogenesis ended with mostly hydrogen present, about 25% of the total mass in the form of helium, about 0.01% as deuterium, and only trace amounts of the berylium and lithium isotopes. Evidently, the heavier elements had to have been formed sometime after the big-bang early episode was over.

It is believed that after about 10^9 years, star formation by gravitational accretion began. Many of these early stars were massive and very hot and hence burned out rapidly, sliding into various stages of old age. At first they burned hydrogen in nuclear reactions leading to the production of more helium. Those that expired through a red giant phase continued to burn helium leading to the production of carbon. In those stellar interiors where temperatures between 2×10^8 K and 3.5×10^9 K existed fusion processes (those nuclear processes where net energy is released) were able to produce nuclei along the stability valley with masses up to the iron isotopes. Beyond iron the build-up of heavier elements requires net energy input and the energy-efficient

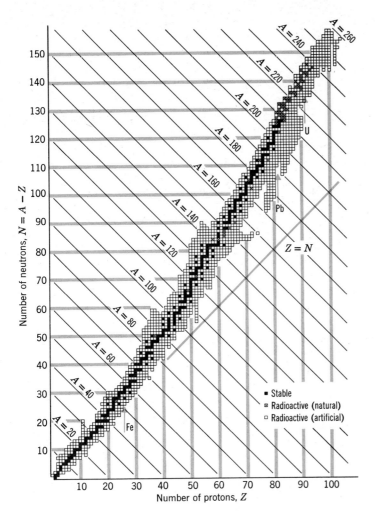

Figure 21.10 Chart of the known nuclei. The arrows indicate the positions of the elements uranium (the heaviest naturally occurring element), lead (the terminating element for most of the heavy radioactive nuclei), and iron (the most abundant element in the region near the maximum of the binding-energy curve). Notice that the N and Z coordinates are on opposite axes compared to Fig. 21.9. The lines of constant mass number are the diagonals marked with the values of A.

fusion processes are no longer possible. To produce the heavier elements required astronomical events of cataclysmic proportions, such as novae and super novae explosions. In these events the necessary energy is available for the heavier-element production processes. All stars, but particularly novae, supernovae, and planetary nebulae eject large fractions of their gaseous constituents into space. The universe, since the big bang, is now believed to be between 12×10^9 and 19.5×10^9 years in age.[3] Our Sun, however, is only 4.5×10^9 years old, and hence is a relatively new star. Thus, our Sun was formed partly out of the debris of earlier stars that have long since disappeared. This accounts for why our solar system contains heavy elements even though the Sun is a young star mostly consisting of hydrogen. The Sun is burning its hydrogen to form helium at a relatively low temperature otherwise incapable of forming any heavier elements directly.

Model calculations based on the nucleogenesis scenario described above lead to abundance predictions that closely resemble those found in nature.

---

[3] The predicated age depends on whether the universe is an open or closed system.

Radiation Exposures of Humans

Every person on Earth is continually exposed to various kinds of radiation from many different sources. Ordinarily, these radiations do us no particular harm. But even the most familiar of radiations—solar radiation—can do damage to the skin or eyes if the exposure is too great. Infrared radiation (from a heating element) or microwave radiation (from a microwave cooking oven) can also cause serious burns if used carelessly. Usually, however, when we speak of "radiation damage," we refer to the effects of high energy radiation, such as X rays from television sets or medical devices and γ rays or particles from radioactive materials. The more energetic X rays and γ rays can penetrate to any point in the human body, whereas ultraviolet radiation is absorbed completely in the skin. Therefore, X and γ radiation can affect the internal organs and nervous system, whereas the effects of UV radiation are generally limited to the exposed areas of the skin. (Overexposure to UV radiation will result in a sunburn, but a long-term effect of repeated overexposure can be the development of skin cancer.)

The high-energy radiation to which the general public is exposed is almost exclusively in the form of X rays or γ rays. Radiation workers, on the other hand, sometimes come into contact with materials that emit α and β particles. All of these radiations can produce biological damage by virtue of their ionizing action in living tissue. The doubly charged, slowly moving α particles from radioactive substances interact very effectively with the atomic electrons in matter and produce a high degree of ionization. The rate at which a single 5-MeV α particle deposits energy through ionization in a medium such as biological tissue is approximately 100 keV/μm. Consequently, in the wake of a moving α particle we find a dense collection of ions and electrons.

The electrons that are emitted in radioactive β decay have energies in the range from a few keV to 1 MeV or so. The corresponding electron velocities are very much greater than those of radioactive α particles. For this reason, an electron passing through matter does not remain near any atom for a time sufficient to interact effectively with the atomic electrons; the degree of ionization that is produced is therefore low. The rate at which a 1-MeV electron deposits energy through ionization in a medium such as tissue is only about 0.25 keV/μm. Consequently, in the wake of a moving electron we find only a diffuse collection of ions and knocked-out electrons.

The much smaller rate of energy loss with distance (or *linear energy transfer*, LET value) for an electron compared with an α particle implies that electrons will penetrate much farther into biological material. In fact, β-decay electrons have ranges in tissue of a few millimeters, whereas radioactive α particles will penetrate only to a depth of about 40 μm (0.04 mm). (The range of a 1-MeV electron in tissue is 4.2 mm, whereas the range of a 5-MeV α particle is 37 μm.)

When γ rays or X rays pass through matter, they interact with the medium via the photoelectric effect or the Compton effect (or by pair production if the photon energy is greater than $2m_ec^2 = 1.02$ MeV) and release energetic electrons. These electrons ionize the surrounding atoms in the same way that β-decay electrons do. Therefore, the characteristics of the ionization produced by β-decay electrons and by γ rays and X rays are the same. The difference between the effects of β-decay electrons and energetic photons is that the latter can penetrate to a substantial depth in matter before the first interaction. For example, the 1.2- and 1.3-MeV γ rays that follow ^{60}Co decay will penetrate about 10 cm of tissue before the incident intensity has been reduced by 50 percent. Consequently, the effect on, for example, the internal organs of the body will be much greater due to an exposure to γ rays or X rays than to an equal exposure to β particles or α particles from an external source. (Of course, if a radioactive material is inhaled or ingested, the effect on the internal organs due to α and β particles can be large.)

Gamma rays and X rays penetrate deeply into matter because they have no electric charge and therefore do not lose energy until they produce photoelectric or Compton electrons. All of the ionization that accompanies the passage of γ rays and X rays through matter is produced by the secondary electrons. Similarly, a neutron does not directly produce any ionization in passing through matter. When the neutron strikes a nucleus, the nucleus recoils as a result of the collision. As the nucleus moves through the surrounding atoms, some of the atomic electrons are stripped away. Thus, the collision produces ionization along the path of the recoiling nucleus. In biological material, which contains a large fraction of hydrogen, neutrons interact primarily with the nuclear protons of the hydrogen atoms. The knocked-on protons are the particles that produce almost all of the ionization in such materials when irradiated with neutrons.

Radiation Units

In order to specify the amount of radioactivity contained in a sample and the amount of radiation absorbed by an object, we make use of two units—the *curie* (Ci) and the *rad*. A curie of radioactivity represents exactly 3.7×10^{10} decay events per second (regardless of the type or energy of the radiation). A clinical source of ^{60}Co might contain several kilocuries (1 kCi = 10^3 Ci), whereas a millicurie (1 mCi = 10^{-3} Ci) of some radioisotope might be administered for internal radiotherapy.

Radiation exposure is measured in terms of a unit called the *rad*, which stands for *radiation absorbed dose*. If 1 kg of material absorbs 0.01 J of radiation energy, the dose is said to be 1 rad:

$$1 \text{ rad} = 0.01 \text{ J/kg} \tag{1}$$

A person standing at a distance of 1 m from a 1-Ci source of ^{60}Co for one hour would receive a dose of approximately 1.2 rad at the front surface of his or her body and a dose of about half this amount at a depth of 10 cm (because of the attenuation of the radiation in the body). It is important to remember that the rad is a measure of the absorbed radiation dose *per kilogram*.

Radiation Exposures

The biological effect of radiation depends not only on the absorbed dose in rads but on several other factors as well. These factors include the LET value of the radiation, the rad distribution within the tissue, as well as certain biological and chemical variables. It has therefore become standard practice to specify the biological damage produced by radiation in terms of a *dose equivalent* measured in *rem*:

$$\text{Dose equivalent in rem} = (\text{absorbed dose in rad}) \times \text{QF} \tag{2}$$

where QF is the *quality factor* of the particular radiation. When a 5-MeV α particle deposits its energy in a dense ionization track through a section of tissue, it does considerably more damage to the tissue than when a number of electrons deposit the same amount of energy in the tissue. Thus, we say that the quality factor of α particles is much greater than that for electrons. QF values can only be approximate because the effectiveness of a particular radiation in producing biological damage depends on many variables. Some working values for the quality factors for different radiations are listed in the following table.*

Radiation	QF (approximate)
X or γ ray	1
Electrons (β particles)	1
α particles	20
Protons	10
Fast neutrons (~MeV)	10
Slow neutrons (~eV)	5

*Notice that *slow neutrons* (neutrons with kinetic energies of a few eV or less) have a rather large QF value even though they can impart very little energy to a nucleus in a collision. However, slow neutrons can produce energetic secondary radiations when they are captured by nuclei and induce nuclear disintegrations.

Thus, if a person receives a 0.2-rad dose of α particles (a substantial dose!), the exposure is measured as (0.2 rad) × (20) = 4 rem. If the exposure is entirely to X and γ radiation or electrons, the dose equivalent in rem is equal to the dose in rad.

The average per capita annual dose for the U.S. population for various sources of radiation is shown in the following table (taken from the *BEIR Report*—see References).

Source	Dose (mrem/y)
Natural radiation	
Cosmic radiation	44
Radionuclides in body	18
External γ radiation	40
Total	102
Man-made radiation	
Medical/dental	73
Fallout	4
Occupational exposure	0.8
Nuclear power (1970)	0.003
Nuclear power (2000)	<1
Total	Approximately 80
Grand total	Approximately 182

(Notice that the values are in *millirem*, mrem.) The figures in this table are *average* values. Variations result because some parts of the country have more natural radioactivity than others; the intensity of cosmic radiation depends on altitude—the resident of Denver receive 50 percent more cosmic radiation than the residents of San Francisco; some wristwatches have luminous dials that contain radium; and so forth. Also, individuals receive widely varying amounts of exposure to medical and dental X radiation.

References

Advisory Committee on the Biological Effects of Ionizing Radiations, Division of Medical Sciences, National Academy of Sciences/National Research Council, *The Effects on Populations of Exposure to Low Levels of Ionizing Radiation*, U.S. Government Printing Office, Washington, D.C., 1974. This is often referred to as the *BEIR Report*.

L. Stanton, *Basic Medical Radiation Physics*, Appleton-Century-Crofts, New York, 1969.

■ Exercises

1 A certain individual receives a whole-body dose of 8 rem of γ radiation, whereas another individual receives a dose of 700 mrad of α particles by inhaling a radioactive material. Which individual will probably suffer the greater biological damage?

2 An accelerator produces a beam of energetic electrons. The beam emerges from the vacuum region of the accelerator through a thin window. Just outside the window the electrons have an energy of 5 MeV. The beam is distributed uniformly over an area of 1 cm² and the beam current is 10 nA (10^{-8} A). A worker in the laboratory accidentally walks close to the beam, exposing his arm to the beam near the window for 1 s. The range of 5-MeV electrons in tissue is 2.2 cm. What radiation dose (in rad) does the exposed tissue receive? (Assume that the tissue has the consistency and properties of

water.) What will be the amount of temperature rise in the tissue (assuming that the energy is deposited uniformly)? Will there by any thermal effects of the exposure? How much energy (in eV) does each molecule receive on average? Is this sufficient to ionize the molecules? What can you conclude about the way that electrons produce biological damage?

21.3 Nuclear Reactions

Neutron Capture

A nuclear *reaction* is a process in which nucleons are added to, removed from, or rearranged within a target nucleus under bombardment by nucleons (that is, protons or neutrons), by groups of nucleons (for example, deuterons or α particles),[4] or by γ radiation. The simplest type of nuclear reaction is that in which a nucleus with mass number A captures a neutron and forms the nucleus $A + 1$ with the accompanying emission of a γ ray. We have already discussed just such a neutron capture process (Section 21.1), namely, the capture of a neutron by a proton to form a deuteron:

$$^1H + n \longrightarrow {}^2H + \gamma$$

It is customary to use a shorthand notation for specifying nuclear reactions. In this notation, the above reaction is written as

$$^1H(n, \gamma)^2H$$

where the first quantity (^1H) specifies the target nucleus and the last quantity (^2H) specifies the final (or *residual*) nucleus; within the parentheses, the first quantity (n) is the incident particle and the second quantity is the outgoing particle (in this case, a γ ray). Neutron capture by ^{12}C, for example, is written as $^{12}C(n, \gamma)^{13}C$.

If the incident neutron has negligibly small kinetic energy, a neutron capture reaction always releases an amount of energy equal to the separation energy of a neutron in the final nucleus. For the case of the deuteron, we found in Section 21.1 that this energy release in 2.224 MeV.

Charged-Particle Reactions

In 1919, Rutherford used α particles from a radioactive source to bombard nitrogen gas and found that occasionally an α particle would react with a nitrogen nucleus to produce a proton. The nuclear reaction he observed (the first such to be discovered) was

$$^{14}N + {}^4He \longrightarrow {}^{17}O + {}^1H$$

or, in the shorthand notation, $^{14}N(\alpha, p)^{17}O$, where we follow custom and write the bombarding and outgoing particles in *nuclear* rather than in *atomic* notation (α for ^4He and p for ^1H). A cloud-chamber[5] photograph of a $^{14}N(\alpha, p)^{17}O$ reaction was first taken in 1925 by P.M.S. Blackett; this photograph is reproduced in Fig. 21.11.

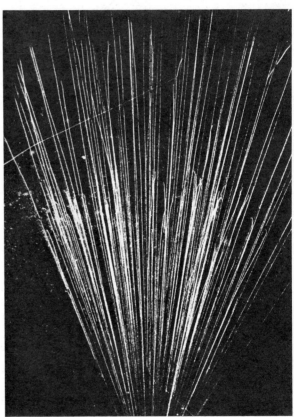

P. M. S. Blackett

Figure 21.11 Cloud-chamber photograph of a $^{14}N(\alpha, p)^{17}O$ reaction amidst the tracks of many α particles that do not induce reactions. A proton moves to the left and downward from the reaction site (near the top of the photograph).

---

[4] Other complex projectiles, such as ^3H, ^3He, ^6Li, ^7Li, ^{12}C, and ^{16}O ions, have also been used to initiate nuclear reactions. High energy electrons and mesons have also been used.

[5] A *cloud chamber* is a device that renders visible the track of a nuclear particle by virtue of the condensation of water droplets on the ions left in the wake of the particle. These devices have now been largely supplanted by *bubble chambers* (see Fig. 13.18).

In all nuclear reactions we must have a balance of protons and neutrons in the initial and final states. For the $^{14}\text{N}(\alpha, p)^{17}\text{O}$ reaction, we have

$$^{14}\text{N} + {}^4\text{He} \longrightarrow {}^{17}\text{O} + {}^1\text{H}$$

No. protons: $7 + 2 = 8 + 1$
No. neutrons: $7 + 2 = 9 + 0$

Using this procedure, we can always identify the fourth nucleus in a reaction if the other three nuclei are known.

● **Example 21.3**

^{10}B is bombarded with neutrons and α particles are observed to be emitted. What is the residual nucleus?

$$^{10}\text{B} + n \longrightarrow (?) + {}^4\text{He}$$

No. protons: $5 + 0 = Z + 2$
No. neutrons: $5 + 1 = N + 2$

Clearly, $Z = 3$ and $N = 4$; therefore, the residual nucleus is ^7Li. ■

The mass-energy balance in a nuclear reaction can be calculated by using exactly the same method as that used in calculating binding energies, namely, we compare the combined mass of the target nucleus and the bombarding particle with the combined mass of the residual nucleus and the outgoing particle. For the $^{14}\text{N}(\alpha, p)^{17}\text{O}$ reaction, we have

$$m(^{14}\text{N}) = 14.003\ 074\ \text{u}$$
$$m(^4\text{He}) = 4.002\ 603\ \text{u}$$
$$\overline{m(^{14}\text{N}) + m(^4\text{He}) = 18.005\ 677\ \text{u}}$$

$$m(^{17}\text{O}) = 16.999\ 133\ \text{u}$$
$$m(^1\text{H}) = 1.007\ 825\ \text{u}$$
$$\overline{m(^{17}\text{O}) + m(^1\text{H}) = 18.006\ 958\ \text{u}}$$

The *energy* difference between the initial and final pairs of particles is

$$\Delta m \times c^2 = \{[m(^{14}\text{N}) + m(^4\text{He})]$$
$$- [m(^{17}\text{O}) + m(^1\text{H})]\} \times c^2$$
$$= (-0.001\ 281\ \text{u}) \times (931.5\ \text{MeV/u})$$
$$= -1.193\ \text{MeV}$$

That is, the mass of $^{14}\text{N} + {}^4\text{He}$ is *less* than the mass of $^{17}\text{O} + {}^1\text{H}$; therefore, an energy of 1.193 MeV must be supplied to the $^{14}\text{N} + {}^4\text{He}$ system in order for the reaction to take place. The requisite energy can be supplied by the kinetic energy of the incident α particle.

The difference in mass between the initial and final states in a nuclear reaction is called the *Q-value* for the reaction and is usually given in energy units. A *positive* Q-value means that energy is *released* in the reaction, and a *negative* Q-value means that energy must be *supplied*. For example,

$$^1\text{H} + n \longrightarrow {}^2\text{D} + \gamma; \qquad Q = +2.224\ \text{MeV}$$
$$^{14}\text{N} + {}^4\text{He} \longrightarrow {}^{17}\text{O} + {}^1\text{H}; \qquad Q = -1.193\ \text{MeV}$$

21.4 Nuclear Fission

The Discovery of Fission

Shortly before his death in 1937, Lord Rutherford stated that "the outlook for gaining useful energy from the atoms by artificial processes of transformation does not look very promising." Although Rutherford's intuition in scientific matters was almost always infallible, within a few years, a series of scientific and technological advances had shown this particular view to be incorrect—incorrect, in fact, to an astonishing degree.

In 1939, the German radio-chemist, Otto Hahn,[6] in collaboration with Fritz Strassman, bombarded uranium with neutrons and performed very careful chemical tests on the resulting radioactive material. They found that among the products of neutron absorption by uranium there was radioactive barium ($Z = 56$) an element much less massive than the original uranium. How could such a light element be formed from uranium? The mystery was soon resolved by Lise Meitner and Otto Frisch, German physicists working then as refugees in Sweden, who suggested that neutron absorption by uranium produced a breakup (or *fission*) of the nucleus into two light fragments:

$$\text{U}(Z = 92) + n \longrightarrow \text{Ba}(Z = 56) + \text{Kr}(Z = 36)$$

This was a startling new type of nuclear reaction. Instead of exchanging only a few nucleons between the incident particle and the target nucleus, as in an (α, p) reaction, this discovery showed that it was possible to split a nucleus into two massive parts.

The Dynamics of Fission

As shown in Fig. 21.2, the binding energy of a heavy nucleus ($A \cong 240$) is approximately 7.5 MeV per nucleon. If such a nucleus were separated into two

---

[6] Otto Hahn (1879–1968), was one of the chief contributors to the discovery of fission. Hahn was awarded the 1944 Nobel Prize in chemistry.

21.4 Nuclear Fission 547

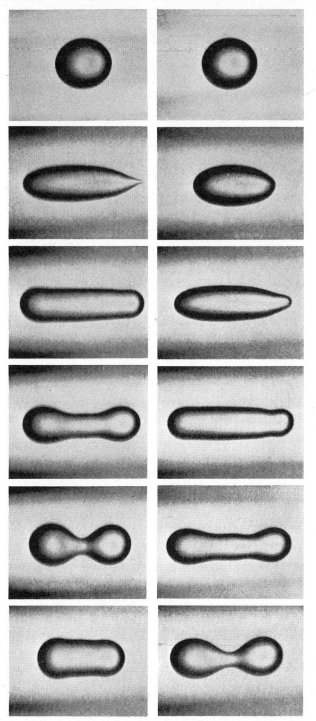

Figure 21.12 Photographs from a motion picture film showing the variation in shapes of an ordinary drop of water suspended in oil when a deformation is induced by a voltage applied across the oil. Each sequence is continued on the following page. In the left-hand sequence, the drop returns to its initial spherical shape without undergoing fission. In the right-hand sequence, the initial deformation is sufficiently large that the drop fissions. In 1939, Niels Bohr and John Wheeler proposed a liquid-drop model of nuclear fission that was successful in explaining the general features of the fission process.

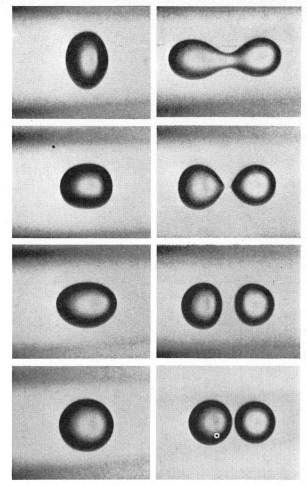

Stanley G. Thompson, Lawrence Radiation Laboratory

parts, each with $A \cong 120$, the binding energy would be *increased* to approximately 8.5 MeV per nucleon. That is, the decrease in binding energy per nucleon with increasing mass number means that a heavy nucleus can break up into two light fragments with the release of a substantial amount of energy. This breakup of a heavy nucleus is somewhat analogous to the splitting apart of a vibrating drop of liquid, as shown in Fig. 21.12. Even though it is energetically favorable for a heavy nucleus to split into two parts, this process is inhibited by the strong attractive nuclear forces. The nucleus may become extended in an effort to fission (as in the left-hand sequence in Fig. 21.12), but it will usually return to and vibrate around its equilibrium shape. The probability of the occurrence of *spontaneous fission* is extremely small, and therefore the corresponding half-life is extremely long ($\sim 10^{17}$ years for ^{235}U). If, however, some additional energy is supplied to the nucleus in the form of the binding energy of a captured neutron, this increase in energy may produce a large nuclear deformation which will be sufficient to permit the relatively easy separation of the nucleus into two fragments; thus, fission can occur (as in the right-hand sequence in Fig. 21.12). For many heavy nuclei, the probability for the occurrence of *neutron-induced fission* is extremely high (half-life about 10^{-21} s).[7]

The fragments that result from the fission of a heavy nucleus do not have equal masses. In fact, the mass numbers for the two fragments are usually quite different. Figure 21.13 shows the distribution of mass numbers observed for the case of the neutron-induced fission of ^{235}U. The most probable pair of mass numbers is 95 + 139. (The sum does not equal 235 + 1 = 236 because, on the average, 2-3 neutrons are released during the fission process.) Two typical fission reactions involving ^{235}U are

$$^{235}\text{U} + n \longrightarrow {}^{139}\text{Ba} + {}^{95}\text{Kr} + 2n \brace \longrightarrow {}^{144}\text{La} + {}^{89}\text{Br} + 3n} \quad (21.6)$$

The isotopes ^{139}Ba, ^{95}Kr, ^{144}La, and ^{89}Br are all on the neutron-rich side of the valley of nuclear stability (Fig. 21.10); consequently, these fission fragments (as do almost all fission fragments) undergo radioactive β^- decay.

The way in which fission energy is distributed, on the average, among the various fission products is indicated in Table 21.7. If the body of the material in which the fission event takes place is sufficiently large that all of the fission products are absorbed (except the neutrinos, which escape), a total of approximately 190 MeV will be converted into heat energy. This is truly an enormous amount of energy. If 1 kg of ^{235}U undergoes fission, approximately 8×10^{13} J of energy is released. This amount of energy is sufficient to raise the temperature of 65,000,000 gallons of water from room temperature to the boiling point.

Chain Reactions

The fact that the fission process releases several neutrons (2.5, on the average, in the case of ^{235}U) makes possible a series or chain of neutron-induced fission events that is self-sustaining. If one neutron from a fission event triggers the fission of another nucleus and one neutron from this event triggers another fission, etc., this series of fission events will sustain itself and will constitute a *chain reaction*. By controlling the environment of the fissioning nuclei

---

[7] The capture of neutron by ^{235}U increases its probability for undergoing fission by a factor of about 10^{45}!

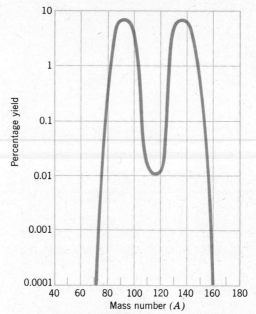

Figure 21.13 The distribution of fission-fragment mass numbers resulting from the neutron-induced fission of ^{235}U. The most probable mass numbers are 95 and 139. The curve represents the average trend of the measured mass fractions.

Table 21.7 Distribution of Energy Among Fission Products

Product	Average Energy (MeV)
Fission fragments (kinetic energy)	168
Fission neutrons (kinetic energy)	5
γ rays (prompt)	5
β^- from fission fragments (kinetic energy)	5
γ rays from fission fragments	7
Neutrinos from β^- decays	10
	200 MeV

it is possible to maintain a condition in which each fission event contributes, on the average, exactly one neutron that triggers another event. In this way, the rate of energy generation (the *power*) is maintained at a constant level. The controlled fission chain reaction (Fig. 21.14) is the principle of the *nuclear reactor*, now widely used in the commercial generation of electric power.

It is also possible to bring together in a small volume a sufficient amount of fissionable material so that fewer of the fission neutrons escape the system and therefore more than a single neutron from each event can trigger a new event. Figure 21.15 shows a series of fission reactions in which each event contributes *two* neutrons toward the next set of events. The rapid multiplication of the number of fissioning nuclei in this uncontrolled situation leads

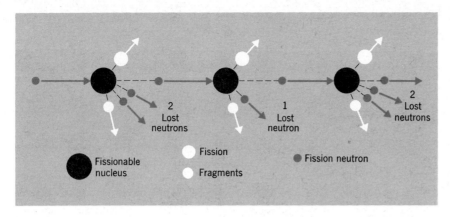

Figure 21.14 A controlled fission chain reaction in which one neutron from each fission event triggers another event. One or two neutrons from each fission event escape the system and are "lost."

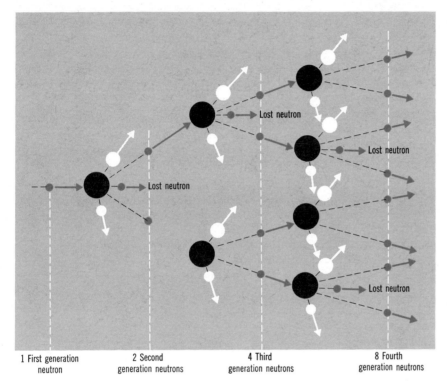

Figure 21.15 An uncontrolled series of fission events. The rapid release of the fission energy in such a system leads to an explosion.

to the explosive release of the fission-generated energy—this is the principle of the atomic bomb (which is, of course, actually a *nuclear* bomb).

In order for the uncontrolled release of fission energy to take place, as many of the fission neutrons as is possible must be kept within the material. Unless the fissionable material has a mass greater than a certain value (called the *critical mass*), too many neutrons will escape the system and the rate of energy release will be too slow for an explosion to occur. The problem in constructing an atomic bomb, therefore, is to bring together into a small volume an amount of fissionable material at least as great as the critical mass. This assembly of the critical mass must be accomplished in an extremely short period of time (about 10^{-3} s), because otherwise a slow, nonexplosive series of fission events will take place. One of the methods devised to overcome this problem is to drive together two or more subcritical masses by means of a conventional (chemical) explosion. The original atomic bombs of 1945 contained several kilograms of fissionable material and were detonated in this way.

Because such a huge amount of energy is released in a very brief time in a localized space, the destructive effect of a nuclear weapon is incredibly large. In 1945 the only two nuclear weapons to be used in warfare devastated the Japanese cities of Hiroshima and Nagasaki. About 100,000 persons (approximately one-quarter of the population of the

21.5 Nuclear Reactors

Basic Components

What are the features that a nuclear reactor must have in order to exploit the release of energy in the fission process? First, there is the matter of the proper fuel. The uranium isotope ^{235}U readily undergoes fission when it absorbs a slowly moving neutron; ^{235}U is a good nuclear fuel. Unfortunately, ^{235}U constitutes only a small part (0.7 percent) of natural uranium. The abundant uranium isotope, ^{238}U, does not undergo fission when struck by a slow neutron and so this isotope is not useful in the types of reactors that employ ^{235}U. Because ^{235}U is such a small fraction of natural uranium (about 1 atom in 150, elaborate measures must be taken to enrich the ^{235}U content of any uranium sample before it becomes useful as a nuclear fuel. The design and construction of an enrichment facility was one of the major problems that had to be solved by the scientists and engineers of the Manhattan Project (the atomic bomb project) during World War II. Several techniques were devised to remove preferentially some of the ^{238}U from a sample, thereby increasing the fraction that consists of ^{235}U. The methods developed at that time are still used today to provide uranium enriched in ^{235}U for nuclear reactors. Atomic weapons and some special types of reactors require uranium that is enriched to 90 percent or more of ^{235}U. However, most commercial power reactors operate with fuel that contains no more than about 3 percent ^{235}U.

The fission of ^{235}U is most efficient when the neutrons that are absorbed have relatively low speeds and correspondingly small energies. Most fission events in a sample of ^{235}U are triggered by neutrons that have energies below 0.1 eV. On the other hand, the neutrons that are released in the fission process have energies of several MeV. Therefore, a nuclear reactor must contain some material that will slow down (or *moderate*) the rapidly moving fission neutrons to the point that they will efficiently trigger additional fission events. What material should be used as a neutron moderator? The maximum amount of energy that can be transferred from a moving particle with mass m occurs when a head-on collision is made with a particle that has the same mass. In such a case *all* of the kinetic energy is transferred to the struck particle and the colliding particle is left at rest. Of course, very few particle-particle collisions are actually head-on, but even in a collision that results in the deflection of the incident particle at some angle, an appreciable fraction of the energy can be transferred. (Refer to Example 7.9.) Therefore, the material that would be most effective in slowing down fast neutrons is *hydrogen*. We need a neutron moderator with a density higher than that provided by a gas, so ordinary water, H_2O, is often used.

Although hydrogen is an efficient moderator, there is a difficulty in using hydrogen in a reactor. In a neutron-proton collision, a capture reaction (Fig. 21.1) will sometimes occur instead of a deflection. A capture reaction removes the neutron from the supply that initiates further fission reactions. Indeed, in order to compensate for losses due to capture reactions, it is necessary that more fissionable nuclei be made available; that is, the uranium fuel must be *enriched*.

One class of nuclear reactors uses deuterated water, D_2O, as the moderator instead of ordinary water, H_2O, because deuterium nuclei do not absorb neutrons as readily as do the nuclei of ordinary hydrogen. The neutron losses are minimized in such reactors and they can operate with natural, unenriched uranium. These reactors are called *heavy-water* reactors; those that use ordinary water are called *light-water* reactors. At present, heavy-water reactors are being built only in Canada. One of the problems with the commercial use of these reactors is the high cost of the heavy-water moderator.

In the United States we have adopted the attitude that we will simply accept the losses due to the capture of neutrons by hydrogen and will make up for these losses by using enriched uranium fuel. All commercial power reactors now in service in the United States are light-water reactors.

Water is used in reactors not only to moderate the fission neutrons but also to transfer the fission energy to the steam loop (Fig. 21.16) that operates the electric generators. When a fission event takes place, the fragments move away from the fission site with high speeds. These fragments collide with and set into motion the atoms along their paths. Thus, the binding energy of the uranium nucleus is first changed into the kinetic energy of the fragments and then into the thermal energy of the system. Water is pumped through the region of the reactor where the fission events are taking place (the reactor *core*) and absorbs some of the heat that is being produced. Water is therefore both the *moderator* and the *coolant* in the reactor. The steam that is pro-

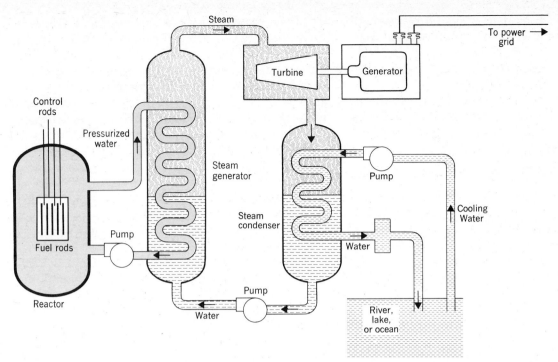

Figure 21.16 Schematic diagram of a nuclear power plant using a pressurized water reactor (PWR).

duced by the heating of the water is used to drive turbine electric generators.

There is one final ingredient that is necessary for a reactor to operate productively and safely. If the power output of a reactor is to remain constant, then we must ensure that exactly one neutron from each fission event triggers a new fission event. If this number (which is called the neutron *multiplication factor*) is slightly less than one, the power output will gradually decrease; similarly, if the multiplication factor is slightly greater than one, the power output will grow. Neither of these situations is desirable, so some control mechanism must be provided to maintain the multiplication factor at exactly one. This is done by introducing into the reactor a number of movable *control rods*. The rods contain an element (usually boron) that readily absorbs neutrons. If the multiplication factor begins to exceed one (rising power level), the control rods are lowered into the reactor core by a small amount. The increased absorption of neutrons by the rods causes the multiplication factor to decrease. Conversely, if the multiplication factor falls below one, the rods are withdrawn slightly and fewer neutrons are absorbed. By inserting the control rods fully into the reactor core, the reactor can be shut down for routine maintenance or in an emergency situation.

Reactor Construction

As we have seen, nuclear reactors consist of four basic parts: the fuel, the moderator, the coolant, and the control rods. In many reactors, the moderator and the coolant are the same. Figure 21.16 shows in a schematic way where these various parts are located in a reactor. First, notice that the uranium fuel is in the form of long, thin rods (about 1 cm in diameter) and that these rods are spaced apart with moderating material between. There are two reasons for using this geometry. When a reactor is operating at any reasonable power level, an enormous amount of heat is being generated by the fission events within the fuel rods. The rods are made thin so that this heat can be transferred quickly to the coolant material (usually water). If the rods were too thick, the heat could not be removed rapidly enough, and the temperature would rise to the point at which the rods would melt. Also, the thinness of the rods permits the fission neutrons to escape into the moderator where they are slowed down. The moderated neutrons then drift back into the fuel rods and trigger new fission events. (Incidently, spacing apart the fuel rods in this way means that too many neutrons escape from the fissionable material for an explosion to take place; a reactor *can never explode* as can an atomic bomb.) In a typical commercial power reactor, there will be about 40,000 fuel rods, each containing about 2 kg of uranium oxide, UO_2. Each fuel rod has a useful life of about three years in the reactor before it must be replaced.

Next, look at the cooling system. Water in a closed system is circulated through the reactor by means of a pump. In the type of reactor illustrated in Fig. 21.16, the cooling water is at high temperature (about 325°C) and at high pressure (about 150 atm). This type of reactor is called a *pressurized water reactor* (PWR). When the high-temperature water passes through the steam generator, heat is transferred to the water in the second closed system and steam is produced. The steam is forced through a turbine which drives an electric generator. Finally, the steam is condensed by cooling water drawn from an external reservoir.

The basic method of producing electric power in a nuclear power plant is the same as that in a conventional oil- or coal-burning plant. In each case, fuel is consumed and heat is produced. This heat converts water into steam and the steam drives an electric generator. A nuclear reactor is simply a device for generating heat!

The Nuclear Power Industry

In the United States, as of 1984, nuclear reactors (approximately 85) produce about 13 percent of our electric power. In certain parts of the country the percentage is considerably higher. For example, in the Chicago area about 30 percent of the electric power is nuclear. It is anticipated that the fraction of the total U.S. electric generating capacity represented by nuclear plants will be about 20 percent in the mid-1990s.

Not only are nuclear power plants expensive to construct, but they have unique safety problems. These safety problems are associated with the fact that the fission process inevitably produces radioactive products and, consequently, intense radioactivity builds up in the reactor fuel rods. There are two important ways in which the general public might become exposed to reactor-produced radioactivity. First, the radioactivity in the fuel rods might become dispersed in some catastrophic accident. Second, radioactive reactor wastes must be stored for long periods of time (because the half-lives of some of the fission products are quite long) and some of this activity might escape into the atmosphere or into the public water supply. (For example, ^{131}I has a half-life of only 8 days, but the half-life of ^{239}Pu is 24,000 years.)

Even though a reactor cannot *explode,* there is the possibility that it could *melt down*. If a reactor's cooling system were to fail, the control rods would automatically shut down the reactor. But there would still be so much heat generated by the radioactivity in the fuel rods that the hot radioactive material could melt its way through the reactor vessel. Even if the vessel were breached, there is a double shielding wall surrounding the reactor (Fig. 21.17). The first of these is a thick concrete shield immediately around the reactor vessel. The second is a very large reinforced concrete dome that is lined with steel plate that encloses the reactor and all of the heat exchange equipment. The primary purpose of this outer containment vessel is to hold any radioactive dust or gas that might be released in an accident. This vessel is strong enough to withstand the impact of a jetliner. In the unlikely event that both of these shielding walls are ruptured, then the radioactive material could escape to the outside world.

Experience with serious reactor accidents is very limited. Studies of the possibility of a catastrophic release of radioactivity in a meltdown have shown that the probability of such an occurrence is extremely remote. The only accident involving a U.S. power reactor ever to release a measurable amount of radioactivity occurred at Three Mile Island in 1979. Only small amounts of radioactivity escaped and no injuries resulted. This accident—which involved both equipment failure and human error—has prompted a thorough reexamination of reactor safety procedures. Also the question of the disposition of reactor wastes is not settled. It appears that burial in certain particularly stable geologic formations will ensure safe disposal but no final decision has yet been made.

Plutonium and Breeder Reactors

When a ^{235}U nucleus absorbs a slow neutron, it undergoes fission. But when the abundant uranium isotope, ^{238}U, absorbs a slow neutron, a capture reaction takes place:

$$^{238}U + n \longrightarrow {}^{239}U$$

The resulting isotope, ^{239}U, is radioactive and undergoes β^- decay to form the transuranic element *neptunium* ($Z = 93$):

$$^{238}U + n \longrightarrow {}^{239}U$$

$$^{239}U \xrightarrow{\beta^-} {}^{239}Np \quad (\tau_{1/2} = 23.5 \text{ min})$$

Figure 21.17 Cutaway view of a nuclear steam supply system. The reactor vessel is in the center and is surrounded by a thick concrete shield. Large pipes connect the reactor vessel with two steam generators on either side. The steam generators are large heat exchangers in which hot, high-pressure water heats a second source of water which boils to steam and leaves through the top to drive a turbine-generator. Note the size comparison of the men on the refueling crane over the reactor vessel. (Photo courtesy of C-E Power Systems, Combustion Engineering, Inc.)

^{239}Np is itself radioactive and forms the element *plutonium* ($Z = 94$) by β^- decay:

$$^{239}\text{Np} \xrightarrow{\beta^-} {}^{239}\text{Pu} \quad (\tau_{1/2} = 2.35 \text{ days})$$

The end product of this sequence, ^{239}Pu, is again a radioisotope, but the half-life is so long (24,000 years) that appreciable quantities can be accumulated. The interesting and important feature of ^{239}Pu is that it undergoes fission upon the absorption of a slow neutron just as ^{235}U does. That is, ^{239}Pu is as good a nuclear fuel as ^{235}U. In fact, one of the two atomic weapons detonated over Japan in 1945 was constructed from ^{239}Pu, and plutonium is used in many of the weapons in the nuclear arsenals around the world today.

When ^{235}U undergoes fission, 2.5 neutrons are produced, on the average, for each fission event. In a nuclear power reactor we want to maintain a constant power output, so one of these neutrons is necessary to trigger a new fission event. If we could arrange for one additional neutron per event to be captured by a nucleus of ^{238}U, then it would be possible to produce one nucleus of ^{239}Pu for each nucleus of ^{235}U consumed. That is, just as much new fuel would be produced as is used. Not only is power generated in the process, but ^{238}U is transformed into a useful fuel at the same rate that ^{235}U disappears. A reactor that is designed to accomplish this is called a *breeder reactor*. The ratio of the amount of fuel produced to the amount consumed is called the *breeding ratio*. In a standard light-water reactor (in which the uranium fuel consists of 3 percent ^{235}U and 97 percent ^{238}U), about one plutonium nucleus is produced for every three ^{235}U nuclei consumed; that is, the breeding ratio is 0.33. A reactor is a true breeder reactor only if it produces more fissionable material than it consumes; that is, the breeding ratio must exceed one.

A breeder reactor is quite different from the light-water, slow-neutron reactors we have been discussing. First of all, in order to decrease the number of neutrons lost by capture in surrounding material and to increase the number that are captured by ^{238}U, the reactor is operated without a moderator. That is, the various events are initiated by *fast* neutrons instead of *slow* neutrons. Because it operates without a moderator, the core of a breeder reactor is much smaller than that of a conventional reactor. With the same amount of power being produced in a much smaller volume, the thermal problems are therefore much more severe in a breeder reactor. Water cannot be used to extract the heat from the core of a breeder reactor because (a) water would act also as a moderator and (b) water is not an efficient medium to use for heat transfer at the high temperatures that exist in a breeder reactor. Molten sodium is the coolant in most of the breeder reactors now operating or under study. These reactors are called *liquid metal fast breeder reactors* (LMFBRs).

Because of the high concentration of power in the core and the necessity of using a liquid metal coolant, a breeder reactor is a very complex device. Nevertheless, breeder reactors have been successfully operated in France, Britain, and the Soviet Union for several years. The performance of these reactors has been exceptionally good, and the construction of large plants is underway.

Biological Effects of Radiation Exposure in Humans

The effects of radiation can be classified as *somatic* (effect on the individual exposed) or *genetic* (effect on the offspring of the individual exposed). In addition, we classify an exposure as *acute* if it is all received within a relatively short time interval (typically, seconds to hours) or *chronic* if it is received over a long time interval (typically, years to decades). The somatic effects for acute whole-body exposures at various radiation dose levels are summarized in the table below.

The delayed somatic effects of acute radiation exposure have been studied by the careful monitoring of the survivors of the Hiroshima and Nagasaki atomic bomb explosions of 1945. This work has been carried out since 1947 by the Radiation Effects Research Foundation (formerly, the Atomic Bomb Casualty Commission). The most notable of the radiation-related disorders detected in this group have been increased occurrences of lenticular opacities (eye cataracts), thyroid tumors, leukemia, chromosome aberrations in blood lymphocytes, and a slight impairment of growth of those who were exposed during early childhood. Recent findings indicate an increased incidence of solid tumors, most apparent in breast and lung carcinoma. It is to be especially noted that an increased risk of cancer still exists in this group more than 30 years after the exposure.

Another effect of radiation exposure is the shortening of lifespan. One estimate places the effect at about 2.5 days/rem for chronic exposure and about 10 days/rem for acute exposure. (Some other estimates are smaller.)

The table and the discussion at the top of the next page refer to *whole-body* exposures. In radiation therapy and in some accidental exposures, only certain or-

γ-Ray Whole-Body Dose (rem)	Effect (acute exposure)	Remarks
0–25	None detectable	
25–100	Some changes in blood but no great discomfort, mild nausea.	Some damage to bone marrow, lymph nodes, and spleen.
100–300	Blood changes, vomiting, fatigue, generally poor feeling.	Complete recovery expected; antibiotic treatment.
300–600	Above effects plus infection, hemorrhaging, temporary sterility.	Treatment involves blood transfusions and antibiotics; severe cases may require bone marrow transplants. Expected recovery about 50 percent at 500 rem.
> 600	Above effects plus damage to central nervous system.	Death inevitable if dose >800 rem.

Organ	Effect	Tolerance dose (rad)[a]	
		For 5% probability of effect in 5 y	For 50% probability of effect in 5 y
Skin	Ulcer, severe fibrosis	5,500	7,000
Stomach	Ulcer, perforation	4,500	5,000
Liver	Failure, ascites	3,500	4,500
Kidney	Nephrosclerosis	2,300	2,800
Heart	Pericarditis, pancarditis	4,000	>10,000
Bone	Necrosis, fracture	6,000	15,000
Eye lens	Cataract	500	1,200
Thyroid	Hypothyroidism	4,500	15,000
Muscle (child)	No development	2,000–3,000	4,000–5,000
Bone marrow	Hypoplasia	200	550
Fetus	Death	200	400

[a]Because medical irradiations are almost always made with X rays, γ rays, or electrons, the rad dose and the rem dose are the same. Usually, doses are specified in rad.

gans or regions of the body are irradiated. The dose that most organs can tolerate is considerably higher than if the entire body is exposed. Some *tolerance doses* for particular organs are listed in the table above (from Rubin and Casarett—see References). It is apparent that very substantial doses must be given to particular organs or parts of organs to produce any effect in radiation therapy procedures.

Genetic effects of human exposure to radiation are much more difficult to identify than are somatic effects. We do know, however, that radiation can produce mutations through unrepaired breaks in the chromosomes or through changes in the sequence of bases in the DNA chain. The radiation level at which mutation production becomes important is not known. In this regard it is interesting to note that extensive studies in Japan have so far failed to detect any genetic effects in the offspring of the exposed population. However, a sufficient time has not yet passed to reveal any damage to recessive genes. A significant number of anomalies would not be expected so soon in the genetically small sample of exposed survivors (about 114,000).

As far as somatic effects are concerned, the biological damage resulting from the accumulated dose due to a chronic exposure is decidedly *less* than if the same dose were received in an acute exposure. The reason for this effect is that the body has an opportunity to repair the damage produced by small doses spaced in time. This may not be true for genetic effects (and probably for some somatic effects such as leukemia susceptibility). Thus, there may be no level of radiation exposure below which there is zero damage to humans. All radiation is probably harmful to some extent.

At the present time, the policy is to acknowledge that there are definite benefits in various operations that involve the production and use of radiation (such as in medical diagnostics and therapeutics and in the generation of electric power by nuclear reactors), so that *some* extra exposure of the general public to radiation is inevitable. Currently, the maximum permissible amount of radiation to which an individual may be exposed is set at 500 mrem per year over and above the dose due to medical and dental irradiations (and, of course, the natural background). However, not every individual can be monitored. This is especially true of persons who do not work with radiation and may therefore be unaware of any radiation in their environment. For this group (the vast majority of the population), the radiation level must be maintained sufficiently low so that no individual is likely to receive a dose greater than 170 mrem/y. It is believed that this level of exposure would not unduly burden the population in terms of increased radiation risks.

There are opponents to this view, however, and they argue that steps should be taken to lower the amount of nonnatural radiation to which the population is exposed. This could be accomplished in a number of ways—for example, by stopping the testing of explosive nuclear devices of any kind, by using diagnostic X rays only when absolutely required, by placing more stringent regulations on the allowable radiation from television sets and microwave cooking ovens, and by halting the proliferation of nuclear power plants. Some of these measures appear to be desirable steps, whereas the population may be unwilling to accept the additional inconvenience and change of life style associated with others.

Nuclear Medicine

X rays have long been used in medicine and dentistry, particularly in the examination of bones and teeth and in the treatment of certain skin conditions. More recently, radioactive isotopes have come into wide usage in specialized diagnostics and therapeutics. One of the primary reasons for the importance of radioisotopes in medicine is the *selectivity* of the body for certain elements. For example, iodine is selectively absorbed by the thyroid gland. Moreover, the isotope ^{131}I has a half-life of 8.05 d. Therefore, if a small quantity of iodine containing ^{131}I is ingested by a patient, the iodine will have ample opportunity to find its way to the thyroid gland but it will decay away within a matter of weeks and will not constitute a long-term radiation hazard.

Radioactive iodine in the thyroid (or another isotope in a different organ) can be located in an effective way by using a scanning device that detects the gamma radiation from the isotope. Such arrangements can be used, for example, to give a brain or body scan. The most advanced type of scanning device is the computerized

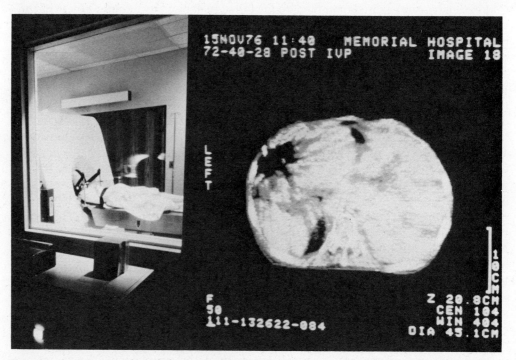

Figure 1 A computerized axial thermography (CAT) scanner in operation at Memorial Sloan-Kettering Hospital, New York City. (Dan McCoy/Black Star)

axial thermography (CAT) scanner. These machines generate cross-sectional views of the body with far greater resolution and accuracy than conventional X ray equipment while using much lower levels of radiation that allow, for example, the examination of pregnant women, which is usually too dangerous if done with X-rays.

CAT scanners provide a tremendous amount of highly detailed information for the diagnostician. They can determine the location and type of a stroke and differentiate among various types of tumors, something not possible with X ray equipment. In addition, CAT scanners are capable of identifying over two thousand tissue densities with the aid of a typewriter-like keyboard computer.

As shown in Fig. 1, the body scan is displayed on a monitor while the procedure is taking place. It is also recorded on videotape so that the scan can be recalled and viewed repeatedly.

In addition to their use as diagnostic tools, radioisotopes are also employed in the treatment of many different types of carcinomas. For example, radioactive iodine taken up by the thyroid is concentrated in the cancerous tissue and the radiation therefore selectively destroys the malignant cells. The table following lists a few of the radioisotopes now in routine use and indicates the diagnostic and therapeutic applications of these radiation sources.

Many types of cancerous growths can be attacked either by surgery or by radiation treatment. In many cases the radiation option has distinct advantages. For example, in cancer of the larynx, surgical removal of the growth is about 80 percent successful, but this procedure invariably affects the functioning of the voice box and leaves the patient totally incapable of speech or with severely impaired speech. Radiation treatment of cancer of the larynx, however, is also about 80 percent successful, and this type of treatment does not affect the ability of the patient to speak.

When deep-seated tumors and other malignancies are given radiation treatment, the radiation is usually from an external source. A typical dose necessary to destroy such a growth is about 6000 rad, usually administered over a period of a month. X rays from conventional machines have a maximum energy of about 150 keV; the penetrating power of such X rays is insufficient to reach deep regions of the body. Radioactive ^{60}Co, which yields γ rays with energies near 1 MeV, is often used in treatments that require deeper penetration than can be achieved with low-energy X radiation. Recently, increasing use is being made of electron beams that are accelerated to energies of 4 MeV in compact linear accelerators designed especially for medical use. These machines provide the most efficient radiation treatments currently available on a routine basis. Probably the best way to irradiate a deep malignancy is to use a high-energy beam of heavy ions, such as carbon, oxygen,

Radioisotope	Half-Life	Use
^{32}P	14.3 d	Treatment of leukemia, other carcinomas
^{51}Cr	27.8 d	Survey of GI bleeding; spleen scans
^{60}Co	5.24 y	External irradiation of carcinomas; implanted in wires or pellets for local irradiations of carcinomas; studies of pernicious anemia
^{75}Se	127 d	Pancreas and parathyroid scans
^{85}Sr	64 d	Bone scans
^{90}Y	64 h	Implanted in wires or pellets for local irradiations of carcinomas
^{99}Tc	6.0 h	Brain, thyroid, kidney, liver, and spleen scans
^{125}I	60 d	Thyroid scans and treatment; brain tumor localization; kidney, liver, and lung studies
^{131}I	8.05 d	
^{198}Au	2.7 d	Treatment of breast cancer

or neon. Studies have shown that large doses of radiation can be delivered by these beams to the region desired while relatively small doses are given to the surrounding healthy tissue. However, because of the high cost of heavy-ion accelerators, routine radiation treatments of this type are far in the future.

References

P. J. Early, M. A. Razzak, and D. B. Sodee, *Textbook of Nuclear Medicine Technology,* Mosby, St. Louis, 1969.

E. J. Hall, *Radiation and Life,* Pergamon, New York, 1976.

P. Rubin and G. W. Casarett, *Clinical Radiation Biology,* Vol. 2, Saunders, Philadelphia, 1968.

E. L. Travis, *Primer of Medical Radiobiology,* Year Book Medical Publishers, Chicago, 1975.

■ Exercise

1. Many elements, when taken into the body, deposit selectively in certain organs or certain regions of the body. Iodine, for example, is concentrated in the thyroid gland and calcium is concentrated in the bones and teeth. Suppose that a certain radiation worker ingests a small amount of a radioisotope that is deposited exclusively and uniformly in the bone marrow. The total mass of the worker's bone marrow is 3 kg. The material emits 5-MeV α particles and the ingested sample has an activity of 0.1 μCi. Assume that none of the sample is eliminated but instead resides permanently in the body. If the particular radioisotope has a long half-life (as does radium, for example), the activity will remain essentially constant for many years. What dose will the individual's bone marrow receive per year? Are any ill effects expected?

21.6 Nuclear Fusion

Another Source of Nuclear Energy

Energy will be released from any group of particles that can be rearranged into a system that has a greater binding energy. Fission, of course, is one example of such a process—the total binding energy of two nuclei such as barium and krypton is greater than the total binding energy of uranium, and so the fission process releases energy. The problem of extracting energy from nuclei can also be approached from the low-mass side of the maximum in the binding energy curve. If we combine two light nuclei to form a tightly bound medium-A nucleus, energy will be released. This process is called *fusion*. For example, if two ^{20}Ne nuclei (binding energy per nucleon approximately 8 MeV; see Fig. 21.2) are combined to form a ^{40}Ca nucleus (binding energy per nucleon approximately 8.5 MeV), there would be a total energy release of 40 × 0.5 MeV = 20 MeV. The difficulty in this particular case, of course, is that a great force would be required to overcome the Coulomb repulsion and to bring the neon nuclei into sufficiently close proximity that the capture process would take place.

The effect of the Coulomb repulsion will be reduced if we use nuclei with small Z. If we bring together two ^{2}H (or ^{2}D) nuclei—deuterons—to form an α particle, the energy release will be almost 24 MeV (see the third entry in Table 21.1) or 6 MeV per nucleon. This amount of energy release per nucleon is more than 6 times greater than for fission (200 MeV for 236 nucleons). But, in fact, when two deuterons combine, it is much more probable that a neutron or a proton, instead of a γ ray, will be emitted. That is, the most probable reactions are

$$\left. \begin{array}{l} ^2\text{D} + {}^2\text{D} \longrightarrow \text{n} + {}^3\text{He} + 3.269 \text{ MeV} \\ ^2\text{D} + {}^2\text{D} \longrightarrow {}^1\text{H} + {}^3\text{H} + 4.033 \text{ MeV} \end{array} \right\} \quad (21.7)$$

Thus, the average energy released in each D + D reaction is approximately 1 MeV per nucleon, comparable with that for the fission of a heavy element.

Because it is relatively easy to separate deuterium from normal hydrogen, there is a vast supply of deuterium available to us in the form of *water,* particularly in the oceans. How can we make use of this enormous reservoir of energy? Coulomb repulsion is an obstacle that must be overcome if fusion energy is to be released. The $^2\text{D}(d, n)^3\text{He}$ and $^2\text{D}(d, p)^3\text{T}$ reactions[8] will take place with an appreciable probability only if the gas sample is heated to about 10^7 K. At such a temperature, some of the deuterons move sufficiently rapidly that they overcome the Coulomb repulsion and closely approach other deuterons so that reactions can occur. Reactions that require these extremely high temperatures are called *thermonuclear reactions.*

One method of achieving a temperature of 10 million degrees or so[9] is by the detonation of a nuclear fission device. In the brief fraction of a second during which the blast takes place, the temperature is sufficiently high to ignite thermonuclear reactions, which then release additional energy and maintain the elevated temperature so that all of the thermonuclear material can "burn." This is, in fact, the principle of the H-bomb.

Although uncontrolled thermonuclear reaction processes have been achieved (in the form of H-bombs), we have not yet succeeded in constructing a device in which the controlled release of fusion energy can be maintained for longer than a small fraction of a second. Experiments are now being carried out with sophisticated devices in several countries (especially the U.S., U.K., and U.S.S.R.) and the hope is that a practical fusion reactor may be constructed before the end of this century.

21.7 Particles and Antiparticles

Positrons

While he was investigating cloud-chamber tracks of cosmic-ray particles in 1932, Carl D. Anderson (1905–   ) observed a track that appeared to be due to an electron. But this track curved the "wrong way" in the magnetic field in which the cloud chamber was located, indicating that the particle carried an electrical charge *opposite* to that of an electron. This was the first observation of a positively charged electron (a *positron*).

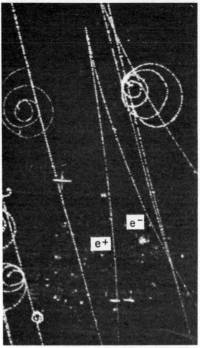

Lawrence Radiation Laboratory

Figure 21.18 Bubble-chamber photograph of the creation of a positron-electron pair. An energetic γ ray enters the chamber from above and interacts with one of the (hydrogen) nuclei in the chamber to produce the pair. The chamber is located in a magnetic field, and so the tracks of the two particles curve in different directions.

Soon after Anderson's discovery, positrons were found to be emitted in the decay of certain artificially produced radioisotopes. Furthermore, it was established that positrons can actually be *created* by the interaction of energetic photons (γ rays) with matter. This creation process, however, always produces a *positron-electron pair* (Fig. 21.18), and therefore does not violate the general principle of charge conservation. In the creation of a positron-electron pair, electromagnetic energy is converted into mass; in order to create two electron masses, the photon energy must be at least $2m_ec^2 = 1.02$ MeV.

Once a positron is created, it interacts via electromagnetic forces with the atomic electrons in its vicinity, eventually losing essentially all of its kinetic energy. As the positron drifts with very low velocity it can encounter and coalesce with an electron. The two particles then *annihilate* one another and the mass-energy of the pair appears in the form of two photons with a total energy of $2m_ec^2$ (Fig. 21.19).[10]

---

[8]The symbol d is used for the deuteron when it is the incident or the emitted particle in a nuclear reaction.

[9]These temperatures occur in the interiors of stars where thermonuclear energy generation takes place.

[10]The annihilation of an e^+-e^- pair at high energy may involve the production of a μ^+-μ^- pair or even a pair of hadrons.

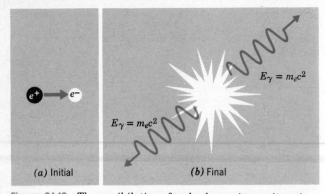

Figure 21.19 The annihilation of a slowly moving positron in an encounter with an electron produces two photons each with energy $\mathcal{E}_\gamma = m_e c^2 = 0.51$ MeV (annihilation radiation). The photons leave the annihilation site "bank-to-back" in order to conserve momentum.

Electrons and positrons are said to be *antiparticles* of one another. The positron is the antiparticle of the electron and *vice versa*, but because the electron is the naturally occurring member of the pair in our world, we usually refer to the electron as the "particle" and to the positron as the "antiparticle."

All elementary particles have antiparticle partners. (The neutral mesons π^0 and η^0, and the photon are in a special category—each of these particles is its *own* antiparticle.) A particle and its antiparticle have exactly the same mass, the same spin quantum number, and, if they are unstable, the particle and the antiparticle decay in the same way with the same half-life. However, the members of a particle-antiparticle pair have *opposite* electrical properties. Thus, the electron carries a negative charge and the positron carries a positive charge.

Antiprotons and Antineutrons

After the positron was discovered, it was natural to wonder whether *antiprotons* and *antineutrons* exist. It requires an energy $2m_e c^2 = 1.02$ MeV to produce an electron-positron pair and, by the same token, it requires an energy $2m_p c^2 = 1876$ MeV to produce a proton-antiproton pair and an energy $2m_n c^2 = 1879$ MeV to produce a neutron-antineutron pair. The concentration of such huge amounts of energy in a single elementary particle that initiates the creation event can be achieved only in the largest accelerators, and such accelerators were not available until the 1950s. In 1955, however, a group working with the 6-GeV accelerator that had recently been constructed at the University of California was successful in producing and identifying antiprotons (symbol: $\bar{p}$); in the following year, the antineutron ($\bar{n}$) was discovered.

Antiprotons are *stable* particles, but a *free* antineutron, if it does not undergo annihilation, will eventually decay into an antiproton and a positron. Because the half-life of the antineutron is so long (presumably, the same as that of the neutron—10.8 min), annihilation will take place before decay occurs; the spontaneous decay of an antineutron has never been observed.

The most complex form of antimatter that has yet been produced and identified is *antihelium* ($2\bar{p} + 2\bar{n}$). Conceivably, *antiatoms*, consisting of antiprotons, antineutrons, and positrons, could be produced; but because they would annihilate immediately on contact with ordinary matter, no such complete antiatom has yet been identified.

21.8 Hadrons: The Strongly Interacting Particles

Mesons

The types of elementary particles that interact via the strong force, collectively called *hadrons*, include mesons, nucleons, and a group of particles (called *hyperons*) that are short lived and have masses greater than the nucleon mass.

In 1935 the Japanese physicist Hideki Yukawa (1907–1981) applied the *field* concept to the strong nuclear force. The highly successful electromagnetic field theory incorporates the idea that the photon is the quantum (or elementary interacting unit) of the field. According to this theory, the electromagnetic force between two interacting electrons derives from the exchange of a virtual photon; refer to Fig. 21.20. Both the emitting and absorbing electrons experience a change in energy and momentum (at the vertices a and b in Fig. 21.20) in exact agreement with the predictions of the laws of electrodynamics. The isolated quantum of this interaction appears as a real photon.

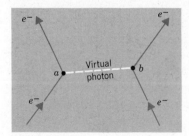

Figure 21.20 The hypothetical path of two interacting electrons. Their scattering derives from the emission of a virtual photon by one electron at vertex a and its absorption by the other electron at vertex b. The changes in path directions at a and b indicate the changes in momentum of the two electrons.

It was Yukawa's hypothesis that the strong nuclear force between nucleons should be represented by a field and that the quantum of this field should be a new elementary particle with mass intermediate between that of an electron and a nucleon. This new particle—called a π meson or *pion*—was finally detected experimentally in 1947. In the following year, pions were first produced artificially in an accelerator and since that time beams of pions have been available for use in detailed studies of their properties and interactions.

There are two types of charged pions, π^+ and π^- (which are the antiparticles of one another), and a neutral pion, π^0 (which is its own antiparticle). The charged pions are slightly more massive than the neutral pion: $m_{\pi^+} = m_{\pi^-} = 140$ MeV/c^2 and $m_{\pi^0} = 135$ MeV/c^2. Thus, the pions are about 270 electron masses.

Pions can be produced in collisions in which an energetic nucleon or photon is incident on a nucleon; for example,

$$p + p \longrightarrow p + n + \pi^+$$
$$n + p \longrightarrow n + p + \pi^0$$
$$\gamma + p \longrightarrow p + \pi^0$$

When particles are produced by high-energy γ radiation, as in the last example, the process is called *photoproduction*.

When an electron-positron annihilation event takes place, the products are the quanta of the electromagnetic field—photons. When a p-$\bar{p}$ or an n-$\bar{n}$ annihilation event takes place, the products are the quanta of the nuclear force field—pions. Figure 21.21 shows the annihilation of an antiproton by a proton in a bubble chamber. In this event, eight charged pions (and probably several neutral pions that leave no tracks) are produced.

The first pion decays to be observed took place in photographic emulsions exposed to cosmic rays (see Fig. 21.22). When a pion comes to rest in an emulsion, it emits a *muon* and a *neutrino* (actually, a *mu* neutrino):

$$\pi^+ \longrightarrow \mu^+ + \nu_\mu$$
$$\pi^- \longrightarrow \mu^- + \bar{\nu}_\mu$$

Muons are weakly interacting elementary particles that have a mass of 106 MeV/c^2 (or about 210 electron masses) and a half-life of 2.2×10^{-6} s. Muon decays are discussed in the next section.

The neutral pion decays predominantly into a pair of γ rays:

$$\pi^0 \longrightarrow \gamma + \gamma$$

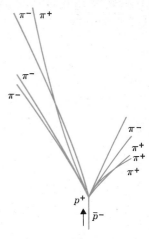

Figure 21.21 Annihilation of an antiproton with a proton in a hydrogen bubble chamber. Eight charged pions (and probably some neutral pions that do not leave tracks in the chamber) were produced.

Figure 21.22 Decay of a pion in a photographic emulsion exposed to cosmic rays. The pion comes to rest in the emulsion (at the bottom of the picture). In the decay process, a muon (μ) is emitted (along with a neutrino which leaves no track in the emulsion). The muon eventually comes to rest in the emulsion (at the top of the picture) and decays with the emission of an electron.

(Can you see why *two* γ rays are produced in this decay? Imagine yourself in the rest frame of the π^0 and apply momentum conservation. Remember, electromagnetic radiation—photons—carry momentum; see Section 15.5.)

When an elementary particle decays, it always does so predominantly through the strongest (and therefore the *fastest*) interaction that is available to the particle. If it is allowed, decay via the strong interaction always dominates. When a proton is supplied with 300 MeV of energy, it forms an "excited state" in much the same way that a hydrogen atom forms its first excited state when it is supplied with 10.2 eV of energy. An excited hydrogen atom returns to its ground state by the emission of a quantum of the electromagnetic field that binds the atom together, namely a *photon*. An excited proton returns to its ground state by the emission of a quantum of the strong interaction field, namely, a *pion*. This process is therefore a *strong decay*, and measurements have shown that the half-life for the decay of an excited proton is about 10^{-23} s. *This time is typical of all strong interaction processes.*

Although pions interact strongly with nucleons and are created in strong interaction processes, pions cannot *decay* via the strong interaction because there are no less massive particles that interact strongly. Therefore, pion decay must involve either the electromagnetic or the weak interaction. Charged pions cannot undergo a purely electromagnetic decay because the electric charge must be carried off by some kind of *particle*. The decay of charged pions is therefore restricted to the weak interaction, and the products of π^+ and π^- decays are muons and neutrinos—weakly interacting particles. Since decays via the weak interaction are slow processes, we expect charged pions to be relatively long-lived particles, and, indeed, the half-life is about 10^{-8} s. Neutral pions decay by gamma rays through the stronger electromagnetic interaction and have a half-life about 10^{-16} s. (Remember, these times are to be compared with the strong-interaction time of 10^{-23} s; therefore, a decay time of 10^{-8} s represents a very slow decay.)

It is now known that, in addition to pions, there are other mesons, called *kaons* and *eta mesons*, that participate in the strong force. Charged kaons (K^+ and K^-) have a mass of 494 MeV/c^2, and the neutral kaon (K^0) has a mass of 498 MeV/c^2. The half-lives of the charged kaons are approximately 10^{-8} s. The neutral eta meson (η^0) has a mass of 549 MeV/c^2 and a half-life less than 10^{-18} s.

Hyperons

The first elementary particles to be found with masses greater than that of the proton were the *lambda particles*. Although these particles (Λ^0 and $\overline{\Lambda^0}$) are electrically neutral, they are readily identified by the V-shaped tracks that the charged decay products leave in emulsions or bubble chambers. Figure 21.23 shows a bubble-chamber photograph of a proton-antiproton collision that results in the production of a $\Lambda^0 - \overline{\Lambda}^0$ pair. Each lambda particle travels only a short distance in the chamber before undergoing decay:

$$\Lambda^0 \longrightarrow \pi^- + p; \qquad \overline{\Lambda^0} \longrightarrow \pi^+ + \overline{p}$$

Following the discovery of the lambda particles, several additional heavy particles were found; these particles bear the labels Σ (sigma particles), Ξ (xi particles,) and Ω (omega particles). Particles in this group are collectively called *hyperons*. The enlarged family, including the hyperons and nucleons, are called *baryons*. They are designated by a new quantum number, the baryon number $B = 1$. All mesons and leptons (Section 21.9) have baryon number $B = 0$.

Hyperons are strongly interacting particles and all undergo decays that lead to nucleons. But these decays take place via the slow-weak interaction and so the half-lives are all long compared to the typical strong-interaction time of 10^{-23} s (Table 21.8).

One striking feature of hyperon production reactions is the fact that they are always produced as pairs (Fig. 21.23) or one in conjunction with a K

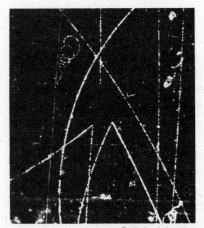
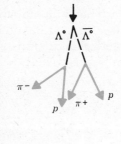

Lawrence Radiation Laboratory

Figure 21.23 Production of a $\Lambda^0 - \overline{\Lambda}^0$ pair by a $p - \overline{p}$ collision in a bubble chamber. The Λ^0 decays into a proton and a π^- meson, and the $\overline{\Lambda}^0$ decays into an antiproton and a π^+ meson.

Table 21.8 Hyperons

Particle	Strangeness (S)	Mass-Energy (MeV)	Half-Life (s)	Anti-Particle[a]
Λ^0	-1	1116	2.6×10^{-10}	$\overline{\Lambda^0}$
Σ^+	-1	1189	8.0×10^{-11}	$\overline{\Sigma^+}$
Σ^0	-1	1193	$<6 \times 10^{-20}$	$\overline{\Sigma^0}$
Σ^-	-1	1197	1.5×10^{-10}	$\overline{\Sigma^-}$
Ξ^0	-2	1315	2.9×10^{-10}	$\overline{\Xi^0}$
Ξ^-	-2	1321	1.6×10^{-10}	$\overline{\Xi^-}$
Ω^-	-3	1672	8.2×10^{-11}	$\overline{\Omega^-}$

[a] The antiparticles all have the opposite sign for the strangeness and $B = -1$.

meson. This necessitates introducing a new quantum number to associate with the hyperons (and the K mesons). This quantum number is called *strangeness*, S.

There are conservation laws associated with the baryon number and strangeness. In all elementary particle reactions the baryon number is conserved, that is the (algebraic) sum of the baryon numbers in the incident state must equal the like sum in the final state. A similar conservation law applies to strangeness but only in strong interactions. For example, consider the reaction

$$\pi^- + p \longrightarrow K^0 + \Lambda^0$$

For the pion $B = 0$, $S = 0$, for the proton $B = 1$, $S = 0$. Thus, on the left-hand side the state has $\Sigma B = 0 + 1 = +1$, $\Sigma S = 0 + 0 = 0$. The kaon K^0 has $B = 0$, $S = +1$ while for the Λ^0, $B = 1$, $S = -1$. The right-hand-side state has $\Sigma B = 0 + 1 = +1$ and $\Sigma S = +1 - 1 = 0$. Thus in this strong interaction both baryon number and strangeness are conserved. Is charge conserved in this reaction? Are charge, baryon number, and strangeness conserved in the $p + \bar{p} = \Lambda^0 + \overline{\Lambda^0}$ reaction shown in Fig. 21.23 if for the proton $B = 1$, $S = 0$ and for the antiproton $B = -1$, $S = 0$?

21.9 Leptons and the Weak Interaction

The Lepton Family

Particles that have small or zero mass and which interact via the weak force[11] are called *leptons*.[12] Electrons (e^+ and e^-) and the neutrinos (ν_e and $\bar{\nu}_e$) associated with electrons in various weak processes constitute members of the family. Muons (μ^+ and μ^-) and the neutrinos (ν_μ and $\bar{\nu}_\mu$) associated with muons in various weak processes constitute another branch of the family. Finally, the recently discovered tau (τ^+ and τ^-) leptons are the heaviest leptons (1784 MeV/c^2) and have the ν_τ and $\bar{\nu}_\tau$ associated neutrinos.[13] All leptons have baryon number $B = 0$.

Electrons and Neutrinos

One of the reasons that Pauli postulated the existence of the neutrino was the necessity to conserve angular momentum in nuclear β decay. For example, if tritium underwent the decay $^3\text{H} \rightarrow {}^3\text{He} + \beta^-$ it would not be possible to balance the intrinsic angular momenta because the spins of ^3H, ^3He, and the electron are all $\frac{1}{2}$, and, according to the rules of quantum processes, there is no way to combine two spin-$\frac{1}{2}$ vectors to produce another spin-$\frac{1}{2}$ vector. This difficulty (as well as the problem of energy balance) disappears if an additional spin-$\frac{1}{2}$ particle is emitted along with the electron. Pauli's neutrino is therefore a *fermion*.

Just as is the case for other fermions (e.g., electrons and nucleons), the neutrino has an antiparticle partner. The particle associated with nuclear β^- decay is called the *antineutrino* ($\bar{\nu}_e$) and that associated with nuclear β^+ decay is called the *neutrino* (ν_e). Nuclear β^- and β^+ decays can be represented as (see Eqs. 21.4).

β^- decay: $n \longrightarrow p + e^- + \bar{\nu}_e$

β^+ decay: $p \longrightarrow n + e^+ + \nu_e$

The Distinguishability of Neutrinos

Several experiments have shown conclusively that the massless particles that are emitted in β^- decay are *not* the same as those emitted in β^+ decay; that is, ν_e and $\bar{\nu}_e$ are *distinguishable particles*.

In what way is a neutrino different from an antineutrino? Because these particles have no mass or charge and always travel with the velocity of light,

---

[11] Of course, a pair of leptons that have charge and mass will interact via the electromagnetic and gravitational force as well; but the essential point is that leptons have *no strong interactions* whatsoever.

[12] From the Greek word *leptos* (meaning "small").

[13] Although the tau meson has almost twice the mass of a nucleon, its properties demand a lepton classification: Yes, Virginia, there are heavy leptons!

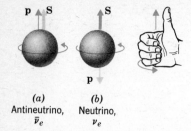

Figure 21.24 The spin vector (**S**) and the linear momentum vector (**p**) have the same direction for $\bar{\nu}_e$ and have opposite directions for ν_e. The arrows on the circles show the directions in which the particles are "spinning" and the direction of the spin vector is then determined by using the right-hand rule.

the distinction between ν and $\bar{\nu}_e$ can depend only on the dynamic properties of the particles; these properties are specified by the intrinsic angular momentum and linear momentum vectors.[14] Experiments have shown that the vectors, **S** and **p**, have the *same* direction for antineutrinos whereas these vectors have *opposite* directions for neutrinos (Fig. 21.24). An antineutrino always advances in the direction in which a right-hand screw advances when turned in the same direction as its "spinning motion." Therefore, we say that an antineutrino is *right-handed*; similarly, a neutrino is said to be *left-handed*. (In modern terminology, we say that an antineutrino has *positive helicity* and that a neutrino has *negative helicity*.)

Heavier Leptons and Neutrinos

When pions decay to muons, neutrinos (or antineutrinos are emitted):

$$\pi^+ \longrightarrow \mu^+ + \nu_\mu$$
$$\pi^- \longrightarrow \mu^- + \bar{\nu}_\mu$$

The muons formed in these decays are also unstable, but muon decay differs in an important respect from pion decay. Unlike the "two-body" pion decay, muon decay is a "three-body" process in which *two* neutrinos are emitted:

$$\mu^+ \longrightarrow e^+ + \bar{\nu}_\mu + \nu_e$$
$$\mu^- \longrightarrow e^- + \nu_\mu + \bar{\nu}_e$$

Again, experiments concerning the production and absorption of muon neutrinos have shown that ν_μ and $\bar{\nu}_\mu$ are distinguishable particles; ν_μ is left-

---

[14] Modern theoretical conjecture would endow neutrinos with a rest mass of about 1 eV/c^2.

handed and $\bar{\nu}_\mu$ is right-handed. Of equal importance for our theories of elementary particles is the fact that electron neutrinos do not participate in weak interaction processes in the same ways that muon neutrinos do. That is, the neutrino pair, $\nu_e, \bar{\nu}_e$ is *not* the same as the neutrino pair $\nu_\mu, \bar{\nu}_\mu$.

The tau meson, likewise, has its own associated distinguishable neutrinos, ν_τ and $\bar{\nu}_\tau$.

What is the Quantum of the Weak Interaction?

The pion (along with kaons and the eta meson) is the mediator of the strong interaction between a pair of nucleons. Does there exist a cousin of the pion that is responsible for mediating the weak interaction between a pair of leptons? The answer to this question was settled in 1983 at the high-energy accelerator at CERN in Geneva when the theoretically predicted W^+, W^-, and Z^0 particles were observed. These particles constitute the mediating agency of the weak interactions.

Quarks

The situation as described above was essentially the status of elementary particle physics 15 to 20 years ago. Since then, over one hundred new particles have been discovered (significantly many more new hadrons than leptons). During the recent phase of development of elementary particle physics, the interaction between theory and experiment, first one leading the other then the other way around, is a fascinating story in scientific progress. Two major achievements since the mid-1960s are the discovery of the fundamental building blocks in the substructure of the hadrons, and the progress towards a unified field theory describing all the basic forces found in nature in terms of a single interaction scheme.

In 1963, M. Gell-Mann and G. Zweig independently postulated an underlying substructure for hadrons that consists of new particles referred to as *quarks*. These bizarre particles had spin angular momentum $\hbar/2$ (i.e., $s = \frac{1}{2}$) but they had fractional charges, $\frac{1}{3}e^\pm$ and $\frac{2}{3}e^\pm$! At first they were three in number: "up" quarks (u), "down" quarks (d), and "strange" quarks (s). Refer to the first three entries in Table 21.9.

All baryons are supposed to be built out of three quarks; for example for the proton:

$$p \cong u + u + d,$$

with charge $\cong \frac{2}{3}e + \frac{2}{3}e - \frac{1}{3}e = +e,$

Table 21.9 Quarks

Particle[a]	Spin	Charge	Strangeness	Charm	Bottomness	Baryon Number
Up (u)	$\frac{1}{2}$	$+\frac{2}{3}e$	0	0	0	$\frac{1}{3}$
Down (d)	$\frac{1}{2}$	$-\frac{1}{3}e$	0	0	0	$\frac{1}{3}$
Strange (s)	$\frac{1}{2}$	$-\frac{1}{3}e$	-1	0	0	$\frac{1}{3}$
Charmed (c)	$\frac{1}{2}$	$+\frac{2}{3}e$	0	$+1$	0	$\frac{1}{3}$
Bottom (b)	$\frac{1}{2}$	$-\frac{1}{3}e$	0	0	$+1$	$\frac{1}{3}$

[a]The antiparticle quarks have opposite signs for charge, strangeness, charm, bottomness, and baryon number.

baryon number $\cong \frac{1}{3} + \frac{1}{3} + \frac{1}{3} = +1$ and
strangeness $\cong 0 + 0 + 0 = 0$

As another example, including strange particles, we have for the Λ^0:

$\Lambda^0 \cong u + d + s$,
with charge $\cong \frac{2}{3}e - \frac{1}{3}e - \frac{1}{3}e = 0$,
baryon number $\cong \frac{1}{3} + \frac{1}{3} + \frac{1}{3} = +1$,
and strangeness $\cong 0 + 0 - 1 = -1$.

All mesons are supposed to consist of a quark and an antiquark. For example the π^+ meson consists of:

$\pi^+ = u + \bar{d}$,
with charge $\cong \frac{2}{3}e + \frac{1}{3}e = e$,
and baryon number $\cong \frac{1}{3} - \frac{1}{3} = 0$.

For the strange meson K^- with strangeness -1, we have

$K^- \cong \bar{u} + s$,
with charge $\cong -\frac{2}{3}e - \frac{1}{3}e = -e$,
baryon number $\cong -\frac{1}{3} + \frac{1}{3} = 0$,
and strangeness $\cong 0 - 1 = -1$.

What would be the two-quark structure for the K^+? (The K^- is the antiparticle of the K^+.)

The above scheme was conjectural for over five years. Then, beginning in 1967 when the Stanford Linear Accelerator Center (SLAC) electron accelerator was completed, a number of landmark experiments involving the scattering of newly available high-energy electrons from protons were started. In many ways the experiments led to results not unlike Rutherford's experiments in the scattering of α-particles from thin foils. A significant number of electrons undergo large-angle scattering from the pointlike quarks within the proton. (See Fig. 21.25.) Analysis of the data not only shows that there are three such centers in the proton but that charges for

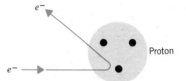

Figure 21.25 Deeply inelastic scattering of 20 GeV electrons from pointlike quarks inside the proton. Compare with Fig. 19.4 referring to Rutherford's experiment.

them of $\frac{2}{3}e$ and $-\frac{1}{3}e$ are involved. However momentum transfer to the three quarks can only account for approximately half of the proton momentum. Something else is also present inside the proton. The something else turns out to be a number of *gluons*, the mediating virtual particles of the strong force that hold the quarks together in the proton.

In 1974 the predicted existence of a new heavy meson called *charmonium*, was experimentally verified with a mass of 3100 MeV/c^2. The substructure of this meson consists of a new quark, c, and its antiquark, $\bar{c}$. A new quantum number is required for the c and $\bar{c}$ quarks; it is called "charm." The properties of the charmed quark are given in Table 21.9. Yet a fifth and also a sixth quark, the "top" and "bottom" quark have been postulated by theorists. An appropriate new meson, mass 9460 MeV/c^2, has been observed experimentally and is believed to consist of a bottom quark and its antiquark.

Virtually all the hadrons consisting of the various possible combinations of three quarks that may be formed without violating the conservation laws, have been observed.

The physical properties listed in Table 21.9 that distinguish the quarks from each other are said to determine the *flavor* of the quarks.[15] A good deal more order and verifiable predictions become possible if in addition to flavor, quarks (and gluons) also

---

[15]By now you have no doubt observed that elementary-particle physicists show an abundant sense of humor by the whimsical names they select to associate with quantities and objects.

possess another property called *color*. The color "charge" (either red, green, or blue) carried by quarks, analogous to electric charge, determines the strong force operating between quarks much the same way as electric charge determines the electromagnetic force acting between electric charges. The color-charge theory for strong interactions between quarks is called *quantum chromodynamics*, QCD.

The alert student has probably noticed that "strong interaction" has been used in two different contexts: as describing the nature of the force between two nucleons and also the force between quarks. How are these strong interactions related? An analogy with electrostatic forces is helpful. The force between the nucleus and the surrounding electrons in an atom is the familiar Coulomb force of Chapter 12. What is the nature of the force between colliding atoms in a gas? The colliding atoms are charge-neutral and hence should have no electric force acting between them. Indeed, when the atoms are a few atomic radii apart, the force between them is vanishingly small. But in close proximity (at a separation between centers on the order of an atomic diameter) there will be a slight force acting between them as the two electron clouds begin to intermix. These so called Van der Waals forces are relatively weaker than the Coulomb force and drop off in strength much faster than an inverse square law for the distance separating the atomic centers. Let us form an analogy: nucleon quark substructure → atomic electron-nucleus substructure and nucleon-nucleon interaction → atom-atom interaction. In QCD theory hadrons are color-charge neutral. Thus, when a proton and a neutron, for example, are a few nucleon radii apart, the color charge force between them is very weak. Only when they are very close (a nucleon diameter apart) does a color-derived force develop that, while strong, is much weaker than the quark-quark force in analogy with the atomic Van der Waals force. If we call the nucleon-nucleon force strong, the quark-quark force is super strong.

Ever since Einstein attempted (and failed) to derive a unified field theory that would account for both the electromagnetic force and the gravitational force by a single interaction scheme, theorists have been attempting the even grander scheme of uniting the gravitational force, electromagnetic force, super strong force, and the weak force in a unified description. The first success came with the *electroweak* theory introduced by Weinberg, Glashow, and Salam uniting the electromagnetic and weak forces. Presently, there are encouraging signs that the electroweak theory and the color-charge super strong forces may be combined in a *grand unified theory* (GUT). One interesting feature of such theories is that if any lepton or quark pair of particles approach each other to within a distance of 10^{-31} m the electromagnetic, weak, and color charge forces would blur into a force with a single strength. A further interesting consequence of the GUT theories is the prediction that the proton is, in fact, unstable in a baryon nonconserving decay such as $p \rightarrow \pi^0 + e^+$, with a half-life of order 10^{32} y. A number of experiments are now in progress attempting to verify the predicted proton decay.

Unifying the GUT theory with gravitation is as yet an elusive goal.

Summary of Important Ideas

Nucleons are bound together in nuclei by the strong nuclear force. The mass of any nucleus is *less* than the mass of the number of free protons and free neutrons that make up that nucleus; this difference in mass-energy is the total *binding energy* of the nucleus.

Nuclei are limited in maximum size by the effects of Coulomb repulsion and the exclusion principle.

Not all groups of nucleons constitute stable nuclei; if there is a less energetic arrangement that is available to the nucleons (i.e., if a configuration of smaller mass is possible), then a *radioactive decay* process will occur, which will transform the original nucleus into a nucleus of smaller mass. Radioactive decay involves emission of an electron, emission of a positron, or emission of an α particle.

The stable nuclei with A less than about 40 contain approximately *equal* numbers of protons and neutrons ($Z \cong N$). For heavier nuclei, the neutron number increases more rapidly than the proton number. The stable elements all derive from primordial hydrogen through nuclear processes mostly taking place in stellar interiors during the evolution of our Universe.

When a nucleon or a nucleus is given a high velocity and is directed toward other nuclei, nuclear *reactions* can take place in which nuclear particles (or photons) are emitted and new nuclei are formed.

Heavy nuclei (such as uranium or plutonium) can absorb a neutron and undergo *fission* by splitting into two fragments of roughly equal mass. Each fission event releases approximately 200 MeV of energy. Energy is also released when two light nuclei combine to form a heavier nucleus; this process is called *fusion*.

All elementary particles have antiparticle partners (e^- and e^+; p and $\bar{p}$; etc.) Some particles (π^0 and photons) are on their *own* antiparticles. The properties of a particle and its antiparticle are the same except that the electrical properties are *opposite*.

Pions are the quanta of the strong internucleon interaction and can be produced copiously in nucleon-nucleon collisions. Pions decay into muons and neutrinos.

The weakly interacting particles (leptons) consist of e^-, μ^-, τ^-, ν_e, ν_μ, ν_τ and their antiparticles. The six types of neutrinos are all different and distinguishable. The $W^\pm$ and Z^0 are the quanta of the weak interaction.

All hadrons are composed of three quarks and all mesons are composed of a quark and antiquark pair. There are five firmly established flavors of quarks and one more probable quark. Gluons are the mediating particle for quark-quark forces. Quarks and gluons carry color charge, an important ingredient in quantum chromodynamics.

Substantial progress exists in a unification of electromagnetic, weak, and super strong (color-charge) forces into a single interaction scheme.

◆ **Questions**

21.1 For mass numbers up to $A = 209$, for only two values of A there are *no* stable nuclei. Use Fig. 21.9 and identify these A values.

21.2 By what processes do you expect the following nuclei to decay: ^{14}O ($Z = 8$), ^{50}Ca ($Z = 20$), ^{67}Cu ($Z = 29$), ^{111}Sn ($Z = 50$)? (Use Fig. 21.10 to determine on which side of the valley of stability each nucleus lies.)

21.3 Discuss the decay history of ^{9}C (see Fig. 21.9). In what form will the original nine nucleons be at the end of the series of decay processes?

21.4 Why are there no pairs of *stable* nuclei with the same value of A but with Z differing by one unit?

21.5 If the nuclei (Z, A) and $(Z + 2, A)$ are both stable, what general statements can be made concerning the nucleus $(Z + 1, A)$? (Is this nucleus radioactive? If so, what type of decay does it undergo?)

21.6 ^{232}Th decays by the following series of emissions: α, β, β, α, α, α, α, β, α, β. Construct a diagram for this case similar to that shown in Fig. 21.8. Identify the intermediate (radioactive) nuclei and the final (stable) nucleus.

21.7 Why is it not reasonable to picture the neutron as a close association of a proton and an electron?

21.8 Neutron capture by a stable target nucleus rarely leads to positron radioactivity? Why?

21.9 An experimenter bombards a natural boron target with deuterons and finds that two different radioactive species are formed, one with a half-life of 20.5 min and the other with a half-life of 0.020 s. What reactions induced these activities? (Refer to Fig. 21.9 and remember that natural boron is a mixture of isotopes.)

21.10 What are the residual nuclei when a (p, α) reaction takes place with the following target nuclei: ^{9}Be, ^{11}B, ^{18}O, and ^{19}F?

21.11 Explain why a (p, n) reaction on a stable target nucleus always has a *negative* Q-value. (If the Q-value were positive, would the target be stable?)

21.12 List some stable targets and incident particles that could be used to produce nuclear reactions that yield ^{13}N + n in the final state.

21.13 A boron target is bombarded with a proton beam. After the beam is turned off, a β-ray detector records 100 counts/s from radioactivity in the target. Forty minutes later, the counting rate has decreased to 25 counts/s. What is the source of the radioactivity and what reaction has taken place? (Use Fig. 21.9).

21.14 List some of the reactions that can take place when ^{9}Be is bombarded with protons.

21.15 List some nuclear reactions that can produce ^{8}Be. (There are at least 16 that involve stable targets and employ bombarding particles with $A \leq 4$.)

21.16 When ^{235}U ($Z = 92$) absorbs a slow neutron, it undergoes fission and releases 2 or 3 neutrons. List 3 or 4 possible pairs of fission-product nuclei (different from those in Eq. 21.6) that could be formed in such a process.

21.17 If a sample of matter containing every element were heated to thermonuclear temperatures in some nuclear cauldron, what element or group of elements would you expect to result from "cooking" this mixture until it is "done?" Explain.

21.18 An electron and a positron can bind together into an "atomic" system called *positronium*. What is "antipositronium?"

21.19 A beam of high energy γ rays strikes a target of ^{3}He. Write down some of the possible photoproduction reactions that produce pions.

★ **Problems**[a]

Section 21.1

21.1 What group of nucleons must be separated from ^{7}Li to form ^{4}He? If these nucleons are removed as a single entity, how much energy is required? (Use the data in Table 21.2.)

21.2 When ^{5}He decays, the range of energy which is shared by the ^{4}He nucleus and the neutron is approximately 0.5 MeV. How long does ^{5}He live on the average? (Use the uncertainty principle, $\Delta\mathscr{E}\,\Delta t \cong h$.)

21.3 How much energy is required to separate a neutron from ^{7}Li? From ^{9}Be? (The neutron separation energy for ^{9}Be is the *lowest* for any stable nucleus.)

21.4 What is the binding energy per nucleon of (a) ^{3}He, (b) ^{6}Li, and (c) ^{7}Li?

21.5 Use the information in Table 21.2 and show that ^{8}Be is unstable and that it can break up *only* into two α particles.

21.6 If two ^{6}Li nuclei were brought together, what nucleus would be formed and how much energy would be released?

21.7 Use Fig. 21.2 and estimate the amount of energy that would be released if 20 protons and 20 neutrons were brought together to form ^{40}Ca.

21.8 The mass of ^{238}U is 238.0508 u. What fraction of the total mass-energy of ^{238}U is its *binding energy*?

21.9 The mass of ^{56}Fe ($Z = 26$) is 55.934 936 u and the mass of ^{56}Co ($Z = 27$) is 55.939 847 u. Which of these nuclei is stable and which decays radioactively into the other? How much energy is available for the decay?

Section 21.2

21.10 What is the maximum energy of electrons emitted in the β decay of tritium?

21.11 What is the available decay energy for the β decay of ^{8}Li?

21.12 Use Fig. 21.7 and estimate the maximum kinetic energy of the electrons emitted by ^{65}Co.

21.13* The available energy in the α decay of ^{210}Po is 5.4 MeV (see Example 21.2). Use energy and momentum conservation and show that this energy is divided between the α particle (5.3 MeV) and the residual nucleus, ^{206}Pb (0.1 MeV).

21.14* The kinetic energy of α particles emitted by ^{226}Ra ($Z = 88$, atomic mass = 226.02536 u) is 4.78 MeV and the recoil energy of the daughter nucleus, ^{222}Rn, is 0.09 MeV. What is the atomic mass of ^{222}Rn ($Z = 86$)?

Section 21.3

21.15 How much kinetic energy will each of the α particles have when a ^{8}Be nucleus breaks up? (Use the data in Table 21.2.)

21.16 What is the Q-value for the ^{4}He(^{3}He, γ)^{7}Be reaction?

21.17 A slow neutron is captured by ^{7}Li and a single γ ray is emitted. What is the energy of the γ ray?

21.18 Calculate the Q-value for the ^{7}Li(d, n)^{8}Be reaction. Explain qualitatively why (d, n) Q-values are generally positive and large.

21.19 When a slow neutron is captured by ^{12}C a 4.95-MeV γ ray is emitted. What is the mass of ^{13}C?

21.20 In an H-bomb the ^{3}T(d, n)^{4}He reaction, instead of the D + D reactions, is used as the primary source of thermonuclear energy. Why? (Compare the Q-values.)

Sections 21.4 and 21.5

21.21 Write down the nuclear equations that represent the conversion of ^{232}Th into ^{233}U by neutron capture and β decay. (The process is another way of breeding useful nuclear fuel—^{233}U—from a nonfissile material—^{232}Th.)

21.22 Assuming the average energy released per fission in a reactor is 200 MeV, how many fissions take place per second in a 1000-MW reactor?

---

[a] For ease of computation in the problems, use the conversion factor, $c^2 = 932$ MeV/u.

21.23 An analysis of nuclear stability to spontaneous fission reveals that when the *fission parameter* Z^2/A for a nucleus exceeds 34, fission with a half-life less than about 10^{20} y will take place. Based on this criterion, which of the following nuclei are unstable to spontaneous fission:

^{209}Po, ^{226}Ra, ^{232}Th, ^{238}U, and ^{254}Cf?

21.24 Determine the energy released in the fission of the nucleus ^{235}U by the thermal capture of a neutron in both of the reactions given in Eq. 21.6. The relevant masses are:

^{235}U = 235.0426; ^{139}Ba = 138.9072;

^{95}Kr = 94.9392; ^{144}La = 143.9179; and

^{89}Br = 88.9270.

Section 21.6

21.25* How much energy would be released if 1 kg of deuterium were completely "burned" in fusion reactions? (Assume that the two reactions in Eq. 21.7 are equally probable and calculate an "average Q-value" for D + D.) Compare this energy with that released by the fission of an equal mass of ^{235}U.

21.26 The fusion cycle converting hydrogen to helium that occurs in hot stars has the net effect, after several separate reaction steps, of

$$4\ ^1\text{H} \rightarrow\ ^4\text{He} + 2\beta^+ + 2\nu + \gamma$$

yielding an energy release of 26.7 MeV. The Sun radiates energy at a rate of approximately 4×10^{26} W. At what rate, kg/s, is hydrogen being consumed in the Sun? What fractional loss of the Sun's mass does this rate correspond to if continued for one year?

Section 21.7

21.27 A positron comes to rest in matter and annihilates with an electron. If three equal-energy photons are produced in the process, what is the energy of each photon and in what relative directions do the photons leave the point of annihilation?

21.28 A p-$\bar{p}$ annihilation takes place at rest and produces 4 charged pions, all of which have the same energy. What is the kinetic energy (in MeV) of each pion?

21.29 What would be the total amount of mass-energy released if an antihydrogen atom annihilated with an ordinary hydrogen atom? (Neglect the atomic binding energy.) What would be the products of the annihilation?

21.30 What would be the total amount of mass-energy released if an antideuterium atom annihilated with an ordinary deuterium atom? (Do not forget the nuclear binding energy but neglect the atomic binding energy.)

Section 21.8

21.31 A negative pion is absorbed (at rest) by a helium nucleus and produces the reaction, $^4\text{He} + \pi^- \rightarrow\ ^3\text{H} + n$. What is the Q-value for this reaction?

21.32 What is the Q-value for the decay of the Σ^+ hyperon: $\Sigma^+ \rightarrow p + \pi^0$?

21.33 What is the minimum photon energy required to produce pions in the reaction $p + \gamma \rightarrow n + \pi^+$?

21.34 What is the total amount of kinetic energy carried by the pions in the decay at rest of a kaon into three pions: $K^+ \rightarrow \pi^+ + \pi^+ + \pi^-$?

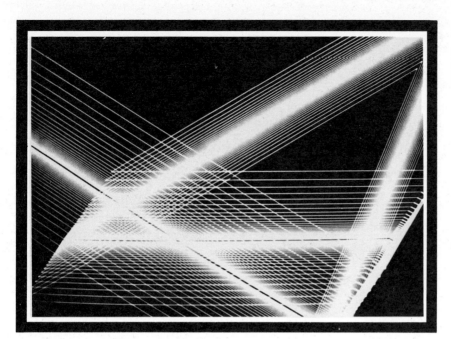

APPENDIX
Mathematical Review

Notations Used

- $\propto$ "is proportional to"
- $\cong$ "is approximately equal to"
- $\approx$ "is very approximately equal to"
- $>$ ($<$)   "is greater (less) than"
- $\geq$ ($\leq$)   "is greater (less) than or equal to"
- $\gg$ ($\ll$)   "is much greater (less) than"

A.1. Powers-of-10 Notation

The number of times that 10 is multiplied together can be expressed as a superscript of 10 (called the *exponent* of 10 or the *power* to which 10 is raised):

$10 \times 10 = 100 = 10^2$
$10 \times 10 \times 10 = 1000 = 10^3$
$10 \times 10 \times 10 \times 10 \times 10 \times 10 = 1,000,000 = 10^6$

Any number can be expressed as a power of 10:

$176,000,000 = 1.76 \times 100,000,000 = 1.76 \times 10^8$

Products of powers of 10 are expressed as

$10^2 \times 10^3 = (10 \times 10) \times (10 \times 10 \times 10)$
$\qquad\quad = 10^5 = 10^{(2+3)}$

That is, in general, the product of 10^n and 10^m is $10^{(n+m)}$:

$$10^n \times 10^m = 10^{(n-m)} \qquad (A.1.1)$$

A power of 10 that appears in the denominator of a fraction is given a negative sign:

$$\frac{1}{10} = 0.1 = 10^{-1}$$

$$\frac{1}{1000} = 0.001 = 10^{-3}$$

That is,

$$\frac{1}{10^m} = 10^{-m} \qquad (A.1.2)$$

Any decimal number can be expressed as a negative power of 10:

$$0.0037 = \frac{37}{10,000} = \frac{3.7}{1000} = 3.7 \times 10^{-3}$$

Calculations involving large or small numbers are made considerably easier (and less subject to error) by using powers-of-10 notation:

$$\frac{640,000}{4,000,000,000} = \frac{6.4 \times 10^5}{4 \times 10^9} = \frac{6.4}{4} \times 10^{(5-9)}$$
$$= 1.6 \times 10^{-4}$$

When a number that is expressed as a power of 10 (or as a power of another number) is itself raised to a power, the result is

$(10^2)^3 = 10^2 \times 10^2 \times 10^2 = 10^6 = 10^{2\times 3}$

$(2^3)^2 = 2^3 \times 2^3 = 8 \times 8 = 64 = 2^6 = 2^{(3\times 2)}$

In general,

$$(a^n)^m = a^{(n \times m)} \quad \text{(A.1.3)}$$

Table A.1.1 Prefixes Equivalent to Powers of 10

Prefix	Symbol	Power of 10
giga-	G	10^{9a}
mega-	M	10^{6a}
kilo-	k	10^3
centi-	c	10^{-2}
milli-	m	10^{-3}
micro-	μ	10^{-6}
nano-	n	10^{-9}
pico-	p	10^{-12}
femto-	f	10^{-15}

[a]$10^6 = 1$ *million*. In the US, $10^9 = 1$ *billion*, but the European convention is that $10^9 = 1000$ million and that 1 billion = 10^{12}; the prefix *giga-* is internationally agreed on to represent 10^9.

A.2. Fractional Exponents

When a number a is multiplied together n times to yield b, we say that a is the *n*th *root* of b; that is,

$$a^n = b \quad \text{means that} \quad a = \sqrt[n]{b} \quad \text{(A.2.1a)}$$

We can express a root as a fractional exponent:

$$a^n = b \quad \text{means that} \quad a = b^{1/n} \quad \text{(A.2.1b)}$$

because

$$a^n = (b^{1/n}) = b^{(1/n)\times n} = b^1 = b$$

and $a^n = b$ was our original hypothesis.

More complicated fractional exponents are handled by combining the various rules for exponents. For example,

$$4^{3/2} = 4^{3\times(1/2)} = (4^3)^{1/2} = (64)^{1/2} = \sqrt{64} = 8$$

or, equivalently,

$$4^{3/2} = 4^{(1/2)\times 3} = (4^{1/2})^3 = (\sqrt{4})^3 = 2^3 = 8$$

A.3. Quadratic Equations

The most complicated equations that we must solve in this book are *quadratic* equations. These equations can always be reduced to the general form:

$$ax^2 + bx + c = 0 \quad \text{(A.3.1)}$$

Sometimes $b = 0$ and we have a particularly simple equation

$$ax^2 + c = 0 \quad \text{(A.3.2)}$$

which has the solution

$$x = \pm\sqrt{-\frac{c}{a}} \quad \text{(A.3.3)}$$

Notice two points concerning this solution. (a) The quantity within the radical is $-c/a$. Now, there exists no real number which, when multiplied by itself, yields a negative number. That is, the square root of a negative number is not a real number. (We will not deal with *imaginary* numbers here.) Therefore, if the solution x in Eq. A.3.3 is to be a real number, the quantity $-c/a$ must be positive; that is, c and a must have opposite signs. (b) We have placed a "plus or minus" sign before the radical in Eq. A.3.3 because either sign for the number $\sqrt{-c/a}$ will produce an acceptable solution. The reason is that the square of a negative number is always a positive number. When solving a problem in physics, the physical situation will dictate which sign is correct. (That is, one sign can be eliminated as unphysical.)

The general solution to Eq. A.3.1 is

$$x = \frac{-b \pm \sqrt{b^2 - 4ac}}{2a} \quad \text{(A.3.4)}$$

Again, the quantity in the radical, $b^2 - 4ac$, must be positive (or zero), and the sign of the radical must be chosen so that the solution is physically meaningful.

A.4. Radian Measure

A convenient measure of angles in scientific problems is in terms of *radians*. The length s of an arc on a circle (Fig. A.4.1) is directly proportional to the radius r of the circle and to the angle θ subtended by the ends of the arc. *One radian* is defined to be the angle subtended when the arc length is exactly equal to the radius. Thus, we can write

$$s = r\theta \quad \text{(A.4.1)}$$

where θ is measured in radians.

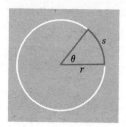

Figure A.4.1 We can write $s = r\theta$ when θ is measured in radians.

If θ is increased until it is equal to 360°, the arc length s becomes the circumference of the circle, $2\pi r$. Then, $s = 2\pi r = r\theta$, so that $\theta = 2\pi$ radians corresponds to $\theta = 360°$. Therefore,

$$1 \text{ rad} = \frac{360°}{2\pi} = 57.2958\cdots° \cong 57.3° \quad (A.4.2)$$

Also,

$$1° = \frac{2\pi}{360°} = 0.01745\cdots \text{ rad} \quad (A.4.3)$$

Notice that the measure of an angle, whether degrees or radians, does not have *physical* dimensions. Although we may carry the unit *rad* throughout calculations to remind us that angles are being measured in radians, this unit does not appear in the final answer. For example, the length of arc s on a circle with $r = 15$ cm that is subtended by an angle $\theta = 0.3$ rad is

$$s = r\theta = (15 \text{ cm}) \times (0.3 \text{ rad}) = 0.45 \text{ cm}$$

A.5. Basic Aspects of Trigonometry

Consider the triangle (called a *right triangle*) shown in Fig. A.5.1, in which the angle $\angle ACB$ is a right angle, that is, 90°. The lengths of the sides opposite the vertices A, B, and C are, respectively, a, b, and c. From the Pythagorean theorem,

$$c^2 = a^2 + b^2 \quad (A.5.1)$$

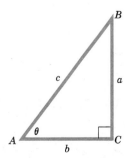

Figure A.5.1 The angle $\angle ACB$ of the triangle is 90°.

We label by θ the angle $\angle BAC$. The ratio of the length of the side opposite θ to the length of the hypotenuse is called the *sine* of the angle θ. We abbreviate this quantity as $\sin \theta$:

$$\sin \theta = \frac{a}{c} \quad (A.5.2)$$

Similarly, we define the ratio b/c to be the *cosine* of θ:

$$\cos \theta = \frac{b}{c} \quad (A.5.3)$$

Another quantity of interest in the *tangent* of θ:

$$\tan \theta = \frac{a}{b} \quad (A.5.4)$$

Graphs of the functions $\sin \theta$, $\cos \theta$, and $\tan \theta$ are shown in Fig. A.5.2 for the range of angles from 0° to 360°. Notice that the tangent function becomes infinitely large and changes sign at angles of 90° and 270°.

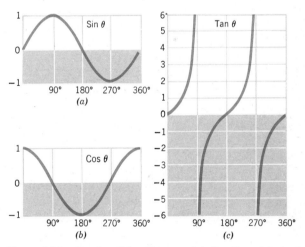

Figure A.5.2 Graphs of the trigonometric functions (a) $\sin \theta$, (b) $\cos \theta$, and (c) $\tan \theta$ for the range of angles from 0° to 360°. Remember, $90° = \pi/2$ rad, $180° = \pi$ rad, $270° = 3\pi/2$ rad, and $360° = 2\pi$ rad.

The tangent function is not independent of the sine and the cosine since

$$\tan \theta = \frac{\sin \theta}{\cos \theta} = \frac{a/c}{b/c} = \frac{a}{b}$$

The relationship between the sine and the cosine can be found by writing, from the defining equations,

$c \sin \theta = a$
$c \cos \theta = b$

If we *square* these equations and add them, we find[1]

$$c^2 \sin^2 \theta = a^2$$
$$c^2 \cos^2 \theta = b^2$$
$$\overline{c^2(\sin^2 \theta + \cos^2 \theta) = a^2 + b^2}$$

Comparing this result with Eq. A.5.1, we see that

$$\sin^2 \theta + \cos^2 \theta = 1 \qquad (A.5.5)$$

Three triangles of particular importance are shown in Fig. A.5.3. Notice the simple relationships among the lengths of the sides of these triangles. (Check that the Pythagorean theorem is verified in each case.) The trigonometric functions for the angles involved in these triangles are easily obtained. For example, referring to the 30°-60°-90° triangle, we see that $\cos 30° = \sqrt{3}/2 = 0.866$.

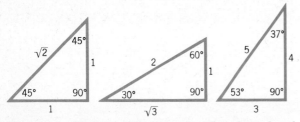

Figure A.5.3 Three important triangles. The relative lengths of the various sides are indicated. The angles given for the 3-4-5 triangle are approximate.

A.6. Small Angle Approximations

We now consider the simplifications that result in the trigonometric functions when the angles involved are small. In Fig. A.6.1 we have a right tri-

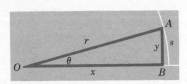

Figure A.6.1 Right triangle with angle $\angle AOB$ much smaller than 90°.

---

[1] It is customary to write, for example, $(\sin \theta)^2$ as $\sin^2 \theta$. Remember that this is just a notational device and do not read into it any extra meaning.

angle in which the side AB is much shorter than the base OB. That is, $y \ll x$ and θ is much less than 90°. Also, $OA \cong OB$ because

$$r = \sqrt{x^2 + y^2} \cong x$$

If we construct a segment of a circle with center at O and radius r, it is evident that AB is approximately equal to the arc length s: $y \cong s$. From the discussion in Section A.4, $s = r\theta$. Therefore, $y \cong r\theta$. Then, the trigonometric functions become, for small θ,

$$\sin \theta = \frac{y}{r} \cong \frac{r\theta}{r} = \theta$$

$$\cos \theta = \frac{x}{r} \cong \frac{r}{r} = 1 \qquad (A.6.1)$$

$$\tan \theta = \frac{y}{x} \cong r\theta r = \theta$$

where θ is expressed in radians.

The following table shows the extent to which the small angle approximation is valid for the sine function. (The percentage error in the tangent is about twice that for the sine, and the percentage error in the cosine is about three times that for the sine.)

	Accuracy of the Small Angle Approximation for $\sin \theta$		
θ	$\sin \theta$ (exact)	$\sin \theta$ ($\cong \theta$)	error (percent)
2°	0.03489	0.03491	0.02
10°	0.17365	0.17453	0.51
20°	0.34202	0.34907	2.06

A.7. Logarithms

The concept of a *logarithm* is a natural extension of the discussion of exponents (Sections A.1 and A.2). We define: The logarithm to the base a of a number x is equal to the exponent y to which the base number a must be raised in order that $x = a^y$. Thus,

$$x = a^y \quad \text{means that} \quad y = \log_a x \qquad (A.7.1)$$

We know that $2^5 = 32$ and $10^3 = 1000$; therefore,

$$5 = \log_2 32 \quad \text{and} \quad 3 = \log_{10} 1000$$

(If the base number is 10, we usually omit the designation from the log; thus, we write 3 = log 1000.)

An important property of logarithms is that *the log of a product is equal to the sum of the logs*. Suppose that

$$A = 10^x \quad \text{and} \quad B = 10^y$$

then,

$$x = \log A \quad \text{and} \quad y = \log B$$

Thus,

$$x + y = \log A + \log B \tag{A.7.2}$$

Also,

$$AB = 10^x \times 10^y = 10^{(x+y)}$$

from which

$$x + y = \log AB \tag{A.7.3}$$

Comparing Eq. A.7.2 with Eq. A.7.3, we conclude that

$$\log AB = \log A + \log B \tag{A.7.4}$$

Similarly, we find

$$\log \frac{A}{B} = \log A - \log B \tag{A.7.5}$$

Using the same reasoning we can easily show that

$$\log a^n = n \log A \tag{A.7.6}$$

These same results hold for logarithms with other base numbers.

Until relatively recently, logarithms to the base 10 were most often used to make easier tedious multiplication and division problems. Now, hand-held electronic calculators perform these tasks accurately and inexpensively. The primary use of base-10 logarithms has now become the graphing of power-law functions. Consider a function of the form

$$y = kx^n \tag{A.7.7}$$

If this function were plotted in the ordinary way (y versus x), we would have difficulty in easily seeing the significance of the exponent n. (n might be some unusual number such as 0.63.) However, if we take logarithms of both sides of Eq. A.7.7, we have, using Eqs. A.7.4 and A.7.6,

$$\log y = \log k + n \log x \tag{A.7.8}$$

Now, let $Y = \log y$, $K = \log k$, and $X = \log x$; then, we can write

$$Y = K + nX \tag{A.7.9}$$

This is a simple linear equations for Y versus X. In fact, the slope of the straight line is n and the intercept is K.

One can compute the logarithms of each (x, y) combination and plot the results to obtain the straight-line relationship of Eq. A.7.9. Or, one can use special graph paper that has a grid with logarithmic spacings. If a logarithmic scale is provided for both axes, this is called *log-log* graph paper. (For other purposes, only one axis is logarithmic; this is called *semilog* graph paper.) Figure A.7.1 shows log-log plots of three power-law equations:

$$y = kx^2 \qquad y = kx^{0.5} \qquad y = kx^{-0.8}$$

In each case the constant k has been adjusted so that the line will pass through the point ($x = 10$, $y = 10$).

Notice, in Fig. A.7.1, that the scales are marked in the units of x and y, whereas the spacing is pro-

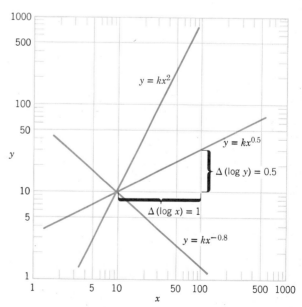

Figure A.7.1 Log-log plots of three power-law equations. Notice how the slope of the line (equal to the exponent of x) is obtained directly from the graph.

portional to the logarithms of these units. This makes easier the plotting of (x, y) combinations, and it means that we can directly measure the slope n of the line to obtain the exponent of x in the power-law equation. In Fig. A.7.1 we show how this is done for the middle line. The value of n is obtained from $n = \Delta Y/\Delta X$, where $Y = \log y$ and $X = \log x$. If we take $\Delta(\log x)$ to be the interval $x = 10$ to $x = 100$, this is 1 unit on the log x scale (one large block on the graph). The corresponding $\Delta(\log y)$ is $\frac{1}{2}$ unit (one-half block on the graph). Therefore, the slope of the line is $\frac{1}{2}$ and the equation is of the form $y \propto x^{0.5}$.

Answers to Odd-Numbered Problems

Chapter 1

- **1.1** 3.281×10^3 ft.
- **1.3** 0.9321
- **1.5** 10.28 s
- **1.7** 1.457×10^6 s or 4.046×10^2 hr
- **1.9** 2.160 min
- **1.11** 3.154×10^7 s
- **1.13** 1.418×10^{17} s
- **1.15** 10^6 n/cm^3
- **1.17** 3.563×10^{-2} kg
- **1.19** 2.499×10^{-2} m
- **1.21** 62.43 lb/ft^3
- **1.23** 0.1250 kg
- **1.25** 3.712×10^{21}
- **1.27** 3.303×10^{-9} m
- **1.29** 63.62 u vs. 63.54 u

Chapter 2

- **2.1** 1.000 m; 1.000 m; 1.600 m
- **2.3** 20.00 m; 31.42 m
- **2.5** 1.100×10^2 mi; 27.50 mi/hr; 2.250 hr; 25.00 mi/hr; 30.00 mi/hr
- **2.7** 29.89 m/s
- **2.9** 50.00 km/hr; 13.89 m/s
- **2.11** 4.000 m/s; 4.000 m/s; $x = 4t - 5$
- **2.13** 0.1732 m/s; 0.000 m/s; -0.1732 m/s
- **2.15** -4.000 (mi/hr)/s; -5.867 ft/s^2
- **2.17** 13.33 m; 16.67 m
- **2.19** 17.72 s; 3.593×10^2 m or 1.179×10^2 ft
- **2.21** 8.722 m/s or 31.40 km/hr
- **2.23** 74.40 m/s
- **2.25** $v_y^2 = v_{y0}^2 + 2gy \rightarrow v = \sqrt{2gh + v_0^2}$
- **2.27** 18.40 s; 2.755×10^2 m
- **2.29** 1.468×10^2 m; -2.682×10^3 m/s^2
- **2.31** 1.429 s
- **2.33** 2.828; 45.00 deg
- **2.35** $\hat{\mathbf{u}} = +0.6000\,\mathbf{i} + 0.800\,\mathbf{j}$
- **2.37** ± 0.4000 m
- **2.39** 42.72; 69.47 deg
- **2.41** 2.000; 1.500×10^2 deg
- **2.43** 5.000×10^2 m/s; $3.000 \times 10^2\,\mathbf{i} + 4.000 \times 10^2\,\mathbf{j}$
- **2.45** No difference
- **2.47** 15.87; 39.38 deg
- **2.49** $4\mathbf{i} - 3\mathbf{j}$; 5.00; 323.1 deg; $0\mathbf{i} + 11\mathbf{j}$; 11.0; 90°
- **2.51** 2.551×10^2 m; 1.786×10^2 m; 96%
- **2.53** (Proof); $\sin(90 + 2\Phi) = \sin(90 - 2\Phi)$
- **2.55** 78.40 m
- **2.57** Yes; 9.381 m; falling
- **2.59** 40.00 m
- **2.61** 4.710×10^3 rad; 1.178×10^3 m
- **2.63** 1.992×10^{-7} rad/s; 7.272×10^{-5} rad/s
- **2.65** 9.390 m/s; 2.007 s
- **2.67** 31.42 rad/s^2; 1.100×10^3 rad/s

2.69 6.283×10^2 rad/s
2.71 3.142×10^2 s; 0 rad net; analogous
2.73 0.5000 s; $a_c = \omega^2 R$; $a = \alpha R$; $\omega = \alpha t$ $t = \sqrt{\frac{1}{2}}$

Chapter 3

3.1 10.00 N; 20.00 N; 2.000 m/s^2
3.3 $a_w = 0.1210$ m/s^2, $a_N = 2.121$ m/s^2
3.5 65.47 N; 92.63°
3.7 7.800×10^2 N; 9.800×10^2 N
3.9 0.6510 s; 11.79 m/s^2
3.11 $a_x = -0.2392$ m/s^2, $a_y = 0.600$ m/s^2
3.13 0.9142 m/s^2
3.15 4.900 m/s^2; 30.00° below horizontal
3.17 $a = 1.032$ m/s^2; M moves up
3.19 0.3571
3.21 1.510 m/s^2 down
3.23 3.240×10^2 N
3.25 $a_{m_1} = a_{m_3} = 2.800$ m/s^2 up; $a_{m_2} = 5.600$ m/s^2 down; $V_{m_2}/V_{m_1} = 2$
3.27 $a = g \tan\theta$; $0.1405\,g$ m/s^2
3.29 2.000×10^3 N
3.31 6.667×10^{-4} s; 6.750×10^3 N
3.33 $v_x = 15.00$ m/s; $v_y = 10.00$ m/s
3.35 $v_{x1} = 3.000$ m/s; $v_{y1} = 1.212$ m/s; $v_{x2} = 3.000$ m/s; $v_{y2} = -1.212$ m/s
3.37 3.819×10^2 m/s
3.39 1.444×10^5 N
3.41 1.033×10^2 m

Chapter 4

4.1 2.250×10^2 mN
4.3 5.208 rad/s
4.5 0.9671
4.7 0.5625 m from ball
4.9 $x_{cm} = 3.000$ m, $y_{cm} = 4.000$ m
4.11 0.2782 rev
4.13 20.00 rad/s^2; 24.49 rad/s; 2.449 m/s
4.15 $a = \frac{2}{3} g$
4.17 2.400 rad/s

Chapter 5

5.1 5.214 N; 85.48° above x-axis
5.3 84.94 N
5.5 $\theta = 63.43°$; $T_1 = 87.64$ N; $T_2 = 1.753 \times 10^2$ N
5.7 $T_{45°} = 25.37$ N; $T_{60°} = 35.87$ N
5.9 88.37 kg
5.11 0.3061
5.13 Yes, since $F_\mu = 39.00$ N $> F_x = 17.32$ N
5.15 $0.6745 < \mu_s < 0.7265$
5.17 25.48 N
5.19 1.416×10^2 N; $\mu > 0.2500$
5.21 $F = +1.663 \times 10^2$ N; $P_x = +1.176 \times 10^2$ N; $P_y + 2.744 \times 10^2$ N
5.23 $T = 3.289 \times 10^{+2}$ N; $F_x = 3.289 \times 10^{+2}$ N; $F_y = 2.450 \times 10^{+2}$ N
5.25 $T_{45°} = 1.690 \times 10^3$ N; $T_{60°} = 2.940 \times 10^2$ N
5.27 $F = 77.61$ N; $O_x = -32.80$ N; $O_y = +8.062$ N
5.29 0.6667
5.31 $0.7500\,L$

Chapter 6

6.1 9.769 m/s^2; 2.442 m/s^2
6.3 3.460×10^8 m from center of earth
6.5 2.739×10^2 m/s^2
6.7 10.02 kg
6.9 6.394×10^{-8} N
6.11 1.732×10^4 s
6.13 3.704×10^{-2} m/s; 2.551×10^{-2} m/s; 9.267×10^{-2} m/s
6.15 Mercury 0.9979; Venus 0.9992; Earth 1.000; Mars 1.004
6.17 $R_{3m} = \frac{1}{3} R_{1m}$; $T_{3m} = T_{1m}$
6.19 1.894×10^{27} kg; 3.170×10^2
6.21 6.424 m/s^2; 7.115×10^3 m/s; 6.959×10^3 s
6.23 8.880×10^6 m/s^2; 8.325×10^3 s; 1.581
6.25 1.536×10^4 s

Chapter 7

7.1 9.800×10^3 J; no, work is independent of path
7.3 5.000×10^2 J
7.5 37.43 N
7.7 3.951×10^5 J
7.9 1.059×10^2 J
7.11 17.60 N; 4.400 N
7.13 9.800×10^{-2} m; 1.000 m/s
7.15 7.840×10^8 J; 1.176×10^9 J; no, bigger potential difference
7.17 24.65 m/s; 14.68 m/s
7.19 4.000×10^5 J; yes; 10.30 m/s
7.21 0.4594 of rated power
7.23 3.600×10^6 J
7.25 13.73 m/s; 27.98°; 4.553×10^5 J
7.27 0.7200 m; 1.280 m
7.29 5.655×10^{28} J; 1.793×10^{13} W
7.31 0.9567 m
7.33 6.288 J
7.35 8.472×10^8 J; 4.058×10^3 m/s
7.37 7.133×10^{15} s; 2.658×10^{41} kg; 1.336×10^{11}; 6.216×10^{40} J; -1.243×10^{41} J
7.39 -6.359×10^{35} J; $+1.590 \times 10^{35}$ J

Chapter 8

8.1 4.900×10^6 Pa
8.3 2.864×10^4 N
8.5 73.09 kg
8.7 4.508×10^4 Pa
8.9 2.940×10^4 Pa; 1.470×10^6 N
8.11 14.69 lbs/in.^{2}
8.13 1.013×10^4 Pa; 1.013 Pa
8.15 12.16 m
8.17 10.34 m; one way valve
8.19 1.004×10^6 Pa
8.21 6.542×10^{-2}
8.23 0.6660 m under water
8.25 11.12 kg
8.27 8.000×10^2 kg/m^3; 7.840×10^3 kg/m^3; iron
8.29 5.598×10^{-2} m
8.31 Water will reach top; it will not squirt out; tension reverses
8.33 -8.674 atms $= -8.788 \times 10^5$ Pa
8.35 56.68 m
8.37 1.952×10^{-3} m^3/s
8.39 Yes; $R_{e1} = 3.000 \times 10^{-5}\,a_1$; $R_{e2} = 1.200 \times 10^{+4}\,a_2$
8.41 6.164×10^{-2} m/s
8.43 2.013 m/s ($R_e = 3.387 \times 10^4$)

Chapter 9

9.1 37.00 °C
9.3 3.596×10^2 °F
9.5 1.000×10^2 °K
9.7 5.000×10^{-6} deg^{-1}
9.9 7.200×10^{-2}%

- 9.11 4.002×10^{-3} m, reading is high; yes
- 9.13 13.20 m
- 9.15 $A = A_0(1 + 2\alpha\Delta T)$
- 9.17 5.235×10^{-4} m^3
- 9.19 5.645×10^{-5} m^3; 5.414×10^{-5} m^3
- 9.21 5.009×10^{-4}; Glycerine
- 9.23 3.000%
- 9.25 1.635×10^3 kg/m^3
- 9.27 4.064 N; 3.829 N
- 9.29 2.081×10^3 J
- 9.31 2.023×10^2 Cal; very poor; 6.743%
- 9.33 2.539 °C
- 9.35 21.53 °C
- 9.37 23.65 J/mol deg; 24.51 J/mol-deg; 26.00 J/mol deg; 25.04 J/mol deg; 26.52 J/mol deg; 25.25 J/mol deg
- 9.39 20.49 °C
- 9.41 23.86 °C; $C_{H_2O} \approx 38 \times C_{Al}$
- 9.43 18.97 °C
- 9.45 46.76 °C or -2.755 °C
- 9.47 0.4253
- 9.49 7.348×10^7 w/m^2; 4.525×10^{26} w
- 9.51 $P = \sigma A e (T_2^4 - T_1^4) \to \sigma A e \times 2T^2 \times 2T(T_2 - T_1) \to C(T_2 - T_1)$

Chapter 10

- 10.1 Hyperbola; PV = constant = 16 atm m^3
- 10.3 0.9094 m; 0.6070 m
- 10.5 0.1170 kg
- 10.7 $P_2 = 3.000 P_1$
- 10.9 4.756 atm; 1.415×10^2 K
- 10.11 2.038 atm
- 10.13 $PV^\gamma = P_0 V_0^\gamma$ also $PV = nRT$, hence $TV^{\gamma-1} = P_0 V_0^\gamma / nR =$ constant
- 10.15 $V/V_0 = T/T_0 \to \Delta V/V_0 = \Delta T/T_0 = \beta \Delta T$
- 10.17 $\frac{1}{2} N_0 m v^2 = \frac{3}{2} RT = \frac{3}{2} PV$; $P = \frac{1}{3} N_0 m/V \, v^2 = \frac{1}{3} \rho v^2$
- 10.19 1.492×10^3 K = 1.219×10^3 °C; 4.000 atm
- 10.21 1.573×10^6 m/s; 1.292×10^4 eV
- 10.23 0.1707 m
- 10.25 2.598×10^2 K; 3.429×10^2 K; 3.454 J; 3.454 J
- 10.27 2.681×10^{-5} Cal/K
- 10.29 3.412%
- 10.31 1.757×10^2 °C
- 10.33 4.754×10^4 J
- 10.35 7.775×10^2 torr
- 10.37 7.600×10^2 torr; 17.50 torr; 7.600×10^2 torr; 7.600×10^2 torr
- 10.39 76.00%
- 10.41 0.1433 Cal
- 10.43 1.420×10^4 Cal
- 10.45 0.4264 Cal
- 10.47 13.33 kg
- 10.49 8.889 °C

Chapter 11

- 11.1 2.121×10^3 N
- 11.3 2.661×10^4 kg
- 11.5 $k = -Y(A/l)$
- 11.7 8.495×10^2 kg
- 11.9 (Proof)
- 11.11 1.440 m
- 11.13 4.000×10^6 N/m^2; 7.407×10^{-4}; 7.407×10^{-5} m
- 11.15 (Proof)
- 11.17 $x = +(2/\pi^2) \cos 4\pi t$; $v = -(8/\pi) \sin 4\pi t$
- 11.19 1.592×10^{-2} m
- 11.21 1.600×10^{-2} J
- 11.23 $x = 0.1 \cos \pi t/2$; $v = -0.05\pi \sin \pi t/2$; $a = -0.025\pi^2 \cos \pi t/2$
- 11.25 $E/m = \frac{1}{2}(k/m) x_0^2 \cos^2 \omega t + \frac{1}{2} \omega^2 x_0^2 \sin^2 \omega t = \frac{1}{2}(k/m) x_0^2 (\cos^2 \omega t + \sin^2 \omega t) = \frac{1}{2}(k/m) x_0^2$
- 11.27 0.4931 m; 3.873 m/s
- 11.29 5.370×10^{-3} m
- 11.31 22.50 s
- 11.33 4.621 s
- 11.35 6.000%
- 11.37 0.2838 s; 2.353×10^{-2} m; 0.5209 m/s

Chapter 12

- 12.1 6.397×10^{-8} N
- 12.3 1.780×10^{-4} C; 1.100×10^{-15}
- 12.5 4.432×10^9 C
- 12.7 $F_x = 0$; $F_y = 56.12$ N on approximate charge
- 12.9 29.62 N
- 12.11 9.216×10^{-8} N
- 12.13 7.500×10^2 J
- 12.15 $w_1 = 49.40$ J; $w_2 = 9.400$ J
- 12.17 9.800×10^6 m/sec
- 12.19 $v_p = 1.860 \times 10^7$ m/sec; $v_u = 1.240 \times 10^7$ m/sec
- 12.21 (Sketch)
- 12.23 3.578×10^{-13} N; $\theta = 63.43°$ from horizontal
- 12.25 3.333×10^3 C; $\sigma = Q/A = 2.653 \times 10^{-4}$ c/m^2
- 12.27 1.440×10^{21} N/C
- 12.29 $F_x = -0.4706$ N; $F_y = +0.1054$ N
- 12.31 -3.840×10^{-15} J; $+3.840 \times 10^{-15}$ J
- 12.33 14.40 volts
- 12.35 6.492×10^6 m/s
- 12.37 1.130×10^5 V/m; 1.130×10^5 V; 2.820×10^4 V/m; 5.650×10^4 V
- 12.39 2.080×10^6 V/m; $E_y = 0$
- 12.41 80 volts
- 12.43 $Q/A = \epsilon_0 E$
- 12.45 Cut sheet and separate by 2.360×10^{-3} m
- 12.47 3.000 μF
- 12.49 0.6590 μF
- 12.51 1.770×10^{-3} μF; 2.830×10^5 N/C; 0.1420 μF; 3.520×10^3 N/C
- 12.53 4.000×10^{-2} m^2
- 12.55 3.150×10^5 m^2; 2.000 m^3; 7.200×10^3; No
- 12.57 2.710×10^{-3} J
- 12.59 4.167×10^{-5} A
- 12.61 $R_{ser} = nR$ Ω; $R_{par} = (R/n)$ Ω
- 12.63 3 Ω
- 12.65 1.450×10^{-3} m^3
- 12.67 5.990×10^{-8} Ω
- 12.69 2.776×10^{-2} Ω
- 12.71 4 volts; 24 watts
- 12.73 $65.12

Chapter 13

- 13.1 0.9483×10^{-11} N; zero—parallel to B
- 13.3 1.034×10^{-2} m up
- 13.5 6.000×10^{-2} N; south $-20°$ from horizontal
- 13.7 $B = 0.1000$ T; 1000 G
- 13.9 6.750×10^{-2} Nm; clockwise
- 13.11 $R = mv/qB$; a) back through same face, b) through bottom c) other side

- 13.13 $B = 5.332 \times 10^{-4}$ T; 5.332 G
- 13.15 Twice (bigger charge has smaller R)
- 13.17 $R = 1.044 \times 10^{19}$ m; 1%
- 13.19 3.573×10^{-11} sec.
- 13.21 0.9132 T
- 13.23 $F = 0$; v is constant
- 13.25 5.660×10^{-3} T
- 13.27 1.421×10^{-8} T
- 13.29 6.898×10^{-3} T
- 13.31 4.000×10^{-2} T
- 13.33 3.200×10^{-5} T
- 13.35 $2 Km (N/R)I$
- 13.37 0.3750 T/A
- 13.39 4.772 V; in direction to maintain field
- 13.41 $E = -B \Delta A/\Delta t = -BLv$; $W = lqvB = qE$; therefore $E = BLv$
- 13.43 1.990×10^{-5} V

Chapter 14
- 14.1 $3.429 \, \Omega$
- 14.3 $2 \, \Omega$; 8 V
- 14.5 0.1111 A; 0.2941 A
- 14.7 3 A through R_1; 3 A through R_2; 0 A through R_3
- 14.9 $\frac{1}{5}$ A through 3 Ω; $+\frac{2}{5}$ A through 12 Ω; $\frac{1}{5}$ A through right 3 Ω; -7.2 Volts
- 14.11 $I_{12\Omega} = 3$ A $\uparrow$; $I_{4\Omega} = 9$ A $\uparrow$; $I_{14\Omega} = 2$ A $\uparrow$; $I_{12V} = 14$ A $\downarrow$
- 14.13 $V = 4.444 \times 10^{-2} \, \Omega$
- 14.15 5.98 V; $V = 0.3 \, \Omega$
- 14.17 1.991×10^{-4} C
- 14.19 $I = 1.730 \times 10^{-2}$
- 14.21 $L = 1.608 \times 10^{-2}$ H; $E = 3.860$ V
- 14.23 Shunt $8.005 \times 10^{-3} \, \Omega$
- 14.25 $R_s = 1.724 \, \Omega$; $R_{s3A} = 0.1672 \, \Omega$; $R_{s10A} = 5.005 \times 10^{-2} \, \Omega$
- 14.27 $I = 10.00$ A
- 14.29 $10.29 \, \Omega$
- 14.31 $N_2 = 60$
- 14.33 $N_1 = 1600$; $N_2 = 160$; $N_3 = 42$; yes (series + parallel)
- 14.35 $L = 2.255$ H
- 14.37 $1.670 \times 10^3 \, \Omega$
- 14.39 $z = 4.258 \times 10^2 \, \Omega$; 1.338×10^2 mA; 4.756×10^2 Hz
- 14.41 $C = 3.016 \times 10^{-4}$ F or 3.328×10^{-3} F
- 14.43 $C = 5.787 \, \mu F$
- 14.45 $I = 0.7296$; $\cos \delta = 0.2146$; Power = 13.31 w; $V_R = 36.48$ V; $V_c = 1.935 \times 10^2$ V; $V_L = 27.51$ V
- 14.47 $\delta = 46.70°$; lags

Chapter 15
- 15.1 40.00 m/s 48.99 m/s
- 15.3 2.500 m
- 15.5 A = 1.5 cm; $\lambda = 10\pi$ meters; v = 20 m/s; to the right ($+ x$)
- 15.7 16 N
- 15.9 3.858×10^2 Hz; 7.716×10^2 Hz; 1.157×10^3 Hz
- 15.11 (Sketch)
- 15.13 7.077 sec; in water
- 15.15 2.580×10^2 m/s
- 15.17 1.006×10^3 m/s
- 15.19 $H_2 = 2.316$ m/sec-deg; Air = 0.6081 m/sec-deg
- 15.21 $v = \sqrt{\gamma p/\rho}$ but $PV = nRT$; therefore v = $\sqrt{\gamma nRT/\rho v} = \sqrt{\gamma RT/M}$; since $M = \rho V/n$
- 15.23 0.9375 meters
- 15.25 M = 1.318
- 15.27 $f_1 = 4.360 \times 10^2$ Hz; (436 or 444); tighten
- 15.29 $I = 5.012 \times 10^{-6}$ W/m^2
- 15.31 $A = 1.088 \times 10^{-5}$ meters
- 15.33 (Sketch)
- 15.35 None
- 15.37 $L_1/L_2 = \frac{4}{3}$
- 15.39 2.176×10^3 Hz; 1.851×10^3 Hz
- 15.41 (Derivation)
- 15.43 (Proof)
- 15.45 $x = (2n + 1) \times 20$ cm; $I = 4I_0$.
- 15.47 0.1 mm
- 15.49 4.2 mm
- 15.51 1.8×10^{-2} m
- 15.53 34.72°
- 15.55 9.59°
- 15.57 12 m; 30 MHz
- 15.59 (Sketch)

Chapter 16
- 16.1 (Sketch)
- 16.3 2.257×10^8 m/s 1.241×10^8 m/s
- 16.5 (Sketch)
- 16.7 40°
- 16.9 $y'/a = 1/\tan \theta_1$ $y/a = 1/\tan \theta$, therefore, $y'/y = \tan \theta_2/\tan \theta_1 = \sin \theta_2/\sin \theta_1 = 1/n$, for θ small
- 16.11 3.682 mm
- 16.13 97.21°
- 16.15 (Sketch)
- 16.17 (Sketch)
- 16.19 -8.571 cm; virtual image
- 16.21 $x_0 = 2f$
- 16.23 -20 cm from second lens; -4 cm (inverted)
- 16.25 20 diopters
- 16.27 37.50 mm; 75.00 mm
- 16.29 $f = 9.77$ cm, 4.29 m
- 16.31 5.438 mm
- 16.33 3.5; 7.143 cm; 2.857 cm
- 16.35 $+f$, virtual; $+3f$, real
- 16.37 $x_i = +30$ cm; $I = 2$ cm; real inverted
- 16.39 Real inverted at center of curvature
- 16.41 Barium, lithium, barium, barium
- 16.43 $\theta_v = 40.52°$; $\theta_r = 41.45°$; $\Delta \theta = 0.914°$
- 16.45 (Proof)
- 16.47 use $l = 18$; $F_0 = 1.200$ cm
- 16.49 $l = 1.620 \times 10^2$; $F = 1.600 \times 10^2$; and $Fe = 2$ cm
- 16.51 1.969×10^{15} m (use 5890); $\alpha = 7.078 \times 10^{-7}$; no
- 16.53 8.356×10^2 m

Chapter 17
- 17.1 4.526 yr; 1.414 yr
- 17.3 1.690×10^{-6} s which is less than 2.2×10^{-6}
- 17.5 3.769×10^{-3} s
- 17.7 0.8660 c or 2.598×10^8 m/s
- 17.9 1.714×10^8 m/s
- 17.11 0.01667% or about 1 part in 6000
- 17.13 $19.48 + 12.5 = 31.98$ sec
- 17.15 6.000 km
- 17.17 a) $\rho = m/l = (m_0/l_0) (1/1 - \beta^2)$
  b) $\rho = m/l = (m_0/\sqrt{1 - \beta^2}) (1/l_0)$
- 17.19 2.828×10^8 m
- 17.21 1 cm; 54 atomic units (Fe)
- 17.23 0.8660 c
- 17.25 3.720×10^{-12} kg
- 17.27 Yes; it would take 50 sec
- 17.29 79.07 kV

17.31 6.328×10^{-8} s (use $\pi^+ = 273 \times 0.511$ MeV)
17.33 $9.000 \times 10^{+8}$ m/sec²; no; 1.089×10^{-13}

Chapter 18
18.1 9.667×10^{-6} m
18.3 1.611×10^3 K
18.5 5.506×10^{-19} J (3.434 eV); 4.580×10^{-19} J (2.857 eV); 4.090×10^{-19} J (2.551 eV); 3.375×10^{-19} J (2.105 eV); 3.030×10^{-19} J (1.890 eV); 2.588×10^{-19} J (1.614 eV)
18.7 1.988×10^{-13} m; gamma ray
18.9 5.035×10^{11} photons/sec; 8.066×10^{-10} Amps
18.11 3.219 eV
18.13 3.712 eV
18.15 1.425 volts; 0.3940 volts
18.17 3.683×10^{-36} kg; 2.476×10^5 photons
18.19 15.10 kev; 0.4375 keV
18.21 58.90 volts
18.23 2.523×10^{-3} sec; 1.374×10^{-6} sec; 3.333×10^{-8} sec
18.25 1.454×10^{-10} m
18.27 6×10^5 oscillations, 0.3000 meters
18.29 5×10^{-11} m or 0.5 Å; 601.7 eV
18.31 2.053×10^6 eV
18.33 $\nu_0 = 0.7046$ Hz; $n \cong 2 \times 10^{29}$
18.35 $2y = 0.1449$ mm
18.37 6.603×10^7 m/sec; 22.76 MeV
18.39 4.136×10^{-7} eV

Chapter 19
19.1 4.293×10^{-14}
19.3 1216 Å; Lyman
19.5 3.085×10^{15} Hz; ultraviolet; no
19.7 (Proof)
19.9 2.187×10^6 m/s or 7.291×10^{-3} C; yes
19.11 2.631×10^4 K
19.13 1.512 eV
19.15 4.054×10^4 Å; 1.459×10^4 Å
19.17 1.094×10^4 Å
19.19 0.2642 Å; 54.40 eV; 304.0 Å; 1641 Å
19.21 (Sketch)
19.23 $\Delta E = 2.122 \times 10^{-3}$ eV
19.25 $N = 2 \Sigma (2l + 1) = 2(1 + 3 + 5 + \cdots) = 2n^2$
19.27 K shell will fill at $Z = 4$ (Be); L shell at $Z = 20$ (Ca)
19.29 21.58 keV
19.31 No spike (65 keV < k shell energy); spike at 59.4 keV (75.1 keV > k shell energy)
19.33 0.100 m; 1.600×10^6 Watts
19.35 3×10^{-4} meters; 4.330×10^2 oscillations; ($\lambda = 6934$)

Chapter 20
20.1 $4.8 - 0.6 = 4.2$ eV ($-4.3 + 3.7 = -0.6$)
20.3 4×10^{-10} meters or 4 Å; $K^+ \gg Na^+$
20.5 ~ 2 eV; energy above $\lambda = 6000$ Å is absorbed

Chapter 21
21.1 2.477 MeV
21.3 7.253 MeV; 1.665 MeV
21.5 95.00 keV < energy required to break up α particle
21.7 340.0 MeV
21.9 $^{56}Co \xrightarrow{\beta^-} {}^{56}Fe$; 4.575 MeV; stable
21.11 16.00 MeV
21.13 (Proof)
21.15 47.61 keV
21.17 2.032 MeV
21.19 13.00 MeV
21.21 $2\beta^-$
21.23 Po = 33.76; Ra = 34.26; Th = 34.91; U = 35.56; Cf = 37.81; all but Po
21.25 3.654 MeV; 8.753×10^{13} J $\gg U^{235}$ fission
21.27 0.3407 MeV; in a plane 120° with respect to each other.
21.29 1877 MeV pions and photons
21.31 119.9 MeV
21.33 140.3 MeV

Answers to Selected Essay Exercises

Chapter 1

Scaling and the Sizes of Things
1. No: $M \propto L^3$ but $A \propto L^2$

Chapter 2

Running Speeds
1. 4.67 m/s^2; 9.18 m; 18.7 s; 20.5 s
2. 10.72 m/s

The Running Long Jump
1. 10.5 m/s, 11.7 m; 1.11 s

Chapter 3

Bone-Breaking Forces in Jumping
1. 5.05 m/s; 3.96×10^{-3} s

Chapter 4

Distribution of Mass in the Human Body
1. 1.06 m
2. 1.15 m

Chapter 5

Traction Systems
1. 93.2 N, 18° below horizontal
2. 100 N, 20° above horizontal

Forces in Muscles and Bones
1. $F_1 = 1570$ N; $F_2 = 1690$ N
3. $F_1 = 2590$ N; $F_2 = 3010$ N

Chapter 6

Artificial Gravity
2. $\theta = \tan^{-1}(\omega^2 r/g)$

Chapter 7

The Energetics of Running
1. $F \cong 2.5w$, each leg

The Energetics of Jumping
2. $\overline{P} = 3430$ W
3. $v = 5.24$ m/s

Chapter 8

The Flow of Blood in the Circulatory System
1. $v = 0.50$ m/s; $\Delta P = 0.985$ torr
2. $\times$ 2.44
4. 1.57 m
6. 1.05 W; 4.20 W

Chapter 9

Energy and the Metabolic Rates of Animals
1. 80 W, 1930 W
2. a factor of 2.8

Metabolic Rates of Humans
1. 38 min^{-1}
2. 20 h, no
4. 5.6 Cal/h

Chapter 10

Diffusion
2. about 2.4×10^{-9} m, 29,000 u
4. 1.4 μm

Osmosis
2. 108 g
3. 8.76 atm
4. 0.059 atm

Chapter 11

Elastic Properties of Biological Materials
1. 1.02×10^5 N
2. 7.7 m
4. about 2000 u

Chapter 12

Membrane Potentials and Nerve Impulses
2. $k = 10$
4. 2.9×10^6 ions; concentration change is 3×10^{-4}

Chapter 13

Electromagnetic Blood Flowmeters
1. 29.5×10^{-6} m³/s
2. 329 A

Chapter 14

Effects of Electric Current in the Human Body
1. 10 A; 4 μF; 50 J

Chapter 15

The Ear and Hearing
1. 5.8×10^{-5} N/m²
2. 0.22×10^{-10} m
5. 92 N/m²

Chapter 16

The Eye and Vision
3. 43.5 to 47.5 diopters
4. myopia, -2.2 diopters

Visual Acuity
1. 2.5×10^{-3} mm
2. 0.6°
3. about 0.6

Chapter 17

(none)

Chapter 18

The Electron Microscope
1. about 5 percent
4. about 6400 Å × 5500 Å; R.P. ≅ 10 Å

Chapter 19

Medical Uses of Lasers
1. 0.015 J, approx. 8×10^4 W/m², 5×10^{16} (for 6600 Å)

Chapter 20

Molecular Spectroscopy
1. 6.36, 3.90, and 1.28 ($\times 10^3$ N/m)
2. 9.20×10^3 N/m
4. Approx. 0.8×10^{-11} m

Chapter 21

Radiation Exposures of Humans
2. 2.3×10^3 rad, 0.0054 deg, 4.25×10^{-6} eV/molecule

Biological Effects of Exposure in Humans
1. 62 rem/y

Chapter Opening Photo Credits

Chapter 1:   Yerkes Observatory Photo/Taurus Photos.

Chapter 2:   Phaneuf/Gurdziel/The Picture Cube.

Chapter 3:   NASA.

Chapter 4:   Museum of Modern Art/Film Stills Archive.

Chapter 5:   Cary Wolinsky/Stock Boston.

Chapter 6:   Chip Maury/Picture Group.

Chapter 7:   Jean–Claude LeJeune/Stock Boston.

Chapter 8:   Christopher S. Johnson/Stock Boston.

Chapter 9:   Michael Gordon/Picture Group.

Chapter 10:   Richard Lodge/Picture Group.

Chapter 11:   Wide World Photos.

Chapter 12:   David S. Strickler/The Picture Cube.

Chapter 13:   Educational Development Center.

Chapter 14:   Franklin Wing/Stock Boston.

Chapter 15:   Michael Gordon/Picture Group.

Chapter 16:   Phyllis Graber Jensen/Stock Boston.

Chapter 17:   Escher Foundation, Gemeentemuseum—The Hague.

Chapter 18:   Escher Foundation, Gemeentemuseum—The Hague.

Chapter 19:   Courtesy of Spectra Physics.

Chapter 20:   Jesse Oroshnik/Illustrator Stock Photos

Chapter 21:   American Cancer Society.

Appendix   Reprinted with permission from ARTIST AND COMPUTER © 1976 by Creative Computing Press, P.O. Box 789-M, Morristown, New Jersey 07960.

Index

Entries in *italics* indicate bioscience essays.

Aberrations:
   chromatic, 408
   spherical, 406
Absolute temperature scale, 181
Absolute zero, 210
Acceleration, 28, 38
   centripetal, 45
   circular motion, 47
   due to gravity, 32
   and force, 56
Acceptors, 520
Achromatic lens, 408
Action potential, *271*
Acuity, visual, *421*
Adiabatic process, 204
Alpha particles, 476, 537
Alpha radioactivity, 537
Alternating current, 335
   heating effect of, 336
Ammeters, 333
Ampere (unit), 277
Anderson, C. D., 559
Aneroid barometer, 158
Angstrom (unit), 4
Ångström, A. J., 476
Angular frequency, 44, 241

Angular velocity, 44, 241
Antiparticles, 559
Apochromatic lens, 408
Archimedes, 159
Archimedes' principle, 158
Astronomical unit (A.U.), 121
Atmosphere (unit), 156
Atom, 13
   masses of, 14
   shell structure of, 487
   sizes of, 16, 475
Atomic bomb, 549
Atomic mass unit, 14
Atomic number, 488
Avogadro, A., 14
Avogadro's number, 14, 205

Ballistic trajectory, 40
Balmer, J., 478
Balmer series, 478
Bardeen, J., 522, 525
Barlow, W., 514
Barometer, 157
Battery, 323
BCS theory, 525
Beats, 361
Becquerel, H., 533

Bel (unit), 363
Bell, A. G., 363
Bernoulli, D., 164
Bernoulli's equation, 165
Beta radioactivity, 534
Beta rays, 533
Big bang, 195, 540
Binding energy, 481, 506
Biological effects of radiation, *541, 554*
Black holes, 444
Blood flowmeter, *299*
Blood pressure, *169*
Bohr, N., 470, 479
Bohr model of hydrogen atoms, 253, 259, 479
Boiling, 224
Boltzmann, L., 205
Boltzmann constant, 205
Born, M., 466, 487
Bose, S. N., 486
Bosons, 486, 525
Boyle, R., 203
Boyle's law, 203
Brahe, Tycho, 119
Brattain, W., 522
Breeder reactors, 552

Bremsstrahlung, 493
Brewster's angle, 416
Broglie, L. V. de, 455
Broglie, de, waves, 455, 466, 482
Bubble chamber, 297
Bulk modulus, 234
Buoyancy, 158

Calorie (unit), 186
Capacitance, 265
   in circuits, 325
Capacitive reactance, 339
Capacitor, 265
   series and parallel combinations, 267
Capillarity, 162
Carbon bonds, 508
Carnot, N. L. S., 220
Carnot cycle, 220
Cavendish, H., 115
Celsius, A., 180
Celsius scale, 179
Center of mass, 83
Centigrade temperature scale, 180
Centripetal acceleration, 45
Chadwick, J., 16
Chain reactions, 548
Charge, 249
   conservation of, 251
   electron, 252, 265
Charles, J., 205
Charles-Gay-Lussac law, 205
Charm, 565
Circuits, 278, 319, 339
Cloud chamber, 545
Coherence, 376, 494
Collisions, 141, 437
Color, 407
Complementarity principle, 470
Compressive stress, 232
Compton, A. H., 463
Compton scattering, 463
Compton wavelength, 463
Conduction, thermal, 191
Conductivity, electric, 280, 516
Conic sections, 119
Conservation laws:
   of angular momentum, 80, 87
   of charge, 251
   of energy, 135
   of linear momentum, 70, 71

   of mass, 10
   of mass-energy, 439
Conservative forces, 133
Convection, 193
Cooper, L., 525
Coulomb (unit), 252
Coulomb, C. A., 251
Coulomb's law, 251, 257
Covalent binding, 506
Cowan, C., 536
Critical mass, 549
Critical point, 225
Cross product, 79
Crystals, 12, 225, 506, 514, 519
Curie (unit), *543*
Curie, M., 533
Current, 276
   alternating, 335
   in human body, *328*
   induced, 307
   loop, 302
Cyclotron, 298

Dalton, J., 13
Damped oscillations, 244
Davisson, C. J., 463
Decibel, 363
Definite proportions, law of, 13
Density, 10, 160
Deuterium, 16
Deuteron, 529, 558
Diamagnetism, 305
Dielectric, 268
Diffraction, 378, 419, 461
Diffraction grating, 379
Diffusion, *211*
Dimensional analysis, 353
Diodes, 342, 521
Diopter, *412*
Dirac, P. A. M., 487
Displacement, 21
DNA, *514*
Domains, magnetic, 306
Donors, 520
Dopants, 519
Doppler, C. J., 366
Doppler effect, 366
Dose:
   equivalent, *543*
   radiation, *543*
Dot product, 128

Ear, *369*

Earth:
   density of, 11
   magnetism of, 290
   mass of, 115
Einstein, A., 429, 453. *See also* Relativity
Elastic collisions, 141
Elasticity, 231
   of bone, *235*
Elastomers, *237*
Electric circuits, 278, 319
   AC, 339
   amplifier, 522
   rectifier, 343
Electric polarization, 269
Electric potential, 258
   in membranes, *271*
Electric resistance, 278, 319
Electromagnet, 292
Electromagnetic force, 296
Electromagnetic induction, 307
Electromagnetic radiation, 381, 455
   spectrum of, 383
Electromagnetic waves, 381
Electromagnetism, 291
Electromotive force, 277
Electron, 15, 252, 563
   charge of, 252, 265
   conduction, 276, 516, 517
   diffraction of, 463
   mass of, 16
   spin of, 484
   valence, 489, 519
   wave description of, 455, 464
Electronics, 342, 520
Electron microscope, *456*
Electronvolt, 256
Electroscope, 251
Electrostatic field, 257
Electrostatic force, 249, 257
Electrostatic force constant, 252
Electrostatic potential energy, 254
Elementary particles, 559, 560, 563
Ellipse, 120
EMF, 277, 310
   of battery, 323
Emissivity, 194
Energy, 127
   bands, 517
   binding, 481, 506
   in capacitor, 270

conservation of, 135, 439
of electromagnetic waves, 383
gravitational, 145
ionization, 489, 493
in jumping, *137*
kinetic, 130
mass-energy, 439
in pole vault, 139
potential, 133
quantization of, 468
rest, 439
rotational, 144
in running, *131*
in SHM, 242
states, 481
thermal, 185
Entropy, 218
Equilibrium, 93
Equivalence principle, 442
Escape velocity, 147
Evaporation, 221
Exclusion principle, 489
Eye, 403, *410, 421*

Fahrenheit, G. D., 180
Fahrenheit scale, 179
Farad (unit), 265
Faraday, M., 266, 308, 327
Faraday's law, 310
Fermi, E., 486
Fermions, 486, 563
Ferromagnetism, 306
Feynman, R. P., 116
Fick, A., *211*
Fick's law, *211*
Field:
 electric, 257
 magnetic, 290, 296
 time-varying, 309
Fission, 441, 546
Fizeau, A., 392
Fluids, 153
 pressure in, 153, 155
Focal length, 398
 of lenses, 399
 of mirrors, 405
Force, 55, *61*, 93
 centripetal, *117*
 conservative, 133
 elastic, 232
 electromagnetic, 296
 electrostatic, 251, 257
 gravitational, 57, 114, 122
 line of, 260

magnetic, 293
in muscles and bones, *103*
normal, 58
strong nuclear, 560
unit of, 59
weak, 563
Foucault, J. B. L., 393
Fraunhofer, J. von, 409
Fraunhofer lines, 409
Frequency, 241
of electromagnetic radiation, 383
fundamental, 358
harmonic, 358
spectrum of, 359, 383
Friction, 35, 56, 66
kinetic, 66
static, 100
Frisch, O., 546
Fundamental frequency, 358
Fusion, 558

Galileo Galilei, 21, 32, 41, 391
Galvani, L., 325
Galvanometer, 325, 333
Gamma rays, 533
Gas constant, 206
Gas laws, 203
Gauss (unit), 293
Gay-Lussac, J. L., 205
Geiger, H., 477
Gell-Mann, M., 564
General relativity, 441
Generators, 335
Germer, L. H., 463
Goudsmit, S. A., 484
Gravitational constant, 115
Gravitational potential energy, 145
Gravity, 113
acceleration due to, 32
artificial, *117*
force of, 57, 114, 122
GUT theory, 566

Hadrons, 560
Hahn, O., 546
Half-life, 533
Harmonics, 358
H-bomb, 559
Heat, 185
transfer, 191
Heat engines, 220
efficiency of, 220

Heat of fusion, 225
Heat of vaporization, 224
Heisenberg, W., 2, 469, 487. *See also* Uncertainty Principle
Helicity, 564
Helium, liquid, 523
Helmholtz, H. von, 127, 186
Henry (unit), 327
Henry, J., 327
Hertz (unit), 241, 451
Hertz, H., 241, 383
Hoff, J. H. van't, *214*
Hooke, R., 130
Hooke's law, 130, 232
Horsepower, 135
Humidity, 223
Huygens, C., 373
Huygens' principle, 373
Hydrogen atom:
 Bohr model of, 479
 quantum theory of, 487
 spectrum of, 478
Hydrogen bond, *513*
Hydrogen molecule, 507
Hyperons, 562

Ideal gas law, 205
Image formation:
 by lenses, 397, 399
 by mirrors, 397, 405
Impedance, 340
Induced current, 307
Inductance, 327
 in circuits, 327
Inductive reactance, 339
Inductor, 327
Inelastic collisions, 141
Inertia, 55
 rotational, 86, 144
Inertial reference frame, 68
Insulators, 518
Integrated circuits, 522
Interference, 356, 374, 465
Ion, 151
Ionic binding, 505
Ionization, 15
 energy, 489, 493
Isothermal process, 203
Isotopes, 16, 532

Jeans, J., 450
Joule (unit), 128

Joule, J. P., 128, 186, 282
Joule heating, 282

Kaons, 562
Kelvin (unit), 181
Kelvin, Lord, *see* Thomson, W. (Lord Kelvin)
Kelvin temperature scale, 181
Kepler, J., 2, 119
Kepler's laws, 119
Kilogram, 4
Kinetic energy, 130
Kinetic theory, 207
Kirchhoff's circuit rules, 321
Kubrick, S., *118*

Lambda particles, 562
Land, E. H., 414
Lasers, 494
   medical uses, *498*
Laue, M. von, 461
Lawrence, E. O., 298
Length, 3
   contraction of, 434
Lens equation, 401
Lenses, 398
Lenz, H., 309
Lenz's law, 309, 327
Leptons, 563
Light, 391
   diffraction of, 419
   focusing of, 397, 404
   polarized, 413
   reflection of, 393
   refraction of, 394
   speed of, 383, 391, 430
Light pipes, 397
Light year (L.Y.), 435
Liquid helium, 523
Livingston, M. S., 298
Lodestone, 289
Lorentz, H. A., 296
Lorentz force, 296
Loudness, 363, *371*

Mach number, 360
McMahon, T., *187*
Magnet, 289
   electromagnets, 292
   superconducting, 525
Magnetic domains, 306
Magnetic field, 290
   due to currents, 290, 291, 302
   strength of, 293

Magnetic flux, 310
Magnetic force, 293
Magnetic force constant, 302
Magnetic materials, 304
Magnetic substates, 484
Magnetism, 289
   atomic, 292
   of Earth, 290
Magnification, 402, 406
   of microscope, 416
   of telescope, 417
Magnus effect, 167
Malus, law of, 415
Manhattan Project, 550
Marsden, E., 477
Mass, 4, 55, 67, 437
   of atoms, 14
   conservation of, 440
   critical, 549
   distribution in body, *85*
   of Earth, 115
   of electron, 16
   of molecules, 14
   of neutron, 16
   of nuclei, 529
   of proton, 16
   in relativity, 437
   rest, 438
   standard of, 4
Mass-energy, 440
Maxwell, J. C., 209, 304, 383
Maxwellian distribution, 209
Mayer, J. R., 186
Meitner, L., 546
Melt down, of reactor, 552
Melting, 225
Mesons, 560
Metabolic rates:
   of animals, *187*
   of humans, *195*
Meter (unit), 3
Metric system, 3
Michelson, A. A., 392
Microscopes, 416
   electron, 456
   resolving power of, 420
Millikan, R. A., 265, 454
Mirrors, 393, 404
Mole, 14, 205
Molecular spectroscopy, *509*
Molecules, 13, 505
   binding of, 505
   hydrogen, 507
   masses of, 14

   organic, 509
   sizes of, 15
   velocities of, 209
Momentum:
   angular, 79, 483
   conservation of, 70, 80, 87
   of electromagnetic waves, 383
   linear, 69
   quantization of, 466, 479
   spin, 484
   total, 485
Moseley, H. G. J., 492
Motion, 21
   circular, 44, 47, 239
   of fluids, 164
   laws of, 56
   in magnetic field, 296
   oscillatory, 235
   planetary, 119
   projectile, 40
   simple harmonic, 238
   two-dimensional, 39
   wave, 351
Motors, electric, 332
Muons, 561

Nernst equation, *271*
Neutrinos, 535, 563
Neutron, 15, 530, 534
Newton (unit), 59
Newton, I., 56, 113, 119, 418
Newton's laws, 56
Normal force, 58
Nuclear force, 560
Nuclear medicine, *556*
Nuclear reactions, 545
   thermonuclear, 559
Nuclei, 16, 476
   fission of, 441
   fusion of, 558
   masses of, 529
   shape of, 532
   size of, 16, 532
   stable, 532
Nucleogenesis, 540
Nucleons, 535
Numerical aperture, 420

Oersted, H. C., 291
Ohm (unit), 278
Ohm, G. S., 278
Ohm's law, 278, 340
Ohm's-law circuits, 319
Onnes, K., 523

Orbits:
   of charged particles in magnetic fields, 296
   helical, 299
   in hydrogen atom, 479
   of planets, 119
   of satellites, 123
Order and disorder, 218
Organic compounds, 509
Oscillations, 235
Osmole, *215*
Osmosis, *214*
Overtones, 358

Paramagnetism, 305
Pascal (unit), 154
Pascal, B., 154
Pascal's principle, 154
Pauli, S., 489
Pauli exclusion principle, 489
Pendulum, simple, 243
Period, 44, 241
   of circular motion, 44
   of planets, 122
   of SHM, 241
Periodic table, 488
Phase:
   change of, 225
   diagram, 225
   relationship, AC, 339
Photoconductors, 519
Photoelectric effect, 451
Photons, 453, 464, 494
Pions, 561
   decay of, 434
   mass of, 561
Planck, M., 450
Planck's constant, 450
Planetary motion, 119
Plasma, 12
Plateau, J., 162
Plutonium, 552
Poiseuille, J. M., *170*
Poiseuille's equation, *170*
Polarization, 413
Positrons, 536, 563
Potential difference, 136, 147
Potential energy, 133, 145
Potentiometer, 324
Power, 134
   electric, 281, 324, 340
   factor, 340
   nuclear, 550
Precession, 122, 442

Pressure, 153
   osmotic, *215*
Probability, 465, 507
Projectile motion, 40
Proton, 15
Proust, J. L., 13
Pulsars, 445
PV diagram, 204

Quanta, 450
Quantum numbers, 480, 485
   angular momentum, 483
   magnetic, 484
   principal, 480
   rotational, *510*
   spin, 484
   total angular momentum, 485
   vibrational, *510*
Quantum theory, 464
   of hydrogen atom, 479
Quarks, 564

Rad (unit), *543*
Radiation, 194, 381, 449
   dose, *543*
   electromagnetic, 381
   exposure, *543*
   nuclear, 533
   stimulated, 494
   thermal, 194, 449
   units, *543*
Radioactivity, 16, 533
   decay chains, 538
Rayleigh, J. W. S., Lord, 450
Reactors, 550
   melt down, 552
Rectifiers, 343, 521
Red shift, 443
Reflection, 393
   internal, 396
Refraction, 393
Refractive index, 394, 407
Reines, F., 536
Relative humidity, 223
Relativity, 429
   general, 441
   length in, 434
   mass-energy in, 439
   mass in, 437
   postulates of, 430
   time in, 432
Rem (unit), *543*
Resistance, 278, 319

Resistivity, 279
Resistors, 278, 319
   parallel connections of, 279, 319
   series connections of, 278, 319
Resolution, 419
Resolving power, 420
   of electron microscope, *457*
Resonance, 244
   in AC circuits, 340
   in air columns, 365
Restitution coefficient, 143
Reynolds number, 168
Right-hand rule:
   for angular momentum, 83
   for electromagnetic waves, 382
   for magnetic field, 292
   for magnetic force, 294
Rocket propulsion, 72
Rotational inertia, 86, 144
Rotational motion, 79, 86, 144
Rutherford, E., Lord, 476
Rydberg constant, 478

Satellites, 123
Scalar, 36
Scalar product, 128
Scaling, 7
Schrieffer, R., 525
Schrödinger, E., 487
Scientific method, 2
Second (unit), 4
Semiconductors, 519
Shawlow, A., 496
Shockley, W., 522
Shock waves, 367
Simple harmonic motion, 238
Simple pendulum, 243
SI units, 3
Snell, W., 394
Snell's law, 394
Sodium-potassium pump, *273*
Solenoid, 303
Sommerfeld, A., 483
Sonic boom, 367
Sound, 355, 358
   and ear, 363, *369*
   intensity level, 363
   musical, 362
   in pipes, 364
   speed of, 358
Specific heat, 189

Spectra, 409, 478
  from beta decay of molecules, *509*
Speed, 22, *30*, 44
Spin, 484
Spring constant, 130
Stefan-Boltzmann law, 195, 450
Stimulated radiation, 494
Strain, 232
Strangeness, 563
Strassman, F., 546
Stress, 232
Strong nuclear force, 560
Sublimation, 13, 226
Superconductors, 281
Superfluidity, 523
Superposition, 257, 356
Surface tension, 161
Synchronous satellites, 123

Telescopes:
  reflecting, 418
  refracting, 417
Temperature, 179, 185
  absolute, 181
  density affected by, 184
  resistivity affected by, 280
Tensile strength, 233
Tensile stress, 233
Terminal velocity, *9*, 35, 173
Tesla (unit), 293
Thermal energy, 185
Thermal expansion, 181
Thermal radiation, 194, 449
Thermodynamics, 216
Thermometer, 179
Thermonuclear reactions, 559
Thomson, G. P., 462
Thomson, J. J., 475
Thomson, W. (Lord Kelvin), 181
Time, 4, 432
Time dilation, 432
Torque, 79, 101
  on current loop, 332

Torr (unit), 157
Torricelli, E., 157
Torricelli's equation, 166
Townes, C., 496
Traction systems, *97*
Transformers, 337
Transistors, 522
Triple point, 225
Tritium, 16
Twin paradox, 435

Uhlenbeck, G. E., 484
Uncertainty principle, 468
Unified atomic mass unit, 14
Uranium, 441, 546, 550

Valence electrons, 489, 519
Valley of nuclear stability, 538
Vaporization, 221
Vapor pressure, 222
Vector, 36
  components of, 37
  unit, 36
Vector product, 79, 295
Velocity, 22, 26, 36
  angular, 44
  electron drift, 277
  escape, 147
  of light, 383, 391, 430
  of molecules, 209
  of sound, 358
  terminal, *9*, 35, 173
Vinci, L. da, 41
Viscosity, 167
Volt (unit), 255, 259
Voltage, 255, 259, 264
  AC, 335
Voltmeters, 333

Watt (unit), 135, 281
Watt, J., 135
Wave, 351, 455
  Broglie, de, 455, 466, 482

  diffraction of, 372
  electromagnetic, 381
  electron, 455, 464
  interference of, 356, 374
  longitudinal, 355
  -particle duality, 466
  plane, 372, 382
  reflection of, 356
  sound, 358
  standing, 357
  transverse, 355, 382
  traveling, 354
Wave function, 467
  hydrogen atom, 487
Wavelength, 354
  Broglie, de, 455, 482
  Compton, 463
  of electromagnetic radiation, 383, 454
  of light, 407
  of standing waves, 357
Wavenumber, *510*
Weak force, 563
Weber (unit), 310
Weber, W. E., 310
Weight, 59, *117*
Weightlessness, 60
Wheatstone bridge, 325
Wheeler, J. A., 471
Wien, W., 450
Wien's law, 450
Wollaston, W., 409
Work, 127
Work function, 453

X rays, 492
  diffraction of, 461

Young, T., 376
Young's double slit, 376
Young's modulus, 233
Yukawa, H., 560

Zweig, G., 564